I0715312

Liquid Rocket Thrust Chambers: Aspects of Modeling, Analysis, and Design

Liquid Rocket Thrust Chambers: Aspects of Modeling, Analysis, and Design

Edited by

Vigor Yang
Pennsylvania State University
University Park, Pennsylvania
Mohammed Habiballah
Office National d'Etudes et de Recherches Aérospatiales
Châtillon, France
James Hulka
Jacobs-Sverdrup Technology, Inc.
Huntsville, Alabama
Michael Popp
Pratt and Whitney Space Propulsion
West Palm Beach, Florida

Volume 200
PROGRESS IN
ASTRONAUTICS AND AERONAUTICS

Paul Zarchan, Editor-in-Chief
MIT Lincoln Laboratory
Lexington, Massachusetts

Published by the
American Institute of Aeronautics and Astronautics, Inc.
1801 Alexander Bell Drive, Reston, Virginia 20191-4344

Table of Contents

Chapter 3   Atomization in Coaxial-Jet Injectors 105

Lucien Vingert and Pierre Gicquel *ONERA, Palaiseau, France*, Michel Ledoux
and Isabelle Caré *CORIA, Université de Rouen, Rouen, France*, and
Michael Micci and Michael Glogowski *Pennsylvania State University,
University Park, Pennsylvania*

Chapter 4   Liquid Bipropellant Injectors 141

William E. Anderson, Matthew R. Long, and Stephen D. Heister *Purdue University, West Lafayette, Indiana*

Chapter 5   Distortion and Disintegration of Liquid Streams 167

William A. Sirignano and Carsten Mehring *University of California, Irvine, California*

Chapter 8 Subcritical/Supercritical Droplet Cluster Behavior in Dense and Dilute Regions of Sprays 323

Josette Bellan *Jet Propulsion Laboratory, California Institute of Technology, Pasadena, California*

Chapter 9 Fundamentals of Supercritical Mixing and Combustion of Cryogenic Propellants. 339

Wolfgang O. H. Mayer *DLR, German Aerospace Research Center, Hardthausen, Germany* and Joshua J. Smith *University of Adelaide, Adelaide, Australia*

Gerald Hagemann and Hans Immich *EADS Space Transportation, Munich, Germany*, Thong Nguyen *GenCorp Aerojet, Sacramento, California*, and Gennady E. Dumnov *NIKA Software, Moscow, Russia*

Patrick Vuillermoz *Centre National d'Etudes Spatiales, Evry, France*, Claus Weiland and Gerald Hagemann *EADS Space Transportation, Ottobrünn, Germany*, Bertrand Aupoix *Office National d'Etudes et de Recherches Aérospatiales, Toulouse, France*, Hervé Grosdemange *SNECMA Moteurs, Vernon, France*, and Mikael Bigert *Volvo Aero Corporation, Trollhattan, Sweden*

Chapter 17   Assessment of Thrust Chamber Performance   601

Douglas E. Coats *Software and Engineering Associates, Inc., Carson City, Nevada*

Chapter 18   Thermodynamic Power Cycles for Pump-Fed Liquid Rocket Engines 621

Randy C. Parsley *Pratt and Whitney, United Technologies Corporation, West Palm Beach, Florida* and Baojiong Zhang *Shanghai Bureau of Astronautics, Shanghai, People's Republic of China*

Chapter 19 Tripropellant Engine Technology for Reusable Launch Vehicles 649

N. S. Gontcharov, V. A. Orlov, V. S. Rachuk, M. A. Rudis, and A. V. Shostak
Chemical Automatics Design Bureau, Voronezh, Russia and R. G. Starke and
J. R. Hulka *Aerojet General Corporation, Sacramento, California*

Chapter 20 Oxidizer-Rich Preburner Technology for Oxygen/Hydrogen Full Flow Cycle Applications 683

Shahram Farhangi, Robert J. Jensen, Ken Hunt, Linda Tuegel, and Tai Yu, *The Boeing Company, Canoga Park, California*

Preface

Liquid-propellant rocket engines have been used as the primary propulsion systems in most launch vehicles and spacecraft since the initial conquest of space. Eight decades after their first fabrication and test, however, many aspects of modeling, analysis, and design of thrust chambers for these engines still present important challenges. Certainly one reason for this continued challenge is the complexity of the problem; although the basic concepts are well established, many of the detailed physiochemical processes of liquid-propellant combustion remain unresolved. Another reason is the difficulty and significant expense of conducting research and development in the harsh and hazardous environments of liquid rocket thrust chambers. Improvements in performance for future rocket engines may require that internal pressures and heat flux, already substantial, continue to increase, rendering the problem even more challenging. Furthermore, any design work addressing performance, life, reliability, and safety for future thrust chamber development will benefit greatly from an improved understanding of the chemical and physical processes and the resulting formulation and use of suitable models that can be incorporated into advanced analysis tools.

Investments in this field have slowed over the past 30 years, and important results and reference publications either have disappeared or have not been made widely or publicly available. The United States and the former Soviet Union have made only sporadic investments since their significant advancements in the 1950s, '60s, and '70s. Europe, China, and Japan, on the other hand, have made major investments in recent decades. At the same time, there have been significant improvements in computational and experimental techniques, and a growing practical experience with full-scale rocket engines worldwide, that have advanced the state of the art. Unfortunately, the smaller scale and narrower dissemination of the literature on this work have limited the widespread utilization of these advances. Existing publications are scattered throughout various journals, technical reviews, conference proceedings, and progress reports.

There is consequently a distinct need for a comprehensive text to organize and make readily available the latest results and developments, and to promote additional insight, research, and collaboration on the topics of modeling, analysis, and design of liquid rocket thrust chambers. There has not been a major publication on this topic since *Liquid Rockets and Propellants*, in 1960. (Volume 2 in the AIAA Progress in Aeronautics and Astronautics Series, edited by L. E. Bollinger, M. Goldsmith, and A. W. Lemmon, Jr.) A volume to capture and present the recent advances and current understanding in the field was therefore considered an important contribution to the research, industrial, and academic communities involved in the design, development, and testing of liquid rocket thrust chambers.

The present volume compiles results from many of the research and development programs conducted over the last several decades across the international community. One of the driving interests behind the production of this volume was to cover the international scope, which was made possible by the significant changes in the political landscape over the past decade.

The original material and impetus for this volume came from the Second International Symposium on Liquid Rocket Propulsion, hosted by l'Office National d'Etudes et de Recherches Aérospatiales, at Châtillon, France, in June of 1995. This meeting was the second in a series of symposia designed to provide open technical exchanges across the international community in the liquid rocket field. At the symposium, technical presentations were made by leading experts from eight countries, including nearly every major governmental and corporate organization. Like its predecessor, this meeting provided a unique opportunity for presentation and discussion of work from the Commonwealth of Independent States and from China that had not been seen previously in the West.

This volume is designed as a reference text, with a balance of fundamental scientific and technological works. It is organized into five subject areas, consistent with a stepwise approach through the design and analysis of thrust chambers, including: 1) injection and atomization processes; 2) combustion and ignition processes; 3) nozzle design and optimization; 4) chamber dynamics, heat transfer, and performance; and 5) influence of engine system design. Twenty chapters are included, all prepared and reviewed by leading experts in this field. The volume can be used and enjoyed by engineers, researchers, and scientists in industry, government, and academia who are involved in rocket propulsion research. The structuring of this volume as a reference also will render it useful to those in other fields involved in the design of combustion devices.

Vigor Yang
Mohammed Habiballah
James Hulka
Michael Popp
November 2004

Acknowledgments

Publication of this volume was made possible through the substantial contributions of a number of individuals and organizations. We would like to first thank the authors of the papers for sharing their time and talent in preparing their manuscripts and carefully revising them. Many other individuals deserve special recognition for reviewing the chapters in this volume; their names are listed here:

D. Ballal	J. Hulka	M. Popp
V. Bazarov	J. Ito	R. Reitz
J. Bellan	R. Jensen	U. Renz
C. Bonniot	T. Jiang	C.-A. Schley
R. Carroll	A. Kumakawa	D. Seymour
Y.-S. Chen	F. Lacas	D. Stepowski
D. Counts	P. Liang	P. Strakey
C. Crowe	W. Mayer	D. Talley
D. Culver	C. Merkle	C. Tropea
C. Dexter	K. Olson	L. Vingert
I. Dubois	M. Moser	T.-S. Wang
S. Fisher	M. Oschwald	J. Wolfrum
K. Gross	S. Pal	P. K. Wu
M. Habiballah	J. Peters	J. Xin
D. Haeseler	J. Pieper	V. Yang

The invaluable assistance of Rodger Williams, Heather Brennan, and Janice Saylor of the AIAA in the preparation of the volume for publication is gratefully acknowledged. We owe a large debt of gratitude to Mary Newby for managing voluminous correspondence among editors, authors, reviewers, and the AIAA. The technical drawing services provided by Danring You and Yanxing Wang deserve special thanks. Last, but by no means least, we wish to thank Anna Creese for her outstanding technical editing expertise.

Financial support for the Second International Symposium on Liquid Rocket Propulsion and the preparation of this volume was provided by the following organizations.

- Office National d'Etudes et de Recherches Aérospatiales (ONERA)
- Agence Spatiale Européenne
- Centre National d'Etudes Spatiales (CNES)
- Snecma Moteurs
- Deutsches Zentrum Für Luft- und Raumfahrt (DLR)
- EADS Space Transportation GmbH
- United States Air Force European Office of Aerospace Research and Development

Their sponsorship is gratefully acknowledged. We wish to express our deep appreciation to Paul Kuentzmann for having hosted the symposium and for his continued support to the international collaboration on space propulsion. He deserves special thanks. We are also indebted to Mitat Birkan and Dieter Preclik for their generous support and encouragement.

Chapter 1

Propellant Injection Systems and Processes

Jackson I. Ito*
GenCorp Aerojet Propulsion Division, Sacramento, California

I. Introduction

$\mathbf{T}$ HE liquid propellant combustion device has always presented design and development risks because of the required harsh operating thermal environment, usually at high pressures, and because of the secondary goals of small packaging, light weight, high performance efficiency, and low cost. The injector design has always been recognized as a key component that often controls the success or failure of the combustion device.

When rocketry was in its infancy, injector designs were developed mainly through a time-consuming and costly process of trial and error. Once a degree of success was achieved, designers attempted to copy previously successful designs. This approach did not always yield the desired results. Eventually, successful engineers recognized that it was not copying the hardware that assured success but rather the proper scaling and control of the combustion process. A design that works well for one application may fail in another because of a subtle difference in operational requirements or system constraints. Analytical tools are now available or are being developed to evaluate these critical combustion processes so that candidate designs can be evaluated and optimized conceptually, thus avoiding or minimizing some of the detailed design, manufacturing, and test cycles historically required. Even when the models may be incompletely understood or when uncertainties exist, it may still be possible to conduct smaller scale, faster, and lower cost experiments to validate necessary assumptions or to plan parallel design approaches for a few high-risk components to increase subsequent probability of success at lower overall development cost.

This chapter addresses key issues that the designer needs to identify so that the technical capabilities provided by the remaining chapters of this volume can be appreciated.

---

*Technical Principal Engineer.

II.   Rocket Application Design Requirements

Before one can expect to achieve success in a combustion device design, it is necessary to determine its functional requirements. It is also helpful to understand what types of development risks are most likely to be encountered and what other constraints are imposed by the system within which it will be expected to operate. This allows prioritization of limited technology resources to ensure solution of the most troublesome problems before committing an entire system design approach. These requirements can be separated into three major categories.

A.   Thrust Level and Operating Pressure

The requirements for thrust level and operating pressure determine the size and weight of the combustion device. Figure 1 illustrates the range of various combustion devices of interest to the international propulsion community.[1-4]

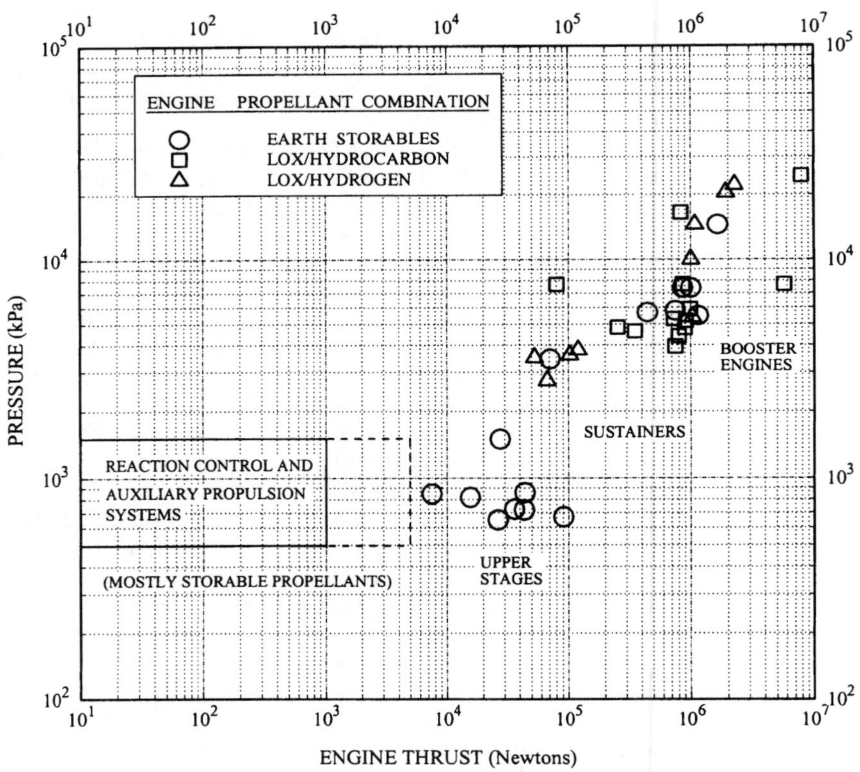

Fig. 1   Liquid rocket engine design applications.

1. Boosters

Boosters are the largest and highest pressure engines. Their high thrust is required to accelerate the gross liftoff weight of the entire vehicle, including sustainer and/or upper stages as well as payload, into orbit. The high pressure is required to accelerate the nozzle exhaust gases against the atmospheric pressure to high Mach numbers, to maximize specific impulse performance. Because the tank volumes are very large, these boosters are pump-fed from lightweight, low-pressure propellant tanks just sufficiently pressurized to suppress pump cavitation.

2. Sustainers or Second-Stage Devices

Sustainers or second-stage vehicle propulsion devices effectively operate outside of the Earth's atmosphere. They can achieve high performance by merely expanding to very high nozzle exit to throat area ratios. They do not need to operate at chamber pressures as high as for boosters and can be pump fed or operated from pressurized propellant tanks.

3. Upper-Stage Devices

Upper stages are smaller versions of sustainers. Their propellant mass fractions relative to total stage weight are less than for lower stages. To save both the weight and costs of a pumping system, they are usually fed from pressurized tanks.

4. Reaction Control and Satellite Propulsion Systems

Reaction control systems or satellite propulsion engines are the smallest rocket thrusters available. These thrusters provide in-flight vehicle guidance or provide in-orbit satellite station keeping functions. They are virtually always pressure fed and operate at low chamber pressures.

B. Propellant Type

Commonly used propellants can be categorized into three major families that differ in their relative volatilities.

1. Cryogenics

Cryogenics remain in liquid form only if kept subcooled below ambient temperature. The most common cryogenic propellant combination is liquid oxygen (LO_2) and liquid hydrogen (LH_2). This pair has the advantage of high specific impulse performance and is environmentally nonpolluting. In most cases the hydrogen, which is an excellent coolant, is used to regeneratively cool the combustion chamber and nozzle. Thus, it is usually in a gaseous state by the time it is injected into the combustion system.

2. Liquid Oxygen/Hydrocarbon

The most commonly used hydrocarbon is kerosene due to its ready availability, low propellant cost, ease of storability and moderate bulk density, which reduces fuel tank structural weight as compared to liquid hydrogen. Liquid oxygen is highly volatile as compared to kerosene. Hence, the combustion chamber length must be designed for the vaporization limited fuel rather than for the oxidizer. This propellant combination offers challenges in the balancing of the design requirements for thermal cooling, high-performance efficiency, and combustion stability, and could benefit greatly from a systematic combustion analysis approach. Some research test firings have been conducted with liquefied natural gas (LNG), liquid methane (CH_4) and liquid propane (C_3H_8). All of the latter are typically stored at temperatures approaching that of LO_2, and are sometimes referred to as space storables.

3. Earth Storable

The oxidizer used with Earth storables is usually a nitric acid (HNO_3) mixture or other oxide of nitrogen such as nitrogen tetroxide (N_2O_4). The most common Earth-storable fuels are amines, a member of the hydrazine (N_2H_4) family or its derivatives. These propellants are liquids at ambient temperature and pressure and are usually hypergolic on contact. Thus, a separate ignition system is not required.

C. Engine Cycle or Feed System

The engine cycle dictates the propellant injection system that the combustion device and injector designers must contend with because it determines propellant states at various component interfaces. A more detailed discussion of liquid rocket engine cycles will be presented in Chapter 19.

1. Pressurized Propellant Tanks

The pressurized propellant tank provides the simplest feed system. The typical tank pressurant is gaseous helium or nitrogen. Helium is usually used in flight because of its lighter weight, whereas nitrogen is usually substituted during use in ground test facilities because of its ready availability and low cost. Ground test facilities capable of operating at high pump-fed system pressures are usually used for initial combustion device development testing to allow parallel development of combustion devices and turbopumps, but for the purposes of the present discussion the injectors will be referred to as pump-fed designs.

To minimize tank structural weight, pressure-fed flight tank pressures are kept low and both combustion device operating pressures and feed system pressure drops are also minimized. To further minimize pressurant storage bottle and gas weight, the propellant tank pressure may only be regulated over the initial portion of its mission and permitted to operate in a blowdown mode to its propellant exhaustion. This requires the combustion device to operate in a throttled (reduced thrust) mode late in its mission.

2. Gas Generator Cycle

The gas generator cycle is the simplest form of the pump-fed engine cycles. A small portion of the main engine propellant is bypassed and burned in a separate combustion device operating at low combustion gas temperature in order to power a turbine that, in turn, drives the propellant pumps. Because the turbine exhaust gases are dumped overboard at low temperature and low pressure, the lower gas generator exhaust gas performance reduces the overall engine system performance. Because the gas generator mass flowrate fraction must increase linearly with the required turbine horsepower, a tradeoff has to be made between increasing the main combustion device performance by increasing operating pressure against an increasing gas generator engine cycle loss. These systems usually optimize performance at moderate pressures.

3. Staged Combustion Cycle

The staged combustion cycle flows all of one propellant and a small fraction of the other to keep combustion gas temperature low enough to permit turbine drive. Then it injects the remaining propellant downstream of the turbine to recover maximum engine performance at high gas temperature. The first combustion device, referred to as a *preburner*, has design criteria similar to those of a gas generator, except that it is usually larger in size. The larger size allows the preburner to accommodate a higher mass flowrate and operate at considerably higher pressure.

The turbine pressure ratio is in series with the main combustion device rather than in parallel as in the gas generator cycle. The turbine mass flowrate available in a staged combustion cycle engine greatly exceeds the mass flowrate available in a gas generator cycle engine. Hence, it can thermodynamically optimize performance at significantly higher operating pressures than could a gas generator cycle engine.

In actual practice, the staged combustion operating pressure is limited from an engine reliability standpoint to a thermal limit to which the combustion device can be cooled. The main combustor in a staged combustion cycle is a gas/liquid injection system, since one propellant circuit has been prevaporized before entering the turbine.

4. Expander Cycle

An expander cycle is somewhat similar to a staged combustion cycle in that no turbine drive gases are exhausted overboard. It has the further simplification that it does not require either a preburner or a gas generator. The turbine drive gases are heated while regeneratively cooling the main combustion chamber and nozzle. In practice, the expander cycle has been developed only for the oxygen/hydrogen propellant combination. Only hydrogen can provide adequate cooling to the regenerative main combustion chamber and still be heated sufficiently to drive the turbine.

Whereas hydrogen is an excellent combustion chamber coolant and delivers high combustion performance, it presents a serious challenge for the fuel turbopump designer. Its low density requires high pump speeds and/or multiple pump stages to raise the pump discharge pressure. This difficult-to-achieve hydrogen pressure is in turn subject to chamber and nozzle coolant pressure losses. It must supply the required turbine pressure ratio and then still have sufficient pressure remaining to meter the flow into the injector and provide chamber pressure. Expander cycle engines therefore operate at much lower pressures than either staged combustion or gas generator cycle engines. This is not a disadvantage for an upper-stage engine operating in space, but it is a serious limitation for a booster engine. Expander cycles also optimize for lower thrust level engines, which have a more favorable exposed heating surface area to engine flowrate ratio, which also makes it ideal for an upper-stage application.

III.   Common Combustion Device Development Risks

The different types of combustion device applications discussed in Section II have different degrees of technical risks. All combustion devices are potentially susceptible, however, to the following primary development problems.

A.   Combustion Instability

Combustion instability has been the single most significant combustion device development problem since the beginning of liquid propellant rockets. An early recognition of the importance and magnitude of the technical concern in the United States is indicated by the broad range of investigators contributing to a systematic sharing of viewpoints compiled in 1972.[5] Likewise, demonstrating that combustion instability is still a major development risk within the propulsion community, the entire First International Symposium on Liquid Rocket Propulsion was devoted to this single subject.[†]

The deadliest form of combustion instability is usually referred to as high frequency combustion instability which is characterized by a coupling between the propellant burning rate with one or more of the transverse combustion chamber acoustic modes. This causes a substantial increase in the forward combustion zone heat flux, and the usual result of a high-frequency combustion instability encounter is immediate catastrophic failure attributed to a burnout of the combustion chamber and/or injector. This problem is most serious for large booster engines and decreases in severity with diminishing engine size. The problem is most common for liquid oxidizer/liquid fuel injectors utilizing the LO_2/hydrocarbon or Earth-storable propellant combinations. It is a lesser problem for the LO_2/H_2 propellant combination, gas/liquid injectors and for small thrusters. Acoustic coupling also occurs with the combustion chamber longitudinal modes between the injector face and nozzle throat plane and is called longitudinal combustion instability, but these modes are generally less damaging.

Low-frequency combustion instability, also called chugging, is characterized by a coupling of the propellant burning rate with the hydraulics of the propellant feed system. This problem is aggravated by low injector pressure drop and selection of injection elements with long atomization and/or vaporization combustion time lags. The combustion device may not be at risk of catastrophic failure as a result of low-frequency combustion instability, but sensitive payloads may incur structural failure, particularly if they possess natural frequencies that could resonate with the chug frequency.

Gas/liquid injection systems could be susceptible to an additional risk from either sufficiently high amplitude feed system coupled instabilities, or longitudinal

---

[†]The First Symposium was held at the Propulsion Engineering Research Center at the Pennsylvania State University, University Park, PA, 18–20 January 1993. The technical papers presented at this symposium were compiled as Vol. 169 of the *AIAA Progress in Astronautics and Aeronautics* series.[6]

acoustic mode combustion instabilities. The rising pressure at the injector face could cause compressibility of the gaseous propellants to result in flow reversal of the combustion gases into the injector manifolds. If the backflowing combustion gases also entrain unvaporized liquid droplets into the opposite gas manifold, they could result in manifold detonation and its structural failure. There have also been occasions when supposedly "non-damaging" forms of combustion instabilities such as longitudinal or chug mode combustion instability perturbations have triggered the *fatal* transverse acoustic mode. Thus, all forms of combustion instabilities should be avoided, even if they appear to be doing no harm at first assessment.

B. Combustion Chamber Overheating and Burnout

Considerable progress has been made in combustion chamber heat flux and wall-cooling predictive technology. This topic will be covered in more detail in Chapters 14 and 15. To achieve higher performance, especially for low-altitude booster engine applications, the first reaction is to increase chamber pressure to expand the exhaust gases to a higher nozzle exit area ratio and thus achieve higher Mach number. Higher heat flux accompanies higher operating pressures.

For a given regeneratively cooled combustion chamber material and wall thickness, a higher heat flux increases the wall temperature differential between the inner coolant wall and outer hot gas wall. The wall thickness can be reduced to limit the maximum hot gas wall temperature. However, the walls must also be designed to withstand a maximum design wall pressure differential, which sometimes occurs during transients. This might be achieved by reducing the cooled wall span, which in turn reduces the coolant passage hydraulic diameter and increases the coolant pressure drop.

From the injector design standpoint, one desirable solution would be to reduce the combustion gas temperature immediately adjacent to the walls by incorporating lower mixture ratio injection elements or pure fuel film cooling injection orifices. Making the cool zone wider than absolutely necessary will reduce engine performance inversely with the engine throat diameter. The combustion chamber thermal design margin is determined by the hottest local streak temperature irrespective of the average gas temperature. Regenerative coolant passage burnout resulting in internal leakage is usually self limiting and seldom results in immediate catastrophic failure. It will, however, cause a loss of engine performance and may cause off-design engine mixture ratio operation to deplete the fuel tank before all of the oxidizer is consumed, compromising mission payload objectives.

An easier solution for sustainer and upper-stage engines is simply to operate at lower chamber pressure and thus reduce heat flux. The only performance penalty of low chamber pressure for an engine in vacuum is the possibility of a minor increase in the nozzle boundary layer and recombination kinetics performance losses.

Reaction control systems and satellite propulsion devices have insufficient propellant consumption rate to regeneratively cool their combustion chambers. Furthermore, these engines are frequently required to fire short, repeated pulses and require a rapid response. Thus, this class of thruster typically uses Earth-storable propellants and depends upon fuel film cooling to provide the required thermal margin.

Fuel film cooling thermal and performance characteristics vary widely, depending upon the type of fuel being used. Hydrogen fuel film cooling is primarily effective when it reduces the local recovery temperature near the design wall temperature. Its high heat capacity still results in relatively high heat flux at moderate temperature ranges. On the other hand, its low molecular weight results in high film coolant specific impulse and low cooling performance losses. The amine fuels undergo monopropellant decomposition and provide relatively stable and predictable heat flux reduction and performance reduction. Hydrocarbon fuels provide very great cooling capacity because of their highly endothermic decomposition, but performance degradation is also high. The best thermal-to-performance trade occurs when the nozzle throat plane recovery temperature is approximately half the stoichiometric temperature. Some trial and error is still required to establish the optimum percentage of fuel film cooling.

C. Injector Face Erosion

Injector face erosion is a potentially mission-compromising failure mode for high-pressure engines, if it results in burn-through into the injector manifold. Such an occurrence will result in loss of engine performance, off-mixture-ratio operation, and premature depletion of one propellant tank before the other, resulting in possible significant reduction in payload terminal velocity.

Face erosion in low-pressure engines is usually limited to superficial erosion that stabilizes after some reduction in local face-plate thickness, absent any combustion instability.

Injector face heat flux models are relatively immature compared with combustion chamber and nozzle thermal models. What has been observed is that high injection velocities tend to aggravate the face heat flux by increasing the recirculation strength. Injector face erosion can be particularly troublesome for the oxygen/hydrogen propellant combination or gas/liquid injection systems if raw oxidizer-rich sprays are allowed to recirculate back to the injector face.

D. Low Thrust Chamber Assembly Performance

Everyone recognizes the importance of high performance. It is an emotional issue. Low combustion-device performance is readily measurable and highly apparent to everyone. Thrust-based specific impulse I_{sp} measurements are most accurate. Chamber-pressure–based characteristic exhaust velocity C^* measurements, although less accurate, can be made using less sophisticated and cheaper test facilities.

The typical reaction to a low performing combustion device by a novice injector designer is to replace the injector with another having more smaller injection orifices, in the belief that more complete combustion will yield higher performance. More often than not, however, the modification can result in combustion instability.

A knowledgeable injector designer understands that low performance can be attributable to any one or more of the following three causes: 1) nonuniform oxidizer to fuel injection distribution across the injector face, 2) inadequate (too large) atomization resulting in incomplete droplet vaporization, or 3) incomplete mixing of fully vaporized combustion products.

E. Unsafe Transients

Relative to the total range of possible combustion device failure modes, too much time and too many resources are spent on studying the steady-state design point, and too little recognition is paid to the possible propellant-type and engine cycle transient operational risks:

1) Propellant-type transient risks are as follows. Liquid/liquid Earth-storable engines have the simplest start transients. Liquid oxygen/hydrocarbons are of intermediate risk. For example, if hydrocarbon contamination were to occur within the oxygen manifold during an engine shutdown transient from a previous test, the subsequent test start transient could be at risk of having a detonation in the oxygen manifold. Cryogenic engines have the most complex start transients because of their severe thermal chill down constraints in addition to the usual pressure variation considerations.

2) Engine cycle transient risks are rated as follows. Pressurized tank feed systems are easiest to operate. The gas-generator cycle has the simplest transient among the pump-fed systems. The staged combustion cycle is considerably more complex. The expander cycle is most difficult to start because of its low turbine power margin and the deep throttling of a low-pressure drop-feed system.

Too often, excessive importance is placed on rapidly achieving steady-state pressures and too little attention is paid to understanding the physics of the slow temperature transient. This is especially true for cryogenic propellants and gas/liquid systems.

Possible negative effects attributable to improper transients are 1) nonignition or nonrestart of cryogenic upper stages (more flight failures have resulted from these than from any other failure mode), 2) delayed ignition, 3) hard starts, 4) combustion gas reversal causing fire within injector manifolds, 5) engine vibration by feed system coupling during deep throttle operation, 6) gas generator or preburner temperature spikes to turbine blades, and 7) rapid, cold cryogenic hydrogen quenching of hot turbine blades and/or hot combustion chamber wall. Some failures can result in immediate termination and mission loss whereas others prematurely limit component cycle life.

IV. Injection System Design Considerations

To simplify this discussion, it will be assumed that the vehicle and system level trades have already been completed. Assume that the combustion device and injector designers have been given the following design requirements: 1) propellant combination, 2) engine thrust, 3) mixture ratio, 4) pressurized tank or pump discharge pressures, and 5) combustion device length and diameter envelopes. The following important design parameters must be taken into consideration and preliminary baseline values (subject to continuing review) should be established. Additionally, the injection element type and injector pattern selection will be discussed separately in Section VI.

A. Engine Pressure Schedule

The total engine pressure available must be allocated between 1) combustion chamber pressure, which is important to maximize for booster applications to obtain high performance, 2) regenerative coolant pressure drop, if applicable, which must be adequate for thermal margin, 3) injector element pressure drop, which must be chug stable at lowest expected flowrates, and 4) propellant distribution system, including propellant lines, valves, and injector manifolding.

B. Nozzle Expansion Ratio

The general procedure is to maximize the nozzle expansion ratio to fill the length envelope for an upper-stage combustion device. Check to verify that payload performance advantage over a lower area ratio, shorter length nozzle merits the weight increase and added complexity. For a booster nozzle, optimize flight trajectory performance from liftoff to second-stage separation. However, the nozzle must also be able to withstand asymmetric separation induced side loads during sea-level firing and during pump-fed start transients.

C. Contraction Ratio

The contraction ratio is the area of subsonic combustion chamber to nozzle throat. Most combustion device contraction ratios are in the 2–4 range. Liquid/liquid boosters are usually within the lower range; staged combustion cycle main injectors and gas/liquid injectors are in the upper range. Fuel-film–cooled reaction control systems and satellite propulsion engines typically have high contraction ratios. The Rayleigh stagnation pressure loss resulting from heat addition at finite Mach number increases rapidly at contraction ratios less than 2.

D. Chamber Length

The chamber length L' from the injector face to the nozzle throat plane needs to be selected together with consideration for probable atomized drop size to achieve high (not necessarily complete) propellant droplet vaporization above the nozzle throat.

E. Injection Element and Pattern

The injection element type and injector pattern selection will be discussed separately in Sec. V.

V. Critical Combustion Processes

Sections II and III describe various liquid propellant rocket engine applications and their combustion device development problems. This section will describe primary physical mechanisms through which the injector designer can

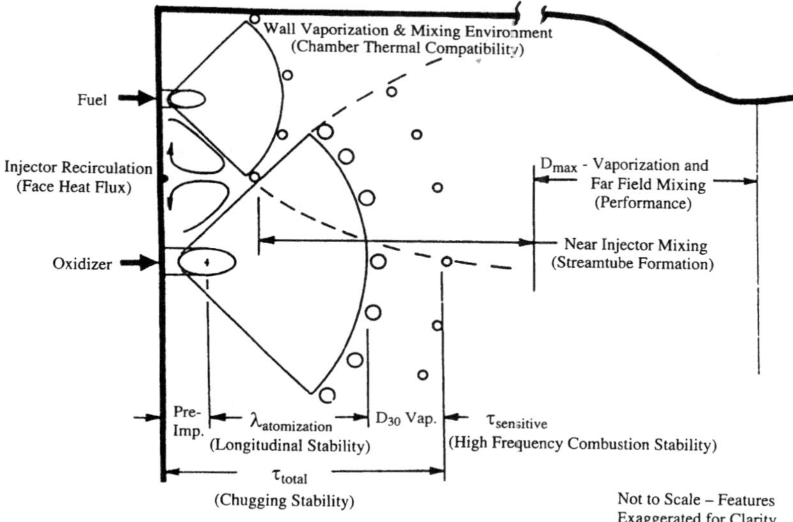

Fig. 2   Combustion processes involved in liquid rocket combustion devices.

establish control to solve these development problems. A schematic showing some of these combustion processes is given in Fig. 2.

A.   Injector Manifold Distribution

The starting point of any injector design is proper distribution of the fuel and oxidizer across the injector face *where you want it!* This requirement is so basic that it should be obvious, but its achievement is often taken for granted and its importance is often overlooked. Uniform mixture ratio distribution across the injector core elements will maximize performance. On the other hand, a uniform mixture ratio at the combustion chamber wall may result in excessive heat flux, which could cause thermal failure or require excessive regenerative coolant circuit pressure drop in a high-pressure engine. In that case, either fuel-film cooling or a barrier mixture ratio bias may be helpful to reduce wall heat flux without reducing chamber pressure. A mass weighted streamtube analysis can provide a quantitative estimate of the effect of mixture ratio maldistribution upon performance penalty. It can account for both intentional cooling bias and unintentional maldistribution performance losses.

Compared with the cost of injector redesign and retesting necessitated by either chamber thermal failure or a disappointingly low injector performance due to injection maldistribution, it would seem prudent to perform simple cold flow hydraulic distribution testing of the injector manifold design before committing the injector design to a specific injector pattern. Injector manifold distribution represents a necessary but not sufficient criterion for design success. That is, a nonuniform injection manifold distribution can present later development problems, but a uniform manifold distribution is only one of the many design requirements for success.

B. Injector Spray Atomization

Many liquid rocket propulsion engineers think that atomization refers to only droplet diameter as it affects subsequent propellant vaporization and performance. That barely scratches the surface of its importance. In fact, it is only the largest droplets that may exhaust through the nozzle throat without being vaporized and degrade vaporization performance. These maximum diameter droplets represent only the largest 10–20% of the total mass distribution.

Everyone acknowledges the critical importance of high-frequency combustion instability. The sensitive time lag is usually approximated by combustion stability analysts with the volume number mean diameter D_{30}, which typically defines the smallest 20% of the cumulative droplet mass distribution. Other drop sizes typically mentioned in the atomization literature refer to the Sauter mean diameter D_{32} and mass median diameter (the diameter corresponding to the median droplet mass). It is of less importance to the injector designer to force a single mean diameter and droplet distribution function to describe the entire spray than it is to understand the mass distributions of small, intermediate, and large drop sizes required by the various combustion process analysis models.

Another critically important consideration that few atomization investigators have recognized is the systematic study of the spatial distribution of spray atomization from the injector face or from the point of jet impingement. The reason spatial atomization distribution is so important to the injector designer is that this breakup distance divided by the injection velocity represents a significant fraction of the combustion dead time. This time lag is needed by the combustion stability analyst to predict the low-frequency feed system or chug stability margin in which a pressure-fed thruster may be required to operate at the end of its tank pressurization blowdown cycle or the intermediate operating point that all pump-fed engines must endure during the start transient before bootstrapping up to full throttle.

An accurate prediction of the *differential* breakup distances from the injector face between the oxidizer and fuel spray fans in a liquid/liquid Earth storable or LO_2/hydrocarbon injector is also critically important to the successful injector designer and thermal analyst. This is especially important for injection elements aligned adjacent to the combustion chamber wall. The atomization distance differential represents whether the fuel or oxidizer spray has a head start, and the relative propellant volatilities determine whether the real vaporized mixture ratio is more fuel rich or more oxidizer rich than the injection mixture ratio at the injector face. The local axial distribution of vaporized wall mixture ratio strongly influences the chamber heat flux and its cooling margin.

Atomization can be approached in a number of ways depending on the resources and preferences of the investigators. It can be measured experimentally and correlated empirically during either cold flow or hot fire testing, as will be described further in Chapters 2–5. It can also be modeled analytically based on first principle theories or inferred from previous experience with similar designs.

To fully reap the benefits of atomization, not only for performance prediction but also for both high-frequency and low-frequency combustion stability analyses as well as for combustion chamber wall and injector face recirculation thermal analyses, a determination of spatial atomization breakup distances as well as drop size distributions is required.

C.   Propellant Droplet Vaporization

As early as the mid-1950s, Priem and Heidmann of the NASA Lewis Research Center had concluded that droplet vaporization could be the rate controlling mechanism in the liquid propellant combustion process.[7] Numerous vaporization and spray combustion models are now available,[5,5,8] and recent advances in this field are discussed in Chapters 7–9.

D.   Bipropellant Mixing

Uniform mixing is essential to the achievement of maximum specific impulse performance. It is also required in gas generators and preburners to achieve uniform turbine inlet gas temperatures and minimize hot streaks, which limit turbine life. On the other hand, it is desirable to minimize mixing between the core and barrier combustion zones to maximize engine performance while minimizing combustion chamber and nozzle heat flux.

Low molecular weight propellant species such as hydrogen have high diffusivity and mix readily. Conversely, high molecular weight propellants such as heavy hydrocarbons mix very slowly. Heavy hydrocarbons have the further disadvantage that they can build up an insulating layer of cooler fuel vapors surrounding the droplet sufficient to retard further droplet vaporization.

Hypergolic propellants, which react spontaneously on contact, can undergo reactive stream separation, also known as blow apart, which retards unlike liquid/liquid propellant mixing. Likewise, gas/gas injectors are notorious for their low mixing efficiencies because of rapid combustion on their mixing interface. Gas/liquid injectors mix not much differently than liquid/liquid systems.

Rupe of the NASA Jet Propulsion Laboratory was one of the earliest investigators to recognize the importance of uniform liquid phase mixing as it related to injection element design parameters, propellant properties, and injection operating conditions.[9] In essence, he reported that optimum unlike mixing could be approached when the propellant jet diameters and injection momentum ratio approached unity.

VI.   Candidate Injectors for Liquid Rocket Applications

References 2 and 10 describe various injection element types that could be beneficially applied to liquid rocket injector designs. Their spray characteristics are depicted schematically in Fig. 3. A cursory discussion of some significant characteristics and some examples of their possible advantages and disadvantages follow.

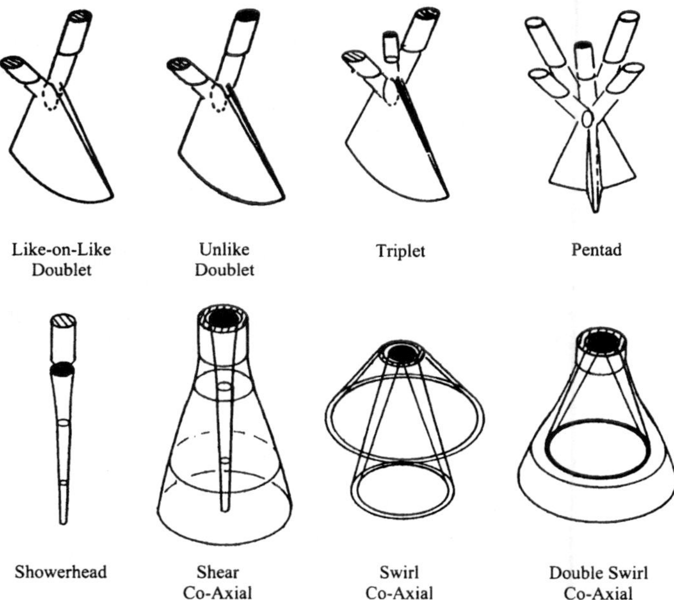

Fig. 3   Schematics of injection element spray patterns.

A.   Coaxial Jet Injectors

This is the single most common element type used for oxygen/hydrogen injectors, which come in two varieties, the shear coaxial and swirl coaxial. Both usually position the hydrogen in the outer annulus and inject the oxygen in the central jet. Because most oxygen/hydrogen thrust chambers operate in the 5–7 mass mixture ratio range, the shear coaxial requires a proportionately higher fuel-to-oxygen injection velocity ratio so that sufficient injection momentum can adequately atomize and mix the liquid oxygen jet.

When there is less hydrogen injection momentum available to adequately shear the oxygen stream, an oxidizer swirl pattern, which can be induced either by inserting a mechanical swirl device to impart rotation or by tangential injection, can help self-atomize the oxygen spray fan either with or without the added assistance of the hydrogen jet. The hydrogen is usually pregasified by regenerative heating in the combustion chamber in a gas generator or expander engine cycle, or precombusted within the pre-burner of a staged combustion cycle engine. Thus, the local vaporized mixture ratio asymptotically approaches the design mixture ratio from the thermally benign fuel-rich side, which benefits both injector face and combustion chamber thermal compatibility. Careful attention must be paid if swirl coaxial injection elements are positioned too close to the chamber wall. Liquid oxygen droplet wall impingement can cause local overheating on the forward chamber wall.

Shear coaxial elements, on the other hand, provide a thermally benign environment on the forward chamber wall. However, this type of injector can cause thermally adverse conditions in the nozzle convergent section if the LO_2 droplets are not completely vaporized by the end of the cylindrical chamber and impinge, shatter, and combust on the convergent throat section. In general, a row of finer elements adjacent to the chamber wall provide better compatibility and higher performance potential. A more detailed discussion of coaxial jet injector atomization will follow in Chapter 3.

B. Impinging Jet Injectors

Many variations on the impinging jet injectors shown in Fig. 3 are used for liquid rocket combustion devices. Some major classifications are like-on-like doublets, quadlet elements, unlike doublets, and unlike triplets.

The like-on-like doublet was one of the earliest injection element concepts used for liquid rocket injectors. Its popularity was generally attributable to its stable combustion characteristics while delivering moderate performance. The like-on-like doublet is composed of both self-impinging fuel doublets and self-impinging oxidizer doublets. The quantities of fuel pairs and oxidizer pairs need not be equal. A functional advantage can be gained by designing more impinging pairs of the less volatile propellant.

Quadlet elements are like-doublet pairs that have been canted toward each other to induce improved unlike propellant mixing. For the same number of impinging pairs and comparable atomization and vaporization efficiencies as like-on-like doublets, quadlet injectors tend to deliver higher performance in mixing limited injectors.

Unlike doublets impinge a single fuel jet upon a single oxidizer jet. This injection element type works best for propellant combinations that have nearly equal fuel and oxidizer injection orifice areas and that also have nearly equal injection momentum ratios.

Unlike triplets impinge two jets of one propellant upon a single jet of the other. Two opposing fuel jets impinging upon an oxidizer is called a F-O-F triplet; whereas two oxidizers impinging upon a single fuel is called an O-F-O triplet. Most liquid/liquid propellant combinations other than oxygen/hydrogen require finer atomization of the less volatile fuel. The F-O-F triplet tends to produce finer fuel droplet atomization for a given total injector element quantity. However, since most propellant injection combinations have higher oxidizer injection momentum ratios, the O-F-O triplet produces better unlike propellant mixing uniformity. The choice between these two triplet orientations depends on whether the propellant combination is more likely to be fuel vaporization limited or mixing performance limited. Special provisions for wall thermal compatibility may be required if the O-F-O triplet is the core element of choice. The *unlike pentad* injector is a variation of the triplet elements except that it impinges 4 on 1 instead of 2 on 1.

Unlike impinging elements tend to produce finer atomization than like impinging elements of similar orifice diameter and pressure drop. They are generally higher performing, but also less combustion stable. A coarser unlike impinging element pattern will exist that produces performance efficiency and combustion stability characteristics comparable to those of a finer like impinging injector. A coarser pattern will probably be cheaper to fabricate but will also provide wider thermal streaks.

C.   Parallel Jet (Showerhead) Injectors

The showerhead injection element is seldom used as a thrust-producing injector because of its poor atomization and mixing characteristics; however, for these very reasons, it is often used as a barrier fuel-film cooling element. It is advantageous to use when the forward chamber can be adequately regeneratively cooled but the throat heat flux is excessive for thermal reliability margin or would otherwise require excessive coolant pressure drop. The coolant jet can be either injected axially parallel to the chamber wall, with a slight impingement angle upon the wall, or with a tangential swirl component for more uniform front end coverage.

D.   Injector Design Synthesis

Historically, the selection criteria for picking a particular injection element to design and develop have been subjective. Either injector designers or liquid rocket companies have favored certain element types and have used them for all applications, disregarding the Rocket Application Design Requirements discussed in Section II or the Common Combustion Device Development Risks in Section III. These choices may have been based on previously successful design experiences, prior design familiarity, or other subjective design considerations.

The analytical design approach adopted at Aerojet since 1966 has been based on the design considerations described in Sections II and III. Atomization breakup distances from the injector face are selected as a design requirement together with a nominal design point pressure drop and injection velocity to determine allowable "combustion dead time" ranges to satisfy feed system combustion stability for transients and required throttle ranges, if applicable.

Characteristic drop sizes for the volume number mean D_{30} can be used to predict allowable sensitive time lags or characteristic high-frequency combustion stability gain relative to the combustion chamber transverse resonance frequencies and combustion damping device margins. Spatial combustion profiles are evaluated or modified to ensure thermal heat flux compatibility at hardware surfaces compared with regenerative cooling flux and wall thermal conductivities. The maximum high end droplet diameters are analyzed parametrically to assess acceptable performance losses caused by unvaporized droplets exhausting through the nozzle throat plane for given chamber lengths. The droplet mass fractions and species (fuel or oxidizer) impinging upon the convergent throat are used to refine the throat heat flux prediction. Note that the average drop size that is the primary focal point of most atomization emphasis was not explicitly mentioned in these functional injector development process models.

The liquid phase or gas/liquid (Rupe) mixing efficiency parameter Em can be used if known to estimate streamtube mixing performance based on distributed mass and mixture ratio distributions.

None of the foregoing design criteria have made any reference thus far to a particular element type. Only *after* the design requirements have been quantitatively defined does the injector designer attempt to evaluate the repertory of available injection element types, orifice diameters, injection velocities, impingement angles, and other design variables to synthesize the injector design that has the highest probability of fulfilling the aforementioned design objectives.

VII. Conclusions and Recommendations

The previous *art of injector design* is maturing and merging with the more systematic *science of combustion device analysis*. The development of injector technology can be based on observation, correlation, experimentation, and ultimately analytical modeling based on basic engineering principles. This methodology is more systematic and far superior to the historical injector design process of trial and error or blindly copying past successes.

The benefit of such an approach is to be able to rank candidate design concepts for relative probability of success or technical risk in all of the important combustion device design requirements and combustion process development risk categories before committing to an engine development program. Even if a single analytical design concept cannot be developed to predict the simultaneous satisfaction of all requirements, a series of risk mitigation, key enabling technologies can be identified for early resolution. Lower cost subscale or laboratory experimentation to demonstrate proof of principle, critical instrumentation requirements, and design discriminating test plans can be developed based on the physical insight provided by these analyses.

The reason this overall procedure may appear intimidating at first is because the development of a large, high-pressure, liquid propellant combustion device itself is a formidable task with many inherent risks. Injector design is a multiple jeopardy problem. There are many individual reasons that any design may become unacceptable; there are considerably fewer combinations of injector designs that satisfy the many demanding design requirements and often contradictory design trades that must be made. *However, the successful seeker will be richly rewarded by long-term cost and schedule benefits.*

References

[1]*Technology Week*, 10th Annual World Missile and Space Encyclopedia Issue, Vol. 19, No. 14, July 25, 1966.

[2]"Liquid Rocket Engine Injectors," NASA Space Vehicle Design Criteria Monograph SP-8089, Cleveland, OH, 1976.

[3]Rowe, J. R., "Liquid Rocket Engines," (A Potential Liquid Rocket Engine Users' Guide), Aerojet Liquid Rocket Co., Sacramento, CA, March 1975.

[4]Isakowitz, S. J., *International Reference Guide to Space Launch Systems*, AIAA, Washington, DC, 1991.

[5]Harrje, D. T., and Reardon, F. H. (eds.), *Liquid Propellant Rocket Combustion Instability*, NASA SP-194, 1972.

[6]Yang, V., and Anderson, W. E. (eds.), *Liquid Rocket Engine Combustion Instability*, Vol. 169, Progress in Astronautics and Aeronautics, AIAA, Washington, DC, 1995.

[7]Priem, R. J., and Heidmann, M. F., "Propellant Vaporization as a Design Criterion for Rocket Engine Combustion Chambers," NASA-TR R-67, 1960.

[8]Cramer, F. B., and Baker, P. D. (eds.), "Combustion Processes in a Bi-Propellant Liquid Rocket Engine (A Critical Review)," Jet Propulsion Lab. Rept. 900-2, Contract NAS7-100, Pasadena, CA, 1967.

[9]Rupe, J. H., "Correlation Between the Dynamic Properties of a Pair of Impinging Streams and the Uniformity of Mixture Ratio Distribution in the Resulting Spray," Jet Propulsion Lab., California Inst. of Technology, Progress Rept. 20-209, Pasadena, CA, 1956.

[10]Penner, S. S., *Chemistry Problems in Jet Propulsion*, Pergamon, New York, 1957, pp. 360–362.

Chapter 2

Design and Dynamics of Jet and Swirl Injectors

Vladimir Bazarov*
Moscow Aviation Institute, Moscow, Russia

and

Vigor Yang[†] and Puneesh Puri[‡]
Pennsylvania State University, University Park, Pennsylvania

Nomenclature

A = geometric characteristic parameter of swirl injector, $A \equiv \bar{R}_{in}/\bar{A}_{in}$; area
a = nondimensional parameter of swirl injector, defined in Eq. (75)
b = nondimensional parameter of swirl injector, defined in Eq. (71)
C = coefficient of nozzle opening, $C \equiv \bar{R}_{in}$
C^* = characteristic velocity
c = specific heat
D = diameter of injector element
d = diameter of substance (drop, spray)
f = functional symbol; frequency
h = liquid film thickness
Im = imaginary part of complex variable
K = momentum-loss coefficient; O/F ratio
k = ratio of specific heats, C_p/C_v
l = length of injector element
M = Mach number
m = mass
$\dot{m}$ = mass flow rate

---

*Professor and Head, Dynamic Processes Division of the Rocket Engines Chair. Member AIAA.
[†]Distinguished Professor, Department of Mechanical Engineering. Fellow AIAA.
[‡]Graduate Student, Department of Mechanical Engineering.

n = number of passages
p = pressure
Δp = pressure drop
Q = volumetric flow rate
$q\,(\lambda)$ = gasdynamic function; tangent of nozzle surface inclination to injector axis
R = gas constant; radius of injector element
Re = Reynolds number
r = radius of jet element; radial location of liquid film
S = surface
Sh = Strouhal number
T = temperature, K
t = time; spacing between injectors
U = velocity
V = volume
x, y, z = coordinates
α = tilt angle of inlet passage; spreading angle of liquid spray
β = tilt angle of wall
Δ = increment
δ = wall thickness, clearance
ε = coefficient of jet contraction
η = pressure-loss coefficient; efficiency; acoustic admittance function
$\aleph$ = polytropic exponent
λ = velocity coefficient; drag coefficient; wavelength
μ = mass flow coefficient
μ_d = dynamic viscosity
ν = kinematic viscosity
Π = response or transfer function
ξ = hydraulic-loss coefficient; fluctuation of liquid film thickness
π = nozzle expansion ratio
ρ = density
σ = surface tension
τ = time interval
Φ = phase angle of individual process
φ = coefficient of passage fullness, i.e., fractional area occupied by liquid in nozzle
Ψ = phase angle of element in the assembly
Ω = amplitude of liquid surface wave
ω = radian frequency

Subscripts

a = axial
c = combustion chamber
eq = equivalent
e = nozzle exit
exp = experimental
ext = external

f = propellant feed system
fl = flow
fr = friction
g = gas
gg = gas generator
gl = gas-liquid
id = ideal
i = injector
in = inlet
j = jet
k = head end of vortex chamber
l = liquid
m = liquid vortex
mix = mixing
N, n = nozzle
out = outlet
r = radial
s = vortex chamber
sp = spray
sw = swirl
T = inlet passage
t = tangential spacing between injectors
th = nozzle throat
u = circumferential
w = wall, wave
v = saturated vapor
vc = vortex chamber
Σ = total
0 = initial conditions
1 = exit conditions
∞ = infinite value

Superscripts

$-$ = dimensionless parameter
$'$ = pulsation component

I. Introduction

MIXTURE formation is one of the most important processes in liquid rocket combustion devices because it determines combustion efficiency, stability, and heat transfer characteristics. This process is implemented through the use of propellant injectors, which not only accomplish their main missions of propellant atomization and combustible mixture formation, but also represent elements of an engine as a complex dynamic system operating under various conditions. Any change in engine operating conditions (such as startup, thrust variation, and shutdown) or flow paths in the feed line and combustion chamber (such as turbulence and pulsations) may lead to a drastically different injector behavior.

The operating conditions of propellant injectors are complicated, and as such the types of injectors and injector assemblies as a whole are numerous. The selection and design of a specific injector depends on the situations of concern and must fulfill the following basic requirements:

1) To provide high combustion efficiency, injectors should ensure high-quality propellant mixing with uniform distributions of mixture ratio and flow intensity over the combustion chamber as much as possible. Minimum consumption of energy is required for propellant atomization and mixing.

2) To protect combustor walls against excessive thermal loading, injectors should provide nonuniform distributions of mixture ratio and flow intensity in regions of concern.

3) To suppress combustion instability in the chamber and to achieve staged combustion, injectors should achieve prescribed distributions of atomization dispersivity, mixture ratio, and flow intensity in the mixture-formation zone. For gas injectors, provisions should be made to remove acoustic energy from the chamber.

4) To suppress flow instabilities, the acoustic conductivity of an injector should be minimized, with a smallest possible response to disturbances arising from variations of the flow rate and other parameters.

5) For more complicated situations such as pulse-triggered instability and thrust transients, injectors must feature prescribed dynamic characteristics within preset limits with fixed steady-state characteristics.

6) For a liquid rocket engine (LRE) operating in a pulse mode, additional requirements of minimizing the volume of injector cavity must be fulfilled. For LRE with a wide range of thrust variation, prescribed mixture-formation parameters should cover the entire operating envelope.

To meet these requirements, injectors should provide pre-specified liquid-sheet thickness, spray-cone angle in the range of 36–120 deg, and dynamic characteristics. In addition, the fabrication procedure should be simplified to achieve reliable designs. Because many of these requirements vary for LRE of different types, a great number of injector types have been developed and implemented. The selection for a specific application is a result of a compromise between the preceding requirements and to a great extent depends on technological expertise, design tradition, and development experience.

A.   Classification of Injectors and Methods of Mixture Formation

Liquid propellant injectors and methods of mixture formation can be classified on the following basis:

1) applications: low-thrust engines, gas generators, medium-thrust engines, and boost engines of launchers;
2) propellants: earth-storable, hypergolic, and cryogenic propellants;
3) pressure drop: high-pressure and low-pressure drops across injectors;
4) design features: dimensions and configurations of flow passages; and
5) propellant mixing: external and internal mixing.

To disperse liquid into droplets and distribute them over the mixture-formation zone, various kinds of energies can be used. The choice, however, is limited for injectors. The classification of propellant injectors from the standpoint of the type of energy used for propellant atomization and mixture formation is discussed here.

1) The most often used concept for an LRE injector is the conversion of potential energy of the liquid in the form of pressure drop across the injector into kinetic energy of a liquid jet, which subsequently produces a spray of droplets. This atomization principle is used for jet, film, and swirl injectors. The energy efficiency of such injectors, however, is rather low. A major part of the energy is consumed in increasing the droplet velocity, rather than overcoming the surface tension. The pressure drop for a commercial LRE typically ranges between 0.29 and 3.75 MPa. The upper limit is determined by chamber stability considerations instead of the required atomization quality.

2) In the case of gas-liquid injectors, the kinetic energy of the gas flow is used to generate wavy motions at the liquid-gas interface, separate wave crests from the liquid core, and to accelerate drops. The pressure drop across the gas passage for achieving required atomization quality is significantly lower than that of liquid injectors. It is mainly determined by the requirements of operation stability and uniformity of propellant distribution between injectors. In the case of increased requirements of atomization quality, for example, in low-thrust engines, gas-liquid injectors are used for liquid propellant atomization with the help of additional high-pressure gas.

3) Thermal energy is often used to heat the liquid being atomized, in order to change its surface tension and viscosity, and in the limiting case (e.g., hydrazine) to evaporate and decompose the liquid to provide gaseous-phase mixing. In most LREs, the heating and evaporation of liquid propellants take place in cooling jackets or in heat exchangers. Only in hydrazine thermal-decomposition reactions, atomizers with developed surface of thermal contact with the liquid are used for initiating reactions.

4) Acoustic energy is used in acoustic and ultrasonic injectors. The resultant flow oscillations promote the formation of surface waves in the liquid stream, which then disintegrates into droplets with a low energy consumption rate.

5) Mechanical energy is used in injectors with reciprocating atomizers or rotors. The former has been employed only in experimental low-thrust engines (up to 1 N) using piezoelectric or magnetostriction vibrators. The latter has found application in low-thrust jet engines and in some designs of liquid propellant rocket engines.[1]

6) Electric energy has been used to activate piezoelectric and magnetostrictive vibrators for injectors of low-thrust engines operating in the pulse mode. In the latter case, the electric energy is directly converted into the potential energy of liquid. In addition, there have been proposals to use electric discharges in non-conducting liquids to obtain pressure pulses, so-called electrohydraulic effect. These methods have not found practical applications in commercial LREs.

7) Combined atomization methods using several different kinds of energy.

Different conversion techniques can be used simultaneously to intensify atomization, such as combined utilization of the potential energy of liquid and kinetic energy of atomizing gas. Application of vibrational energy during pneumatic and hydraulic atomization produces the most significant effect. This combination allows considerable savings of the energy spent for atomization.

B. Liquid Injectors

1. Jet and Slit Injectors

The jet injector is the simplest device for converting the potential energy of liquid in the form of pressures drop to the kinetic energy of a jet. The injector also represents a local contraction connecting the propellant manifold to the atomization zone. The ideal-liquid exhaust velocity U_{id} is determined by Bernoulli's equation:

$$U_{id} = \sqrt{\frac{2}{\rho}\Delta P_i}$$ (1)

where ΔP_i is the pressure drop across the injector, and ρ is the liquid density. The effect of viscous loss is usually taken into account by using either empirical coefficients or expressions derived from the boundary-layer theory. The liquid flow rate is determined by the following equation:

$$m_{i.id} = \mu_n A_N \sqrt{2\rho\Delta P_i}$$ (2)

where A_N is the injector cross-sectional area. The discharge coefficient μ_N mainly depends on the injector shape.

The slit injector has a flow passage formed by either flat or concentric surfaces. The hydraulics of slit injectors are well understood and reliable design methods are available.[2] The atomization mechanism of a liquid stream involves disintegration into drops due to the loss of flow stability under the effects of aerodynamic forces on the liquid surface.[3]

The jet and slit injectors are exclusively used for bipropellant applications, with their sprays formed by the intersection of liquid streams. Figure 1 shows an injector with five intersecting jets (four oxidizer jets and one fuel jet). It

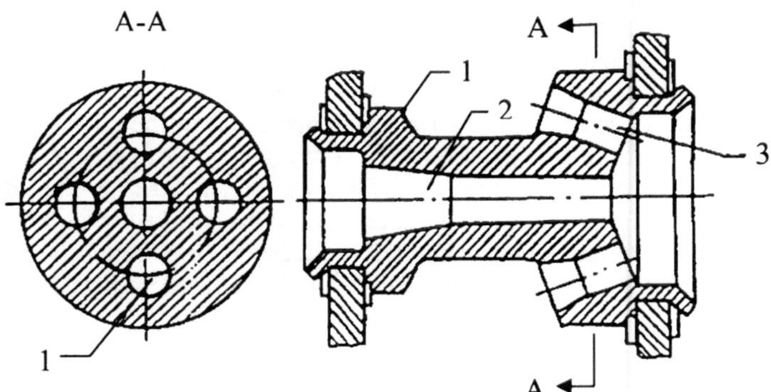

Fig. 1 Bipropellant spray injector with intersecting fuel and oxidizer jets; 1-casing, 2-fuel passage, 3-oxidizer passage.

contains a casing (1) brazed into the assembly bottom, an axial hole (2), and four axially symmetric holes (3) for the other propellant tilted towards its axis. The number of holes (3) may vary from two to four. In case two holes are employed, no axial hole is needed, and one of the holes (3) should be connected to the manifold of the other propellant. This situation is referred to as a doublet injector.[4] As the propellants flow out, a thin flat bipropellant lobe providing high-quality mixing and sufficiently fine dispersion of propellants is formed, as illustrated in Fig. 2. Injectors of the "triplet" type (with two lateral holes for one of the propellant) have the same scheme of liquid-propellant atomization and mixing. In other cases, a sharp bipropellant spray cone with rather coarse atomization is formed at the place of jet intersection.

The advantages of injectors with intersecting jets are:

1) simplicity of design and reliability of hydraulic analysis;
2) satisfactory atomization quality and uniform propellant mixing for doublet and triplet injectors; and
3) short flow residence time in the injector and short ignition delay.

The disadvantages of jet injectors are:

1) stingy requirements imposed on the fabrication technology. A small deviation in dimensions may considerably change the atomization and mixing quality;
2) significant non-reproducibility of the resultant spray property;
3) considerable differences of propellant-mixing pattern between flow tests for model liquids and real propellants in case of hypergolic propellants. This is attributed to the liquid sheet repulsion and deterioration of the mixing conditions, caused by gaseous products produced in liquid-phase reactions between the intersecting propellant flows;
4) the spray fan formed in the propellant-flow intersection region is extremely sensitive to local flow fluctuations, especially in the transverse direction. Injector assemblies consisting of doublet injectors are very susceptible to high-frequency transverse instabilities.

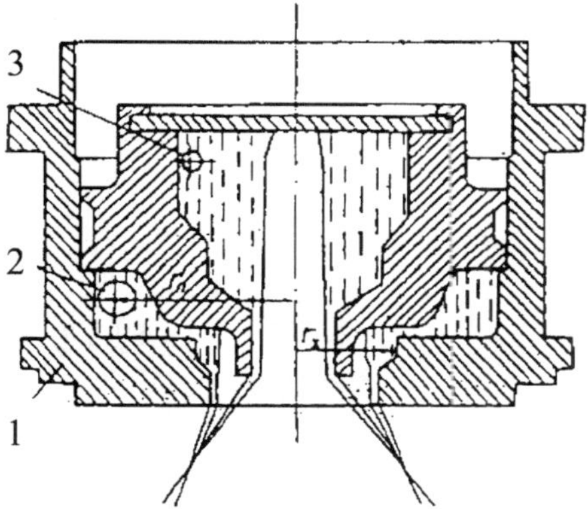

Fig. 2 Formation of fine dispersion of propellants in a doublet swirl injector; 1-casing, 2-fuel channel, 3-oxidizer channel.

To reduce the effect of fabrication errors, slit injectors with flat holes producing intersecting sheets are made by spark erosion machining. For example, the platelet manufacturing technology was used for the impinge jet multi-injector assembly for the Space Shuttle Orbit Maneuver System (OMS),[5] as shown in Fig. 3. Although the vacuum bonding of thin sheets of metal results in injector passages of stepwise form with enhanced hydraulic losses, the propellant mixing efficiency is considerably improved, especially for hypergolic propellants. Injectors with three or five jets forming a narrow long spray (see Fig. 1) were placed on the injector assembly as a means to increase the combustion zone length and thus to decrease its response to pressure fluctuations.

Application of injectors with intersecting jets has practically ceased for high- and medium-thrust engines with staged-combustion cycles, except for gas generators in which high-frequency instabilities were not nearly so critical due to the lower power intensity and longer operation processes. Injectors with intersecting coaxial sheets have found applications in so-called pintle injectors[6] for throttable LRE, as shown in Fig. 4. The disadvantages common to all of the designs of combined slit injectors with variable passage cross sections are stingy requirements on manufacturing accuracy. The generally permitted misalignment $\sim$0.03 mm is too large for this injector. For example, in the case of nominal liquid-sheet thickness of 1 mm for a 10-fold decrease of thrust, the misalignment amounts to $\pm$30% of the average width of the injector passage cross section. The ensuing deviation in the mixture ratio reduces the engine efficiency. The presence of a movable element in the immediate vicinity of the heat release zone poses another serious challenge in ensuring reliable operation of injector assemblies, and consequently restricts the design to single-injector assemblies.

2. Swirl Injectors

Swirl injectors are predominantly used in Russian LRE gas generators and in combustion chambers with pressurized feed systems or gas-generator cycles.

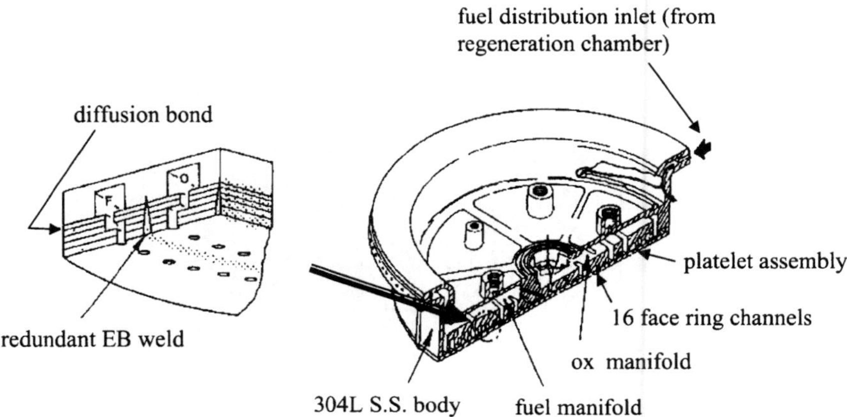

Fig. 3   Space Shuttle OMS engine injector.

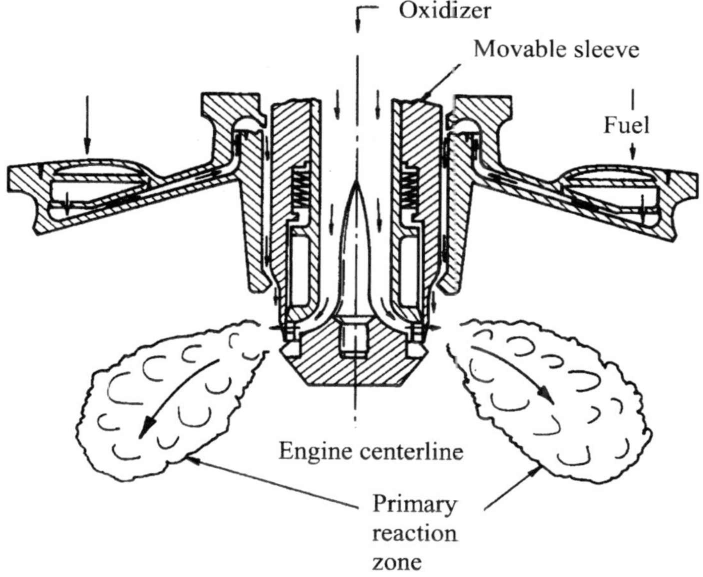

Fig. 4   Pintle injectors of throttable LRE.[6]

Monopropellant swirl injectors of screw-conveyer or tangential-channel type were used in early LRE, as shown in Fig. 5. The former employs a screw conveyer as swirler and are predominantly used in gas generators with moderate combustion temperatures. Each injector has a casing (1) brazed into the bottom of the assembly. The casing has a nozzle (2) and an axisymmetric cavity (3) connected to the liquid manifold through tangential passages (4). The bottom (5) is flared out and welded in the casing (1) forming a vortex chamber (6). In recent injector designs, the bottom (5) is made profiled to optimize the shape of the passage cross section of the vortex chamber (6). In screw-conveyer injectors (Fig. 5a), a screw-conveyer swirler (7) whose external passages (8) serve as tangential passages (4) is fitted into the casing (1).

Important geometric parameters determining swirl-injector characteristics are:

1) nozzle radius R_n;
2) cross-sectional area of the inlet flow passage A_{in};
3) swirling arm, i.e., the distance from the axis of the tangential passage to the injector axis, R_{in}.

These parameters form a dimensionless number known as the geometric characteristic parameter of a swirl injector:

$$A = \frac{A_n}{A_{in}} \cdot \frac{R_{in}}{R_n} \tag{3}$$

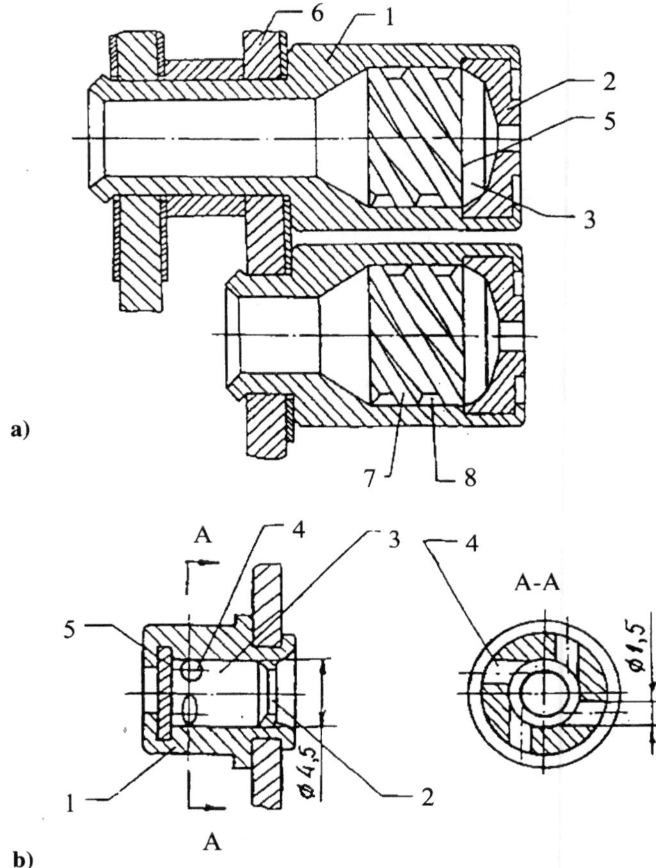

Fig. 5 Monopropellant swirl injectors with a) screw conveyer and b) tangential passages.

It determines the injector flow coefficient μ, the nozzle filling coefficient φ, the spray cone angle at the cylindrical nozzle exit, and other output parameters. A table of expressions for calculating these parameters is given in Ref. 7. In addition, there are some secondary parameters, which are of importance in determining the liquid flow residence time and viscous losses in injectors. These include:

1) the diameter and length of the vortex chamber;
2) the nozzle length and the convergence angle of the vortex-chamber wall adjacent to the nozzle.

When fed through the tangential (4) or screw passages (8), the liquid is set in rotary motion in the chamber (3), and forms a liquid vortex with a free internal surface whose radius smoothly changes from the minimum at the bottom

(5) $r_{mk} = R_N\sqrt{a}$ to $r_{mN} = R_N\sqrt{1 - \varphi}$ in the nozzle, where the flow area ratio $\varphi = A_L/A_N$ is the liquid-occupied fraction of the injector nozzle section, and $a = 2(1 - \varphi)^2/(2 - \varphi)$. At the nozzle exit, the spreading angle of the liquid sheet slightly increases due to the conversion of the centrifugal pressure produced by the rotating liquid sheet to the axial velocity component, i.e., the Skobelkin effect.[2] The pressure on the liquid surface is equal to the pressure in the combustion chamber. For an ideal fluid, the velocity at the liquid surface can be written as

$$U = \sqrt{U_a^2 + U_u^2 + U_r^2} = \text{const} = \sqrt{\frac{2}{\rho}\Delta P_i} \qquad (4)$$

where ΔP_i is the pressure drop across the entire injector. Any change in the velocity components U_i is unequivocally associated with a change in the surface radius r_m. The liquid flows out of the nozzle exit as a thin sheet whose thickness and spreading angle are independent of the pressure drop in a wide range of its variations, and the initial section has the shape of a single-cavity hyperboloid. The thickness of the sheet decreases farther downstream. It loses its stability and disintegrates into droplets according to the mechanism described in Ref. 3. It should be noted that under typical chamber conditions of a modern LRE, the liquid sheet from a swirl injector does not have enough time to get thin and to lose its stability since the aerodynamic effects of the surrounding high-pressure turbulent flow tends to disintegrate the liquid sheet more effectively. Thus, one has to analyze atomization for the startup and steady operating conditions of the same injector using two different methods.

The following features of swirl injectors, which determine their predominant applications in Russian LREs, are noteworthy:

1) For the same pressure drop and liquid flow rate, the average median diameter of droplets is 2.2 to 2.5-fold smaller than that of jet injectors. This advantage prevails for high flow rates and decreases when the counter pressure (i.e., the sum of the combustion chamber pressure and the centrifugal pressure created by liquid swirling motion) grows.

2) Compared with jet injectors, swirl injectors are not so sensitive to manufacturing errors such as deviation from prescribed diameter and surface misalignment.

3) The flow passage areas of swirl injectors are much larger than those of jet injectors with the same flow rates, and consequently they are less susceptible to choking or cavitation.

4) The pressure drop across a swirl injector is shared between the tangential channels ΔP_T and the vortex chamber ΔP_{vc}. Under steady-state conditions, the relation between ΔP_T and ΔP_{vc} can be easily defined, with the latter much higher than the former in most operational injectors. During the engine startup, when the vortex chamber is initially empty, the entire pressure drop is applied to the tangential channels and the liquid velocity is much higher than its steady-state value. The vortex chamber begins to be filled with high-speed rotating liquid. The ensuing increases in the centrifugal pressure and viscous losses then decrease the pressure drop across the inlet passage and subsequently the

mass flow rate prior to ignition. This self-tuning capability with variable flow resistance under transient conditions improves the engine startup operation.

Swirl injectors also feature wide spray-cone angles (in the range of 36 and 140 deg) and sharp non-uniformity of flow distribution. The liquid is practically absent near the injector axis. These features can be considered as a disadvantage when flow non-uniformity is detrimental for the process, for example, irrigation of catalyst grains for hydrazine decomposition. They are, however, advantageous for most applications in LRE since hot combustion products are recirculated to the injection region to stabilize the flame. The formation of a thin uniform liquid sheet also protects the injector face plate against excessive heat transfer from the high-temperature region.

Extensive effort has been applied to develop theories of swirl-injector hydraulics in Russia since as early as the 18th century (e.g., Leonard Euler, a member of the St. Petersburg Academy of Sciences). Among them, the principle of maximum flow rate postulated by N. Abramovich[8] in 1944 (Diploma of Discovery No. 49) is most noteworthy, which essentially laid the foundation of modern development of swirl injectors. The work was supplemented by the studies of L. A. Klyachko[2] and A. M. Prakhov,[3] as well as empirical data of Y. I. Khavkin[9] and many other scientists. It now gives reasonably reliable calculation results. Investigation of the dynamics of swirl injectors[7,10] has made it possible to design injectors with prescribed dynamic properties and use them as a means of suppressing various mechanisms of high-frequency instabilities. Work was conducted to study injectors with flow modulation capabilities[11] for modulating spray cone angles and flow rates with prescribed degrees of atomization.

The disadvantages of swirl injectors are:

1) the internal-cavity volume is significantly larger than that of a jet injector and has a longer startup transient time, which restricts their application in low-thrust LREs with pulse operations;

2) complex configuration and heavy weight.

The types of swirl injectors used in LREs are diversified because of specific requirements for different applications.

3.   Monopropellant Swirl Injectors

Monopropellant injectors (see Fig. 5) were widely used in open-loop engines. As a rule, they are arranged in assemblies in the chess-board or honeycomb patterns with alternating fuel and oxidizer injectors. In welded and brazed assemblies, oxidizer injectors with elongated casings are attached to the assembly bottoms and thus serve as structural elements. They often have two- to four-run screw-conveyer swirlers with cylindrical configurations. Conic screw-conveyer injectors, which are technologically less effective and widely used in jet engines, are seldom used in LREs. They are used only in low-flow injectors to minimize the vortex-chamber volume and to eliminate nonswirling liquid leaking through the gap between the screw conveyer and casing.

Contemporary monopropellant swirl injectors have tangential passages. As a rule, the number of passages is three or four. A smaller number of passages increases the nonuniformity of flow distribution along the spray cone circle. Two passages are used only in highly closed injectors or in very long injectors. One passage is used for special purposes, when flow nonuniformity is required along the circumference, such as in near-wall regions. At present, monopropellant swirl injectors are mainly used in gas generators to provide efficient propellant delivery or as near-wall injectors in combustion chambers. Their relatively low combustion efficiency (~97%) and high sensitivity of fuel and oxidizer mixing increases the susceptibility to combustion-driven flow oscillations.

4. *Bipropellant Swirl Injectors*

Since the mid-1960s, bipropellant swirl injectors have been used most often in Russian LREs of various types and applications. Figure 6a shows a typical design of such injector elements. It has a hollow casing (1) with a nozzle (2). The hollow insert (3) with the nozzle (4) and flared-cut bottom (5) are brazed into

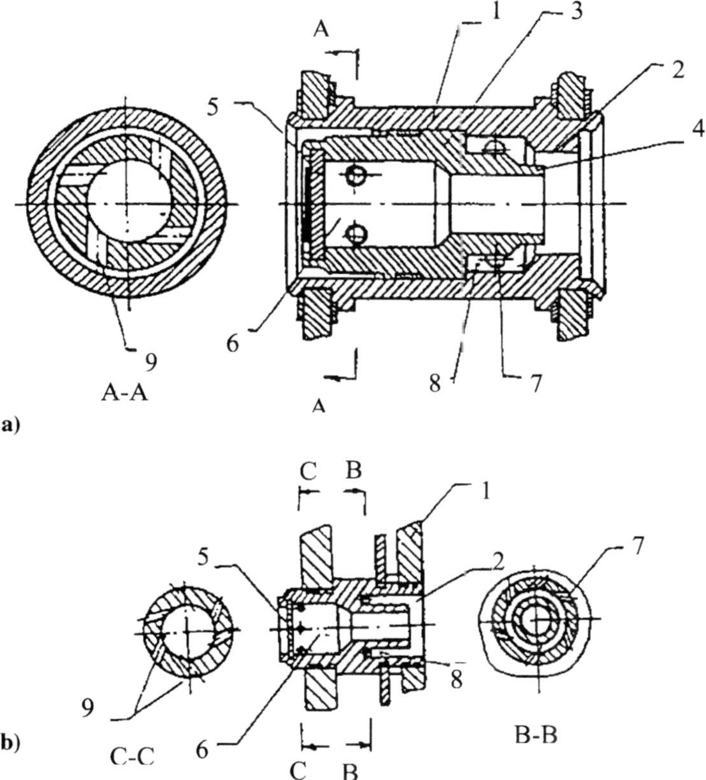

Fig. 6   Two different designs of bipropellant swirl injectors; 1-casing; 2-casing nozzle; 3-insert; 4-insert nozzle; 5-bottom; 6-central vortex chamber; 7–9 tangential passages; 8-peripheral vortex chamber.

it. The bottom and the insert form the vortex chamber (6), which is connected to the propellant-delivery manifold through tangential passages (9). The insert (3) and casing (1) form a peripheral vortex chamber (8) for the other propellant connected to its delivery manifold through tangential passages (7).

A more advanced and compact design of bipropellant injectors is given in Fig. 6b. It is beam welded to the fire face. For hypergolic propellants, the edge of the nozzle (4) is usually buried in the nozzle (2) of the peripheral injector by 0.7–2.2 mm. The specific dimension depends on the propellant type and injector geometry. Oxidizer injectors can be used as peripheral injectors (so-called direct scheme). Such an arrangement is more preferential from the standpoint of mixing hydrodynamics since it will produce sheets of the same or approximately the same thickness. In more recent engines, however, the reverse scheme with fuel delivery through the peripheral stage has found applications, due to the ease of arrangement with the cooling jacket and the fire bottom of the injector assembly. Injectors of the direct scheme have either propellant mixing outside the injector (in this case, the spreading angle of the external spray should be less than its counterpart of the internal spray), or mixing at the external-nozzle edge. Injectors of the reverse scheme, whose peripheral fuel stage have a considerably wider spray cone angle than that of the internal oxidizer injector, may have propellant mixing at the nozzle edge alone. Internal mixing of hypergolic propellants should be avoided in such injectors since they are not safe against ingress of one of the propellants to the cavity of the other one. Cryogenic propellants of the oxygen-kerosene type can be mixed directly in the vortex chamber of the peripheral oxygen stage, in which the swirling liquid-oxygen layer provides the cooling of the injector walls.

5. Combined Jet-Swirl Injectors

To increase the length of the combustion zone and to decrease the combustion response to chamber flow fluctuations, combined jet and swirl injectors have been widely used since the 1950s. Figure 7 shows sectors of the injector assemblies equipped by such injector elements. Each or some of the screw swirlers inside the vortex chambers are drilled to form a passage for a liquid jet inside the surrounding conical spray. A combination of impinging jets and such a type of injector element was used in place of baffles in several Chinese booster LREs (see Fig. 8) to suppress high frequency combustion instabilities. One should be careful with this design because such an injector could be a source of LRE unsteadiness if improperly designed,[12] and may lead to operations in two different manners, even at the same pressure drop. There may be a situation with separate outflows of jet and swirling liquid or with a joint outflow. The central jet could also disappear and the swirling spray could have a narrow spreading angle and become poorly atomized. Figure 9 schematically shows a sector of a combustion chamber injector assembly with alternating bipropellant swirl and impinged jet injectors.

C. Gas-Liquid Injectors

Gas-liquid injectors utilize the kinetic energy of a gas flow for liquid atomization and are mainly used in combustion chambers of closed-loop LREs and in

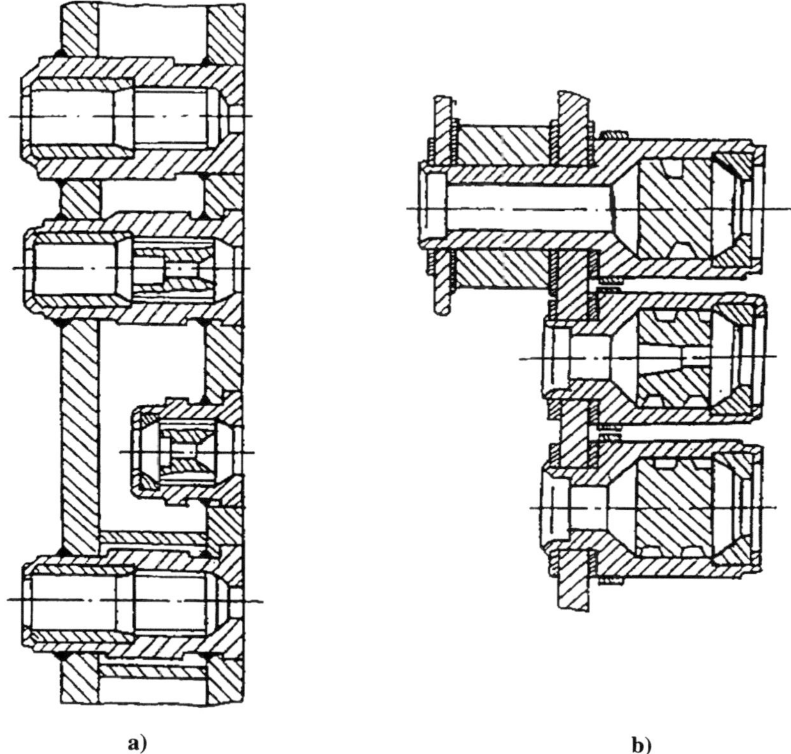

a)   b)

Fig. 7   Injector assemblies with alternating swirl and spray-swirl injectors.

some open-loop LREs having special operating conditions such as large thrust control while using prevaporized, decomposed, or gasified propellant, with preburning of a small amount of the propellant.

Figure 10 shows the main design schemes of gas injectors used in LREs. They can be classified into three basic categories: injectors with peripheral liquid and central gas delivery (a, d); injectors with peripheral gas and central liquid delivery (b, c, and e); and injectors with two-sided atomizing-gas delivery to the liquid sheet being atomized (f).

1.   Jet Injectors

The gas-liquid jet injector (Fig. 10a) is one of the designs most commonly used in both hypergolic and hydrocarbon-oxygen LREs. It has a tubular casing (1) with an axial gas passage (2) and holes (3) for liquid propellant delivery. Passages can be arranged along the length of the casing (1) in different ways, depending on the propellants used. Passages (3) are predominantly located in the vicinity of the exhaust edge of the casing (1). The axes of passages (3) were made intersecting with the casing axis at an angle of 45 to 60 deg. Recently, it has become

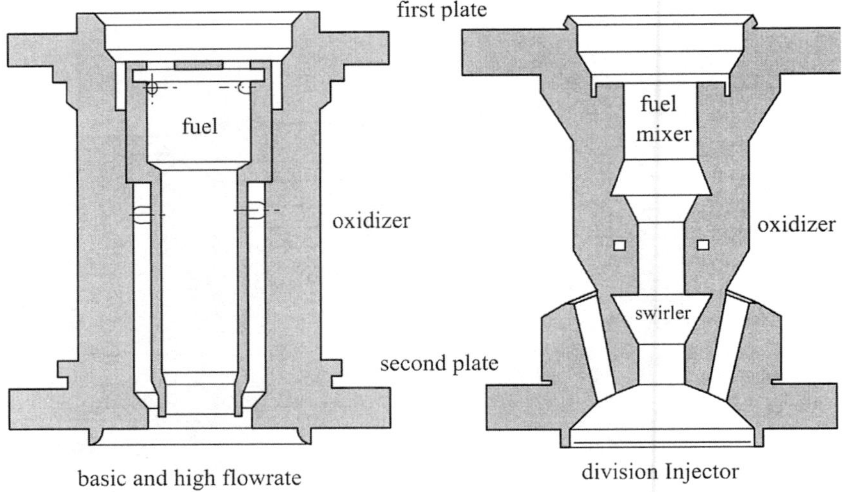

Fig. 8   High flow rate and division injection elements for Chinese YF-1 engine.

preferential to make passages (3) chordal, which increases the uniformity of propellant mixing.

The diameter of passage (2) usually varies between 6 and 18 mm, depending on the engine thrust requirement. The length of passage (2) is chosen to maximize the

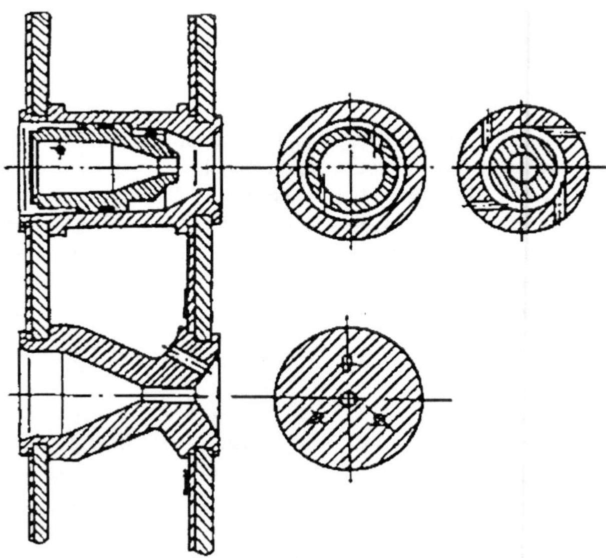

Fig. 9   Combustion chamber injector assembly with adjacent bipropellant jet and swirl injectors.

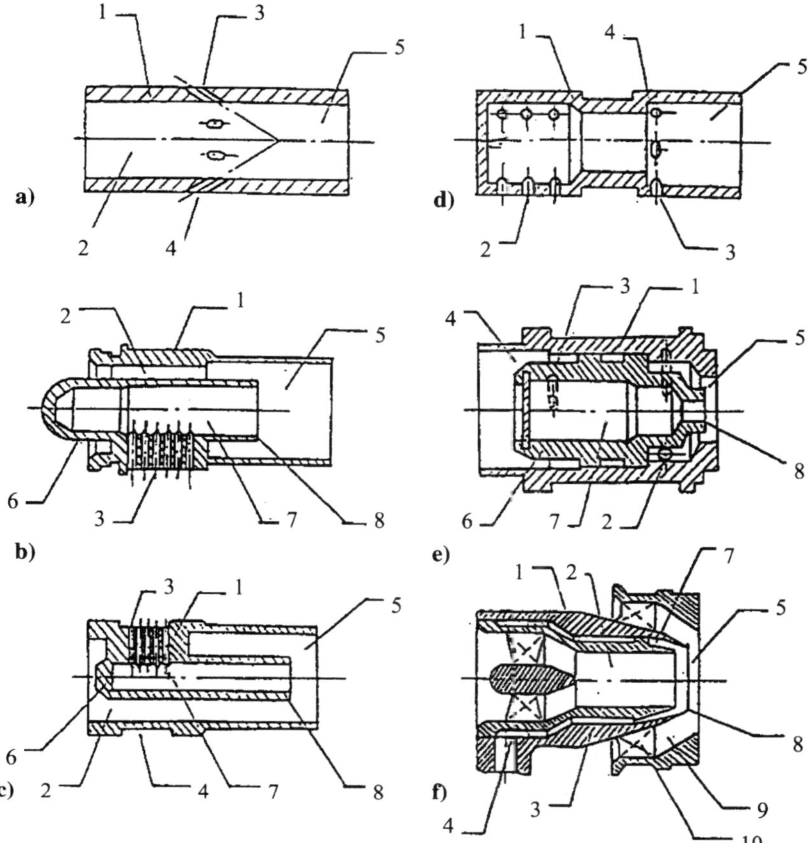

Fig. 10   Pneumatic injectors used in main combustion chambers and gas generators of LRE: a) jet-jet injector; b), c) jet-swirl injector; d), e) swirl-swirl injector; f) slit injector with alternating liquid and gas injection; 1-casing; 2, 9 gas passage; 3-liquid passage; 4-liquid manifold; 5-mixer; 6-bottom; 7-vortex chamber; 8-nozzle; 10-gas swirler.

removal of acoustic energy from the combustion chamber by treating the injector as a half-wave acoustic resonator. Figure 11 shows a gas-liquid spray injector assembly with a swirl fuel injector at the periphery, which protects the fire bottom of the assembly against overheating. Liquid propellant jets injected into the mixer (5) are atomized by the gas flow from passage (2). The combustion process occurs immediately downstream of the injected jets since the flame is stabilized on the walls of casing (1) behind holes (3). Simplicity of design and fabrication and high-quality mixing are the advantages of injectors of the type shown in Fig. 10a. The disadvantage is the relatively high sensitivity to pressure pulsations, which, however, can be compensated by changing the injector geometry to remove acoustic energy at prescribed frequencies from the combustion chamber.

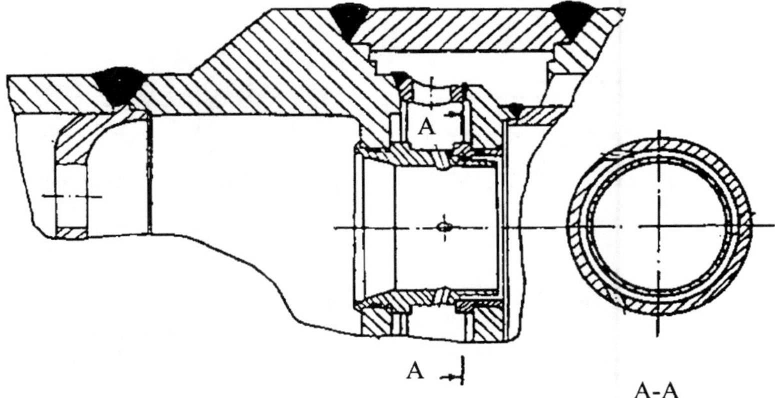

Fig. 11 Injector assembly with gas-liquid jet-jet injector equipped with a peripheral liquid swirl stage.

2. Swirl Injectors

In the aforementioned case, if holes (3) in the casing wall are made tangential, blocked from the entrance face, a swirl injector is formed with peripheral liquid propellant delivery and central delivery of a hot swirl gas flow (Fig. 10d). This design features a contraction between passages (2) and (3) and a shoulder in front of passages (3) to ensure the formation of a swirling liquid sheet on the internal walls of the mixer (5). The gaseous propellant arriving through the tangential passages (2) forms a swirling flow in the internal cavity of casing (1), travels around the liquid sheet running along the walls of the mixer (5), forms surface waves on it, and finally blows away the crests and disperses the liquid into droplets. Such injectors are widely used in high-power LREs with hypergolic propellants. The main advantage is a wide stability margin for operation over a broad range of mass flow rates. This feature may be attributed to the low sensitivity of the atomization and mixing processes stabilized by the mixer wall to pressure fluctuations and flow-velocity variations. Positioning liquid passages (4) at the periphery of the mixer (5) allows cooling of the fire bottom of the assembly by means of a cone-shaped screen formed by the evaporating-liquid sheet.

High-quality atomization achieved by swirling gas and liquid flows allows manufacturing of large injectors with diameters of the mixer (5) in a range of 50–60 mm and flow rates of several kilograms per second. To suppress high-frequency instability in the combustion chamber, unique acoustic properties of injectors such as acoustic impedance of the mixer (5) and acoustic resonance of the gaseous-propellant vortex chamber are used. Important results are obtained when the acoustic properties of the mixer and vortex chamber are combined.

The disadvantage of swirl injectors is the non-uniformity of flow intensity and mixture composition along the spray-cone radius, which does not provide adequate combustion efficiency. Figure 12 shows an example of a full-scale injector with central delivery of swirling gas. The gas resonance cavity minimizing flow

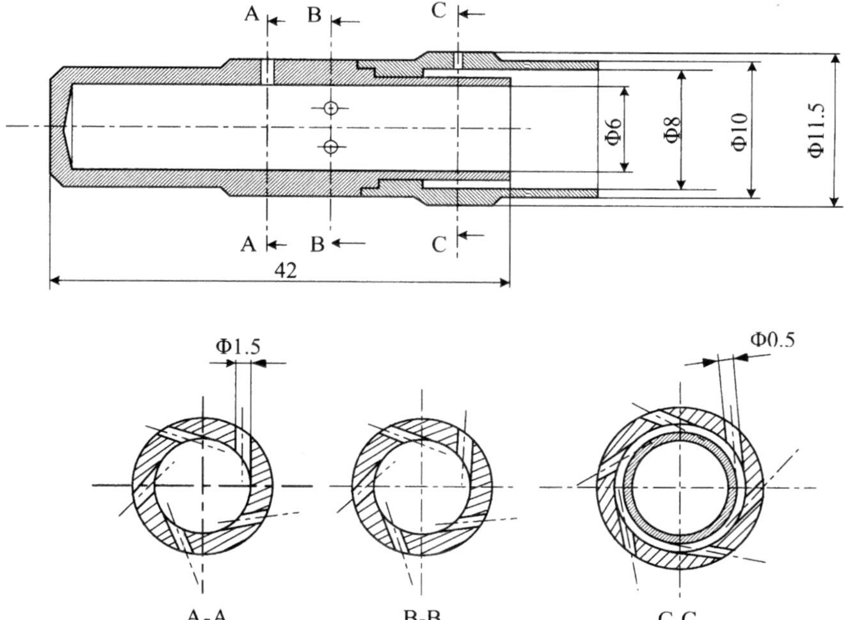

Fig. 12 Injector with central injection of swirling gas and acoustically tuned gas stage.

fluctuations is clearly visible. In low mass flow-rate injector elements with exterior swirling liquid flows, the central gas stage could be made of a non-swirling stepped channel, as shown in Fig. 13.

3.   Coaxial Injectors with Central Liquid Stage

This type of design includes jet, jet-swirl, and swirl-swirl injectors, as shown in Figs. 10b, c, and e. The injector has a tubular casing (1) with passages (2) for gaseous propellant delivery [usually between the pylons having passages (3) for liquid propellant delivery]. The pipe (6) with passage (7) and nozzle (8) is usually buried in the tubular casing (1) and mounted rigidly on the pylons. When the liquid stage (6) is made as a jet injector, passages (3) are radial and passage (7) is elongated. In this case, injectors are used for mixing liquid oxygen and hydrogen-enriched gas. When the central stage is made as a swirl injector, passages (3) are of either the tangential or screw-conveyer type (the latter is mainly used in hydro-gen-oxygen gas generators). In this case, a swirling liquid sheet flowing out of the nozzle (8) is in the form of a cone-shaped film interacting with the coaxial gas flow. The quality of propellant atomization and uniformity of mixing are signifi-cantly enhanced, which allows manufacturing of larger injectors (with nozzle diameters of 8–12.5 mm) with propellant flow rates of up to 2 kg/s for each injector element. The presence of a hollow liquid vortex inside the passage (7)

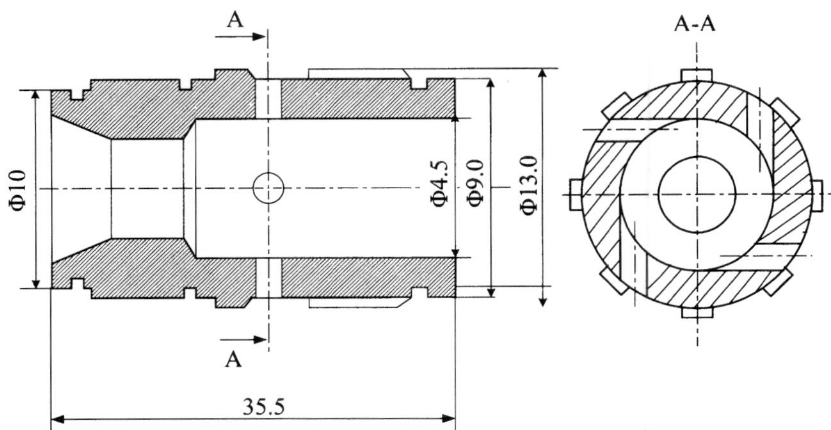

Fig. 13   Oxygen-kerosene injector with tangential injection of liquid oxygen.

makes it possible to use the liquid stage of the injector as a quarter-wave resonator to absorb the acoustic energy in the combustion chamber, for which purpose the passage (2) should be further elongated (see Fig. 10b), and in some cases it is necessary to make additional holes in the walls of the acoustic damper (6) or to equip the liquid stage of the injector with an additional Helmholtz transmission resonator.

The main advantage of such injectors is high-quality atomization and mixing of propellants. Their disadvantage lies in the complexity of manufacturing and adjustment. The possibility of self-oscillations of the liquid sheet in the coaxial gas flow in injectors of this type, which were first noted in Ref. 13, is an additional problem. Testing of full-scale engines revealed the presence of self-oscillations at the frequency of $\sim$5500 Hz, leading to pressure pulsations in the oxidizer passage with an amplitude of up to 0.9 MPa accompanied by subsequent breakdown of pipe connections with no significant pressure fluctuations in the combustion chamber.

Investigations of these injectors using models and actual propellants made it possible to establish the mechanism of self-oscillations described in Ref. 14 and to develop methods for their suppression. The presence of self-oscillations may sharply intensify propellant atomization and mixing, especially for injectors with low pressure drops. It has been proved both theoretically and experimentally that such induced high-frequency fluctuations are not related to intrinsic oscillations in the combustion chamber and are safe and even beneficial for combustion stability and efficiency. The self-oscillation regime is sometimes displaced into the engine operating regime requiring throttling (below 75% thrust), and injectors are equipped with devices dampening pressure oscillations arising in the combustion chamber.

The principle of the swirl-swirl injector (see Fig. 10e) is similar to that of the liquid-liquid injector shown in Fig. 6a. The only difference lies in the cross-sectional area of the gaseous propellant passage. Because the gaseous propellant comes through tangential passages (2) from the periphery of the casing (1), the mixer is not made long. This, however, may lead to overheating and burnout of the casing. Additional means of cooling are thus required to prevent overheating of the fire bottom. Similar measures should also be introduced in jet injectors (see Fig. 10a), usually by putting an additional low-flow swirl injector at the periphery of the mixer nozzle (see Fig. 11). The designs of coaxial swirl-swirl injectors are highly diversified. As an example, Figs. 14 and 15 show swirl-swirl injectors of gaseous oxygen-kerosene engines with gaseous oxygen-cooled walls. Figure 16 shows injectors of oxygen-hydrogen gas generators.

4.   *Slit Injectors*

Figure 10f shows an example of the injector design to provide atomization of a liquid sheet (or a vast number of fine jets) with the gas flow on both sides. This design has found applications in both jet engines and LREs. It contains a hollow

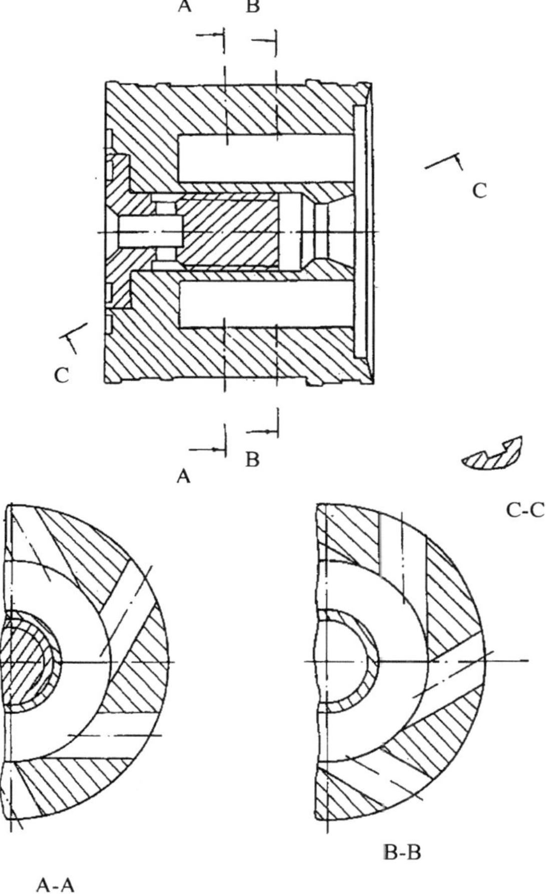

Fig. 14   Swirl-swirl injectors of gaseous oxygen and liquid hydrocarbon.

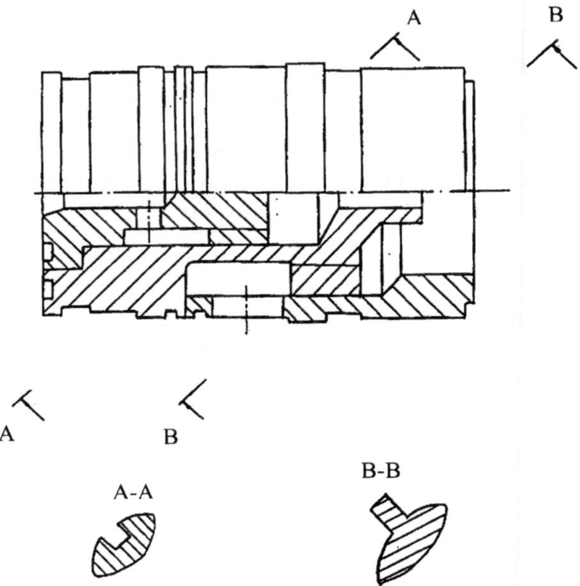

Fig. 15 Modified oxygen-kerosene injector with screw conveyer swirler.

casing (1) with an insert having axial gas passage (2) and slit passage (3) formed
by the insert, and a casing (1) for the liquid propellant. The passage (3) is connected to the feed line (4). The casing (1) is mounted on an external yoke (9)
by means of a gas-flow swirler (10) and forms a vortex chamber (7) for the
liquid, which ends in a nozzle (8). The mixer (5) consists of the nozzle (8) and
yoke (9). The gas-flow swirler (11) or an additional low-flow liquid injector
(not shown in the figure) can be placed into the passage (2). Several alternating
liquid (3) and gas (2) circumferential passages can be made. The advantages of
this design are the most uniform and best-quality propellant atomization and
mixing. As a consequence, one engine with such a design has the combustion
efficiency of around 0.995. The disadvantages include increased production
complexity and cost and insufficient understanding of its dynamic characteristics,
which forbid its application in high-thrust engines.

5. Jet-Swirl Injectors

Injector assemblies using jet and swirl elements in the fuel or oxidizer lines are
shown schematically in Fig. 11. Application of a liquid swirl injector in combination
with a jet injector improves the cooling conditions of the fire bottom. The design has
allowed the development of tripropellant injectors (see Fig. 17), which represent in
essence a combination of well-developed designs shown in Figs. 10a and 10b. It
consists of a casing (1) with a cavity (2) and a chamber (3) with a nozzle (4) coaxi-
ally mounted in the casing, which forms a cavity (5) connected to the liquid propellant delivery manifold by tangential or radial passages (6). Passages (7) are formed

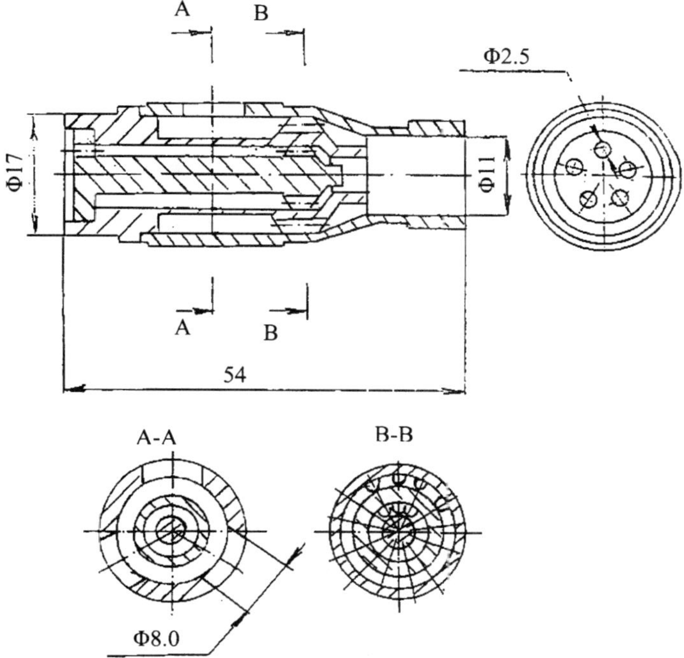

Fig. 16   Injector of oxygen-hydrogen gas generator.

between the chamber (3), the pylons with tangential passages (6) and casing (1). The mixer (8) is formed between the nozzle (4) and the casing (1). Chordal (9) and radial (10) passages for additional propellant delivery are made in the walls of the casing (1). This injector is designed for two applications: reducing and oxidizing generator gases. These versions differ in the dimensions of the passage cross sections alone. In case reducing gas is fed to the injector through passages (7), liquid oxidizer comes to the chamber (3) through passages (6) (similar to the injector in Fig. 10b), and additional fuel (hydrocarbon of the liquefied methane or propane type) is fed through passages (10). In the case of operation according to the scheme with afterburning of the oxidizing generator gas fed via passages (7), less volatile liquid fuel (kerosene) is fed to the chamber (3) through passages (6) and more volatile fuel (hydrogen) via passages (9) and (10). Under this condition, there exists a possibility of throttling of kerosene passage (6) up to its complete disengagement. The injector will operate in the regime characteristic of the injector in Fig. 10b. The results of injector development and testing are given in Refs. 14 and 15.

D.   Intensification of Propellant Atomization and Mixing in Liquid Injectors

The efficiency of liquid injectors is relatively low because much energy is expended for liquid acceleration rather than for atomization. Injectors possessing

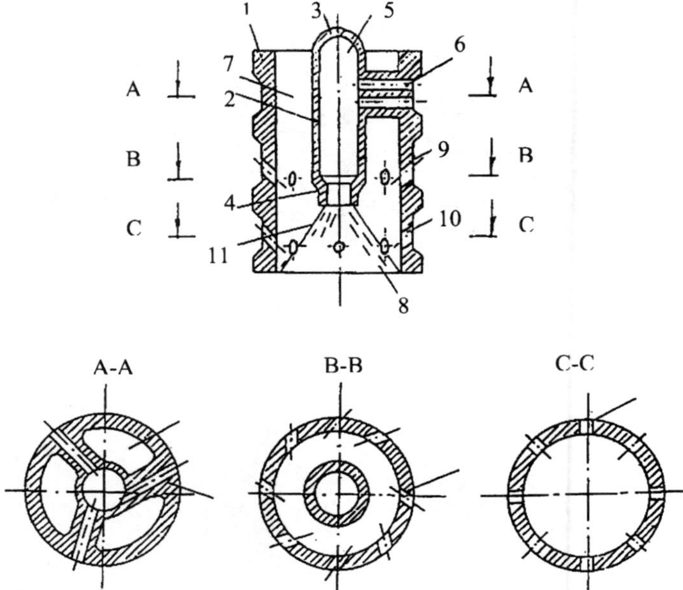

Fig. 17 Design of a tripropellant injector.

higher atomization and mixing characteristics have been developed. The principal means of improving injector efficiency involve reducing viscous losses of the liquid flow and applying nonstationary processes to intensify propellant atomization and mixing. In jet injectors, the former can be accomplished by increasing the quality of the liquid surface and reducing its area (e.g., by shortening injectors). Investigations of viscous-loss mechanisms in swirl injectors have shown the following two contributing processes: interlayer liquid friction in the vortex chamber and friction of the external liquid layers on the walls of the vortex chamber and nozzle. To reduce viscous losses, it is advisable to shorten the lengths of all of the elements, in particular the nozzle, and to reduce the liquid flow velocity by increasing the swirling arm of the liquid, R_{in}. Optimization of the injector shape to meet these requirements has led to the development of short and flat platelet swirl injector assemblies consisting of a stack of plates welded in vacuum with holes, grooves, and saw-cuts to form nozzles, vortex chambers, and passages. Figure 18 shows an injector assembly with stacked plate arrangement having two passages for propellant delivery. The assembly has a casing (1) with an annular manifold (2) to deliver the first-stage flow and disks (3) and (4) fitted in the casing with separation (5), tangential passages (6), vortex chambers (7), and nozzles (8). The casing has an additional manifold (9) for the second-stage propellant, and the disk (4) has additional supply (10) and tangential (11) passages. To decrease the dimensions of the assembly and the distance between the injectors, propellant distribution between them is provided with disks (12) and (13) having grooves (14) and

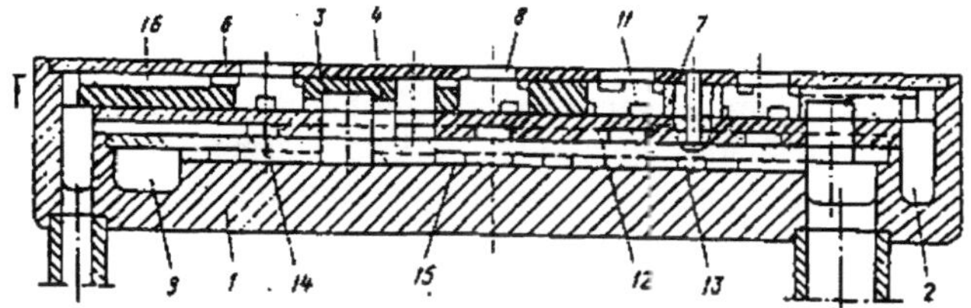

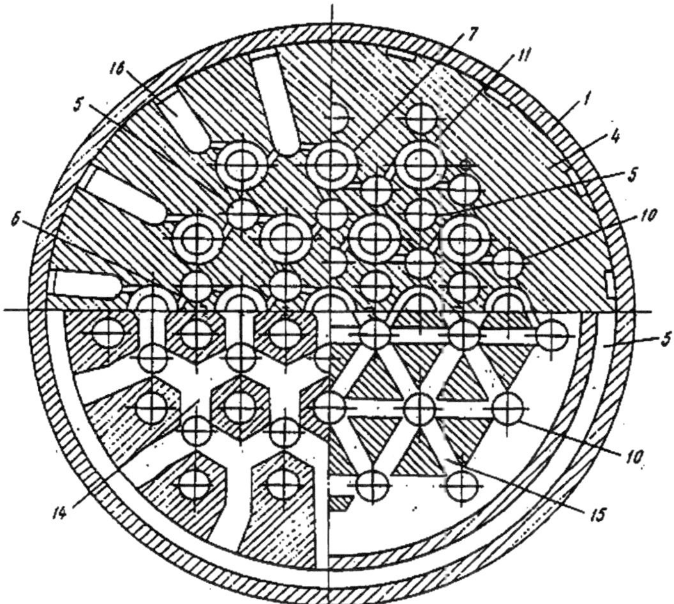

Fig. 18   Design of a stacked bottom injector assembly; 1-casing; 2, 9 propellant manifolds; 3-fire bottom; 4-injector bottom; 5, 10, 14, 16 supply passages; 6, 11-swirl passages; 7-injector vortex chambers; 8-injector nozzles; 12–13 separating disks.

(15) for separation. The fire bottom (3) has grooves (16), which not only deliver part of the propellant to the injectors but also cool it. When accommodated in the casing, disks (3), (4), (12), and (13) are diffusion welded to it in vacuum, after which propellant delivery pipes are welded to the casing.

When fed to the manifolds (2) and (9), the propellant enters tangential passages (6) and (11) through grooves (14), (15), and (16) and delivery passages (5) and (10), gains rotating motion in the vortex chambers (7) and exits from the nozzles (8) in the form of thin circumferential sheets. To decrease viscous losses inside the vortex chamber when using viscous, adhesive or non-Newtonian liquids such as jells and metal particle suspensions, the vortex chamber and nozzle can be made of a porous material. Figure 19 schematically shows a porous swirl injector in a mandrel for cold-flow tests. The injector has an insert mounted on a branch pipe (1). The insert is made by diffusion-welded porous bottoms (2) and casing (3) fixed by a coupling nut (4). The casing has a heat-resistant coating on the combustion-zone side. The annular manifold formed by the casing and coupling nut (4) can be connected by holes to the branch pipe, as shown in Fig. 18, or to an independent manifold, to deliver low-viscous liquid that penetrates through the porous material of the casing and forms a boundary layer that separates main propellant from the surfaces of vortex chamber and the nozzle of the injector.[16]

When fed through the branch pipe, the liquid enters the casing cavity through the tangential passages, forms a liquid vortex in the vortex chamber, and leaves the nozzle as a circumferential liquid sheet. Simultaneously, part of the liquid leaks through the porous bottom and casing and reaches the boundary layer flowing around the internal surface of the vortex chamber, thereby decreasing

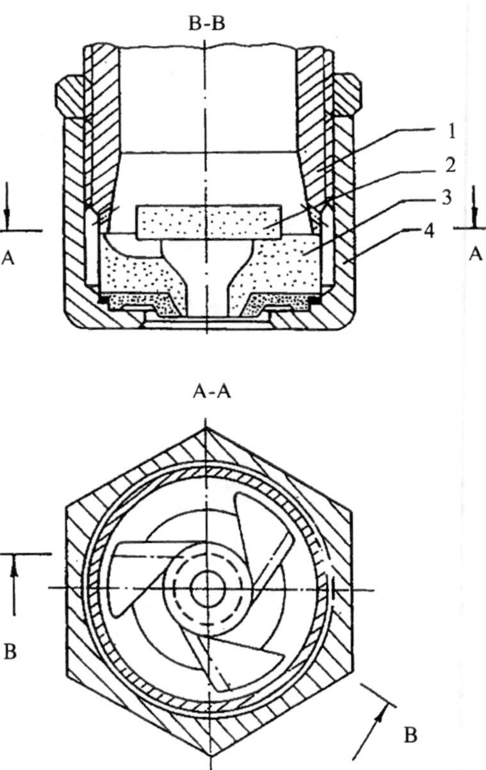

Fig. 19 Swirl injector with porous ceramic-metal swirler; 1-mandrel; 2-bottom; 3-porous casing with tangential grooves; 4-coupling unit.

the gradient of the circumferential velocity and hence the viscous losses. In the case of increased propellant viscosity, a small amount of low-viscosity liquid such as kerosene is fed through the porous walls from an independent source.

In liquid swirl injectors of the classical design, the exhaust process is stable and not accompanied by regular self-oscillations. However, in the presence of some destabilizing factors such as exhaust of a boiling liquid and bubbled media, or with specially profiled vortex chambers, self-oscillations at frequencies from several Hz to several hundred Hz can be produced and used to intensify the process of mixture formation.[17] As for vibrational liquid injectors with magneto-striction or piezoelectric actuators, driven by additional sources of electrical energy, they were used only in experimental designs for research purposes.[18]

E.   Intensification of Propellant Atomization and Mixing in Gas-Liquid Injectors

Because gas-liquid injectors use generator gas produced mainly by one of the propellants, the engine is not in short supply of the atomizing flow. The problem of improving the pneumatic atomization process is not so critical. The limiting factors are length of the combustion zone permitted in chambers of limited dimensions, high flow intensity, maximum flow uniformity and mixture ratio (which is especially important for low-thrust systems), and sensitivity to acoustic instability. In such chambers, methods for intensifying the mixture formation process are:

1) decreasing injector dimensions;
2) increasing surface area wetted by the liquid and flown around by the atomizing gas;
3) emulsification of the liquid, i.e., formation of gaseous bubbles in the liquid flow.

II.   Theory and Design of Liquid Monopropellant Jet Injectors
A.   Flow Characteristics

The mass flow rate of a jet (spray) injector, shown schematically in Fig. 20, can be determined using Bernoulli's theorem:

$$p_{01} = p_1 + \frac{\rho U_1^2}{2} = p_2 + \frac{\rho U_2^2}{2} + \Delta p_{1-2} \tag{5}$$

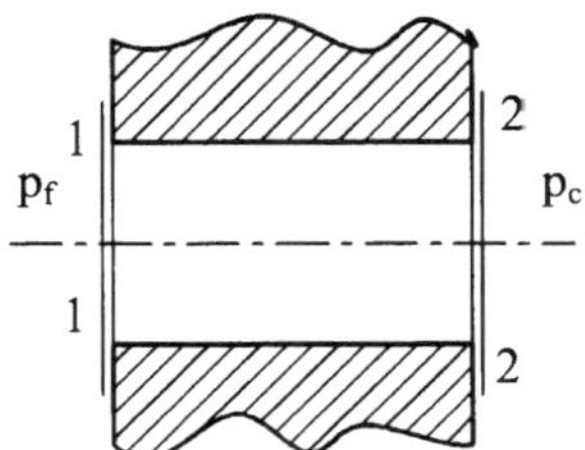

Fig. 20   Schematic diagram of jet injector.

where Δp_{1-2} represents the pressure losses in the injector passage and p_{01} the total pressure at the entrance. From mass continuity,

$$\dot{m} = \rho U_1 A_1 = \rho U_2 A_2 \tag{6}$$

The flow area at the injection exit A_2 is usually smaller than the injector cross-sectional area A_n. Thus, $A_2 < A_n$, and $A_2 = \varepsilon A_n$, with ε (≤ 1.0) being the coefficient of jet contraction. For continuous exhaust, $A_2 \equiv A_n$ and $\varepsilon = 1.0$. If the propellant is delivered from a large manifold with $A_1 \gg A_2$, the inlet velocity U_1 can be neglected and $p_1 = p_{01}$. The pressure drop across the injector Δp_i can be obtained from Eq. (5):

$$\Delta p_i = p_f - p_c \equiv p_{01} - p_2 = \frac{\rho U_2^2}{2}(1 + \xi_i) \tag{7}$$

where $\xi_i = \Delta p_{1-2}/(\rho U_2^2/2)$ is the hydraulic-loss coefficient. The subscripts f and c refer to the conditions in the propellant manifold and combustion chamber, respectively. The exit velocity U_2 takes the form

$$U_2 = \frac{1}{\sqrt{1 + \xi_i}} \sqrt{\frac{2\Delta p_i}{\rho}} \tag{8}$$

Substitution of U_2 from this expression into the continuity equation gives the mass flow rate of the injector:

$$\dot{m}_i = \frac{\varepsilon}{\sqrt{1 + \xi_i}} A_n \sqrt{2\rho \Delta p_i} \tag{9}$$

For an ideal liquid flow with $\xi_i = 0$ and $\varepsilon = 1.0$, we have

$$\dot{m}_{i.\mathrm{id}} = A_n \sqrt{2\rho \Delta p_i} \tag{10}$$

Comparison of Eqs. (9) and (10) leads to the expression of the flow coefficient of an injector μ:

$$\frac{\dot{m}_i}{\dot{m}_{i.\mathrm{id}}} = \frac{\varepsilon}{\sqrt{1 + \xi_i}} = \mu \tag{11}$$

Since the hydraulic loss coefficient ξ_i is always positive, the injector flow coefficient μ is less than unity in all cases. In practical designs, ξ_i and ε are determined experimentally, as functions of injector length l_i and cross-sectional area A_n. The mass flow rate can be calculated from Eq. (9) for given pressure drop, fluid density, and injector area.

B. Effect of Injector Configuration

Figure 21 schematically shows the liquid flow in a real injector passage. When entering the injector, the liquid flow separates from the sharp edge at the entrance

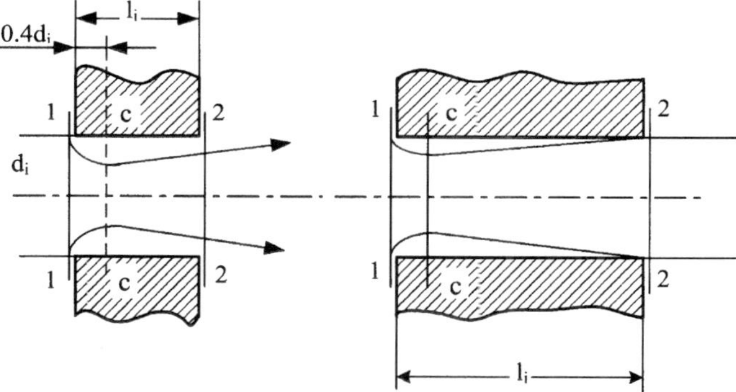

Fig. 21   Schematic diagram of liquid flow in jet injector.

due to the inertia force, and contracts as it proceeds. The largest contraction takes place at a distance of 0.4 d_i from the inlet. The flow then expands and exhausts from the injector. Two different modes of injector operation have been observed:

1) If the passage is short with the aspect ratio of l_i/d_i less than 1.5, the expanding jet is unable to reach the nozzle wall (see Fig. 21a) with the jet contraction coefficient ε less than unity, and an annular space is formed between the jet and the injector wall, which is in communication with the combustion chamber. The exhaust liquid flow is unstable and the mass flow rate is reduced. For this reason, short injectors are seldom used in operational engines.

2) If the injector passage is sufficiently long with l_i/d_i greater than 1.5, the expanding jet reaches the injector wall (see Fig. 21b) and the contraction ratio ε equals to unity. A stagnant zone, however, forms between the flow boundary and injector wall, and a vortex-type liquid flow reaches its steady state there. At high flow velocities, cavitation may occur in the contraction section, and the stagnant zone becomes a cavity. It is clear that each mode of injector operation affects the flow coefficient μ in its own way.

C.   Flow Coefficient

The flow coefficient μ can be determined experimentally based on either the measured mass flow rate or the hydraulic loss coefficient. The total pressure losses $\Delta p_{1-2} = \xi_i \rho U_2^2/2$ for a jet injector can be represented as follows:

$$\Delta p_{1-2} = \Delta p_{1-c} + \Delta p_{c-2} + \Delta p_{\mathrm{fr}} = (\xi_{1-c} + \xi_{c-2} + \xi_{\mathrm{fr}})\frac{\rho U_2^2}{2} = \xi_i \frac{\rho U_2^2}{2} \qquad (12)$$

where Δp_{1-c} and Δp_{2-c} are the losses in the $1 - c$ and $c - 2$ sections, respectively, and Δp_{fr} the friction loss on the passage wall. Here,

$$\xi_i = \xi_{1-c} + \xi_{c-2} + \xi_{\mathrm{fr}}; \quad \mu = \frac{\varepsilon}{\sqrt{1 + \xi_{1-c} + \xi_{c-2} + \xi_{\mathrm{fr}}}} \qquad (13)$$

The coefficient ξ_{1-c} represents the energy loss associated with the vortex generation when the liquid flows into the injector passage. Figure 22 shows the

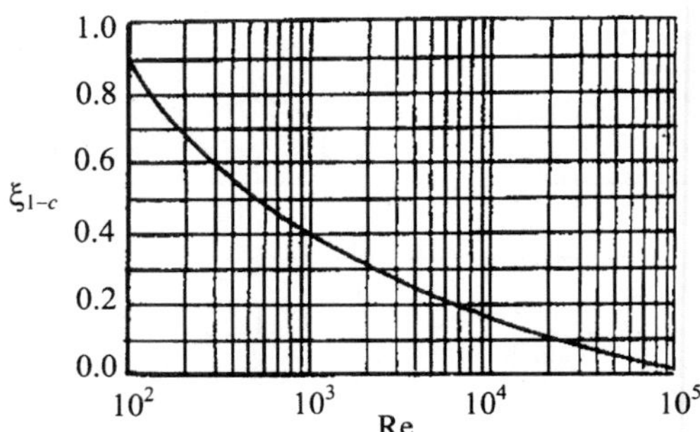

Fig. 22   Effect of Reynolds number on coefficient ξ_{1-c}.

measured ξ_{1-c} vs the Reynolds number ($Re = U_i d_i / v$) on a logarithmic scale. It decreases with increasing Reynolds number. The coefficient ξ_{c-2} represents the energy losses associated with the vortices formed when the flow expands after the contraction at the $c - c$ station. The explicit expression for ξ_{c-2} can be derived from the following equations:

Bernoulli's equation

$$p_n + \frac{\rho U_n^2}{2} = p_2 + \frac{\rho U_2^2}{2} + \Delta p_{c-2} \tag{14}$$

where the subscript n denotes the flow properties for an ideal injector without flow contraction.

Momentum conservation

$$(p_2 - p_n)A_2 = \rho U_2 A_2 (U_n - U_2) \tag{15}$$

Mass continuity

$$\rho U_2 A_2 = \rho U_n A_n \tag{16}$$

A simple manipulation of Eqs. (14)–(16) gives the expression for the total-pressure loss:

$$\Delta p_{c-2} = \frac{\rho U_2^2}{2} \left(\frac{U_n}{U_2} - 1 \right)^2 \tag{17}$$

which is also known as the Borda-Carnot theorem. It follows from this equation that

$$\xi_{c-2} = \left(\frac{U_n}{U_2} - 1 \right)^2 \tag{18}$$

As is shown by experiments, this coefficient depends not only on the degree of flow contraction but also on the Reynolds number in the narrow passage section. However, for $Re > 10^5$, which, in fact, is always the case with LRE injectors, viscosity practically exerts no effect on ξ_{c-2}. It is solely determined by the inlet configuration that affects flow contraction, i.e., $\xi_{c-2} \equiv \xi_{in}$. Figures 23 and 24 show such effects on ξ_{in} for two different designs.[19] As the inlet convergence angle β increases, the value of ξ_{in} initially decreases, reaches a minimum at about 50 deg, and then increases. This optimum value decreases with an increase in the ratio of l_{in}/d_i. Figure 25 shows the ξ_{in} behavior when the injector passage has sharp inlet edges and is tilted with respect to the inlet plane.[19]

The coefficient ξ_{fr} represents the friction losses on the passage wall, and can be expressed as follows:

$$\xi_{fr} = \lambda l_i/d_i \tag{19}$$

where λ is the drag coefficient. For hydraulically smooth pipes in the turbulent regime ($Re > 4 \times 10^3$), λ can be expressed as $\lambda = 0.3164 Re^{-0.25}$. A passage is considered hydraulically smooth if the wall-surface roughness, which depends on the quality of injector production, is less than $0.007 d_i$.

The hydraulic-loss coefficient of a jet injector takes the form

$$\xi_i = \xi_{1-c} + \xi_{in} + \lambda l_i/d_i \tag{20}$$

D.   Design Procedure

Once the theory behind the operation of jet injectors is established, the following steps can be used to design a jet injector based on the equations derived in the preceding section. These are eight equations and one experimental correlation for 11 variables (i.e., Δp_i, $\dot{m}_i$, l_i, Re, U_i, d_i, λ, ξ_{fr}, μ, ξ_{in}, and ξ_{1-c}). The system can be easily solved using an iterative scheme if two of these parameters such as Δp_i and $\dot{m}_i$ are known:

1) Specify initial data of Δp_i, m_i, l_i, ρ, and ν for the expected injector design.
2) Initiate the calculation with the flow coefficient $\mu = 1/\sqrt{1 + \xi_{in}}$ as first approximation.

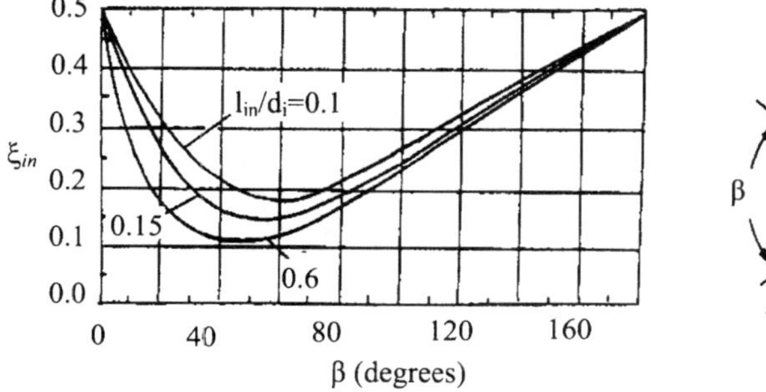
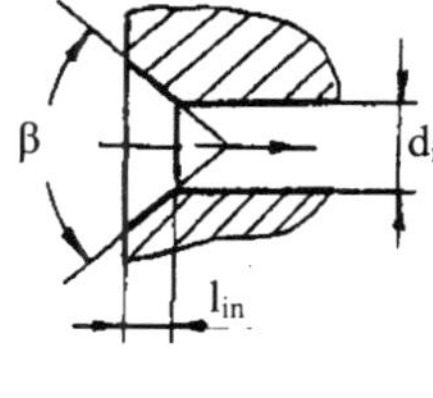

Fig. 23   Effect of inlet-edge contraction on coefficient ξ_{in} in jet injectors.

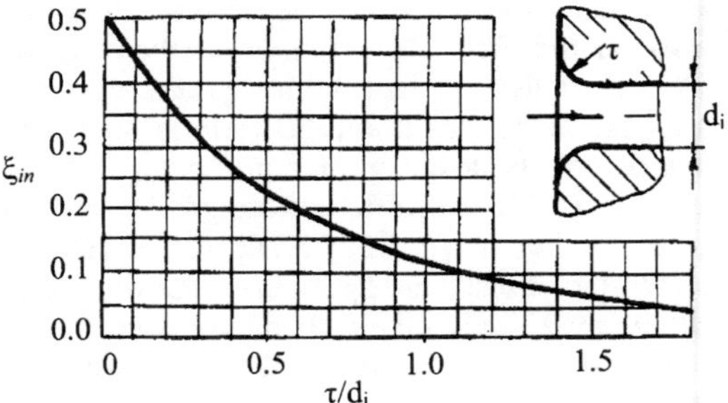

Fig. 24 **Effect of inlet-edge rounding on coefficient ξ_{in} in jet injectors.**

3) Calculate the diameter of the injector passage, $d_i = 0.95\dot{m}_i^{0.5}(\mu)^{-0.5} \times (\rho\Delta p_i)^{-.25}$.

4) Determine the velocity and Reynolds number based on the values obtained in the previous steps with $U_i = 1.273\dot{m}_i\rho^{-1}(d_i)^{-2}$.

5) Determine λ and ξ_{fr} from their respective relations, $\lambda = 0.3164Re^{-0.25}$ and $\xi_{fr} = \lambda l_i/d_i$.

6) Determine ξ_{1-c} from Fig. 22 and calculate $\xi_i = \xi_{1-c} + \xi_{in} + \xi_{fr}$.

7) Update the flow coefficient with $\mu = 1/\sqrt{1 + \xi_i}$ and repeat steps 3–7 until the calculated coefficient μ converges.

The injector diameter d_i usually falls in the range of 0.8–2.5 mm because a small diameter is susceptible to clogging and a large diameter gives rise to poor atomization quality and increased spray length. If d_i falls outside the aforementioned limits, some changes should be introduced into the mixture formation process and the mass flow rate $\dot{m}_i$ should either be decreased or increased.

The possibility of occurrence of cavitation in the injector passage should be checked. Assuming that $p_2 \equiv p_s$ and $\xi_{c-2} \equiv \xi_{in}$, from Eq. (14) we have

$$p_s - p_n = \frac{\rho U_2^2}{2}\left(\frac{U_n^2}{U_2^2} - 1\right) - \frac{\rho U_2^2}{2}\left(\frac{U_n}{U_2} - 1\right)^2 = 2\sqrt{\xi_{in}}\frac{\rho U_2^2}{2} \tag{21}$$

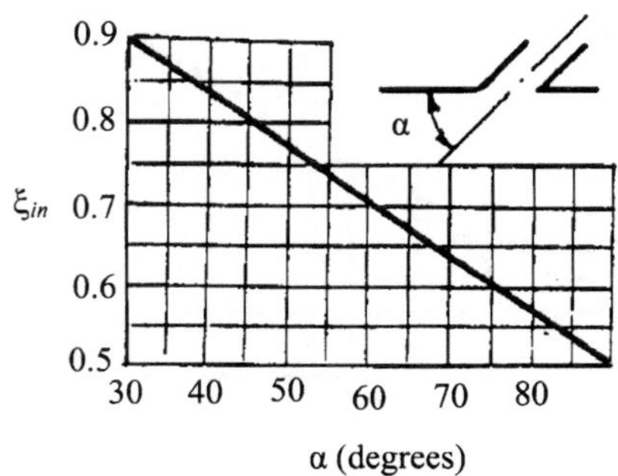

Fig. 25 **Effect of tilt angle of injector passage on coefficient ξ_{in}.**

Then,

$$\frac{\Delta p_i}{p_s - p_n} = \frac{1 + \xi_{1-c} + \xi_{in} + \lambda l_i/d_i}{2\sqrt{\xi_{in}}} = \eta \tag{22}$$

The condition for preventing cavitation becomes

$$p_n = p_s - \frac{\Delta p_i}{\eta} \geq p_v \tag{23}$$

where p_v is the vapor pressure of saturated propellant.

III.   Theory and Design of Gaseous Monopropellant Jet Injectors

A.   Flow Characteristics

The delivery of gaseous combustible mixtures produced in the gas generator into the main combustion chamber is achieved with the aid of jet injectors. The general theory of gas injectors is based on the conservation of mass and energy and the equation of state. The mass flow rate of an injector is expressed as

$$\dot{m}_i = \mu \rho_2 U_2 A_2 \tag{24}$$

where μ is the flow coefficient defined in Eq. (11). The gas density at the injector exit is obtained by assuming an isentropic process through the flow passage:

$$\rho_2 = \rho_{01} \left[\frac{p_2}{p_{01}} \right]^{1/k} \tag{25}$$

The ideal exit velocity U_2 becomes

$$U_2 = \sqrt{2 \frac{k}{k-1} R T_{01} \left[1 - \left(\frac{p_c}{p_{01}} \right)^{\frac{k-1}{k}} \right]} \tag{26}$$

where R denotes the gas constant and k the ratio of specific heats. The total pressure of the gas flow p_{01} is the sum of the chamber pressure p_c and the pressure drop across the injector Δp_i, i.e., $p_{01} = p_c + \Delta p_i$. The total temperature T_{01} equals the gas temperature in the manifold. The exit velocity reaches its maximum U_{th}, in the limit of $p_c = 0$. We define the velocity coefficient λ as

$$\lambda_2 = \frac{U_2}{U_{th}} = \frac{U_2}{\sqrt{2(k/k+1)RT_{01}}} = \sqrt{\frac{k+1}{k-1} \left[1 - \left(\frac{p_c}{p_{01}} \right)^{\frac{k-1}{k}} \right]} \tag{27}$$

Substitution of Eq. (27) into Eq. (24) gives the injector mass flow rate in terms of λ:

$$\dot{m}_i = \mu \frac{(p_c + \Delta p_i)A_n}{C^*} q(\lambda_2) \tag{28}$$

where C^* is the characteristic velocity[20] defined as

$$C^* = \frac{\sqrt{kRT_{01}}}{k\sqrt{[2/(k+1)]^{(k+1)/(k-1)}}} \tag{29}$$

The gasdynamic coefficient function $q(\lambda_2)$ takes the form

$$q(\lambda_2) = \left(\frac{k+1}{2}\right)^{1/(k-1)} \lambda_2 \left(1 - \frac{k-1}{k+1}\lambda_2^2\right)^{1/(k-1)} \tag{30}$$

The flow coefficient μ of a gas injector for incompressible fluids can be written as

$$\mu = \frac{1}{\sqrt{1+\xi_i}} \tag{31}$$

In evaluating the flow loss coefficient ξ_i, the high permeability of the gas injector assembly should be taken into account. The ratio of the area occupied by the gas flow in the $c - c$ cross section, A_{gas}, to the total injector assembly area, A_{total}, is rather high compared with liquid injector assemblies. This highly packed situation tends to straighten the flow at the injector entrance and reduce the subsequent flow contraction in the injector passage. To account for the effect of neighboring injectors, the following empirical relationship is used:

$$\xi_{c-2} = \xi_{\text{in}}(1 - d_i^2/d_1^2) \tag{32}$$

where d_1 is the diameter of the gas flow before entering the injector (i.e., at the $1-1$ cross section). It is taken numerically to be the average distance between the axes of the neighboring injectors. In addition, as a consequence of high-intensity turbulence, viscosity exerts no influence on the local hydraulic losses, i.e., $\xi_{1-c} = 0$. Thus,

$$\xi_i = \xi_{\text{in}}(1 - d_i^2/d_1^2) + \lambda l_i/d_i \tag{33}$$

B. Design Procedure

Several parameters are usually specified at the engine system design stage. These include the mass flow rate of the injector assembly and associated flow conditions, diameter of the injector assembly, pressure in the combustion chamber, and pressure drop across injectors. The type of injectors and their surmised number are also provided. The problem can be formalized to determine the injector passage diameter for the prescribed conditions. The numbers of unknown parameters and equations used to close the formulation are the same as those for liquid injectors. There are, however, three additional unknowns, λ_2, C^*, and $q(\lambda_2)$, which can be solved using Eqs. (27), (29), and (30), respectively.

The injector design proceeds in the following steps:

1) Assume a value for the flow coefficient μ.

2) Calculate $p_{01} = p_c + \Delta p_i$ and determine the velocity coefficient λ_2 from Eq. (27).

3) Calculate $q(\lambda_2)$ using Eq. (30).

4) Calculate C^* from Eq. (29).

5) Determine $A_n = \dot{m}_i C^*/\mu p_{01} q(\lambda_2)$ and $d_i = 1.128\sqrt{[\dot{m}_i C^*/\mu p_{01} q(\lambda_2)]}$.

6) Calculate ξ_i from Eq. (33) and determine $\mu = 1/\sqrt{1 + \xi_i}$.

7) Repeat steps 3–6 until the calculated d_i converges.

If calculations show the lack of space to accommodate the intended number of injectors with the calculated passage diameter d_i on the injector assembly (e.g., the diameter d_i is too large), changes should be made in the engine design by increasing the pressure drop Δp_i since d_i is inversely proportional to Δp_i. If, on the contrary, the injectors with the diameter d_i prove to underutilize the area of the injector assembly (i.e., low permeability of the injector assembly due to small d_i), then a lower value of Δp_i should be implemented. The injector design is considered to be completed when the parameters in Eq. (28) are correlated not only with each other, but also with the engine parameters of the propulsion system and the design of the injector assembly.

IV.   Theory and Design of Gas-Liquid Jet Injectors

The main objective of a gas-liquid mixing element is to provide a uniform initial distribution of liquid propellant through the gas flow. This can be achieved by introducing thin liquid jets into the gas flow. Numerous versions of injectors of this type have been designed and fall into the three categories shown schematically in Fig. 26. Slit injectors with gas passages made in the form of concentric slots in the injector bottom represent a derivative of the preceding designs. The principle behind slit injectors and gas-liquid spray injectors is identical. As shown by practice, slit injectors having high permeability are less susceptible to high-frequency instability.

Keeping all of the other factors the same, injectors with external propellant mixing (see Fig. 26a) lead to lower combustion efficiency and therefore have

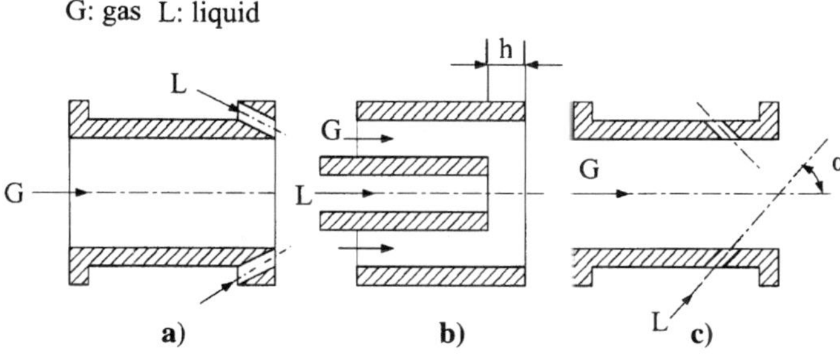

Fig. 26   Major designs of gas-liquid jet-jet injectors.

not found wide applications. Liquid propellant jets can be introduced into the gas flow over a range of injection angle of $\alpha = 0$ to 90 deg. The value of α influences the depth of liquid jet penetration into the gas flow, distribution of droplets, and eventually combustion efficiency. An injector in which the liquid-propellant jet is introduced along the gas-flow axis ($\alpha = 0$ deg) is referred to as coaxial (see Fig. 26b). Injectors of this type have been implemented in liquid oxygen/hydrogen engines, where liquid oxygen is fed through the central passage and gasified hydrogen is delivered through the circumferential passage. The gas-flow velocity should far exceed the liquid-jet velocity to achieve effective atomization. The recess distance between the central post and outer tube also plays an important role in determining the efficiency and stability of injector operation.

As a specific example illustrating the injector design, we consider the following injector geometry and operating conditions (the corresponding nomenclature is given in Fig. 27):

1) injector geometry (taken based on recommendations)

$$d_g = 20 \, \text{mm} \quad d_l = 2.2 \, \text{mm} \quad \delta = 4 \, \text{mm}$$
$$\Delta l = 15 \, \text{mm} \quad l_{\text{in}} = 12 \, \text{mm} \quad l_i = 45 \, \text{mm}$$
$$\alpha = 45 \, \text{deg} \quad \beta = 30 \, \text{deg} \quad t_i = 30 \, \text{mm}$$

2) operating parameters (obtained from the engine design and preliminary analysis of mixture formation)

$$\dot{m}_{i,g} = 2.21 \, \text{kg/s}; \quad \dot{m}_{i,l} = 0.516 \, \text{kg/s}; \quad p_c = 150 \cdot 10^5 \, \text{N/m}^2$$
$$(RT)_g = 180{,}000 \, \text{J/kg}; \quad \rho_l = 785 \, \text{kg/m}^3$$

The objectives are

1) to construct the liquid jet trajectory in the injector passage;
2) to calculate the flow coefficients of the gas and liquid passages;
3) to evaluate the pressure drops of $\Delta p_{i,g}$ and $\Delta p_{i,l}$, and the pressures of p_g and p_l required to provide the prescribed gas and liquid mass flow rates through the injector.

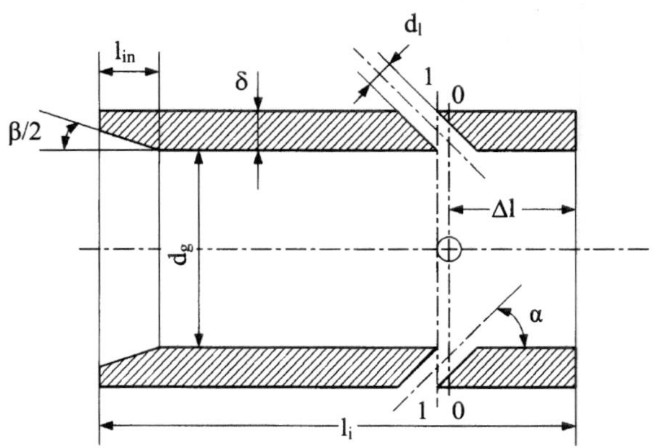

Fig. 27 Schematic diagram of gas-liquid injectors.

Liquid jet trajectory. The analysis of the liquid trajectory in the gas flow is based on the results of experimental and theoretical studies of the jet shape in a cross flow. The gas flow in the passage tends to bend the injected liquid jet more severely as compared to the situation with an unbounded flow. The following expression is usually used for the external boundary of the jet:

$$\bar{x}_{\text{ext}} = 1.1\,\bar{y}^2 \rho_g U_g^2 / \rho_l U_l^2 + \bar{y}\tan(90 - \alpha_j) \tag{34}$$

where $\bar{x}_{\text{ext}} = x_{\text{ext}}/d_l$ is the longitudinal coordinate of the external jet boundary reckoned from the 1-1 section, $\bar{y} = y/d_l$, the radial coordinate from the internal surface of the gas passage, and α_j the angle cf the jet exhausted from the passage. If $\delta/d_l > 1.0$, the jet direction at the exit can be assumed to coincide with the passage direction, i.e., $\alpha_j = \alpha$. As expected, the momentum ratio of the gas to the liquid flow, $\rho_g U_g^2 / \rho_l U_l^2$, and the injection angle α determine the jet trajectory.

Flow coefficient of gas passage. A gas-liquid injector is usually designed using the results of cold-flow tests. The flow coefficient of the gas passage can be written in the following form:

$$\mu_g = \frac{1}{\sqrt{1 + \xi_{\text{in}}(1 - d_g^2/l_i^2) + \xi_{l,j}}} \tag{35}$$

where $\xi_{\text{in}}(1 - d_g^2/l_i^2)$ is the hydraulic-loss coefficient at the inlet to the gas passage (see Fig. 23). It depends on the ratio between the gas and liquid heads, the angle of the jet entering the flow, and the number and diameters of jets. Figure 28 plots $\xi_{l,j}$ vs $\rho_g U_g^2 / \rho_l U_l^2$ and α for the case of four jets.

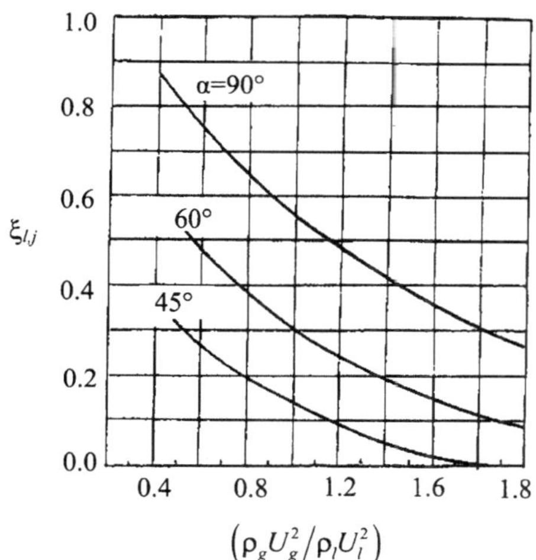

Fig. 28   Effect of momentum ratio on μ_g for gas-liquid injectors.

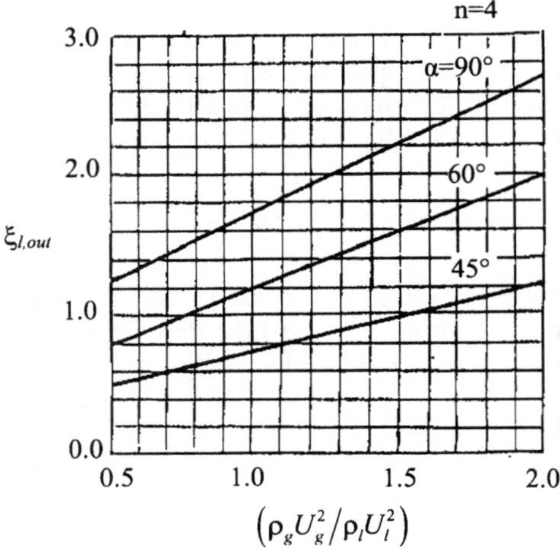

Fig. 29   Effect of momentum ratio on μ_l for gas-liquid injectors.

Flow coefficient of liquid passage.   In calculating the flow coefficient for the liquid passage μ_l, an increase in the pressure loss due to tilted entry of the liquid into the cylindrical gas passage should be taken into consideration. Furthermore, μ_l is also influenced by the conditions outside the passage, which depend on the energy of the gas flow and the injection angle α. If we ignore these factors, μ_l becomes

$$\mu_l = \frac{1}{\sqrt{1 + \xi_{in} + \xi_{l.out}}} \tag{36}$$

The coefficient ξ_{in} represents the hydraulic loss at the inlet to the cylindrical tilting passage with sharp edges (see Fig. 25), and $\xi_{l.out}$ is the coefficient characterizing the additional losses when the tilted jet is introduced into the gas flow. Figure 29 shows the effects of the momentum ratio and injection angle on $\xi_{l.out}$, based on the data of cold-flow tests. In calculating μ_l, the length of the tilted passage is assumed to be greater than its diameter, i.e., $\delta > d_l$. If the gas injector wall is thinner, the effect of wall thickness on the flow coefficient should be taken into account.[17,19]

V.   Theory and Design of Liquid Monopropellant Swirl Injectors

Although the fundamentals of swirling flow dynamics were established more than 60 years ago by G. N. Abramovich[8] in 1944 and independently by Taylor[12] in 1947, the hydraulic characteristics of a liquid swirl injector remains a complicated problem since fluid properties and injector geometric parameters have pronounced and sometimes conflicting effects on injector characteristics. For example, the mass flow through a swirl injector increases with an increase in liquid viscosity, while the situation is reversed in jet injectors, despite the fact that the general trend of the two types of injectors is identical for ideal fluids.

Publications on the effect of injector geometric and operating parameters on spray characteristics often report contradictory results. Numerous and quite different empirical expressions[16] used to calculate the flow properties of swirl injectors do not allow evaluation of the effect of the primary parameters, and can only be applied to the specific configuration of concern or those similar to it.

A.   Flow Characteristics of Ideal Swirl Injector

To gain a fundamental understanding of the flow development in a swirl injector and various underlying parameters, we first consider an ideal situation by ignoring the flow non-uniformity associated with the discrete tangential passages. Figure 30 shows a typical swirl injector consisting of a hollow casing (1), an axisymmetric vortex chamber (2), a nozzle (3), and tangential passages (4) connected upstream with the propellant feed system. Liquid propellant enters the vortex chamber at the velocity U_{in}, forms a circumferential swirling flow, exhausts at the axial velocity U_{an} through the nozzle, and finally establish a near-conic sheet in the mixture-formation zone. Ideally, the liquid sheet has the shape of a hyperboloid of revolution of one nappe. The spreading angle is determined by the ratio between the circumferential and axial components of the liquid velocity near the injector exit.

In such a swirl injector, the pressure all over the internal surface of the liquid vortex is equal to the pressure in the combustion chamber. The liquid potential energy in the form the pressure drop across the injector is fully converted to the kinetic energy. Thus, the liquid flow velocity on the surface becomes

$$U_{\Sigma} = \sqrt{\frac{2}{\rho}\Delta p_i} = \sqrt{U_u^2 + U_r^2 + U_a^2} \qquad (37)$$

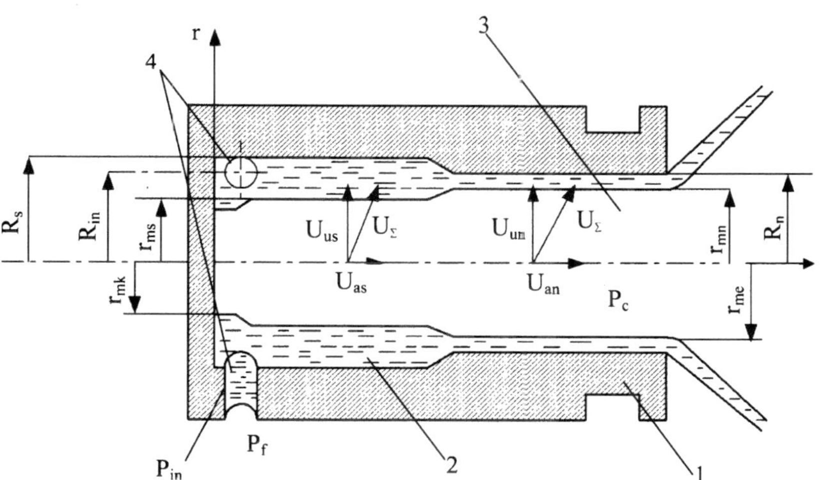

Fig. 30   Schematic diagram of liquid flow in swirl injector; 1-injector casing; 2-vortex chamber; 3-nozzle passage; 4-tangential passages.

On the injector wall $U_r = 0$, and

$$U_\Sigma = \sqrt{U_{us}^2 + U_{as}^2} = \sqrt{U_{un}^2 + U_{an}^2} = U_{uk} \qquad (38)$$

The subscripts k, s, and n denote the conditions at the injector head end, vortex chamber, and nozzle, respectively. At the injector head end, $U_a = 0$. The circumferential component of the liquid velocity U_{uk} is maximum and the radius of the liquid-vortex surface, on the contrary, is minimum. In the vortex chamber, the axial velocity U_{as} is positive and the circumferential velocity U_{us} is smaller than U_{uk}, giving $r_{ms} > r_{mk}$. In the nozzle, the smaller liquid passage area leads to an increase of the axial velocity U_{an} and a decrease of U_{un}, giving $r_{mn} > r_{ms}$. Finally, at the nozzle exit, the centrifugal force arising from the swirling motion acts as a velocity head, leading to an additional increase of the axial velocity and subsequently an increase of the liquid-surface radius r_{me}. The swirling-liquid flow in the field of centrifugal force bears a resemblance to the liquid flow through a dam in the field of gravitational force of the Earth. According to N. E. Zhukovsky,[21] the longitudinal velocity along the dam cannot exceed the velocity of surface-wave propagation, much as the velocity of a gas flow in a pipe of constant cross section cannot exceed the sound velocity. The concept of a critical liquid flow in a swirl-injector nozzle results from the principle of maximum flow postulated by G. N. Abramovich[8] and later proved by L. A. Klyachko[2] in 1962. It serves as the basis of modern theories of swirl injectors. The whole theory of an ideal swirl injector is based on three principles, namely, Bernoulli's equation, conservation of mass energy, and conservation of angular momentum.

The swirl injector design involves 20 main parameters listed in Table 1. The variable r with subscript stands for the radius of the liquid film, and R with subscript the radial dimension of the injector.

All of these parameters can be related to each other through Bernoulli's equation and conservation of mass, energy, and angular momentum. The total velocity can be determined in terms of the pressure drop $(p_f - p_c)$ using Bernoulli's equation:

$$U_\Sigma = \sqrt{2(p_f - p_c)/\rho} \qquad (39)$$

where p_f is the pressure in the propellant feed system for tangential channels and p_c the chamber pressure. The mass flow rate can be expressed in terms of the total velocity U_Σ and the nozzle area A_n as well as the mass flow coefficient:

$$\dot{m} = \mu A_n \sqrt{2\rho(p_f - p_c)} \qquad (40)$$

The total velocity is the vectorial sum of its three components:

$$U_\Sigma = \sqrt{U_{un}^2 + U_{an}^2 + U_{rn}^2} \qquad (41)$$

Table 1   Parameters involved in swirl injector design

Parameter	Definition
$\dot{m}$	Mass flow rate
p_f	Pressure in propellant feed system
p_c	Combustion chamber pressure
p_{in}	Inlet pressure at tangential channel
α	Spray cone angle
μ	Mass flow coefficient
φ	Fractional area occupied by liquid in nozzle
h	Liquid film thickness
U_Σ	Total velocity
U_{un}	Swirl velocity in nozzle
U_{an}	Axial velocity in nozzle
U_{rn}	Radial velocity in nozzle
U_{uk}	Swirl velocity at head end of vortex chamber
U_{rk}	Radial velocity at head end of vortex chamber
U_{in}	Velocity at entrance of vortex chamber
r_{mk}	Radius of liquid film at head end of vortex chamber
r_{mn}	Radius of liquid film in nozzle
A_n	Area of nozzle
R_{in}	Radial location of center of tangential channel
A	Geometrical characteristic parameter

Application of Bernoulli's equation at the tangential entry gives

$$U_{in} = \sqrt{2(p_f - p_{in})/\rho} \tag{42}$$

where p_{in} is the tangential inlet pressure. It has been assumed in ideal injector theory that the radial of velocity in the liquid film is zero in the vortex chamber and nozzle:

$$U_{rk} = U_{rn} = 0 \tag{43}$$

According to the conservation of angular momentum, the azimuthal velocity satisfies the following relations:

$$U_{um}r_m = U_{uk}r_{mk} = U_{in}R_{in} = U_{rn}r_{mn} \tag{44}$$

Since U_a and U_r are zero at the liquid surface at the head end of the vortex chamber, the total velocity equals the circumferential velocity:

$$U_\Sigma = U_{uk} \tag{45}$$

Equation (44) implies that the swirl velocity becomes infinity as the radius approaches zero. Since the angular velocity cannot be infinite, a gas core must be present, and the liquid will not fully occupy the entire injector. We can thus define a parameter φ, known as the coefficient of passage fullness, that relates the area filled by the liquid to the nozzle area:

$$\varphi = \frac{\pi(R_n^2 - r_{mn}^2)}{\pi R_n^2} = 1 - \frac{r_{mn}^2}{R_n^2} \tag{46}$$

where R_n is the radius of the nozzle and r_{mn} is the radius of liquid film in the nozzle. Similarly, a non-dimensional parameter μ, known as the mass flow coefficient, is defined that relates the actual flow rate to the maximum possible flow rate through the nozzle. The two parameters can also be related to each other.

$$\mu = \frac{\rho U_{an} A_n \varphi}{\rho U_\Sigma A_n} = \frac{U_{an}\varphi}{U_\Sigma}$$

where U_{an} is the axial velocity in the nozzle. Thus,

$$\mu = \varphi\sqrt{\frac{U_\Sigma^2 - U_{un}^2}{U_\Sigma^2}} = \varphi\sqrt{1 - \frac{r_{mk}^2}{r_{mn}^2}} \tag{47}$$

Various geometrical parameters can be correlated to form a non-dimensional geometrical characteristic parameter defined by

$$A = A_n R_{in}/A_{in} R_n \tag{48}$$

where R_n is the nozzle radius, R_{in} the radial location of the center of the inlet passage, and A_{in} is the total area of inlet passages. Finally, the spreading angle of the liquid sheet at the nozzle exit, α, can be expressed in terms of the velocity components:

$$\tan\alpha = U_{un}/U_{an} \tag{49}$$

Parameters such as $\dot{m}, p_f, p_c, p_{in}$, and α are design specifications that are chosen at the engine system design stage. Thus, there are in total 15 unknown parameters involved in the injector analysis. Since there are only 14 equations, Eqs. (39)–(49), an additional equation is required, which can be found using the principle of maximum flow or an alternative differential volume approach as shown next. Introducing $P_t = P_f - P_c$ for this derivation and equating the pressure and centrifugal forces on a liquid element of radius r with width dr, length $rd\Phi$, and unit thickness, we have

$$rd\Phi dP = dm\frac{U_u^2}{r} \tag{50}$$

where U_u is the circumferential velocity at radius r inside the liquid film, and $dm = \rho r dr d\Phi$. The conservation of angular momentum,

$$U_u = U_{um} r_m/r \tag{51}$$

leads to

$$dP = \rho U_{um}^2 r_m^2 \frac{dr}{r^3} \tag{52}$$

where U_{um} is the circumferential velocity at the liquid film radius r_m. Integrating the preceding equation and applying $P = 0$ at $r = r_m$, we have

$$P = \frac{\rho}{2}(U_{um}^2 - U_u^2) \tag{53}$$

From Bernoulli's equation,

$$\frac{\rho}{2}(U_a^2 + U_u^2) + P = P_t \tag{54}$$

Substituting the value of P from Eq. (53) into Eq. (54), and rearranging the result, we have

$$U_a = \sqrt{\frac{2P_t}{\rho} - U_{um}^2} \tag{55}$$

The conservation of angular momentum gives

$$U_{um} = \frac{U_{in}R_{in}}{r_m} \tag{56}$$

The total volumetric flow rate can be expressed as the product of the inlet passage area and velocity U_{in}:

$$Q = n\pi r_{in}^2 U_{in} \tag{57}$$

where n is the number of tangential inlet passages and r_{in} the radius of tangential inlet passage. Substituting the value U_{um} from Eq. (56) in Eq. (55) and replacing the value of U_{in} from Eq. (56), we obtain

$$U_a = \sqrt{\frac{2P_t}{\rho} - \frac{R_{in}^2 Q^2}{n^2 \pi^2 r_{in}^4 r_m^2}} \tag{58}$$

The definition of the coefficient of passage fullness, φ, gives

$$Q = \varphi(\pi R_n^2 U_a) \tag{59}$$

With the preceding three equations, we may eliminate U_a and substitute for the parameter A to obtain the following expression for the volumetric flow rate Q:

$$Q = \frac{1}{\sqrt{\dfrac{A^2}{1-\varphi} + \dfrac{1}{\varphi^2}}} \pi R_n^2 \sqrt{\frac{2P_t}{\rho}} \tag{60}$$

Since $\sqrt{2P_t/\rho}$ represents the total velocity, $\pi R_n^2 \sqrt{2P_t/\rho}$ is the total volumetric flow rate possible through the nozzle. By substituting the definition of the flow coefficient μ, Eq. (60) can be rearranged as follows:

$$\mu = \frac{1}{\sqrt{\dfrac{A^2}{1-\varphi} + \dfrac{1}{\varphi^2}}} \tag{61}$$

The flow discharge coefficient depends on the injector geometric parameter A and the coefficient of passage fullness φ. If φ is decreased, then the decrease in the equivalent flow area is faster than the increase in the axial velocity, and so the mass flow rate decreases. Similarly, for an increase in φ, the decrease in the axial velocity is faster than the increase in the equivalent flow area, and so the mass flow rate decreases. Thus there exists an optimum maximum mass flow rate (or the discharge coefficient). Figure 31 shows the mass flow coefficient as a function of the φ for various values of the geometric characteristic A. A maximum value of μ for each value of A is observed, indicating the existence of the maximum flow rate for a given value of A. Application of the condition $d\mu/d\varphi = 0$ gives the optimum value of the discharge coefficient:

$$\mu = \varphi \sqrt{\frac{\varphi}{2-\varphi}} \tag{62}$$

Equation (62) is the final equation required to close the formulation for the injector analysis. The axial velocity in the cylindrical nozzle is

$$U_{an} = \frac{\mu}{\varphi} \sqrt{\frac{2}{\rho}(p_f - p_c)} \tag{63}$$

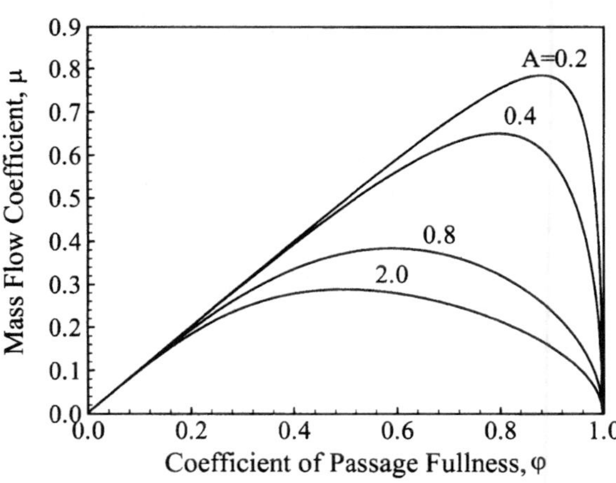

Fig. 31   Effect of coefficient of passage fullness on mass flow coefficient for various geometric characteristic parameters.

According to the flow continuity condition, the liquid mass flow rates in the nozzle and tangential passages are equated:

$$\mu A_n \sqrt{2\rho(p_f - p_c)} = \mu_t n \pi R_{\text{in}}^2 \sqrt{2\rho(p_f - p_{\text{in}})} \tag{64}$$

We now determine the azimuthal velocity U_{um} at some arbitrary point of the liquid vortex at the radius r_m. The conservation of angular momentum gives

$$U_{um} = \frac{R_{\text{in}}}{r_m} U_{\text{in}} = \frac{r_{mk}}{r_m} U_\Sigma = \frac{R_{\text{in}}}{r_m} \sqrt{\frac{2}{\rho}(p_f - p_{\text{in}})} \tag{65}$$

where U_Σ is the idealized total liquid velocity. Application of Bernoulli's theorem to the cylindrical part of the nozzle passage leads to

$$p_f = p_c + \frac{\rho}{2}(U_{un}^2 + U_{cn}^2) \tag{66}$$

Substitution of Eqs. (63) and (65) into Eq. (66) gives

$$p_f - p_c = \frac{\mu^2}{\varphi^2}(p_f - p_c) + \frac{R_{\text{in}}^2}{r_{mn}^2}(p_f - p_{\text{in}}) \tag{67}$$

A simple manipulation yields

$$\frac{p_f - p_{\text{in}}}{p_f - p_c} = \frac{\Delta p_{\text{in}}}{\Delta p_i} = \frac{1 - \mu^2/\varphi^2}{(R_{\text{in}}/r_{mn})^2} \tag{68}$$

With the aid of the principle of maximum flow, $\mu = \sqrt{\varphi^3/(2 - \varphi)}$, and $r_{mn} = R_n\sqrt{1 - \varphi}$, Eq. (68) results in the ratio of the pressure drops across the tangential passage and the injector as a whole:

$$\frac{\Delta p_{\text{in}}}{\Delta p_i} = \frac{2(1 - \varphi)^2/(2 - \varphi)}{(R_{\text{in}}/R_n)^2} \tag{69}$$

The preceding equation is valid only for injectors having $R_{\text{in}}/R_n > 1$; otherwise, it may give a physically unrealistic solution with $\Delta p_{\text{in}}/\Delta p_i > 1$, which makes no sense. Indirectly, this means that for such injectors the principle of flow maximum does not hold.

We now normalize all of the radii with respect to R_n and all of the velocities with respect to U_Σ to express injector parameters in terms of three non-dimensional parameters φ, μ, and A. To simplify notation, all the non-dimensional quantities are expressed with a bar over them. Thus the radius of the liquid film at the head end of the vortex chamber becomes

$$\bar{r}_{mk} = \sqrt{2(1 - \varphi)^2/(2 - \varphi)} \equiv \sqrt{a}$$

This quantity will be hereinafter included in many expressions. It is convenient to use because of its sole dependence on the injector geometric characteristic parameter A. Figure 32 shows the effects of the geometric characteristic parameter, A on various commonly used injector parameters. The azimuthal velocities U_{un} and U_{ue} increase with A, whereas the axial velocities U_{an} and U_{ae} decrease with increasing A. The coefficient of passage fullness φ and the mass flow coefficient μ also show a decrease with an increase in A.

At the head of the vortex chamber, $U_a = 0$, and

$$\frac{\Delta p_{in}}{\Delta p_i} = \frac{U_{in}^2}{U_{uk}^2} = \left(\frac{r_{mk}}{R_{in}}\right)^2 = \frac{(r_{mk}/R_n)^2}{(R_{in}/R_n)^2} \tag{70}$$

With Eq. (70) in mind, the physical meaning of the parameter $a \equiv \bar{r}_{mk}^2$ is evident. The ratio of the liquid film radius at the head end to that in the nozzle becomes

$$\bar{r}_{mk}^2/\bar{r}_{mn}^2 = a/(1 - \varphi) = 2(1 - \varphi)/(2 - \varphi) \equiv b \tag{71}$$

Equations (70) and (71) are employed to derive the components of the steady-state velocity in the injector element. Applying the conservation of angular momentum, $\bar{U}_{un}\bar{r}_{mn} = \bar{U}_{\Sigma}\bar{r}_{mk}$, we obtain the circumferential velocity in the nozzle:

$$\bar{U}_{un} = \bar{r}_{mk}/\bar{r}_{mn} = \sqrt{2(1 - \varphi)/(2 - \varphi)} \tag{72}$$

The axial velocity in the nozzle becomes

$$\bar{U}_{an} = \sqrt{1 - \bar{U}_{un}^2} = \sqrt{1 - 2(1 - \varphi)/(2 - \varphi)} = \sqrt{\varphi/(2 - \varphi)} \tag{73}$$

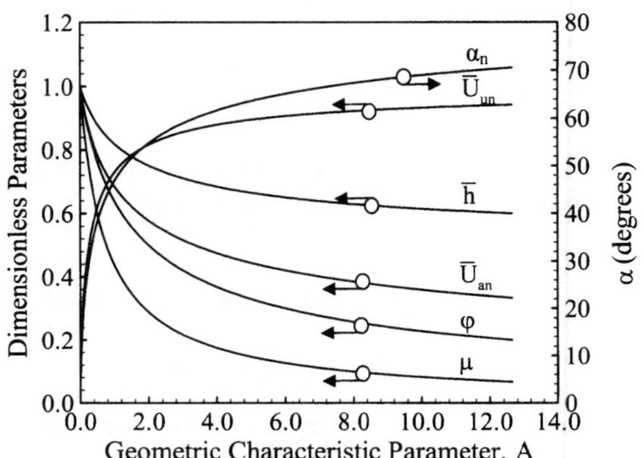

Fig. 32 Effects of geometric characteristic parameter A on other injector design and flow parameters.

Combining Eqs. (72) and (73) determines the spreading angle of the liquid sheet at the injector exit:

$$\alpha_n = \tan^{-1}(U_{un}/U_{an}) = \tan^{-1}(\bar{U}_{un}/\bar{U}_{an}) = \tan^{-1}\sqrt{2(1-\varphi)/\varphi} \tag{74}$$

Owing to the effect of centrifugal force, the liquid surface radius at the nozzle exit r_{me} is larger than r_{mn}, and the circumferential velocity decreases as well.[2] According to Eq. (66), the circumferential velocity downstream of the nozzle exit is

$$\bar{U}_{ue} = r_{mk}/R_n = \sqrt{2(1-\varphi)^2/(2-\varphi)} = \sqrt{a} \tag{75}$$

The axial velocity becomes

$$\bar{U}_{ae} = \sqrt{1-a} \tag{76}$$

Combining Eqs. (75) and (76) gives the spreading angle of the liquid sheet downstream of the injector exit:

$$\alpha_e = \tan^{-1}\sqrt{a/(1-a)} \tag{77}$$

It should be noted from Eq. (76) that the axial velocity at the nozzle exit,

$$U_{ae} = U_\Sigma\sqrt{1-a} = U_\Sigma\sqrt{(3-2\varphi)\varphi/(2-\varphi)} \tag{78}$$

exceeds that in the nozzle section. The ratio of the two velocities depends solely on the coefficient of passage fullness φ:

$$\lambda_e = \bar{U}_{ae}/\bar{U}_{an} = \sqrt{(1-a)(2-\varphi)/\varphi} = \sqrt{3-2\varphi} \tag{79}$$

In the limit of $\varphi \to 0$ (i.e., infinitesimally thin liquid film),

$$\bar{U}_{ae} = \sqrt{3}\,\bar{U}_{an} \tag{80}$$

The liquid velocity exceeds the critical velocity at the nozzle exit. Hence, the nozzle throat is offset by some distance from the exit. The relation between r_{me} and injector parameters can be expressed by the following transcendental equation[2]:

$$\mu = \sqrt{1-\mu^2 A^2} = \bar{r}_{me}\sqrt{\bar{r}_{me}-\mu^2 A^2} - \mu^2 A^2 \ln\frac{1+\sqrt{1-\mu^2 A^2}}{\bar{r}_{me}+\sqrt{\bar{r}_{me}-\mu^2 A^2}} \tag{81}$$

A more accurate evaluation of the liquid spreading angle α_e can be made by substituting r_{me} from Eq. (81) for a.

The velocity in the tangential passage, U_{in}, can be determined from Eq. (44):

$$\bar{U}_{in}\bar{R}_{in} = \bar{r}_{mk}$$

This implies

$$\bar{U}_{in} = \sqrt{a}/\bar{R}_{in} \tag{82}$$

We determine the velocity components on the liquid surface in the vortex chamber. In the root section of the vortex chamber, U_{am} is calculated from the condition of flow continuity:

$$\bar{U}_{am} = \bar{U}_{an}(\bar{R}_n^2 - \bar{r}_{mn}^2)/(\bar{R}_m^2 - \bar{r}_m^2) \tag{83}$$

Note that $\pi(\bar{R}_n^2 - \bar{r}_{mn}^2)$ is the flow area in the nozzle and $\pi(\bar{R}_m^2 - \bar{r}_m^2)$ is the flow area in the vortex chamber. Equation (83) can be written as

$$\bar{U}_{am} = \bar{U}_{an}\varphi/(\bar{R}_m^2 - \bar{r}_m^2) = \mu/(\bar{R}_m^2 - \bar{r}_m^2) \tag{84}$$

Applying the conservation of angular momentum, $U_m r_m = U_\Sigma r_{mk}$, and noting $U_{am}^2 + U_{um}^2 = U^2$ and Eq. (71), we obtain:

$$\bar{r}_m^2 = a/(1 - \bar{U}_{am}^2) \tag{85}$$

By substituting r_m from Eq. (85) into Eq. (84), U_{am} and r_m can be obtained by the successive approximation method.

B. Flow Characteristics of Real Swirl Injectors

The flow process in a real swirl injector can be described by taking into account viscous effects with the Navier-Stokes equations.[14] No analytical solutions are available for general cases, and the use of numerical calculations is inevitable.[26] In practice, the real conditions can be approximately taken into account by introducing the hydraulic loss coefficient, ξ_i, which characterizes the total-pressure loss in the injector, and the angular-momentum loss coefficient K:

$$p_f = p_c + \frac{\rho U_{un}^2}{2} + \frac{\rho U_{an}^2}{2} + \xi_i \frac{\rho U_{in}^2}{2}$$

Thus,

$$U_{an} = \sqrt{\frac{2}{\rho}\Delta p_i - \xi_i U_{in}^2 - U_{un}^2} \tag{86}$$

where $\Delta p_i = p_f - p_c$ is the pressure drop across the injector. The actual mass flow rate through the injector nozzle can be represented in the following form:

$$\dot{m}_i = \varphi\pi R_n^2 \rho U_{an} = \varphi\pi R_n^2\sqrt{2\rho\Delta p_i - \xi_i \rho^2 U_{in}^2 - \rho^2 U_{un}^2} \tag{87}$$

where

$$U_{\text{in}} = \frac{\dot{m}_i}{n\pi r_{\text{in}}^2 \rho} \quad \text{and} \quad U_{un} = \frac{KR_{\text{in}} U_{\text{in}}}{r_{mn}}$$

where K represents the loss of angular momentum. Substitution of the definitions of the geometric parameter A and the coefficient of nozzle opening, $\bar{R}_{\text{in}} = R_{\text{in}}/R_n$, gives

$$U_{\text{in}} = \frac{\dot{m}_i A}{\pi R_n^2 \rho \bar{R}_{\text{in}}} \tag{88}$$

$$U_{un} = \frac{KR_{\text{in}} U_{\text{in}}}{R_n\sqrt{1-\varphi}} = \frac{KR_{\text{in}}\dot{m}_i}{R_n\sqrt{1-\varphi}\,n\pi r_{\text{in}}^2 \rho} = \frac{K\dot{m}_i}{\rho\pi R_n^2\sqrt{1-\varphi}}A \tag{89}$$

Substitution of Eqs. (88) and (89) into Eq. (87) and rearrangement of the result yields an explicit expression for the flow characteristic parameter A that accounts for viscous losses:

$$\dot{m}_i = \mu_i \pi R_n^2 \sqrt{2\rho\Delta p_i} \tag{90}$$

where the mass flow coefficient μ_i takes the form

$$\mu_i \equiv \frac{1}{\sqrt{\dfrac{1}{\varphi^2} + \dfrac{A^2 K^2}{1-\varphi} + \xi_i \dfrac{A^2}{\bar{R}_{\text{in}}^2}}} \tag{91}$$

As is evident, the flow coefficient depends on the flow area ratio, φ, combination of the geometric dimensions, A and R_{in}, hydraulic losses, ξ_i, and angular momentum losses, K. With the use of the principle of maximum flow,

$$AK = \frac{(1-\varphi)\sqrt{2}}{\varphi\sqrt{\varphi}} \tag{92}$$

we have

$$\mu = \frac{1}{\sqrt{\dfrac{2-\varphi}{\varphi^3} + \xi_i \dfrac{A^2}{R_{\text{in}}^2}}} \tag{93}$$

The spray-cone angle is determined from the ratio between the circumferential and total velocities in the nozzle exit section: $\sin\alpha = U_{un}/U_{\Sigma n}$. Application of Eqs. (88) and (90) gives

$$U_{un} = \frac{R_n}{r_{mn}} \mu_i A K \sqrt{\frac{2}{\rho}\Delta p_i} \tag{94}$$

and the total velocity becomes

$$U_{\Sigma n} = \sqrt{U_u^2 + U_a^2 + U_r^2 - \xi_i U_{in}^2} = \sqrt{1 - \xi_i \mu_i^2 \frac{A^2}{\bar{R}_{in}^2}} \sqrt{\frac{2}{\rho} \Delta p_i} \qquad (95)$$

Equations (94) and (95) lead to the following equation for the liquid-sheet spreading angle:

$$\sin \alpha = \frac{R_n}{r_{mn}} \mu_i A \frac{K}{\sqrt{1 - \xi_i \mu_i^2 \frac{A^2}{\bar{R}_{in}^2}}} \qquad (96)$$

Equation (96) suggests that the spreading angle is different for liquid particles located at various distances r from the axis ($\sin \alpha \sim 1/r$). In calculations, with some inaccuracy assumed, the spray cone angle α corresponding to the average radius is used:

$$r_{av} = \frac{R_n + r_{mn}}{2} = \frac{R_n}{2} \left(1 + \sqrt{1 - \varphi}\right) \qquad (97)$$

Thus,

$$\sin \alpha = \frac{2\mu_i A K}{(1 + \sqrt{1 - \varphi})\sqrt{1 - \xi_i \mu_i^2 \frac{A^2}{\bar{R}_{in}^2}}} \qquad (98)$$

For ideal liquid ($\xi_i = 0$ and $K = 1$), with the neglect of the radial component of the liquid velocity, the spray cone angle 2α and flow parameters μ and φ are determined by the injector geometric characteristics alone and can be calculated as a function of A.

C. Effect of Viscosity on Injector Operation

Propellant viscosity and the ensuing friction losses affect the injector characteristics in terms of μ, α, and φ. The momentum loss measured by the coefficient K is first considered. For simplicity, the hydraulic losses are neglected with $\xi_i = 0$. The injector performance can be conveniently evaluated by the equivalent-injector characteristic parameter A_{eq} from Eq. (92):

$$A_{eq} \equiv AK = \frac{(1 - \varphi_{eq})\sqrt{2}}{\varphi_{eq}\sqrt{\varphi_{eq}}}$$

Consequently,

$$\mu_{eq} = \frac{\varphi_{eq}\sqrt{\varphi_{eq}}}{\sqrt{2 - \varphi_{eq}}}; \quad \text{and} \quad \mu_i = \frac{\mu_{eq}}{\sqrt{1 + \xi_i \mu_{eq}^2 \frac{A^2}{\bar{R}_{in}^2}}} \qquad (99)$$

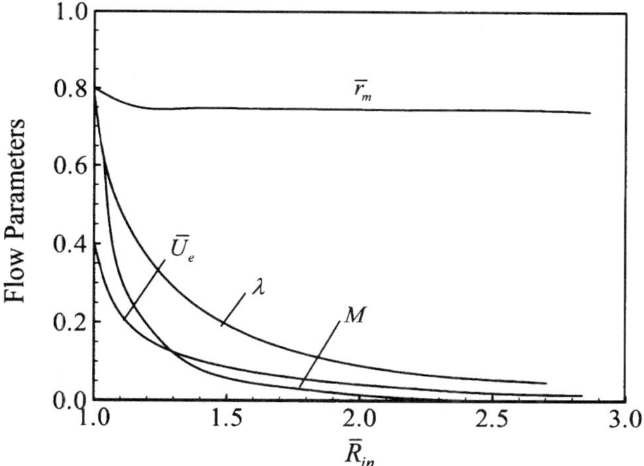

Fig. 33   Effect of nozzle opening on flow parameters of swirl-injector vortex chamber.

This method is convenient because the numerical relations obtained for an ideal injector and displayed graphically in Fig. 32 remain valid for the equivalent injector by replacing A with A_{eq}. Since $K < 1.0$ and $A_{eq} < A$ for the same injector under the effect of viscosity, it is evident from Fig. 33 that the momentum losses lead to increase of the mass flow (μ_{eq}) and passage fullness (φ_{eq}) coefficients and a decrease of the spray cone angle $2\alpha_{eq}$.

To evaluate A_{eq} for the suggested injector geometry ($A = R_{in}R_n/nr_{in}^2$) and to find real values of μ, α, and φ, the following expression can be used[23]:

$$A_{eq} = \frac{R_{in}R_n}{nr_{in}^2 + \dfrac{\lambda}{2}R_{in}(R_{in} - R_n)} \tag{100}$$

where

$$\lambda = 0.3164/(Re_{in})^{0.25} \quad \text{and} \quad Re_{in} = \frac{U_{in}r_{in}2\sqrt{n}}{\nu} = \frac{2\dot{m}_i}{\pi\sqrt{n}r_{in}\rho\nu} \tag{101}$$

Equation (100) characterizes the behavior of an open-type injector ($C = \bar{R}_{in} = 1.0$) with $R_{in} = R_n$, $A_{eq} \equiv A$. Unlike the momentum losses K, the total-pressure losses ξ_i decrease the flow coefficient. The main total-pressure losses occur in the inlet passages. For most designs, we can assume that

$$\xi_i = \xi_{in} + \lambda\frac{l_{in}}{d_{in}} \tag{102}$$

D.   Design Procedure

When designing an injector, the mass flow rate $\dot{m}_i$, pressure drop Δp_i, and propellant properties are usually known, and we need to evaluate the actual flow coefficient μ_i and the injector dimensions. The problem is reduced to correlating the parameters in Eq. (99). The calculation proceeds as follows:

1) Prescribe the spray cone angle based on the injector operating conditions (usually between 90 and 120 deg, lower values may be used for special cases).[1] The geometric characteristic parameter A and the flow coefficient μ_i are then determined from the plots in Fig. 32.

2) Determine the nozzle radius using

$$R_n = 0.475 \sqrt{\frac{\dot{m}_i}{\mu \sqrt{\rho \Delta p_i}}} \tag{103}$$

3) Specify the number of inlet passages (usually between two and four) and the coefficient of injector opening, based on structural considerations. Then, the radius of the inlet passage is obtained:

$$r_{\text{in}} = \sqrt{\frac{R_{\text{in}} R_n}{nA}} \tag{104}$$

4) Revise the following injector parameters:
    a) length of the tangential passages, usually $l_{\text{in}} = (3-6)r_{\text{in}}$;
    b) nozzle length, $l_n = (0.5-2)R_n$, vortex-chamber length $(l_s > 2R_{\text{in}})$ and vortex chamber radius $(R_s = R_{\text{in}} + r_{\text{in}})$

5) Find the Reynolds number in the inlet passages using

$$Re_{\text{in}} = 0.637 \frac{\dot{m}_i}{\sqrt{n} r_{\text{in}} \rho \nu}$$

and the friction coefficient using $\lambda = 0.3164/(Re_{\text{in}})^{0.25}$.

6) Determine A_{eq} using Eq. (100), and find μ_{eq} and α_{eq} from the plots of Fig. 32.

7) Calculate the hydraulic-loss coefficient in the tangential passages using

$$\xi = \xi_{\text{in}} + \lambda \frac{l_{\text{in}}}{2r_{\text{in}}}$$

The coefficient ξ_{in} is determined from the plots of Fig. 25 with allowance for the tilting angle of the tangential passage relative to the external surface of the vortex chamber, which can be calculated using the following formula:

$$\alpha = 90\,\text{deg} - \tan^{-1}\frac{R_s}{l_{\text{in}}} \tag{105}$$

8) Determine the actual flow coefficient μ_i using Eq. (99).

9) Calculate the nozzle radius using the new approximation

$$R_n = 0.475 \sqrt{\frac{\dot{m}_i}{\mu_i \sqrt{\rho \Delta p_i}}}$$

10) Calculate the geometric parameter A in the new approximation

$$A^{(1)} = \frac{R_{in}R_n^{(1)}}{nr_{in}^2}$$

11) Repeat steps 1–10 until the calculated injector parameters converge.

Another method of designing swirl injectors is also possible. It is based on the results of model experiments shown in Figs. 34 and 35. The advantage of this approach is its simplicity and adequate accuracy. The limited amount of experimental data is the major limitation. The design procedure is as follows:

1) Prescribe the spray cone angle and l_n/D_n from the plots in Fig. 34. Find the value of A, and then obtain $\mu_{exp} \equiv \mu_{in}$, using the plots in Fig. 34.
2) Calculate the nozzle radius using Eq. (103).
3) Prescribe the number of inlet passages and the coefficient of injector opening R_{in} based on structural considerations. Then, the radius of the inlet passage is obtained:

$$r_{in} = \sqrt{\frac{R_{in}R_n}{nA}}$$

4) Determine the Reynolds number Re_{in} in the inlet passages from Eq. (101), and use this result for design if $Re_{in} > 10^4$.
5) Determine the other injector parameters such as l_{in}, l_n, l_s, and R_s, find the relative liquid vortex radius $\bar{r}_m$ from the plots in Fig. 35, and then r_m from $r_m = \bar{r}_m R_n$.

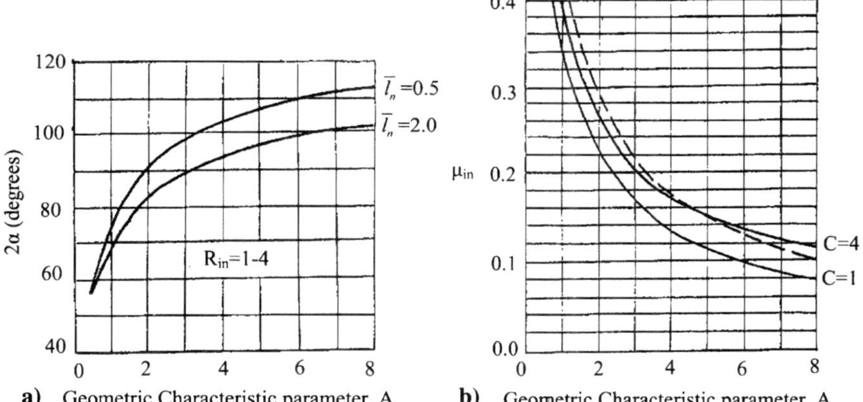

a)   Geometric Characteristic parameter, A   b)   Geometric Characteristic parameter, A

Fig. 34   Experimental plots of a) spray cone angle 2α and b) flow coefficient as functions of geometric parameter of swirl injector.

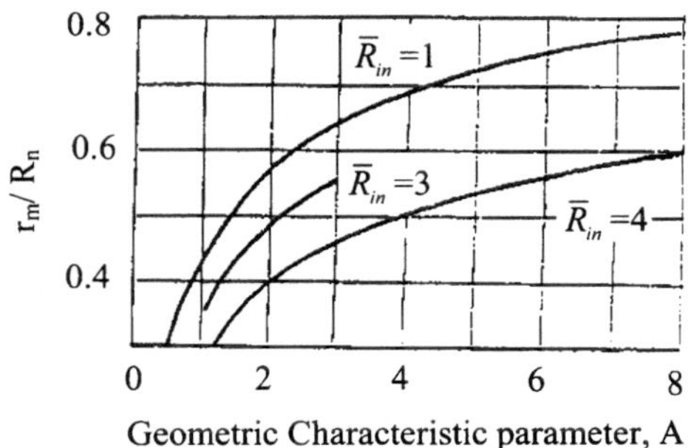

Fig. 35 Experimental plot of relative liquid vortex radius in vortex chamber as function of geometric parameter of swirl injector.

VI. Theory and Design of Liquid Bipropellant Swirl Injectors

In bipropellant injectors, the liquid-phase mixing occurs even before the liquid sheet starts disintegrating. Both designs with internal and external mixing have been implemented.

A. Injectors with External Mixing

Figure 36 shows three different configurations of injectors with external mixing. Two swirl injectors are structurally connected in such a way that the nozzle of stage 1 is located concentrically inside the nozzle of stage 2. The exit sections of both injectors are located at the same plane. Sometimes, to decrease the overall dimension of the injector, stage 2 or both stages are made completely open.

A basic design requirement for injectors with external mixing is that the spray-cone angle of stage 1 should be larger than its counterpart of stage 2, such that the fuel and oxidizer sheets intersect and mix outside the injector even before they start disintegrating into droplets. The injector designs can be further classified into two categories. If one of the nozzles is inside the other one, two injector designs are possible. In version 1, the nozzle of stage 1 is accommodated by the gas vortex of stage 2. Both injectors are hydraulically independent of each other and can be designed using the procedure described in Section II. In version 2, the nozzle of stage 1 is submerged in the liquid stream in stage 2. This design is usually associated with the quest for increased flow capacity of stage 1 without increasing the pressure drop Δp_i and decreasing the spray cone angle by increasing the nozzle dimension ($A_{n1} = \pi R_{n1}^2$). The operation of stage 2, in particular its flow coefficient μ_2, depends on the ratio $R = R_{n2}/R_1$ where R_{n2} is the radius of the nozzle of stage 2 and R_1 is the external radius of the nozzle of stage 1. Figures 37 and 38 show the theoretical results of the spray cone angle and flow coefficient of stage 2 as functions of $\Delta \bar{l}_n$ and A_2, respectively.

1. Design Procedure for Version 1

To initiate the injector design, the pressure drop Δp_i, mass flow rate $\dot{m}_i$, and propellant properties for each injector stage are prespecified. In addition, the following parameters are provided from structural considerations:

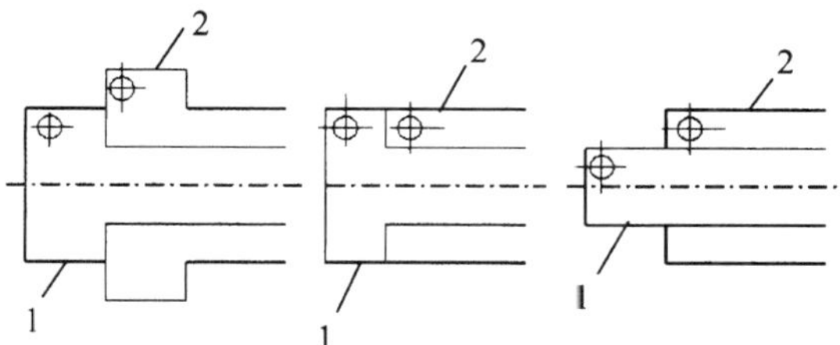

Fig. 36   Versions of bipropellant swirl injectors with external mixing.

1) $\bar{R}_{in1}$ and $\bar{R}_{in2}$ are the coefficients of nozzle opening;
2) $\bar{l}_{n1}$ and $\bar{l}_{n2}$ are the relative nozzle lengths $\bar{l}_n = l_n/2R_n$;
3) n_1 and n_2 are the number of inlet passages.

The design is based on the results of model experiments (see Section II) and is carried out in accordance with the following procedure:

1) Prescribe the spray cone angles $2\alpha_2$ and $2\alpha_1$, according to the empirical condition $2\alpha_1 - 2\alpha_2 = 10$ to 15 deg based on injector operating conditions. With these values and the correlation given in Fig. 34a, find the geometric characteristic parameters, A_1 and A_2. The flow coefficients of stages 1 and 2, μ_1 and μ_2, are then determined from Fig. 34b.

2) Calculate the nozzle radii R_{n1} and R_{n2} from Eq. (103), and determine the tangential-entry radii r_{in1} and r_{in2} from Eq. (104).

3) Determine the Reynolds numbers Re_{in1} and Re_{in2} using Eq. (101). The design is completed if $Re_{in} > 10^4$, and the injector dimensions and flow parameters are calculated.

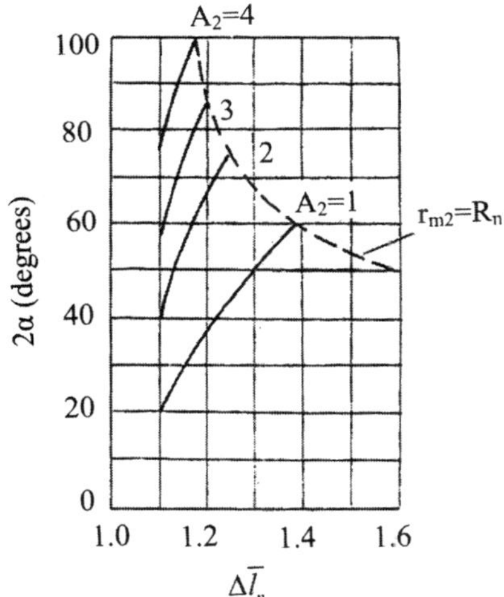

Fig. 37   Spray cone angle as function of relative radial spacing between nozzles in bipropellant swirl injectors.

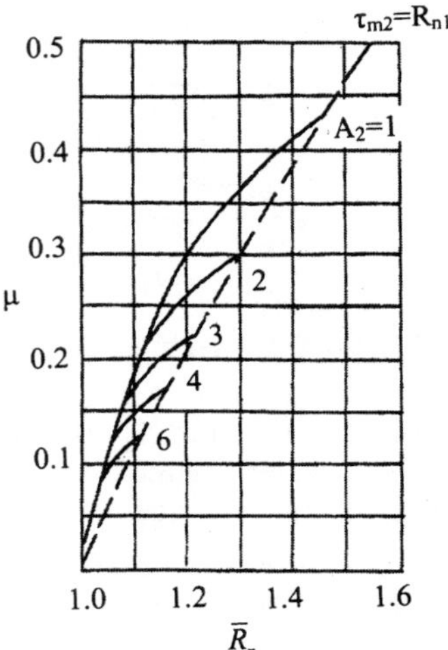

Fig. 38   **Flow coefficient as function of relative radial spacing between nozzles in bipropellant swirl injectors.**

There are cases that the initial requirements are not satisfied. For example, the nozzle of stage 1 is not accommodated inside the gas vortex of stage 2, a situation frequently observed when oxidizer is fed through stage 1. In this case, version 2 of the injector design should be chosen, with the nozzle of stage 1 submerged in the liquid stream in stage 2.

2.   *Calculation Procedure for Version 2*

The initial injector requirements are the same as those in the previous version. With the spray cone angles $2\alpha_2$ and $2\alpha_1$ prescribed, the injector of stage 1 is designed following the same procedure for version 1. The calculation proceeds as follows:

1) Specify the thickness of the nozzle wall δ_w and determine the external radius of the nozzle of stage 1, $R_1 = R_{n1} + \delta_{w1}$.

2) Specify the spacing between the nozzles Δr (no less than 0.3 mm) and calculate the nozzle radius $R_{n2} = R_1 + \Delta r$ of stage 2.

3) Determine the geometric parameter A_2 from Fig. 37 and find μ_2 from Fig. 38.

4) Calculate the inlet-passage radius $r_{in2} = \sqrt{R_{in2}R_{n2}/n_2A_2}$.

5) Determine the required pressure drop across stage 2 following the standard formula for the mass flow rate in terms of the injector pressure drop:

$$\Delta p_{i2}^{(1)} = 0.05 \frac{\dot{m}_{i2}^2}{\mu_2^2 \rho_2 R_{n2}^4} [N/m^2]$$

6) Repeat steps 1–5 using another R_1 until the calculated Δp_{i2} matches its prespecified value.

B. Injectors with Internal Mixing

Figure 39 shows three different versions of injectors with internal mixing. The inner injector (stage 1) is recessed from the exit of stage 2, to achieve stable and efficient mixing of propellants on the internal surface of the nozzle of stage 2. This part of the nozzle of stage 2 is referred to as the injector mixer, whose length can be varied to provide the desired propellant flow residence time in the mixer, τ_i. If τ_i is too long (e.g., 1.5–10 ms), burnouts and explosions may occur in injectors. Conversely, if τ_i is too short (e.g., less than 0.1 ms), poor mixing of the propellants may take place leading to degraded combustion efficiency. The optimal value of τ_i depends on propellant properties, injector flow rate, and several factors whose effects are still not clearly understood. Provisionally, $\tau_i = 0.1$ ms is recommended for hypergolic propellants, and $\tau_i = 0.2$ ms for non-hypergolic propellants with the total propellant flow rate $m_{i1} + m_{i2}$ in the range of 0.2–1.0 kg/s. The final τ_i value (and hence, the recess length Δl_n) is determined during the engine development.

The spray cone angle, when both stages operate simultaneously, depends on many factors. It is generally assumed that the total angle $2\alpha_2$ is 30–40 deg, smaller than the spray cone angle of an isolated stage 2 without the inclusion of stage 1. During the design of an injector, hydraulically independent operation of each stage should be provided, namely,

1) the gas-column radius of stage 2 should exceed the external radius of the nozzle of stage 1, with $r_{m2} - r_{m1} = 0.2–0.3$ mm.

2) the spray cone angle of stage 1 should be such that the propellant arrives at the mixer wall 2–3 mm downstream of the tangential entries of stage 2.

The preceding conditions prevent the ingress of propellant from one of the stages into the other.

To initiate the injector design, propellant properties, pressure drops, Δp_{i1} and Δp_{i2}, and mass flow rates, $\dot{m}_{i1}$ and $\dot{m}_{i2}$, for both stages should be prespecified as basic requirements. In addition, the following parameters should be given for each stage based on structural considerations:

1) $\bar{R}_{in1}$ and $\bar{R}_{in2}$: coefficients of nozzle opening; $\bar{R}_{in} = 3$ for closed stages and $\bar{R}_{in} = 0.7–0.8$ for open ones;

2) $\bar{l}_{n1}$ and $\bar{l}_{n2}$: relative nozzle lengths, $\bar{l}_{n1} = 1.0$;

3) n_1 and n_2: number of inlet passages (2–6);

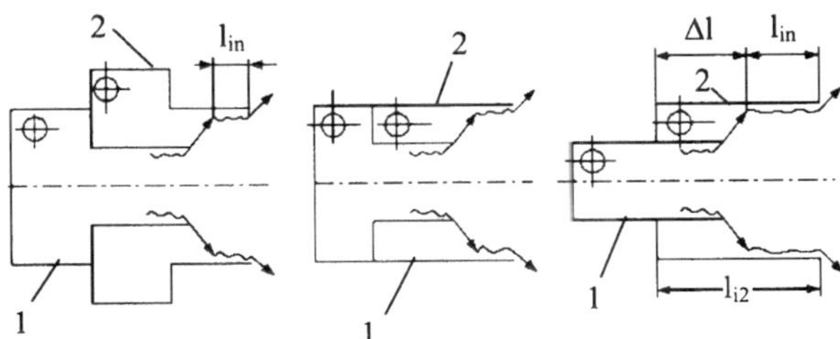

Fig. 39   Versions of bipropellant swirl injectors with internal mixing.

4) $2\alpha_1$: spray cone angle of stage 1, 60–80 deg

The injector design proceeds in the following steps.

1. Design of Stage 1

1) From the experimental correlations given in Fig. 34, determine the non-dimensional parameter a, coefficient of passage fullness φ, geometric characteristic parameter A_1, and mass flow coefficient μ_1.

2) Calculate the nozzle radius R_{n1} from the relation

$$R_{n1} = 0.475\sqrt{\dot{m}_1/(\mu_1\sqrt{\rho_1\Delta p_{i1}})}$$

radius $R_{\text{in}1}$ using $R_{\text{in}1} = \bar{R}_{\text{in}}R_{n1}$, and the radius of the inlet passages,

$$r_{\text{in}1} = \sqrt{R_{\text{in}1}R_{n1}/(n_1 A_1)}$$

3) Determine the Reynolds number in the inlet passages from Eq. (101). If $Re > 10^4$, consider the design of stage 1 completed and calculate the other parameters of stage 1.

4) Calculate the length of the tangential passages using $l_{\text{in}1} = (3\text{–}4)\,r_{\text{in}1}$, the length of the nozzle using $l_{n1} = 2R_{n1}$, and the length of the vortex chamber using $l_{s1} = (2\text{–}3)\,R_{\text{in}1}$.

5) Calculate the external radius of the nozzle, $R_1 = R_{n1} + \delta_w$, where the nozzle wall thickness is $\delta_w = 0.2\text{–}0.8$ mm.

The relative vortex radius r_m is found from Fig. 35, and the vortex radius r_{m1} is calculated.

2. Design of Stage 2

1) Determine the permitted gas-vortex radius, $r_{m2} = R_1 + 0.3$ mm.

2) Assume $r_{m2} = R_{n2}$ to first approximation, and calculate μ using $\mu^{(I)} = 0.225\dot{m}_{i2}/(R_{n2})^2\sqrt{\rho_2\Delta p_{i2}}$, where the superscript (I) denotes the initially guessed value of the mass flow coefficient.

3) Determine A_2 from experimental correlations in Fig. 34, and then the relative vortex radius r_{m2} from Fig. 35.

4) Determine the nozzle radius $R_{n2}^{(II)} = r_{m2}/r_{m2}^{(1)}$ based on the known values of $r_{m2}^{(1)}$ and r_{m2}, and calculate $\mu_2^{(II)}$ using $\mu_2^{(II)} = 0.225\dot{m}_{i2}/(R_{n2}^{(II)})^2\sqrt{\rho_2\Delta p_{i2}}$ with the updated value of $R_{n2}^{(II)}$, where the superscript (II) denotes the iteration step. Repeat steps 3 and 4 until the calculated $R_{n2}^{(i)}$ converges, and update the values of A_2, R_{n2}, and r_{m2}.

5) Calculate $R_{\text{in}2}$ with $R_{\text{in}2} = \bar{R}_{\text{in}2}R_{n2}$ and $r_{\text{in}2}$ with $r_{\text{in}2} = \sqrt{R_{\text{in}2}R_{n2}/n_2 A_2}$.

6) Determine Re_{in} from Eq. (101). If $Re_{\text{in}} > 10^4$, consider the design of stage 2 completed, and calculate the other parameters of stage 2. Determine the spray cone angle $2\alpha_2$ from Fig. 34 with stage 1 being idle and assume the total spray cone angle of the injector 2α to be $2\alpha_2 - 35$ deg. Using the prescribed value of $\tau_i = 0.1\text{–}0.2$ ms, calculate the length of propellant mixing using the

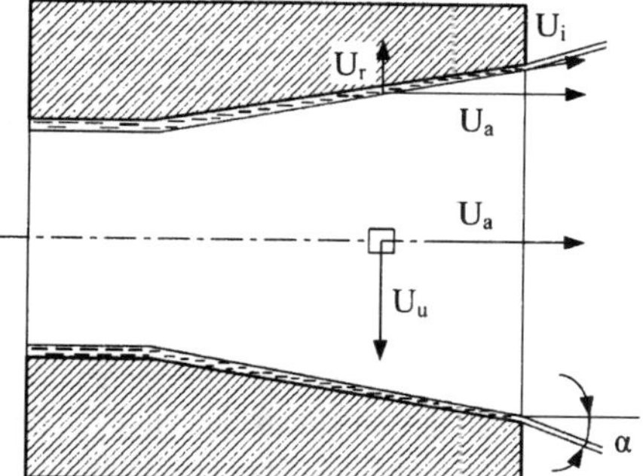

Fig. 40   Schematic diagram of liquid flow along swirl injector with divergent nozzle.

following equation:

$$l_{\text{mix}} = \sqrt{2}\,\tau_i \left(\frac{K_m \mu_2}{(K_m + 1)\varphi_2} \sqrt{\frac{\Delta p_{i2}}{\rho_2}} + \frac{\mu_1}{(K_m + 1)\varphi_1} \sqrt{\frac{\Delta p_{i1}}{\rho_1}} \right)$$

where K_m is the propellant ratio and φ_1 and φ_2 are the coefficients of stage passage fullness ($\varphi = 1 - \bar{r}_m^2$).

7) Calculate the nozzle length $l_{n2} = 2\bar{l}_{n2}R_{n2}$, compare it with l_{mix}, and finally obtain $l_{\text{mix}} + \Delta l_n = l_{n2}$. Determine the final values of τ_i and 2α during the experimental development.

VII.   Modulation of Liquid Spray Characteristics of Swirl Injectors

Theoretically, the spray of a swirl injector with an ideal liquid resembles a hyperboloid of revolution of one nappe. The spray cone angle is confined by the asymptotes of the hyperbolas bounding it and is determined solely by the geometric characteristic parameter A. For real injectors, as a result of viscous losses, the spray shape varies from a tulip-like to a near-conic configuration, depending on the pressure drop. In practice, there are often cases in which the spray cone angle needs to be changed without affecting the geometric parameter and mass flow of the injector. A notable example is the requirement for the spray of a coaxial bipropellant liquid injector to intersect in the mixture formation zone, especially when hypergolic propellants are used. Theoretically, it is impossible to vary the spray cone angle without changing A. Designers are thus forced to make the sheets intersect on the wall of the peripheral-injector nozzle, as described in Section VI. The potential disadvantages of such propellant mixing are:

1) nozzle erosion due to decreased distance between the combustion zone and the injector; and

2) ingress of one propellant into the vortex chamber of the other and ensuing explosions of the bipropellant mixture during engine restarts. The other important

requirement imposed on flow-controlled injectors is minimum changes in the spray cone angle with respect to variations of the geometric parameter.[1]

Methods of modulating the spray cone angle independently of the geometric parameter have been developed by Bazarov at the Moscow Aviation Institute. The techniques do not require any moving parts in injectors. According to classic theories of ideal swirl injectors,[2] the spray cone angle of an injector is unambiguously determined by its geometric characteristic A, which is related to the flow coefficient μ_i. However, as shown in Ref. 1, the flow coefficient μ_i and the spray cone angle can be changed independently. One can either act upon the liquid sheet in the injector nozzle once it has passed through the throat section where the liquid axial velocity U_a equals its surface wave velocity U_w, or use means leading to the violation of the principle of maximum flow.

Several methods may be used to control the spray cone angle independently of the liquid flow, as shown in Fig. 40. The simplest method is profiling of the post-throat section of a swirl injector nozzle. When the degree of nozzle opening increases, the circumferential velocity decreases, thereby leading to an increase in both the axial and radial components. The relationship between the liquid circumferential velocity and the nozzle radius is based on the conservation of angular momentum:

$$U_u r_i = U_\Sigma r_{mk} \tag{106}$$

where r_i is the radius of the passage section of the injector nozzle; r_{mk} is the radial location of liquid film at the head end of the vortex chamber. The nozzle profile determines the ratio between the axial and radial velocities:

$$U_r/U_a = dr_i/dz \tag{107}$$

where z is the coordinate along the nozzle length and dr_i/dz represents the slope at the nozzle.

According to gas-hydraulic analogy between the free-surface liquid flow and gas flow through a pipe proposed by N. E. Zhukovsky, liquid velocity is critical in the narrowest section and is equal to the velocity of disturbance (long waves) propagation over the surface. This conclusion refers not only to spillways, but also, as shown by L. A. Klyachko,[2] to other potential liquid flows with free-surface swirling motions in the field of gravitational forces in particular.

The inference about the impossibility of reaching a supercritical velocity,[2] however, does not hold for swirling liquid flows in the divergent part of the nozzle. It is reasonable to suggest that with the gas-hydraulic analogy proposed by N. E. Zhukovsky, the free-surface liquid flow may become supercritical in the divergent nozzle with its velocity exceeding the surface wave propagation speed.

Consider an ideal swirling liquid flow along an axisymmetric divergent nozzle (see Fig. 40). With all of the velocity components normalized by the total velocity of the liquid exhausted from the nozzle, U_Σ, and the geometric dimensions by R_n, Eqs. (41), (106), and (107) can be written in the dimensionless form:

$$\bar{U}_a^2 + \bar{U}_u^2 + \bar{U}_r^2 = 1; \quad \bar{U}_u = \bar{r}_{mk}/\bar{r}_i; \quad \bar{U}_r/\bar{U}_a = d\bar{r}_i/d\bar{z} = q \tag{108}$$

where q is the tangent of the nozzle-profile angle in reference to the injector axis, and r_{mk} is the liquid-vortex radius at the head end of the vortex chamber. From the theory of ideal swirl injectors, the normalized circumferential velocity at the nozzle throat is

$$\bar{U}_{u,\text{th}} = \bar{r}_{mk} \tag{109}$$

where r_{mk} depends solely on the geometric characteristic parameter A and can be expressed in terms of the coefficient of passage fullness φ. With $r_i = f(z)$ available, the velocity components in each nozzle section can be calculated from Eq. (108), and hence the spray cone angle is obtained.

Substitution of $U_r = qU_a$ in Eq. (108), along with the use of Eq. (75) and the conservation of angular momentum, gives

$$\bar{U}_u^2 + \bar{U}_a^2 + q^2\bar{U}_a^2 = 1; \quad \bar{U}_a = \sqrt{(1 - a/\bar{r}_i^2)/(1 + q^2)} \tag{110}$$

The spray cone angle becomes

$$\tan\alpha = \sqrt{a(1 + q^2)/(\bar{r}_i^2 - a)} \tag{111}$$

The desired spray angle can thus be achieved by varying q. For a cylindrical nozzle exit $q = 0$ and

$$\tan\alpha = \sqrt{a/(\bar{r}_i^2 - a)} \tag{112}$$

For an axisymmetric divergent nozzle with $A = 2$, $a = 1/3$, $\bar{r}_i = 2$, and $\tan\alpha \approx 0.301$. The resultant spray cone angle becomes $\alpha = 16.7$ deg, which is 12 deg less than that for a cylindrical nozzle with $A = 2$.

The liquid velocity along the wall is

$$\bar{U}_L = \sqrt{\bar{U}_a^2 + \bar{U}_r^2} = \sqrt{1 - \bar{U}_u^2} \tag{113}$$

Substitution of $\bar{U}_u$ from Eq. (108) into Eq. (113) gives

$$\bar{U}_L = \sqrt{1 - a/\bar{r}_i^2} \tag{114}$$

The value of $\bar{U}_L$ in the divergent nozzle exit section $(\bar{r}_i > 1)$ exceeds the velocity of wave propagation in the throat section given by $\bar{U}_w = \bar{U}_e = \sqrt{1 - a}$.

The ratio between the liquid velocity along the nozzle wall U_L and the velocity of wave propagation in the nozzle throat section U_{th}, Eq. (73), can be found in a manner similar to that for gas flow, and is denoted as λ_i:

$$\lambda_i = U_L/U_{\text{th}} = \sqrt{(1 - a/\bar{r}_i^2)(2 - \varphi)/\varphi} \tag{115}$$

Substitution of the a value in terms of φ, Eq. (75), for the cylindrical nozzle exit into Eq. (115) gives the expression for λ_i in the cylindrical nozzle with allowance for the Skobelkin effect described in Ref. 2:

$$\lambda_i = \sqrt{3 - 2\varphi} \tag{116}$$

Since φ lies between zero and unity, λ_i is always greater than unity.

The liquid velocity U_L increases in the remaining length of the nozzle as r_i increases according to Eq. (115). As $r_i \to \infty$, $\lambda_{i.\max} = \sqrt{(2 - \varphi)/\varphi}$. There exists a limiting λ_i value for each geometric characteristic parameter A that can be achieved by the liquid.

The coefficient of passage fullness at the exit of a profiled nozzle, $\varphi_{i.\text{exit}}$, can be determined as follows. Application of conservation of mass gives

$$\mu_i A_{\text{th}} U_\Sigma = \varphi_{i.\text{exit}} A_i U_a \tag{117}$$

Substitution of U_a from Eq. (110) to Eq. (117) yields

$$\varphi_{i.\text{exit}} = \frac{\varphi\sqrt{\varphi}\sqrt{1 + q^2}}{\sqrt{2 - \varphi}\,\bar{A}_i\sqrt{1 - a/\bar{r}_i^2}}$$

For a cylindrical nozzle ($q = 0$ and $\bar{A}_i = 1$), the coefficient of passage fullness is obtained by considering for the Skobelkin effect:

$$\varphi_{i.\text{exit}} = \frac{\varphi\sqrt{\varphi}}{\sqrt{2 - \varphi}\sqrt{1 - 2(1 - \varphi)^2/(2 - \varphi)}} = \frac{\varphi}{\sqrt{3 - 2\varphi}} \tag{118}$$

The liquid-vortex radius r_v in the profiled nozzle is readily calculated

$$\pi \varphi_{i,\text{exit}} r_i^2 = \pi (r_i^2 - r_v^2) \tag{119}$$

and

$$\bar{r}_v = \bar{r}_i \sqrt{1 - \varphi_{i,\text{exit}}} \tag{120}$$

where φ_i is the coefficient of passage fullness of the injector. Since $r_i - r_v = h$, with h being the liquid-layer thickness, we have

$$\bar{h} = 1 - \sqrt{1 - \varphi_i} \tag{121}$$

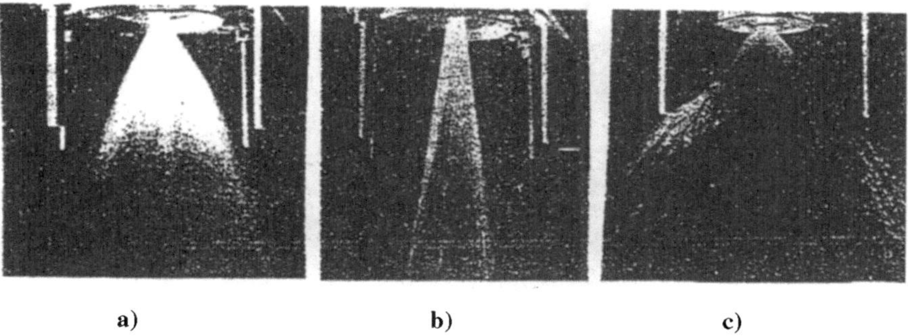

a)   b)   c)

Fig. 41   Liquid spray cone of swirl injector with $R_s = 7$ mm, $r_{in} = 1$ mm, $n = 2$: a) cylindrical nozzle with $d_i = 2$ mm; b) divergent nozzle with $d_{i.th} = 2$ mm and $d_{i.exit} = 6$ mm; c) cylindrical nozzle with $d_i = 6$ mm.

Equation (111) shows that a profiled divergent extension of a nozzle whose length is of the order of its diameter is an effective means to provide spray intersection in coaxial bipropellant injectors. Figure 41 shows photographs of swirl-injector sprays for various nozzle configurations. The divergent nozzle extension decreases the spray cone angle and the range of its variations. It does not ensure a constant spray cone angle. Furthermore, such a divergence piece increases frictional losses at the wall and, consequently, leads to deterioration of atomization quality.

A decrease in the flow velocity by increasing the injector area A_i causes a simultaneous decrease in the spray cone angle. On the other hand, a decrease in the flow velocity by increasing μ_i increases the spray cone angle. Thus, the spray cone angle can be fixed within a prescribed narrow region by approximately adjusting these two factors. Figure 42 schematically shows the liquid flow along a combined dual-orifice injector with its spray cone angle varied using a hydrolock. When fed to the chamber (3) through passages (5) and (6) and to the nozzle (2) from the manifold through the slot (11), the liquid is swirled and flows out from the nozzle edge as a hollow near-conic spray (12). The liquid flow running out through the slot is bent by the main swirling flow and forms a liquid bulkhead (13) in the nozzle, through which the main swirling flow exhausts in a manner similar to a profiled nozzle. With the throttle valve (10) open, the flow passage in the nozzle (2) is reduced, thereby decreasing the main liquid flow and the spray cone angle. With the throttle valve (8) closed, the flow decreases and the spray cone angle increases. Modulating the flow areas of throttle valves (8) and (10) makes it possible to vary either the spray cone angle or the flow rate, without affecting the other parameters. The previously described method of maintaining the prespecified spray cone angle can also be used in combination with a bypass injector. To do this, passage (6) is connected to the throttle valve (8) in line (4) and the liquid from the vortex chamber is throttled.

The passage area of the nozzle can be changed using other methods. For dual-orifice injectors, application of partition cowlings with nozzles between the

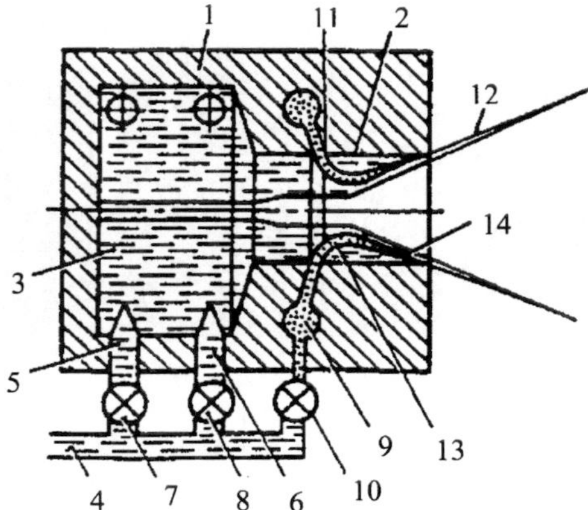

Fig. 42 Liquid flows in combined dual-orifice injector with its spray cone angle varied using a hydrolock; 1-casing; 2-nozzle; 3-vortex chamber; 4-feed line; 5, 6 tangential passages; 7, 8, 10 throttles; 9-ring collector; 11-ring slot; 12-liquid spray; 13-liquid bulkhead; 14-liquid film.

high- and low-flow parts of the vortex chamber is very effective. If the internal-nozzle diameter is smaller than the external-nozzle diameter, liquid exhaust from such an injector is close to that along a divergent extension piece. The possibility of controlling the spray cone angle by compressing the liquid flow in the area between the vortex chamber and the nozzle of the swirl injector is also considered. The classic theory of swirl injectors neglects the radial velocity component in the vortex chamber. However, as shown by A. M. Prakhov,[3] the presence of the radial velocity component in the area between the vortex chamber and nozzle is responsible for the failure of the principle of maximum flow, since the axial velocity in the nozzle is higher than the velocity of disturbance propagation along the liquid surface. In other words, a "supercritical" liquid flow appears with a corresponding decrease of the sheet thickness in the nozzle. The conditions of "supercritical flow" migrate downstream from the vortex chamber to the nozzle.

Both the axial and the radial velocity components in the transition area between the vortex chamber and the nozzle can be increased by introducing a central body whose diameter is larger than the diameter of the liquid-vortex free surface into the nozzle, or by means of axial contraction between the vortex chamber and nozzle. Although the former measure is undesirable because of increased losses in the kinetic energy of the liquid flow, it is often unavoidable in practice due to structural considerations in the case of coaxial arrangement of multiple-orifice injectors. On the other hand, axial contraction does not cause any increase in losses and is applicable to swirl injectors of any type.

Consider the liquid flow under the condition of axial contraction. Part of the pressure is consumed in the annular radial slot connecting the vortex chamber with the passages. The resultant decrease in the liquid flow velocity reduces the circumferential velocity of the liquid sheet velocity in the nozzle and the spray cone angle. By changing the cross-sectional areas of the tangential inlet passages and the width of the radial slot, the prescribed value of the flow coefficient μ_i and a spray cone angle smaller than the one based on the principle of maximum flow can be obtained.

This contraction decreases the liquid velocity in the tangential passages and can be expressed as $U_{in}^* = \zeta U_{in0}$. In this case, r_{mk} decreases by a factor of ζ according to the conservation of angular momentum:

$$r_{mk}^* = \zeta r_{mk} \tag{122}$$

Physically, this decrease of r_{mk} is caused by the presence of the radial velocity U_r in the annular slot and can be determined from the conservation of energy,

$$\Delta p/\rho + U_r^2/2 + U_u^2/2 = \text{const} \tag{123}$$

and the mass continuity equation,

$$2\pi r l U_r = U_{in} A_{in} \tag{124}$$

The spray cone angle for ideal liquid is determined from the formula[14]

$$\alpha = 2\tan^{-1}\sqrt{\bar{r}_{mk}^*/(1 - \bar{r}_{mk}^*)} \tag{125}$$

Substitution of r_{mk}^* from Eq. (122) into Eq. (125) gives

$$\alpha^* = 2\tan^{-1}\sqrt{\zeta\bar{r}_{mk}/(1 - \zeta\bar{r}_{mk})} \tag{126}$$

Figure 43 shows schematically an experimental injector, and Fig. 44 presents the experimental data of the spray cone angle as a function of the slot widths h_1 and h_2. The flow testing results agree well with the calculations. Axial contraction is an effective method of modulating the spray cone angle without changing the injector geometric characteristic parameter A. The decrease in the flow rate due to the velocity drop in the inlet passages U_{in} can be compensated by increasing the passage flow area. Such a measure decreases the geometric characteristic parameter A, and thus provides an additional decrease of the spray cone angle because not only ζ but also $\bar{r}_{mk} = (1 - \varphi)\sqrt{2}/\sqrt{2 - \varphi}$ decrease in this case.

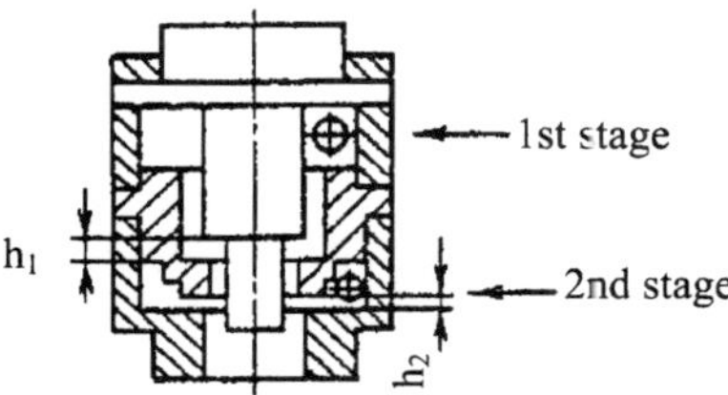

Fig. 43   Experimental swirl injector with axial contraction between vortex chamber and nozzle.

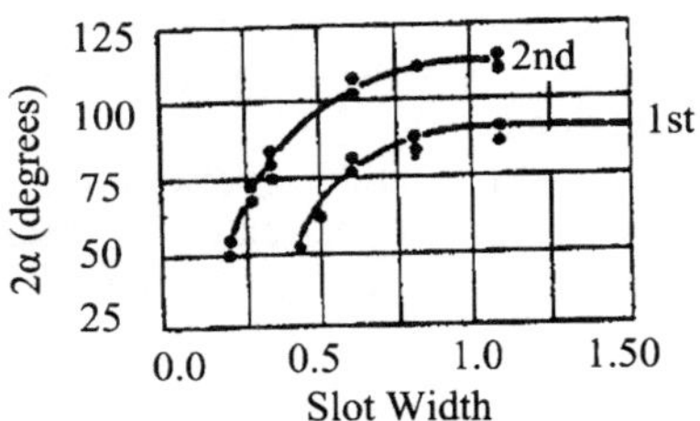

Fig. 44   Effect of slot width on spray cone angle for an experimental swirl injector with axial contraction.

The spray cone angle of the injector with axial contraction can be determined from the formula

$$\tan^2 \alpha = \frac{1}{(1 + 1/\tan^2 \alpha_0)(1 + 1/\zeta^2) - 1} \tag{127}$$

where α_0 is the spray cone angle without flow contraction.

Figure 2 shows a bipropellant swirl injector [A.C. 792023 (USSR)] with an annular slot made in its peripheral region between the vortex chamber and the nozzle, to provide guaranteed intersection between sprays 1 and 2. This design is especially attractive when peripheral fuel delivery is used, and the sprays can hardly be intersected in space using conventional methods due to the lower propellant flows and greater geometric characteristic parameter of the peripheral injector. As discussed in earlier sections, to provide contact for hypergolic ignition, the central nozzle should be deepened into the peripheral one for propellants to mix at the edge of the external nozzle. Such a design leads to propellant ingress to the pre-injector cavity of the other propellant, and causes explosions when the engine start cyclogram is disrupted.

To reduce the effect of flow compression by either decreasing the flow rate or increasing the injector geometric characteristic parameter is a disadvantage of the preceding method of modulating the spray cone angle as applied to LREs. Therefore, spray flow compression in multimode injectors with fixed elements is applied only to provide intersection of sprays in a prescribed range of modes. In the process of thrust control, such injectors have a wider range of spray cone angle variations than conventional ones. For swirl injectors with moving elements, especially with moving screw conveyers, the radial clearance can be changed simultaneously with the tangential passages area, and thus, to achieve constant spray cone angle.

Consider the effect of liquid swirl U_u/U_a in the vortex chamber on the spray cone angle. According to the conservation of momentum, a change in U_u leads to a change in U_u/U_a. This can be achieved, for example, by turning the injection flow passage to the extent that the liquid swirl ceases completely and the swirl injector becomes a jet one. Consequently not only the spray cone angle is decreased but also the flow rate is increased in accordance with the change in the nozzle flow coefficient. In this design, with the spray cone angle controlled, the flow rate can be maintained fixed by correspondingly decreasing the inlet passage section area.[1]

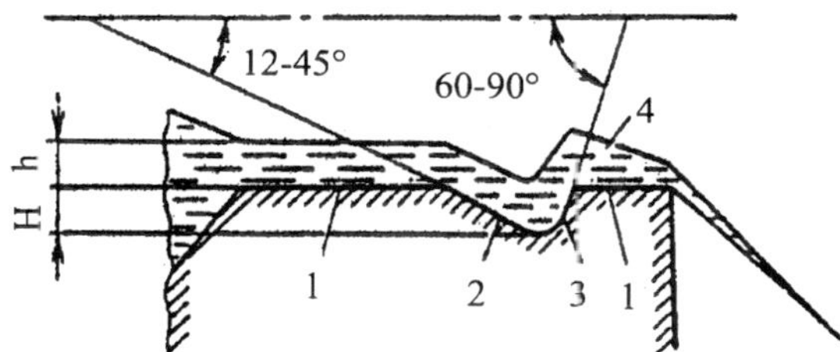

Fig. 45   Profiled nozzle of swirl injector with jump in liquid level.

Stabilization of the spray cone angle can be achieved by using dual-orifice injectors for each propellant. As for the flow rate and spray cone angle, each injector should operate independently and be designed to fulfill the prescribed requirements. In the transition mode, the spray cone angle remains unchanged and the atomization quality is ensured by intersecting the external throttled spray with the central spray operating under excess pressure drop. Stabilization of the spray cone angle can also be obtained by varying the degree of swirling through the use of low flow rate swirling passages with smaller swirling radii.

Consider the possibility of stabilizing the spray cone angle by introducing a liquid jump in the nozzle. Following the gas-hydraulic analogy of N. E. Zhukovsky,[21] we obtain a liquid flow with a jump of the liquid level in a swirl injector similar to the compression jump in a supersonic gas flow. As shown schematically in Fig. 45, a supercritical liquid flow should first be achieved by making a divergent section (2) in the nozzle (1) and a subsequent contraction (3). The liquid travels along the nozzle and forms a standing annular wave (4) whose radius is no less than r_{mk}. Because of flow anisotropy in the jump, the circumferential velocity component does not reach U_{uk}. During subsequent decrease of the liquid level and the flow exhaust from the nozzle to the mixing zone, the spray cone angle of a swirl injector with a profiled groove in the nozzle becomes smaller than its theoretical counterpart for the case without a jump. The effect of the groove decreases with the decreasing ratio between its depth H and the liquid sheet thickness h, and completely disappears with $H/h < 0.3$. This allows the use of a liquid jump in the nozzle to stabilize the spray cone angle of a swirl injector with variable flow coefficient μ_i and coefficient of passage fullness φ. In the mode of maximum flow, a groove with depth $H < 0.3h$ exerts little influence on the spray. As μ_i and φ decrease, the liquid sheet thickness h decreases, and the effect of groove on the spray increases. The increase of the spray cone angle associated with the decrease of μ_i is partially or completely counterbalanced.

VIII.   Design of Gas Swirl Injectors

A.   Design Procedure

A gas injector, shown schematically in Fig. 46, can be designed by determining its basic dimensions D_n, d_{in}, R_{in}, and n, on which the prescribed flow and spray properties are dependent. The design proceeds in the following steps:

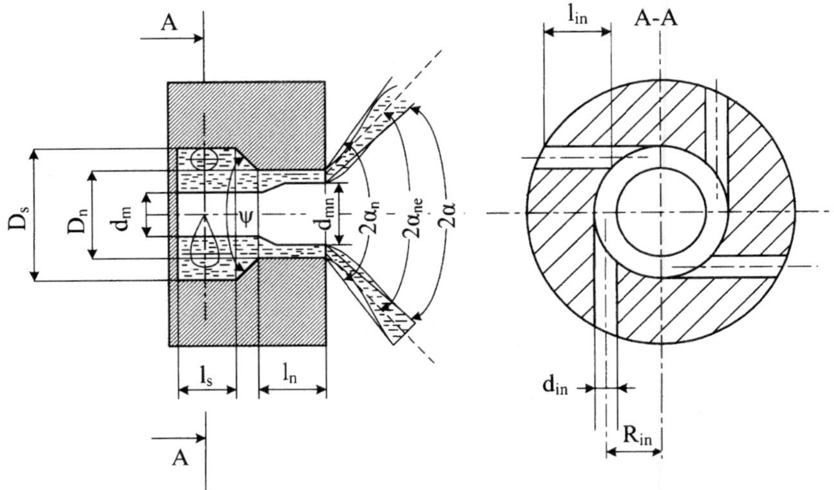

Fig. 46 Schematic diagram of gas swirl injector.

1) Choose the nozzle expansion ratio π_{i}, the geometric characteristic parameter A, and the degree of nozzle opening $\bar{R}_{in}$ to accommodate the specific features of the injector under consideration.

2) Determine the flow coefficient μ from Fig. 47. For intermediate $\bar{R}_{in}$ values in the range between 0.2 and 1, μ is calculated from the following formula:

$$\mu = \mu_{ref}\bar{\mu} \tag{128}$$

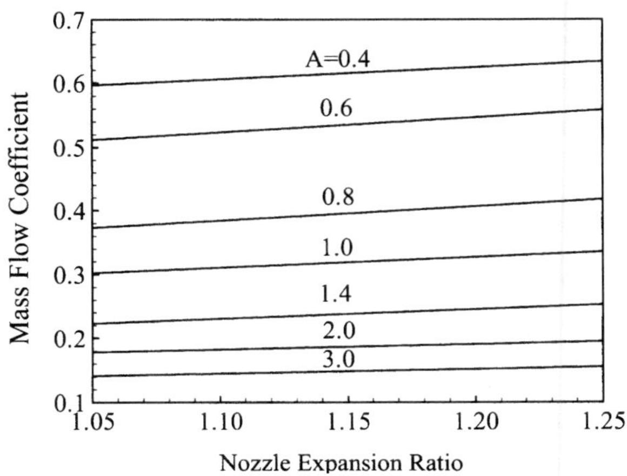

Fig. 47 Flow coefficient μ_{ref} vs A and nozzle expansion ratio π_i for $\bar{R}_{in} = 0.75$ of gas swirl injector.

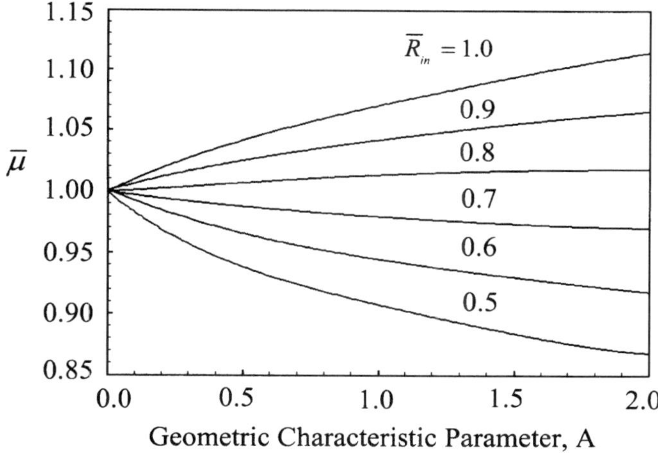

Fig. 48   Correction of flow coefficient based on the degree of nozzle opening.

where μ_{ref} is the initial flow coefficient for $\bar{R}_{in} = 0.75$ and $\bar{\mu}$ is a correction for the nozzle opening determined from Fig. 48. While making calculations, one must meet the following conditions under which the experimental data given by Figs. 47 and 48 were obtained: $Re_{in} > 3000$; $\bar{l}_{in} = 1\text{--}1.2$; $\bar{l}_n = 0.2\text{--}1$ and $\bar{l}_s = 0.2\text{--}0.3$.

3) Calculate the Reynolds number for the inlet passages Re_{in} using

$$Re_{in} = 1413 \frac{\mu d_n p_2 \varphi}{\mu_d \sqrt{RT_1}} \sqrt{\frac{A}{R_{in}}} \tag{129}$$

where μ_d is the dynamic viscosity of the gas, $Pa \cdot s$.

4) Calculate the nozzle diameter in mm using

$$d_n = 0.948 \sqrt{\frac{\dot{m}\sqrt{RT_1}}{\mu p_2 \varphi}} \tag{130}$$

where T_1 is the gas temperature at the inlet to the injector, K; p_2 is the gas pressure downstream of the injector, MPa; R is the gas constant, $J/kg\,K$; $\dot{m}$ is the mass flowrate through the injector, kg/s; and φ is calculated from the formula:

$$\varphi = \pi_\iota \sqrt{\frac{\aleph}{\aleph - 1}\left(\frac{1}{\pi_i^{\frac{2}{\aleph}}} - \frac{1}{\pi_i^{\frac{\aleph+1}{\aleph}}}\right)} \tag{131}$$

where $\aleph$ is the polytropic exponent of gas expansion.

5) Calculate the gas flow rate $\dot{m}$ using the prescribed nozzle diameter,

$$\dot{m} = 1.11 \frac{d_n^2 \mu p_2 \varphi}{\sqrt{RT_1}} \tag{132}$$

6) Calculate the total gas velocity at the nozzle exit, U_Σ, in m/s using

$$U_\Sigma = 1.413 \frac{\mu \varphi \sqrt{RT_1}}{\left(1 - \bar{d}_m^2\right) \pi_i^{\frac{\aleph-1}{\aleph}} \cos \alpha_{\text{exit}}} \tag{133}$$

Calculate the axial velocity component using

$$U_a = U_\Sigma \cos \alpha_{\text{exit}} \tag{134}$$

and the tangential component using

$$U_u = U_\Sigma \sin \alpha_{\text{exit}} \tag{135}$$

7) Calculate the inlet passage diameter d_{in} in mm using

$$d_{\text{in}} = d_n \frac{R_{\text{in}}}{A n_{\text{in}}} \tag{136}$$

Determine the inlet passage diameter for a fully open injector using Fig. 49.

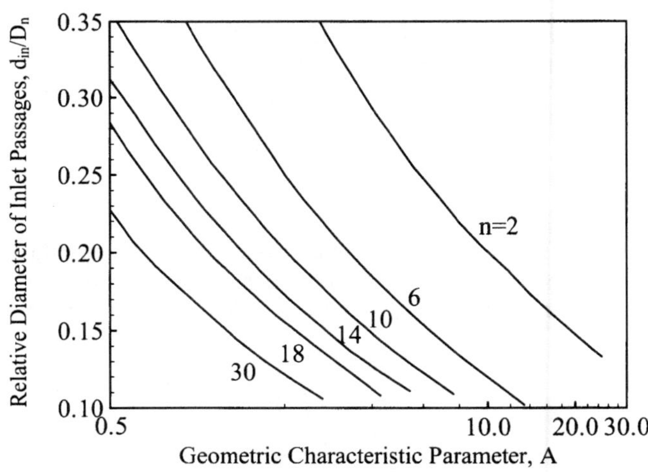

Fig. 49   Relative diameter of inlet passage d_{in}/D_n vs A and n for fully open injector $(D_s = D_n)$.

8) Calculate the inlet radius R_{in} and the vortex-chamber diameter D_s using the following formulae:

$$R_{in} = \bar{R}_{in} \frac{D_n}{2} \qquad (137)$$

$$D_s = 2R_{in} + d_{in} \qquad (138)$$

B.   Selection of Geometric Dimensions and Flow Parameters

This section summarizes the common practice used in the design of gas injectors in Russia:

1) The relative length of the inlet passages $\bar{l}_{in}$ is chosen between 1.0 and 1.5.

2) The relative length of the vortex chamber $\bar{l}_s = l_s/D_s$ is chosen between 0.1 and 0.3. If $\bar{l}_s$ is greater than 0.3, μ is calculated from the formula

$$\mu = \mu_1 + b\mu_2 \qquad (139)$$

where μ_1 can be found from Figs. 47 and 48 for prescribed values of A, $\bar{R}_{in}$ and π_i and μ_2 for $\pi_i = 1.05$. The parameter b is determined form the empirical formula

$$b = 0.0225A(\bar{R}_{in} - 0.3)(\bar{l}_s - 0.3) \qquad (140)$$

The formula is applicable for $A \leq 3$, $\bar{R}_i \leq 4$, and $\bar{l}_s \leq 4$.

3) The number of inlet passages n is chosen based on the condition that the required coefficient of gas-distribution non-uniformity K is obtained from Figs. 50 and 51.

4) The convergence angle at the nozzle entrance Ψ is chosen between 90 and 120 deg.

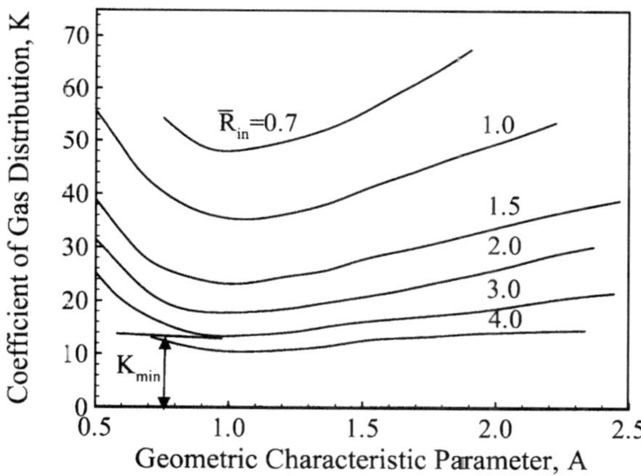

Fig. 50   Effect of geometric characteristic parameter A on coefficient of gas-distribution non-uniformity in mixing layer for different $\bar{R}_{in}$ ($n = 4$).

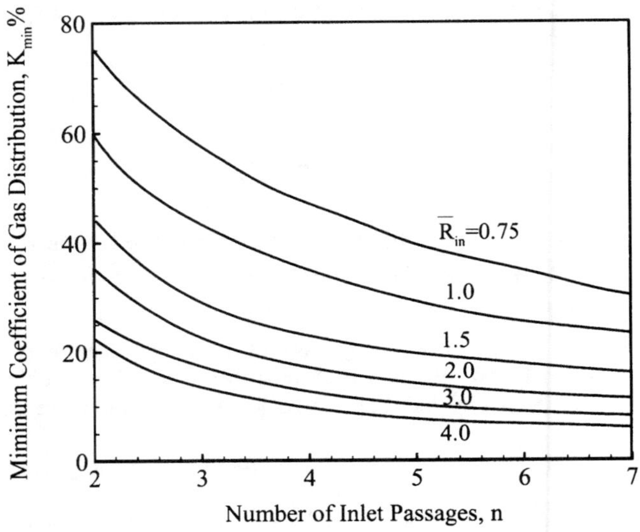

Fig. 51 Effect of number of inlet passages on minimum coefficient of gas-distribution non-uniformity in mixing layer.

5) With the relative nozzle length $\bar{l}_n$ between 1 and 10, the gas flow coefficient μ remains constant.

6) The tolerances for the basic dimensions are chosen according to the manufacturing specifications. The error for the injector flow coefficient should be less than 10% for $D_n < 10$ mm and 16% for $D_n > 10$ mm. The final geometric dimensions are determined in accordance with the results of injector flow tests.

7) The surface roughness should be $Rz < 40$ μm for the inlet passages, $Rz < 20$ μm for the vortex chamber, and $Rz < 2.2$ μm for the cylindrical and end surfaces of the nozzle. No burrs are permitted at the nozzle edge and in the inlet passages. The radius of the blunting chamber is 0.05–0.2 mm.

IX. Dynamics of Liquid Rocket Injectors

A liquid rocket engine, as shown schematically in Fig. 52, contains various sources of intense pressure fluctuations caused by turbulent flows in the feed line, fluttering of pump wheel blades, vibrations of control valves, and unsteady motions in the combustion chamber and gas generator. As a consequence, the actual process of mixture formation in injector elements typically occurs in the presence of highly developed fluctuations, as the feedback coupling (loop 1 in Fig. 53) affects the processes occurring in the combustion chamber and forms a self-oscillating circuit.[23,27] All conceivable mechanisms of intrachamber instability are included here. Additionally, the chamber pressure fluctuation P'_c directly affects the liquid stage of the injector, L, forming another feedback

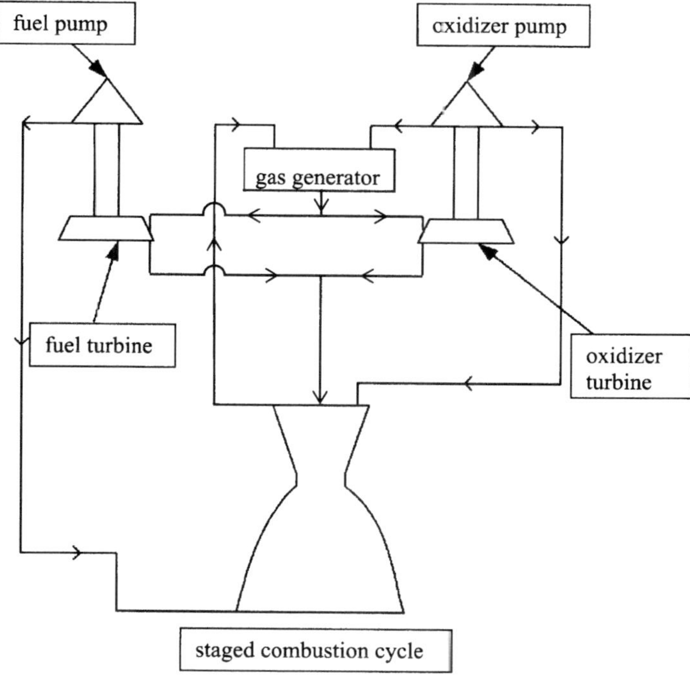

Fig. 52   Schematic diagram of liquid rocket engine with staged combustion.

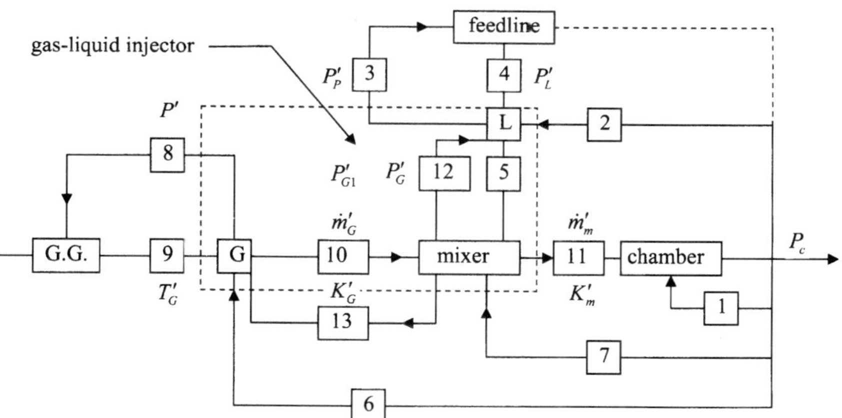

Fig. 53   Interactions of dynamic processes in liquid rocket engine with staged combustion.

coupling (loop 2). Acting as an oscillator in the feed system with the feedback coupling 3, the injector excites pressure fluctuations P'_L which then affect the injector response through the direct coupling 4. The ensuing fluctuation in pressure drop across the liquid injector, P'_L, causes liquid-flow fluctuations $\dot{m}'_L$ in the nozzle exit. In parallel, the chamber pressure fluctuations P'_c affect the gas stage of the injector, G, by means of the feedback coupling 6, and consequently, the gas generator ($G.G.$) through the feedback coupling 8. For a gas-liquid injector with internal mixing, the pressure fluctuations P'_c affect the mixer through the feedback coupling 7, and consequently, the liquid (12) and gas (13) stages of the injectors. The gas generator responds to the disturbances of the flow rate $\dot{m}'_G$, temperature T'_G, composition K'_G, exhaust velocity U'_G, and pressure P'_G of the generator gas which, when passing through the mixer, results in fluctuations of droplet mass and size distributions of the combustible mixture spray. The feedback couplings 12 and 13 can form their own self-oscillating circuits, causing fluctuations in the propellant flow at the injector exit as well as changes in the main mixture-formation parameters, including the atomized droplet-size distribution, spray angle, and uniformity of mixture composition.[14]

In LRE systems, injection is a key process since through it all feedback couplings of the combustion chamber with other engine components are realized. In addition to its main function of preparing a combustible mixture, an injector acts as a sensitive element that may generate and modify flow oscillations.[24] This section summarizes various important aspects of injector dynamics. The mechanisms of driving self-pulsations in both liquid and gas-liquid injectors are addressed systematically.

A. Linear Dynamics of Jet Injectors

For a short injector whose length is much less than the wavelength of oscillation, the equation of motion for inviscid liquid takes the form

$$\frac{d\tilde{U}}{dt} + \frac{\tilde{U}^2}{2L_i} = \frac{\tilde{P}_f - \tilde{P}_c}{\rho L_i} \equiv \frac{\Delta\tilde{P}}{\rho L_i} \tag{141}$$

where $\sim$ stands for the instantaneous quantity and L_i for the injector length. Each flow property may be decomposed into mean and fluctuating parts:

$$\Delta\tilde{P} = \Delta P + \Delta P' \quad \text{and} \quad \tilde{U} = U + U' \tag{142}$$

For time-harmonic oscillations,

$$\Delta P' = |\Delta P'|e^{i\omega t} \quad \text{and} \quad U' = |U'|e^{i\omega t} \tag{143}$$

Substitute Eq. (142) into Eq. (141) and linearize the result to get

$$\frac{dU'}{dt} + \frac{U}{L_i}U' = \frac{|\Delta P'|}{\rho L_i}e^{i\omega t} \tag{144}$$

The solution to the preceding equation is

$$U' = \frac{\Delta P'}{\rho U + i\omega\rho L_i} \tag{145}$$

A transfer function relating the fluctuating velocity U' and pressure drop $\Delta P'$ is obtained as follows:

$$\Pi_j = \frac{U'/U}{\Delta P'/\Delta P} \equiv \frac{\bar{U}'_j}{\Delta \bar{P}'_j} = \frac{1}{2} \cdot \frac{1 - i\omega L_i/U}{1 + (\omega L_i/U)^2} = \frac{1}{2} \cdot \frac{1 - iSh_j}{1 + Sh_j^2} \tag{146}$$

where the overbar denotes a dimensionless quantity. The Strouhal number of the jet injector, Sh_j, is defined as $Sh_j \equiv \omega L_i/U$.

Figure 54 shows the amplitude-phase diagram of the transfer function Π_j for a short jet injector. The normalized pressure-drop fluctuation $\Delta \bar{P}'_j$ is taken to be unity, and the phase angle between $\bar{U}'_j$ and $\Delta \bar{P}'_j$ is Φ_j. The locus is obtained by increasing the Strouhal number (or oscillation frequency) in Eq. (146). For practical injector dimensions and oscillation frequencies commonly observed in LREs, a jet injector can be considered as a single inertial element in which the amplitude of flow oscillation U' decreases smoothly as the Strouhal number increases, and the phase angle Φ_j increases asymptotically to $\pi/2$. The transfer functions Π_j for step-shaped and other shapes of jet passages can be calculated as a synthesis of several passages connected in series. In the case of long liquid injectors and coaxial gas-liquid injectors, resonance at multiple frequencies may occur when the injector length becomes comparable to the wavelength of the fluctuation. The influence of injector length should be taken into account

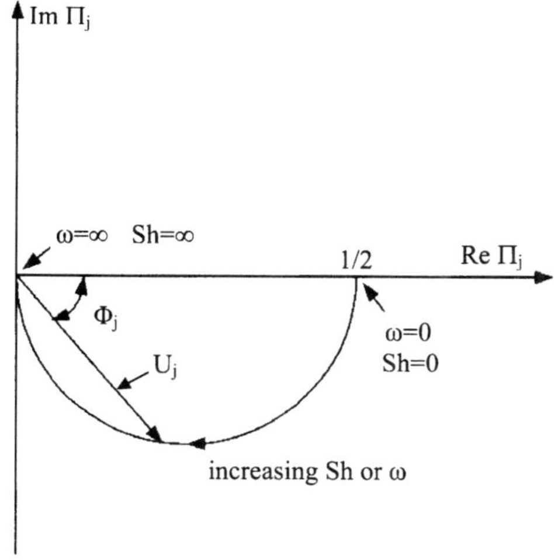

Fig. 54   Amplitude-phase diagram of response function of a short jet injector.

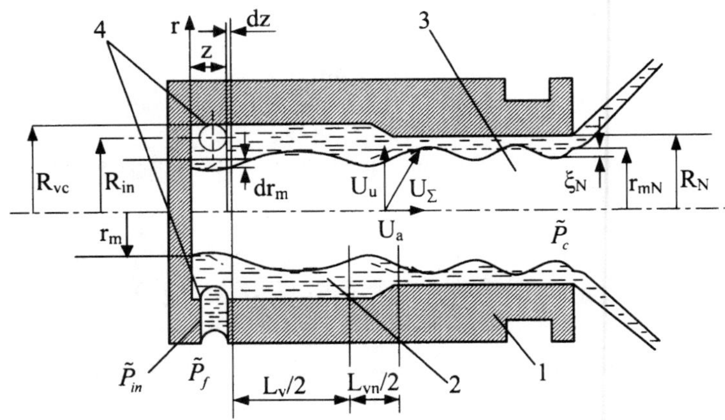

Fig. 55 Schematic of liquid swirl injector; 1-casing; 2-vortex chamber; 3-nozzle; 4-tangential passage.

for cryogenic liquids since the sound speed is relatively low due to the existence of gas bubbles.

B. Linear Dynamics of Swirl Injectors

Figure 55 shows schematically a swirl injector with liquid flow. The liquid is fed to the injector through tangential passages (station 4), and forms a liquid layer in the vortex chamber (station 2) with a free internal surface shown by the dashed line for the stationary case. The liquid is exhausted from the nozzle (station 3) in the form of a thin, near-conical sheet that then breaks up into fine droplets. Compared with a jet injector of the same flow rate, the flow passage of a swirl injector is much larger, and as such any manufacturing inaccuracy exerts a much weaker effect on its atomization characteristics. The resultant droplets are finer and have higher uniformity, thereby motivating the predominant application of swirl injectors in Russian LREs. From the dynamics standpoint, a swirl injector is a much more complicated element than a jet injector. The liquid residence time is longer than that of a jet injector; its axial velocity component U_a is smaller for the same pressure drop, and the speed of disturbance propagation is lower due to the existence of a central gas-filled cavity in the liquid vortex. A swirl injector contains an inertial element (i.e., tangential passage), an energy capacitor (i.e., vortex chamber partially filled with rotating fluid), and a transport element (i.e., nozzle). Each of these elements can be described with rather simple relationships.

The unsteady behavior of the tangential passage can be determined following the same analysis as that for a jet injector, Eq. (146). The dynamics of the liquid layer inside the vortex chamber and the nozzle can be modeled by means of a wave equation that takes into account the disturbance propagation in the liquid with centrifugal force. If we ignore the liquid-layer thickness and radial velocity, and follow the approach given in Ref. 2, a wave equation characterizing the flow oscillation in the liquid layer is obtained:

$$\frac{\partial^2 \xi}{\partial t^2} = \frac{1}{r_m^4} U_{in}^2 R_{in}^2 \left(\frac{R_{vc}^2 - r_m^2}{2} \right) \frac{\partial^2 \xi}{\partial z^2} \tag{147}$$

Here ξ denotes the fluctuation of the liquid-layer thickness, and r_m the radius of the liquid surface. The surface-wave propagation speed U_w is

$$U_w = \sqrt{ \left(\frac{U_{in}^2 R_{in}^2}{r_m^3} \right) \left(\frac{R_{vc}^2 - r_m^2}{2 r_m} \right) } = \frac{U_{in} R_{in}}{r_m^2} \sqrt{ \frac{R_{vc}^2 - r_m^2}{2} } \tag{148}$$

The first parenthesized term in the square root represents the centrifugal acceleration, and the second parenthesized term the effective thickness of the liquid layer. The expression for the wave speed is analogous to that for shallow-water wave propagation. The solution to Eq. (147) for a semi-infinite vortex is

$$\xi = \Omega e^{i\omega(t - z/U_w)} \tag{149}$$

where Ω represents the amplitude of the liquid surface wave. For an axisymmetric rotating flow with a free interior surface, linearization of the equations of motion leads to a relationship between the fluctuations of the liquid surface and axial velocity:

$$\frac{\partial U_a'}{\partial t} = \frac{U_{in}^2 R_{in}^2}{r_m^3} \frac{\partial \xi}{\partial z} \tag{150}$$

The amplitude of the axial velocity fluctuation U_a' is

$$|U_a'| = \Omega U_{in}^2 R_{in}^2 / (U_w r_m^3) \tag{151}$$

In a non-dimensional form, the liquid surface-wave velocity inside the vortex chamber can be determined from Eq. (148):

$$(\bar{U}_w)_{vc} \equiv \frac{U_{w,vc}}{U_\Sigma} = \sqrt{ \frac{1}{2} \left(\frac{\bar{R}_{vc}^2}{a} - 1 \right) } \tag{152}$$

Here U_Σ is the liquid velocity at the head end of the vortex chamber. The non-dimensional parameters a and $\bar{R}_{vc}$ are defined respectively as

$$a \equiv \left(\frac{r_{mk}}{R_n} \right)^2 = A^2 \mu^2; \quad \bar{R}_{vc} \equiv \frac{R_{vc}}{R_n} \tag{153}$$

The subscript k denotes the head end of the vortex chamber, and A is the geometric characteristic parameter.

In accordance with the principle of maximum flow,[2] the axial velocity of the liquid flow inside the injector nozzle is the same as the surface wave speed, analogous to the gas flow in a choked nozzle. Thus,

$$(\bar{U}_w)_n = (\bar{U}_a)_n \equiv \frac{U_{an}}{U_\Sigma} = \sqrt{\frac{\varphi}{2 - \varphi}} \tag{154}$$

where $\varphi \equiv A_L/A_n$ is the coefficient of passage fullness, representing the ratio of the cross-sectional area occupied by the liquid to that of the entire nozzle.

As the wave propagation speed varies in the injector, the unsteady liquid flow rate also changes with the flow. To quantify the dynamic characteristics of the vortex chamber subject to flow disturbances, a reflection coefficient of the surface wave at the nozzle entrance (or the exit of the vortex chamber), β, is defined according to the fluctuation of the liquid flow rate. A simple analysis based on mass conservation leads to the following relation between the flow-rate oscillations in the vortex chamber and the nozzle:

$$\frac{Q'_n}{Q'_{vc}} = \frac{U_{wn}}{U_{w,vc}} \cdot \frac{\Omega_n}{\Omega_{vc}} \cdot \frac{r_{mn}}{r_{m,vc}} \tag{155}$$

After some straightforward manipulations, a reflection coefficient β characterizing the nozzle dynamics is obtained:

$$\beta = \frac{Q'_{vc} - Q'_n}{Q'_{vc}} = 1 - \frac{2\sqrt{\varphi}}{\sqrt{\bar{R}_{vc}^2 - a}} \tag{156}$$

The amplitude of the surface wave in an infinitely long vortex chamber (i.e., no wave reflection) is

$$\bar{\Omega}_\infty \equiv \frac{\Omega_\infty}{r_{mk}} = \frac{1}{A\sqrt{2\left(\bar{R}_{in}^2 - a\right)}} \frac{U'_{in}}{U_{in}} \tag{157}$$

For a vortex chamber with zero length, the surface-wave amplitude can be determined by the acoustic conductivity of its nozzle:

$$\bar{\Omega}_o = \frac{\bar{\Omega}_\infty \sqrt{\bar{R}_{in}^2 - a}}{2\sqrt{\varphi}} = \frac{\varphi \bar{R}_{in}}{4\sqrt{1 - \varphi}} \frac{U'_{in}}{U_{in}} \tag{158}$$

which is much greater than $\bar{\Omega}_\infty$. For an intermediate case, the surface-wave amplitude depends on the reflection coefficients at the head end and the exit of the vortex chamber, as well as on liquid viscosity. The surface wave causes

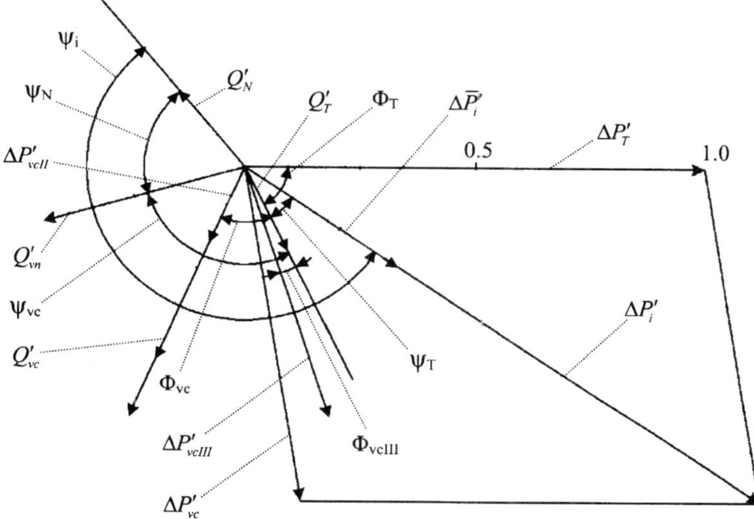

Fig. 56 **Amplitude-phase diagram of response function of a swirl injector.**

pulsations of the circumferential velocity U'_u in the radial direction according to the conservation of angular momentum ($U_u r = $ const) and consequently gives rise to pulsations of centrifugal pressure. The non-dimensional amplitude of centrifugal-pressure pulsations $\Delta\bar{P}'_{vc}$ (defined as the pressure difference between the tangential entry and the liquid free surface) caused by surface wave motions is not high and is equal to the non-dimensional amplitude of surface waves in the swirl chamber.

The main difference between a swirl and a jet injector as a dynamic element lies in the different mechanisms of disturbance propagation between the combustion chamber and the feed system. For conventional injector dimensions that are significantly smaller than disturbance wavelengths in the gas and liquid, pressure oscillations arising in the combustion chamber propagate through the liquid vortex layer almost instantaneously. This results in fluctuations of pressure drop across the tangential entries, $\Delta P'_T$, as shown in the amplitude-phase diagram in Fig. 56, where $\Delta P'_T$ is set to unity on the abscissa. Similar to a jet injector, these fluctuations lead to oscillations of the liquid flow rate, Q'_T, which subsequently produce surface waves in the vortex chamber propagating back and forth. Their amplitudes and phase angles ψ_{vcII} with respect to the pressure oscillations depend on the resonance properties of the liquid vortex in the injector, and can be determined by the reflection coefficients based on Eq. (156). When disturbances occur, part of these waves pass through the injector nozzle and cause fluctuations of the flow rate Q'_N and spray angle at the exit. Concurrently, the fluctuation Q'_T gives rise to oscillations of the circumferential velocity U'_u in the vortex chamber that propagate with the liquid flow and produce centrifugal-pressure fluctuations $\Delta P'_{vcIII}$ on the vortex chamber wall.

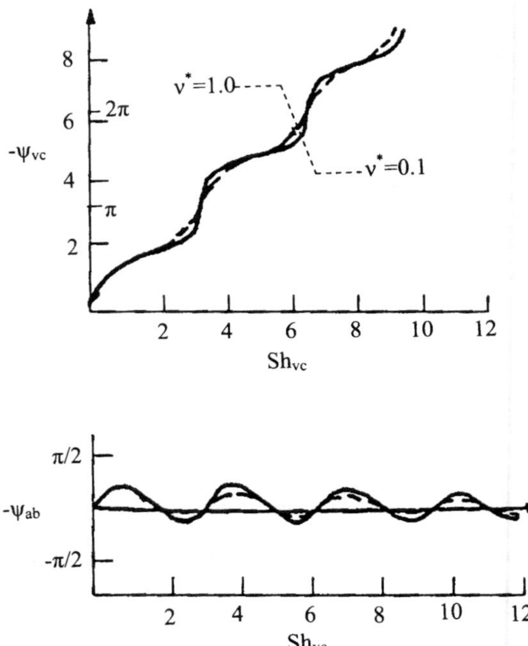

Fig. 57   Phase angle of pressure pulsation in vortex chamber as function of Strouhal number.

This secondary fluctuation $\Delta P'_{vcIII}$ can be vectorially summed with the original pressure-drop fluctuation in the liquid vortex $\Delta P'_{vcII}$ under the action of the surface wave to obtain the total pressure-drop fluctuation $\Delta P'_{vc}$. Finally, the vector sum of $\Delta P'_{vc}$ and $\Delta P'_{T}$ forms the dynamic pressure drop in the injector $\Delta P'_{i}$ relative to which the flow-rate fluctuation in the tangential entry is phase-shifted by an angle ψ_{T}. If the tangential-velocity disturbance is not damped by viscous losses, it will reach the injector nozzle exit considerably later than the surface wave. The resultant fluctuations of the flow rate and other properties must be determined by their vector sum.

Compared to the unsteady flow in the tangential inlet channel, the characteristic time of circumferential-velocity pulsations in the liquid vortex layer is much shorter, and as such any disturbance in the liquid layer is rapidly transmitted in the radial direction. Measurements of centrifugal-pressure pulsations at the outer wall of a typical vortex chamber at frequencies of hundreds of Hz reveal that the liquid vortex layer responds in a quasi-steady-state manner to radial disturbances. Their amplitudes are only a few percent lower than the quasi-stationary variations for given changes of the circumferential velocity. Figure 57 shows the phase angle between the pressure pulsations in the feed line and the vortex chamber, where ψ_{vc} is the phase difference between oscillations at the head end and exit of the vortex chamber, and ψ_{ab} between the

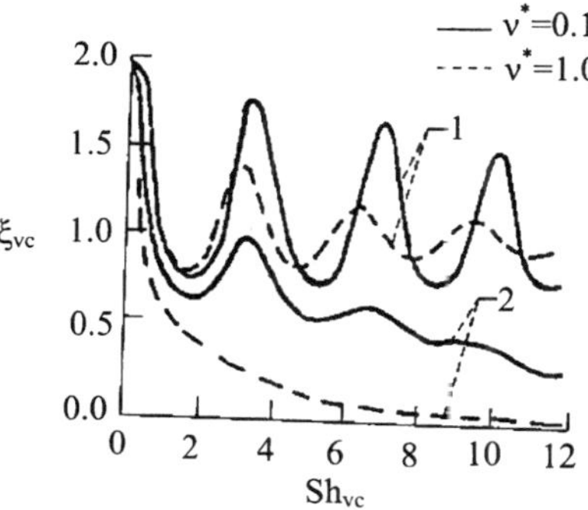

Fig. 58 Amplitude of liquid surface wave in vortex chamber as function of Strouhal number; 1-head end, 2-exit.

exit of the tangential entry and the liquid free surface at the vortex-chamber head end. The Strouhal number Sh_{vc} is defined as $\omega L_{vc}/U_w$. Two different liquids with dimensionless viscosities of $\nu^* = 0.1$ and 1.0 are considered here. For reference, $\nu^* = 0.08$ for water at room conditions. Figure 58 shows the amplitude of the liquid surface wave in the vortex chamber, where station 1 corresponds to the head end and 2 to the exit.

Theoretical and experimental studies on the effect of the velocity pulsations in the tangential entries U'_{in} on the liquid swirling flow suggest at least two different mechanisms of disturbance propagation in the vortex chamber. First, U'_{in} pulsations cause fluctuations of the liquid free surface r'_{mk} which then propagate at the speed U_w according to Eq. (148). Second, U'_{in} pulsations result in an energy disturbance in the form of circumferential velocity fluctuations that propagate throughout the entire liquid layer in both the radial and axial directions. Analogous energy waves in gaseous flows are observed and generally referred to as entropy waves.[24] When propagating along the axis of the swirler, the lengths of these waves decrease but the amplitudes grow in accordance with the conservation of angular momentum.

The pressure variation in the radial direction is obtained by integrating the centrifugal force across the liquid vortex layer:

$$\Delta \tilde{P}_{sc} = \rho \int_{R_{in}}^{r_m} \frac{\tilde{U}_u^2}{r}\,dr \tag{159}$$

In dimensionless form, Eq. (159) becomes

$$\Delta \bar{P}'_{sc} = 2\rho U_{in} U'_{in} \int_{1}^{\sqrt{a}/\bar{R}_{in}} \frac{\arg \overline{U'_u}}{\bar{r}^3}\,d\bar{r} \tag{160}$$

where $\arg \bar{U}'_u \equiv U'_u/U_{in}$ is the deviation from the stationary dependence of U_u per dimensionless radius $\bar{r}$. As an example, for an infinitely long vortex

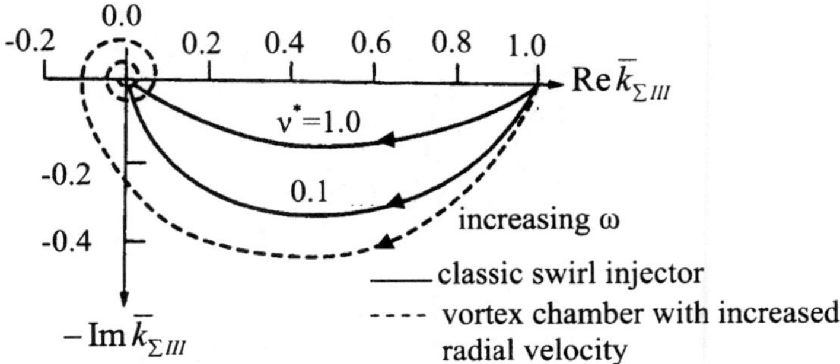

Fig. 59 Amplitude-phase diagram of amplification coefficient of a typical liquid vortex chamber.

chamber (i.e., no wave reflection from the nozzle), we have from Ref. 14

$$\arg \bar{U}_u = \tan\left[\frac{\pi}{2} \cdot \frac{\bar{R}_{vc}(1 - \bar{r})}{\bar{R}_{vc} - \sqrt{a}}\right] \cdot \exp\left[iw\left\{\tau - \frac{R_{in}}{U_\Sigma}\frac{\bar{R}_{vc}^2}{\mu}(1 - \bar{r})\right\}\right] \tag{161}$$

Figure 59 shows the amplitude-phase diagram of the liquid swirler response to incoming pressure pulsations for different values of liquid viscosity. The subscript *III* stands for the results obtained from the third model. At zero frequency, the amplification coefficient k (defined as $k \equiv U'_{in}/\Delta P'_{sc}$) has its stationary value of unity. The coefficient decreases rapidly with increasing frequency, but the phase angle grows from 0 to $\pi/2$. Thus, the swirling flow movement is stable if the amplification coefficient is placed in the fourth quadrant of this complex plane.

For small disturbances, surface and entropy (or energy) waves behave independently and the net effect can be represented by their vectorial sum. At high frequencies, the entropy wave and its influence on centrifugal pressure can be ignored due to the high inertia of the liquid vortex layer. In contrast, the influences of surface waves become negligible at low frequencies because of the rapid decrease of their amplitudes. Entropy waves prevail in this situation. Calculations have shown that centrifugal-pressure pulsations resulting from circumferential velocity fluctuations may exist for several periods of pulsation, but at the same time surface waves may propagate throughout the liquid almost instantaneously with their high wave speed. This effect, known as the memory effect of a swirling flow, may suppress these fluctuations if they are out of phase.

The overall response function of a swirl injector Π_{sw} can be obtained in terms of the transfer characteristics of each individual element of the injector. The transfer function between the fluctuating velocity in the tangential inlet passage and the pressure drop $\Delta P'_T$ across the inlet is defined as

$$\Pi_T = \frac{U'_{in}/U_{in}}{\Delta P'_T/\Delta P_T} = \frac{Q'_T/Q_T}{\Delta P'_T/\Delta P_T} \tag{162}$$

The centrifugal-pressure pulsations caused by liquid surface wave motions and circumferential velocity fluctuations can be characterized by the following transfer functions, respectively:

$$\Pi_{vcII} = \frac{\Delta P'_{vcII}/\Delta P_T}{2U'_{in}/U_{in}} = \frac{\Delta P'_{vcII}/\Delta P_T}{2Q'_T/Q_T} \tag{163}$$

$$\Pi_{vcIII} = \frac{\Delta P'_{vcIII}/\Delta P_T}{2U'_{in}/U_{in}} = \frac{\Delta P'_{vcIII}/\Delta P_T}{2Q'_T/Q_T} \tag{164}$$

Note that all of the variables in the preceding equations are complex to account for the phase differences between the various processes of concern. The fluctuation of the total pressure drop across the entire injector $\Delta P'_i$ is the vector sum of $\Delta P'_T$ and P'_{vc}, with the latter being $P'_{vcII} + P'_{vcIII}$. Thus,

$$\Delta P'_i = \Delta P'_T + \Delta P'_{vc} = \Delta P'_T + \Delta P'_{vcII} + \Delta P'_{vcIII} \tag{165}$$

Substitution of Eqs. (162–164) into (165) and rearrangement of the result give rise to the transfer function between Q'_T and $\Delta P'_i$:

$$\Pi_{in} = \frac{Q'_T/Q_T}{\Delta P'_i/\Delta P_i} = \frac{\Pi_T}{1 + 2\,(\Pi_{vcII} + \Pi_{vcIII})\Pi_T}\left(\frac{\Delta P_i}{\Delta P_T}\right) \tag{166}$$

The fluctuating flow rate at the exit of the vortex chamber (or the entrance of nozzle) Q'_{vc} can be expressed as

$$\frac{Q'_{vc}}{Q_{vc}} = \Pi_{vc}\frac{Q'_T}{Q_T} \tag{167}$$

Similarly, the fluctuating flow rate at the nozzle exit becomes

$$\frac{Q'_n}{Q_n} = \Pi_n\frac{Q'_{vc}}{Q_{vc}} \tag{168}$$

Since the mean flow rate through all elements of the injector is identical,

$$Q_T = Q_{vc} = Q_n \tag{169}$$

Substituting from Eqs. (167) and (168) and using Eq. (166), we obtain the mass transfer function for the entire injector.

The overall response function of a swirl injector Π_{sw} can be obtained by combining Eqs. (166–169)

$$\Pi_{sw} = \frac{\bar{Q}'_n}{\Delta \bar{P}'_i} = \frac{\Delta P_i}{\Delta P_T}\cdot\frac{\Pi_T\Pi_n\Pi_{vc}}{2\Pi_T\Pi_{vc} + 1} \tag{170}$$

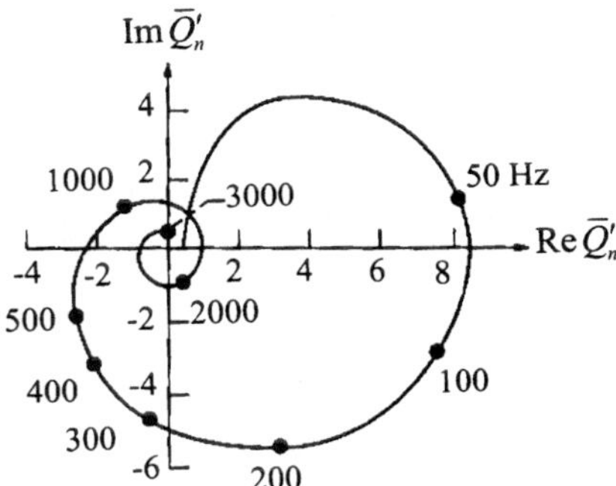

Fig. 60 Amplitude-phase diagram of response function of a typical liquid swirl injector.

Here $\Delta P_i\,(\equiv P_f - P_c)$ denotes the pressure drop across the entire injector. The pressure drop ratio can also be expressed in terms of the geometrical parameters of the injector. The intricate dynamics in a swirl injector produce complicated amplitude-phase characteristics of its overall response function, as shown in Fig. 60 where $\Delta \bar{P}'_i$ is set to unity for simplicity. This diagram allows one, under practical design limitations, to obtain any desired pulsation characteristic by either suppressing or amplifying flow oscillations. Thus, it becomes possible to control the engine combustion dynamics by changing the injector dynamics alone without modifying the other parts of the combustion device.

Acknowledgments

The authors wish to express their sincere thanks to Piyush Thakre for helping to prepare the figures.

References

[1]Mikhailov, V. V., and Bazarov, V. G., *Throttlable Liquid Rocket Engines*, Mashinostroenie Pub., Moscow, Russia, 1985.

[2]Dityakin, Y. F., Klyachko, L. A., and Jagodkin, *Atomization of Liquids*, Mashinostroenie Pub., Moscow, Russia, 1977.

[3]Pazhi, D. G., and Prakhov, A. M., *Liquid Atomizers*, Khimiya, Moscow, Russia, 1979.

[4]Rollbuhler, H. J., "Experimental Investigation of Reaction Control, Storable Bipropellant Thrusters," NASA TND 4416, 1976.

[5]"Space Shuttle Orbital Maneuvering Subsystem Rocket Engine Design Features," Aerojet Liquid Rocket Co., Rept. N6673:271, PRA/SA, Rockwell, July 1976.

[6]Elverum, G., Jr., Staudhammer, P., Miller, J., Hoffman, A., and Rockow, R., "The Descent Engine for the Lunar Module," AIAA Paper 1967-521, 1967.

[7]Bazarov, V. G., *Dynamics of Liquid Injectors*, Mashinostroenie Pub., Moscow, Russia, 1979.

[8]Abramovich, G. N., *Applied Gas Dynamics*, Nauka, Moscow, Russia, 1976.

[9]Khavkin, Y. I., *Swirl Injectors*, Mashinostroenie Pub., Moscow, Russia, 1976.

[10]Andreev, A. V., Bazarov, V. G., Marchukov, E. Y., and Zhdariov, V. I., "Conditions of Hydrodynamic Instability Occurrence in Liquid Swirl Injectors," *Energetika*, 1985, pp. 6–10.

[11]Lyul'ka, L. A., and Bazarov, V. G., "Investigations of the Self-Oscillation Mode of Liquid Sheets in a Coaxial Air Flow," *Aviatsionnaya Tekhnika*, No. 3, 1978, pp. 19–24.

[12]Taylor, G., "The Mechanism of Swirl Atomizers," *Proceedings of 7th International Congress for Applied Mechanics*, Vol. 2, London, 1948.

[13]*Liquid Rocket Engine Injectors*, NASA SP-8089, 1976.

[14]Andreev, A. V., Bazarov, V. G., Dushkin, A. L., Ggrigoriev, S. S., and Lul'ka, L. A., *Dynamics of Gas-Liquid Injectors*, Mashinostroenie Pub., Moscow, Russia, 1991.

[15]Bazarov, V. G., "Injectors for Three-Propellant Liquid Rocket Engine with Smooth Thrust Variation," IAF-95-S.1.03, *Proceedings of 46th International Astronautical Congress*, Oslo, Norway, 1995.

[16]Baywel, L., and Orzechovski, Z., *Liquid Atomization*, Taylor & Francis, Washington, DC, 1993.

[17]Bazarov, V. G., "The Effect of Injector Characteristics on Combustion Efficiency and Stability," IAF Paper 92-0645, *Proceedings of 43rd Space Conference*, Washington, DC, 1991.

[18]Dressler, L., and Jackson, T., "Acoustically Driven Liquid Sheet Breakup," *Proceedings of 4th ILASS American Conference*, Hartford, CT, Vol. 4, 1990, pp. 132–141.

[19]Idelchik, I. E., *Handbook on Hydraulic Resistances*, Mashinostroenie Pub., Moscow, Russia, 1975.

[20]Kudriavtzev, V. M., *Basics of Theory and Design of Liquid Rocket Engines*, 4th ed., Visshaya Shkola, Moscow, Russia, 1993.

[21]Zhukovski, N. E., *Hydraulics*, Vol. 7, ONTI-NKTP, 1937 (in Russian).

[22]Bazarov, V. G., "Self-Pulsations in Coaxial Injectors with Central Swirl Liquid Stage," AIAA Paper 1995-2358, 1995.

[23]Bazarov, V. G., *Fluid Injectors Dynamics*, Mashinostroenie Pub., Moscow, Russia, 1979.

[24]Glickman, B. F., *Dynamics of Pneumo-hydraulic Liquid Rocket Engine Systems*, Mashinostroenie Pub., Moscow, Russia, 1983.

[25]Ditiakin, Y., Kliatchko, L., Novikov, B., and Yagodicin, V., *Atomization of Liquids*, 3rd ed., Mashinostroenie Pub., Moscow, Russia, 1987.

[26]Zong, N., and Yang, V., "Dynamics of Simplex Swirl Injectors for Cryogenic Propellants at Supercritical Conditions," AIAA Paper 2004-1332, 2004.

[27]Yang, V., and Bazarov, V., "Propellant Rocket Engine Injector Dynamics," *Journal of Propulsion and Power*, Vol. 14, No. 5, 1998, pp. 797–806.

Chapter 3

Atomization in Coaxial-Jet Injectors

Lucien Vingert* and Pierre Gicquel*
ONERA, Palaiseau, France

Michel Ledoux[†] and Isabelle Caré[‡]
CORIA, Université de Rouen, Rouen, France

and

Michael Micci[§] and Michael Glogowski[¶]
Pennsylvania State University, University Park, Pennsylvania

Nomenclature

a = radius of atomizer
C = model constant
d = liquid nozzle diameter
d_i = drop diameter
D_0 = initial diameter of vaporizing drop
D_{10} = arithmetic mean diameter
h = thickness of annular air nozzle
L_R = length of recess
L = length of combustor
M_f = mixture ratio
M_r = momentum flux ratio
n = number of droplets produced per unit time
N_i = number of droplets
P_l = pressure of liquid upstream in injector

*Aerospace Research Engineer.
[†]Professor.
[‡]Graduate Research Assistant.
[§]Professor, Aerospace Engineering, Propulsion Engineering Research Center.
[¶]Graduate Research Assistant, Propulsion Engineering Research Center.

P_v = vapor pressure
q_m = mass flow rate of atomized fluid at interface
r = radial distance from spray axis
R_d, R_h = Reynolds numbers
SMD or D_{32} = Sauter mean diameter
t_e = vaporization time
t_s = residence time
U_{air}, U_g = mean velocities of air and gas flow, respectively
ΔU = difference of velocity between air and liquid at the interface
U_l = mean velocity of liquid flow
V = velocity
We = Weber number
x = distance from injector outlet
λ_e = vaporization constant
Λ = wavelength of most unstable mode
σ = surface tension
ρ_g, ρ_l = densities of gas and liquid, respectively
Ω = growth rate of most unstable mode
η = perturbation of interface
τ_r = nondimensionalized breakup time of a ligament

I. Introduction

A COMBUSTION chamber using coaxial atomizers is a complex system in which numerous elementary processes such as atomization, mixing, evaporation, and chemical reaction take place with a broad range of spatial and time scales. This chapter focuses on the atomization of a shear coaxial injector. After a brief description of the basic phenomena and a review of the available literature in Section II, three sets of experimental investigations will be presented in Section III, focusing on different aspects with increasing complexity.

II. Phenomenological Description and Literature Review

Several reviews of experimental and theoretical works on coaxial atomizers are available (see, for example, Ferrenberg et al.,[1] Lefebvre,[2] and Vingert et al.[3]). The present survey is based on this last work, with emphasis placed on the fundamental aspects.

A. General Scheme of Jet Disintegration and Drop Formation

The aim of atomization is to substantially increase the liquid surface area to enhance vaporization, mixing, and burning. Although this surface-area increase may be achieved in various ways, the end result is that the liquid jet becomes unstable, leading to disintegration of the liquid surface into droplets. In the case of a shear coaxial injector, commonly used in cryogenic-propellant engines, the liquid jet is atomized by a coflowing high-velocity gas stream (gaseous hydrogen in cryogenic engines), as illustrated in Fig. 1. Although this is true for any type of injector, the breakup of the liquid jet is a result of the

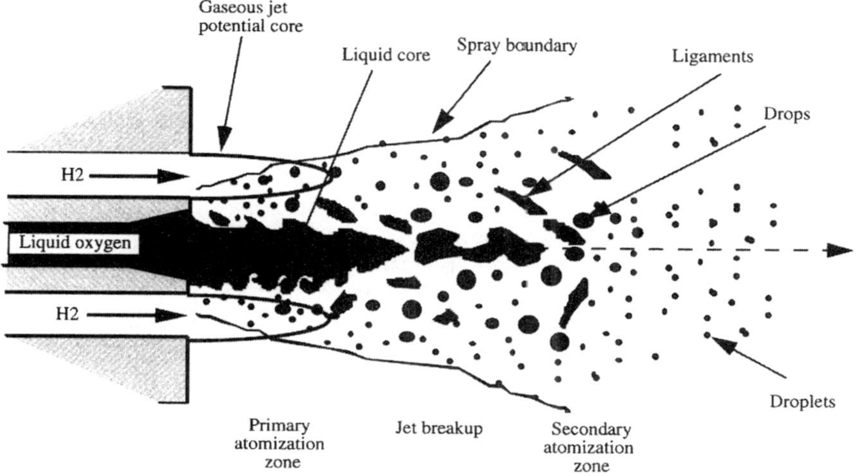

Fig. 1   General scheme of coaxial jet disintegration.

complex interaction among inertial, viscous, and surface tension forces, which can be summarized as follows. Liquid viscosity has a damping effect on the growth of disturbances on the surface. Aerodynamic forces, on the contrary, tend to promote them. Surface tension tends to pull the liquid together. Pressure oscillations and turbulence in the injected fluids and their surrounding gases also affect the dynamics of atomization. The overall process can be characterized by several nondimensional parameters defined as

$$R_d = \frac{U_l d}{\nu_l} \qquad R_h = \frac{U_l h}{\nu_l} \tag{1}$$

$$We = \frac{\rho_g U_l^2 d}{\sigma} \qquad We_m = \frac{\rho_l U_l^2 d}{\sigma} \qquad We_h = \frac{\rho_g U_l^2 h}{\sigma} \tag{2}$$

$$M_f = \frac{\dot{m}_l}{\dot{m}_g} \qquad M_r = \frac{\rho_g U_g^2}{\rho_l U_l^2} \tag{3}$$

The Reynolds numbers R_d and R_h measure the ratios of viscous to aerodynamic forces. The Weber numbers defined by Eq. (2) relate aerodynamic forces to surface tension. M_f is the mixture ratio and M_r is a ratio of local momentum fluxes; $\dot{m}_l$ is the liquid mass flow rate and $\dot{m}_g$ is the gaseous mass flow rate. Another important number can be obtained by combining We_m and R_d (the Ohnesorge number) and depends on both the liquid properties and the nozzle diameter:

$$O_h = \frac{\mu_l}{\sqrt{\rho_l \sigma d}} \tag{4}$$

By plotting the Ohnesorge number vs the Reynolds number, one can distinguish between the low-velocity region, where the jet breakup is due to the action of surface tension forces (i.e., the Rayleigh regime described in Section II.B), and the high-velocity regions, where the influence of aerodynamic (inertial) forces increases exponentially with R_d at constant O_h, so that it rapidly dominates the process.

It is generally agreed (see, for example, Farago and Chigier,[4] Mayer and Krülle,[5] Gomi,[6] and Hopfinger and Lasheras[7]) that the disintegration of the liquid jet proceeds roughly as follows:

1) In the near field of the nozzle (typically between the nozzle exit and a few nozzle diameters downstream, i.e., in the gaseous stream potential core), the difference of velocity between the gas and liquid leads to a surface instability and a stripping of filaments or drops from the jet surface. This is the primary atomization zone.

2) In the far field of the nozzle (further downstream, but before the end of the liquid core), the gas velocity decreases because of mixing with the external atmosphere, and instability scales grow. Different breakup modes may occur in this region. Simultaneously, in the spray surrounding the liquid jet, the so-called secondary atomization process takes place. The largest droplets and the ligaments produced in the primary atomization zone, as well as in the jet breakup zone, break up further during their flight into smaller and more stable droplets, depending on the local Weber number. The breakup time of these droplets can be expressed as a function of the local relative velocity and the ratio of the gas to liquid density. This breakup time, together with the initial droplet velocity, is very important in combustion applications because it gives the distance from the injectors where the secondary atomization takes place (as compared) to the flame position.

Farago and Chigier[4] carried out an extensive experimental study, using a spark photograph technique, to visualize the rupture of round liquid jets under conditions with and without a coflowing gas stream. They observed in the far field: 1) a Rayleigh type breakup, which can be divided into two subgroups, axisymmetric breakup ($We < 15$) and nonaxisymmetric breakup ($15 < We < 25$); 2) a membrane-type breakup ($25 < We < 70$), in which the round jet develops into a thin sheet (membrane), which forms Kelvin-Helmholtz waves and breaks up into drops; and 3) a fiber-type breakup ($100 < We < 500$), which can be further divided into pulsating, which is the normal submode, and superpulsating jet disruption.

Gomi,[6] based on previous experimental works on coaxial injectors and his own results, classified the breakup regimes of such atomizers into three categories, depending on the mixture ratio M_f defined earlier: 1) $M_f < 0.2$, the relative velocity determines the drop size; 2) $0.2 < M_f < 1$, the relative velocity and the mixture ratio determine the drop size; and 3) $M_f > 1$, many parameters affect drop sizes. The drop size depends on the potential core length of the annular gas jet relative to the length of the atomizing region.

Hopfinger and Lasheras,[7] using water and air as simulant fluids, found several breakup regimes based on the Weber number We, but in contrast to Farago and Chigier.[4] A membrane-type breakup exists for We up to 150. They also investigated the importance of the local gas/liquid momentum ratio M_r. The liquid core may disintegrate completely for M_r greater than 10. If M_r is greater than 20, a recirculation of gas near the nozzle leading to liquid core disintegration could explain the superpulsating mode of Farago and Chigier.[4] Le Visage[8] and Carreau et al.[9] investigated the two-phase flow in a spray using an optical fiber probe. Le Visage pointed out the importance of the development of the hydrodynamic regime in the nozzle tube. He also showed that both the momentum and density ratios determine the breakup length of the liquid core.

Vingert[10] presented the results of measurements performed in a pressurized vessel (up to 30 bar), with a Malvern particle sizer, and gave a correlation expression applicable to European Vulcain engine injectors. The same injectors were tested at atmospheric pressure by Gaic et al.[11] They showed that, at atmospheric pressure, good atomization can be obtained only when the gas-to-liquid momentum ratio approaches its nominal value.

There are also papers dealing with points that are often neglected, such as the influence of large amplitude pressure waves interacting with the spray, real cryogenic fluids, supercritical conditions, and combustion (Huynh et al.,[12] Beisler et al.,[13] Schick et al.,[14] Tamura and Mayer,[15] Goix et al.[16]).

B. Studies of Elementary Processes

Experimental information relevant to the far field is found in the studies of more basic flows: round jet, plane or annular sheets, and disintegration of a jet by a normal gas flow.

For round jets, Reitz[17] summarized the conclusions of several previous investigators and distinguished four breakup regimes with increasing Reynolds number at constant Ohnesorge number: 1) Rayleigh jet breakup: surface tension force, which induces axisymmetric oscillations of the jet surface, is mainly responsible for breakup; drops with diameter exceeding that of the jet are pinched off from the end of the jet; 2) the first wind-induced regime: aerodynamic forces start playing a role; unstable waves at long wavelengths grow along the jet and drops with diameter on the order of the jet diameter are pinched off from the end of the jet; 3) the second wind-induced regime: drops that are smaller than the jet diameter produced by the unstable growth of short wavelength surface waves are torn off from the jet surface; and 4) the atomization regime: the jet surface appears to break up immediately at the nozzle exit into very small drops.

A classical "stability curve" is found very often in the literature; it describes the evolution of the breakup length vs the jet mean velocity. It shows a maximum at the transition between the Rayleigh and first wind-induced zones. A common opinion is that this maximum velocity is determined by aerodynamic forces, as shown by the theory of Weber.[18] Grant and Middleman,[19] at atmospheric pressure, and Fenn and Middleman,[20] in subatmospheric pressures, showed that the Weber theory did not predict correctly this transition. Sterling and Sleicher[21] then provided a correction to Weber's theory. Experiments at high pressures, up to 80 bar, showed that breakup could be attributed to two kinds of phenomena, depending on the properties of the fluids involved, the ambient pressure, the Ohnesorge number, and the ratio of the length of the nozzle to its diameter, i.e., the velocity profile at the exit (Leroux et al.[22]).

The classical linear theory examines the stability of a liquid jet surface due to an infinitesimal axisymmetric disturbance with a Fourier form: $X = X_0 e^{ikz+\alpha t}$, where k is the wave number and α the wave amplitude exponential growth rate, α is governed by the dispersion equation, obtained from the linearized Navier–Stokes equations by considering the aerodynamic effects with some simplifying assumptions, and using the linear stability theory. The result is that α depends on the local relative velocity, on the gas and liquid densities, on the kinematic viscosity, and on the wave number.

The problem of the destabilization of an annular sheet by two coflowing, possibly swirling, airflows is basic for aircraft atomizers. This rather general case induced several developments of the linear theory.[23–27] The different models vary as to the hypotheses made on the relative values of sheet thickness, radius, and instability wave length. Camatte and Ledoux[28] removed most of these hypotheses and demonstrated the limitations of this kind of model by comparing its results with the visualization of a specifically built sheet generator.[29] For liquid velocity values below 1 m/s, the wavelengths observed are spanwise and well predicted by linear theory. When the liquid velocity is above 1 m/s, unpredictable streamwise instabilities appear. This phenomenon may be compared to the streamwise structures observed by Bernal and Roshko[30] or Breidenthal[31] in gas/gas or liquid/liquid mixing layers. It is attributed to the development of a (lateral) three-dimensional instability of the initially linear vortex line constituting the interface. Some analyses have been proposed in the area of turbulent jet modeling.[32] Lozano et al.[33,34] proposed a numerical approach to this phenomenon for a liquid/gas planar interface and planar liquid sheet. This phenomenon could be important in the breakup of jets into filaments; this may be observed in diesel sprays, for example, because the velocity of the round jet is sufficiently high.

Near-nozzle surface stripping is generally attributed to Kelvin-Helmholtz type instabilities. Wu et al. and Faeth et al.,[35–37] however, claimed in a series of papers that surface deformation and the stripping of the interface are due to turbulence energy and to the presence of vorticity produced by boundary layers at the nozzle wall. This opinion is supported by a model in which scales are deduced from considerations based on turbulent scales and energies. These scales compare fairly well to experimental Sauter mean diameter (SMD) values. In an amazing experiment they showed, by removing the external part of the liquid jet with a small annulus or "cutter," which leaves the jet without any boundary layer, that the first part of the interface near the nozzle remains smooth. Because the Kelvin-Helmholtz theory does fit many experimental data in the near-nozzle field, there remain unanswered questions here.

Lefebvre[38] distinguished two atomization mechanisms in airblast atomizers, classical atomization, where the droplet formation is due to wave instability on the liquid surface, and prompt atomization, which occurs if the gas stream impacts the liquid jet at high angles of incidence, or if the gas velocity is unusually high. This may be related to the so-called membrane-type breakup of Farago and Chigier. Most of the studies of jets destabilized by normal gaseous flow have been conducted under supersonic conditions.[39–41] For subsonic conditions, a classification of the different aspects observed, including membrane type, is given by Vich and Ledoux[42] in terms of M_r and Reynolds number R_d.

C. Numerical Simulations of the Atomization Process

Available linear theories, developed for round or planar jets, may be convenient for users, but they are limited to small amplitudes of a sinusoidal single mode. They cannot predict the real shape of ligaments near breakup. A ligament's deviation from sine shape can be interpreted in terms of the appearance of harmonics. The modal analysis of linear theory turns then to a study of nonlinear interaction of modes, and it is necessary to perform nonlinear calculations. There are several different approaches: direct solution of the Navier–Stokes equations or discretized vortex sheet theories. The Navier–Stokes calculations may take into account the viscosity of fluids, i.e., loss of vorticity from the interface, but they involve the solution of the difficult problem of the definition of interface deformation. Various solutions have been proposed (use of adaptive meshes, or volume of fluid method).[43]

The discretized vortex sheet method[44–46] can be applied only to ideal (nonviscous) fluids but avoids the problem of surface localization, because the positions of vortex wires constituting the interface are directly calculated.

For simpler as well as less extensive calculations, a method is proposed in Section III for the SMD of the drops formed at the interface in the near-nozzle zone. Section III also points out the difficulty of estimating the skimming kinetics (mass flux of drops torn off from the surface per unit time).

D. Derivation of Droplet Size Distribution Functions

Neither linear nor nonlinear methods for the determination of the scales of instability can deal with the complex reality of jet breakup. The use of information theory was proposed by Sellens and Brzustowski[47] and by Li and Tankin.[48] This is the so-called maximum entropy formalism. A distribution of drop sizes is inferred from the minimization of Shannon entropy. The physical principles of conservation are applied by these authors, producing source terms that are generally not well known. Sangakiri and Ruff[49] tried to address this problem. A new procedure that avoids the use of source terms was proposed by Cousin et al.[50,51] It was applied with success to the prediction of drop sizes produced by pressure swirl atomizers, automobile fuel injectors, and ultrasonic atomizers.

III. Investigations of Atomization in Shear Coaxial Injectors

A. Experimental and Theoretical Investigation at Atmospheric Pressure with Simulants

Atomization processes in coaxial atomizers are complex. They cannot be reduced to only one type of drop formation. Among these processes, the "skimming" of the liquid core through a nonlinear Kelvin-Helmholtz type instability produces a spray of fine droplets. Although this process may not be the major atomization process in terms of mass production, it can be credited with a decrease of the liquid core diameter along, for instance, a length on the order of five jet diameters. Furthermore, in many cases, the fine spray formed can be of major importance, as far as ignition or flame stabilization by recirculation of the smallest drops is concerned. The present section focuses on this specific aspect of the formation of drops.

An experimental study is presented as well as a model of two aspects of the primary atomization, namely the prediction of the Sauter mean diameter through a linear/nonlinear model and the determination of the skimming kinetics. Experiments provide data on the radial distribution of drop sizes in the spray and give a value for a constant C needed for the skimming kinetics model. The SMD calculations are validated from these results.

1. Experimental Study of Primary Atomization Near the Nozzle

a. Apparatus. The atomization of a liquid cylindrical flow by a coaxial external annular airflow was achieved by means of a device shown in Fig. 2. Special attention was paid to axisymmetry of both (nonswirling) flows. Liquid and air flows issued from a large settling chamber, equipped with a porous plate and honeycombs in the path of the airflow. The radius of the liquid nozzle is $a = 1.95$ mm; the thickness of the annular air nozzle is $h = 0.54$ mm. Results presented here are for a recess length L_R equal to 0.

Water or kerosene and air were used as substitute fluids for oxygen and hydrogen, respectively. For all experiments, the velocity of the liquid was fixed at 1.4 m/s, whereas the air velocity varied from 100 to 220 m/s at the nozzle exit. Because this part of the study was focused on Kelvin-Helmholtz instabilities in the near-nozzle field, the conservation of local ratios of specific momenta ρU^2 (where ρ is the density and U the velocity of each phase) of the fluids was chosen as a similarity criterion. For the far-nozzle zone, the ratio of the total momenta of liquid and air jets should be preferred (as shown, for example, by Hopfinger and Lasheras[7]). Table 1 summarizes the experimental parameters, as well as the main

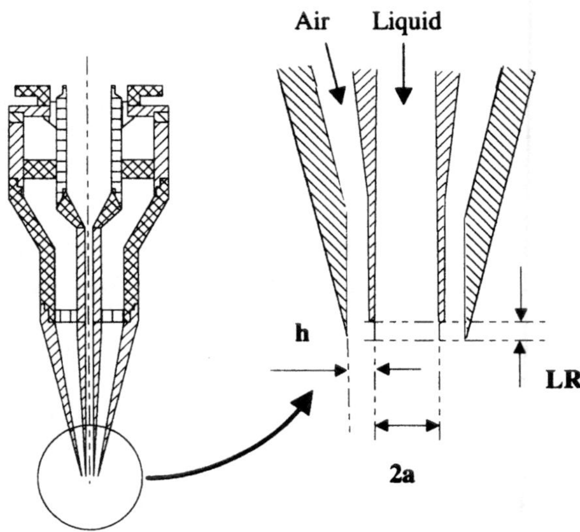

Fig. 2 Atomizer: $a = 1.95$ mm, $h = 0.54$ mm.

Table 1   Operating conditions for water/air tests ($m_l = 0.0167$ kg/s, $U_a = 1.4$ m/s, $R_d = 5460$)

U_{air}, m/s	m_{air}, kg/s	$M_f = m_l/m_{air}$	$M_r = \rho_{air}U_{air}/\rho_l U_l$	W_e
140	0.0014	12	13	1380
170	0.00165	10.1	19.2	2035
200	0.00196	8.5	26.5	2816
230	0.00225	6.7	35.1	3725

corresponding nondimensional numbers defined in Section II, Weber number W_e, ratio of liquid to gas mass flow M_f, and ratio of local momenta M_r.

A short time-exposure photographic technique was performed, using a strobe lamp (10 μs flash duration). Movies were also obtained by use of a continuous polychromatic source. The structure of the spray described in the following was then observed. Different lighting techniques (short time and continuous) were used to distinguish between continuous liquid flow (liquid core) and fine spray.

The structures of the flow observed downstream of the nozzle are similar to those observed previously and described in Section II. It can be summarized as follows (full details are given in Ref. 52):

1) A liquid core: short time exposure (Fig. 3a) shows a liquid jet with large lateral structures, and lateral movement with an amplitude on the order of half a nozzle diameter; with long time exposure (Fig. 3b), a conical shape is easier to observe. The dimensions of this conical shape are reproducible, and this information will be used for skimming kinetics measurements.

2) A dispersed phase around this liquid flow (Fig. 3c): this phase consists of a fine spray entrained by the annular air jet.

According to the mapping in the plane W_e/R_d of Farago and Chigier,[4] these experiments should lie in the fiber-type jet breakup zone. According to Gomi,[6] we are in the zone $M_l > 1$, where many parameters influence the granulometry, not just the relative velocities of the liquid and gas. This remark is not of major importance because we are interested in only the near-nozzle zone. According to Hopfinger and Lasheras,[7] the cone should disintegrate completely ($M_r > 10$) and, for air velocities above 170 m/s ($M_r > 20$), recirculation should be present. This may explain the observed beating of the jet. The appearance of recirculation may depend, however, on the external geometry of the nozzle. The external geometry of our atomizer (Fig. 2) is conical, while the nozzle of Hopfinger and Lasheras discharges through a plate perpendicular to the jet axis. This latter geometry may more easily induce recirculating flows.

The skimming process is not sufficient to provide complete atomization. Because of the presence of mixing layers between the thin annular airflow and the external atmosphere, the relative velocity between air and water decreases strongly along the jet axis. Therefore, atomization does not continue some centimeters downstream of the nozzle. A falling liquid jet of several millimeters in diameter is observed. The present study is confined to the first atomization zone. Air velocities were measured through a laser Doppler velocimetry LDV

L. VINGERT ET AL.

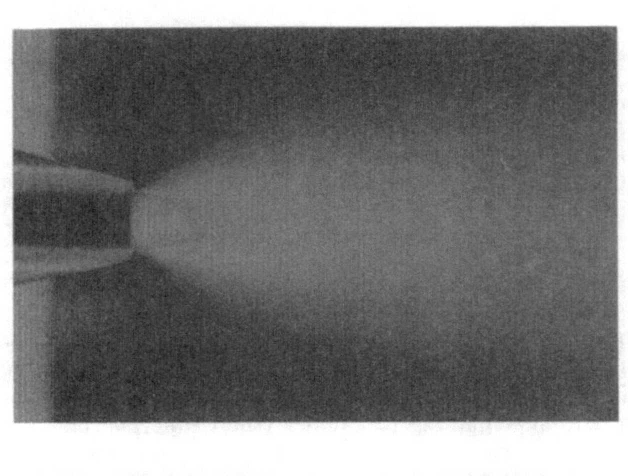

c) Long time exposure, light source at 90°

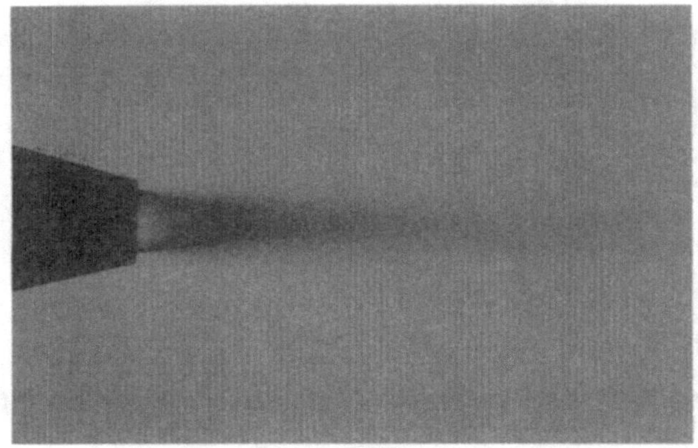

b) Long time exposure (1 s), light source at 180°

a) Short time exposure (10 μs), light source at 180°

Fig. 3 Stroboscopic visualizations of the atomizing jet.

system. Drop size measurements were obtained through a 2600 diffraction particle sizer. Because the investigated spray was annular and relatively thin, the diameter of the laser beam, which is 7 mm for the commercial apparatus, was reduced to 2 mm. The diffraction particle sizer is a line-of-sight device. Because measurements were made through a fraction of an annular dispersed flow, an inversion technique, similar to the one used by Gomi,[6] was employed.[53] The main assumption made at this point is that the spray has a symmetry of revolution; however, the computed corrections were not superior to some microns.

b. Experimental results. In addition to the photographic investigation of the liquid cone using short and long exposures, the results of which have been summarized, the following studies were performed:

1) Radial profiles of drop sizes were established. They are presented here in terms of Sauter mean diameter.

2) Air velocity fields were measured around a slightly conical solid cylinder simulating the liquid cone. This allowed determination of the aerodynamics in a dry atmosphere around an ideal model of the liquid core. Use of this device prevented influence of the measurements by the lateral velocities occurring from the beating of the liquid jet. Of course, the possible effects of the liquid on the momentum of the annular airflow could not be evaluated.

Gas velocity U_{air}, surface tension σ (of water and kerosene), airflow thickness h, and recess L_R were varied. Some of the results are shown in Figs. 4–7. As in Gomi's[6] results, one can observe an increase of SMD values radially from the liquid core surface to the external part of the spray (Fig. 4). Because of lateral movements of the jet, measurements of drop size at distances of less than

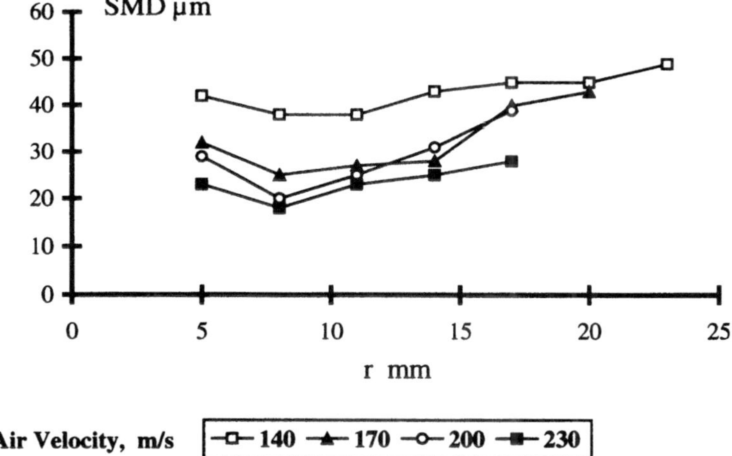

Fig. 4 Radial profiles of SMD, 30 mm downstream of the nozzle, for different air velocities.

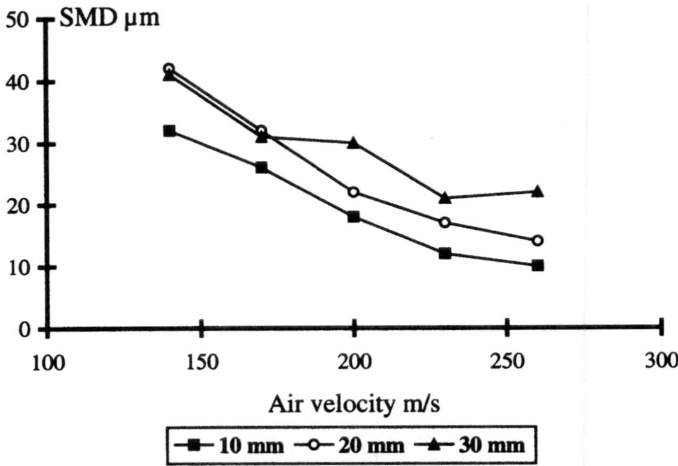

Fig. 5 Influence of air velocity, at 10, 20, and 30 mm downstream of the nozzle.

5 mm from the mean position of the interface were not found to be reliable. This can be explained through inertial effects of drops. The largest droplets have a much greater Stokes response time (varying as d^2) than the others. They will preserve their radial and axial velocity components over greater distance along the axis, while the smallest drops are immediately entrained by the airflow, resulting in a concentration of large drops away from the liquid jet.

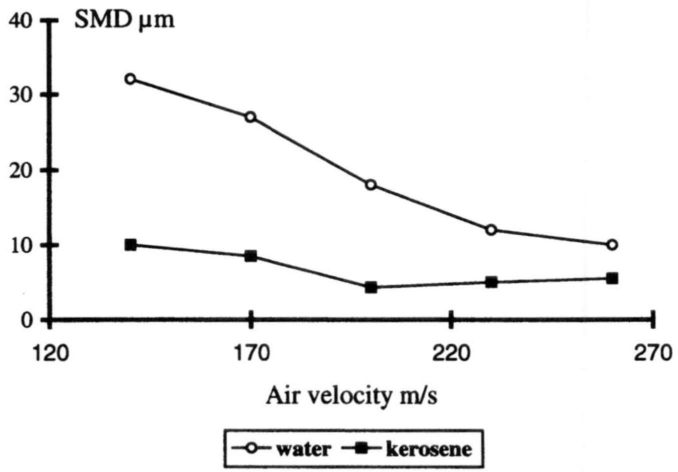

Fig. 6 Effect of surface tension on drop size.

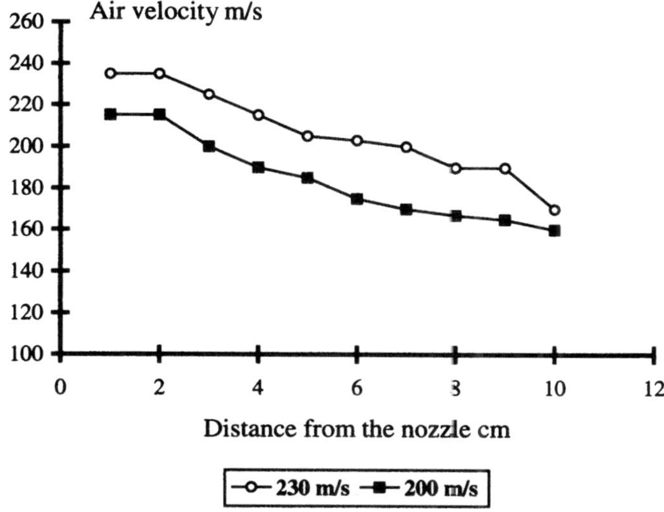

Fig. 7   Variation in atomizing air velocity along the liquid jet.

According to the following atomization theory, an increase of ΔU, the difference between liquid and air velocities, leads to smaller drops. The destabilizing effect of aerodynamic forces is thus demonstrated (Fig. 5). A smaller surface tension leads to smaller drops (Fig. 6). The stabilizing effect of surface tension is then confirmed. Figure 7 shows that an atomization process can be continued for some distance downstream of the nozzle.

2.   Droplet Sizing in the Downstream Region

To extend the experiments just described, phase Doppler particle analyzer (PDPA) measurements were performed at atmospheric pressure, with water and air serving as substitutes for liquid oxygen and gaseous hydrogen, to examine the influence of the atomizer geometry, especially the effects of recessing and tapering the liquid oxygen (LOX) post. Although water/air experiments cannot provide absolute drop size and velocity values that can be assumed to be valid in a high-pressure liquid oxygen/gaseous hydrogen (LOX/GH$_2$) rocket engine, trends produced by changes in injector design may be comparable and, because of the ease of conducting experiments, greater amounts of data can be obtained. The baseline dimensions of the shear coaxial injector are shown in Fig. 8. They were derived from the dimensions of the prototype injector element of the fuel preburner of the space shuttle main engine (SSME). A second LOX post was fabricated with a 7 deg taper at the tip to match the shape of the production injector element. Mass flow regulation of both air and water was performed by means of calibrated, variable area rotameters. Quantitative data on the spray characteristics were obtained with a phase Doppler particle analyzer, which allowed simultaneous

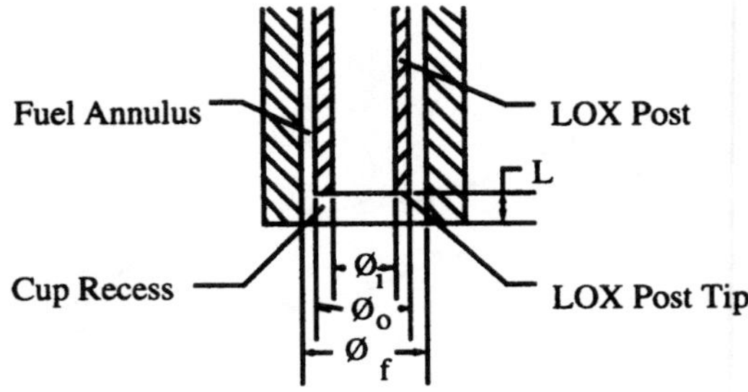

Baseline Dimensions

Fuel Annulus Diameter, $\varnothing_f$	5.03 mm
LOX Post Outer Diameter, $\varnothing_o$	3.76 mm
LOX Post Inner Diameter, $\varnothing_i$	2.26 mm
Post Tip Land Width	0.76 mm
Recess Length, L	2.54 mm

Fig. 8   Shear coaxial injector schematic with baseline dimensions.

measurement of droplet size and velocity at spatially resolved positions within the spray. The optical configuration for these tests included a helium neon laser emitting at 632.8 nm, a diffraction grating based transmitter, and a receiver oriented at either 30 or 90 deg from the optical axis of the transmitter. The transmitter was fitted with a 500 mm focal length (f.l.) lens and the receiver with a 495 mm f.l. collection lens. Doppler signal filtering and acquisition were performed by a counter-based processor provided by the manufacturer of the device. After completion of data sampling, dedicated software computed the mean values of droplet size, which includes the arithmetic mean and Sauter mean diameters, and the mean and rms values of the droplet velocity.

Figures 9 and 10 illustrate the radial distribution of droplet mean axial velocity and size for the straight and tapered post positioned flush with the injector exit. Mass flow rates for water and air were maintained at 3.12 g/s and 1.40 g/s, respectively. Data were taken at a single axial location of 50 mm from the injector exit and at several radial positions from the injector centerline to 20 mm in the spray. As can be seen in Fig. 9, the two post tip shapes have yielded similar velocity profiles. The magnitude of droplet velocities tended to be less for the tapered post at the injector centerline, but approached that of the straight post moving outward in the spray. The distribution of drop sizes, as depicted in Fig. 10, reveals larger drop sizes for the tapered post than for the straight post.

The effect of recessing the tapered post on the drop size and velocity distributions is depicted in Figs. 11 and 12. The recessed post tended to produce a less steep velocity profile than the nonrecessed post and lower droplet velocities close to the injector centerline. There was little to no variation in drop size between the recessed and nonrecessed post within approximately 10 mm of the centerline. Beyond this distance the recessing of the post did produce smaller

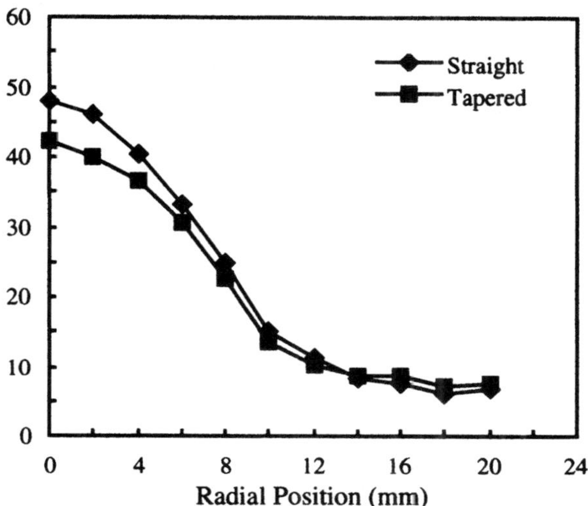

Fig. 9   Radial variation in mean axial velocity at 50 mm downstream of injector operating with tapered and straight post.

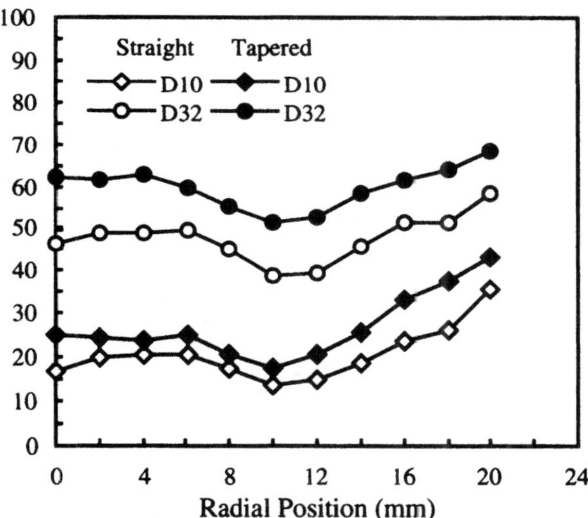

Fig. 10   Radial variation in mean drop size at 50 mm downstream of injector operating with tapered and straight post.

                    L. VINGERT ET AL.

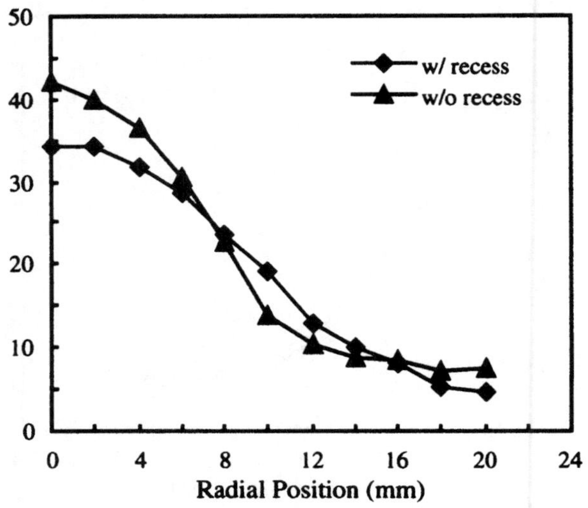

Fig. 11   Radial variation in mean axial velocity at 50 mm downstream of injector with tapered post not recessed and recessed to 2.54 mm.

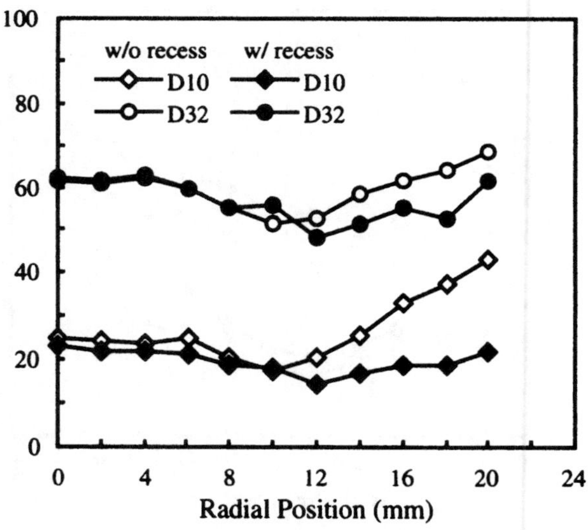

Fig. 12   Radial variation in mean drop size at 50 mm downstream of injector with tapered post not recessed and recessed to 2.54 mm.

drop sizes. The same behavior was found when one moved closer to the injector face and into the dense spray region. Measurements were made as close as 1.27 mm to the LOX post tip, but no reverse flow, which would be an indication of a recirculation region, was found.

3. Modeling of the Primary Atomization Process

a. Prediction of drop sizes.   Drop sizes away from the surface ($r > 5$ mm) may be influenced by several processes, such as the aerodynamic effects quoted in the preceding discussion or secondary atomization due to breakup of drops. The present calculations are concerned only with primary atomization, i.e., the formation of drops at the liquid jet surface. A second and quite different model has to be developed to predict ballistic effects and secondary breakup.

The authors have previously developed a complete linear model of the instability of a swirling liquid jet in a swirling coflowing air current.[52] For high relative liquid-air velocities, this model was found to be asymptotic to the classical Kelvin-Helmholtz theory for capillary waves at a plane interface. The dominant wavelength Λ (wavelength of the instability mode showing the fastest growth) and its growth rate Ω (in linear temporal theory the perturbation is growing as $\exp \Omega t$) are expressed by

$$\Lambda = \frac{3\pi\sigma}{\rho_g \Delta U} \qquad \Omega = \frac{2\rho_g \Delta U^2}{3\sqrt{3}\sigma}\left(\frac{\rho_g}{\rho_l}\right)^{1/2} \tag{5}$$

where σ is the surface tension, ρ_g and ρ_l the gas and liquid densities, and ΔU the difference of velocities between liquid and gas.

The propagation velocity c of this wave is then

$$c = \frac{\rho_g U_g + \rho_l U_l}{\rho_g + \rho_l U_l} \tag{6}$$

From Λ and Ω one can deduce different parameters of the atomizing zone: SMD, skimming kinetics, and even initial cone angle.[53]

The development of the interfacial wave is linear only in its initial stage, i.e., for short times. For longer times, the authors developed a nonlinear method of calculation based on the vortex sheet technique. This method has been used to investigate the nonlinear interaction between different modes by Cousin et al.[46] For the present case, it was considered sufficient to calculate the value of the dominant wavelength through linear theory and to apply a simple description of the nonlinear subsequent evolution of the wave, which starts from a sinusoidal shape and evolves to a filament-like shape. The wave is initially sinusoidal, given by a normal mode analysis:

$$\eta(x, t) = \eta_0 \exp \Omega t \exp i(kx - \omega_r t) \tag{7}$$

where $\omega_r = kc$, k is the wave number of the dominant mode $k = 2\pi/\Lambda$, η_0 is the initial disturbance amplitude, $\eta(x, t)$ is its value at time t and abscissa x, Ω is the growth coefficient of the most unstable mode, and Λ is the wavelength (scale) of this mode. The velocity of the crest is given by a Lagrangian derivative $d\eta/dt = \Omega\eta$. Here it is assumed that, near the breakup, the wave has the shape of a cylindrical ligament of diameter $\Lambda/2$ and length $A\Lambda$. A is a constant that will be estimated in the following. This shape agrees with photographs and calculations by different authors.

In reality, atomization results in a spectrum of drop diameters $N_i(d_i)$, with N_i being the number of drops of diameter d_i formed. One can write the conservation of mass for the breakup of a ligament:

$$\sum N_i d_i^3 = 3A\Lambda^{3/8} \tag{8}$$

Surface tension energy is also conserved, leading to

$$\sum N_i d_i^2 = A\Lambda/2 \tag{9}$$

By dividing these two equations, and owing to the definition of the Sauter mean diameter, the unknown constant A disappears, and the effect of the dispersion is statistically eliminated from the definition of SMD,

$$\text{SMD} = \frac{3\Lambda}{4} \tag{10}$$

The mean number of drops formed from one ligament is on the order of A ($n = 8/9A$).

A comparison is made in Fig. 13 between calculated drop sizes and SMDs measured 10 mm downstream of the nozzle at 5 mm from the jet axis for water and kerosene sprays. In this figure, for each liquid, each point corresponds to a different air velocity. The comparison appears to be satisfactory.

b. Skimming kinetics. This parameter has, to our knowledge, not yet been accurately measured, although it is extremely important in practical applications. An attempt was made[52,53] to determine it experimentally and then produce a model predicting the mass flow rate of liquid torn off from the surface by capillary instabilities (skimming kinetics). The density of atomized liquid (mass of liquid torn off from the surface per meter^2 and second), q_m, was expressed as

$$q_m = C\Lambda\Omega \tag{11}$$

This formula is based on simple considerations. The number of drops formed per unit time is inversely proportional to the time needed for a ligament to grow and break up; this number will then be proportional to the growth factor Ω. The number of ligaments appearing at the same time per surface unit is inversely proportional to the area of the section of one ligament, which is $\pi\Lambda^2$. The mass flow rate per second and meter^2 of interface is proportional to the number of filaments on a surface unit, inverse to the time needed for breakup of a filament, and proportional to the mass of the filament, itself proportional to Λ^3. C was

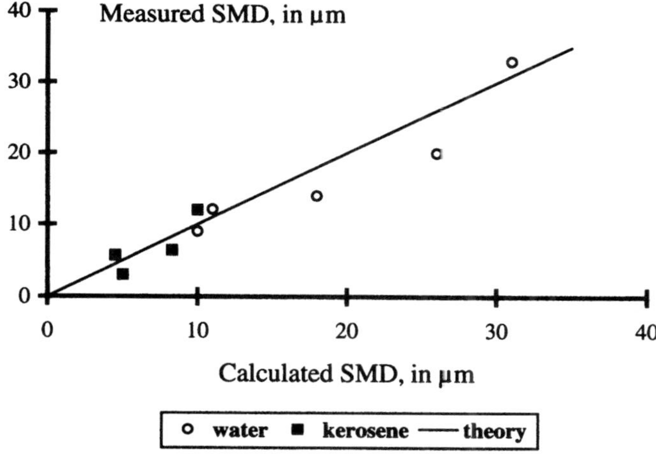

Fig. 13 Comparison between predicted SMD and experimental values. For each fluid, each point corresponds to a different air velocity.

experimentally deduced from a mass balance in the liquid cone. The velocity U_L was assumed to be constant along the axis of the liquid cone (which is debatable, because the drag by rapid airflow is neglected). The momentum of the airflow is four times that of the liquid.

Experimental determination of C under these conditions can only be a rough estimate (but worth attempting, because of the scarcity of literature on the subject). Table 2 shows experimental values for water and kerosene sprays. A mean value of 1.8×10^{-2} was found, with a standard deviation of 5×10^{-3}.

The nonlinear nature of instability at breakup must be taken into account. Ω is the growth coefficient of the linear theory; it is associated with an exponential growth of the amplitude. It has been shown (for cylinders and plane interfaces, for example) that the nonlinear part of the growth is no longer exponential. The use of Ω could be considered irrelevant from this point of view. In the following we shall thus consider Eq. (11) as an empirical expression defining C (which is correct).

**Table 2 Values of C for different liquids
and air velocities**

U_{air}, m/s	Water	Kerosene
123	0.022	0.029
147	0.014	0.023
170	0.018	0.014
195	0.015	0.017
220	0.015	0.015

We now consider that, reasonably, the number of drops n produced by unit surface and unit time can be expressed as

$$\dot{n} = K \frac{n}{t_r \Lambda} = K \frac{n \Delta U}{\tau_r \Lambda^3} \tag{12}$$

New parameters appear: n is the number of drops formed from one ligament, and τ_r is a nondimensional time defined as $t_r \, \Delta U / \Lambda$, where t_r is the time needed to form drops from each ligament (breakup time of the ligament).

Defining B as a constant such that the drop diameter d can be expressed by $d = B\Lambda$, the density of the mass flow rate q_m skimmed from the liquid surface can be expressed as

$$q_m = \rho_l \dot{n} G \frac{\pi d^3}{6} = 0.52 K B^3 \frac{n \Delta U}{\tau_r} \tag{13}$$

A drawback of this new formulation is that τ_r and n are not known immediately. Mayer[54] attempted to estimate this time from considerations derived from the linear theory. He used his model to derive a drop distribution but did not give an expression for the skimmed mass flow rate. With a nonlinear technique such as the vortex sheet method, it is now possible to determine these two parameters. In particular, the vortex sheet method gives the true shape of the ligaments for each value of τ_r. The advantage of the present procedure is then twofold: the model is more correct and does not suffer from the limitations of the linear theory, and the nonlinear vortex sheet method can account for the ambient pressure so that this formula can be used for most practical cases.

The product of constants $0.52 B^3 K$ can be easily derived from C. A value of 0.0234 ± 0.006 is thus determined. Notice that the physical meaning of K is clear: it is the ratio of the liquid surface occupied by ligaments to the total surface of the atomizing zone (i.e., the skimmed interface). A reasonable value should be between 0.1 and 1. With B estimated at 0.5, its value is found here to be equal to 0.36.

Finally, two methods of calculation of q_m could be proposed. The first one is straighforward, by use of the formula:

$$q_m = 1.8 \, 10^{-2} \Lambda \Omega \tag{14}$$

or, after calculation of n and τ_r, by the more recommended expression:

$$q_m = 0.023 \frac{n \Delta U}{\tau_r} \tag{15}$$

4.   Conclusion

The experimental study of a coaxial airblast atomizer showed a complex structure. The model presented here represents only primary atomization. An attempt was made to predict the mass flow rate of the atomized liquid in this zone. Two formulas are given. One is for a rough but rapid and easy estimate. The other, based on more physical grounds, requires information from a nonlinear model presented elsewhere. The estimation of Sauter mean diameter from primary atomization is validated satisfactorily with droplet size measurements in the near field of the nozzle. The action of turbulent structures is often advocated as the main process for interfacial atomization. This was discussed in Section II.A. More experimental data are needed to assess the prediction of skimming kinetics, but some indications are given here for a rough estimate of the efficiency of the atomizing process.

B. Photographic Studies Using Cryogenic Fluids (Liquid/Gaseous Nitrogen)

The preceding study was a simplified case, at atmospheric pressure and with water or kerosene and air as simulants. A second step toward actual rocket engines is to take into account the cryogenic nature of the liquid, where atomization and vaporization may be strongly coupled phenomena. This was done in a series of tests performed with liquid nitrogen serving as a simulant to liquid oxygen. In addition to this, the main objective of the liquid nitrogen/gaseous nitrogen (LN_2/GN_2) tests was to characterize the effects of chamber pressure and liquid-to-gas mixture ratio on atomization and vaporization. The baseline dimensions of the shear coaxial injector are shown in Fig. 8. Pressure and temperature were measured in the fuel plenum region upstream of the fuel annulus and in the cryogen supply line at the entrance to the LOX post.

For the elevated pressures required for LN_2 testing, a pressure chamber was constructed to operate at pressures up to 10.0 MPa and to provide optical access. The chamber has a square cross section with sides measuring 10.2 cm. Four ports drilled into the injector end of the chamber serve to provide additional gas for chamber pressurization and to suppress recirculating flow. A range of chamber pressures P_c can be investigated for specific propellant mass flow rates by the replacement of the nozzle plate. Optical access to the spray is through rectangular Plexiglas® windows, measuring 3.2 cm wide by 15.2 cm long by 2.5 cm thick, located on both sides of the chamber. Mean chamber pressure and internal gas temperature are measured at the center of the chamber. Flow visualization was performed with the nontapered LOX post at chamber pressures between 2.2 and 3.3 MPa and at two liquid-to-gas mixture ratios of 0.80 and 1.58. To determine whether any observable differences in liquid atomization exists between the two LOX post designs, several stroboscope tests were conducted with both LOX posts operating at the same mixture ratio and at a range of chamber pressures from 2.2 to 4.1 MPa. Table 3 summarizes the run conditions for the flow visualization tests. Because cavitation due to heat transfer to the cryogen is a significant concern in LN_2 atomization experiments at subcritical conditions, a criterion was established based on the work by Nurick[55] to determine whether a test was influenced by LN_2 cavitation. The criterion for the tests was

$$\frac{P_l - P_v}{P_l - P_c} > 1.64 \tag{16}$$

Table 3   Operating conditions for LN_2/GN_2 flow visualization tests
($P_c = 3.4$ MPa, $T_c = 126$ K)

Test No.	P_c, MPa	m_l, kg/s	m_g, kg/s	m_l/n_g, kg/s	U_g/U_l	$U_g - U_l$, m/s	ρ_g/ρ_l
A1	2.93	0.050	0.062	0.81	8.99	164.8	0.061
A2	2.60	0.049	0.062	0.79	10.06	184.3	0.056
A3	2.32	0.046	0.063	0.74	11.58	203.9	0.052
A4	2.79	0.052	0.033	1.58	4.82	82.3	0.060
A5	2.63	0.052	0.033	1.58	5.06	87.7	0.057
A6	2.34	0.050	0.033	1.54	5.87	103.9	0.051
A7	2.94	0.076	0.094	0.82	8.68	244.0	0.065
A8	3.97	0.076	0.094	0.81	6.99	183.6	0.081

which states that the difference between the LOX post upstream pressure P_l, and vapor pressure P_v must exceed the pressure drop through the LOX post by a factor of 1.64. The value of 1.64 was determined from the entrance geometry (an extremely large $L/D > 50$) and discharge coefficient of the LOX post (a sharp-edged exit orifice). Although the SSME rocket engine operates in the supercritical regime, the dynamics of cavitation at subcritical pressures could yield spray structures, drop size distributions, and overall propellant mixing quality not concurrent with the operation of other subcritical LOX/GH_2 rocket engines. The liquid nitrogen injection temperatures were always subcritical, as is the case for LOX injection in cryogenic rocket engines, and sufficiently low that Eq. (16) was always satisfied. The gaseous nitrogen was injected at ambient temperature.

To gain qualitative insight into the overall morphology of a shear coaxial injector spray, the spray was examined by imaging the scattered light from a laser sheet passing through the axis of the spray and, in a separate test series, from a stroboscope illuminating the back side of the spray. For the laser sheet imaging tests, a frequency doubled Nd : YAG laser (532 nm), pulsed at 10 Hz with a pulse duration of 10 ns, was used with a spherical and cylindrical lens combination to form a laser sheet with a thickness of 0.15 mm and negligible divergence at the injector centerline. The scattered light from the spray was visualized perpendicular to the sheet with a 35-mm camera. A 1-nm bandpass filter centered at 532 nm was installed in front of the camera lens to eliminate ambient light. In an effort to ascertain whether a more extensive droplet flowfield exists beyond the observable limits of the laser sheet images, a modified approach to spray visualization was undertaken with the stroboscope. Based on previous flow visualization experience with water/air sprays at atmospheric pressure, the strobe light was directed toward the back side of the spray, off center from the camera and toward the injector face. With this orientation both the dispersed droplet region and the structure of the liquid core may be imaged with the 35-mm camera. The flash rate of the strobe was set at 10 Hz with a flash duration of 3 μs.

The first series of photographs in Fig. 14 corresponds to the run conditions of tests A1−A3 of Table 3 and illustrates the effect of chamber pressure on the

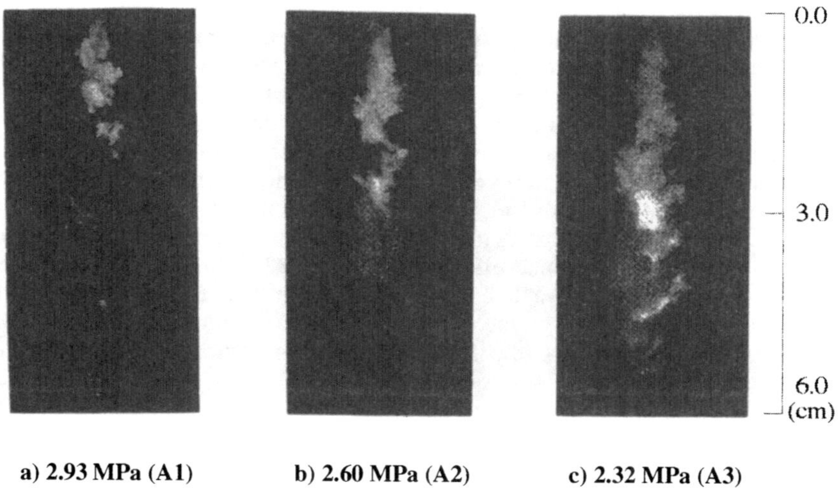

a) 2.93 MPa (A1)   b) 2.60 MPa (A2)   c) 2.32 MPa (A3)

Fig. 14   Effect of chamber pressure on LN$_2$/GN$_2$ spray (mean mixture ratio: 0.78).

overall spray morphology of the coaxial injector using the nontapered LOX post. The photographs were imaged with the laser sheet and have been digitally reproduced from color photographs to improve image quality for photoduplication and to accentuate noticeable structures within the spray. Figure 14 depicts an overall contraction of the large-scale liquid structures (liquid core and detached ligaments) and droplet flow regions with increasing ambient pressure. Regions of droplet flow are most evident downstream of the liquid structures, whereas very little droplet production is observed along the liquid column close to the injector exit. According to the classical description of shear coaxial injector atomization, liquid ligaments and droplets are produced from the jet as a result of unstable surface wave growths induced by liquid turbulence and a shearing force between the surrounding gas and liquid. These ligaments and droplets in turn undergo secondary atomization through surface deformation and breakup by aerodynamic forces resulting in a spray of polydispersed droplets around the liquid jet. With rising chamber pressure the increased gas density enhances the breakup of the jet,[5] as can be observed in Fig. 14, but reduces secondary atomization due to a decrease in the slip velocity between the gas and the droplet. As a consequence, larger droplets pervade the spray of a nonevaporating liquid at higher chamber pressures,[56] although shorter liquid breakup lengths may also be present. Based on the diminishing presence of droplets at higher chamber pressures in Fig. 10, evaporation has become an important mechanism in the evolutionary behavior of the LN$_2$ spray. The large gas mass flux appears to have rapidly vaporized droplets surrounding the core except for those in the shadow of the LN$_2$ jet, a region characterized by lower gas temperatures and velocities. With increasing chamber pressure the rate of liquid vaporization is further enhanced by the reduction to zero of both the latent heat of vaporization and the liquid surface tension at the critical pressure.

In tests A4–A6 of Table 3 the gas mass flow rate was decreased to approximately half of the value for tests A1–A3, while the chamber pressure was varied from 2.34 to 2.79 MPa. The photographic results of these tests were obtained with the laser sheet visualization technique and are digitally reproduced in Fig. 15. Overall, the reduction in gas mass flow rate results in a larger, more dispersed spray, indicating a strong effect of mixture ratio on the atomization and vaporization of liquid nitrogen. The behavior of the spray with respect to increasing chamber pressure follows the same trend as tests A1–A3 in that the LN_2 spray contracts in size as the chamber pressure increases. In terms of droplet sampling with the phase Doppler interferometric device, these reduced gas flow conditions are advantageous in that the observable dilute spray region is spread over a larger volume. The droplet flow regions of tests A1–A3 appear to be confined close to the liquid core, which may create potential difficulties in droplet sampling.

Tests A7 and A8 were performed with the tapered LOX post using the stroboscope to visualize the spray. Test A7 depicted in Fig. 16a matches closely the chamber pressure and mixture ratio of test A1 but was conducted with substantially higher liquid and gas mass flow rates. Regardless of this difference, the lengths of the intact liquid core appear very similar between the two tests. It is

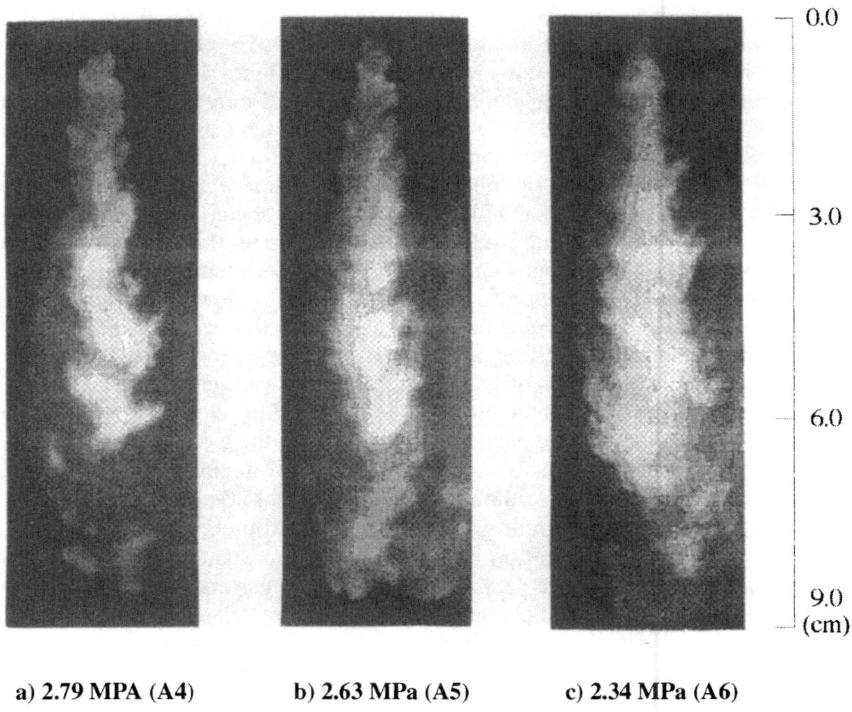

a) 2.79 MPA (A4)   b) 2.63 MPa (A5)   c) 2.34 MPa (A6)

Fig. 15   Effect of chamber pressure on LN_2/GN_2 spray (mean mixture ratio: 1.57).

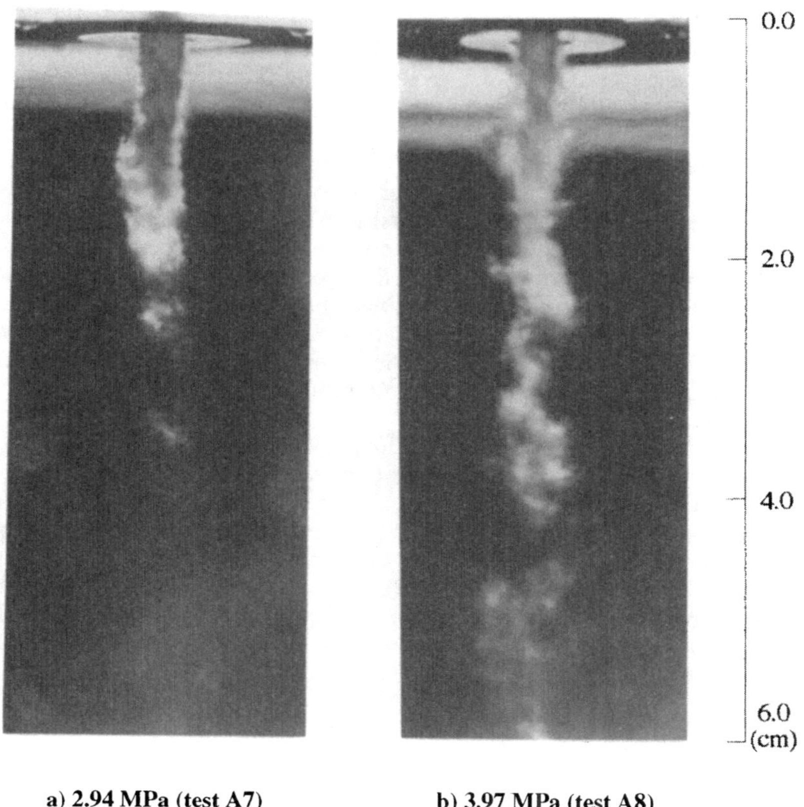

a) 2.94 MPa (test A7)   b) 3.97 MPa (test A8)

Fig. 16   Stroboscope results for tapered LOX post at sub- and supercritical pressures (mean mixture ratio: 0.81).

evident, however, that the stroboscopic image provides more detail on the structure of the liquid core and detached ligaments than the laser sheet images but less information on the droplet flow region. Figure 16b shows the structure of the jet issuing from the injector at a supercritical pressure of 3.97 MPa. Contradictory to the observed contraction of the spray with increasing pressure, the liquid core breakup length has significantly increased for a chamber pressure greater than the critical pressure (3.4 MPa). This behavior may be attributed to the drop in relative velocity between the gas and liquid at very high chamber pressures, which may surpass any enhancements to liquid breakup due to increased gas density, and may be related as well to the increase in droplet evaporation times observed in the supercritical region.[57] Stroboscope tests with the nontapered LOX post under similar operating conditions revealed the same behavior for pressures above and below the critical point but indicated little differences in spray morphology between the two LOX post designs.

C. LOX Spray Combustion

In addition to the photographic studies just described, a first series of drop size measurements in a cryogenic spray produced by a rocket engine coaxial injector was performed at the cryogenic test facility of the University of Liège, Belgium.[58] These tests provided some results on droplet sizes and velocities under pressures up to 1 MPa, but still under cold-flow conditions. A further step toward actual rocket engine conditions is to include combustion in the investigation. Therefore, a new set of measurements was performed on the Mascotte test bench, mainly focused on the LOX droplet size analysis under hot-fire conditions. Mascotte is a cryogenic test facility developed to study the elementary processes that are involved in the combustion of cryogenic propellants (LOX and H_2). It is aimed at feeding a single-element combustor with actual propellants at chamber pressures up to 8 MPa in the present version. For a more detailed description of the bench and of the combustor, refer to Chapter 18 of the present volume.

To avoid cavitation in the LOX injector, the LOX flow crosses a liquid nitrogen bath to reach a temperature of 80 K in the feed line. At these conditions, the vapor pressure is $P_v = 0.03$ MPa. If we assume a classical value of 0.7 for the LOX post discharge coefficient, then the pressure drop through the injector is $P_l - P_c = 0.006$ MPa. If the chamber pressure is atmospheric, which is the most critical case, we have $K = (P_l - P_v)/(P_l - P_c) = 12.6$ while, with the considered geometry, the cavitation criterion of Eq. (16) states that cavitation does not occur if $K > 1.27$.

1. Flow Visualization

As a first step, it was decided to provide preliminary information on the fluid dynamics of the LOX core breakup process to complete the available cold-flow conditions visualizations and to localize the atomization zone where phase Doppler measurements could be achieved. The visualization technique used for this purpose was based on a stroboscopic laser sheet synchronized with a high-speed movie camera. The laser sheet was formed by a stroboscopic copper laser beam (at a wavelength of 510.6 nm: green) focused through a cylindrical and a spherical lens. The floodlit zone was 70 mm long by 50 mm high, while the sheet thickness was roughly 1 mm near the chamber axis. The sheet was introduced into the chamber through fused-silica windows (see Chapter 18). The strobe frequency was 2 kHz, and the pulse time duration was about 40 ns. A high-speed 16 mm camera was used in conjunction with black and white high-speed sensitivity movie film (500 ISO) to record good quality images of the scattered light of the LOX structures.

The four different operating conditions investigated in this test program are summarized in Table 4. Two chamber pressures (atmospheric and 1 MPa) and two gas H_2 mass flow rates were used, while the LOX mass flow rate was maintained constant at 50 g/s.

Two types of images were recorded during these tests. First, the phenomena were recorded without any filter to eliminate the flame emission (Fig. 17). Those images permitted the structures of both the luminous flame and the liquid inside the reaction zone to be recorded. In a second configuration, a bandpass filter,

Table 4 Operating conditions for LOX spray combustion tests

Pressure, MPa		V_{LOX}, m/s	V_{GH_2}, m/s	$\dot{m}_{\text{LOX}}$, g/s	$\dot{m}_{\text{GH}_2}$, g/s	Mixture ratio	We, 10^5	Re, 10^5
Point A	0.1	2.21	893	50	15.0	3.3	0.598	0.67
	1	2.23	308	50	23.7	2.1	0.20	0.67
Point C	0.1	2.21	628	50	10.0	5.0	0.304	0.67
	1	2.23	207	50	15.8	3.2	0.09	0.67

centered at the wavelength of the laser beam around 510 nm, was placed in front of the camera lens. Thus, only the scattered light of the liquid structure was visualized (Fig. 18). Notice that only 70 mm were illuminated by the laser sheet in the axial direction, and so what appears as the sharp end of the spray in Figs. 17 and 18 is actually the end of the laser sheet. Liquid objects do exist far downstream of that location. Note that the time resolution is different for the liquid structures illuminated by the laser for 40 ns and the flame emission, which is recorded for 1/6000 s, related to the cadence of the camera. The spatial resolution is also different. The liquid is visible in the plane of the laser, whereas the flame emission is detected throughout the combustor.

The recorded movies were analyzed with the objective of determining where the LOX core was too dense to allow any measurements with the phase Doppler particle analyzer. However, the most interesting information provided by these images is the difference in flame emission between 0.1 and 1 MPa. At atmospheric pressure, the light emitted by the flame is not intense enough to be detected; the images are the same with and without filter. Analyzing the spectral response of the entire detection arrangement allows us to state that the observed light may be in the visible or in the near UV range (350–400 nm). This is only qualitative information, and spectroscopic measurements are needed to obtain complementary information on the spectral emission of the flame. Numerical processing of a series of images also permitted the mean velocity of liquid objects to be obtained, as in PDPA velocity measurements.

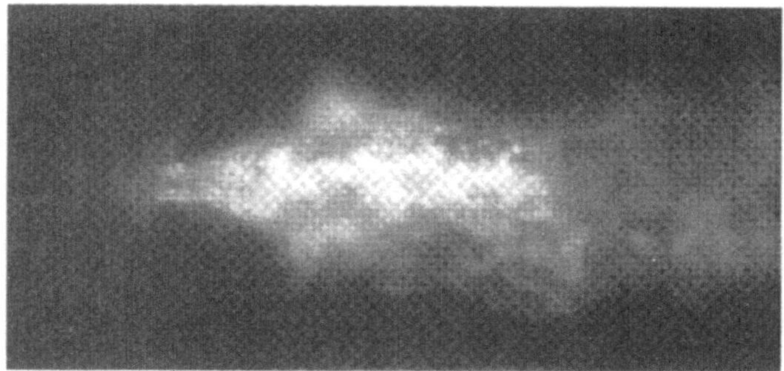

Fig. 17 Visualization of flame and spray (without filter).

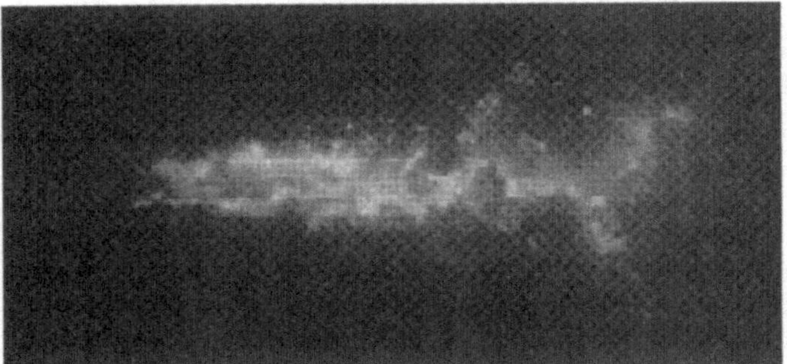

Fig. 18   Visualization of spray only (with green filter).

2.   *Drop Size and Velocity Measurements*

After the initial visualization tests, the main objective of this study consisted of measuring LOX droplet sizes and velocities under combusting conditions using the PDPA in a commercial one-component configuration and describing the atomization and vaporization zone by a fine mapping of those measurements around the LOX core. The laser was a 15-mW helium/neon laser. It was focused into the test chamber by a 1000 mm transmitter lens. The receiver system was situated in the forward direction, at 30 deg off axis, to collect the scattered light with a 500-mm standard lens. A red bandpass filter was placed in front of the three detectors behind the receiver lens. In our experiment, due to limited access, it would have been very difficult to have a good collecting angle through the windows without the following specific optical arrangement: the transmitter and the receiver were placed on opposite sides of the combustor, at an angle of 15 deg. The LOX refractive index is 1.221.

Tests were performed at the operating points A and C as presented in Table 4, precisely 55 runs at point A and 52 at point C. Seven radial profiles were explored along the flow axis, the first one as close as possible to the LOX post tip. Those profiles ranged from $x/d = 4$ out of the axis (due to the presence of the LOX core) to $x/d = 36$. An additional measurement location was also placed outside of the combustion chamber at $x/d = 72.4$. This measurement was successful: LOX droplets were detected and measured at this relatively far distance from the injector. The radial variations of measured arithmetic and Sauter mean diameters, mean axial velocity, and validation rate are plotted in Fig. 19. Because of the basic principle of the phase Doppler technique, the detected particles are only measured if they are spherical and have a diameter between the specified minimum and maximum diameters. All nonspherical as well as out-of-range particles are rejected by the software, which nevertheless indicates the validation rate, defined as the ratio of the actually measured (validated) particles to the total number of particles detected in the probe volume during the test run time. A histogram of the different reasons for rejection (nonsphericity, too large, too small) is also saved. Looking at this histogram after the test, it was determined that most rejected particles were nonspherical, whereas only a few were rejected because of the size criterion. This indicates that, for the validated particles,

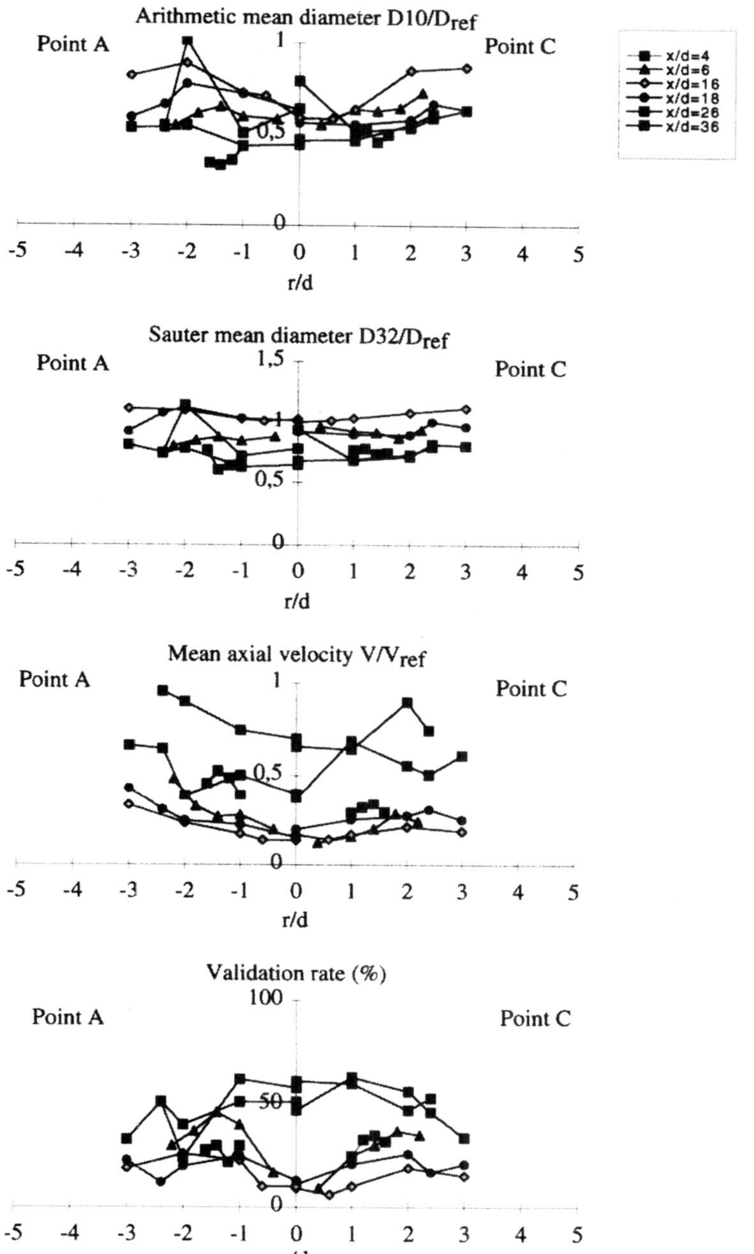

Fig. 19 Radial variations in droplet mean diameters, axial velocity, and validation rate.

the size measurement range was correct, but also that the atomization was not complete in the investigated zone and that many nonspherical objects (clusters, ligaments, deformed droplets) were present in addition to the droplets.

Based on mean sizes and velocities observation, several remarks can be made. The atomization seems slightly better for condition A than condition C, although the overall atomization process looks very similar for both cases, which are both situated in the fiber-type breakup region defined by Farago and Chigier.[4] The extremely high experimental Weber numbers are due to very high hydrogen injection velocities, together with very low LOX surface tension. In each case, we saw a primary atomization zone near the injector exit, probably due to the skimming mechanism described in Section II.A, where a spray of rather small droplets coexists with the liquid core (full liquid jet) flowing out of the LOX post. This zone extends from $x/d = 0$ to almost $x/d = 8$. Droplets are measured away from the centerline ($1 < r/d < 2$), while no measurement is possible for $r/d < 1$. No liquid objects are detected for $r/d > 2$, which permits localization of the boundary of the spray. Downstream of this first zone, the spray looks like a region of large ligaments and droplets where the validation rate is very low (less than 30%). This zone extends from $x/d = 10$ to $x/d \geq 20$. Further downstream, we find an area where the atomization is more complete (in terms of validation $> 60\%$), but where the particles are relatively large (SMD = 80 μm).

Concerning the particle velocity in cases A and C, from $x/d = 0$ to $x/d = 15$, the mean velocity decreases because of both the very low validation rate and the increasing droplet sizes (factor 2 on SMD between $x/d = 4$ and $x/d = 20$). After this zone of clusters and ligaments, the velocity increases again, because of dragging by the accelerating combustion products.

Measurements were also performed at 1 MPa: a first radial profile very close to the injector, at $x/d = 4$ and another at $x/d = 16$. At this pressure, a measurement outside of the combustion chamber after the throat showed that no droplets were present at a distance of $x/d = 83.6$.

Comparisons between results at atmospheric pressure and at 1 MPa in two cases (A and C) show that the droplet sizes are similar. This result seems to show that the most important similarity parameter for spray atomization is the momentum ratio, because it is the only nondimensional number that was kept constant, while the pressure in the combustor was increased.

Chin and Lefebvre[59] define an effective value of evaporation constant λ_e, which takes into account and combines several effects, including the initial heating phase of the droplet. The total vaporization time of the droplet t_e is then written as $t_e = (D_0)^2/\lambda_e$, where D_0 is the initial diameter of the droplet. This time may be compared to the average residence time of the droplet in the combustor defined as $t_s = L/V$, where L is the length of the combustion chamber downstream of the ligament zone (from $x/d = 20$ to the exit) and V is the measured droplet average axial velocity. For an initial droplet diameter of 200 μm and $\lambda_e = 6.44$ mm^2/s, we obtain for condition A, $t_s = 5.8$ s and $t_e = 6.2$ s. Droplets larger than 200 μm may reach the combustor exit before they vanish.

The measured droplet sizes may also be compared to empirical correlations obtained from the available literature on coaxial injector investigations. Many of these correlations are listed in Refs. 1 and 2. Let us choose those of Refs. 60–62, for instance (Fig. 20). Applying the equation of Ref. 60, which

Atomizer	Geometry	Fluids (nature and physical properties)	Operating range	Correlation equation
Nukiyama - Tanasawa 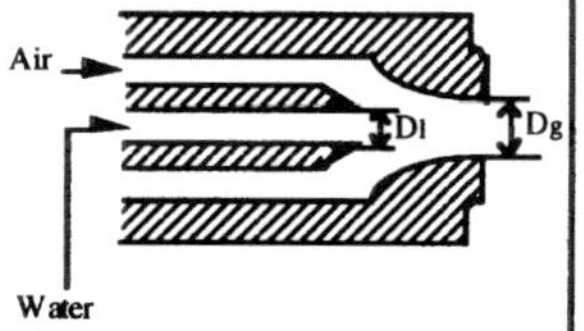	$D_l = 0.2 - 1.0\ 10^{-3}$ m $D_g = 2.0 - 5.0\ 10^{-3}$ m	Water, Glycerin, Alcohol / Air $\mu_l = 1 - 50\ 10^{-3}$ Pa.s $\sigma_l = 19 - 73\ 10^{-3}$ N/m $\rho_l = 700 - 1200$ kg/m^3 $\rho_g = 1.2$ kg/m^3	$V_l = 1 - 12$ m/s $V_g = 50$ m/s - sonic $M_f = 0.07 - 2$	$D_{32} = 0.585\left[\dfrac{\sigma}{\rho_l\,\Delta V^2}\right]^{0.5} + 53\left[\dfrac{\mu_l^2}{\sigma\rho_l}\right]^{0.225}\left[\dfrac{V_l}{V}\right]^{1.5}$
Lorenzetto - Lefebvre 	$D_l = 0.4 - 1.6\ 10^{-3}$ m $D_g = 6. - 25.\ 10^{-3}$ m	Water, Kerosene, Alcohol / Air $\mu_l = 1.3 - 76\ 10^{-3}$ Pa.s $\sigma_l = 24 - 74\ 10^{-3}$ N/m $\rho_l = 812 - 2180$ kg/m^3 $\rho_g = 1.2$ kg/m^3	$V_l = 0.5 - 12$ m/s $V_g = 61 - 108$ m/s $M_f = 0.1 - 1$	$D_{32} = \left\{0.95\,\dfrac{(\sigma\dot{m}_l)^{0.33}}{\Delta V\rho_l^{0.37}\rho^{0.30}} + 0.13\mu_l\left[\dfrac{d_l}{\sigma\rho_l}\right]^{0.5}\right\}\left[1 + \dfrac{\dot{m}_l}{\dot{m}}\right]^{1.7}$
Kim - Marshall 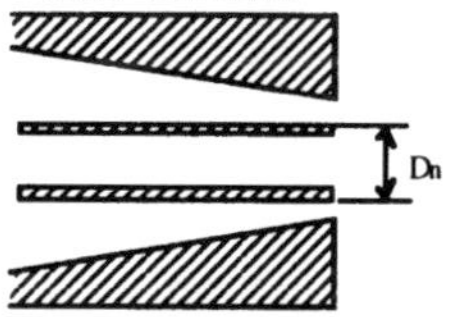	$D_l = 1.4 - 5.6\ 10^{-3}$ m $D_n = 1.8 - 6.4\ 10^{-3}$ m $D_g = 3.3 - 6.9\ 10^{-3}$ m	Wax / Air $\mu_l = 9 - 49\ 10^{-3}$ Pa.s $\sigma_l = 30 - 31\ 10^{-3}$ N/m $\rho_l = 780 - 830$ kg/m^3 $\rho_g = 0.9$ kg/m^3	$V_l = 0.001 - 9$ m/s $V_g = 76$ m/s - sonic $M_f = 0.02 - 17$	$D_{v,0.5} = 5.36\ 10^{-3}\,\dfrac{\sigma^{0.41}\mu_l^{0.32}}{\rho\Delta V^{2\,0.57}\,A^{0.36}\,\rho^{0.16}}$ $+ 3.44\ 10^{-3}\left[\dfrac{\mu_l^2}{\sigma\rho_l}\right]^{0.17}\dfrac{\dot{m}_l^{\,\alpha}}{\dot{m}}\dfrac{1}{\Delta V^{0.54}}$ with $\alpha=1$ for $\dfrac{\dot{m}_l}{\dot{m}} > 1/3$ and $\alpha=0.5$ for $\dfrac{\dot{m}_l}{\dot{m}} < 1/3$

Fig. 20   Empirical correlation equations for coaxial injectors.

was obtained for water and air, we have, respectively, for case A and C, $SMD_A = 45$ μm and $SMD_C = 71$ μm. The equation of Ref. 61, obtained in another water and air coaxial configuration, gives $SMD_A = 151$ μm and $SMD_C = 210$ μm. The equation of Ref. 62, established with liquid wax and air gives $SMD_A = 16$ μm and $SMD_C = 21$ μm. Finally, applying the model of Section II.A gives $SMD = 24$ μm.

Each of the three results obtained with the phase Doppler system (near-injector region, ligament zone, downstream region) may be predicted with one of the preceding correlations.

IV. Conclusions

In the past, twin fluid coaxial injectors have been used extensively in cryogenic rocket engines, in general providing high propulsive efficiency and good combustion stability behavior. The basic atomization phenomena that convert the primary liquid jets into droplets are not yet fully understood and not yet well modeled, however. This is required if a numerical simulation of the complete engine is to be achieved.

This chapter attempts to provide a deeper insight into the state of the art of coaxial injection element characterization in cryogenic rocket engines. Relevant literature on theoretical aspects and experimental work was reviewed and discussed. The main conclusion of this review is that no unified theory is currently available and that experimental investigations remain the best way to characterize a given injection element.

Furthermore, the specific features of the coaxial design make the measurements quite impossible in engine operation conditions and still difficult in simulation. Problems have to be uncoupled, and several aspects have to be investigated separately. This chapter presents several test programs following that strategy, namely 1) cold-flow tests at atmospheric pressure using water and air as simulants in the primary atomization zone, for which a model was derived, and in the far-field zone, 2) higher pressure tests with cryogenic fluids, and 3) LOX spray combustion. General features of the spray are qualitatively presented, but it appears that quantitative comparison to theory or even other empirical results remains questionable. Measured drop sizes vary in space, and when comparing experiments of different authors, one has to take care not only to be using similar operating conditions but also to be investigating the same region of the spray.

Acknowledgment

Work performed at CORIA and at ONERA was supported by the Centre National d'Etudes Spatiale and the Société Européenne de Propulsion.

References

[1]Ferrenberg, A., Hunt, K., and Duesberg, J., "Atomization and Mixing Study," NASA-CR-178751, 1985.

[2]Lefebvre, A. H., "Atomization and Sprays," *Combustion: An International Series*, Hemisphere, 1989.

[3]Vingert, L., Gicquel, P., Lourme, D., and Ménoret L., "Coaxial Injector Atomization," Progress in Astronautics and Aeronautics, Vol. 169, AIAA, Washington, DC, 1994, pp. 145–189.

[4]Farago, Z., and Chigier, N., "Morphological Classification of Disintegration of Round Liquid Jets," *Atomization and Sprays*, Vol. 2, No. 2, 1992, pp. 137–153.

[5]Mayer, W., and Krülle, G., "Rocket Engine Coaxial Injector. Liquid/Gas Interface Flow Phenomena," AIAA Paper 92-3389, July 1992.

[6]Gomi, H., "Pneumatic Atomization with Coaxial Injectors: Measurements of Drop Sizes by the Diffraction Method and Liquid Phase Fraction by the Attenuation of Light Method," Ph.D. Thesis, Univ. of Sheffield, Sheffield, England, Dec. 1984.

[7]Hopfinger, E. J., and Lasheras, J. C., "Breakup of a Water Jet in High Velocity Co-Flowing Air," *Proceedings of ICLASS 94*, Begell House, Rouen, France, July 18–22, 1994.

[8]Le Visage, D., "Pulverisation d'un jet issu d'un injecteur coaxial assisté: Géométrie de l'injecteur, Modélisation et Approche Cryogénique," Ph.D. Thesis, Univ. of Poitiers, Poitiers, France, Jan. 1996.

[9]Carreau, J. L., Le Visage, D., Monote, G., Gicquel, P., and Roger, F., "Characterisation of the Near-Injector Region of Coaxial Jets," *Proceedings of ICLASS 94*, Begell House, Rouen, France, July 18–22, 1994.

[10]Vingert, L., "Coaxial Injector Spray Characterization for the Ariane 5 Vulcain Engine," Proceedings of 6th Annual Conference of ILASS Europe, Pisa, Italy, July 4–6, 1990.

[11]Gaic, P., Burnage, H., Yoon, S. J., and Lourme, D., "Distribution of Mean Concentration Drop Size and Velocity and Size-Velocity Correlations in the Spray of an Air-Assist Nozzle," *International Journal of Turbo and Jet Engines*, Vol. 3, 1986, p. 105.

[12]Huynh, C., Ghafourian, A., Mahalingam, S., and Daily, J. W., "Combustor Design for Atomization Study in Liquid Rocket Engines," AIAA Paper 92-0465, Jan. 1992.

[13]Beisler, M. A., Pal, S., Moser, M. D., and Santoro, R. J., "Shear Caoaxial Injector Atomization in a LOX/GH$_2$ Propellant Rocket," AIAA Paper 94-2775, June 1994.

[14]Schick, A., Schneider, G., and Krülle, G., "Experimental Simulation of Supercritical Injection in Liquid Rocket Engines," Proceedings of 11th Annual Conference of ILASS Europe, Nürnberg, Germany, March 21–23, 1995.

[15]Tamura, H., and Mayer, W., "Observations of a Single-Shear Coaxial Injector Spray in a LOX/GH$_2$ Rocket Chamber," *Proceedings of the 20th Internaional Symposium on Space Technology and Science*, Gifu, Japan, May 19–25, 1996.

[16]Goix, P. J., Edwards, C. F., Cessou, A., Dunsky, C. M., and Stepowski, D., "Structure of a Methanol-Air Coaxial Reacting Spray near the Stabilization Region," *Combustion and Flame*, Vol. 98, 1994, pp. 205–219.

[17]Reitz, R. D., "Breakup Regimes of a Single Liquid Jet," AMS Rept. 1262, 1976.

[18]Weber, C., "Disintegration of Liquid Jets," *Zeitschrift Angew. Mathematical Mechanics*, Vol. 11, No. 2, 1931, pp. 136–159.

[19]Grant, R. P., and Middleman, S., "Newtonian Jet Stability," *AIChE Journal*, Vol. 12, No. 4, 1966, pp. 669–678.

[20]Fenn, R. W., and Middleman, S., "Newtonian Jet Stability: the Role of Air Resistance," *AIChE Journal*, Vol. 15, No. 3, 1969, pp. 379–383.

[21]Sterling A. M., and Sleicher C. A., "The Instability of Capillary Jets," *Journal of Fluid Mechanics*, Vol. 68, No. 3, 1975, pp. 497–502.

[22]Leroux, S., Dumouchel, C., and Ledoux, M., "The Stability Curve of Newtonian Liquid Jets," *Atomization and Sprays* (to be published).

[23]Crapper, G. D., Dombrowski, N., and Pyott, G. A. D., "Kelvin-Helmholtz Wave Growth on Cylindrical Sheets," *Journal of Fluid Mechanics*, Vol. 68, No. 3, 1975, pp. 497–502.

[24]Lee, C. P., and Wang, T. G., "A Theoretical Model for the Annular Jet Instability," *Physics of Fluids*, Vol. 29, No. 7, 1986, pp. 2076–2085.

[25]Kendall, J. M., "Experiments on Annular Liquid Jet Iinstability and on the Formation of Liquid Shells," *Physics of Fluids*, Vol. 29, No. 7, 1986, pp. 2086–2094.

[26]Meyer, J., and Weihs, D., "Capillary Instabillity of an Annular Liquid Jet," *Journal of Fluid Mechanics*, Vol. 179, 1987, pp. 531–545.

[27]Hashimoto, H., Suzuki, T., and Matsuya, T., "Interaction between two thin Liquid Jets with Disintegration by Gas Flow," *Atomization and Sprays*, Vol. 1, 1991, pp. 47–61.

[28]Camatte, P., and Ledoux, M., "Airblast Atomization. Instability of an Annular Sheet Surrounded by Two Air Flows of Different Velocities," *Proceedings of ICLASS 91*, Gaithersburg, MD, 1991.

[29]Camatte, P., "Etude des instabilités d'une nappe liquide de géométrie annulaire en pulvérisation assistée. Modèle. Expérience (in French)," Ph.D. Thesis, Univ. of Rouen, Rouen, France, May 1992.

[30]Bernal, L. P., and Roshko, A., "Streamise Vortex Structure in Plane Mixing Layers," *Journal of Fluid Mechanics*, Vol. 170, 1986, pp. 499–525.

[31]Breidenthal, R., "Structure in Turbulent Mixing Layers and Wakes Using a Chemical Reaction," *Journal of Fluid Mechanics*, Vol. 109, 1981, pp. 1–24.

[32]Jang, P. S., Benney, D. J., and Gran, R. L., "On the Origin of Streamwise Vortices in a Turbulent Boundary Layer," *Journal of Fluid Mechanics*, Vol. 169, 1986, pp. 109–123.

[33]Lozano, A., Call, C. J., and Dopazo, C., "An Experimental and Numerical Study of the Atomization of a Planar Liquid Sheet," *Proceedings of ICLASS 94*, Begell House, Rouen, France, July 18–22, 1994.

[34]Lozano, A., Garcia-Olivares, A., Dopazo, C., Barreras F., and Lopez, E., "The Instability Growth Leading to a Liquid Sheet Breakup," Proceedings of 12th Annual Conference of ILASS Europe, Lundt, Sweden, June 1996.

[35]Wu, P. K., Tseng, L. K., and Faeth, G. M., "Primary Breakup in Gas/Liquid Mixing Layers for Turbulent Liquids," *Atomization and Sprays*, Vol. 2, 1992, pp. 295–317.

[36]Faeth, G. M., Hsiang, L. P., and Wu, P. K., "Structure and Breakup Properties of Sprays," *International Journal of Multiphase Flow*, Vol. 21, Supp., 1995, pp. 99–127.

[37]Wu, P. K., Tseng, L. K., and Faeth, G. M., "Primary Breakup in Gas/Liquid Mixing Layers for Turbulent Liquids," AIAA Paper 92-0462, Jan. 1992.

[38]Lefebvre, A. H., "Twin-Fluid Atomization: Factors Influencing Mean Drop Size," *Atomization and Sprays*, Vol. 2, No. 2, 1992, pp. 101–119.

[39]Shetz, J. A., Kush, Jr., and Joshi, P. B., "Wave Phenomena in Liquid Jet Breakup in a Supersonic Crossflow," *AIAA Journal*, Vol. 18, No. 7, 1980, pp. 774–778.

[40]Less, D. M., and Shetz, J. A., "Transient Behavior of Liquid Jets Injected Normal to a High Velocity Gas Stream," *AIAA Journal*, Vol. 24, No. 12, 1986, pp. 1979–1986.

[41]Shetz, J. A., and Padhye, A., "Penetration and Breakup of Liquid Fuel Jets in High Subsonic Speed Air Streams," AIAA Paper 77-201, Jan. 1977.

[42]Vich, G., and Ledoux, M., "Destabilization of a Liquid Jet by an Air Flow at Normal Incidence," Proceedings of 11th Annual Conference of ILASS Europe, Nürnberg, Germany, March 21–33, 1995.

[43]Keller, F. X., Li, J., Vallet, A., Vandromme, D., and Zaleski, S., "Direct Numerical Simulation of the Interface Breakup and Atomization," *Proceedings of ICLASS 94*, Begell House, Rouen, France, July 18–22, 1994.

[44]Zalosh, R. G., "Discretized Simulation of Vortex Sheet Evolution with Buoyancy and Surface Tension Effects," *AIAA Journal*, Vol. 14, No. 11, 1976, pp. 1517–1523.

[45]Rangel, R. H., and Sirignano, W. A., "The Linear and Nonlinear Shear Instability of a Fluid Sheet," *Physics of Fluids*, A3, No. 10, 1991, pp. 2392–2400.

[46]Cousin, J., Camatte, P., and Ledoux, M., "Instability of a Gas/Liquid Interface: Nonlinear Effects," *Proceedings of ICLASS 94*, Begell House, Rouen, France, July 18–22, 1994.

[47]Sellens, R. W., and Brzustowski, T. A., "A Prediction of the Drop Size Distribution in a Spray from First Principles," *Atomization and Spray Technology*, Vol. 1, 1985, pp. 89–102.

[48]Li, X., and Tankin, R. S., "Prediction of Droplet Size and Velocity Distributions in Sprays Using Maximum Entropy Principle," *Combustion Science and Technology*, Vol. 68, 1989, pp. 147–155.

[49]Sangakiri, N., and Ruff, G. A., "Extension of Spray Nozzle Correlations to the Prediction of Drop Size Distributions Using Priciples of Maximum Entropy," AIAA Paper 93-0899, Jan. 1993.

[50]Cousin, J., Dumouchel, C., and Ledoux, M., "Prediction of Drop Size Distributions of Sprays Produced by Low Pressure Car Injectors," *Proceedings of 12th Annual Conference of ILASS Europe*, Lundt, Sweden, June 1996.

[51]Cousin, J., Yoon, S. J., and Dumouchel, C., "Coupling of Classical Linear Theory and Maximum Entropy Formalism for Prediction of Drop Size Distribution in Sprays. Application to Pressure Swirl Atomizers," *Atomization and Sprays* (to be published).

[52]Care, I., and Ledoux, M., "Study of an Airblast Coaxial Atomizer. Experiments and Modelisation," *Proceedings of the 5th ICLASS*, Gaithersburg, MD, July 1991, pp. 763–768.

[53]Care, I., "Etude d'un injecteur coaxial assisté (in French)," Ph.D. Thesis, Univ. of Rouen, Rouen, France, Dec. 1990.

[54]Mayer, E., "Capillary Mechanism of Liquid Atomization in High-Velocity Gas Streams," *Proceedings of 12th International Astronautical Congress*, Washington, DC, 1961, pp. 731–740.

[55]Nurick, W. H., "Orifice Cavitation and Its Effect on Spray Mixing," *Journal of Fluids Engineering*, Vol. 98, Dec. 1976, pp. 681–687.

[56]Sankar, S. V., Wang, G., Brena de la Rosa, A., Rudoff, R. C., Isakovic, A., and Bachalo, W. D., "Characterization of Coaxial Rocket Injector Sprays Under High Pressure Environments," AIAA Paper 92-0228, Jan. 1992.

[57]Faeth, G. M., Dominicis, D. P., Tulpinsky, J. F., and Olson, D. R., "Supercritical Bipropellant Droplet Combustion," *Twelfth Symposium (International) on Combustion*, Combustion Institute, 1969, pp. 9–18.

[58]Gicquel, P., Vingert, L., and Millan, P., "Faisabilité de mesures granulométriques dans des sprays cryogéniques d'injecteurs coaxiaux (in French)," 3ème Congrès Francophone de Vélocimétrie Laser, Toulouse, France, Sept. 1992.

[59]Chin, J. S., and Lefebvre, A. H., "Effective Values of Evaporation Constant for Hydrocarbon Fuel Drops," *Proceedings of the 20th Automotive Technology Development Contractor Coordination Meeting*, 1982, pp. 325–331.

[60]Nukiyama, S., and Tanasawa, Y., "Experiments on the Atomization of Liquids in an Air Stream," *Transactions of the Society of Mechanical Engineers Japan*, Vol. 5, No. 18, 1939, pp. 62–67.

[61]Lorenzetto, E. G., and Lefebvre, A. H., "Measurements of Drops Size on a Plain Jet Airblast Atomizer," *AIAA Journal*, Vol. 2, No. 3, 1964, pp. 583–585.

[62]Kim, K. Y., and Marshal, W. R., "Drop Size Distributions from Pneumatic Atomizers," *Journal of American Institute of Chemical Engineering*, Vol. 17, No. 3, 1971, pp. 575–584.

Chapter 4

Liquid Bipropellant Injectors

William E. Anderson,* Matthew R. Long,⁻ and Stephen D. Heister‡
Purdue University, West Lafayette, Indiana

Nomenclature

A = empirical constant, or area, m^2
α = spray cone half-angle
C_d = discharge coefficient
d = diameter, m
d_{32} = Sauter mean diameter, m
f = frequency, s^{-1}
E = efficiency
F = fuel
h = sheet thickness, m
K_1 = non-dimensional swirl atomizer constant, $A_i/r_s r_o$
L = length, m
$\dot{m}$ = flow rate, kg/s
m = streamtube mass collected with $MF > mf$
$\bar{m}$ = streamtube mass collected with $MF < mf$
mf = streamtube mass fraction $m_2/(m_1 + m_2)$ collected with $MF > mf$
$\bar{mf}$ = streamtube mass fraction $\bar{m}_2/(\bar{m}_1 + \bar{m}_2)$ collected with $MF < mf$
M = total flow of injected simulants
MF = injection mass fraction, $M_2/(M_1 + M_2)$
n = number of streamtubes with $MF > mf$
$\bar{n}$ = number of streamtubes with $MF < mf$
O = oxidizer

---

*Assistant Professor, School of Aeronautics and Astronautics. Member AIAA.
†Graduate Student, School of Aeronautics and Astronautics. Member AIAA.
‡Professor, School of Aeronautics and Astronautics. Associate Fellow AIAA.

p = pressure, MPa
Q = volumetric flow rate, m^3/s
r = radius, m
Re = Reynolds number, $\rho UL/\mu$
s = density ratio
t = time, s
U = velocity, m/s
u = axial velocity, m/s
We = Weber number, $\rho U^2 d/\sigma$
w = tangential or swirl velocity, m/s
x = axial distance, m
X = ratio of gas core area to orifice area, A_g/A_o
Δ = delta
ϕ = azimuthal angle, deg
η = efficiency
λ = wavelength, m
μ = viscosity, kg/m/s
θ = impingement angle, deg
ρ = density, kg/m^3
σ = surface tension, N/m
v = radial velocity, m/s
ω = angular frequency, s^{-1}

Subscripts

1 = propellant simulant of type one
2 = propellant simulant of type two
amb = ambient
b = breakup
c = chamber
CL = centerline
D = drop
g = gas
i = inlet
j = jet
l = liquid
L = ligament
m = mixing
o = orifice
p = pintle
r = radial
s = swirl chamber
z = axial

I.   Introduction

THE most critical part of a rocket engine combustor is the injector. The injector includes the injector elements that deliver propellant to the combustor and a manifold system that distributes the propellant to the injector elements. Because the design of the injector, along with the injection pressure drop and the propellant properties, determines the propellant mass distribution and the spray drop size distribution for liquid propellants, the injector determines the maximum achievable combustion efficiency, the heat transfer rates to the combustion chamber walls, and whether or not high frequency combustion instabilities will occur.

The choice of injector element type and its specific design is dependent on a number of factors: the propellant combination and the state of the propellants at the injector inlet, the oxidizer-to-fuel mass ratio, the chamber pressure, the injection pressure drop, the chamber diameter, the performance and life requirements, and the experience of the engine manufacturer. All of these factors are usually determined long before the processes of design, analysis, and testing begin. An optimal injector design is one that meets the performance and life requirements of the propulsion system, is stable from combustion instabilities throughout its operating range, provides a compatible environment around the chamber walls, and can be fabricated inexpensively and reliably.

In rocket engines that use liquid/liquid propellant combinations (e.g., liquid oxygen/kerosene or nitrogen tetroxide/monomethyl hydrazine), impinging jet, bicentrifugal swirl, and pintle injectors have typically been used. Liquid/liquid injectors are used in both pressure-fed and gas generator cycle main chamber combustors, and in typical gas generators and preburners. Through the year 2000, impinging jet injectors have been the most commonly used type in the United States,[1] while the bicentrifugal swirl injector, where both liquid propellants are swirled, has been used extensively in Russia.[2] The pintle injector is a component somewhat unique to a single engine manufacturer, but has been used for injecting both storable and cryogenic liquid oxygen/liquid hydrogen (LOX/LH_2) propellants.[3] This chapter provides a survey of these three types of injector elements. The important design features are described, results from previous studies are summarized, and a discussion is provided on the controlling physical processes.

II.   Impinging Jet Injector

A.   General Description

Liquid jet impingement is a very simple method of injecting and distributing liquid propellants into a rocket engine combustion chamber. It is conducive to the traditional manufacturing practices of milling, lathing, and twist-drilling. It has a high efficacy of atomization and mixing because of the direct use of the dynamic head of the propellant stream to disperse the propellant.

A number of combinations of impinging jet elements have been used, including the like doublets (two streams of the same propellant impinge on each other), unlike doublets (one stream of oxidizer impinges on one stream of fuel), triplets (two streams of one propellant impinge on one central stream of the opposite propellant), quadlets (two streams of oxidizer and two streams of fuel impinge concurrently), and pentads (four streams of one type of propellant impinge on a central stream of the opposite propellant). An injector element is the orifices that make up the individual doublet, triplet, etc. Detailed descriptions of all of these injector element types along with design and manufacturing guidelines can be found in the NASA special publication *Liquid Rocket Injectors.*[1]

The like doublet (LD) and unlike doublet (UD) are the most common type of impinging jet injector elements and are also the most studied. Both doublet types have the advantage that they are easy to manifold. The injector body is typically built out of concentric rings that are welded or brazed together. Radial and axial downcomers feed machined channels in the rings that distribute propellants to the injector elements. The propellants flow through the orifices in the injector face-plate at a dynamic head typically about 20% of the chamber pressure (to preclude injection-coupled combustion instability) at steady-state operation.

The UD element promotes rapid mixing and combustion, whereas the LD element provides a more distributed combustion zone along the combustor axis. For this reason, the UD element is typically used more often in small thruster applications, whereas the LD element is used more often in large thrusters where combustion instability can be a problem.

B. Applications and Design Guidelines

Geometric and operating characteristics of a variety of impinging jet injectors used in production rocket engines are given in Table 1. In all of the engines listed in Table 1, both oxidizer and fuel enter the combustion chamber liquids. The simplest and most common type of impinging jet injector is the LD injector (Fig. 1), where equal diameter liquid jets impinge to form a fan-like elliptical liquid sheet. The angle between the jets, 2θ, is generally about 60 deg. The impingement point is typically about five length-to-diameter ratios (L_j/d_o) downstream of the injector face. The fan may be inclined at a small angle α toward an adjacent fan of the opposite propellant to promote interpropellant mixing. The angular coordinate in the plane of the sheet is defined as ϕ. For the case of directly opposed jets, a circular disc would be formed after impingement.

Injector designers mostly rely on empirical correlations[4,5] for mixing and drop size for input to combustor performance and stability models. Most of the design correlations used to predict drop size are dependent on propellant properties (density and viscosity), ambient pressure, and injector element design parameters (injector pressure drop, orifice diameter, and impingement angle).

High efficiency injectors must promote intra-element mixing. The predominantly axial, high-speed gas flow does not provide much mixing after the combustion products and unburnt propellants are accelerated downstream of the injector faceplate. Thus, most mixing must be accomplished near the faceplate, preferably while the propellants are in the liquid state.

Mixing correlations are often based on uni-element mixing efficiencies $E_{m,\text{uni}}$, and corrected for inter-element mixing by adding a term based on element density. Uni-element mixing efficiencies, defined as

$$E_{m,\text{uni}} = 1 - \left[\sum_{0}^{n} \frac{m(MF - mf)}{M \cdot MF} + \sum_{0}^{\bar{n}} \frac{\bar{m}(MF - \bar{m}\bar{f})}{M(MF - 1)} \right] \tag{1}$$

Table 1   Engines using impinging jet injectors

Engine	Propellants	O/F	P_c, MPa	Thrust, MN	Chamber diameter, m	Number and type of elements	Orifice diameter, mm, oxygen/ fuel	Injector pressure drop, $\Delta P/P_c$
Gemini First Stage	NTO/A-50	2.00	5.41	0.956[a]	0.55	O:[c] 568 LD F:[c] 516 LD	3.05/2.03	
Gemini Second Stage	NTO/A-50	1.80	5.54	0.445[b]	0.37	Quadlets O: 1319 F: 818	1.27/1.02	O: 0.17 F: 0.17
Apollo LEMDE	NTO/A-50	1.60	1.03	0.047[b]	0.29	165 FOF	1.96/1.24	0.26
Apollo LEMA	NTO/A-50	1.60	0.83	0.016[b]	0.20	177 UD	0.050/0.040 0.036/0.032	
Titan II Booster	NTO/A-50	1.93	5.40	0.954[a]	0.54	Quadlets O: 1319 F: 818	3.02/2.08	
AGENA	HDA/UDMH	2.69	3.69	0.075[b]	0.27	Triplets: O: 88 F: 176	2.82/1.24	
Long March 3, FY-20	NTO/UDMH	2.21	7.38	0.697[a]	0.40	O: 607 LD F: 605 LD	2.70/2.30	O: 0.16 F: 0.14
Long March 4, YF-40	NTO/UDMH	2.20	4.5	0.005[b]	0.15	O: 100 LD F: 105 LD	1.35/0.9	O: 0.31 F: 0.28
Ariane, Viking V	NTO/UH25	1.85	5.35	0.68[a]	0.61	O: 216 LD F: 216 LD	4.3/2.9	O: 0.30 F: 0.25

(*Continued*)

Table 1 Engines using impinging jet injectors (continued)

Engine	Propellants	O/F	P_c, MPa	Thrust, MN	Chamber diameter, m	Number and type of elements	Orifice diameter, mm, oxygen/ fuel	Injector pressure drop, $\Delta P/P_c$
Space Shuttle OME	NTO/MMH	1.65	0.86	0.027[b]	0.21	O: 272 LD F: 272 LD	0.81/0.71	O: 0.44 F: 0.48
Redstone A-7	LOX/EtOH	1.35	2.17	0.347[a]	0.55	355 LD	2.87/2.58	
Titan I Booster	LOX/RP-1	2.25	4.39	0.801[a]	0.55	O: 560 LD F: 610 LD		
Saturn 1B H-1	LOX/RP-1	2.23	4.86	0.91[a]	0.52	O: 365 LD F: 612 TR	3.05/2.08	
Saturn 1C F-1	LOX/RP-1	2.27	7.87	6.73[a]	1.02	O: 714 LD F: 702 LD	6.15/7.14	O: 0.122 F: 0.015

[a]Sea-level value. [b]Altitude value. [c]O = oxygen; F = fuel.

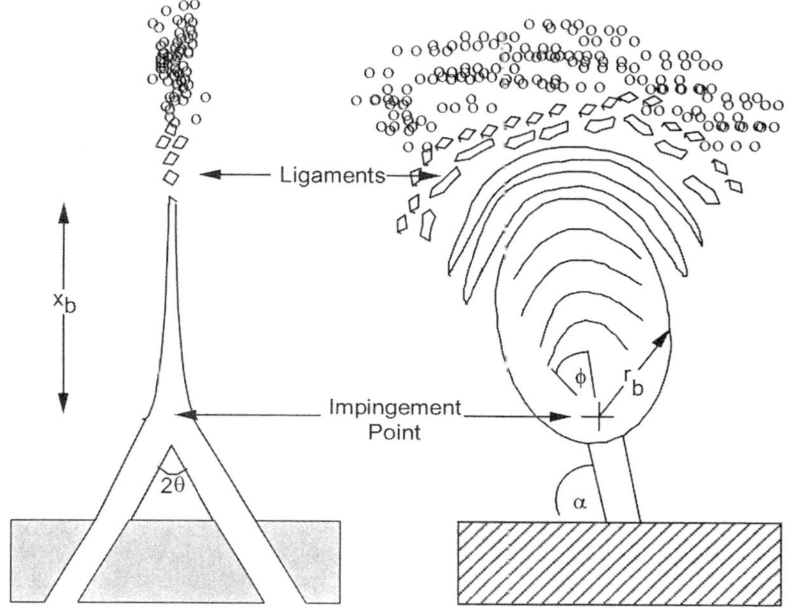

Fig. 1   Schematic of typical like-doublet impinging jet injector element.

are obtained usually by patternator measurements and typically range from 0.5 to 0.85.[5] The mixing efficiency represents the tightness of the mixture ratio distribution about the local injected average. Higher efficiencies are found for configurations where the ratios of the product of the velocity head and orifice diameters are on the order of unity (Rupe's criterion).[5] A mixing efficiency of zero represents a completely unmixed state, and a mixing efficiency of unity indicates complete mixing.

The effects of mixing on performance efficiency are often calculated through a streamtube analysis, whereby the performance parameters of a finite number of oxidizer-rich streamtubes distributed about an average value ($E_m^{-1} \times O/F$) and fuel-rich streamtubes distributed about an average fuel-rich value ($E_m \times O/F$) are mass-averaged to yield performance values for the imperfectly mixed gas. If the combustor uses film-cooling to reduce wall temperatures, a near-wall streamtube is considered to account for the performance decrement associated with the near-wall layer. Entrainment models are used to calculate the near-wall mixture ratio. If the combustor is not long enough for the propellant drops to completely vaporize, the streamtube mixture ratios are adjusted to account for incomplete vaporization.

The combustion stability of impinging jet injectors is another design aspect that must be considered. Three modes of combustion instability are typically encountered—non-acoustic bulk mode instabilities (chug), acoustic injection-coupled modes, which often present themselves as longitudinal modes, and acoustic intrinsic instability modes, for which the physical mechanisms are still being investigated.[6] Chug and injection-coupled instabilities can be modeled using time-lag models, but the intrinsic instabilities still resist prediction. Chug and injection-coupled instabilities can typically be avoided by having a sufficiently high injector pressure drop. A design guideline for chug elimination is to maintain nominal propellant injector drops above 20% of chamber pressure. Intrinsic instabilities in combustors that use impinging jet injectors are typically avoided by increasing orifice size or decreasing injection velocity, both of which lead to a decrement in performance efficiency.

C. Mechanistic Study

Figure 2 shows images of the impinging jet injector. Based on the images taken at lower ambient pressure conditions, a three-step process of impingement and impact wave formation, primary atomization of the fan into ligaments, and finally complete atomization is evident. The initial fragmentation process appears to be periodic and the ligament shedding frequency is controlled by the impact waves. The diameter of the ligaments is dependent on the thickness of the fan at breakup, the distance between the impact waves, and the angle away from the plane containing the jet axes. The fan thickness at breakup is dependent on the geometry of the impinging jets and the axial distance from the impingement

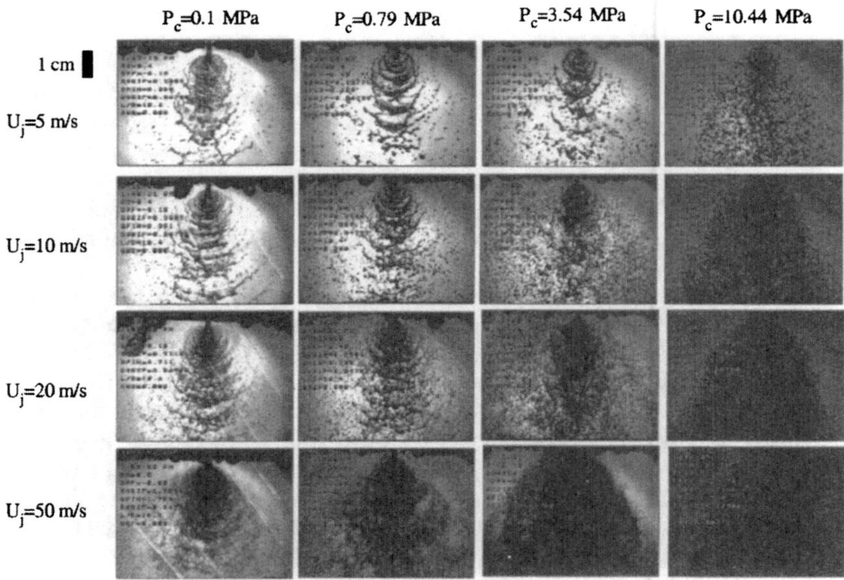

Fig. 2   Instantaneous images of spray formed by impinging jets at 0.1 to 10.44 MPa (left to right) and 5 to 50 m/s (top to bottom); $d_o = 1.194$ mm, $L/d_o = 18.4$, $\theta = 60$ deg.[14]

point where incipient fan breakup occurs.[7] Far downstream of the initial impingement, atomization is complete when the drop size distribution achieves dynamic equilibrium with the surrounding gas flowfield. At very high ambient pressure conditions, early atomization is enhanced by aerodynamic forces.

The obvious wave nature of the early stages of breakup has been studied to elucidate the incipient mechanisms of atomization. Heidmann et al.[8] were the first to report periodic waves of drops projecting from the point of impingement. They reported that this "wave frequency" was linearly proportional to resultant sheet velocity and that orifice diameter and pre-impingement length had minor effects. A single, well-defined characteristic frequency could be determined only over a limited distance with respect to the impingement point, and this distance varied as operating conditions were changed.

These observations led to later studies by Dombrowski and Hooper[9] of the disintegration of the sheet formed by turbulent and laminar impinging water jets. They concluded that one of three mechanisms was operative in the disintegration of the liquid fan, primarily depending on the relative jet inertia and whether the jets were turbulent or laminar. For turbulent jets, impact wave effects were seen at velocities as low as 7 m/s for large impingement angles. Drop size decreased with increasing jet velocity and increasing impingement angle. The transition to impact wave-controlled breakup for the turbulent jet case was determined to occur for relative jet Weber numbers, $\rho_l U_j^2 d_j \sin^2 \theta / \sigma$, greater than 100.

Yatsuyanagi[10] studied the effects of ambient pressure on the breakup of liquid fans formed by impinging jets and determined that increases in ambient pressure led to increases in measured wave frequency. A correlation was obtained for the wave frequency in the form of $f = A We_s^{0.29}$, where We_s is the sheet Weber number based on measured sheet length, and A is an empirical constant equal to 274. The other major factor that affected the wave frequency was the liquid injection velocity. At atmospheric pressure, measured values of wave frequency were about one-third of the wave frequencies reported by Heidmann et al.[8] This difference may be attributed to the measurement technique that was employed; Heidmann et al. reported a time-averaged frequency based on the obscuration of a light beam caused by waves of drops, while Yatsuyanagi used high-speed cinematography.

Ryan et al.[11] made measurements of the impact wavelength and breakup length of impinging water jets at a variety of conditions and determined that the characteristic length of the impact waves is linearly dependent on jet diameter and independent of jet velocity, impingement angle, or ambient pressure. It was postulated that these waves, first seen on the sheet within a few diameters downstream of the impingement point with a characteristic length of about one jet diameter, have their origin near the impingement point and are the results of large pressure and momentum fluctuations.[11]

Measurements of the length, frequency, and propagation speed of the impact waves on a fan were made by Anderson.[12] Directly opposed turbulent water jets with a velocity U_j of 5.2 m/s were studied at atmospheric conditions. Measured disturbance wavelengths varied from about one jet diameter near the impingement point to about three jet diameters near the breakup point of the sheet. The large disturbances merged as they moved downstream, indicating a dispersion in disturbance wavelength and resulting in increased wavelengths with distance from the impingement point. These results explained the differences in wavelength reported by previous researchers in Refs. 9, 10, and 11. The average speed of the impact waves was close to the jet velocity, whereas the variation in impact wave speeds was attributed to fluctuating stagnation conditions.

Ashgriz et al.[13] studied the effect of jet structure and impact conditions on propellant mixing using a patternator. They observed that the impact waves were not symmetric and suggested that the "zig-zag" wave pattern on the sheet and spray fan was due to helical disturbances on the pre-impingement jets. The helical disturbances caused transient, periodic partial misalignment between the impinging jets, leading to some propellant crossover from one side to the other. This "transmitive" process was a mixing loss mechanism that became worse at high jet velocity.

The mode of atomization is determined largely by the density of the ambient medium and the relative velocity between it and the liquid phase. The highest ambient density that is achieved in the combustors of rocket propulsion systems occurs in the preburners of staged combustion cycle engines, where the operating pressure is high and the gas temperature is relatively low. For reference, assuming a well-mixed combustion mixture at nominal overall mixture ratio, the ambient gas density in the fuel-rich preburner of the space shuttle main engine is about $15 \, kg/m^3$. Strakey and Talley[14] measured the drop size distribution and the breakup length of impinging jets at ambient density conditions ranging up to $125 \, kg/m^3$ using high-pressure nitrogen as the ambient fluid. The combined effects of injection velocity and ambient pressure are shown in Fig. 2. Above ambient pressures above about 3 MPa, breakup length apparently became independent of injector operating conditions, and it was concluded that aerodynamic effects are most important in this high-pressure regime. Secondary atomization was reported to be dominant at these conditions, and impact waves were not apparent due to the extremely short fan lengths prior to breakup.

D. Modeling Approaches

The various breakup regimes of a liquid fan formed by impinging liquid jets at atmospheric ambient conditions were characterized by Heidmann et al.[8] At low We_j (<350) and low ambient pressure conditions, the atomization process of a sheet formed by the impingement of dynamically similar and symmetrical jets is described well by Taylor's model[15] based on the balance between tensile forces pulling inward and inertial forces pushing outward. As the relative velocity between the ambient fluid and the liquid increases, Kelvin-Helmholtz waves may appear to cause breakup, and most previous mechanistic modeling efforts (e.g., Refs. 16–21) have used linear stability-based analyses of sheet breakup to predict breakup and drop size. These models consider the aerodynamic amplification of small periodic disturbances on the sheet formed by the impinging jets and have generally been shown to be accurate for relatively low-speed laminar impinging jets.

At rocket engine operating conditions, however, the atomization process is dominated by liquid inertial forces. The jet Reynolds number ($Re_j = \rho_l U_j d_o / \mu_l$) is on the order of 10^5 to 10^6. The jet Weber number ($We_j = \rho_l U_j^2 d_o / \sigma$), characterizing the balance between liquid inertial forces that tend to fragment the propellant and surface tension forces that tend to maintain the integrity of the liquid mass, is on the order of 10^5, and injection pressure drops range from about 0.5 to 3 MPa. The liquid inertial forces, which are focused at the stagnation point, are at least an order of magnitude higher than any of the other forces. For jet diameters on the order of 1 mm, the pressure gradients near the impingement zone can easily exceed 1000 MPa/m. The jet-to-jet interaction dynamics dictate the atomization characteristics, as do most certainly the condition of the jet before impingement.

A phenomenological model that accounts for the processes of impact wave formation, sheet breakup, and ligament formation was developed by Anderson.[12] Cold flow data[11] were used to develop empirical correlations for the impact wave length λ at the point on the fan at which it breaks up, and the length of the fan at the breakup point, x_b. The correlations normalized to injector orifice d_o are

$$\frac{\lambda}{d_o} = 0.687 + 0.1019\frac{x_b}{d_o} \tag{2}$$

$$\frac{x_b}{d_o} = 13.56(We_j s^2)^{-0.102} \tag{3}$$

Finally, the diameter of the ligament shed from the fan leading edge d_L was obtained by assuming it contracts into a cylindrical form. From continuity, at the fan centerline:

$$\left.\frac{d_L}{d_o}\right|_{\text{CL}} \approx 0.22\sqrt{\frac{f(\theta)}{\pi}}[(We_j s^2)^{0.102} + 2] \tag{4}$$

By using existing correlations[22] for the Sauter mean diameter of the drops, a correlation for the process of drop formation from the ligaments was determined as

$$\frac{d_{32}}{d_L} = 1.01(We_{g,L})^{-0.156} \tag{5}$$

Although impinging jets have been used since the beginning of modern liquid propellant rocketry, there are still important unknown aspects. The atomization process at high liquid inertial and high ambient pressures, the effects of practical injector geometries, and the physical mechanisms that lead to intrinsic combustion instability are chief among them. The controlling physical processes at high pressures and high velocities representative of rocket conditions, a determination of the source of impact waves and identification of methods for their control, a characterization of the effects of noncircular orifices and manifold cross flows, and the heat transfer to subcritical temperature jets at supercritical pressures and its effect on atomization are some of the important areas that need to be investigated further.

III.   Bicentrifugal Swirl Injector

A.   General Description

Although their use has been limited in the United States primarily to gas-liquid propellants, liquid-liquid swirl injectors have remained an interesting alternative because of their potential for less stringent manufacturing tolerances and relatively high flow rate per element. These advantages are significant. The high mass flow rate per element is possible because the element efficiently provides high interfacial areas for interpropellant mixing. Furthermore, because the liquid cone that is formed by the injector expands and thins as it swirls outward, small drops are formed and dispersed. These smaller drops tend to mix, vaporize, and react with the opposite propellant more rapidly. The more rapid mixing and combustion allows the use of smaller combustion chambers, which results in the possibility for lower weight propulsion systems. The fact that higher flow-rate-per-element injectors can be used translates directly to geometrically larger elements, which results in relaxed manufacturing tolerances and lower costs. Furthermore, with the initial mixing plane recessed upstream of the injector faceplate, the susceptibility of various processes to transverse velocity oscillations in the combustion chamber is reduced.

A single swirl injector is often referred to as a simplex or pressure-swirl atomizer. There are two types of simplex nozzles: solid-cone and hollow-cone spray simplex atomizers. The solid-cone atomizer atomizes a liquid by swirling it, while permitting most of the flow to go through a middle cylindrical hole, providing spray drops at the center of the spray pattern. Solid-cone atomizers tend to suffer from coarse atomization with the drops in the center of the spray pattern being larger than the drops near the periphery. Hollow-cone atomizers provide better atomization and radial liquid distribution, rendering them the preferred configuration for combustion applications.[23]

The hollow-cone simplex atomizer creates angular momentum by injecting the liquid tangentially into a vortex chamber. Typically two to four tangential inlets are used. A gas core forms inside the swirl chamber. The swirl chamber usually contracts near its exit, and then straightens out again to form a finite-length orifice. The flow emerges from the atomizer to form a hollow conical sheet, the angle of which is determined by the axial and tangential velocities.[23]

A bicentrifugal swirl injector consists of an inner simplex atomizer and an outer swirler. Figure 3 provides schematics for the simplex and bicentrifugal swirl atomizers. In the bicentrifugal swirler, fluid is injected tangentially into both the inner and outer swirl chambers. The swirling fluids proceed down the length of the chamber, are slung out into the chamber, and then coalesce into a single hollow cone if their trajectories intersect. The differences between the inner and outer swirler are a higher wetted surface area (increased viscous effects) and the absence of a gas core for the outer swirler. Flow in the inner swirl chamber is more complicated because of the presence of the gas core.

B.   Applications and Design Guidelines

Bicentrifugal swirl injectors are used extensively in the Russian rocket program.[2] In addition to the possibility of high thrust per element, these injector types feature a wider range of throttleability, an increased margin from

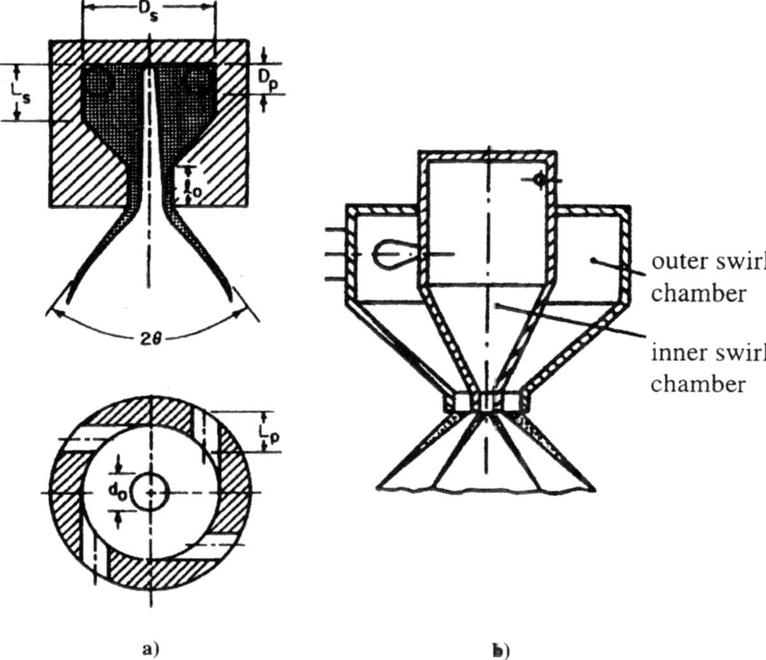

Fig. 3   Schematics of swirl injectors: a) simplex injector[23] and b) bicentrifugal swirl injector.[24]

combustion instability,[25] and reduced pressure drop through the injector,[2] which can result in a decreased demand on rocket turbomachinery. The Russian approach to injector design has been empirical and relies heavily, as it does in the United States, on "know-how." Whereas limited technical details are available in the open literature, two references that yield some clues toward a successful design philosophy for bicentrifugal swirl injectors are provided in the following.

Addressing the combustion instability of the Russian RD-0110 engine (third stage of the Soyuz space vehicle), Rubinsky[26] detailed the bicentrifugal injector scheme shown in Fig. 4. The RD-0110 is a LOX-kerosene gas generator cycle rocket engine. In the injector, six tangential inlets are used to inject the oxidizer and the fuel into the central and peripheral swirl chambers, respectively. The propellants are injected so that they swirl in the same direction. A slight recess is shown, and the exit chamber of the oxidizer circuit is contracted, with an exit area about half that of the swirl chamber.

Detailed laboratory-scale experiments conducted to assess the feasibility of upgrading the space shuttle with nontoxic propellants for its orbital maneuvering

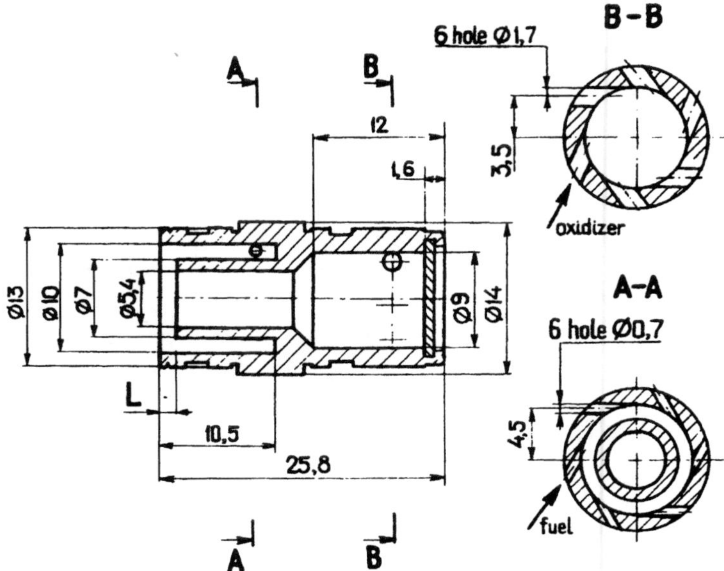

Fig. 4   Injector schematic of the RD-0110 engine.[26] Dimensions are in mm.

system and reaction control system are discussed in another paper by Bazarov.[27] The propellants were LOX-ethanol, and both bicentrifugal and pintle injectors were investigated because of the presumed throttleability capabilities of these injector types. The bicentrifugal swirl injectors were of Russian heritage and based on experience with earth storable, hypergolic propellants. In contrast to the RD-0110 engine, the bicentrifugal swirl injector was ethanol centered (both swirling counterclockwise), with LOX on the outside to provide cooling to the chamber and the injector face.

C.   Modeling Approaches

Theoretical investigations of the swirling flows in atomizers include those by Taylor[28] and a few years later by Giffen and Muraszew.[29] Lefebvre,[23] Bayvel and Orzechowski,[24] and Yule and Chinn[30] revisited swirl injector theory in the late 1980s and early 1990s to compare the results from Taylor's original theory with experimental data. Doumas and Laster[31] conducted an experimental study on swirl atomizers, and used Novikov's theoretical treatment[32] as a basis for their experiments. The applied studies all used empirical corrections to the theory to obtain better agreement between the inviscid theory and experiment, which of course included real viscous effects.

Yule and Chinn[30] disputed a key assumption made by Taylor[28] and Novikov[32] regarding the maximum allowable flow rate through the orifice of a swirl chamber and presented a more complete inviscid theory based on the Bernoulli equation and axial momentum conservation. Benjamin et al.[33] compared experimental data with results from viscous flow simulations, inviscid analyses, and empirical correlations, and noted that all the inviscid models tended to underestimate the film thickness and overestimate the spray angle.

The goal of the swirl atomizer flow model is to relate the spray angle and film thickness to the swirl atomizer geometry. As with impinging jet injectors, predictions of drop size are primarily empirical. It has been shown that average drop size is proportional to the film thickness at the exit of the discharge orifice $(d_D \sim Ah^{0.4})$.[34] The following model has been adapted from Giffen and Muraszew[29] to the outer swirl atomizer. The notation used is for both inner and outer atomizers and is consistent with that used in the discussion of the theory presented in Yule and Chinn.[30] Because many of the same principles apply to both inner and outer swirl atomizers, the simpler flow physics of the outer atomizer are examined first.

From the conservation of angular momentum of the swirling flow,

$$wr = \text{const} \tag{6}$$

At the inlet, the velocity of the fluid is purely tangential, and thus

$$w_s(r_s - r_i) = wr = \text{const} \tag{7}$$

where from continuity

$$w_s = \frac{Q}{A_i} \tag{8}$$

The tangential velocity can therefore be determined at any radius r in the swirl atomizer. The spray cone half-angle can be calculated from the ratio of the tangential velocity at the midpoint of the annulus, $\bar{w}$, to the total velocity of the fluid, U:

$$\sin \alpha = \frac{\bar{w}}{U} \tag{9}$$

With the presence of a gas core, the flow physics of the inner swirl atomizer are more complex than the outer atomizer. The classical inviscid theory as presented by Taylor[28] and Giffen and Muraszew[29] is briefly summarized here. Again, one starts with the conservation of angular momentum:

$$wr = \text{const} = w_i R_s \tag{10}$$

Because the swirl velocity at the chamber inlet is purely tangential, from continuity

$$w_i = \frac{\dot{m}_l}{\rho_l A_i} \tag{11}$$

If viscous losses are ignored, the total pressure at any point in the liquid is

$$P = p + 0.5\rho_l u^2 + 0.5\rho_l w^2 \tag{12}$$

At the interface with the gas core, the static pressure equals the back pressure of the ambient gas, which is constant and can be arbitrarily set to zero, and thus

$$P = 0.5\rho_L u_g^2 + 0.5\rho_L w_g^2 \tag{13}$$

The axial velocity remains constant in the radial direction, giving $u = u_g$. From continuity the axial velocity is

$$u = \frac{\dot{m}_l}{\rho_l(A_o - A_g)} \tag{14}$$

where A_o is the orifice outlet area and A_g is the gas core area. From Eqs. (7) and (11), the tangential velocity at the liquid-gas interface becomes

$$w = \frac{\dot{m}_l R_S}{\rho_l A_i r_g} \tag{15}$$

where r_g is the radius of the gas core in the swirl chamber. Substitution of Eqs. (14) and (15) into Eq. (13) yields

$$p = 0.5\rho_l \left(\frac{\dot{m}_l}{\rho_l(A_o - A_g)}\right)^2 + 0.5\rho_l \left(\frac{\dot{m}_l R_S}{\rho_l A_i r_g}\right)^2 \tag{16}$$

Combination with the mass continuity equation leads to

$$\dot{m}_l = C_d A_o (2\rho_l P)^{0.5} \tag{17}$$

Equations (15) and (16) can be properly manipulated to give the key non-dimensional equation that relates the injector geometry to the size of the gas core and the injector discharge coefficient:

$$\frac{1}{C_d^2} = \frac{1}{K_i^2 X} + \frac{1}{(1 - X)^2} \tag{18}$$

To obtain the additional equation necessary to solve Eq. (18), the assumption is made[28-31] that for any given injection pressure the radius of the gas core in the outlet r_g will adjust itself so that the volumetric flow rate will be a maximum—the so-called principle of maximum flow—so that $\partial Q/\partial r_g = 0$ at the exit. Doumas and Laster[31] made measurements of the axial reaction force exerted on the atomizer (as well as photographing the flow at the orifice exit) and concluded that the gas flow area at the orifice was independent of flow rate, thereby demonstrating the principle of maximum flow. Yule and Chinn[30] disputed this assumption and instead approached the problem from a perspective of the conservation of axial momentum.

According to the principle of maximum flow, the discharge coefficient and atomizer constant can be written explicitly as follows:

$$C_d = \sqrt{\frac{(1-X)^3}{(1+X)}} \tag{19}$$

$$K_1 = \frac{(1-X)^{3/2}}{\sqrt{2}X} \tag{20}$$

Once the cone angles for the inner and outer swirlers are determined, the total cone angle can be determined from the momentum balance:

$$\cos \alpha_T = \frac{U_1 \dot{m}_1 \cos \alpha_1 + U_2 \dot{m}_2 \cos \alpha_2}{U_1 \dot{m}_1 + U_2 \dot{m}_2} \tag{21}$$

The need still exists for detailed measurements within the swirl chamber that provide information on the coupling between the swirl flow mechanics and the ambient conditions. The disagreement that exists regarding the universal application of the principle of maximum flow has not been conclusively resolved. Viscous effects can be important and are not accurately accounted for in the models. Finally, it should also be noted that the bicentrifugal injector design methodology may be drastically different according to whether hypergolic or cryogenic propellants are used.

IV. Pintle Injector

A. General Description

The pintle injector is unique among the various injector options that have been successfully used in liquid rocket engines. In a bipropellant engine, one of the propellants flows down the inside of the pintle and is ejected radially through a series of holes or slots near the tip of the pintle. The other propellant leaves the manifold through an annular sheet around the base of the pintle. Vigorous mixing and atomization of the propellants result from the collision of the radial jets with the thin liquid sheet. Figure 5 is a series of photographs of water flow tests on a single pintle injector, looking back toward the injector element. Figure 5a shows characteristic flow for the outer, annular injection; Figure 5b shows a wider-angle view of the inner passage flow being injected as a radial sheet; and Fig. 5c shows the spray fan resulting from the combined flows. This resultant flowfield yields a curved combustion zone that is substantially different from those formed by "flat face type" injectors.

Pintle injectors enjoy several advantages over other types of liquid-liquid injectors. First, the design is inherently simpler in the sense that only a single injector element is required. This issue is somewhat misleading in that the "single element" can have multiple radial holes, but in any case pintle engines have inherently lower number of injection sites than the face-type injectors.

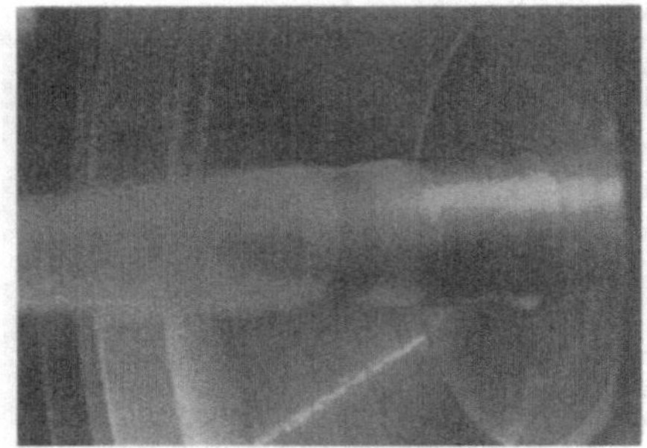

a) Outer flow only

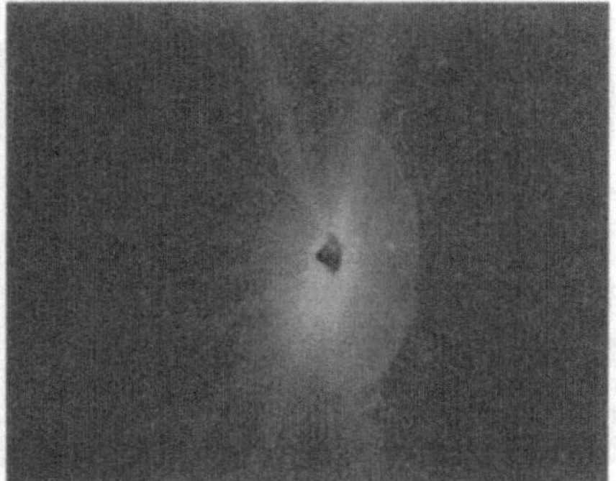

b) Inner flow only

c) Combined flows

Fig. 5   Water flows in pintle injector.

For this reason, pintle engines are inherently simpler to build; all machining operations involve placing holes normal to the local surface and the precise alignment issues associated with impinging element injectors are not present.

The second advantage is the inherent combustion stability afforded by this injector. There has never been an instance of combustion instability associated with pintle engines[3]; this factor reduces risk and eliminates the need for stability aids such as baffles and acoustic liners/cavities. The curved combustion zone developed as a result of this unique injection geometry leads to energy release that, compared to uniformly distributed flat face injectors, is away from the pressure antinodes of the chamber. Also the radial flow component of the injected propellant ensures that there is always a significant mean velocity difference between the gas flow and the propellant drops.

The third attractive feature of the pintle injector involves its throttleability. By using a translating sleeve, the flow areas in both the annulus and the holes/slots can be adjusted to provide deep throttling and/or face shutoff of the propellant flows. Throttling ratios of $10-20:1$ have been demonstrated with hypergolic propellants using this capability.[3] Pulsing applications demanding a face shutoff can also realize significant advantage using this injector design concept. The small dribble volume and face shutoff features associated with this injector type make for rapid pulsing capabilities; 2-ms pulses have been demonstrated using this methodology.[35]

B. Applications and Design Guidelines

Prior applications of pintle injector technology are summarized in Refs. 3 and 35–50. The pintle injector design concept was initially developed as a standardized technique to test hypergolicity of storable propellant combinations under development in the late 1950s.[3] The first rocket engine applications were developed in the early 1960s.[36] Since that time, pintle engines have been successfully fired using 25 different propellant combinations in thrust levels from 5 to $650 \text{ k} \cdot \text{lb}_\text{f}$.[3] Over 60 different engine designs have been developed over the past 40 years since the genesis of the concept.

The lunar module descent engine (LMDE) was one of the more notable achievements of the pintle injector technology. This engine used a translating sleeve to control both annular and slot flow areas to provide the $10:1$ throttling ratio demanded of this application.[37,38] Since this time, the TR201 liquid apogee engine was developed as a fixed-thrust version of the LMDE technology; this engine has flown nearly 80 times without a failure as the second-stage engine of the Delta launch vehicle. Gas/liquid injection has recently been demonstrated in an 870-lbf engine using gaseous oxygen/ethanol propellants.[46] Austin et al.[50] describe experiments with small engines ($100-200 \text{ lb}_\text{f}$ thrust) using nontoxic hypergolic propellants based on hydrogen peroxide and methanol.

Figure 6 highlights some of the major design features of a pintle injector. The most important design variable is the total momentum ratio (TMR), defined as the ratio of radial-to-axial stream momentum:

$$\text{TMR} = \frac{(\dot{m}U)_r}{(\dot{m}U)_z} \tag{22}$$

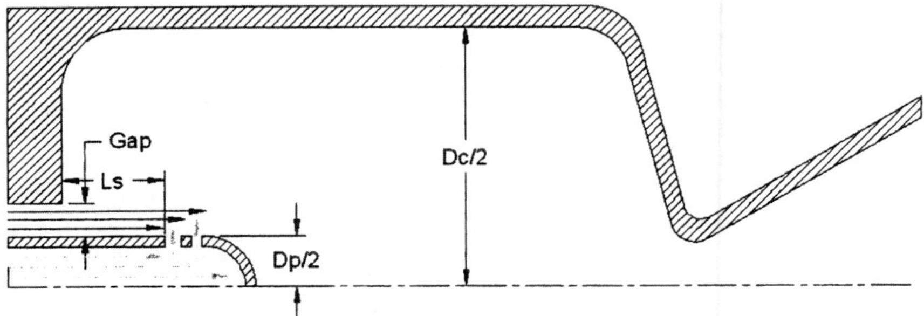

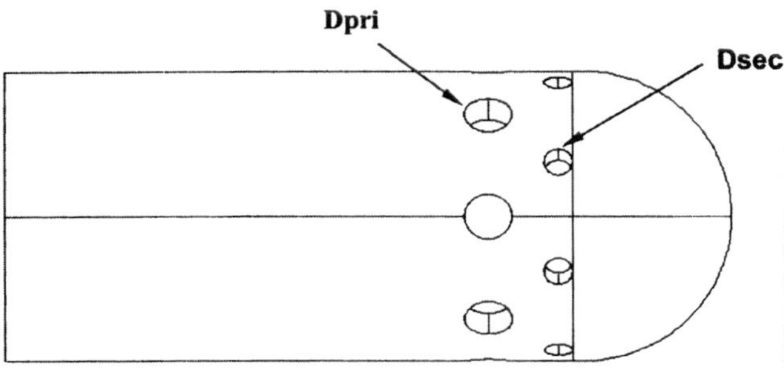

Fig. 6 Key design variables for a pintle injector.

The cone angle formed by the propellant spray increases with TMR; design experience shows that TMR values near unity provide optimal performance. The cone angle of the spray scales as $\mathrm{TMR}^{0.5}$, similar to the scaling of jet penetration in a crossflow with momentum ratio.[48] Typical injection velocities for both axial and radial streams range from 10 to 50 m/s in most cases.

The blockage factor (BF), defined as the ratio of the total hole/slot circumferential length divided by the circumference of the pintle, is another important design variable:

$$\mathrm{BF} = \frac{Nd_o}{\pi d_p} \tag{23}$$

where N is the number of holes/slots in the pintle tip (typically 20–36). In some designs, a set of secondary holes is placed just downstream of the primary holes in the pintle. These holes are placed circumferentially to lie in the gaps formed by the primary holes; they generally are smaller in size than the primary holes.

Another important dimensionless variable is the ratio of chamber-to-pintle diameters, d_c/d_p. Typical values for this quantity range from 3 to 5.[35] Finally, the skip distance is defined as the length that the annular flow must travel before impacting the radial holes divided by the pintle diameter, L_s/d_p.

A typical value for this parameter is around 1; larger skip distances are subject to substantial deceleration of the liquid because of friction against the pintle post, whereas very short skip distances may lead to spray impingement on the head end of the combustion chamber.

Pintle flows are classified as fuel centered or oxidizer centered, depending on which propellant flows inside the pintle tip. Typically, smaller engines utilize fuel-centered configurations and larger engines are oxidizer centered.[35] When using a fuel-centered design, it is possible to increase TMR such that fuel jets actually penetrate the axial oxidizer annulus and impact on the wall to provide film cooling. Oxidizer-centered designs would operate best at lower TMR values to insure that oxidizer does not impact chamber walls and cause damage or compatibility issues.

In general, the combustion chambers that use pintle injectors tend to have a higher contraction ratio than face-type injectors to accomodate the radial flows induced by this injection scheme. However, the combustor lengths may actually be shorter because of this same phenomenon; the overall chamber characteristic length values are not dramatically different between the two designs. Pintle thrusters have demonstrated combustion efficiencies in the 96–99% range for most of the engines that have been developed for flight programs. Because of the limited number of holes in the pintle, the thrust per element (thrust per hole) can be very large for high-thrust engine designs, and the combustion efficiency tends to be a bit lower on large engines for this reason. Of course, this difference in efficiency must be weighed against the inherent simplicity and cost of the injector.

The aforementioned design guidelines along with required propellant flow rates allow one to size the dimensions shown in Fig. 6. Design changes are very easy to implement by simply machining another pintle tip. In general, it is difficult to manufacture very small thrust injectors (as in the case of all bipropellant engines) because both the orifice sizes and the annular gap become very small. Tight tolerances are required on the pintle outer surface to insure a uniform propellant gap under these conditions.

C. Mechanistic Study

The physics associated with pintle injector flows and combustion have been much less studied than those of the flat-face injector types. To date, the database for design has largely been developed empirically. Design analysis codes exist within industry, but these are not generally available to the public. Limited academic studies have been reported.[35,49] Even the very fundamental problem of multiple radial jets impinging on a flat liquid sheet has received little attention at this point in time. When hypergolic fluids are used, the "blowapart" phenomenon attributed to the gas evolution at the contact surface has not been addressed or quantified in this injector scheme. Even basic waterflow data on the drop sizes produced for various designs are not readily available in open literature.

The issues that tend to complicate the development of these injectors include:

1) Manufacturing issues are associated with maintaining a small annular gap around the pintle for injection of the axial stream.

2) Heat affected regions at the tip of the pintle must be considered. The pintle tip lies in a recirculation zone and is subject to high heat flux. There is some

limited evidence that use of a rounded tip tends to somewhat alleviate this effect.

3) Heat affected regions near axial/radial stream impact points must be considered. In engines using hypergolic fluids, small recirculation zones near the base of the radial jets can lead to pintle damage because of local combustion in this region. Careful design of the pintle surface can help alleviate this problem by creating shapes that minimize the size of recirculation zones.

4) As with any bipropellant engine, wall heat flux problems can also be present, but these can normally be alleviated through variation of TMR.

V. Summary and Conclusions

Liquid/liquid injectors are an important type of injector used in rocket combustors ranging from less than a few to over a million pounds of thrust and with virtually every propellant combination there is. To design these injectors most efficiently, it is important to understand the processes of atomization and spray dispersion. In the past, designers have relied heavily on empirical guidelines and extrapolating results from cold-flow studies. Much work needs to be done to understand the controlling design parameters in a high-pressure combusting environment.

Even though the history of impinging jet atomizers extends back to the 1950s, our understanding of the atomization process is far from complete, and a general model that accounts for all the critical processes still does not exist. Based on laboratory experiments at reduced-flow conditions, impact waves seem to dominate atomization. These impact waves are a major factor in the resulting drop size distribution and the atomization frequency, which may play a major role in combustion instability. To design high-performing impinging jet injectors that are not prone to combustion instability, an understanding of how these impact waves are formed in a practical injector is needed. The importance of impact waves at very high ambient density such as in a high-pressure preburner has not been established.

Bicentrifugal swirl injectors offer the advantages of throttleability and high thrust per element. Reasonable design models exist, but the physics of the gas core swirl injector need to be better understood to fully realize the benefits of these injectors. The interactions between oscillations in the ambient gas phase and the instabilities on the swirling surface have been speculated to play a role in combustion instability, and this needs to be investigated. Also the interactions and mixing between the swirling cones as they exit the swirl injector is an area where little work has been done and increased understanding will yield higher performing and more robust designs.

Pintle injectors also offer simple high-thrust-per-element designs. Very little fundamental work on pintle injectors is available in the open literature, and the design process is highly empirical. More than the other injectors, the design of the pintle injector must be combined with the design of the combustion chamber to yield good results. Better understanding of the interactions between the pintle injector flow and the combustor could help reduce the combustor size.

References

[1]*Liquid Rocket Engine Injectors*, NASA SP-8089, 1976.

[2]Bazarov, V. G., "Dynamics of Liquid Rocket Injectors," Second International Symposium on Liquid Rocket Propulsion, Chatillon, FRANCE, June 1995.

[3]Dressler, G. A., and Bauer, J. M., "TRW Pintle Engine Heritage and Performance Characteristics," AIAA Paper 2000-3871, July 2000.

[4]Ferrenberg, A., Hunt, K., and Duesberg, J., "Atomization and Mixing Study," NASA Contract NAS8-34504, 1985.

[5]Rupe, J. H., "The Liquid-Phase Mixing of a Pair of Impinging Streams," Jet Propulsion Lab., Progress Rept. 20-195, Pasadena, CA, 1953.

[6]Anderson, W. E., Miller, K. L., Ryan, H. M., Pal, S., Santoro, R. J., and Dressler, J. L., "Effects of Periodic Atomization on Combustion Instability in Liquid-Fueled Propulsion Systems," *Journal of Propulsion and Power*, Vol. 14, No. 11, 1998, pp. 818–825.

[7]Hasson, D., and Peck, R. E., "Thickness Distribution in a Sheet Formed by Impinging Jets," *AIChE Journal*, Sept. 1964, pp. 752–754.

[8]Heidmann, M. F., Priem, R. J., and Humphrey, J. C., "A Study of Sprays Formed by Two Impinging Jets," NACA TN 3835, 1957.

[9]Dombrowski, N., and Hooper, P. C., "A Study of the Sprays Formed by Impinging Jets in Laminar and Turbulent Flow," *Journal of Fluid Mechanics*, Vol. 18, Part 3, 1964, 392–400.

[10]Yatsuyanagi, N., "The Effect of Environmental Pressure on Spray Formed by Two Impinging Jets," National Aerospace Lab., NAL TM-268, Japan, 1974.

[11]Ryan, H. M., Anderson, W. E., Pal, S., and Santoro, R. J., "Atomization Characteristics of Impinging Liquid Jets," *Journal of Propulsion and Power*, Vol. 11, No. 1, 1995, pp. 135–145.

[12]Anderson, W. E., "The Effects of Atomization on Combustion Stability," Ph.D. Dissertation, Pennsylvania State Univ., 1996.

[13]Ashgriz, N., Brocklehurst, W., and Talley, D. G., "On the Mixing Mechanisms in a Pair of Impinging Jets," AIAA Paper 95-2421, July 1995.

[14]Strakey, P. A., and Talley, D. G., "Spray Characteristics of Impinging Jet Injectors at High Back-Pressure," Eighth International Conference on Liquid Atomization and Spray Systems, Pasadena, CA, July 2000.

[15]Taylor, G. I., "The Dynamics of Thin Sheets of Fluid II. Waves on Fluid Sheets," *Proceedings of the Royal Society of London, Series A: Mathematical and Physical Sciences*, Vol. 253, 1959, pp. 296–312.

[16]Squire, H. B., "Investigation of the Instability of a Moving Liquid Film," *British Journal of Applied Physics*, Vol. 4, 1953, pp. 167–169.

[17]Dombrowski, N., and Johns, W. R., "The Aerodynamic Instability and Disintegration of Viscous Liquid Sheets," *Chemical Engineering Science*, Vol. 18, 1963, pp. 203–214.

[18]Clark, C. J., and Dombrowski, N., "Aerodynamic Instability and Disintegration of Inviscid Liquid Sheets," *Proceedings of the Royal Society of London, Series A: Mathematical and Physical Sciences*, Vol. 329, 1972, pp. 467–478.

[19]Lin, S. P., Lian, Z. W., and Creighton, B. J., "Absolute and Convective Instability of a Liquid Sheet," *Journal of Fluid Mechanics*, Vol. 220, 1990, pp. 673–689.

[20]Chuech, S. G., "A Comparative Study of Temporal and Spatial Instabilities of a Two-Dimensional Liquid Sheet," *Heat and Mass Transfer in Spray Systems*, HTD-Vol. 187, ASME, 1991, pp. 19–25.

[21]Ibrahim, E. A., and Przekwas, A. J., "Impinging Jets Atomization," *Physics of Fluids A*, Vol. 3, No. 12, 1991, pp. 2981–2987.

[22]Dombrowski, N., and Hooper, P. C., "The Effect of Ambient Density on Drop Formation in Sprays," *Chemical Engineering Science*, Vol. 17, 1962, pp. 291–305.

[23]Lefebvre, A. H., *Atomization and Sprays*, Hemisphere, 1989.

[24]Bayvel, L., and Orzechowski, Z., *Liquid Atomization*, Taylor and Francis, 1993.

[25]Bazarov, V., "Self-Pulsations in Coaxial Injectors with Central Swirl Liquid Stage," AIAA Paper 95-2358, July 1995.

[26]Rubinsky, V. R., "Combustion Instability in the RD-0110 Engine," *Liquid Rocket Engine Combustion Instability*, edited by V. Yang and W. Anderson, Progress in Astronautics and Aeronautics, Vol. 169, AIAA Washington, DC, 1995, pp. 89–112.

[27]Bazarov, V., "Injector Research for Shuttle OMS Upgrade Using LOX/Ethanol Propellants," AIAA Paper 98-3816, July 1998.

[28]Taylor, G. I., "The Mechanics of Swirl Atomizers," 7th Int. Congress of Applied Mechanics, V.2., Pt. 1. (S.M. 6845.220), Sept. 1948.

[29]Giffen, E., and Muraszew, A., *Atomization of Liquid Fuels*, Chapman and Hall, London, 1953.

[30]Yule, A. J., and Chinn, J. J., "Swirl Atomizer Flow: Classical Inviscid Theory Revisited," ICLASS-94 Rouen, France, July 1994.

[31]Doumas, M., and Laster, R., "Liquid-Film Properties for Centrifugal Spray Nozzles," *Chemical Engineering Progress*, Vol. 49, No. 10, 1953, pp. 518–526.

[32]Novikov, I. I., *Journal of Technical Physics*, Vol. 18, No. 3, 1948, p. 345.

[33]Benjamin, M. A., et al., "Comparison of Simplex Atomizer Correlations with Detailed CFD and Experimental Data," Proc. ILASS-Americas '97, NRC Ottawa, Ont., 1997.

[34]Rizk, N. K., and Lefebvre, A. H., "Influence of Liquid Film Thickness on Airblast Atomization," *Transactions of ASME, Journal of Engineering for Power*, Vol. 102, 1980, pp. 706–710.

[35]Escher, D. W., "Design and Preliminary Hot Fire and Cold Flow Testing of Pintle Injectors," M.S. Thesis, Mechanical Engineering Dept., Pennsylvania State Univ., Dec. 1996.

[36]Siegel, B., "Research of Low-Thrust Bipropellant Engines," Space Technology Lab., Independent Research Program Annual Rept. No. 9990-6020-RU-000, Redondo Beach, CA, 1961, pp. VI-59–VI-73.

[37]Elverum, G., Staudhammer, P., Miller, J., Hoffman, A., and Rockow, R., "The Descent Engine for the Lunar Module," AIAA Paper 67-521, July 1967.

[38]Gilroy, R., and Sackheim, R., "The Lunar Module Descent Engine—A Historical Perspective," AIAA Paper 89-2385, 1989.

[39]Fritz, D., Dressler, G., Mayer, N., and Johnson, L., "Development and Flight Qualification of the Propulsion and Reaction Control System for ERIS," AIAA Paper 92-3663, July 1992.

[40]Fritz, D., and Gavitt, K., "Gel Propulsion for the Fourth Generation Escape System," 1993 SAFE Symposium, Las Vegas, NV, Nov. 1993.

[41]Dressler, G., Stoddard, F., Gavitt, K., and Klem, M., "Test Results from a Simple, Low-Cost Pressure-Fed Liquid Hydrogen/Liquid Oxygen Rocket Combustor," Chemical Propulsion Information Agency, Pub. 602, Vol. II, 1993, pp. 51–67.

[42]Hodge, K., Crofoot, T., and Nelson, S., "Gelled Propellants for Tactical Missile Applications," AIAA Paper 99-2976, June 1999.

[43]Gavitt, K., and Mueller, T., "TRW LCPE 650 Klbf LOX/LH2 Test Results," AIAA Paper 2000-3853, July 2000.

[44]Klem, M. D., and Stoddard, F. J., "Results of 178 kN (40,00 lbf) Thrust LOX/LH$_2$ Pintle Injector Engine Tests," JANNAF CS/PSH/EPTS/SPIRITS Joint Meeting, Huntsville, AL, Oct. 1995.

[45]Klem, M. D., Jankovsky, A. L., and Stoddard, F. J., "Results of LOX/RP-1 Pintle Injector Engine Tests," JANNAF CS/PSH/EPTS/SPIRITS Joint Meeting, Huntsville, AL, Oct. 1995.

[46]Calvignac, J., "Design and Testing of a Non-Toxic 870-LBf Engine," AIAA Paper 2000-3851, July 2000.

[47]Mueller, T., and Dressler, G., "TRW 40 KLBf LOX/RP-1 Low Cost Pintle Engine Test Results," AIAA Paper 2000-3863, July 2000.

[48]Heister, S. D., Nguyen, T. T., and Karagozian, A. R., "Modeling of Liquid Jets Injected Transversely into a Supersonic Crossflow," *AIAA Journal*, Vol. 27, No. 12, 1989, pp. 1727–1734.

[49]Austin, B. J., "Development of a Pintle-Based Engine for Nontoxic Hypergolic Propellants," 38th AIAA/ASME/SAE/ASEE Joint Propulsion Conference, July 2002.

[50]Austin, B., Matthews, J., and Heister, S. D., "Engine/Injector Development for New Nontoxic, Storable, Hypergolic Bipropellants," 13th Propulsion Engineering Research Center Annual Symposium on Propulsion, Huntsville, AL, 2001.

Chapter 5

Distortion and Disintegration of Liquid Streams

William A. Sirignano* and Carsten Mehring[†]
University of California, Irvine, Irvine, California

Nomenclature

a = half of undisturbed sheet thickness
A = area, constant
B = constant
B_0 = Bond number
c = nondimensional wave speed
C = coth kh for symmetric mode, tanh kh for antisymmetric mode, constant
e_1, e_2 = constants defined in Eqs. (86) and (87)
f_1, f_2, f_3 = functions defined in Eqs. (56)
$F_{\pm}$ = function defined in Eq. (84)
Fr = Froude number
g = gravitational acceleration
G = Green's function
$G_{\pm}$ = function defined in Eq. (85)
h = nondimensional film thickness perturbation
I_n = modified Bessel function of the first kind
k = wave number
l_c = characteristic length for pinch-off
l = wave number
L = characteristic length
m = growth rate, transverse sinuous mode wave number
n = azimuthal wave number, transverse dilational mode wave number
$\boldsymbol{n}$ = unit normal vector

---

*Professor, Department of Mechanical and Aerospace Engineering. Fellow AIAA.
†Research Associate, Department of Mechanical and Aerospace Engineering. Member AIAA.

Oh = Ohnesorge number
p = pressure
p_{ref} = reference pressure
P = separated pressure perturbation function
q = gradient at surface
r = radial position
r' = nondimensional annular radius perturbation
R = radius of curvature, droplet radius
Re = Reynolds number
s = nondimensional growth rate, arc length
s^* = dimensional growth rate
S = jet or droplet surface area
St = Strouhal number
t = time coordinate
t_c = characteristic time for pinch-off
T_{ex} = reference time for exponential function
T_p = reference time for trigonometric function
u = velocity component
u_o = mean jet velocity
U = separated velocity perturbation function, wave velocity
$\mathbf{v}$ = velocity vector
v = velocity component
v = volume
V = characteristic velocity
V_o = applied voltage
W = work, also $W = [2\pi\sigma/\lambda(\Delta u)^2][(1/\rho_1) + (1/\rho_2)]$, a parameter used in
    Ref. 69
We = Weber number
w = velocity component
x = stream coordinate in planar case
y = transverse coordinate in plane
$\bar{y}$ = local sheet centerline position
$\tilde{y}$ = local sheet thickness
Y = nondimensional sheet centerline position
z = downstream or axial position in cylindrical coordinates, spanwise
    coordinate in Cartesian coordinates

Greek and Arabic

α = wave phase, growth rate, node constant in Eq. (70)
β = nondimensional wave number, growth time constant
δ = differential quantity
Δ = change in a quantity
$\epsilon = \sqrt{1/(2We)}$
ϵ_o = permittivity of free space
ε = perturbation amplitude
ϕ = velocity potential angle corresponding to arc length on surface
ϕ_e = electrostatic potential

Γ = normalized electrostatic pressure
η = coefficient of kinematic viscosity, disturbance amplitude
κ = curvature
λ = wavelength
μ = coefficient of dynamic viscosity
θ = azimuthal position, ± 1 or ∓ 1
ρ = density
σ = surface tension
$\boldsymbol{\tau}$ = viscous stress tensor
τ = nondimensional time, sometimes in coordinate system moving with fluid
ω = frequency
ξ = spacial coordinate in reference frame moving with fluid

Subscripts

g = gas
k = integer index indicating particular fluid
l = liquid
r, θ, z = components in r, θ, z directions
$\pm$ = liquid/gas interface location

Superscript

$'$ = perturbation quantity

I. Introduction

S PRAYS are important to the economy and to the health and well-being of society and have many practical applications. In many cases, the detailed characteristics of the spray are consequential in the application. The droplet diameter and velocity of distributions affect the spray penetration (and, thereby, the mixture-ratio profiles) and vaporization rates. As noted by Sirignano et al.,[1] droplet vaporization is the slowest of the combustion processes and therefore rate-controlling in a wide range of liquid-propellant-rocket applications. In this chapter, the aspects of spray formation most important to liquid-propellant technology will be discussed. These aspects will also be relevant to many other applications.

A major challenge occurs because of the need to replace existing empiricism with theoretical foundations for the prediction of the spray characteristics that emerge from the liquid injection and atomization process. Empirical knowledge pertains to specific injector systems, and therefore predictive capability for the initial spray character does not extend beyond an envelope related to the particular design configuration. Improved theory would allow both a more universal methodology for prediction and a more highly optimized design. It is intended herein to present the current status of the fundamental understanding about the liquid atomization processes for various injection configurations. The limitations of the theory and the need for future work will be made apparent. This chapter is not intended to be a guide for the practicing engineer; the current state-of-the-art is based upon empirical approaches[2,3] that will not be discussed herein. Rather, this chapter will review theoretical research that should eventually lead to improved design methodology and design tools for liquid atomization systems. Other overviews of the theory can be found in Refs. 2, 3, and 4.

The atomization problem could be divided according to three subdomains of the fluid mechanical field. The upstream subdomain lies within the liquid-supply piping, plenum chamber, and orifice (nozzle) of the injector hardware. More than the mass flow and average velocity from the orifice into the combustion chamber are important here; velocity and pressure fluctuations in the liquid due to turbulence, cavitation, supply pressure unsteadiness, and/or active-control devices are critical in affecting the temporal and spatial variation of the liquid flow over the orifice exit cross section. A small amount of research has been performed on this subject.

The second subdomain involves the liquid stream from the orifice exit to the downstream point where disintegration of the stream begins. The neighboring gas flow (or gas and droplet flow) is part of this subdomain. This region involves severe distortion of the liquid stream that typically occurs in a wave-like manner. We will focus our discussion in this chapter on this subdomain. This is the subdomain in which most of the research work has been performed. The nonequal treatment of the three subdomains is a reflection of the amount of research performed rather than of their relative importance.

The third subdomain in the spray formation phenomenon involves the cascading of the ligaments that occur in the earliest disintegration to smaller and smaller elements, resulting eventually in the essentially spherical droplets of the spray. This is the most challenging portion of the theoretical problem. No theory has been developed to explain the details of the dynamics in this cascade process. Some statistical theories[1,2] have attempted to describe the droplets that are yielded from this process.

There are many types of injectors (sometimes called atomization systems) that inject liquid streams with sufficient energy to provide for breakup or atomization into droplets. Energy can be provided through various mechanisms: liquid pressure, air pressure, rotation of cups or disks, vibration or acoustics, and electric fields. Discussions of practical injectors can be found in Refs. 2 and 3.

There are some important geometrical characteristics of injected streams: 1) curvature of the gas-liquid interface and 2) divergence of the liquid stream. The interface can have a mean or average curvature due to basic geometry; round jets, annular jets, coaxial jets, and conical jets share this character. A planar jet does not have a mean curvature. Any liquid jet or stream with a free surface can have curvature because of surface waves such as capillary waves. Curvature leads to a pressure jump across the surface because of surface tension. With wave motion at the surface, the pressure jump will vary with time and position along the surface, thereby leading to pressure gradients parallel to the surface and subsequent accelerations. These induced motions can become unstable, causing the disintegration of the stream.

Divergence of the stream occurs in conical streams and fan jets (such as result from two impinging round jets). For hollow cone and fan jet injections, this results in a thinning of the stream and affects the eventual droplet size. Of course, the divergence increases the number of interesting dimensions for the phenomenon.

Important liquid properties for atomization are density ρ_l, surface tension coefficient σ, and dynamic viscosity μ_l. For common liquids, density is of the order of $1000 \ kg/m^3$. The surface tension coefficient is of the order of $10^{-2} \ N/m$ for many liquids. Dynamic viscosity varies from the order of 10^{-4} to the order of $1 \ kg/ms$. Liquid density is very weakly dependent on pressure and temperature. Surface tension and viscosity are weakly dependent on pressure and will decrease as temperature increases.

There are four forces acting on the liquid that are typically important in atomization (not including electromagnetic forces acting on charged liquids). They are 1) gravity force, $\rho_l L^3 g$; 2) inertia, $\rho_l L^2 V^2$; 3) surface tension force, σL, and 4) viscous force, $\mu_l L V$. Note that g can also represent any acceleration of the liquid that can be formulated as a reversed D'Alembert acceleration in a noninertial frame of reference. From these four forces three independent, well-known nondimensional groupings result:

$$Re = \frac{\rho_l L V}{\mu_l}$$

$$We = \frac{\rho_l L V^2}{\sigma}$$

$$Fr = \frac{V^2}{gL}$$

Another number in common use is the Ohnesorge number:

$$Oh = \frac{We^{0.5}}{Re} = \frac{\mu_l}{(\rho_l \sigma L)^{0.5}}$$

In some cases, the Bond number replaces the Froude number:

$$Bo = \frac{\rho_l L^2 g}{\sigma}$$

When a periodic disturbance of frequency ω is imposed on the liquid, it may be convenient to use the Strouhal number:

$$St = \frac{\omega L}{V}$$

The characteristic velocity and length typically are based on the phenomenon to be observed. For discharging jets, the average stream velocity and the diameter (for a round cylindrical jet) or the thickness (for a liquid sheet-like flow) are typically employed. For the temporal analysis of periodically disturbed streams, the disturbance wave number and a characteristic velocity other than the average sheet velocity might be employed. The nondimensional groupings indicate that $\sqrt{gL}$ and $\sqrt{\sigma/\rho_l L}$ are natural characteristic velocities related to gravity waves and capillary waves, respectively. It is seen that gravity wave speeds dominate on large length scales, whereas capillary wave speeds will dominate on short length scales. With the small sizes typically associated with injected liquid streams, gravity waves are not interesting, but capillary waves are critical in the distortion and disintegration process.

Later in our discussion of the effects of electric fields, a fifth type of force on the liquid will be examined. It occurs when the liquid can conduct electricity and an electric field is applied. Its magnitude is characterized by $\epsilon_o V_o^2$. The pressure on the liquid surface due to the electrostatic force can be normalized by the capillary pressure to give another nondimensional grouping, the normalized electrostatic pressure

$$\Gamma = \epsilon_o V_o^2 / \sigma L$$

In addition, there are viscous forces, gravity force, and inertia for the surrounding gas. This will result in four more nondimensional parameters: gas-phase Weber number, gas-phase Reynolds number, and the gas-to-liquid density and viscosity ratios. These Weber and Reynolds numbers will be based on a gas characteristic velocity and gas properties.

Instabilities that involve only the liquid inertia, liquid viscous forces and body forces, and the surface tension are described as capillary instabilities. The inertia and the viscosity of the gas might modify the instability in a quantitative manner; however, they are not essential to the capillary mechanism. An instability that involves a parallel flow of gas with the liquid stream can also occur. If a continuous variation of the parallel velocity occurs, this hydrodynamic instability is named the Rayleigh instability in the inviscid case and the Orr–Sommerfeld instability in the viscous case.[5] (Note that this Rayleigh hydrodynamic instability is distinct from the Rayleigh capillary instability to be discussed later.) When the variation of the parallel velocity involves a discontinuity of that velocity component at the liquid-gas interface forming a vortex sheet there, the phenomenon is identified as Kelvin–Helmholtz instability. When there is instability associated with an acceleration or body force in a normal direction to the gas-liquid interface, we identify the behavior as a Rayleigh–Taylor instability. Any of these instabilities can result in a disintegration of the liquid stream and the formation of smaller liquid ligaments and droplets.

Generally two types of instabilities for liquid streams have been analyzed: temporal instabilities and spatially developing instabilities, both of which will be discussed here. Temporal stability analyses consider solutions that are periodic in space and can be oscillatory or exponential in time. The spatial periodicity implies infinite length for the liquid stream. Spatial stability analyses consider solutions for a semi-infinite stream flowing from a nozzle. Periodicity in time is considered whereas the behavior in space can be oscillatory or exponential.

The earliest analyses were temporal, but spatial stability is of greater practical interest. Note that perturbations of liquid jets in open-flow problems might evolve both in space and time, and therefore temporal or spatial analyses need not provide the whole picture even in linear stability. Convective and absolute instabilities can become important here.

Two general modes of wave phenomena can be found for either temporal or spatially developing instabilities in most injection configurations. The dilational mode involves primarily a pulsing of the stream width of thickness. This mode has also been named a "sausage" mode. For a planar sheet, the pulsing is symmetric, so that this mode is also identified as the symmetric mode. The other general mode is named the sinuous mode because the primary distortion involves a wavy ribbon-like motion of the stream with only secondary effects on the stream thickness. Thus the center of the stream cross section follows a wavy path in the sinuous distortion while it is not significantly affected in the dilational mode. The sinuous mode has an antisymmetric distortion of the two liquid-gas interfaces in the planar configuration and has also been named the antisymmetric mode.

The discussion of liquid stream stability will be separated into four general categories based on the geometrical configuration: round jets, planar sheets, annular sheets, and conical sheets. They are addressed in Sections III, IV, V, and VI, respectively. Coaxial or compound jets will be discussed in the section on round jets, and fan jets will be discussed in the section on planar sheets. The terms sheet and free film will be used interchangeably to denote a liquid stream whose thickness is much smaller than its other dimensions. In the atomization process, thin sheets or free films are often formed as a first step toward obtaining a spray with small droplets in a relatively fast manner. The term conical is used casually here following industrial practice; the conical surfaces discussed here are not necessarily generated by straight lines emerging from a common point. Rather the generating lines are typically curved so that bell-shaped surfaces will be included in this category. The form categories are chosen arbitrarily, and several of them can be portrayed as limiting cases of some other configurations. A common formulation of the governing equations is given in Section II. Concluding remarks are presented in Section VII.

II. Formulation of Governing Equations

The general equations governing the motion of a liquid stream or bulk of liquid in another immiscible fluid and under the influence of gravity are presented for the specific case of an annular liquid stream. Three limiting cases are possible for this configuration. The round jet in an ambient gas stream is recovered as the radius for the inner interface decreases to zero. As both the inner and outer annular radii grow infinitely large, the planar sheet configuration is recovered with the sheet thickness given by the difference of inner and outer annular radius. Finally, for finite values of the inner radius and infinitely large outer radius, one obtains the case of a submerged gas jet. Our discussion will address the round jet, planar sheet, and annular sheet.

Boundary conditions must be applied at the liquid interface. One condition is the kinematic condition that a particle of fluid on the surface moves with the surface so as to remain on the surface. In other words, the velocity component normal to the interface is continuous across the interface, i.e., $v_1 \cdot n = v_2 \cdot n$, where $v_{1,2}$ are the local velocity vectors on both sides of the interface of fluid 1 and fluid 2, and n is the local unit normal vector of the interface.

The second condition considers the balance between the surface stresses on both sides of an interface between fluid 1 and fluid 2, including the pressure jump across the interface due to surface tension. The dynamic boundary condition is $(p_1 - p_2 + \sigma \kappa)n = (\tau_1 - \tau_2) \cdot n$, where σ is the surface tension coefficient of fluid 1 in fluid 2 (in units of force per unit length), p_k is the pressure and τ_k is the viscous stress tensor, n is the unit normal (into fluid 2) at the interface, and κ is the local surface curvature, $R_1^{-1} + R_2^{-1}$, where R_1 and R_2 are the principal radii of curvature of the surface.

For the motion of a swirling viscous annular (or conical) liquid sheet, it is convenient to formulate the equations in cylindrical coordinates. Assuming Newtonian viscous incompressible fluids only and gravity as the only body force, one obtains

$$\frac{\partial v_{z,k}}{\partial z} + \frac{\partial v_{r,k}}{\partial r} + \frac{v_{r,k}}{r} + \frac{1}{r}\frac{\partial v_{\theta,k}}{\partial \theta} = 0 \tag{1}$$

$$\frac{\partial v_{z,k}}{\partial t} + v_{z,k}\frac{\partial v_{z,k}}{\partial z} + v_{r,k}\frac{\partial v_{z,k}}{\partial r} + \frac{1}{r}v_{\theta,k}\frac{\partial v_{z,k}}{\partial \theta} = -\frac{1}{\rho_k}\frac{\partial p}{\partial z} + g_z + \eta_k(\nabla^2 v_{z,k}) \tag{2}$$

$$\frac{\partial v_{r,k}}{\partial t} + v_{z,k}\frac{\partial v_{r,k}}{\partial z} + v_{r,k}\frac{\partial v_{r,k}}{\partial r} + \frac{1}{r}v_{\theta,k}\frac{\partial v_{r,k}}{\partial \theta} - \frac{v_{\theta,k}^2}{r}$$

$$= -\frac{1}{\rho_k}\frac{\partial p}{\partial r} + g_r + \eta_k\left(\nabla^2 v_{r,k} - \frac{2}{r^2}\frac{\partial v_{r,k}}{\partial \theta} - \frac{v_{r,k}}{r^2}\right) \tag{3}$$

$$\frac{\partial v_{\theta,k}}{\partial t} + v_{z,k}\frac{\partial v_{\theta,k}}{\partial z} + v_{r,k}\frac{\partial v_{\theta,k}}{\partial r} + \frac{1}{r}v_{\theta,k}\frac{\partial v_{\theta,k}}{\partial \theta} + \frac{v_{r,k}v_{\theta,k}}{r}$$

$$= -\frac{1}{\rho_k r}\frac{\partial p_k}{\partial \theta} + g_\theta + \eta_k\left(\nabla^2 v_{\theta,k} + \frac{2}{r^2}\frac{\partial v_{r,k}}{\partial \theta} - \frac{v_{\theta,k}}{r^2}\right) \tag{4}$$

where the subscript $k = 1, 2, 3$ now refers to either the fluid within the liquid annulus ($k = 1$), the fluid within the core of the annulus ($k = 2$), or the fluid surrounding the annulus ($k = 3$) (see Fig. 1).

Denoting the annular liquid jet's inner (subscript$-$) and outer (subscript$+$) interfaces by $r_-(t, z, \theta)$ and $r_+(t, z, \theta)$, respectively, the kinematic boundary

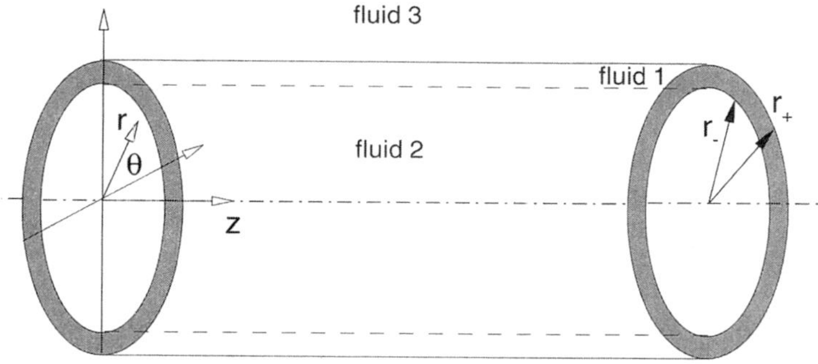

Fig. 1   Annular fluid film configuration under consideration.

conditions at these interfaces are given by

$$v_{r,k}(r_-, \theta, z, t) = \frac{\partial r_-}{\partial t} + v_{z,k}(r_-, \theta, z, t)\frac{\partial r_-}{\partial z}$$

$$+ \frac{1}{r_-} v_{\theta,k}(r_-, \theta, z, t)\frac{\partial r_-}{\partial \theta}, \qquad k = 1,2 \tag{5}$$

$$v_{r,k}(r_+, \theta, z, t) = \frac{\partial r_+}{\partial t} + v_{z,k}(r_+, \theta, z, t)\frac{\partial r_+}{\partial z}$$

$$+ \frac{1}{r_+} v_{\theta,k}(r_+, \theta, z, t)\frac{\partial r_+}{\partial \theta}, \qquad k = 1,3 \tag{6}$$

These equations establish that the jet's surfaces are material surfaces. The dynamic boundary conditions establish that the shear stress is continuous, and the jump in normal stresses across the interfaces is balanced by surface tension. Assuming inviscid fluids only, the viscous shear stress at both interfaces r_+ and r_- vanishes, and the dynamic boundary conditions at the two interfaces reduce to a condition for the normal stress only, i.e.,

$$p_1(r_-, \theta, z, t) = p_2(r_-, \theta, z, t) - \sigma_- \kappa_- \tag{7}$$

$$p_1(r_+, \theta, z, t) = p_3(r_+, \theta, z, t) + \sigma_+ \kappa_+ \tag{8}$$

where $\sigma_\pm$ denote the constant surface tension coefficient between fluid 1 and fluid 2 (subscript$-$) or fluid 1 and fluid 3 (subscript$+$), respectively. κ_- and κ_+ refer to the local curvature of the inner and outer interface, respectively.

The pressure jump across the interface due to surface tension is obtained from the local curvature $\kappa_{(\pm)}$, i.e.,

$$\kappa = \nabla \cdot \boldsymbol{n} = \frac{1}{R_1} + \frac{1}{R_2} \qquad (9)$$

where $\boldsymbol{n}$ is the local normal unit vector at the considered interface.

For an inviscid flow without vorticity, it is possible to set the velocity equal to the gradient of a scalar velocity potential:

$$v = \nabla \phi$$

Substitution of this relation into Eq. (1) yields Laplace's equation $\nabla^2 \phi = 0$.

The unsteady Bernoulli equation can be developed by integrating the inviscid vector form of the momentum equation (2), (3), and (4) along a path that is locally tangent to the velocity vector (not necessarily the particle path in an unsteady case). We can obtain

$$\rho_l \frac{\partial \phi}{\partial t} + \frac{\rho_l}{2} \nabla \phi \cdot \nabla \phi + p_l = p_{\text{ref}} \qquad (10)$$

where p_{ref} is a constant reference pressure. Equation (10) is especially useful when applied along the liquid interface where it can relate velocity potential to pressure. An important nonlinear effect is introduced through the Bernoulli equation.

In the limit of very large values for the radial locations of the two interfaces $r_+(z, \theta, t)$ and $r_-(z, \theta, t)$, we recover the equations for the three-dimensional distortion of an incompressible inviscid planar fluid sheet separating two (different) inviscid incompressible fluids. Also, for finite values of r_+ but zero value for r_-, the governing equations for a round jet in an ambient fluid are obtained. In the limit of very large values for r_+, the equations for a submerged gas jet are obtained, if $\rho_2 \ll \rho_1$.

III.   Round Jet Analyses

The round liquid jet is the simplest injection configuration and has received the most attention. It is well known that four distinct regimes can be found on a plot of Weber number vs Reynolds number (or on a plot of Weber number vs Ohnesorge number). These four regimes, shown in Fig. 2, are separated on a log-log plot by three straight lines of negative slope.[4] At the lowest values of *We* and *Re*, the Rayleigh capillary mechanism prevails. Here aerodynamics interaction with the surrounding gas is not important, and axisymmetric dilational oscillations occur. At higher values of *We* and *Re*, the first-wind-induced regime appears where aerodynamics becomes important and nonaxisymmetric sinuous oscillations appear. In these first two regimes, the resulting droplet diameters after breakup are of the same order of magnitude as the diameter of the original jet. At still higher *We* and *Re*, the second-wind-induced regime appears, which is characterized by smaller droplets after breakup. At the

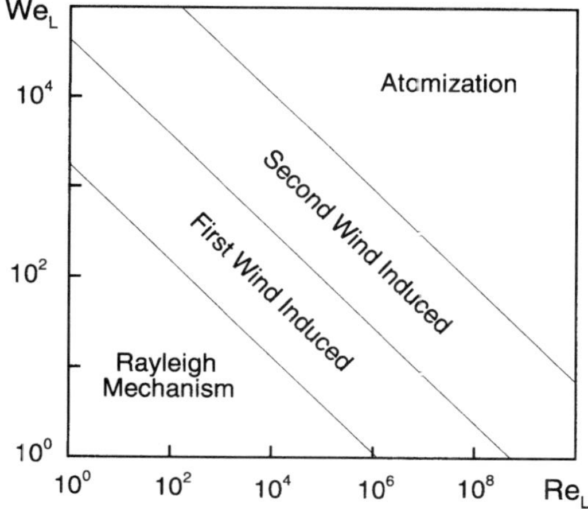

Fig. 2   Modes of disintegration on a liquid Reynolds number and Weber number plane.

highest *We* and *Re* values, the atomization regime appears, wherein breakup occurs very close to the nozzle exit with the smallest droplets as a result.

Consider an initially cylindrical liquid jet undergoing the capillary oscillation. Assume no gravitational or viscous effect. At low pressures, the density of the surrounding gas is negligible compared to the liquid density, so that gas inertia can be neglected. Thereby, the liquid jet is essentially flowing in a passive medium, implying very small or zero values for the gas-phase Weber number. The governing continuity and momentum equations for the liquid jet are now given by Eq. (1) through Eq. (4) for $k = 1$ with $\eta = 0$ and $g_z = g_r = g_\theta = 0$. For the round jet only one free surface exists, i.e., $R = r_+(z, \theta, t)$ and $r_- = 0$; and because the surrounding medium has been assumed to be passive, field equations for fluid 3 ($k = 3$) do not have to be solved.

Accordingly, the only kinematic and dynamic boundary conditions that apply here are Eq. (6) for $k = 1$ and Eq. (8) with $p_3 = 0$, respectively.

A.   Temporal Stability Analysis

Rayleigh[5] made an early contribution to the understanding of the formation of droplets from a liquid stream by capillary effects. In a temporal stability situation, a Galilean transformation can be made to the infinite stream so that stream velocity is eliminated. (In the spatial stability case, that transformation is not useful because the nozzle exit position should remain stationary in the frame of reference.) Small magnitudes of disturbances to the initially cylindrical column of liquid allow a linearized analysis. Two approaches are possible. One employs separation of variables to solve the governing Laplace partial differential equation constrained by certain interface conditions and predicts spatial and temporal behavior. Another approach seeks a final configuration with minimum surface energy.

Rayleigh sought solutions of the form

$$p' = P(r)e^{in\theta}e^{ikz}e^{i\omega t}$$

$$u'_r = U(r)e^{in\theta}e^{ikz}e^{i\omega t}$$

$$R' = ce^{in\theta}e^{ikz}e^{i\omega t}$$

for the perturbations of pressure, radial velocity, and jet radius, n and k are wave numbers in the circumferential and axial directions, ω is an angular frequency, and c is a constant. The function P is governed by a modified Bessel equation. It is found that the growth or decay rate is given by

$$\omega^2 = \frac{\sigma}{R_0^3 \rho_l} \frac{kR_0 I'_n(kR_0)}{I_n(kR_0)} [k^2 R_0^2 + n^2 - 1] \tag{11}$$

$I_n(kr)$ is a modified Bessel function of the first kind and R_0 is the undisturbed radius of the liquid cylinder. Because the right-hand side of Eq. (11) is real, ω^2 is real, and because the term outside of the bracket is positive, ω^2 takes the sign of $[k^2 R_0^2 + n^2 - 1]$. Because n is an integer or zero, ω^2 is always positive when n is a nonzero integer, resulting in a stable oscillatory behavior. When $n = 0$, oscillations with growth or decay can result when $k^2 R_0^2 - 1 < 0$. For $kR_0 > 1$, neutral oscillations can result. Note that $n = 0$ occurs for axisymmetric behavior. Thus, only axisymmetric modes can grow in amplitude. Note that the condition for growth is that $kR_0 < 1$, which implies that the wavelength $\lambda = 2\pi/k$ is longer than the circumference of the jet. The nonaxisymmetric modes with $|n| \geq 1$ do not grow or decay. The maximum growth rate occurs when the wavelength to undisturbed radius ratio is $\lambda/R_0 = 9.016$. The axisymmetric linear-theoretical behavior results in a dilational or varicose waviness of the interface. The amplitude growth continues until the stream is severed at positions one wavelength apart. Eventually one spherical droplet forms for each original wavelength along the cylindrical liquid. The droplet radius is given by

$$R_{\text{drop}} = R_0 \left(\frac{3\pi}{2kR_0} \right)^{1/3} = R_0 \left(\frac{3\lambda}{4R_0} \right)^{1/3} \tag{12}$$

Rayleigh showed that the perturbations of surface area and of surface energy during temporal instability take the sign of

$$k^2 R_0^2 + n^2 - 1$$

which is the factor that emerged from Eq. (11). Thus when $|n| \geq 1$, surface energy would increase. Because this is not allowed by the energy conservation principle, nonaxisymmetric modes do not occur. Only with $n = 0$ (axisymmetric behavior) and $kR_0 < 1$ (wavelength longer than circumference) is a decreasing surface energy solution obtained. This implies that a more stable configuration is being found. The modal analysis and the surface energy analysis therefore lead to identical conclusions.

The surface energy per unit length for a round jet of radius R is $2\pi R\sigma$ while the kinetic energy per unit length is $\rho_l \pi R^2 V^2/2$. For stream breakup to occur, the kinetic energy must exceed the surface energy; this implies that $We = \rho_l V^2 R/\sigma > 4$. For a sheet of flowing liquid, breakup according to this simplistic concept requires that again $We > 4$, but now the Weber number is based upon the sheet thickness. The same concept applied to spherical droplets yields $We > 6$ with the droplet radius as the characteristic length.

B.   Surface Energy

For hydrostatic capillary phenomena in a mechanical system in equilibrium, the total system energy is unvaried under arbitrary virtual displacements that are consistent with the constraints of that system. The surface energy of the system is modified by the pressure work done on the surface. Consider the pressure in the liquid at the surface to be

$$p = p_a + \sigma\left(\frac{1}{R_1} + \frac{1}{R_2}\right) \tag{13}$$

where p_a is the gas pressure at the interface. If δz is a displacement locally normal to the surface (that is positive in the direction toward the liquid) and dA is a local increment of surface area, then the work done on the liquid is given by

$$\delta W = \int_A p_a \delta z\, dA + \int_A \left(\frac{1}{R_1} + \frac{1}{R_2}\right)\delta z\, dA \tag{14}$$

The element of surface is given by

$$dA = ds_1 \cdot ds_2 = R_1 d\phi_1 \cdot R_2 d\phi_2 \tag{15}$$

where ds_1 and ds_2 are infinitesimal arc lengths and $d\phi_1$ and $d\phi_2$ are the corresponding angle increments. The surface displacement will scale as follows:

$$\frac{\delta z}{R_1} = -\frac{\delta(ds_1)}{ds_1}; \quad \frac{\delta z}{R_2} = -\frac{\delta(ds_2)}{ds_2} \tag{16}$$

where $\delta(ds_1)$ and $\delta(ds_2)$ are the changes in the infinitesimal arc lengths due to the normal displacement δz. As a result by neglecting higher order terms,

$$\delta(dA) = \delta z\, dA\left(\frac{1}{R_1} + \frac{1}{R_2}\right) \tag{17}$$

Now we have

$$\delta W = \int_A p_a \delta z \, dA + \sigma \int_A \delta(dA)$$

$$= \int_A p_a \delta z \, dA + \sigma \, (\Delta A) \qquad (18)$$

It follows that $\sigma \Delta A$ is the surface energy. The analysis based on virtual displacements is valid for both hydrodynamic and hydrostatic cases because the interface has zero thickness mass and kinetic energy. It is noteworthy here that several recent analyses of the final droplet diameter and droplet velocity joint distributions based on the maximum entropy principle assume certain constraints on the surface energy. Some workers[6-9] assume that the surface energy is conserved so that the original liquid stream and the final droplet collection have the same surface energy. Another group[10-13] assumes that the sum of kinetic energy and surface energy is conserved. As will be shown, these constraints do not apply to the simplest configuration that we understand, namely the Rayleigh capillary breakup mechanism. Therefore, we have no reason to accept these constraints for more complex situations. In the Rayleigh breakup, the velocities of the jet stream and of the droplets are identical, so that the kinetic energy is constant there. (No aerodynamic interaction with the gas is considered.)

As the surface of a jet distorts to form droplets, the surface area will decrease during the process. For example, consider a sinusoidal disturbance of the surface of an initially cylindrical jet of radius R_0. As a higher order effect, the mean radius R_m of the disturbed jet will be smaller than the original radius R_0. If the local radius of the surface of the round jet is given by $R = R_m + \varepsilon \cos(2\pi x/\lambda)$, it follows that to conserve volume $R_0^2 = R_m^2 + \varepsilon^2/2$. Therefore, $R_m < R_0$ if $|\varepsilon| > 0$. The surface area of the disturbed liquid is $S = 2\pi R_m \lambda$ for a segment that extends one wavelength. This area is smaller by a factor of R_m/R_0 than an area segment of the same length for the undisturbed liquid surface. As the amplitude ε increases, the disturbed surface area becomes still smaller. This can be counterintuitive because, if the change in mean radius is not considered, one might expect increased surface area. The decrease in surface area results in a proportional decrease in surface energy.

Equations (14–18) show that as the surface area decreases, work is done that converts surface energy into some other form. As the surface moves, the liquid acquires a kinetic energy of oscillation that is distinct from the translational kinetic energy. The presence of viscous dissipation will cause the oscillation to cease. The dissipation will slightly increase the liquid temperature by a negligible amount. Although the prescribed kinetic energy and viscous-dissipative heating are insignificant compared to the internal energy, they are not negligible compared with the surface energy. For example, if we consider the fastest growing temporal instability for the Rayleigh mechanism, the wavelength-to-initial-radius ratio of 9.016 together with Eq. (12) indicates that the surface area (and surface energy) ratio between the final drop and the initial cylinder is

$$\frac{S_{\text{drop}}}{S_0} = 2\left(\frac{R_{\text{drop}}}{R_0}\right)^2\left(\frac{\lambda}{R_0}\right)^{-1} = 0.7933 \tag{19}$$

This means that about 21% of the original surface energy changed to kinetic energy of oscillatory motion and then dissipative heat. For ideal inviscid fluids, energy dissipation is absent, and the droplets resulting from stream disintegration would be subject to continuous oscillation/distortion. Note that Eq. (13) can be used to demonstrate that the pressure difference between the ambient value and the liquid volume increases by a factor of

$$\frac{2R_0}{R_{\text{drop}}} = 2\left(\frac{4}{3}\frac{R_0}{\lambda}\right)^{1/3} \tag{20}$$

For the fastest growing Rayleigh wavelength, this factor is 1.058. This pressure increase indicates a slight enthalpy increase that can be significant for the energy balance in the flowing system in the spatially developing instability. For example, a cylindrical jet flowing at velocity u_0 has a surface energy flux of $2\pi\sigma R_0 u_0$ and a pressure work term exactly one-half of that value. Thus, this 6% increase in the pressure contribution to the enthalpy requires only a 3% change in the surface energy. We expect, therefore, viscous dissipation to be a dominant factor for the spatially developing instability as well as for the temporal instability.

When aerodynamic effects are present, the kinetic energy associated with the relative motion of the surrounding gas can be transferred to the liquid. In this case, the resulting spray can have more energy than the original liquid jet. In the case in which the gas stream causes the liquid stream to disintegrate into many droplets with diameters smaller than the original transverse dimension of the liquid stream, the surface energy per unit volume of droplets will be much greater than that value for the initial stream. In the case in which the entrainment of the liquid and of the droplets by the gas stream tends to accelerate the liquid, an increase in liquid kinetic energy results. These transfers of energy from the gas to the liquid are not possible in the capillary limit, in which gas density is neglected. In the aerodynamic case in which gas density is finite, meaningful energy conservation principles can be written only for the liquid and gas as a combined system.

C. Spatial Stability Analysis

Consider now a round jet flowing at mean velocity u_0 from an orifice. Keller et al.[14] found the dispersion relation

$$(\omega + ku_0)^2 = \frac{\sigma}{\rho_l R_0^3}\frac{kR_0 I_n'(kR_0)}{I_n(kR_0)}(k^2R_0^2 + n^2 - 1) \tag{21}$$

from a three-dimensional inviscid analysis. Here, we consider ω to be real and to be imposed through modulation at the orifice exit. The wave number k can be complex, whereby the imaginary part implies exponential growth or decay. When the wave number n (which is only real and represents azimuthal waves) takes a nonzero integer value, k cannot become complex. Thus asymmetric modes are neutrally stable. For axisymmetric behavior (i.e., $n = 0$), neutral stability still occurs when the wavelength is shorter than the circumference ($k^2 R_0^2 - 1 > 0$), whereas exponential behavior occurs when the wavelength exceeds the circumference ($k^2 R_0^2 - 1 < 0$). The latter observation can also be made by employing simplified reduced one-dimensional model equations.[15]

The Keller et al. dispersion relation yields four values of k, one of which shows a long wavelength oscillation with exponential growth of amplitude. However, that was never found experimentally. This inconsistency was later explained by Bogy,[16] who introduced the use of group velocities to the linear analysis of capillary instability problems on a one-dimensional inviscid jet.

It is well known that, for conservation of wave number in multidimensional unsteady cases,

$$\frac{\partial k_i}{\partial t} + \frac{\partial \omega}{\partial x_i} = 0 \tag{22}$$

When $\omega = \omega(k_x, k_y, k_z) = \omega(\boldsymbol{k})$ as given by the dispersion relationship, we have

$$\frac{\partial k_i}{\partial t} + \frac{\partial \omega}{\partial k_j}\frac{\partial k_j}{\partial x_i} = \frac{\partial k_i}{\partial t} + C_j(\boldsymbol{k})\frac{\partial k_j}{\partial x_i} = \frac{\partial k_i}{\partial t} + C_j(\boldsymbol{k})\frac{\partial k_i}{\partial x_j} = 0 \tag{23}$$

where $C_i(\boldsymbol{k})$ is the group velocity that can be derived by differentiation from a dispersion relation such as Eq. (21). Because k_i is the gradient of a phase α, it follows that

$$\frac{\partial k_j}{\partial x_i} = \frac{\partial k_i}{\partial x_j} = \frac{\partial^2 \alpha}{\partial x_i \partial x_j}$$

In a one-dimensional situation, the wave number k becomes a scalar, and the gradient becomes a spatial derivative. For a linear system, the group velocity gives the rate at which information/energy is propagated, i.e., group velocity and energy propagation velocity are identical.[17] Each of the four branches for k has its own group velocity. When C is positive, information for that branch is propagated in the downstream direction, so that a boundary condition is required at $x = 0$ to provide that information. A negative value of C means that information is propagated in the upstream direction. That branch requires a boundary condition at infinity. As a consequence, Bogy showed that all four boundary conditions should not be placed at the orifice exit as had been done by previous investigators. An unstable mode with non-zero-valued group velocity is named a convective instability, whereas an unstable mode with zero-valued group velocity is an absolute or nonconvective instability. With an absolute instability, the growth is confined to some fixed region where the group velocity is zero.

It can be assumed that no energy is added to the stream oscillation at infinity. All modulation occurs at the nozzle exit, $x = 0$. The branch that yields exponential growth in the downstream direction has negative group velocity so that in the absence of energy addition or modulation downstream, it is never excited. This explains why the exponential growth term predicted by theory is not observed experimentally. Following this observation, Pimbley[18] and Bogy[16,19,20] correctly analyzed the jet problem for large Weber numbers. However, the authors failed to notice the absolute instability later found by Leib and Goldstein[21] for small jet velocities. The latter authors also showed that neutral waves with upstream (i.e., negative) group velocity can be found on the jet even downstream from their source (i.e., the nozzle exit). Mehring and Sirignano have also made the similar observation for spatially developing annular liquid sheets.[22]

D.   Nonlinear Effects

The nonlinear terms in the governing equations for the round jet have profound effects. They predict the deviation from sinusoidal wave distortion and the subsequent deviation from a monodisperse spray. For example, satellite droplet formation has been explained by a second-order modulation amplitude perturbation analysis.

The simplest way to explore nonlinear effects in jet breakup is to extend Rayleigh's linear analysis to higher order in the perturbation amplitude ε. Such an expansion was attempted by Yuen,[23] who used the method of strained coordinates in a third-order perturbation expansion. Yuen's analysis provided some indication of the nonuniform capillary breakup of an inviscid liquid jet and also predicted a shift of the stability limit to larger wave numbers than observed within linear theory (i.e., $k = 1/R_0$). Nayfeh[24] showed the invalidity of Yuen's result near the stability limit, $k = R_0$. Using the method of multiple scales, he obtained two second-order expansions: one valid away from $k = 1/R_0$ and one valid near $k = 1/R_0$. The latter has an even larger instability region than predicted by Yuen with a cutoff wave number of $k = [1 + 3\varepsilon^2/(4R_0^2)]/R_0$, vs $k = [1 + 9\varepsilon^2/(16R_0^2)]/R_0$ given by Yuen,[23] where ε denotes the amplitude of the imposed disturbance of wavelength k. The validity of Nayfeh's result was demonstrated by Eggers,[25] who compared Nayfeh's solution with those obtained from nonlinear numerical simulations of the one-dimensional lubrication equations. Using the method of strained coordinates, Chaudhary and Redekopp[26] developed a third-order solution for an infinite jet with initial velocity disturbance consisting of a fundamental and one harmonic component. Chaudhary and Maxworthy[27,28] conducted experiments on the breakup of a discharging jet with axial velocity modulations at the nozzle exit and obtained satisfactory agreement with the theoretical results by Chaudhary and Redekopp.[26]

Pimbley and Lee[29] performed a second-order analysis with nozzle boundary conditions that were later recognized as inappropriate by Bogy.[16] They also conducted an experiment. Some significant differences from the temporal analysis were found. In the temporal case, symmetry existed around the position of maximum amplitude. This symmetry is no longer present in the spatial stability case. The relationship between a large droplet and the satellite droplet immediately downstream of it will, in general, be different from the relationship with the satellite droplet immediately upstream. In the flowing spatially unstable round jet cases, Pimbley and Lee found that the nonlinear effects produce higher harmonics in the diameter as a function of downstream position and time. The second harmonic amplitude is typically smaller than the fundamental amplitude but causes two local maxima per wavelength and leads to a satellite droplet. Thus, every wavelength produces one parent droplet and one satellite droplet in this theory.

Various authors addressed shortcomings of the linear and nonlinear one-dimensional models by Bogy[19] and Pimbley and Lee.[29] Using a formal perturbation expansion, Schulkes[30] showed that Lee's inviscid one-dimensional equations[31] are inconsistent because terms that have been retained in the boundary conditions should have been rejected according to the approximations made for the momentum equations.

A systematic derivation of inviscid and viscous one-dimensional equations starting from the full Navier–Stokes equations was performed by Bechtel et al.,[32] Eggers and Dupont,[33] and Garcia and Castellanos.[34]

The discussion on self-consistent higher order one-dimensional models was continued by Bechtel et al.,[15] identifying Eggers' "regularized" model applied in Refs. 33 and 35–37 as consistent with exception of the surface tension/curvature term, which appears, to all orders, coupled to the leading-order approximations of all other physical effects. On the other hand, the same authors note that the models of Refs. 32, 38, and 39 for the same physical geometry maintain the slenderness approximation consistently in all terms. Based on this understanding, the authors produced one-dimensional nonlinear slender Newtonian jet equations for leading-order behavior and higher order corrections, incorporating the effects of surface tension, inertia, viscosity, and gravity. The models by Garcia and Castellanos[34] and Bechtel and coworkers[15,32] were contrasted with each other by Eggers.[25] Using essentially a Galerkin approximation of the equations of motion, Eggers[25] illustrated the appearance of the full curvature term from surface-tension forces even at leading order and providing a consistent strong argument for the "ad-hoc" consideration of the full curvature term in otherwise lower order one-dimensional models. As Eggers notes, a consistent leading-order system (i.e., without full curvature term) is unstable against short wavelength "noise," making it ill-suited for numerical simulations. Because the leading-order equations are unaffected by higher order terms, this problem remains even if higher order perturbation terms are included.[25]

The use of the slender-jet approximation, employed by many authors to predict nonlinear jet distortion away from breakup and until breakup, will lead to increasing errors as contributions of short wavelength disturbances increase during jet pinch-off. Even more severe limitations apply to inviscid one-dimensional or slender-jet models, for which the development of finite time-singularities is observed even before jet breakup. For a more detailed discussion on the limitations of one-dimensional models, we refer to Schulkes[40] and Eggers.[25]

The capillary temporal stability of a round viscous jet in a vacuum has been studied by Ashgriz and Mashayek[41] by a Galerkin finite element numerical solution of the Navier–Stokes equations together with the free-surface conditions. A Reynolds number is defined to be

$$Re = \frac{1}{\nu}\left(\frac{\sigma R}{\rho_l}\right)^{1/2} \tag{24}$$

which is actually a reciprocal Ohnesorge number. With $Re \leq O(1)$, imposed modulations are overdamped and no oscillations occur. Higher Re values result in oscillations. Above a certain value of Re dependent on the wave number, satellite droplets are formed with satellite size increasing with Re; clearly, viscosity is found to inhibit satellite formation and, for sufficiently large viscosity, any droplet formation is inhibited. As the modulation amplitude is increased, the threshold Re value for satellite formation increases and satellite size decreases. The results confirm that the second-harmonic component is responsible for the satellite formation.

Ashgriz and Mashayek[41] found that growth rates agree well with linear theory for low values of Re, but linear theory overpredicts the growth rates at higher Re values. Breakup time decreases exponentially with increasing amplitude. The cutoff wave number value (below which the jet is unstable) increases with initial amplitude of the oscillation.

Heister and coworkers[42-45] and Lundgren and Mansour[46] have performed an interesting set of round-jet calculations employing a boundary element method (BEM). Hilbing and Heister[44] showed that variations in Weber number, modulation frequency (and thereby wavelength), and amplitude can be used to control the eventual droplet size (see Fig. 3). Satellite droplet velocities are found to be less than the main droplet velocity, indicating the possibility of recombination. Only inviscid axisymmetric behavior was considered with the neglect of aerodynamic interactions. Mansour and Lundgren studied the Rayleigh capillary instability; neglecting viscosity and air inertia, satellite droplets were predicted at all unstable wave numbers. Comparison of the nonlinear calculations with classical linear analysis indicated that the linear treatment predicts well the early jet deformation, although the formation of the satellites is not predicted. The technique represents the flow by singular dipole solutions of Laplace's equation distributed over the surface. The authors claim that this is equivalent to distributed vortex rings of varying strength determined by local surface distortion.

Spangler et al.[42] extended the work of Lundgren and Mansour to consider the inertia of the surrounding air. Now the Kelvin–Helmholtz and capillary instability mechanisms can appear in an integrated fashion. Aerodynamic interactions were important even at low relative velocities, and the nonlinear effects were significant for initial deflections as small as 1% of the undisturbed radius. Transition from the Rayleigh regime to the first wind-induced regime and then to the second wind-induced domain occurred approximately at gas-phase Weber numbers of 1 and 2, respectively. Nonlinear and aerodynamic effects tended to cause broad troughs and narrow peaks in the waveform with two points of minimum radius for each wavelength. One satellite drop for each main drop was formed in the first wind-induced regime. At higher Weber numbers in the second wind-induced regime, a spiked peak occurs in the waveform with the pinching off of

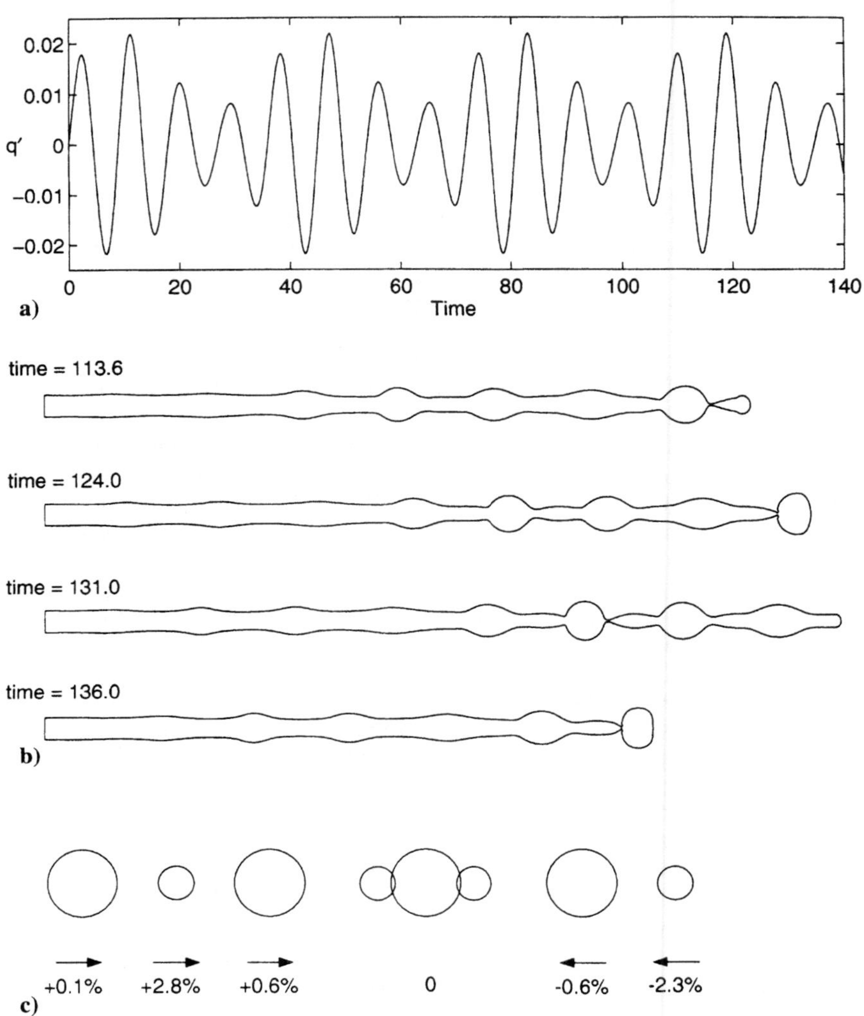

Fig. 3 Jet breakup with an amplitude-modulated inflow velocity: a) inflow disturbance waveform, b) jet profiles at four successive pinch events, and c) relative droplet velocities.[43]

a fluid ring at the peak location for each wavelength. This indicates that small droplets will be formed when this ring of fluid disintegrates. Of course, axisymmetric calculations cannot capture this breakup.

Hilbing and Heister[45] analyzed the unsteady nonlinear evolution of a high-speed viscous liquid jet issuing from a circular orifice. The authors employed an integral method for a thin viscous region at the jet periphery, while a boundary-element method was used for the inviscid "core" flow. Under certain conditions, jet "swelling" was observed, and the boundary layer thinned out to a shear layer over a length of about half an orifice radius.

E. Viscous Effects

Weber[47] considered the viscous effects in a one-dimensional temporal instability analysis of the axisymmetric case. His results yielded that the growth rate is given by

$$\alpha = -\frac{3}{2}\frac{\mu}{\rho_l}k^2 \pm \left[\left(\frac{3}{2}\frac{\mu}{\rho_l}k^2\right) + \frac{\sigma R_0^2 k^2}{2\rho_l R_0^3}(1 - k^2 R_0^2)\right]^{1/2} \tag{25}$$

When the viscosity $\mu = 0$, this agrees with the result of Bogy's one-dimensional analysis[16] in the temporal limit where $u_0 = 0$. Nondimensionalization of Eq. (25) yields an equation containing the Ohnesorge number Oh with $L = R_0$.

Viscosity tends to damp any oscillation, causing an exponential decay. The decay rate increases (because of increasing velocity gradients) as the wavelength increases with increasing viscosity. The breakup length and time also increase.

Leib and Goldstein[21] considered the differences for spatially developing instability between flat (plug-flow) velocity profiles and parabolic (Hagen-Poiseuille) velocity profiles. Their formulation, however, was inviscid, so that the viscous problem was not completely simulated. They found that the flat profiles were more unstable. The implication is that, after a flow with an initially parabolic profile exits an orifice, it becomes less stable or more unstable as the profile relaxes.

The local behavior of the liquid stream at the point in space and time where the thickness goes to zero thickness is an important matter of current research. The large fractional change in thickness results in large velocity gradients. Viscosity becomes important locally; by means of the viscous forces, singularities that can appear in the inviscid solution may not manifest.

Eggers[25] presents a thorough review of the pinch-off behavior of a round capillary jet. He presents the local characteristic time and length as

$$t_c = \rho_l^2 \eta_l^3 / \sigma^2 \tag{26}$$

and

$$l_c = \rho_l \eta_l^2 / \sigma \tag{27}$$

where η_l is the kinematic viscosity of the liquid. Nondimensionalization in terms of these quantities would aid in the development of self-similar solutions. These characteristic dimensions result from a balance of surface tension, viscous, and inertial forces. In particular, the Reynolds number and Weber number based on l_c can each be set to unity. The characteristic velocity in these two numbers can be defined as l_c/t_c. As the liquid stream comes closer to the zero thickness point, we can expect that continuum theory will become invalid and intermolecular forces will increase in importance. The final stream pinch-off can be predicted only if intermolecular forces are properly taken into account. Although Eggers' analysis[25] was developed for the round jet, the discussed concepts should be applicable to the liquid sheet as well.

F.   Spray Control via Electric Fields

Electrohydrodynamic spraying is of interest because it can produce very fine droplets. Furthermore, the droplets will be charged and become amenable to guided trajectories through the use of an applied electric field. The atomization system typically will have electrically conducting injection hardware serving as one electrode and a nearby counterelectrode with a potential difference of several thousand volts. The liquid should be an electrical conductor so that it acquires a charge in the injector and is accelerated by the applied electric field between the two electrodes. With this electric field, lower upstream pressures are required to achieve the desired mass flow and droplet size. Electrospraying has not yet been applied to practical systems with high mass flows. It is, however, an interesting candidate for future utilization in combustion applications.

This atomization process always yields highly charged droplets. If that is not desirable, it might be necessary to employ a neutralization process downstream.

At very low pressure-difference and potential-difference across the injector, a dripping mode and a cone-jet mode can appear. The cone jet is a result of a certain balance between electrostatic pressure and capillary pressure shown by Taylor.[48] When the balance is exact, a conical meniscus with no through flow results. A higher applied potential causes a jet to emerge from the apex of the cone. Thus the flow emerges from the round orifice and the freejet has two distinct portions: an approximately conical convergent flow followed by an approximately cylindrical flow. The latter region can undergo hydrodynamic and/or capillary instabilities that result in droplet formation. When the applied voltage is increased, the conical portion of the jet can deform and the meniscus can form multiple cusps that result in multiple jets. Cloupeau and Prunet-Foch[49] give a review of the dripping and cone-jet modes. Here we are interested in applications that require higher applied voltages and produce smaller droplets.

Under the influence of the higher voltage, the primary liquid jet can break into many fine jets that stream off from the axially flowing primary jet with relatively large radial-velocity components. Of course, these fine jets produce fine droplets.

Some interesting work[50,51] on charged liquid column stability is relevant to the round jet injection problem. Setiawan and Heister[50] considered an inviscid, axisymmetric electrified jet of infinite length. The gas density and, therefore, the inertia of the gas were taken to be negligible. The liquid was assumed to be a perfect electrical conductor so that the gas-liquid interface achieved a uniform electrostatic potential (voltage). Space change in the gas surrounding the jet was neglected. The electrostatic potential is set to zero at a concentric cylindrical surface (the electrode) at some distance from the jet. Now Laplace's equation describes two fields: the velocity potential ϕ in the liquid and the electrostatic potential ϕ_e in the field between the liquid surface and the surrounding cylindrical surface of the electrode. Equations (6) and (8) will also apply at the liquid surface. The unsteady Bernoulli equation (10) is also applied at the surface to relate pressure and velocity potential.

The pressure jump across the interface is due to both capillary and electrostatic effects. That is, Eq. (8) is modified to yield

$$p_1 - p = \kappa\sigma - \frac{\epsilon_o}{2}\left(\frac{\partial\phi_e}{\partial r}\right)^2 \tag{28}$$

where ϵ_o is the permittivity of free space and n is the local normal direction at the liquid surface. The capillary effect pulls the surface inward toward the liquid so that a force balance at the liquid surface requires a higher liquid pressure than external gas pressure. With a positive (negative) potential on the liquid relative to the distant electrode, the positive (negative) charges on the liquid are attracted outward toward the oppositely charged distant electrode. This outward electrostatic force reduces the pressure at the liquid surface. The electric field can be seen as a means of reducing and thereby controlling the surface tension effect.

If V_o is the voltage drop between the two electrodes, and R_o is defined as the undisturbed liquid jet radius, a meaningful nondimensional parameter $\Gamma = \epsilon_o V_o^2/R_o\sigma$ appears. This normalized electrostatic pressure Γ is sometimes called a Taylor number. $\Gamma = 0$ reduces the phenomenon to the classical Rayleigh case.

Nonlinear calculations[50] have been performed with a boundary-element method, and the time-dependent evolution has been predicted. The distortion predicted by nonlinear theory[50] is substantially different from the results found through linear theory. At Γ values of 10 and below, a modification of the nonlinear capillary instability was found where a main droplet and a satellite droplet were formed for every wavelength of the disturbance. At higher Γ values of 15 and above, it was found that droplets no longer form by pinching off at the centerline; rather, at the local regions of maximum jet width, droplets are pinched off at the periphery. Linear theory unreliably predicts that breakup time would increase with increasing Γ, whereas nonlinear theory shows that the opposite result actually occurs. Linear theory[50] shows that the electrostatic effect will increase the wave number domain to extend to larger wave numbers where the wavelength can be less than the circumference of the undisturbed jet. The upper (lower) limit of the wave number (wavelength) increases (decreases) as Γ increases. The growth rate of the disturbance at lower wave numbers will decrease as Γ increases.

Yoon et al.[51] extended the boundary-element method of Setiawan and Heister[50] to consider nonaxisymmetric distortions of a round liquid jet or column. Only variations with t, r, and θ were considered; properties remained constant as z varied. Deviations of the initial distortions of the radius from the undisturbed radius were proportional to $\cos n\theta$. Values of $n = 2$, 3, and 4 were considered. There is a minimum distance between the undisturbed liquid jet surface and the electrode below which the distortion of the surface is unstable. This minimum distance increases as the applied voltage increases. For a sufficiently large value of Γ, instability is always found for the $n = 2$ mode.

At higher values of Γ, instabilities in the $n = 3$ and $n = 4$ modes were found. At Γ values above the critical value, the liquid jet no longer oscillates but rather distortion occurs monotonically with two spikes eventually forming for an $n = 2$ case. There is a tendency for liquid to be pinched off near the spikes on the periphery. For lower values of Γ where oscillations occur, the frequency decreases with increasing values of the amplitude.

Yoon et al.[51] also examined the case in which multiple jets are formed at the region where the stream exits the origin. They noted that linear theory applied to a single jet indicates that the axisymmetric disturbance grows fastest and therefore does not predict the formation of the multiple jet. Thus a nonlinear explanation is required. The authors represented the nonlinear three-dimensional phenomena with a two-dimensional model. Using the boundary-element method, they determined the ratio of electrostatic surface energy to capillary surface energy as a function of the applied voltage for a given number n of the fine jets. The critical value where electrostatic surface energy first equals capillary surface energy does not predict well the experimental observation. The supposition that the multiple jets form when the two energies are of the same order of magnitude does not explain the observations. The observed applied voltages are an order of magnitude larger than predicted. Therefore, we still lack a model that predicts the nonlinear, time-dependent behavior of the multijet mode. Some other interesting research on electrospraying can be found in Refs. 52–55.

G.   Coaxial Jets

In many applications, two liquids are injected in concentric cylindrical streams with a round jet of the inner fluid surrounded by an annual stream of the second fluid. Although these are described as coaxial jets in the liquid-propellant-rocket literature, the scientific literature on capillary and aerodynamic jet instabilities describes them as compound jets. Coaxial jets can be formed by injection of one liquid through a round orifice that is surrounded by an annular orifice through which another liquid flows. Another method of formation involves a high momentum jet of the primary liquid penetrating a layer of another liquid and entering the ambient gas field while entraining some of the secondary liquid to form an annular sheath around the primary liquid. Generally the jets break up before substantial molecular mixing of the two liquids occurs so that the liquids are considered to be immiscible for practical purposes.

Hertz and Hermanrud[56] found experimentally with two immiscible liquids in a coaxial configuration that the inner and outer liquids can break into drops at different rates. Figure 4 shows that the inner liquid in a particular case forms droplets within the outer liquid before the outer liquid breaks into droplets. In this case, the inner jet is composed of 80% water with 20% glycerol plus dye with a surface tension of 72 dynes/cm. The outer jet is silicon fluid with a surface tension of 20 dynes/cm. The interfacial tension is 52 dynes/cm. The overall distance from the nozzle exit to the point of drop formation is about 25 mm.

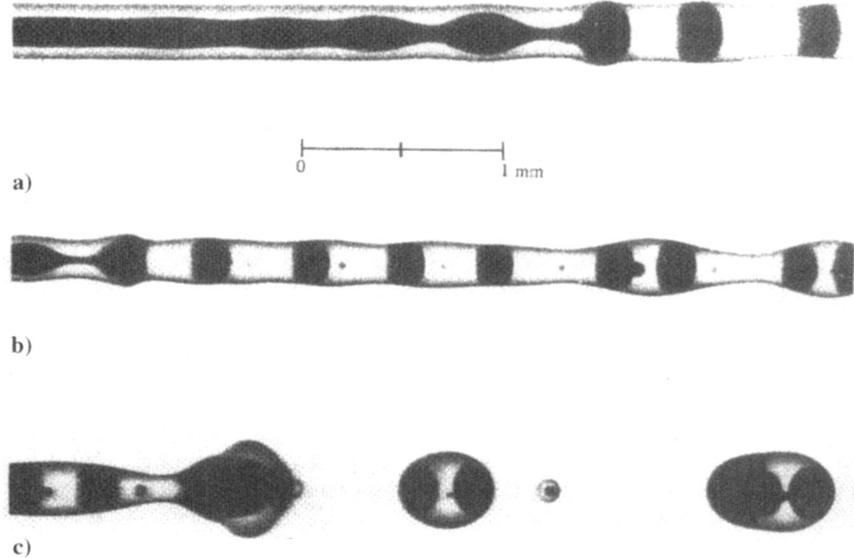

Fig. 4 Drop formation for a coaxial jet[56]: a) base region, b) mid-region, and c) region of final drop formation.

Some linear inviscid and viscous analyses of compound jets are presented in the literature. The analyses by Crapper et al.[57] and Shen and Li[58] included the effects of a gas inside and around an annular liquid flow. The surrounding gas is not included in the analyses of compound jets presented by Chauhan et al.[59] and Shkadov and Sisoev.[60] These authors analyzed the capillary instability of axisymmeteric two-layer capillary liquid layer or annular liquid jet that is surrounded by a void. Both liquids were considered to be immiscible viscous fluids. As with the analysis by Shen and Li, the flowfield within the annular second liquid layer is part of the solution and might in fact determine the stability of the overall system. The boundary-layer approximation to the full Navier–Stokes equations was applied to determine the stationary axisymmetric jet flow developing downstream from the location where initial or boundary conditions were prescribed. Here, the velocity of the two-layer jet flow (under zero gravity) was found to tend asymptotically to a uniform distribution across the jet with growing downstream distance and independently of the initial velocity profile at the jet exit.

Using the assumption of a locally parallel main flow, a linear stability analysis was performed for various cross sections along the jet, i.e., for various axial velocity profiles, by assuming a solution of the form $f(\eta)\exp[i\beta(\xi - c\tau)]$, where c and β are the nondimensional wave velocity and wave number, ξ and η denote nondimensional axial and radial coordinates, and τ is the nondimensional time coordinate. The characteristic length and velocity for the nondimensionalization are the local radius of the surface interface and the average velocity determined from the volume flow rate. Two linearly independent solutions are determined in the regions of both layers. The stability problem is defined by six parameters, i.e., the density ratio ρ_2/ρ_1, the ratio of the kinematic viscosities pertaining to the two fluids η_2/η_1, the ratio of the surface tension coefficients pertaining to the inner and outer interfaces σ_2/σ_1, the Reynolds number $Re = U_c R_2/\eta_1$, and the Weber number $We = \rho_1 U_c^2 R_2/\sigma_1$. Here, the average velocity U_c is determined from the volume flow rate and denotes the local radius of the surface interface pertaining to the steady solution. The subscripts 1 and 2 refer to the values of the inner and outer layers, or the fluid and surface interfaces, respectively. For the case of a uniform velocity profile across the jet, two types of instability disturbances are described: one relating to the interface between the layers and one relating to the jet surface. No consideration is made of the relative motion between the outer (surface) and inner (fluid) interfaces. Comparison of the growth rate predictions with those made within the compound jet analysis by Chauhan et al.[59] (see the following discussion) suggests that the surface instability is related to a sinuous deformation (stretching) of the outer layer and the interface disturbance to a dilational deformation (squeezing) of the same layer. It is noted, however, that in Ref. 59 "stretching" and "squeezing modes" are dominated by destabilizing effects of the inner and outer interfaces, respectively.

Shkadov and Sisoev[60] note that, as already observed by other authors for inviscid and viscous jets, the instability related to the jet surface dominates for large Weber numbers We. Parameter studies have been considered by varying one of the governing parameters from a base-case parameter set given by $\rho_2/\rho_1 = 0.752$, $R_1/R_2 = 0.5$, $\eta_2/\eta_1 = 0.709$, $\rho_2/\rho_1 = 0.752$, $We = 20$, $\sigma_2/\sigma_1 = 1.5$. Variations were considered within $1.33 < We < 133$, $0.1 < \sigma_2/\sigma_1 < 10$, $0.1 < R_1/R_2 < 0.9$, $100 < Re < 200$, $0.1 < \rho_2/\rho_1 < 1.4$, $0.1 < \eta_2/\eta_1 < 1.4$. For the considered We range, predominance of the surface disturbance is observed. The maximum amplitude of the unstable growth rates for both types of instability disturbances increases with increasing the σ_2/σ_1 ratio, with surface disturbances being more unstable than interface disturbances. Within the examined parameter range of R_1/R_2, the predominance of the surface instability over the interface instability is observed; for a thinner outer layer, i.e., larger R_1/R_2 ratio, the difference in the maximum growth rate becomes more significant. Also, the range of unstable surface disturbances is decreased with increasing value R_1/R_2, whereas changing R_1/R_2 from 0.1 to 0.9 does not significantly influence the stability range for unstable interface disturbances.

An increase in unstable growth rates for both types of instability is observed with increasing Reynolds number values, as well as for decreasing η_2/η_1 values, illustrating a stabilizing effect of viscosity in the considered parameter range. Also, an increase in ρ_2/ρ_1 (i.e., increasing density of the outer layer with fixed density of the inner jet) reduces the growth rates of both surface and interface instabilities. Similarly to Crapper et al.,[57] the wave velocity for both disturbances is given to be the same as the jet velocity.

Shkadov and Sisoev[60] also analyzed the stability of compound jets with nonuniform velocity profile of the steady-state solution, which allowed the analysis of the instability mechanism related to the discontinuity of the velocity gradient at the interface between the layers and its relative importance with respect to the prescribed capillary instability. For the case of a constant axial velocity across the jet (as discussed here), the authors report agreement with results found in other papers in the Russian literature.

In Ref. 59, the authors presented both linear temporal and linear spatial stability analyses, the latter addressing both convective and absolute instabilities. The compound jet was assumed to be inviscid and moving in a vacuum. Because capillary instabilities were the focus of the analysis, the jet velocity was assumed to be constant with the same value in both liquids. As noted by the authors, no slip at the liquid interface prevents the Kelvin–Helmholtz type of instability and allows focus on the capillary driven instability phenomena. The linear spatial and temporal stability analyses were conducted by employing a Fourier–Laplace transform of the problem and investigating the transform inversions by considering the poles in the s-plane and the corresponding poles in the k-plane. Here, s and k denote the transformed time and space coordinates. The temporal analysis showed that the dispersion relation is found to be quadratic in s^2, so that for each wave number k there are four roots s whereby two of them are the negative of the other two. It was shown that s^2 must be real.

For all the modes as $k \to 0$, s was found to be linear in k, so that the growth rate of both modes goes to zero as the disturbance wavelength becomes very large. One unstable mode (referred to as the stretching mode) is unstable for all $k < 1/R_1$, whereby the interfaces are in phase and the other mode (later referred to as squeezing mode) is unstable for all $k < 1/R_2$ with the interfaces moving out of phase; both modes are purely imaginary beyond their cutoff. In the stable region, the movement of the interfaces with respect to each other is just inverse for the two modes; for the mode that grows for $k < 1/R_1$, the interfaces are out of phase for the stable case, but in phase for the mode that grows for $k < 1/R_2$. As before, R_1 and R_2 refer to the annular radii of the inner and outer interface, respectively. The stretching mode is reported to always dominate the squeezing mode. (This has been illustrated for thin films and also for the case $\rho_2/\rho_1 = 1$, $\sigma_2/\sigma_1 = 2$, and $R_2/R_1 = 2$.) The relative importance of the two modes (causing film rupture or core rupture) with respect to the initial conditions was investigated.

The effects of changes in σ_2/σ_1, ρ_2/ρ_1, and R_2/R_1 have also been described. For the squeezing mode the growth rate increases with increasing values for σ_2/σ_1, whereas the growth rate for the stretching mode increases for $k < 1/R_2$ but decreases for $k > 1/R_2$. Recall that the stability limit for the stretching mode is $1/R_1$ and $R_1 < R_2$. As reported, an increase in the density of the outer layer leads to a decrease of the unstable growth rate for both modes. Also, a decrease in the thickness of the outer layer leads to an increase in the maximum growth rate for the squeezing mode. This same behavior is found for the stretching mode if the sheet is thin enough. Note that Chauhan et al.[59] and Shkadov and Sisoev[60] present results only in a narrow parameter band. The base case analyzed by Chauhan et al. is given by $\sigma_2/\sigma_1 = 2$, $\rho_2/\rho_1 = 1$, and $R_2/R_1 = 2$. Parameter variations for σ_2/σ_1 and R_2/R_1 ranged (for the temporal case) from $0.5 < \sigma_2/\sigma_1 < 10$ and $1 < R_2/R_1 < 2, 10$. No results were illustrated for $\rho_2/\rho_1 = 1$.

To analyze the case of a thin annular second layer, Chauhan et al.[59] approximated R_2/R_1 by $1 + \varepsilon$, which provides the growth rates $s = s_0\varepsilon^0$ or $s = s_{1/2}\varepsilon^{1/2}$ for the stretching or squeezing mode to the lowest order in ε, with

$$s_{1/2} = \pm\sqrt{\frac{k^2(1 - k^2)}{\rho_2/\rho_1}\frac{\sigma_2/\sigma_1}{1 + \sigma_2/\sigma_1}} \tag{29}$$

$$s_0 = \pm\sqrt{(1 + \sigma_2/\sigma_1)(1 - k^2)k\frac{I_1(k)}{I_0(k)}} \tag{30}$$

Here, wavelike solutions of the form e^{ikz+st} were assumed. Note that the dimensional growth rate s^* is given by $s^* = \sqrt{(\sigma_1/\rho_1 R_1^3)}s$. The magnitude of pressure and velocity components in both liquid film and liquid core are described for both modes in terms of their order in ε. For the squeezing mode, the solution for the second (thin) fluid layer is independent of the core. For the stretching mode, the liquid core behaves like a single jet with an effective surface tension coefficient of $\sigma = \sigma_1 + \sigma_2$. For both modes the pressure does not vary in the radial direction.

The spatial analysis presented by Chauhan et al.[59] follows the same strategy as their temporal analysis. Spatial stability is analyzed by investigating the movement of poles in the k-plane (transformed space coordinate), with respect to the contour chosen for the Fourier inversion, due to motions in the s-plane (transformed time coordinate) along the contour chosen for the inversion of the Laplace integral (i.e., Bromwich contour). Hereby, absolute instability of the sheet is identified as the merging of roots from inside and outside the Fourier inversion contour on the contour itself as σ is varied along the contour chosen for the Laplace inversion. Only jet stability is analyzed and the back-transformation of the problem into physical space is made only in principle. No solution is presented for specific initial conditions. Compound jet stability due to harmonic forcing at the source is analyzed for a particular set of parameters (i.e., $\rho_2/\rho_1 = 1$, $\sigma_2/\sigma_1 = 2$, $R_2/R_1 = 2$, $We = 4$).

Two potentially unstable modes are found. The modes are unstable below critical Strouhal numbers of about 1 or R_1/R_2, whereby the exact values depend, as stated, on the Weber number. The authors note that Leib and Goldstein[21] found similar results for the single inviscid jet. The axial growth rates of both modes increases with a decrease in jet velocity. At high jet velocities, the two modes become the stretching and squeezing mode of the temporal analysis. The critical value below which the jet becomes absolutely unstable is different for both modes. Good agreement between theoretical predictions of the spatial analysis with experimental observations is demonstrated. For thin sheets the surface instability described by Shkadov and Sisoev[60] corresponds to the stretching mode identified by Chauhan et al.[59] The latter authors showed that the stretching mode (i.e., sinuous deformation of the outer liquid layer or film) is the dominant instability for a compound jet in the case of a thin outer layer or film. The instability in this case is determined by single jet behavior of the inner fluid with surface tension given by the sum of the surface tension of the two fluids. This behavior is very different from the one observed for a thin annular liquid sheet with constant gas core pressure considered here and previously analyzed by Dumbleton and Hermans.[51] In particular, the sinuous distorting annular films analyzed here are characterized by a linear pressure profile across the sheet. However, for the stretching (and squeezing) mode analyzed by Chauhan et al.,[59] the pressure profile across both fluids is constant. Constant pressure across the compound jet has also been assumed within the one-dimensional linear stability analysis of this configuration presented by Sanz and Meseguer.[62] Shkadov and Sisoev[60] note that the analysis by Sanz and Meseguer corresponds to their own analysis in the inviscid limit and for the case of a constant axial velocity profile across the sheet.

Note that neither of the two described analyses of the compound jet[59,60] apply in the limit of large values for the ratio between the densities of the outer and inner fluid. That limit would lead to a configuration similar to that analyzed by Crapper et al.[57] and Shen and Li.[58] Shkadov and Sisoev[60] do consider slight variations in the density ratio of the two viscous fluid layers, i.e., between 0.752 and 1.4, whereas the results illustrated by Chauhan et al.[59] for two inviscid layers refer to the case with density ratio 1 only.

The maximum amplitude of the unstable growth rates for both types of instability disturbances increases with increasing the σ_2/σ_1 ratio, with surface disturbances being more unstable than interface disturbances. Within the examined parameter range of the R_1/R_2 ratio, the difference in the maximum growth rate becomes more significant. Also, the range of unstable surface disturbances is decreased with increasing value R_1/R_2, the predominance of the surface instability over the interface instability is observed. For a thinner outer layer, i.e., larger R_1/R_2 ratio, the difference in the maximum growth rate becomes more significant.

IV. Planar Sheet Analyses

The dynamics of sheets of fluid was studied as long ago as 1833 by Savart,[63] who produced and analyzed bell-like or flat axisymmetric sheets, created by a disk-shaped obstruction in the path of a cylindrical water jet, or by impingement of two jets. Dorman,[64] Fraser and Eisenklam,[65] and later Dombrowski and Fraser[66] were the first to describe the breakup and drop formation of plane fan sheets. Three modes of sheet disintegration, referred to as rim, wave, and perforated-sheet disintegration, were identified. In the rim mode, forces created by surface tension cause the free edge of a liquid sheet to contract into a thick rim, which then breaks up by a mechanism corresponding to the disintegration of a freejet. This mode of disintegration is most prominent where the viscosity and surface tension of the liquid are both high. In the perforated-sheet disintegration mode, holes appear in the sheet and are delineated by rims formed from the liquid that was initially included inside. These holes grow rapidly in size until the rims of adjacent holes coalesce to produce ligaments of irregular shape that finally break up into drops of varying size. Disintegration through the generation of wave motion on the sheet and in the absence of perforations is referred to as wave disintegration, whereby areas of the sheet, corresponding to half or full wavelengths of the oscillation, are torn away before the leading edge is reached. The relative importance of the different modes can greatly influence both the mean drop size and the drop size distribution.[2] Rim sheet disintegration is found to be important at low relative velocities, whereas wave sheet disintegration due to aerodynamic interaction with the surrounding gas dominates at most injection pressures used in practice. Our discussions will focus on the wave disintegration mode.

A. Linear Theory

Hagerty and Shea[67] and Squire[68] first considered temporal (spatially periodic) behavior on an infinite liquid sheet at low gas-to-liquid-density ratio. Rangel and Sirignano[69] and Sirignano and Mehring[4] extended the analysis to high density ratio. In the limit of large radius r ($r \to \infty$), Eqs. (1–9) yield the governing equations for a planar liquid sheet (fluid 1) in a surrounding ambient gas (fluid 2 = fluid 3) in Cartesian coordinates. Considering two-dimensional disturbances only (i.e., $\partial .../\partial\theta = 0$) and neglecting the effects of viscosity and gravity in both the liquid and the gas phase, linearized treatment[4] yields the velocity potential:

$$\phi = (A \cosh ky + B \sinh ky)e^{i(kx-\omega t)} \tag{31}$$

for the liquid sheet of thickness $2a$, where $A = 0$ for the antisymmetric mode and $B = 0$ for the symmetric mode, and where y and x denote the directions perpendicular and parallel to the main flow direction of the sheet, respectively. Furthermore,

$$\phi = A_1 e^{\pm ky} e^{i(kx-\omega t)} \tag{32}$$

where A_1 assumes a different value for the gas region $y < -h$ below the liquid than for the gas region $y > h$ above the liquid. The plus sign is used in the exponent if $y < -h$ while the minus sign applies if $y > h$.

The dispersion relation can be determined using the above relations plus the interface conditions:

$$\frac{\omega}{ku_0} = \frac{1}{2}\frac{\rho - C}{\rho + C} \pm \frac{[2\pi/We_1(C + \rho) - C\rho]^{1/2}}{\rho + C}$$

$$= \frac{1}{2}\frac{\rho - C}{\rho + C} \pm \frac{[kh/We_2(C + \rho) - C\rho]^{1/2}}{\rho + C} \tag{33}$$

with $We_1 = \rho_l u_0^2 \lambda/\sigma$ and $We_2 = \rho_l u_0^2 h/\sigma$. $C = \coth kh$ for the symmetric mode, and $C = \tanh kh$ for the antisymmetric mode. Here u_0 is the relative velocity between the liquid and the gas, ω will have an imaginary part for sufficiently low σ, sufficiently high $\rho = \rho_g/\rho_l$, and sufficiently high relative velocity. When $\sigma = 0$, we have a pure Kelvin–Helmholtz behavior, while for $u_0 = 0$, or $\rho = \rho_g/\rho_l = 0$, we have pure capillary waves.

The two solutions obtained from Eq. (33) for ω/k denote two different waves traveling in opposite directions along the sheet. From Eq. (31), note the fluid sheet is unstable if ω/k in Eq. (33) becomes complex, i.e., when the Weber number We_1 or We_2 exceeds its critical value:

$$We_{1,c} = 2\pi\left[\frac{1}{\rho} + \frac{1}{C}\right], \quad We_{2,c} = kh\left[\frac{1}{\rho} + \frac{1}{C}\right] \tag{34}$$

The growth rate of an initial disturbance is given by the imaginary part of the disturbance frequency ω provided by Eq. (33), i.e.,

$$\text{Im}[\omega/(ku_0)] = \pm\sqrt{\frac{C\rho}{(C+\rho)^2} - \frac{kh}{We_2}\frac{1}{C+\rho}} \tag{35}$$

where, according to Eq. (31), the $+$ sign denotes the solution with unstable behavior and Im means "the imaginary part of." The results obtained from the prescribed linear analysis can be summarized as follows:

1) Only two "principal" modes of sheet distortion, i.e., sinuous or varicose (dilational) modes, develop on planar sheets. The varicose waves are dispersive while the sinuous waves are generally dispersive but become nondispersive, as kh and ρ both tend toward zero.

2) The stability and growth rates of any sinuous or varicose disturbance depend on the Weber number $We_2 = \rho_l u_0^2 h/\sigma$ and the density ratio $\rho = \rho_g/\rho_l$. See the results for growth rate in Fig. 5.

3) For all density ratios, the growth rates for both sinuous and dilational waves increases as the Weber number We_2 is increased. The maximum growth rate for the sinuous disturbances does not significantly change with changes in the density ratio ρ. However, the maximum growth rate for the dilational case increases significantly as ρ is increased. For low-density ratios, the maximum growth rate for the sinuous case is always higher than for dilational waves. As ρ is increased beyond a certain value, the maximum growth rate for dilational waves eventually overcomes the value for sinuous growth.

4) For all density ratios, there exists a region of wave numbers, in which dilational waves are more unstable than the sinuous ones; the latter might even be stable in that region.

5) The disturbance wavelength at the maximum growth rate decreases as the density ratio is increased for either sinuous or dilational waves.

6) For intermediate Weber numbers and small thickness-to-wavelength ratios, sinuous waves are more unstable than dilational ones with the instability occurring at Weber numbers larger than unity and with the most unstable wave number given by

$$k \approx \rho_g u_0^2/2\sigma$$

with a growth rate of

$$\omega_{\text{max}} = \frac{1}{2}\frac{\rho_g}{\rho_l}\frac{u_0}{h}We^{1/2} \tag{36}$$

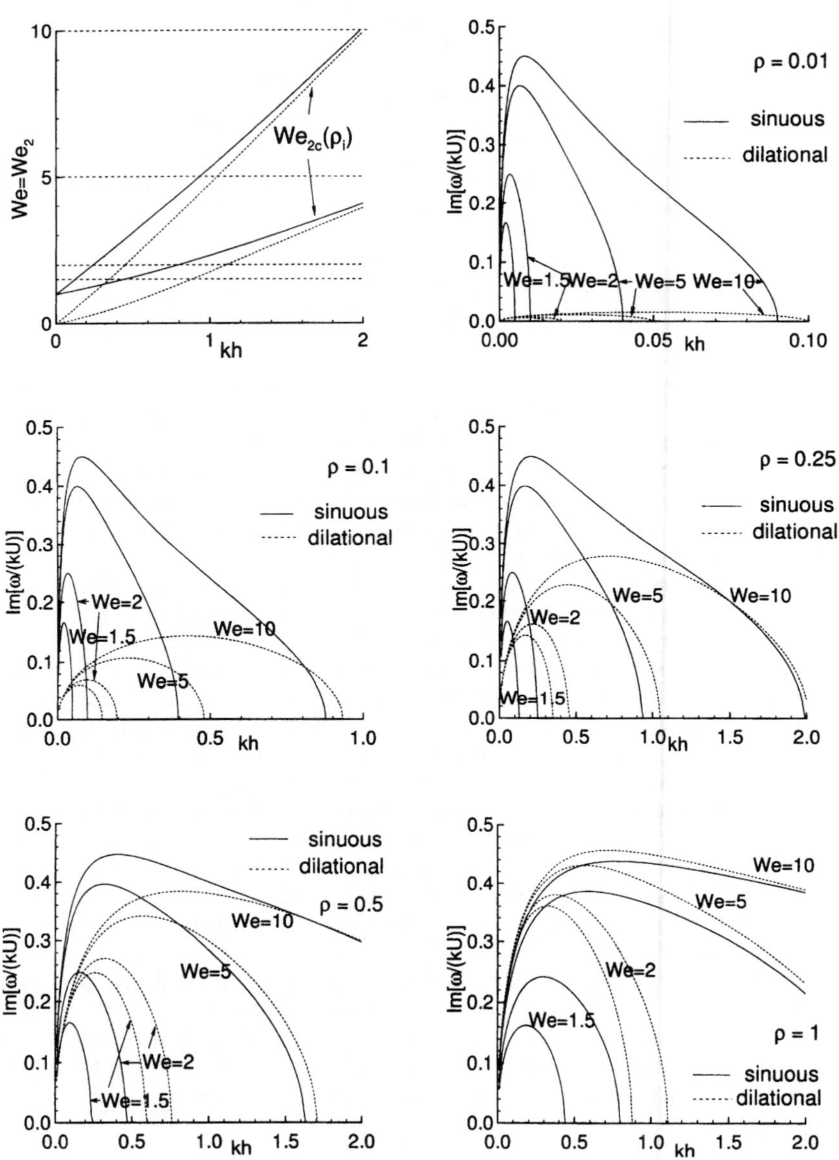

Fig. 5   Dimensionless growth rate $\mathrm{Im}[\omega/(ku_0)]$ as a function of $kh = 2\pi h/\lambda$ for different values of $We = We_2$ and $\rho = \rho_g/\rho_l$.

At low Weber numbers (near or less than one), the varicose or dilational mode is more unstable, and at very large Weber numbers, the instability of both modes is equally important. In fact, for $We < 1$, sinuous waves even become neutrally stable, so that in that case only varicose waves can be aerodynamically unstable. The Weber number at which the instability of sinuous waves starts dominating the varicose instability depends on the gas-to-liquid-density ratio ρ_g/ρ_l and increases with increasing values of ρ_g/ρ_l. At low gas density, the sinuous wave dominates over a wide range of Weber numbers. However, at high ambient gas densities (e.g., $\rho_g \geq 0.25\rho_l$), the dilational mode has the faster growth rate over a very wide range of Weber numbers.

Temporal and spatial stability analyses show that liquid viscosity plays a dual role in the stability of liquid sheets. At low Weber numbers, viscosity introduces an additional mode of instability, which (under certain conditions) can grow faster than the aerodynamic instability. However, at high Weber numbers, linear theory shows that aerodynamic instability always dominates and liquid viscosity always reduces the disturbance growth rates and shifts the dominant disturbances to longer wavelengths.[70,71]

Differences between predictions for the most unstable wavelength, i.e., wavelength with the maximum growth rate, on two-dimensional viscous sheets and observed dominant waves on plane and fan sheets were explained by disturbances forced onto the sheet internally by flow disturbances even at subcritical Reynolds numbers[72] or externally by nozzle vibrations at some natural frequency.[73,74] Sheet thinning in fan sheets and conical sheets, and the magnitude of initial disturbance amplitudes of unstable waves, were also used to argue that disturbances at the optimum wavelength, i.e., the wavelength with maximum growth rate, as predicted by linear two-dimensional theories, do not necessarily dominate the instability or breakup of practical liquid sheets.

B. Fan Sheets

A fan sheet or fan jet involves a thin liquid sheet that, in the undisturbed state, has a plane as the mid surface between the two gas-liquid interfaces of the sheet. Furthermore, the streamlines in this plane are straight lines diverging from a common point; the name fan sheet provides a description of the geometry. The velocity along these radial streamlines is found to be approximately constant with the distance from the origin.[75] The curvature of the nearly planar gas-liquid interfaces is insufficient to cause pressure gradients and velocity change in the steady case. Thus, it can be assumed from continuity that, in the steady state, the product of thickness and distance from the theoretical point of origin of the streamlines is a constant. Therefore, the sheet thickness decreases as it flows downstream. In practice, of course, the fan sheet does not emerge from a point; it can emerge from a properly designed nozzle or be the result of two identical round jets impinging at an angle with the center plane of the fan sheet coinciding with the symmetry plane of the two-jet configuration. These impinging jets are common in liquid rocket engines. The antisymmetric mode has been found experimentally only at atmospheric pressure. Thus, analysis has focused on that mode. However, it would not be surprising to see the importance of the symmetric dilational mode increase with increases in pressure based on other

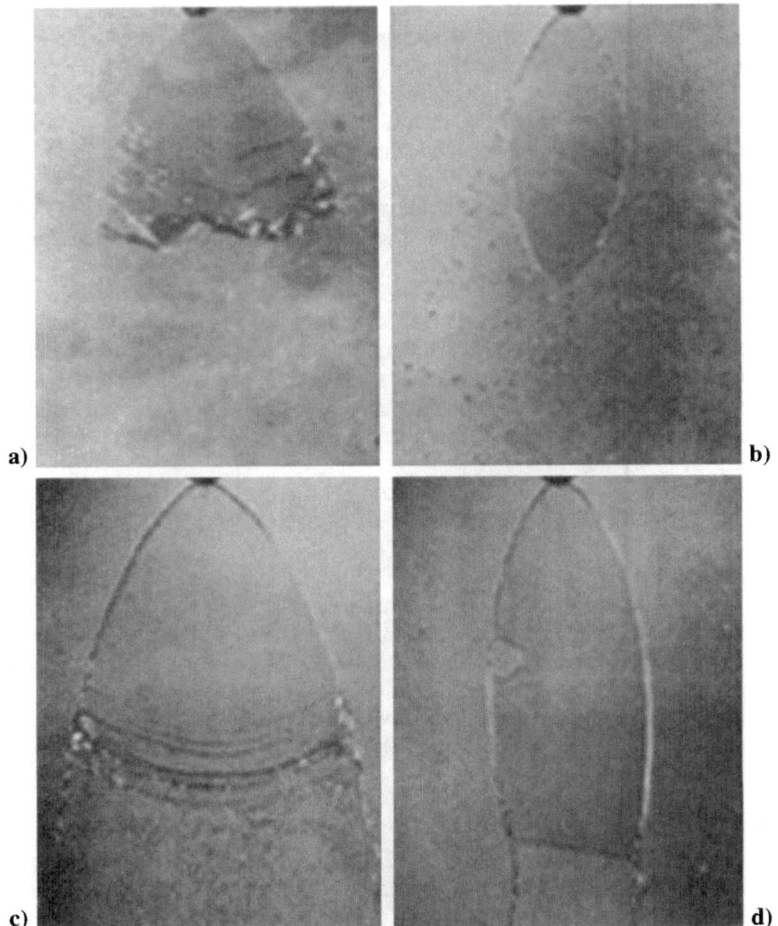

Fig. 6 Sheet formation from fan spray nozzles showing the effect of surface tension and viscosity[76]: a) ethyl acetate, b) water at 55°C, c) isobutyl alcohol, and d) 47.5% glycerine in water.

experiences with planar sheets. The surface tension and viscosity effects for fan spray sheets are shown in Fig. 6.[76] Here, surface tension is increasing from left to right while liquid viscosity increases from top to bottom.

The thickness variation of fan sheets can be described by $t = K/x$, where t, x, and K are the local sheet thickness at distance x from the nozzle, and K is a thickness parameter depending on fluid viscosity, surface tension, and liquid injection pressure.[77]

Dombrowski and Johns[78] found that, for a viscous fluid in an attenuating sheet, the wave with maximum growth is not necessarily one with the maximum growth rate, whereas in the inviscid case, the wave of maximum growth has also the maximum growth rate through the sheet. Here, maximum growth was determined by integrating the growth rate equations over time. An analytical linear solution for radially moving attenuating sheets was obtained by Weihs[79] in terms of hypergeometric functions, leading to similar observations as those described before; critical wavelengths were found to depend on the original location of the disturbance, and the wave number for maximum instability was found to increase with distance from the nozzle. For cases of practical interest, predictions deviated from those obtained from parallel flow models by a factor of five.

Weihs[79] considered a viscous liquid fan sheet interacting aerodynamically with a gaseous potential flow resulting in wavy interfaces of the antisymmetric type. He showed that both amplitude and waveform can change with downstream distance that is consistent with experimental findings.[73,80] In fact, a critical radial distance was predicted below which instability of the wave was possible and beyond which stability was found.

C.   Simplified Breakup Theories

Two-dimensional linear analyses of infinite (spatially periodic) and semi-infinite (time periodic) liquid sheets break down as the disturbance amplitudes become large and as three-dimensional effects appear.[80,81] They also fail to predict sheet breakup due to sinuous disturbances, and they do not account for the finite (lateral) dimensions of practical liquid sheets. However, they can be used to estimate breakup lengths and resulting drop size for practical sheet atomizers, if combined with other theories, e.g., Rayleigh breakup mechanism for ligament or thread disintegration, and empirical information (see, for example, Refs. 78 and 82–85).

Some simple concepts can often be applied to make rough predictions of the breakup of liquid jets, drops, and ligaments. The atomization of a liquid jet consists of the airstream stripping away filaments of liquid that subsequently break up into droplets. The maximum size drop that can remain stable within a gas stream is dependent on the ratio of the dynamic pressure force to the surface tension force, known as the Weber number

$$We = \frac{\rho_g V^2 d}{\sigma}$$

where ρ_g is the density of the gas stream, V is the relative velocity between liquid/gas, d is the droplet diameter, and σ is the surface tension coefficient of the liquid.

A liquid drop that is stripped from a jet inside a moving gas is exposed to the aerodynamic pressure effect that depends on the relative velocity of the drop with respect to the free flow. If this pressure is large enough to overcome the restoring force of the surface tension, the drop will disintegrate into smaller droplets. The forced deformation of a liquid droplet, under the influence of the pressure distribution caused by a turbulent gas flow around the drop, has been investigated by Hinze.[86] He considered two cases: suddenly applied steady velocity and gradually increasing velocity. He concluded that a drop will shatter if the surface deformation at the stagnation point is roughly equal to the droplet radius. This occurs when the Weber number exceeds a value ranging from 6 to 10 depending on the viscosity of the liquid. The critical Weber number We_{crit} has been determined[87] to be between 5 and 10 with the higher values for the case of free falling droplets. The drop breakup time for a low viscosity drop can be estimated by

$$t_b = \frac{d}{2V}\sqrt{\rho_l/\rho_g}$$

Some experimental studies[88,89] showed that the drop breakup time is of the same form as that given by the preceding equation within one order of magnitude.

The maximum stable drop size for drops exposed to an airstream is then estimated by

$$d_{max} = \frac{\sigma We_{crit}}{\rho_g V^2}$$

Obviously in the case of a large mass of liquid fuel moving in a gas stream, the bulk of the liquid does not encounter the full gas stream velocity instantaneously. The fuel filaments closer to the outer edge are expected to break up into drops earlier than the liquid close to the center of the fuel bulk. Therefore, an entire spectrum of drop sizes is expected.

Consider that the surrounding surface of a large liquid fuel volume is subjected to the relative gas flow around it. To calculate an estimate of the stripping rate of a large liquid mass, the following simplified analysis can be followed. This analysis is a variation of an approach presented by Sirignano et al. in a report[90] of restricted circulation. It treats ligaments of liquid that have been created by the first-stage breakup of liquid streams as large droplets. The macroscopic spherical character of the liquid mass is maintained while the surrounding flow raises small capillary waves of wavelength λ of maximum growth rate on the liquid-gas interface. These waves form ring structures around the sphere. It is also assumed that as these waves crest, circular torroidal rings with an estimated characteristic cross-sectional diameter of λ are eventually formed. These rings are exposed to high velocities and eventually break to form new droplets of diameter 1.89λ. This estimate of the new droplet size follows if the most unstable wavelength according to the Rayleigh mechanism is applied to the cylindrical tube of liquid. Thus from the discussion following Eq. (11), we conclude that the most unstable wavelength on the cylindrical tube is $9.016 \, (\lambda/2)$ and the radius of the new drop is $\lambda/1.058$, which means the new drop diameter is 1.89λ.

A linear model is employed based on the Kelvin–Helmholtz instability on the surface of the liquid volume. By assuming that only the transversal disturbances (those that result in transverse ligaments) are important and by neglecting the effect of curvature (essentially assuming that the wavelength of the disturbance is much smaller than the diameter of the large mass of liquid), the wavelength λ_{opt} of the most unstable wave on the surface of the liquid is found following Rangel and Sirignano[91] to be

$$\lambda_{\text{opt}} = \frac{3\pi\sigma}{\Delta u^2}\left(\frac{1}{\rho_g} + \frac{1}{\rho_l}\right) \tag{37}$$

where Δu is the velocity between the gas and the liquid. In the analysis, the interface is considered to move at a velocity $\Delta u/2$ relative to the liquid and at a velocity $-\Delta u/2$ relative to the gas. The subscript "opt" implies the optimum value for early film rupture, i.e., maximum growth rate.

Linear theory predicts that the growth rate of these disturbances is exponential in time according to

$$\eta = \eta_0 e^{\beta t}$$

where η is the dimensional amplitude of the disturbance and β the growth time constant. The growth time is immediately obtained from this last equation as

$$t = \frac{1}{\beta}\ln\left(\frac{\eta}{\eta_0}\right) \tag{38}$$

The growth time constant is provided by the linear analysis as

$$\beta = 0.385C \tag{39}$$

where the constant C is given by

$$C = \frac{\Delta u^3 \rho_g}{\sigma}\frac{(\rho_g/\rho_l)^{0.5}}{(1 + \rho_g/\rho_l)^2} \tag{40}$$

To obtain a characteristic critical time for the breakup of a ring-shaped ligament from the liquid surface, one must assume the critical size of the amplitude and the initial size of the amplitude. It is reasonable to assume that the critical size is of the order of the wavelength. The initial amplitude is somewhat more arbitrary because it depends on the particular source of the disturbance. We note that, from Eq. (38), a 10^n change in the amplitude ratio results in a variation of the critical time that is of the order of n only. To obtain a growth rate with a reasonable correctness for its order of magnitude, we choose an initial amplitude equal to $1/10$ of the disturbance wavelength; thus from Eq. (38)

$$t_{\text{crit}} = \frac{2.30}{\beta}$$

and using the value of β given above through Eqs. (39) and (40), we obtain

$$t_{\text{crit}} = \frac{5.97\sigma\,(1 + \rho_g/\rho_l)^2}{\rho_g \Delta u^3 (\rho_g/\rho_l)^{0.5}} \tag{41}$$

Based on the linear theory of Rangel and Sirignano,[91] it can be shown that the wave velocity along the surface relative to the liquid can be approximated by

$$U_{\text{wave}} = \left(\frac{\Delta u}{1 + \rho_g/\rho_l}\right) \tag{42}$$

This is strictly valid for planar wave propagation, and so it is assumed here that the wavelength is very small compared to the mean radius of curvature of the surface. The length along the surface that a wave crest travels during the period L_{crit} can be determined as

$$L_{\text{crit}} = U_{\text{wave}}\, t_{\text{crit}} \tag{43}$$

It follows from Eqs. (41–43) that the critical Weber number is given by

$$We_{\text{crit}} = \frac{\rho_g (\Delta u)^2 L_{\text{crit}}}{\sigma} = 5.97 \left(\frac{\rho_l}{\rho_g}\right)^{0.5} \left(1 + \frac{\rho_g}{\rho_l}\right)$$

Clearly, the proposed mechanism can work only for masses of liquid with dimensions larger than L_{crit}. For most applications, ρ_l is much larger than ρ_g, so that the critical Weber number is much larger than unity. Thus the mechanism requires a high Weber number.

Because the wavelength of the disturbance is much smaller than the radius R of the large mass, the resulting ring-shaped ligaments will also have a small cross section compared to the large mass radius. The volume of a ligament will be approximately given by

$$\delta v = \frac{\pi \lambda^2}{4} 2\pi R \sin\theta \tag{44}$$

where θ is the polar angle, and the diameter of the cross section of the ligament is λ.

The angle θ is also the ratio of the arc length to the radius, and so $\theta = L_{crit}/R$. Furthermore, the ligament is shed at the L_{crit} position once every cycle. The volumetric shedding rate is given by

$$\frac{U_{wave}}{\lambda_{opt}} \delta v = \frac{\pi^2}{2} \frac{\lambda_{opt} R \Delta u}{1 + \rho_g/\rho_l} \sin \frac{L_{crit}}{R} \tag{45}$$

The shedding rate of ligaments equals the rate of change of droplet volume, so that

$$\frac{dR}{dt} = \frac{3\pi^2}{8} \frac{\sigma}{\rho_g R \Delta u} \sin \frac{L_{crit}}{R} \tag{46}$$

The toroidal ligaments that form can be considered as columns of liquid because the cross-sectional diameter is small compared with the radius of curvature of the torus. Then a Rayleigh capillary mechanism could break the torus into many spherical ligaments. These small spheres can, if the Weber number is sufficiently large, undergo the same Kelvin–Helmholtz instability previously discussed for the larger parent droplet. A cascade process can continue until the droplets are so small that their Weber numbers are too small for substantial distortion to occur.

D. Nonlinear Theory

More insight on the sheet breakup mechanism for the sinuous mode was provided by Clark and Dombrowski.[75] They considered a second-order temporal analysis of the aerodynamic growth of sinuous waves on nonattenuating inviscid sheets and predicted the appearance of the first harmonic dilational mode due to energy transfer from the fundamental sinuous mode, which subsequently leads to sheet breakup at half-wavelengths of the fundamental sinuous mode. The same observation was also made within the third-order stability analysis by Jazayeri and Li[92] and within the nonlinear discrete-vortex method simulations of the two-dimensional sheet by Rangel and Sirignano.[69] In both analyses, the nonlinear sinuous growth rates were found to be less than predicted by linear theory. Also, as shown in the following, the prescribed nonlinear coupling between sinuous and dilational capillary waves does not depend on the presence of a surrounding gas flow. Rangel and Sirignano,[69] using a vortex dynamics method, performed a two-dimensional, nonlinear, inviscid analysis of the temporal stability of planar sheets. The liquid–gas interfaces are themselves vortex sheets in the inviscid limit. The strength of the vortex sheet can be modified because of surface tension and density discontinuity across the interface.

Rangel and Sirignano[69] assumed periodic spatial behavior on an infinitely long liquid stream and calculated the temporal behavior for both the sinuous and dilational modes. The method is powerful in that no restrictions on the magnitude of the gas-density-to-liquid-density ratio, on the sheet-thickness-to-

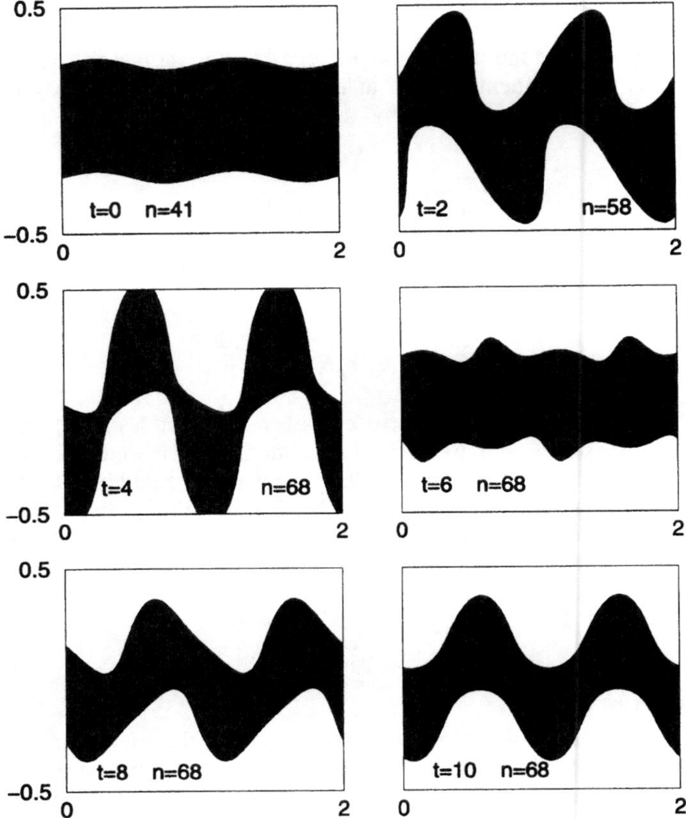

Fig. 7   Time evolution of sinuous sheet disturbance for $\rho = 1$, $W = 0.67$, and $h/\lambda = 0.25$.[69]

wavelength ratio, or on the amplitude of the sheet distortion are required. However, the method cannot readily be extended to semi-infinite sheet configurations or to three-dimensional and axisymmetric cases. Also, the numerical method requires great care when the sheet thickness locally becomes comparable to the arc distance between discrete vortex elements.

Figures 7 and 8 give the results of Rangel and Sirignano[69] for the sinuous and dilational modes, respectively, in the case in which gas and liquid densities are equal and the initial sheet thickness is one quarter of the wavelength of the disturbance. The sinuous case displays half-wave thinning. The theory cannot predict stream breakup because numerical errors occur as the sheet thickness approaches the length of discretization. However, the possibility of the stream breaking at half-wavelengths is indicated. At the larger gas-to-liquid-density

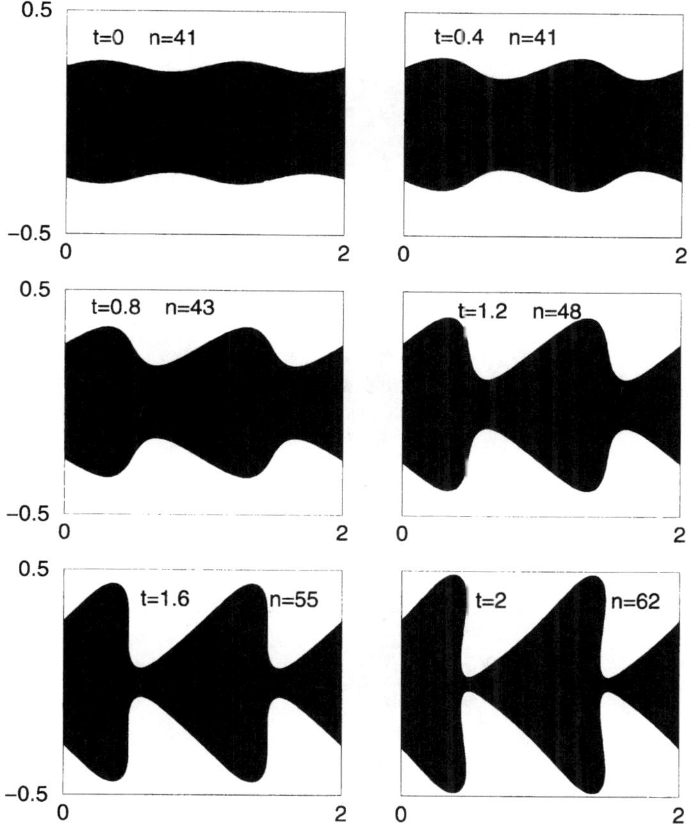

Fig. 8   Time evolution of dilational sheet disturbance for $\rho = 1$, $W = 0.67$, and $h/\lambda = 0.25$.[69]

ratio, an oscillation in the waveform occurred; the sinuous shape disappears and reappears. The dilational waves can assume a "heart" shape with thinning at wavelength intervals.

1.   Thin Sheet Analysis

A thin planar liquid sheet infinitely or semi-infinitely long in the flow direction (x) is considered as shown in Fig. 9. The liquid sheet is initially injected into a gas of negligible density compared to the liquid density with the undisturbed velocity u_0 and the undisturbed semithickness a. The liquid is assumed inviscid, incompressible, and free of gravity force. Then, following Kim and Sirignano[93] in

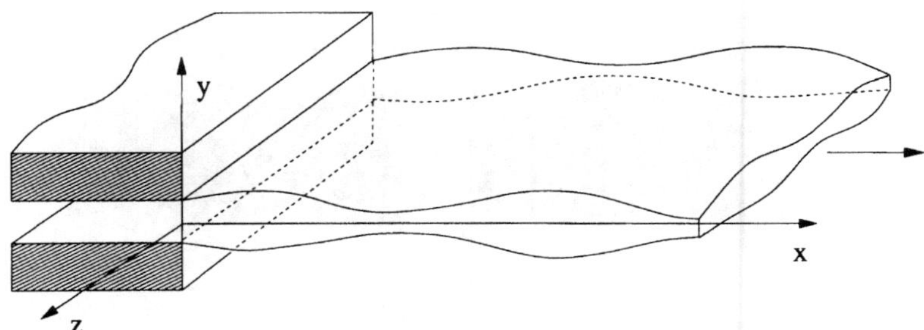

Fig. 9　Flow geometry and coordinates.

the limit of infinite radius of curvature, Eqs. (1–4) become

$$\frac{\partial u}{\partial x} + \frac{\partial v}{\partial y} + \frac{\partial w}{\partial z} = 0 \tag{47}$$

$$\frac{\partial u}{\partial t} + \frac{\partial (u^2)}{\partial x} + \frac{\partial (uv)}{\partial y} + \frac{\partial (uw)}{\partial z} + \frac{1}{\rho}\frac{\partial p}{\partial x} = 0 \tag{48}$$

$$\frac{\partial w}{\partial t} + \frac{\partial (uw)}{\partial x} + \frac{\partial (vw)}{\partial y} + \frac{\partial (w^2)}{\partial z} + \frac{1}{\rho}\frac{\partial p}{\partial z} = 0 \tag{49}$$

$$\frac{\partial v}{\partial t} + \frac{\partial (uv)}{\partial x} + \frac{\partial (v^2)}{\partial y} + \frac{\partial (vw)}{\partial z} + \frac{1}{\rho}\frac{\partial p}{\partial y} = 0 \tag{50}$$

The upper boundary of the liquid sheet is given by $y_+(x, z, t)$, and the lower boundary is described as $y_-(x, z, t)$. The boundary conditions (5) through (8) at the upper and lower surfaces become

$$v_{\pm} = \frac{\partial y_{\pm}}{\partial t} + u_{\pm}\frac{\partial y_{\pm}}{\partial x} + w_{\pm}\frac{\partial y_{\pm}}{\partial z} \tag{51}$$

$$p_{\pm} = \sigma\left(\frac{1}{R_{1\pm}} + \frac{1}{R_{2\pm}}\right)$$

$$= \frac{\pm\sigma}{(1 + y_x^2 + y_z^2)^{3/2}}\left[y_{xx}(1 + y_z^2) + y_{zz}(1 + y_x^2) - 2y_x y_z y_{xz}\right] \tag{52}$$

where the subscripts $+$ and $-$ denote values at the upper and lower sheet surfaces, respectively, and $R_{1\pm}$ and $R_{2\pm}$ are the principal radii of curvature of the surface. Note that the subscripts x and z imply partial differentiation with respect to that coordinate.

Define the sheet thickness,

$$\tilde{y}(x, z, t) = y_+ - y_-, \qquad \bar{y}(x, z, t) = (y_+ + y_-)/2 \tag{53}$$

Also define Δp and $\bar{p}$ in a similar fashion. They are related to $\tilde{y}$ and $\bar{y}$ by using Eqs. (52) and (53) to yield

$$\Delta p = p_+ - p_-$$

$$= -\sigma\left[(f_{1+} + f_{1-})\frac{\partial^2 \bar{y}}{\partial x^2} + \tfrac{1}{2}(f_{1+} - f_{1-})\frac{\partial^2 \tilde{y}}{\partial x^2} + (f_{2+} + f_{2-})\frac{\partial^2 \bar{y}}{\partial z^2}\right.$$

$$\left. + \tfrac{1}{2}(f_{2+} - f_{2-})\frac{\partial^2 \tilde{y}}{\partial z^2} + (f_{3+} + f_{3-})\frac{\partial^2 \bar{y}}{\partial x\,\partial z} + \tfrac{1}{2}(f_{3+} - f_{3-})\frac{\partial^2 \tilde{y}}{\partial x\,\partial z}\right] \quad (54)$$

$$\bar{p} = (p_+ + p_-)/2$$

$$= -\frac{\sigma}{2}\left[(f_{1+} - f_{1-})\frac{\partial^2 \bar{y}}{\partial x^2} + \tfrac{1}{2}(f_{1+} + f_{1-})\frac{\partial^2 \tilde{y}}{\partial x^2} + (f_{2+} - f_{2-})\frac{\partial^2 \bar{y}}{\partial z^2}\right.$$

$$\left. + \tfrac{1}{2}(f_{2+} + f_{2-})\frac{\partial^2 \tilde{y}}{\partial z^2} + (f_{3+} - f_{3-})\frac{\partial^2 \bar{y}}{\partial x\,\partial z} + \tfrac{1}{2}(f_{3+} + f_{3-})\frac{\partial^2 \tilde{y}}{\partial x\,\partial z}\right] \quad (55)$$

where $f_{1+}, f_{1-}, f_{2+}, f_{2-}, f_{3+}$, and f_{3-} are defined as

$$f_{1\pm} = \frac{1 + \left(\bar{y}_z \pm \tfrac{1}{2}\tilde{y}_z\right)^2}{\left[1 + \left(\bar{y}_x \pm \tfrac{1}{2}\tilde{y}_x\right)^2 + \left(\bar{y}_z \pm \tfrac{1}{2}\tilde{y}_z\right)^2\right]^{3/2}}$$

$$f_{2\pm} = \frac{1 + \left(\bar{y}_x \pm \tfrac{1}{2}\tilde{y}_x\right)^2}{\left[1 + \left(\bar{y}_x \pm \tfrac{1}{2}\tilde{y}_x\right)^2 + \left(\bar{y}_z \pm \tfrac{1}{2}\tilde{y}_z\right)^2\right]^{3/2}}$$

$$f_{3\pm} = \frac{-2\left(\bar{y}_x \pm \tfrac{1}{2}\tilde{y}_x\right)\left(\bar{y}_z \pm \tfrac{1}{2}\tilde{y}_z\right)}{\left[1 + \left(\bar{y}_x \pm \tfrac{1}{2}\tilde{y}_x\right)^2 + \left(\bar{y}_z \pm \tfrac{1}{2}\tilde{y}_z\right)^2\right]^{3/2}} \quad (56)$$

For a sheet whose thickness is small compared with the wavelength of a disturbance, we consider u, $\partial v/\partial y$, w, and $\partial p/\partial y$ to be nearly constant with variation of y. For the two-dimensional disturbance, Mehring and Sirignano[94] showed that these behaviors were predicted as the leading behavior in an asymptotic representation for long wavelengths. The problem can therefore be reduced to a two-dimensional, unsteady formulation. We define average velocities

$\bar{u}(x, z, t)$, $\bar{v}(x, z, t)$, and $\bar{w}(x, z, t)$ to be

$$\bar{u}(x, z, t) = \frac{1}{\tilde{y}} \int_{y_-}^{y_+} u \, dy, \quad \bar{v}(x, z, t) = \frac{1}{\tilde{y}} \int_{y_-}^{y_+} v \, dy, \quad \bar{w}(x, z, t) = \frac{1}{\tilde{y}} \int_{y_-}^{y_+} w \, dy \quad (57)$$

Average pressure $\bar{p}(x, z, t)$ is defined in a similar manner. Equations (47–50) can be integrated term by term from y_- to y_+ and incorporated with the kinematic and dynamic boundary conditions [Eqs. (51) and (52)] and the preceding definitions. The results are

$$\frac{\partial \tilde{y}}{\partial t} + \frac{\partial \tilde{y}\bar{u}}{\partial x} + \frac{\partial \tilde{y}\bar{w}}{\partial z} = 0 \tag{58}$$

$$\frac{\partial \bar{u}}{\partial t} + \bar{u}\frac{\partial \bar{u}}{\partial x} + \bar{w}\frac{\partial \bar{u}}{\partial z} = -\frac{1}{\rho}\left(\frac{\partial \bar{p}}{\partial x} - \frac{\Delta p}{\tilde{y}}\frac{\partial \tilde{y}}{\partial x}\right) \tag{59}$$

$$\frac{\partial \bar{w}}{\partial t} + \bar{u}\frac{\partial \bar{w}}{\partial x} + \bar{w}\frac{\partial \bar{w}}{\partial z} = -\frac{1}{\rho}\left(\frac{\partial \bar{p}}{\partial z} - \frac{\Delta p}{\tilde{y}}\frac{\partial \tilde{y}}{\partial z}\right) \tag{60}$$

$$\frac{\partial \bar{v}}{\partial t} + \bar{u}\frac{\partial \bar{v}}{\partial x} + \bar{w}\frac{\partial \bar{v}}{\partial z} = -\frac{1}{\rho}\frac{\Delta p}{\tilde{y}} \tag{61}$$

Equations (58–61) show that the number of unknowns is five ($\bar{y}$, $\tilde{y}$, $\bar{u}$, $\bar{v}$, and $\bar{w}$), but the number of equations is four. An additional equation is obtained by combining the kinematic boundary conditions for v_+ and v_- and by using $\bar{v} = (v_+ + v_-)/2$. Mehring and Sirignano[94] showed that v can be expressed by a polynomial expansion in terms of y or $(y - \bar{y})$. As a consequence, v can be expressed as a linear function of y by the first-order approximation. Thus, the expression $v = (v_+ + v_-)/2$ is consistent with Eq. (57) by the first-order approximation:

$$\bar{v} = \frac{\partial \bar{y}}{\partial t} + \bar{u}\frac{\partial \bar{y}}{\partial x} + \bar{w}\frac{\partial \bar{y}}{\partial z} \tag{62}$$

Equations (58–62) represent the "lubrication approximation" or "slender jet" assumption. For liquid sheets (or jets) with large amplitude distortions, such as observed close to pinch-off, this assumption needs careful reevaluation, e.g., comparison with solutions to the full system of equations.[94] For two-dimensional planar sheet distortion, Eqs. (58–62) agree with the results of Mehring and Sirignano,[94] whereby $\partial(\ )/\partial z = 0$ and $\bar{w} = 0$. The system of equations has been solved by finite-difference computations using the Richtmyer splitting of the Lax-Wendroff method.

Considering only two-dimensional disturbances, the reduced-dimension approach yields, after linearization, two partial differential equations governing small amplitude dilational and sinuous capillary waves on thin two-dimensional inviscid ($\eta_1 = 0$) planar sheets, i.e.,

$$\frac{\partial^2 h}{\partial \tau^2} + \frac{\sigma a}{\rho_1} \frac{\partial^4 h}{\partial \xi^4} = 0 \tag{63}$$

$$\frac{\partial^2 Y}{\partial \tau^2} - \frac{\sigma}{\rho_1 a} \frac{\partial^2 Y}{\partial \xi^2} = 0 \tag{64}$$

Equation (63) describes thickness variations h (dilational waves), and Eq. (64) describes variations in the sheet-centerline position Y while the sheet thickness remains constant (sinuous waves). τ and ξ denote time and space variables in a coordinate system moving with the undisturbed liquid stream. The previous equations readily show that linear sinuous and dilational capillary waves are decoupled. A modal analysis also quickly reveals that dilational waves are dispersive, whereas sinuous waves are nondispersive. Equations (63) and (64) are both well-known equations in mechanics, governing the transverse vibration of a uniform beam and vibrations on a taut string, respectively.[95] Within the analysis of thin liquid sheets, Taylor[96] first derived and analyzed Eqs. (63) and (64), the latter in its steady form and for radially expanding planar sheets.

A similar linear analysis of spatially developing dilational and sinuous capillary waves on thin planar sheets shows that there are four dilational ($k_{1,\dots,4}$) and two sinuous ($l_{1,2}$) waves generated if a harmonic disturbance is locally forced onto the moving sheet at a frequency ω, i.e.,

$$k_{1,2} = \frac{1 \pm (1 - 8a\epsilon\omega/u_c)^{1/2}}{4a\epsilon}$$

$$k_{3,4} = \frac{-1 \pm (1 + 8a\epsilon\omega/u_o)^{1/2}}{4a\epsilon} \tag{65}$$

$$l_{1,2} = \frac{\omega}{2a(1 \pm 2\epsilon)}$$

with $\epsilon = 1/\sqrt{2We}$ and $We = 2\rho a u_o^2/\sigma$, where $2a$ is the undisturbed sheet thickness, and u_o is the undisturbed base flow velocity of the sheet exiting from the orifice. Figure 10 illustrates the dilational mode wave numbers $k_i(i = 1, \dots, 4)$ in dependence on the time period $T = 2\pi/\omega$ of the imposed sheet modulation.

Group velocity arguments analogous to those employed by Bogy[16,20] for slender cylindrical liquid jets show that, of the four dilational waves generated by a harmonic dilational forcing at the nozzle exit, only two (i.e., k_1 and k_3) will appear downstream from the nozzle, resulting in the superposition of two waves of similar wavelengths or a single traveling wave superimposed onto a wave with an exponentially decaying envelope in the downstream direction. For sinuous sheet modulations, only two wave numbers are generated, both of which will have positive group velocities relative to the nozzle exit if $\epsilon < 0.5$ or $We > 2$, resulting in a beat in the envelope of two sinuous waves traveling in the downstream direction. However, for $We < 2$, only wave number $l_1 = \omega/[2a(1 + 2\epsilon)]$ is expected to appear downstream from the nozzle. The appearance of dilational

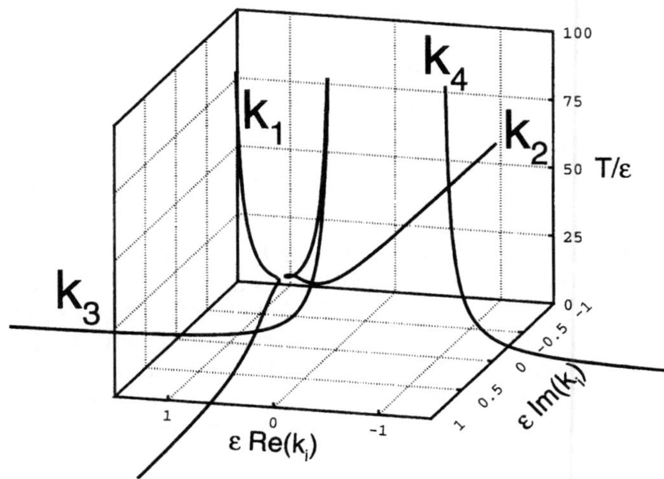

Fig. 10   The dependence of wave numbers k_i for modulated symmetric distorting semi-infinite sheets on forcing period $T = 2\pi/\omega$.[94]

and sinuous capillary wave on planar sheets is independent of the imposed modulation frequency, because the group velocity for each wave number has either positive or negative values for any forcing frequency. Linear analysis also shows that there exists no forcing frequency resulting in exponential growth of sinuous or dilational sheet disturbances forced at the nozzle exit.

The general nonlinear dimensionally reduced system of equations for sinuous and dilational capillary waves on thin planar two-dimensional (inviscid) liquid sheets in a passive ambient gas or void is obtained as a special case of Eqs. (58–62) and is given by

$$\frac{\partial \tilde{y}}{\partial t} + \frac{\partial}{\partial x}(\bar{u}\tilde{y}) = 0 \tag{66}$$

$$\frac{\partial \bar{y}}{\partial t} + \bar{u}\frac{\partial \bar{y}}{\partial x} = \bar{v} \tag{67}$$

$$\frac{\partial \bar{u}}{\partial t} + \bar{u}\frac{\partial \bar{u}}{\partial x} = \frac{\sigma}{2\rho_l}\left(\frac{\partial^3 \tilde{y}}{\partial x^3} - \frac{4}{\tilde{y}}\frac{\partial \tilde{y}}{\partial x}\frac{\partial^2 \tilde{y}}{\partial x^2}\right) \tag{68}$$

$$\frac{\partial \bar{v}}{\partial t} + \bar{u}\frac{\partial \bar{v}}{\partial x} = \frac{2\sigma}{\rho_l}\frac{1}{\tilde{y}}\frac{\partial^2 \bar{y}}{\partial x^2} \tag{69}$$

These equations govern the average axial and transverse velocity components $\bar{u}$ and $\bar{v}$, the sheet thickness $\tilde{y}$, and the deviation of the sheet-centerline location from its undisturbed position $\bar{y}$. It has been assumed in Eqs. (68) and (69) that the variations in $f_\pm$ defined in Eq. (56) are higher order. Linearization and Galilean transformation of these equations yields Eqs. (63) and (64). As already mentioned, sinuous and dilational modes are decoupled in linear analysis; however, the second term on the right-hand side of the $\bar{u}$-momentum equation yields a nonlinear coupling between the two modes and leads to the prescribed excitation of the dilational mode, once the sheet is subjected to sufficiently large sinuous disturbances. Apart from the prescribed nonlinear coupling effect, temporal analysis also reveals that the transverse oscillation of nonlinear sinuous traveling waves already observed by Rangel and Sirignano[69] for the case of finite density air, also appears for the zero ambient density case. Both vortex-dynamics simulations[69] and reduced-dimension analysis[93,94] for thin periodically disturbed sheets show that, for the nonlinear dilational traveling wave, a temporal steepening and de-steepening of the wave (wobbling) occurs. The latter analysis also reveals the possibility of sheet instability (due to capillary effects) triggered by a small amplitude dilational sheet disturbance with wavelength λ_s superimposed onto a dilational traveling wave with shorter wavelengths and, in particular, for $\lambda = \lambda_s/n$, where n has an integer value. This type of instability has first been observed by Matsuuchi.[97,98]

Equations (66–69) can also be used to analyze the nonlinear distortion of thin planar liquid sheets, which are modulated at the nozzle exit. See Figs. 11 and 12 for modulations of the transverse and axial velocity component, respectively. The characteristic envelope in variations of the sheet-thickness or the sheet-centerline location in the downstream direction along the dilational or sinuous modulated sheet predicted by linear theory and described earlier is altered somewhat by nonlinear effects. As within the temporal analysis, the modulated sinuous sheet distortion excites the dilational mode as the sheet propagates downstream, leading to fluid agglomeration in the maximum deflection region of the sinuous distorting sheet, as already observed experimentally by Hashimoto and Suzuki.[99]

For both sinuous and dilational modulations, nonlinear effects result in the accumulation of fluid into lumps, connected by threads of fluid that show increased thinning and that eventually break up at points close to these blobs of fluid. This observation is analogous to the case of a cylindrical jet, where nonlinear effects are responsible for the satellite droplets between the larger droplets. Drop-size predictions from these nonlinear two-dimensional results can again be obtained by employing the simplified assumption that the larger two-dimensional liquid columns and the thin threads between them will disintegrate according to Rayleigh's theory, assuming that no further breakup of these ligaments occurs in the longitudinal direction, i.e., in the downstream direction. Clearly, to predict accurately the overall sheet breakup process, a three-dimensional theory is needed.

2. *Nonlinear Aerodynamic Effects*

The nonlinear evolution of a thin planar liquid sheet under the influence of capillary and aerodynamic effects and with disturbances in the main flow direction (x-direction) only has been modeled[100] by employing the reduced-dimension approach[94] to describe the thin planar sheet and a BEM formulation[43] for

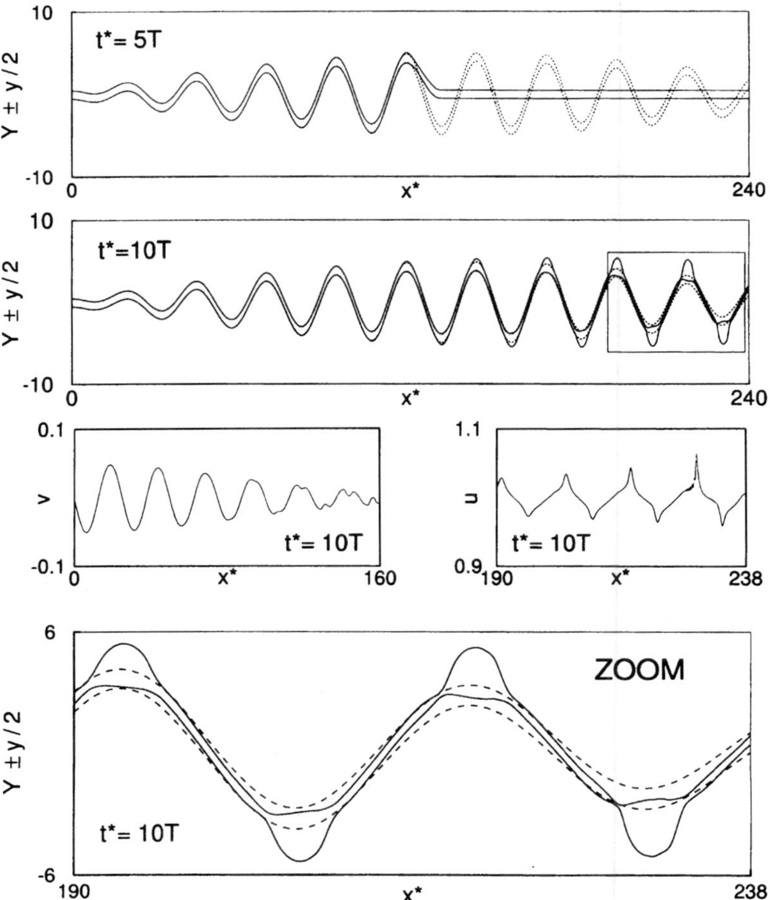

Fig. 11   Sinuous distortion of semi-infinite planar liquid sheet harmonically forced at $x = 0$ with forcing frequency ω; $We = 1000$, $T = 2\pi/\omega = 25$; – – – linear limit-cycle solution at $\tau = nT$, where n is an integer; —— nonlinear transient solution.[94]

the inviscid incompressible gas phase. The boundary-element method in combination with the unsteady Bernoulli equation (10) is used to determine the instantaneous gas pressure at the liquid-gas interfaces, which is needed within the reduced-dimension equations governing the liquid phase. The use of a discrete boundary-element method for the gas phase allows, in principle, the consideration of more practical applications in which sheets of liquid are injected into a gaseous flowfield with its own physical constraints or boundary conditions. The latter is of particular importance with regard to the complicated flowfields

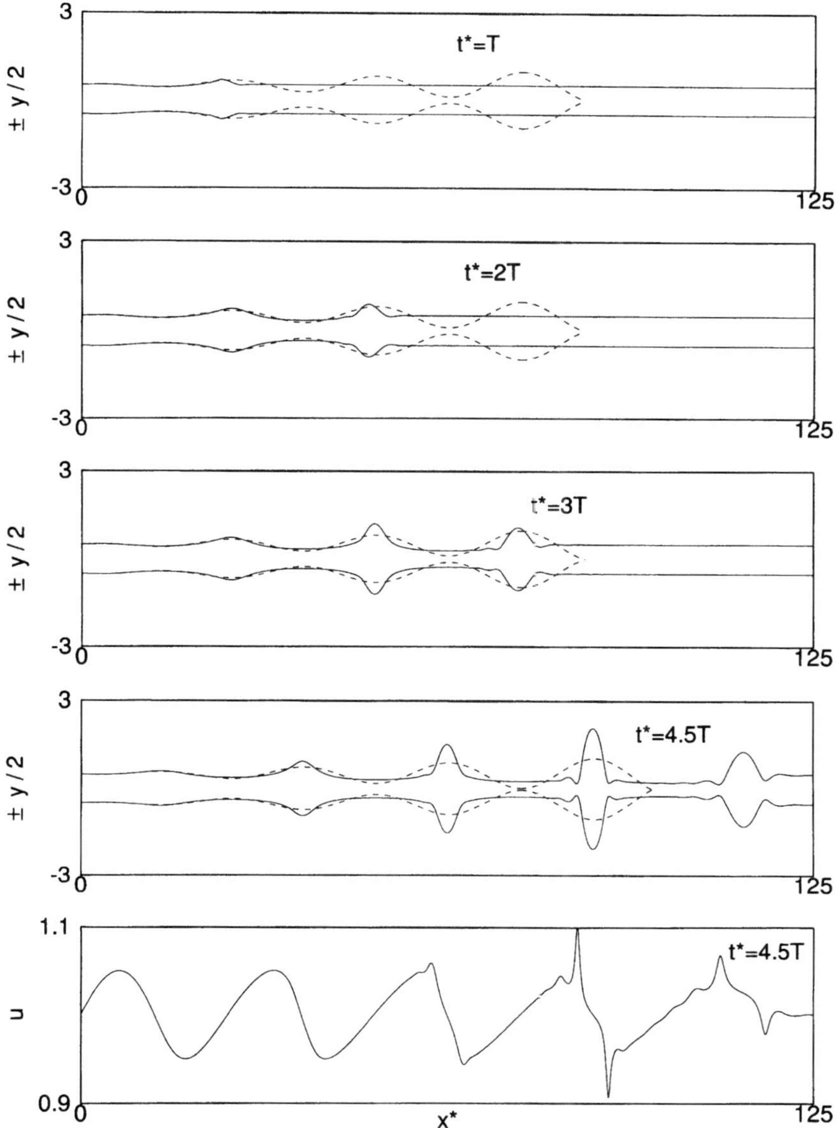

Fig. 12 Dilational distortion of semi-infinite planar liquid sheet harmonically forced at $x = 0$ with forcing frequency ω; $We = 500$, $T = 2\pi/\omega = 25$; – – – linear limit-cycle solutions at $\tau = nT$, where n is an integer; ⸺ nonlinear transient solution.[94]

within and/or around fuel injection elements or atomizers used for spray combustion purposes, e.g., prefilming airblast atomizers.

The proposed model is applicable for both dilational and sinuous modes. However, only dilational sheet distortions are discussed here.[100] The solution procedure for the liquid phase assumes that the sheet is only disturbed locally and that the disturbance is prescribed by the initial conditions. There is no time-dependent forcing imposed onto the sheet, and the sheet remains undisturbed at the boundaries of the computational domain. However, consideration of spatially developing semi-infinite sheets modulated at the nozzle exit and impacted by a surrounding gas flow is possible, if the solution algorithm for the liquid phase is replaced by the one previously employed for (capillary waves on) modulated semi-infinite sheets[94] and if appropriate boundary conditions are specified within the BEM procedure for the flowfield solution in the gas phase.[101,102]

The computational domains (and associated boundary conditions) for the considered flow problem are illustrated in Fig. 13 together with the appropriate governing equations within these domains, i.e., for gas and liquid phase. The coordinates in the axial and transverse flow direction are denoted by x and y, and the velocity potential in the gas phase is $\phi = \phi_g$. The coordinate system is fixed to the liquid sheet, and the gas phase is moving from left to right with velocity U relative to the fluid. Note that we are considering symmetric (dilational) disturbances only,[100] so that only the lower half of the sheet is examined here; the axis of symmetry is given by $y = 0$. Accordingly, this flow problem corresponds also to the flow of a thin inviscid planar liquid sheet along a wall (with slip boundary

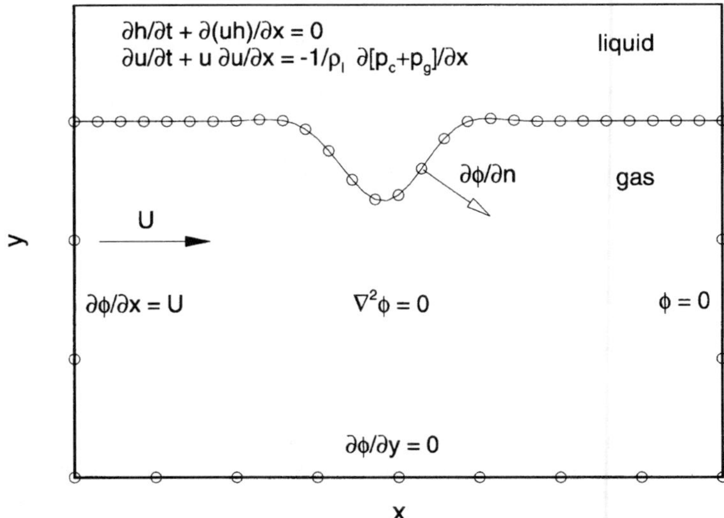

Fig. 13 Computational domain and gas-phase boundary conditions for infinite planar sheet with local sheet disturbance and coflowing gas streams.[101]

conditions) and impacted by an inviscid coflowing gas stream; this configuration is of potential interest in context with the previously mentioned prefilming airblast atomizers.

The integral representation of Laplace's equation may be written as

$$\alpha\phi_g(\vec{r}_i) + \int_S \left(\phi_g \frac{\partial G}{\partial n} - qG \right) dS = 0 \tag{70}$$

where $\phi_g(\vec{r}_i)$ is the value of the potential at a point $\vec{r}_i$, S denotes the boundary of the domain, $q = \partial\phi_g/\partial n$ is the gradient of ϕ_g on S normal to the boundary, α is a constant (for a given node), and G is the free-space Green's function corresponding to the governing equation. Because the preceding equation involves integration only around the boundary, we need not discretize the entire domain. It is presumed that either ϕ_g or q is specified at each node on the boundary, and the other quantity is returned as part of the solution. Coupling between the gas and liquid phase is described through kinematic and dynamic boundary conditions at the phase interface. The specification of the instantaneous interface location and the normal velocity of the interface represent kinematic conditions; this determines the solution for the velocity potential in the gas phase. On the other hand, the flowfield in the gas phase impacts the liquid-sheet distortion through the dynamic condition at the interface, requiring that the pressure inside the liquid balances with the pressure on the gas side of the interface combined with the capillary pressure.

The equations governing the nonlinear dilational sheet distortion are obtained by setting $\bar{y} = \bar{v} = 0$, adding the gas-phase pressure gradient term $-(1/\rho_l)\partial p_g/\partial x$ to the right-hand side of Eq. (68), and then using Eq. (66) and the modified Eq. (68).

The equations have to be integrated in time, which also includes the solution of Laplace's equation for ϕ_g at each time step. The procedure begins with the solution of Laplace's equation for ϕ_g by using the current interface location and the velocity normal to this interface, i.e., $\partial\phi_g/\partial n$. The resulting solution gives the values of the velocity potential ϕ_g along the interface. This information permits the updating of the gas pressure using Eq. (10). The newly determined gas pressure can now be used with the two-dimensional form of Eqs. (58–62) to update the interface location (including the location for the BEM surface nodes) and the velocity component normal to the interface. By repeating this procedure, the surface shape can be determined at all times prior to sheet pinch-off. The prescribed solution procedure closely follows the procedure employed by Spangler et al.,[42] who analyzed the nonlinear evolution of an axisymmetric liquid jet by using a BEM for the solution of Laplace's equation in both gas and liquid phases and by employing the unsteady Bernoulli equation in both phases along the interface.

Figure 14 illustrates two typical results generated using the new model. Both solutions are for the case $We_g = 2$ with zero liquid stream velocity and a local thinning or thickening of the sheet initially, i.e.,

$$y_\pm(x, t = 0) = \pm 0.5 \quad \text{for} \quad \left[x > x_0 - \frac{\Delta x}{2} \right]$$

$$y_\pm(x, t = 0) = \pm 0.5 + 0.35\theta\left\{ 1 - \cos\left[\frac{2\pi}{\Delta x}(x - \Delta x) \right] \right\}$$

$$\text{for} \quad \left[x_0 - \frac{\Delta x}{2} < x < x_0 + \frac{\Delta x}{2} \right]$$

$$y_\pm(x, t = 0) = \pm 0.5 \quad \text{for} \quad \left[x < x_0 + \frac{\Delta x}{2} \right]$$

and $u(x, t = 0) = 0$. Here $x_0 = 50$ and $\theta = \pm 1$ for local increasing of the sheet thickness and $\theta = \mp 1$ for local thinning of the sheet. As seen from Fig. 14, the gas stream flowing along the disturbed liquid-gas interface causes gas-pressure fluctuations, which result in significant sheet distortion and which eventually cause the liquid stream to break. Steepening of the originally symmetric disturbances is observed on the "valley" or "hillside" opposite to the direction from which the gas flow impacts the disturbances. For the case with locally increased sheet thickness, the breakup time is shorter than predicted for the similar case but with locally reduced sheet thickness. In the latter case, the

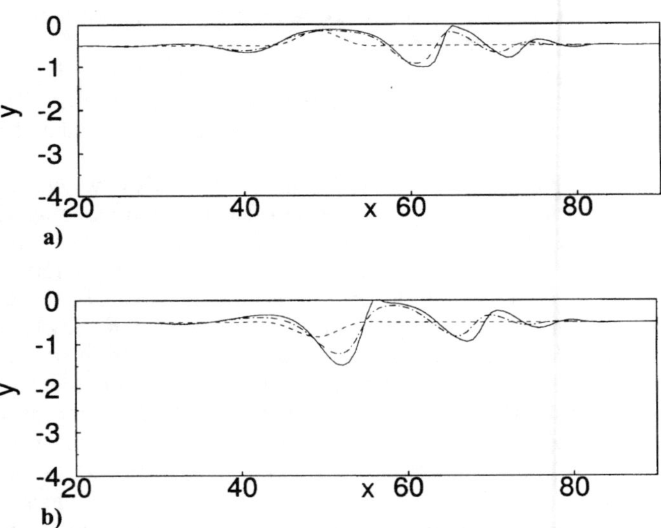

Fig. 14 Dilational sheet distortion due to parallel gas flow and local sheet thickness disturbance: a) local sheet depression and b) local sheet thickness increase. Symmetry axis at $y = 0$; $We_g = 2$, $\rho_g/\rho_l = 1.2 \cdot 10^{-3}$; – – – $\tau = 0$; –·–· $\tau = 24$; —— $\tau = 28$; $[\tau = t/t_{\text{ref}}, t_{\text{ref}} = h\sqrt{\sigma/(2\rho_l h)}]$. [101]

sheet tends to break "simultaneously" at two locations (creating a free liquid column), whereas in the former case the sheet only breaks once.

The significance of gas-pressure fluctuations for the distortion and breakup of the liquid sheet depends greatly on the gas-phase Weber number. This importance is illustrated in Figs. 15 and 16 for the case with a positive cos-hill variation in the sheet thickness for the same case shown in Fig. 14b.

Without surrounding gas flow, i.e., in a surrounding gas of zero density, the gas pressure does not appear in the governing equations for the liquid phase, so that the subsequent sheet distortion is determined by the dispersive nature of capillary waves on symmetric distorting sheets. In other words, the original local sheet thickness variation (composed of a wide variety of wave numbers) will disperse along the sheet. This dispersion is symmetrical around $x = x_0$ because the initial conditions along the sheet (i.e., for sheet thickness and sheet velocity) are symmetric around this location (see Fig. 15a).

Figure 15b shows the same case as in Fig. 15a but for a value of $We_g = 1$. Here one clearly observes the influence of the gas-pressure fluctuations on the sheet deformation. The dispersive nature of the capillary waves initially present is still visible, resulting in wave propagation in both directions of the sheet; however, now the surrounding gas heavily impacts the sheet distortion flow, causing asymmetric (with respect to $x = x_0$) sheet distortion and an increase in the maximum disturbance amplitude of the sheet.

The continuous decrease of breakup time with increases in We_g can be observed from Fig. 14b and Fig. 16. In Fig. 16, the similar case is analyzed as in Fig. 14b, but now for $We_g = 2.5$. Similar observations as those described for

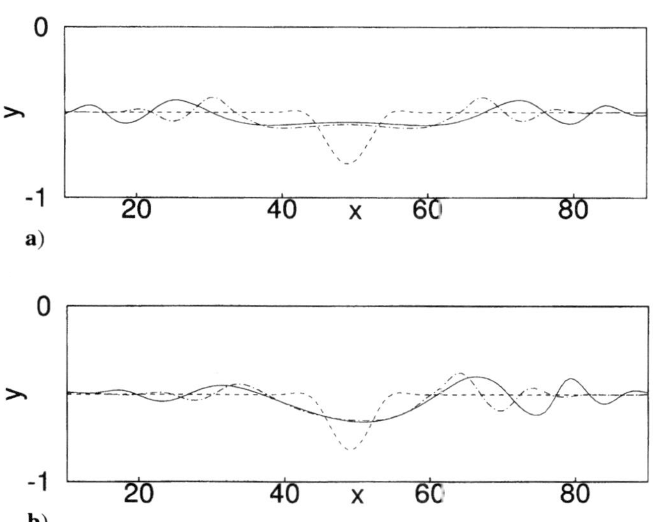

Fig. 15 Dilationally distorting sheet due to initial local sheet thickness increase: a) $We_g = 0$ and b) $We_g = 1$. Symmetry axis at $y = 0$; $\rho_g/\rho_l = 1.2 \cdot 10^{-3}$; − − − $\tau = 0$; −·−·− $\tau = 36$; $[\tau = t/t_{\text{ref}}, t_{\text{ref}} = h\sqrt{\sigma/(2\rho_l h)}]$.[101]

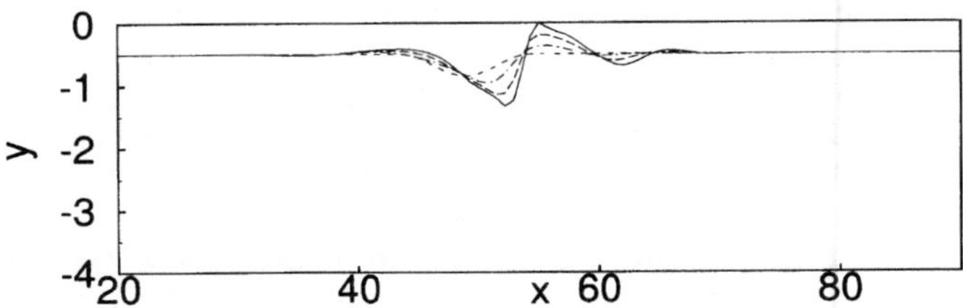

Fig. 16 Dilational distorting sheet due to initial local sheet thickness increase. Symmetry axis at $y = 0$; $We_g = 2.5$, $\rho_g/\rho_l = 1.2 \cdot 10^{-3}$; $- - -$ $\tau = 0$; $- \cdot - \cdot$ $\tau = 8$; $- - \tau = 12$; ——— $\tau = 14.5$; $[\tau = t/t_{\rm ref}, t_{\rm ref} = h\sqrt{\sigma/(2\rho_l h)}]$.[101]

variations in We_g can be made for variations in the density ratio ρ_g/ρ_l at fixed We_g (not illustrated). Comparison shows that with increasing gas phase influence (due to increasing values of We_g or ρ_g/ρ_l) capillary effects, which are stabilizing and cause the prescribed dispersive effects, become less important, and the initial local sheet disturbance remains concentrated to a more and more confined region of the sheet.

Note that for large values of We_g destabilizing aerodynamic effects dominate over stabilizing capillary effects, and numerical error associated with the solution of Laplace's equation and the unsteady Bernoulli equation in the gas phase becomes more important. More accurate predictions of the sheet disintegration process at larger values of We_g require a larger number of nodes for the BEM solution of Laplace's equation in the gas phase as well as a more accurate time-integration scheme.

3. Three-Dimensional Theory

Although liquid sheet breakup is an inherently three-dimensional problem, only a few three-dimensional or quasi-three-dimensional analyses on planar liquid sheets have been reported so far.[103–105] Ibrahim and Akpan[104] presented a fully three-dimensional linear analysis of a plane viscous liquid sheet in an inviscid gas medium. At low Weber numbers, two-dimensional disturbances were found to always dominate the instability of symmetric and antisymmetric waves. Furthermore, when the Weber number is high, long-wave three-dimensional symmetric disturbances were found to have larger growth rates than their two-dimensional counterparts, whereas the opposite was true for antisymmetric disturbances. For short waves, both two- and three-dimensional disturbances grow at approximately the same rate. Increasing the gas-to-liquid-density ratio or decreasing the Ohnesorge number was found to enhance the departure in the growth rates of two- and three-dimensional symmetric disturbances of long wavelength. Both the maximum growth rate and the dominant wave number were predicted to increase with Weber number and density ratio but decrease with Ohnesorge number. Using the reduced-dimension approach previously employed for capillary waves on a two-dimensional planar sheet,[105] a linear temporal analysis provides the following dispersion relations in a coordinate system moving with the undisturbed (uniform) sheet velocity:

$$\omega = \frac{1}{2}\frac{\sigma}{\rho_l a}(k^2 + n^2)$$

for the dilational mode and

$$\omega = \frac{\sigma}{\rho_l a}\sqrt{l^2 + m^2}$$

for the sinuous mode, where k and l, (n, and m) denote dilational and sinuous mode wave numbers in the axial and transverse direction, respectively. This illustrates that, in contrast to planar sheets, both sinuous and dilational capillary waves are dispersive in the presence of disturbances parallel and transverse to the main flow direction. Nonlinear temporal and spatially reduced-dimension analyses of three-dimensional distorting planar sheets are able to predict sheet breakup into droplets without further simplifications. They also provide information on the nonlinear interaction between transverse and longitudinal waves, both of which are observed experimentally.

Three-dimensional nonlinear sinuous and dilational oscillations on a thin planar sheet were considered by Kim and Sirignano.[93] They extended the Mehring and Sirignano[94] analysis by using the reduced-dimension approach to yield an approximate system of two-dimensional, unsteady equations. They examined both temporal spatially periodic waves with prescribed initial conditions and spatially developing instabilities on sheets injected from slot orifices with exit velocity modulation. Standing waves in the lateral direction were assumed while waves traveled in the mainstream direction. A linear analysis was also performed to have a basis for comparison with the numerical nonlinear results. This linear analysis was also useful in setting the boundary conditions for the spatially developing instability by determining the sign of the group velocities. In both the linear analysis and the numerical analysis, the viscous effects and the inertia of the surrounding gas were neglected.

Interesting effects occur when the streamwise and lateral wave numbers are close to each other. For the dilational temporal mode, the kinetic energy and the surface energy oscillate on both a long-time scale and a short-time scale. The surface energy and the kinetic energy can each be divided into two parts, one part for lateral wave motion and the other for streamwise wave motion. On the long-time scale, there is no oscillation of the total kinetic energy or the total surface energy. However, energy is continually transferred from lateral motion to streamwise motion and back to lateral motion. Over the long-time scale, lateral waviness in the thickness disappears while higher harmonics appear in the streamwise thickness wave. Then, the higher harmonics disappear and the lateral waviness reappears. This continues in a cyclic manner on the long period. Similar results occur for spatially developing instabilities at low Weber numbers with small differences between the two wave numbers. At high Weber numbers, with similar wave numbers for each direction, fluid lumps are formed in the dilational mode at wavelength intervals in each direction. These fluid lumps will be the first ligaments in a capillary disintegration. Sinuous three-dimensional oscillations are found to be dispersive, unlike two-dimensional oscillations. Similar to the two-dimensional case, the three-dimensional sinuous waves involve the appearance of dilational waves through nonlinear effects. When the lateral and streamwise wave numbers are close in value, fluid lumps are formed at half-wavelength intervals in both directions for the temporal oscillation. This same phenomenon occurs at high Weber numbers in the spatially developing case.

For the sinuous case with lower Weber numbers, two waves of very different speed and wavelength will propagate differently. Because one wave propagates much faster than the other, there is a domain where only one wave exists.

V.   Annular Free Films

A free film (sheet) is annular when, in the undisturbed state, its inner and outer surfaces are cylindrical and concentric. A conical film (sheet), as noted earlier, does not necessarily have a conical surface but rather a curved axisymmetric surface that can have an approximate cone or bell shape. For both annular and conical films, the curvatures of the outer surface and of the inner surface usually result in positive pressure jumps across the surfaces in the direction of decreasing radial position in cylindrical coordinates aligned with the axis of the flow. If the gas pressure outside of the annular or conical film equals the pressure in the gaseous core surrounded by the film, the capillary pressure causes a radial pressure gradient in the liquid that will tend to collapse the sheet toward the axis of symmetry. This can be stabilized in one of two ways: pressurization of the gaseous core or swirl of the liquid. If the swirl and/or pressurization balance the mean capillary pressure, an annular cylindrical sheet forms. If the swirl and/or pressurization is more than sufficient to balance the capillary pressure, the sheet radius will increase as it flows downstream, causing a conical film to form.

Both the annular and conical films have a surface curvature in the undisturbed state. There is one radius of curvature for a surface in the annular case in the undisturbed state. That radius of curvature is uniform over the surface. On the contrary, the conical film surface has two radii of curvature, each of which can vary with position on the surface. The conical film also differs because of the stream divergence as the flow proceeds downstream. As a consequence, the mean local thickness of the conical film decreases with downstream distance while the thickness of the undisturbed annular film is uniform.

From a practical point of view, annular liquid films or conical films are of greater interest than the planar film. From a theoretical point of view, an annular film is a general two-dimensional geometry, and other common geometries, such as the cylindrical jet and the planar film, can be treated as special cases.[105-109] For annular films or jets, a pure dilational or sinuous mode does not exist, and dominantly sinuous or dilational film disturbances are also often referred to as para-sinuous or para-dilational disturbances.[108,110]

A.   Linear Theory

A linear analysis of an inviscid annular liquid film subject to aerodynamic forces and without the thin film assumption was studied by Crapper et al.,[80] recovering Squire's result[68] for the limiting planar case. Crapper et al.[80] present the general dispersion relation for waves of wave number k on an inviscid thin annular film moving at constant velocity through a quiescent gas. Numerical solutions for the growth rate ω and the wave velocity ω/k are obtained from the dispersion relation (which is of fourth order in ω) for a range of values for the film velocity, film thickness, and cylinder radius. The authors do not address the simultaneous appearance of both modes, but merely note that, in the limit of planar films, the dispersion relation factors into two quadratic equations in ω. The growth rate and range of unstable symmetric (dilational) and antisymmetric (sinuous) modes increase with decreasing annular radius. Also, for symmetric (aerodynamically unstable) waves, the optimum frequency is independent of the film thickness, but for antisymmetric waves, the optimum value (i.e., wave number of maximum growth) is affected by both the radius and film thickness. The velocities of symmetric waves relative to the liquid are the same as the film velocity for all frequencies and are unaffected by the film curvature and thickness. On the other hand, antisymmetric wave velocities are always less than the film velocity and increase with increasing radius and diminishing thickness, whereby the differences between wave and film velocity decrease with increasing frequency. The increase in growth rates with increasing film curvature might be a reason for the observed shortness of conical films produced from swirl spray nozzles, compared with flat films. Note that the waves analyzed by Crapper et al.[80] are dominated by aerodynamic effects.

A linear temporal analysis of the capillary instability for thin axisymmetric annular films as shown in Fig. 1, stabilized in their undisturbed configuration by a constant pressure difference between the inner and outer fluids surrounding the film (i.e., pressure stabilization), yields[22,111]

$$\frac{\partial^2 h}{\partial \tau^2} + \epsilon^2 \left\{ \frac{\partial^4 h}{\partial \xi^4} - C \left[2R \frac{\partial^2 r'}{\partial \xi^2} - (R^2 + 1/4) \frac{\partial^2 h}{\partial \xi^2} \right] \right\}$$

$$= -\frac{1}{R} \epsilon^2 \left\{ 4 \frac{\partial^2 r'}{\partial \xi^2} + 2C[(2R^2 + 1/2)r' - Rh] \right\} \tag{71}$$

$$\frac{\partial^2 r'}{\partial \tau^2} = \epsilon^2 \left\{ 4 \frac{\partial^2 r'}{\partial \xi^2} + 2C[(2R^2 + 1/2)r' - Rh] \right\} \tag{72}$$

where h denotes the nondimensional disturbance of the film thickness, r' is the non-dimensional disturbance of the annular radius, and R denotes its nondimensional undisturbed value. Here, pressure stabilization of the annulus refers to the stabilization of the film in its undisturbed configuration by a constant pressure difference between the inside and outside of the annular film. The undisturbed film thickness $2a$ is used as characteristic length in the nondimensionalization process, and τ and ξ denote the nondimensional time and space variables in a coordinate system moving with the undisturbed film velocity. Also, $C = (R^2 - 1/4)^{-2}$ in the preceding equations, and $\epsilon^2 = 1/(2We) = 1$, if $0.5\sqrt{\sigma/(\rho_l a)}$ is chosen as the characteristic velocity in the nondimensionalization process. Neglecting terms of order R^{-3}, the previous equations reduce to the same equations governing linear waves along a thin rod under prestress or along a string on an elastic foundation; however, the "spring constant" in Eq. (72) is negative. Assuming solutions of the form $e^{i(\omega\tau+k\xi)}$, the corresponding nondimensional growth rates of dilational and sinuous disturbances are decoupled and given by

$$\omega_d = \pm\, \epsilon k\sqrt{k^2 - R^{-2}}; \qquad \omega_s = \pm 2\epsilon\sqrt{k^2 - R^{-2}} \tag{73}$$

In other words, if terms of order R^{-3} can be neglected, then a dilational distortion does not couple with a sinuous deformation, and the sinuous mode leads to variations in the film thickness only through the conservation of mass equation. Also note that the previous analysis assumes that the gas-core pressure is constant and unaffected by perturbations of the liquid interfaces. Equations (73) indicate that instability occurs if the disturbance wave number is smaller than the reciprocal of the undisturbed annular radius, i.e., $k < 1/R$.

In contrast with the planar geometry, not only dilational but both sinuous and dilational waves are dispersive. Linear theory also shows that, in the absence of swirl, disturbances that are not cylindrically symmetric, i.e., circumferential or azimuthal disturbances, are always stable.[61] For the annular geometry, linear coupling between sinuous and dilational waves occurs if terms of order R^{-3} are considered. Note that R corresponds to the nondimensional ratio between annular radius and thickness of the undisturbed film. Similarly to thin planar films,[97,98] a nonlinear reduced-dimension analysis for thin annular films also demonstrates the presence of a nonlinear capillary instability for the dilational (and sinuous) mode in the presence of subharmonic dilational film disturbances.

Linear reduced-dimension spatial analysis shows that, for annular films modulated at the nozzle exit, capillary dilational wave propagation is similar to that observed on thin planar films. Figures 17 and 18 illustrate the dependence of the real parts of sinuous and dilational mode wave numbers, l and k, on modulated thin annular films on the forcing frequency ω. Note that information and energy

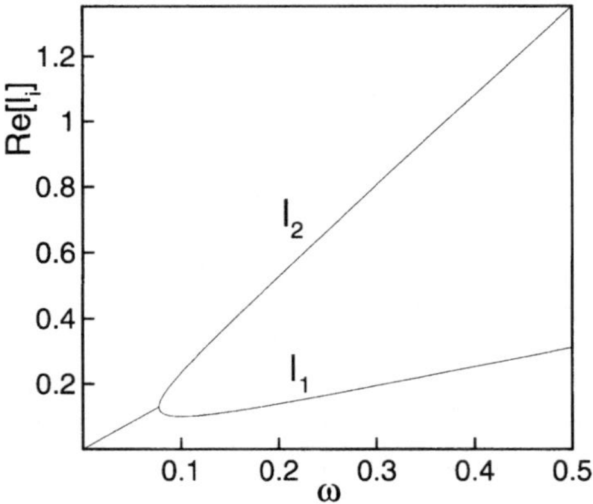

Fig. 17   Real values of sinuous mode wave numbers l_i, i.e., Re[l_i], with dependence on ω for $R = 10$ and $We = 5$.[22]

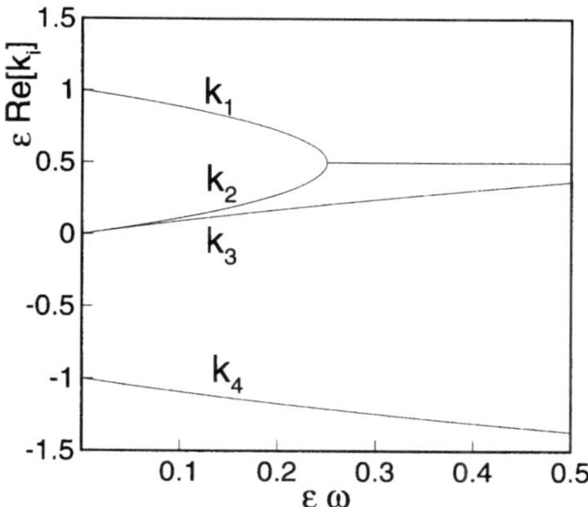

Fig. 18   Real values of dilational mode wave numbers k_i, i.e., $\epsilon\,\mathrm{Re}[k_i]$, with dependence on $\epsilon\omega$ for $(\epsilon/R)^2 = 2\cdot 10^{-5}$.[22]

propagate with the group velocity C given by $C = d\omega/d\mathrm{Re}[l]$ or $C = d\omega/d\mathrm{Re}[k]$, where Re means "real part of."

Only two of the four dilational waves obtained from the dispersion relation, i.e., $k_{2,3}$, will appear on the film for any forcing frequency. For $We > 2$, both sinuous mode waves that are solutions to the dispersion relation will appear even though one of them has negative group velocity in some range of the forcing frequency; for $We < 2$ linear theory predicts that only one of the sinuous mode waves appears downstream from the nozzle. (The former can be shown by numerical simulations of the transient initial and boundary value problem.[22]) For sufficiently large Weber numbers, i.e., $We > 2$, and stable films, the film distortion with a beat in the envelope of the film-centerline location or the film thickness (depending on whether sinuous or dilational modulations are forced onto the film) is analogous to the one observed on modulated planar films. Sheet breakup or collapse might occur if the beat amplitude in the variation of film thickness or annular centerline location is sufficiently large. However, in contrast to planar films and because of the nonzero second radius of curvature, there are now ranges of forcing frequencies for which dilational and sinuous film modulations lead to exponential growth in the film thickness disturbance or the film-centerline disturbance downstream from the nozzle. The corresponding critical frequency for the sinuous mode is given by

$$\omega_{s,\mathrm{cr}} = \frac{1}{R}\sqrt{1 - 4\epsilon^2} \qquad (74)$$

which is accurate through order R^{-2}. For the dilational mode, the critical frequency $\omega_{d,\mathrm{cr}}$ is given by the solution of

$$1 + F - \frac{1+F}{3} + \frac{a}{3b^{1/3}} + \frac{b^{1/3}}{3} = 0 \qquad (75)$$

with

$$F = (\epsilon/R)^2, \qquad a = 1 + 2F + F^2 - 12(\epsilon\omega_{d,\mathrm{cr}})^2$$

$$b = -1 - 3F - 3F^2 - F^3 + (18 - 36F)(\epsilon\omega_{d,\mathrm{cr}})^2 + 6\sqrt{3}c\epsilon\omega_{d,c}$$

$$c = F + 3F^2 + 3F^3 + F^4 - \epsilon\omega_{d,\mathrm{cr}}[1 + 20F - 8F^2 - 16(\epsilon\omega_{d,\mathrm{cr}})^2]$$

which is also accurate through order R^{-2}. Note that for this spatial analysis, the undisturbed film velocity (with respect to the nozzle exit) is chosen as the characteristic length for the nondimensionalization, and $\epsilon^2 \neq 1$.

The growth-rate predictions obtained from the temporal analysis agree with the results of Dumbleton and Hermans[61] in the limit of very thin films. They presented a linear two-dimensional temporal analysis of the capillary instability observed in axisymmetric inviscid annular films of finite thickness. The capillary stability of liquid films has also been considered in context with the stability analysis of compound jets, in which a liquid jet (inner layer, fluid 2 in Fig. 1) is surrounded by a second (outer) liquid layer (fluid 1 in Fig. 1) or annular liquid jet that is surrounded by a void (fluid 3 in Fig. 1). The flowfield within the second annular liquid layer is part of the solution and might in fact determine the stability of the overall system[110] (see Section III.G).

Dumbleton and Hermans[61] considered an inviscid annular liquid film in a surrounding void and a gaseous core of constant pressure. As later done by Chauhan et al.[112] Dumbleton and Hermans analyzed the axisymmetric case after observing that, for disturbances that are not cylindrically symmetric, all axial wave numbers are stable. Linear modal analysis, with disturbances of the inner (subscript a) or outer interfaces (subscript b) given by $\epsilon_{a,b}\exp[i(kz + mt)]$, provided a second-order equation for m^2, whereby m denotes the growth rate and m^2 is purely real. The two identified modes given by m_1^2 and m_2^2 essentially correspond to the squeezing and stretching modes later mentioned by Chauhan et al.[112] The mode pertaining to m_1^2 is shown to become unstable if $k < 1/R_2$, and the one related to m_2^2 is shown to be unstable for $k < 1/R_1$. Also, for the case in which both modes are unstable, i.e., $m_{1,2}^2 < 0$, $-m_2^2$ is found to be significantly larger than $-m_1^2$. This has been illustrated by tabulating the values of $m_{1,2}^2$ as functions of k for three different film thicknesses. The analysis by Dumbleton and Hermans also showed $-m_2^2$ is larger for $R_2/R_1 = 2$ than for $R_2/R_1 = 10$. The authors also state that, by proper choice of initial conditions, it is possible to suppress either one of the modes. This implies that if one of the two solutions of m^2 is negative and the other one positive (for a given parameter set), the initial conditions determine if the film is stable or unstable.

From an analysis of the two modes pertaining to m_1^2 and m_2^2 for thin annular films, Dumbleton and Herman find the same behavior of the amplitude disturbance ratio ϵ_a/ϵ_b for each mode around its particular stability limit, as described by Chauhan et al.[112] for the compound jet. (Compare Fig. 2 of Ref. 61 and Fig. 6 in Ref. 112.) For the first mode, i.e., m_1^2, the inner and outer interface are in phase (symmetric distortion) for $k > 1/R_2$ (and $\epsilon_a/\epsilon_b = 1$ for $k > 1/R_1$ in particular), whereas for $k < 1/R_2$ (unstable mode) the interfaces move out of phase (antisymmetric distortion, $\epsilon_a/\epsilon_b < 0$). For the second mode, i.e., m_2^2, the interfaces move out of phase in the stable region, i.e., for $k > 1/R_1$, but in phase in the unstable region $k < 1/R_1$ (in particular $\epsilon_a/\epsilon_b = 1$ if $k < 1/R_2$). In their examination of the interfacial shapes, Dumbleton and Hermans and Chauhan et al. both report that for m_1^2 (squeezing mode) the inner interface is undisturbed at $k = 1/R_2$. The similar neutral mode is found for m_2^2 (stretching mode), whereas the outer interface remains undisturbed for $k = 1/R_1$.

Even though many similarities between the two prescribed analyses have been found, there is also one significant difference. The growth rate[112] for both modes approaches zero as the wave number approaches zero. However, in the analysis presented by Dumbleton and Hermans, this is the case only for the squeezing mode or m_1^2. The stretching mode represented by m_2^2 is found to have a finite value in the limit of very large wavelengths. These differences can be explained by the fact that, in contrast to Chauhan et al., Dumbleton and Hermans considered the pressure variation across the thin annular film. These pressure variations are not relevant for the compound jet because the liquid core jet determines the behavior of the overall jet. However, one observes that, as the density of the liquid core approaches zero, the dynamics of the thin outer layer become important for the stretching mode. In that case, the driving force for the film distortion is the pressure difference across the film and needs to be considered within the analysis (see Dumbleton and Hermans[61]).

B.   Nonlinear Theory

Nonlinear analyses of annular liquid films were presented by Lee and Wang[113–115] and Ramos,[116] who considered the formation of closed cylindrical liquid shells through the collapse of annular liquid films. The analyses assumed the liquid layer to be a structureless sheet and neglected axial pressure gradients, omitting the possibility for dilational waves. Time-independent boundary conditions were specified at the nozzle exit. Boundary conditions were specified for the film thickness, the annular radius, and the axial and transverse velocity components at the nozzle exit. Wave propagation properties at this boundary were not taken into consideration. Ramos concluded that the convergence length and the volume enclosed by annular films increase as the Froude and Weber numbers, the pressure coefficient (i.e., the nondimensional pressure difference across the film), and the nozzle exit angle are increased, but decrease as the thickness-to-radius ratio at the nozzle is increased. Depending on the balance between pressure forces and surface tension forces, the annular liquid jets remain cylindrical (if the nozzle exit angle is zero) and can converge or not converge.

The thin sheet assumption leads to a reduced-dimension approach for axisymmetric annular and conical (swirling or nonswirling) sheets as shown in the following.[22,111,117,118] The problem is cast in cylindrical coordinates.

We can assume an analytical behavior of the governing equations away from $r = 0$, as a function of r. Also, assuming that the sheet thickness is small compared with the streamwise disturbance wavelength, it is consistent to consider v_z and v_θ to be nearly constant and v_r and p to be linearly varying with r. It is convenient, therefore, to reduce the problem to a one-dimensional, unsteady formulation by integrating Eqs. (1–4) over the sheet thickness. This is done by using Leibnitz's rule and by considering the previously mentioned velocity and pressure profile approximations. Introducing averaged quantities $\bar{\phi}$, defined by

$$\bar{\phi} \equiv \frac{\int_{r-}^{r+} (\phi 2\pi r)\, dr}{\int_{r-}^{r+} 2\pi r\, dr} = \frac{\int_{r-}^{r+} (\phi r)\, dr}{\bar{r}\,\Delta r} \tag{76}$$

one obtains

$$\frac{\partial \Delta r}{\partial t} + \frac{\partial(\bar{v}_z \Delta r)}{\partial z} = -\frac{\Delta r}{\bar{r}}\,\bar{v}_r \tag{77}$$

$$\frac{\partial \bar{v}_z}{\partial t} + \bar{v}_z \frac{\partial \bar{v}_z}{\partial z} = -\frac{1}{\rho_l}\left[\frac{\partial \bar{p}}{\partial z} - \Delta p\left(\frac{1}{\Delta r}\frac{\partial \bar{r}}{\partial z} + \frac{1}{4\bar{r}}\frac{\partial \Delta r}{\partial z}\right)\right] \tag{78}$$

$$\frac{\partial \bar{v}_r}{\partial t} + \bar{v}_z \frac{\partial \bar{v}_r}{\partial z} = -\frac{1}{\rho_l}\frac{\Delta p}{\Delta r} + \frac{\bar{v}_\theta^2}{\bar{r}} \tag{79}$$

$$\frac{\partial \bar{v}_\theta}{\partial t} + \bar{v}_z \frac{\partial \bar{v}_\theta}{\partial z} = -\frac{\bar{v}_r \bar{v}_\theta}{\bar{r}} \tag{80}$$

$$\frac{\partial \bar{r}}{\partial t} + \bar{v}_z \frac{\partial \bar{r}}{\partial z} = \bar{v}_r \tag{81}$$

with $\Delta p = p_+ - p_-$ and $\bar{p} = (p_+ + p_-)/2$ given by

$$\bar{p} = -\frac{\sigma}{2}\left\{\frac{G_+ + G_-}{2}\frac{\partial^2 \Delta r}{\partial z^2} + (G_+ - G_-)\frac{\partial^2 \bar{r}}{\partial z^2}\right.$$
$$\left. + \frac{1}{D}\left[\frac{F_+ + F_-}{2}\Delta r - (F_+ - F_-)\bar{r}\right]\right\} \tag{82}$$

$$\Delta p = -\sigma\left\{(G_+ + G_-)\frac{\partial^2 \bar{r}}{\partial z^2} + \frac{G_+ - G_-}{2}\frac{\partial^2 \Delta r}{\partial z^2}\right.$$
$$\left. - \frac{1}{D}\left[(F_+ - F_-)\bar{r} - \frac{F_+ - F_-}{2}\Delta r\right]\right\} \tag{83}$$

where $D = \bar{r}^2 - \Delta r^2/4$, and $F_\pm$ and $G_\pm$ are given by

$$F_\pm = G_\pm^{1/3} \tag{84}$$

$$G_\pm = \left[1 + \left(\frac{\partial \bar{r}}{\partial z}\right)^2 \pm \frac{\partial \bar{r}}{\partial z}\frac{\partial \Delta r}{\partial z} + \frac{1}{4}\left(\frac{\partial \Delta r}{\partial z}\right)^2\right]^{-3/2} \tag{85}$$

Equations (77–85) form a closed system of equations that, together with appropriate boundary and initial conditions, govern the nonlinear distortion of thin axisymmetric liquid sheets exiting from a nozzle or atomizer into a void. Sheet breakup occurs when the independent variable for the sheet thickness reaches zero value locally. For the subsequent analyses, the preceding equations have been nondimensionalized by using the sheet thickness and axial velocity component at the nozzle exit, i.e., Δr_0 and $\bar{v}_{z,0}$, as characteristic length and velocity. The Weber number in the resulting nondimensional equations is then given by $We = \rho_l \bar{v}_{z,0}^2 \Delta r_0 / \sigma$.

It has previously been shown[94] that, for the case of a planar sheet, the employed assumptions for the velocity and pressure profiles across the sheet agree with the lowest-order expansion of the full two-dimensional problem in terms of $(y - \bar{y})/\lambda$, where y denotes the direction perpendicular to the undisturbed sheet, $\bar{y}(x, t)$ is the instantaneous location of the sheet centerline in the y-direction, and λ is the wavelength of a disturbance in the x-direction. In other words, for the planar case the reduced-dimension equations are exact in the limit where the ratio between sheet thickness and λ equals zero.

For the steady-state annular liquid membranes with sinuous distortions, the analogy between integral (control-volume) formulation and rigorous Taylor series expansions has been demonstrated by Ramos.[119] The integral representation employed by the authors for nonswirling annular sheets[22,111] follows, as mentioned earlier, Ramos's analysis[116] for the steady version of the configuration but extends it to the unsteady cases of both sinuous and dilational sheet distortions.

The limitations of the employed formulation, in particular during pinch-off (when short wavelength disturbances cannot be neglected), have been addressed.[22,111] It is noted, however, that despite these constraints, the usefulness of the approach has been demonstrated conclusively for planar sheets, by comparison with accurate two-dimensional vortex-dynamics simulations.[94]

Using this reduced-dimension analysis, Mehring and Sirignano[22,111] provided time-dependent solutions for nonlinearly distorting, initially undisturbed axisymmetric annular liquid films modulated at the nozzle exit. Boundary conditions were specified according to wave propagation information obtained from the corresponding linear boundary-value problem. Linearly stable or unstable dilationally and sinuoidally modulated films were also found to be nonlinearly stable or unstable, respectively. For stable dilationally modulated films, nonlinear effects lead to an increase in breakup time and breakup length (Fig. 19).

The same observation is made for linearly stable dilationally modulated films for which large beat amplitudes result in film breakup. Strong nonlinear coupling between the dilational and sinuous modes might also lead to the formation of large gas bubbles between the fluid rings if the gas core is assumed to have constant pressure (Fig. 20).

The constant gas-core pressure assumption implies that pressure disturbances generated within the gas core by the distorting annulus are adjusted instantaneously by the undisturbed part of the semi-infinitely long gas-core column. Thus, the speed of sound of the gas within the annulus is assumed to be infinite or significantly larger than the velocity of the investigated capillary waves. Note that the ratio of disturbed and undisturbed gas-core volumes is 1 in the case of locally modulated semi-infinite films, but the ratio has a finite value for periodically disturbed infinite films. In the latter case, the time-dependent (but spatially uniform) gas-core pressure might introduce "elastic" phenomena such as those observed in bubble dynamics.

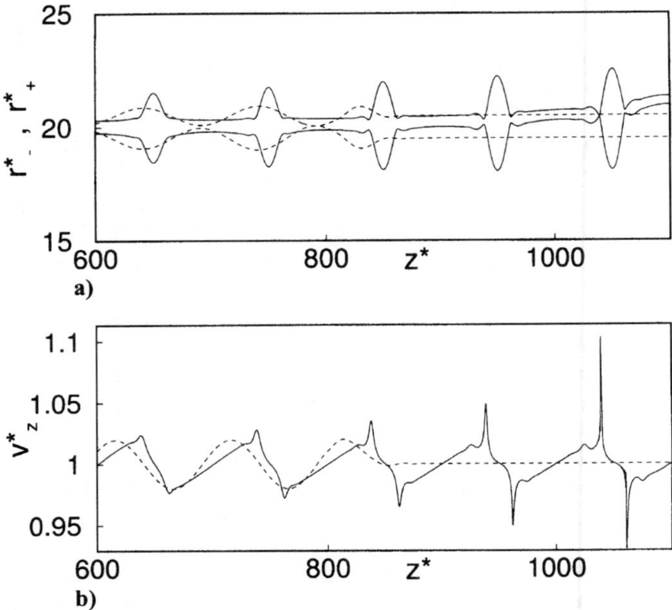

Fig. 19   Linear (– – –) and nonlinear (——) distortion of dilationally modulated and linearly stable film at breakup: a) interface location and b) nondimensional axial velocity component [$We = 150$, $R = 20$, $T = 2\pi/\omega = 100$]. Nozzle exit at $z^* = 0$.

Also, analogous to the planar case, at the point of film rupture (see Fig. 19), the presence of short wavelength disturbances will lead to increasing uncertainties in the accuracy of the employed long-wavelength approximation. The same is true for the bubble formation process shown in Fig. 20. Here, large slopes in the film centerline with respect to the x-axis lead to significant errors within the employed long-wavelength theory, which uses essentially lower order approximations for the velocity and pressure profiles within the film in the direction perpendicular to the x-axis.

Similar to the capillary breakup characteristics of cylindrical liquid jets and symmetric distorting planar liquid films, nonlinear effects on dilationally modulated films result in the formation of fluid rings connected by thin annular liquid shells or smaller satellite rings. The same observation was also made by Panchagnula et al.[120] Film breakup occurs when the prescribed shells or smaller rings thin out even more in regions close to the larger liquid rings.

For linearly unstable sinuous modulated annular films, nonlinear effects always decrease film collapse lengths and times (Fig. 21). For linearly stable sinuous modulations, they result in the agglomeration of fluid into liquid rings that for small values of R might have considerably different cross-sectional areas and circumferential lengths along the film. Depending on the breakup length and the wavelength of the beat envelope in the variation of the annular

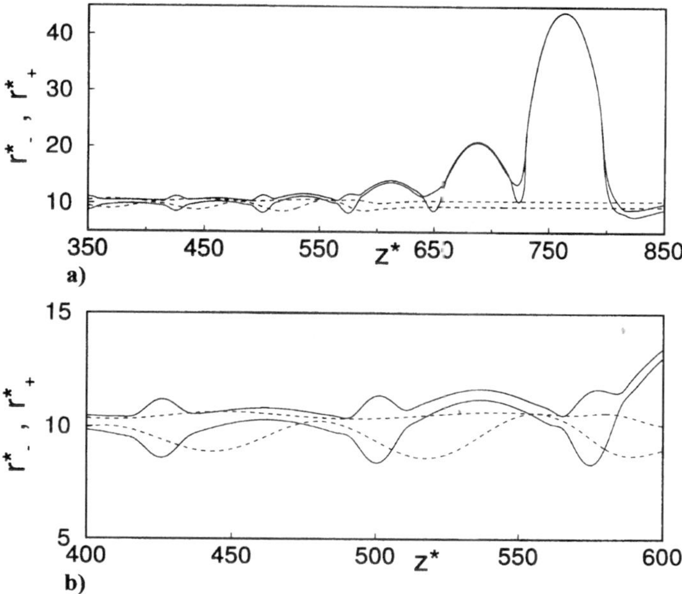

Fig. 20   Linear (– – –) and nonlinear (——) distortion of dilationally modulated and linearly unstable film at breakup: [$We = 150$, $R = 10$, $T = 2\pi/\omega = 75$].[22] Nozzle exit at $z^* = 0$. a) Overview and b) magnified details.

radius, the disintegrating annulus will show the presence of satellite rings between the larger liquid rings. The variation of film breakup or collapse length and time with a variation of Weber number We, annular radius R, forcing period T, or disturbance amplitude A depends on the location in parameter space (We, R, T, A). It is noted here that the linear results shown in Figs. 19–21 include terms of order R^{-3}.

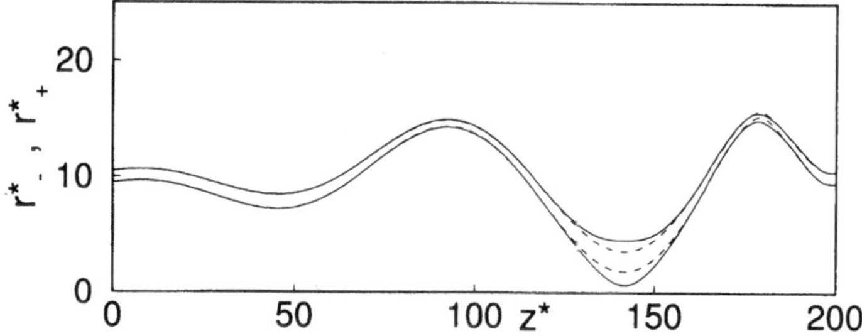

Fig. 21   Linear (– – –) and nonlinear (——) distortion of sinusoidally modulated and linearly unstable film at $t^* = 190$ [$We = 1000$, $R = 10$, $T = 2\pi/\omega = 100$].[22] Nozzle exit at $z^* = 0$.

C. Effect of Swirl

A linear temporal analysis of three-dimensional swirling inviscid annular films with inner and outer gas flows of different velocities was conducted by Panchagnula et al.[121] For small Weber numbers, inner gas flows lead to significantly faster growing instability modes than outer gas flows at the corresponding Weber number. For large Weber numbers inner and outer gas flow have the same effects. In the absence of swirl, the disturbance mode with maximum growth rate always has a circumferential mode number of zero. For nonzero axial Weber numbers, an increase in swirl increases the range of unstable axial and circumferential modes as well as their growth rates. It also increases the axial wave number with maximum growth. For large enough swirl Weber numbers, the fastest growing disturbance was predicted to have nonzero circumferential but zero axial wave number.

An extension of the prescribed reduced-dimension analysis for annular nonswirling films to swirling annular films allows for the study of the influence of the previously mentioned constant gas-core pressure assumption on the overall film distortion; it is assumed that the undisturbed annular film is now stabilized by a constant swirl. It also allows for the study of film divergence if the angular momentum at the nozzle exit exceeds the amount required to stabilize the undisturbed annular configuration.

Neglecting terms of order R^{-3}, the linear spatial stability analysis for capillary waves on swirling annular films shows that the stability limits for the forcing frequency ω are larger than in the corresponding pressure-stabilized cases. Also, dispersion relations for dilational and sinuous waves cannot be obtained separately, and sheet distortion is characterized by predominantly dilational or predominantly sinuous waves. For the predominantly sinuous mode, no spatial instability is observed if the film is swirl stabilized rather than pressure stabilized. The argument here is that the stabilizing effect of the centrifugal forces due to swirl exceeds the destabilizing capillary forces if variations of the annular radius occur. This is in contrast to the pressure-stabilized annular film where both capillary forces and pressure forces, exerted on the film by the gas core, destabilize the thin sinusoidally deformed annular film. Spatial stability analysis shows that, for sinuous modulated swirl-stabilized films, exponential solutions exist only for $We < 2$. However, the group velocities of the corresponding waves are negative, so that these waves are not expected to appear downstream from the nozzle. Pressure-stabilized films with modulated sinuous mode disturbances might be unstable only if $We > 2$ (see Figs. 22 and 23).

A nonlinear analysis of swirling thin annular films is obtained analogously to the nonswirling pressure-stabilized case.[117] For films with radial displacement only and without disturbances in the axial direction, the governing equations are given by

$$\Delta r\bar{r} = e_1 \tag{86}$$

$$\bar{v}_\theta \bar{r} = e_2 \tag{87}$$

$$\frac{\mathrm{d}^2\bar{r}}{\mathrm{d}t^2} = -\frac{2\sigma}{\rho_l e_1} + \frac{e_2^2}{\bar{r}^3} \tag{88}$$

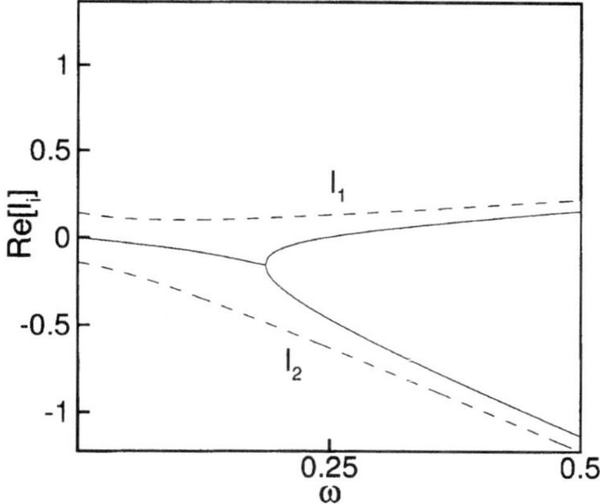

Fig. 22   Real values of wave numbers l_i, i.e., Re[l_i], with dependence on ω for $R = 10$ and $We = 1$ [solid: swirl-stabilized case; dashed: pressure-stabilized case].[117]

where Δr and $\bar{r}$ are the film thickness and the radial film-centerline location, $\bar{v}_\theta$ is the circumferential velocity component, and the constant values e_1 and e_2 are determined by the initial geometry of the cylindrical film and the angular momentum or swirl imposed at $t = 0$. Equation (88) shows that, depending on the initial swirl prescribed by e_2, the initial configuration prescribed by e_1, and the liquid properties (i.e., surface tension coefficient σ and density ρ_l), film collapse is possible, whereas continuous film blowout is not possible. An oscillatory behavior for the temporal variation of $\bar{r}$ occurs around the equilibrium location $\bar{r}_s = We_x^{1/2}\bar{r}_0$ (assuming $d\bar{r}/dt = 0$ initially), whereby the swirl Weber number We_s is formed by $\bar{v}_\theta(t = 0)$, $\Delta r/2(t = 0)$, σ and ρ_l. If $We_s < 1$, radial acceleration of the film is

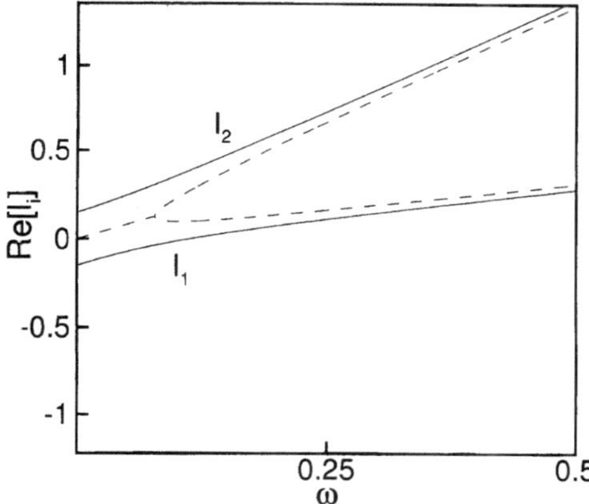

Fig. 23   Real values of wave numbers l_i, i.e., Re[l_i], with dependence on ω for $R = 10$ and $We = 5$ [solid: swirl-stabilized case; dashed: pressure-stabilized case].[117]

initially inward (toward $\bar{r}_s$); if $We_s > 1$, the film accelerates radially outward toward its equilibrium position at $\bar{r}_s$. In the former case, the sheet might collapse into a "full" cylindrical jet. In a swirl-stabilized situation with only material and capillary forces, sheet blowout is not possible. However, for a pressure-stabilized annular sheet, either collapse or blowout is possible under certain circumstances.

For swirling annular films with inertia and surface tension only, energy is conserved with energy exchange between potential (surface) energy and kinetic energy (due to angular and radial momentum). However, with an imposed pressure difference, energy exchange with the ambient gas takes place, which might render the film unstable. For the latter case, one should note the similarity to the observations for a steady-state spatially developing free-falling liquid curtain with pressure difference across the film made by Finnicum et al.[122] for the planar case and by Ramos[123] for the axisymmetric annular case.

VI. Conical Free Films

The spatial development of films exiting from an annular nozzle or atomizer, with excess swirl, into a void, is characterized by film divergence close to the nozzle exit and subsequent oscillatory variations of the film radius in the downstream direction. Referring to the diverging part of the film directly at the nozzle exit, a spatial analysis of swirling conical films can be provided.

In Ref. 117, steady-state nonlinear solutions for the pure initial-value problem of a spatially developing swirling axisymmetric annular film exiting from a nozzle or atomizer (into a passive ambient gas) are presented for various sets of boundary conditions at the nozzle exit. Similar steady-state analyses were also presented by Yarin[124] for the case of swirling liquid membranes and by Ramos[116] for nonswirling thin liquid films. As within the described analysis,[117] the latter authors[116,124] specified boundary conditions only at the location where the film exits the nozzle or atomizer.

For the steady-state problem, the specification of the first- and second-order derivatives of the film thickness and the film-centerline location at the nozzle exit is a delicate matter, because it is not possible to determine a priori the fate of the film downstream, i.e., with respect to the encounter of singular points that render the existence of a steady-state solution impossible for the specified combination of nozzle exit conditions.

For the unsteady transient or time-periodic problem, in which capillary waves are generated on the film due to some forcing conditions imposed along the film, constraints that apply to the propagation of these waves and their energy facilitate the specification of boundary conditions. For example, for the time-periodic linear problem, the number of boundary conditions to be prescribed at the inflow and outflow planes of the considered spatial domain will depend, as already discussed for planar films, on the group velocities or energy propagation characteristics associated with the (capillary) waves traveling along the film. For example, if energy is imposed onto the semi-infinite film only at the nozzle exit, then capillary waves associated with energy transport upstream, i.e., with group velocities directed in the upstream direction, have to be excluded from the solution of the linear problem. The rejection of (linear) solutions associated with such wave numbers effectively results from the use of the Sommerfeld radiation (boundary) condition at infinity, i.e., the requirement that energy has to be propagated downstream (or outward) at infinity.

Richer in content and of more practical interest than time-periodic solutions are transient solutions to the initial- and boundary-value problem. Transient linear solutions can be obtained numerically or analytically by employing Fourier–Laplace transforms to obtain the dispersion relation in terms of complex wave number and frequencies. Numerical solution also allows ready access to the solution of the nonlinear problem. Here, the specification of boundary conditions is even more problematic as wave reflection at the nozzle exit might have to be considered. The latter occurs, for example, in the development of an absolute film instability triggered by some downstream disturbance on the liquid film at low Weber numbers. It remains a challenge to specify boundary conditions (in number and type) that result in an accurate representation of wave reflection within a simplified mathematical model for the overall film dynamics and provide reasonable agreement with solutions obtained by more elaborate models or with experimental observation.

Steady-state or time-periodic solutions can be obtained by transient simulations starting from some reasonable initial conditions.[125] Accordingly, the selection of boundary conditions for the linear time-periodic problem, based on the group velocity of linear capillary waves (as discussed earlier), can in many cases be validated by linear transient numerical solutions employing these boundary conditions. Similarly, nonlinear steady-state solutions can be obtained by nonlinear transient numerical simulations. If disturbances in the flow quantities are not too large at the nozzle exit, and nonlinear effects remain contained further downstream, then it is reasonable to consider the same nozzle exit conditions for both linear and nonlinear transient simulations.

Depending on the employed solution procedure for the transient problem, more boundary conditions might have to be specified at the nozzle than suggested by modal analysis of the time-periodic problem. These numerical boundary conditions need to be specified such that they do not significantly influence the behavior of the system. Also, the stability of these additional numerical conditions has to be assured.[126]

In an effort to determine physically reasonable boundary conditions for simplified liquid stream distortion models, the full governing equations at the nozzle were considered.[127,128] The understanding gained from an analysis considering the wide variety of physical phenomena present at the transition point from a wall-bounded stream to a freestream could be used to reformulate the problem by considering inner and outer solutions to the spatially developing liquid stream.[127–129]

For the nonlinear transient simulation of a semi-infinitely long swirling axisymmetric annular film,[117] the number of nozzle exit boundary conditions (excluding numerical conditions) was chosen according to the number of waves with positive group velocities for the corresponding linear time-periodic problem. Furthermore, the specification of downstream boundary conditions was omitted in Ref. 117 by choosing the computational domain large enough, preventing any modulation-generated waves from reaching the domain boundary. A more detailed description regarding the specification of boundary conditions (for inviscid flows) and selection of relevant wave numbers by means of group velocity arguments for the possible wave systems can be found elsewhere.[22,94]

Figures 24a–d, taken from Ref. 117, illustrate nonlinear steady-state solutions for $We = 1000$ and swirl number (defined in the following) $k = 10, 2, 0.1$, and 1, respectively. The remaining boundary conditions[117] are identical for cases a–c but differ for case d. $G_\pm = F_\pm = 1$.

If the angular momentum at the nozzle exit is smaller than the amount needed to stabilize the undisturbed annular film with radius $\bar{r}_0$ and thickness Δr_0, the film collapses onto the symmetry axis (see Fig. 24c). In this case, the centrifugal forces cannot compensate for the capillary pressure, which promotes a minimization of the surface area of the swirling film. If the angular velocity at the nozzle exit exceeds the critical value that stabilizes the annular configuration, i.e.,

$$\bar{v}^*_{\theta,0} = \sqrt{2/\{We[1 - \Delta r_0^2/(4\bar{r}_0^2)]\}}$$

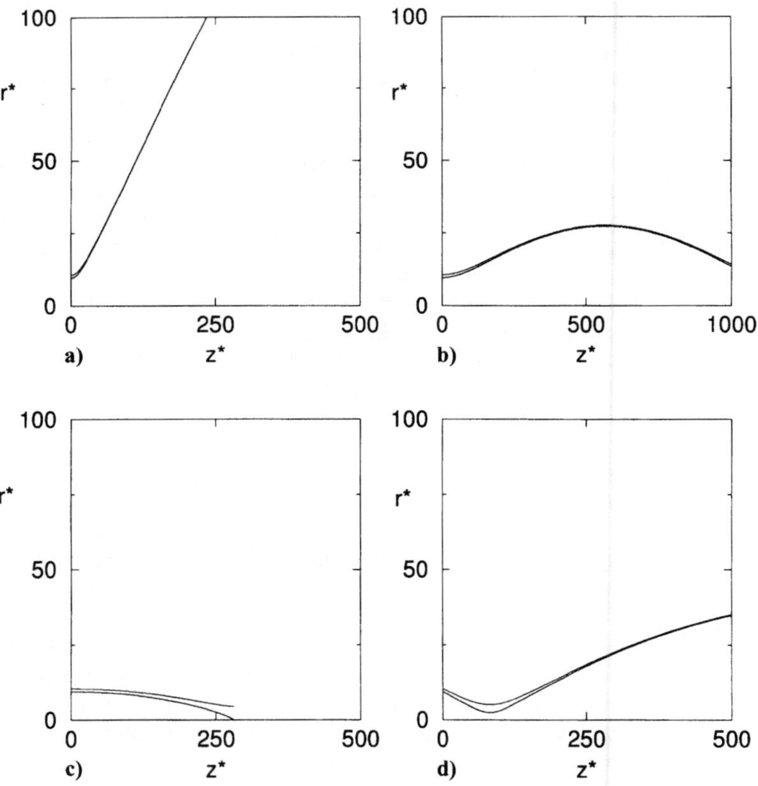

Fig. 24   **Steady-state solutions for conical free film under various conditions.**[117]

the radial centerline position of the film oscillates in the downstream direction of the nozzle according to the energy transfer between surface energy and energy stored within the swirling motion of the film (see Figs. 24a,b,d). These films are often described as conical in the engineering practice, although they can deviate substantially from a true conical shape. For the large Weber number cases illustrated in Fig. 24, steady-state solutions were readily obtained. However, for low Weber numbers, the steady-state equations might become singular at some location downstream from the nozzle. (See the equation for $ds/dz^* = d^2\bar{r}^*/dz^{*2}$ in Ref. 117.) The singularity is removable by appropriate choices for the amount of swirl imposed at the nozzle exit and/or by changing the boundary conditions for $d\bar{r}^*/dz^*$ and $d^2\bar{r}^*/dz^{*2}$ at the nozzle exit.

Steady-state equations for nonswirling axisymmetric annular liquid membranes and their singularities have been analyzed by Ramos,[123] who included gravity effects into the steady-state version of the model by Lee and Wang.[113] Similar to the preceding discussion, the singularities identified by Ramos at the nozzle exit or downstream from the nozzle were removable by imposing some restrictions on the slope and the curvature for the annular membrane at the nozzle exit.

The nonlinear analyses[117] on conical films with modulations of the axial or transverse velocity component at the nozzle exit show that film thinning due to film divergence does not fundamentally change the characteristics of the non-linear capillary breakup observed for swirling but nondiverging annular films modulated at the nozzle exit (see, for example, Fig. 25). However, an increase in the thinning rate or cone angle of diverging annular films, resulting from an increase in the angular momentum or swirl at the nozzle exit, leads to smaller disturbance amplitudes and longer breakup lengths for the same modulation at the nozzle exit. These breakup lengths and times might be significantly larger than those observed for the corresponding modulated planar films. For dilational film modulations, the effect of Weber number changes on film breakup length l_b and breakup time t_b is determined by the relative importance of the radius of curvature in the main flow direction R_1, the radius of curvature in the corresponding perpendicular direction R_2, and the relative changes of the thinning rate with changes in We at constant swirl number. Here, the term swirl number refers to the ratio k of swirl c_2 imposed at the nozzle exit to the amount of swirl $c_{2,0}$ necessary to stabilize the film in its annular configuration, i.e., $c^2 = kc_{2,0}$. For sinuous film modulations, film breakup occurs due to nonlinear coupling with the dilational mode. Here, the dependency of nonlinear breakup length and time on Weber number is greatly influenced by the nonlinear mode coupling and the linear and nonlinear dilational mode behavior. The former decreases with increasing Weber number (and annular radius). A description of the dependencies for both modes is given in Ref. 117.

Comparison between thin annular swirl- or pressure-stabilized films shows that swirl stabilization for annular liquid films can significantly reduce nonlinear and linear breakup lengths for thin dilationally modulated films (see Fig. 26). The comparison also shows that because of the constant gas-core pressure, bubble formation between fluid rings observed for pressure-stabilized films is not found on swirl-stabilized films.

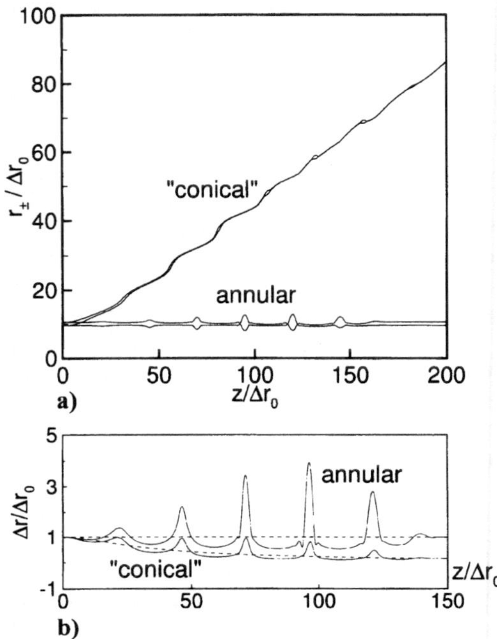

Fig. 25　Annular and conical swirl-stabilized liquid film dilationally modulated at the nozzle exit $z = 0$, r_+ = radius of the outer $(+)$ and inner $(-)$ interface, Δr = local film thickness (subscript 0 denotes its value at $z = 0$).[117]

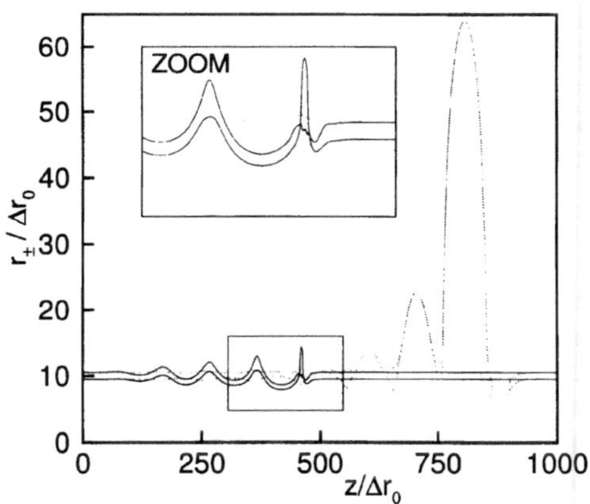

Fig. 26　Annular dilationally modulated liquid film stabilized at breakup with constant gas-core pressure and no swirl (dotted), or with constant swirl but zero gas-core pressure (solid). [$We = 150$, $T = 2\pi/\omega = 100$].[117]

The analysis of axisymmetric conical films that are modulated at the nozzle exit shows that an increase in the thinning rate results in smaller absolute amplitude disturbances downstream for the same amplitude at the nozzle. Consequently, the appearance of nonlinear effects (i.e., higher harmonic disturbances) is typically delayed, thus resulting in longer breakup times and larger breakup lengths. This is particularly true for sinuous modulated films (i.e., modulation of the transverse velocity at the nozzle exit). Also, for large enough thinning rates (i.e., swirl numbers), film breakup might not occur, even though breakup was predicted at lower thinning rates under the same forcing conditions. The results[117] show that the breakup lengths and times for diverging films (with excess swirl at the nozzle exit) are generally significantly greater than for the similar planar case. This is found to be particularly true for larger swirl numbers k and sinuous mode simulations. On the other hand, breakup lengths and times for swirl-stabilized modulated annular films are in general smaller than those predicted for the corresponding planar cases. Planar films have been approximated[117] by simulating swirl-stabilized annular films with an annular radius of $R = \bar{r}^* = 10^4$ at the nozzle exit.

Following the reduced-dimension approach, Mehring and Sirignano[118,130] have recently extended their analysis of swirling conical films to three-dimensional disturbances. Initial conditions for the nonlinear numerical simulations were obtained from the solution to the corresponding nonlinear steady-state problem without modulation imposed at the nozzle exit. The similar boundary conditions as previously employed for axisymmetric swirling conical films were utilized, with the exception that now modulations were imposed onto the film thickness and film-centerline location rather than onto the axial and transverse film velocity. Also now, the modulations varied spatially in the circumferential direction at the nozzle exit.

Figure 27 shows the outer film surface location for the case of a swirling conical sheet generated by liquid that exits an annular slit nozzle with 10 times the amount of swirl needed to stabilize the film in its annular configuration. Initial conditions for this case were obtained by solving the nonlinear steady-state axisymmetric equations. The film interface location of this steady state is illustrated in Fig. 24a. The liquid film is modulated sinusoidally at the nozzle ($z = 0$) according to

$$v_0 = C \exp(-t^*/T_{\text{ex}})\sin(2\pi\tau^*/T_p)\cos(n\theta)$$

with $C = 0.1$ and $n = 5$ (i.e., 5 standing waves imposed at the nozzle in the circumferential direction). The Weber number for this case is $We = 1000$, $R = 10$, and $T_{\text{ex}} = T_p = 10$. For the same case and the same time, Fig. 28 shows the instantaneous film thickness distribution. Areas of the film, with a local maximum of the film thickness at the nozzle exit, develop into areas with two local maxima in the film thickness. The latter is not found for the similar swirling annular sheet and results from sheet divergence in the conical case. Sheet divergence due to excess swirl also causes the regular cellular structure observed at the nozzle to stretch, forming a net-like structure of thicker fluid ligaments imposed onto a thinner layer of liquid (see Fig. 27).

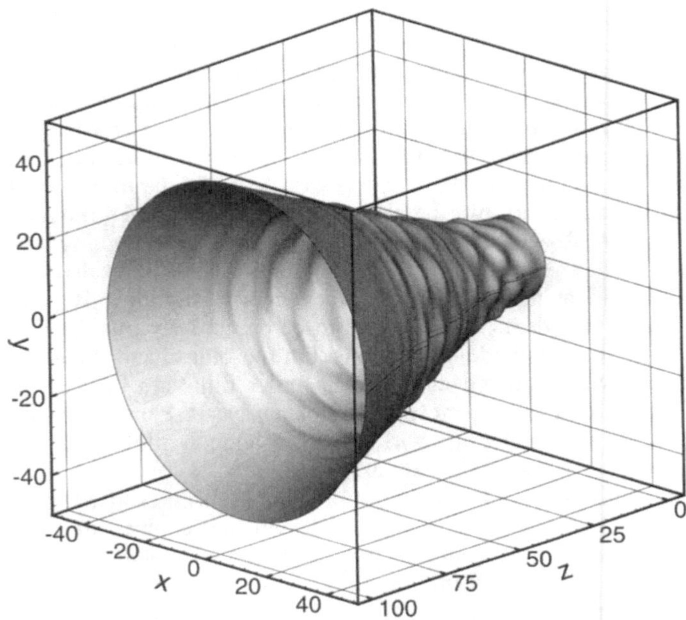

Fig. 27   Instantaneous location of the outer interface for swirling conical liquid film sinusoidally modulated at the nozzle $z = 0$.[130]

For conical films, increased film divergence leads to increased film thinning and smaller droplets once the thin ligaments detached from the film do disintegrate further (e.g., according to Rayleigh's mechanism for jet breakup). However, for purely capillary film breakup and diverging films, the decreasing effect of the increasing second radius of curvature and the prescribed decrease in disturbance amplitude renders the film more stable. This contrasts to the case in which aerodynamic effects dominate the film breakup. Here, film divergence leads to an increase in surface area onto which the surrounding gas can impose forces or transfer energy. Within the present analysis[118,130] of capillary waves on diverging annular films, energy is transferred onto the film only at the nozzle exit. Nonlinear analyses considering the aerodynamic stability of annular and conical films have not been published so far.

The reduced one-dimensional models, discussed here and employed to analyze the nonlinear distortion and disintegration of thin films subject to long-wavelength disturbances on infinite periodic or semi-infinite modulated annular or conical liquid films, provide important qualitative and quantitative information on the stability and on the breakup of liquid films used in practical applications. As already mentioned, they allow rough predictions of film breakup length and resulting mean drop size. More details regarding the analyses presented in Refs. 22, 100–102, 111, and 117 and discussed here can be found in Ref. 131.

A nonlinear numerical analysis of free liquid films by using a general-purpose multifluid method has been reported by Poo and Ashgriz.[132] The accuracy of

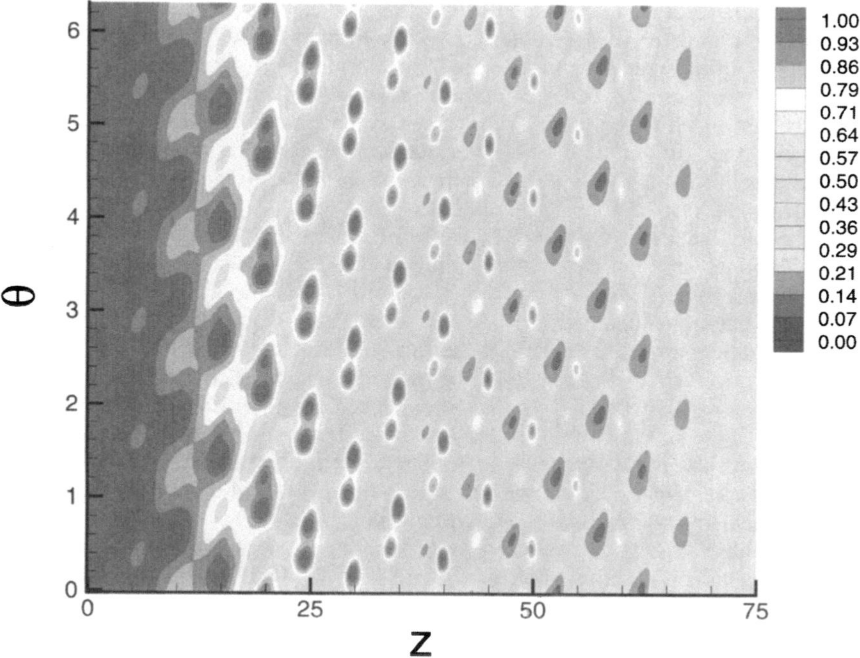

Fig. 28   Instantaneous nondimensional film thickness distribution for the swirling conical liquid film of Fig. 27.[130]

these predominantly Eulerian techniques is difficult to maintain at an acceptable level when thin films with considerable nonlinear distortion are considered. More accurate results can be expected by Lagrangian methods, e.g., boundary-element techniques or discrete-vortex methods as well as the prescribed Eulerian methods with dynamic regridding at the location of the fluid interfaces. However, such simulations are computationally very intensive and do not typically yield the same insights as the prescribed simplified analyses, e.g., with respect to the propagation properties of capillary waves and their interaction between each other.

VII.   Concluding Remarks

Some linear and nonlinear analyses have been performed for round jets and planar sheets and for the annular and conical domains with and without swirl. Aerodynamic effects have been considered for the round jets and planar sheet but remain to be studied for nonlinear annular and linear and nonlinear conical films. There are clear indications that nonlinear effects are important, especially in the latter stages of deformation. Substantial deviations from sinusoidal-shaped waveforms of linear theory have been demonstrated in various configurations due to nonlinear effects. Viscous effects are critical in the pinch-off process. In addition, the relative importance of viscous and capillary forces will determine distortion characteristics and rupture of the stretching and/or contracting fluid ligaments initially shed from the continuous discharging stream.[133] In the final stages of pinch-off, the role of intermolecular forces can be important.[133]

The setting of the boundary conditions in the spatially developing wave case must be done with care. The required number of upstream boundary conditions can change from one regime to another. Group velocity analysis is very useful in setting the proper boundary conditions.

The fan sheet stability should be examined for high-pressure effect in both the dilational and sinuous modes. Nonlinear effects for fan sheets also remain to be determined.

Little theoretical analysis exists on the coupling of the wave phenomenon with behavior upstream in the orifice. The effects of generated turbulence, nonuniform velocity profiles, cavitation, and upstream pressure disturbances should be studied further. The impact of gas streams colliding with (rather than flowing parallel to) the liquid stream should be studied further. Similarly, acoustic interactions with the liquid stream should be analyzed. More theoretical work on the interaction of electric fields with charged liquid streams should be performed. More three-dimensional analyses are needed.

The interesting work[25] on the process of stream pinch-off must be extended. Molecular dynamics methodology offers some hope for that evaluation.[134] The breakup of the ligaments formed from pinch-off into smaller ligaments and droplets is not well understood. Some simple theories exist that model the process of cascading from large ligaments to small droplets. There is substantial need for refinement and further understanding of the physical behavior. A relatively large amount of information is known about the fluid dynamics, trajectories, secondary atomization, heating, and vaporization of the droplets and spray.[135]

This article has emphasized atomization processes that are initiated by capillary instabilities, Kelvin–Helmholtz instabilities, or a combination of these two mechanisms. Other interesting mechanisms exist. The Rayleigh–Taylor mechanism can be important in situations with gravitational and/or inertial accelerations. It plays a role in secondary atomization of droplets in many spray flow configurations. Another important mechanism[135–138] relates to the stripping of mass from a viscous liquid boundary layer at the interface in cases where a gas motion occurs relative to the liquid, thereby causing shear. The gas flow over a nearly spherical droplet or the gas cross flow over a liquid jet will tend to separate at moderately high Reynolds numbers or greater. The separating gas flow can remove liquid from the surface of the droplet or jet. Some analysis of this effect is presented by Delplanque and Sirignano[136] and Ranger and Nicholls.[137,138] Delplanque and Sirignano[135,136] show how this mechanism can be especially important at near-critical conditions and can explain the stability of hydrogen/oxygen rocket engines.

Acknowledgment

This research has been supported by the U.S. Army Research Office through Grant/Contract DAAH 04-96-1-0055, with David Mann as the program manager.

References

[1]Sirignano, W. A., Delplanque, J. -P., Chiang, C. H., and Bhatia, R., "Liquid Propellant Droplet Vaporization: A Rate-Controlling Process for Combustion Instability," edited by V. Yang, Progress in Astronautics and Aeronautics, Vol. 169, AIAA, Washington, DC, 1995, pp. 307–343.

[2]Lefebvre, A. H., *Atomization and Sprays*, Hemisphere, New York, 1989.

[3]Bayvel, L., and Orzechowski, Z., *Liquid Atomization*, Taylor and Francis, Washington, DC, 1993.

[4]Sirignano, W. A., and Mehring, C., "Review of Theory of Distortion and Disintegration of Liquid Streams," *Progress in Energy and Combustion Science*, Vol. 26, 2000, pp. 609–655.

[5]Drazin, P. G., and Reid, W. H., *Hydrodynamic Stability*, Cambridge Univ. Press, New York, 1981.

[6]Sellens, R. W., and Brzustowski, T. A., "A Prediction of the Drop Size and Velocity Distribution in a Spray and First Principles," *Atomization and Spray Technology*, Vol. 1, 1985, pp. 85–102.

[7]Sellens, R. W., and Brzustowski, T. A., "A Simplified Prediction of Droplet Velocity Distributions in a Spray," *Combustion and Flame*, Vol. 65, 1986, pp. 273–279.

[8]Sellens, R. W., "Prediction of the Drop Size and Velocity Distribution in a Spray Based on the Maximum Entropy Formalism," *Part. Part. Syst. Charact.*, Vol. 6, 1989, pp. 17–27.

[9]Ahmadi, M., and Sellens, R. W., "A Simplified Maximum-Entropy-Based Drop Size Distribution," *Atomization and Sprays*, Vol. 3, 1993, pp. 291–310.

[10]Mitra, S. K., and Li, X., "A Prediction Model for Droplet Size Distribution in Sprays," *Atomization and Sprays*, Vol. 9, 1999, pp. 29–50.

[11]Li, X., and Tankin, R. S., "On the Prediction of Droplet Size and Velocity Distribution in Sprays Through Maximum Entropy," *Part. Part. Syst. Charact.*, Vol. 9, 1992, pp. 195–201.

[12]Li, X., Tankin, R. S., and Renksizbulut, M., "Calculated Characteristics of Droplet Size and Velocity Distributions in Liquid Sprays," *Part. Part. Syst. Charact.*, Vol. 7, 1990, pp. 54–59.

[13]Li, X., and Tankin, R. S., "Droplet Size Distribution: A Derivation of a Nukiyama-Tanasowa Type Distribution Function," *Combustion Science and Technology*, Vol. 56, 1988, pp. 65–76.

[14]Keller, J. B., Rubinow, S. I., and Tu, Y. O., "Spatial Instability of a Jet," *Physics of Fluids*, Vol. 16, No. 12, 1973, pp. 2052–2055.

[15]Bechtel, S. E., Carlson, S. D., and Forest M. G., "Recovery of the Rayleigh Capillary Instability from Slender 1-D Inviscid and Viscous Models," *Physics of Fluids*, Vol. 7, No. 12, 1995, pp. 2956–2971.

[16]Bogy, D. B., "Wave Propagation and Instability in a Circular Semi-Infinite Liquid Jet Harmonically Forced at the Nozzle," *Journal of Applied Mechanics*, Vol. 45, 1978, pp. 469–474.

[17]Lighthill, J., *Waves in Fluids*, Cambridge Univ. Press, Cambridge, England, UK, 1978.

[18]Pimbley, W. T., "Drop Formation from a Liquid Jet: A Linear One-Dimensional Analysis Considered as a Boundary Value Problem," *IBM Journal of Research and Development*, Vol. 20, No. 2, 1976, pp. 148–156.

[19]Bogy, D. B., "Use of One-Dimensional Cosserat Theory to Study Instability in a Viscous Liquid Jet," *Physics of Fluids*, Vol. 21, 1978, pp. 190–197.

[20]Bogy, D. B., "Drop Formation in a Circular Liquid Jet," *Annual Review of Fluid Mechanics*, Vol. 11, 1979, pp. 207–228.

[21]Leib, S. J., and Goldstein, M. E., "The Generation of Capillary Instabilities on a Liquid Jet," *Journal of Fluid Mechanics*, Vol. 168, 1986, pp. 479–500.

[22]Mehring, C., and Sirignano, W. A., "Axisymmetric Capillary Waves on Thin Annular Liquid Sheets. Part II: Spatial Development," *Physics of Fluids*, Vol. 12, 2000, pp. 1440–1460.

[23]Yuen, M. C., "Non-Linear Capillary Instability of a Liquid Jet," *Journal of Fluid Mechanics*, Vol. 33, No. 1, 1968, pp. 151–163.

[24]Nayfeh, A. H., "Nonlinear Stability of a Liquid Jet," *Physics of Fluids*, Vol. 13, No. 4, 1970, pp. 841–847.

[25]Eggers, J., "Nonlinear Dynamics and Breakup of Free-Surface Flows," *Reviews of Modern Physics*, Vol. 69, No. 3, 1997, pp. 865–929.

[26]Chaudhary, K. C., and Redekopp, L., "The Nonlinear Capillary Instability of a Liquid Jet: Part 1. Theory," *Journal of Fluid Mechanics*, Vol. 96, 1980, pp. 257–274.

[27]Chaudhary, K. C., and Maxworthy, T., "The Nonlinear Capillary Instability of a Liquid Jet: Part 2. Experiments on Jet Behavior Before Droplet Formation," *Journal of Fluid Mechanics*, Vol. 96, 1980, pp. 275–286.

[28]Chaudhary, K. C., and Maxworthy, T., "The Nonlinear Capillary Instability of a Liquid Jet: Part 3. Experiments on Satellite Drop Formation and Control," *Journal of Fluid Mechanics*, Vol. 96, 1980, pp. 275–286.

[29]Pimbley, W. T., and Lee, H. C., "Satellite Droplet Formation in a Liquid Jet," *IBM Journal of Research and Development*, Vol. 21, No. 1, 1977, pp. 21–30.

[30]Schulkes, R. M. S. M., "Dynamics of Liquid Jets Revisited," *Journal of Fluid Mechanics*, Vol. 250, 1993, pp. 635–650.

[31]Lee, H. C., "Drop Formation in a Liquid Jet," *IBM Journal of Research and Development*, Vol. 18, 1974, pp. 364–369.

[32]Bechtel, S. E., Forest, M. G., and Lin, K. J., "Closure to All Orders in 1-D Models for Slender Viscoelastic Jets: An Integrated Theory of Axisymmetric, Torsionless Flow," *Stability Appl. Anal. Continuous Media*, Vol. 2, 1992, p. 59.

[33]Eggers, J., and Dupont, T. F., "Drop Formation in a One-Dimensional Approximation of the Navier–Stokes Equations," *Journal of Fluid Mechanics*, Vol. 262, 1994, pp. 205–221.

[34]Garcia, F. J., and Castellanos, A., "One-Dimensional Models for Slender Axisymmetric Viscous Liquid Jets," *Physics of Fluids*, Vol. 6, No. 8, 1994, pp. 2676–2689.

[35]Shi, X. D., Brenner, M. P., and Nagel, S. R., "A Cascade of Structure in a Drop Falling from a Faucet," *Science*, Vol. 265, 1994, pp. 219–222.

[36]Brenner, M. P., Shi, X. D., and Nagel, S. R., "Iterated Instability During Droplet Fission," *Physical Review Letters*, Vol. 73, No. 25, 1994, pp. 3391–3394.

[37]Eggers, J., "Theory of Drop Formation," *Physics of Fluids*, Vol. 7, No. 5, 1995, pp. 941–953.

[38]Ting, L., and Keller, J. B., "Slender Jets and Thin Sheets with Surface Tension," *SIAM Journal on Applied Mathematics*, Vol. 50, No. 6, 1990, pp. 1533–1546.

[39]Forest, M. G., and Wang, Q., "Change-of-Type Behavior in Viscoelastic Slender Jet Models," *J. Theor. Comp. Fluid Dyn.*, Vol. 2, No. 1, 1990, pp. 1–25.

[40]Schulkes, R. M. S. M., "Nonlinear Dynamics of Liquid Columns: A Comparative Study," *Phys. Fluids A (Fluid Dyn.)*, Vol. 5, No. 9, 1993, pp. 2121–2130.

[41]Ashgriz, N., and Mashayek, F., "Temporal Analysis of Capillary Jet Breakup," *Journal of Fluid Mechanics*, Vol. 219, 1995, pp. 163–190.

[42]Spangler, C. A., Hilbing, J. H., and Heister, S. D., "Nonlinear Modeling of Jet Atomization in the Wind-Induced Regime," *Physics of Fluids*, Vol. 7, 1995, pp. 964–971.

[43]Hilbing, J. H., Heister, S. D., and Spangler, C. A., "A Boundary-Element Method for Atomization of a Finite Liquid Jet," *Atomization and Sprays*, Vol. 5, 1995, pp. 621–638.

[44]Hilbing, J. H., and Heister, S. D., "Droplet Size Control in Liquid Jet Breakup," *Physics of Fluids*, Vol. 8, 1996, pp. 1574–1581.

[45]Hilbing, J. H., and Heister, S. D., "Nonlinear Simulation of a High-Speed, Viscous Liquid Jet," *Atomization and Sprays*, Vol. 8, 1997, pp 155–178.

[46]Lundgren, T. S., and Masour, N. N., "Computation of Drop Pinch-Off and Oscillation," *Proc. 3rd Int. Coll. On Drops and Bubbles*, edited by T. G. Wang, Monterey, CA, 1988, pp. 208–215.

[47]Weber, C., "Zum Zerfall eines Flussigkeitsstrahles. Z.Angew," *Math. Mech.*, Vol. 11, 1931, p. 136.

[48]Taylor, G. I., "Disintegration of Water Drops in Electric Field," *Proceedings of the Royal Society of London, Series A: Mathematical and Physical Sciences*, Vol. 280, 1964, pp. 383–397.

[49]Cloupeau, M., and Prunet-Foch, B., "Electrohydrodynamic Spraying Functioning Modes: A Critical Review," *J. Aerosol. Sci.*, Vol. 25, 1994, pp. 1021–1036.

[50]Setiawan, E. R., and Heister, S. D., "Nonlinear Modeling of an Infinite Electrified Jet," *Journal of Electrostatics*, Vol. 42, 1997, pp. 243–257.

[51]Yoon, S. S., Heister, S. D., Epperson, J. T., and Sojka, P. E., "Modeling Multi-jet Mode Electrostatic Atomization Using Boundary Element Methods," *Journal of Electrostatics*, Vol. 50, 2001, pp. 91–108.

[52]Kelly, A. J., "On the Statistical, Quantum and Practical Mechanics of Electrostatic Atomization," *J. Aerosol. Sci.*, Vol. 25, 1994, pp. 1159–1177.

[53]Shtern, V., and Barrero, A., "Striking Features of Fluid Flows in Taylor Cones Related to Electrosprays," *J. Aerosol. Sci.*, Vol. 25, 1994, pp. 1049–1063.

[54]Ganan-Calvo, A. M., Lasheras, J. C., Davila, J., and Barrero, A., "The Electrostatic Spray Emitted from an Electrified Conical Meniscus," *J. Aerosol. Sci.*, Vol. 25, 1994, pp. 1121–1142.

[55]Shrimpton, J. S., "Space Charge Transport in Dielectric Fluid Flows," *Proceedings of Eighth International Conference on Liquid Atomization and Spray Systems*, Institute for Liquid Atomization and Spray Systems, July 2000.

[56]Hertz, C. H., and Hermanrud, B., "A Liquid Compound Jet," *Journal of Fluid Mechanics*, Vol. 131, 1983, p. 271–287.

[57]Crapper, G. D., Dombrowski, N., and Pyott, G. A. D., "Kelvin–Helmholtz Wave Growth on Cylindrical Sheets," *Journal of Fluid Mechanics*, Vol. 68, 1975, pp. 497–502.

[58]Shen, J., and Li, X., "Instability of an Annular Viscous Liquid Jet," *Acta Mech.*, Vol. 114, 1996, p. 167.

[59]Chauhan, A., Maldrelli, C., Rumschitzki, D. S., and Papageorgiu, D. T., "Temporal and Spatial Instability of an Inviscid Compound Jet," *Rheol. Acta*, Vol. 35, 1996, p. 567.

[60]Shkadov, V. Y., and Sisoev, G. M., "Instability of a Two-Layer Capillary Jet," *International Journal of Multiphase Flow*, Vol. 22, 1996, pp. 363–377.

[61]Dumbleton, J. H., and Hermans, J. J., "Capillary Stability of a Hollow Inviscid Cylinder," *Physical of Fluids*, Vol. 13, 1970, pp. 12–17.

[62]Sanz, A., and Meseguer, J., "One-Dimensional Linear Analysis of the Compound Jet," *Journal of Fluid Mechanics*, Vol. 159, 1985, p. 55.

[63]Savart, F., "Suite du Memoire Sur Le Cho d'une Veine Liquide Lancee Contre un Plan Circulaire," *Annales de Chimie et de Physique*, Vol. 54, 1833, pp. 113–165.

[64]Dorman, R. G., "The Atomization of Liquid in a Flat Spray," *British Journal of Applied Physics*, Vol. 3, 1952, pp. 189–192.

[65]Fraser, R. P., and Eisenklam, P., "Research into the Performance of Atomizers of Liquids," *Imp. Coll. Chem. Eng. Soc. J.*, Vol. 7, 1953, pp. 52–68.

[66]Dombrowski, N., and Fraser, R. P., "A Photographic Investigation into the Disintegration of Liquid Sheets," *Philosophical Transactions of the Royal Society of London, Series A: Mathematical and Physical Sciences*, Vol. 247, No. 924, 1954, pp. 101–130.

[67]Hagerty, W. W., and Shea, J. F., "A Study of the Stability of Plane Fluid Sheets," *Journal of Applied Mechanics*, Vol. 22, No. 4, 1955, pp. 509–514.

[68]Squire, H. B., "Investigation of the Instability of a Moving Liquid Film," *British Journal of Applied Physics*, Vol. 4, 1953, pp. 167–169.

[69]Rangel, R. H., and Sirignano, W. A., "The Linear and Nonlinear Shear Instability of a Fluid Sheet," *Physics of Fluids*, Vol. 3, No. 10, 1991, pp. 2392–2400.

[70]Li, X., and Tankin, R. S., "On the Temporal Instability of a Two-Dimensional Viscous Liquid Sheet," *Journal of Fluid Mechanics*, Vol. 226, 1991, pp. 425–443.

[71]Li, X., "Spatial Instability of Plane Liquid Sheets," *Chemical Engineering Science*, Vol. 48, 1993, pp. 2973–2981.

[72]Hashimoto, H., and Suzuki, T., "Experimental and Theoretical Study of Fine Interfacial Waves on Thin Liquid Sheet," *JSME Int. J., Ser. II*, Vol. 34, 1991, pp. 277–283.

[73]Crapper, G. D., Dombrowski, N., Jepson, W. P., and Pyott, G. A. D., "A Note on the Growth of Kelvin–Helmholtz Waves on Thin Liquid Sheets," *Journal of Fluid Mechanics*, Vol. 57, No. 4, 1973, pp. 671–672.

[74]Crapper, G. D., and Dombrowski, N., "A Note on the Effect of Forced Disturbances on the Stability of Thin Liquid Sheets and on the Resulting Drop Size," *International Journal of Multiphase Flow*, Vol. 10, No. 6, 1984, pp. 731–736.

[75]Clark, C. J., and Dombrowski, N., "Aerodynamic Instability and Disintegration of Inviscid Liquid Sheets," *Proceedings of the Royal Society of London, Series A: Mathematical and Physical Sciences*, Vol. 329, No. 1579, 1972, pp. 467–478.

[76]Dombrowski, N., and Munday, G. I., "Spray Drying," *Biochemical and Biological Engineering Science*, Academic International Press, London, 1968, Chap. 16.

[77]Dombrowski, N., Hasson, D., and Ward, D. E., "Some Aspects on Liquid Flow Through Fan Spray Nozzles," *Chemical Engineering Science*, Vol. 12, 1960, pp. 35–50.

[78]Dombrowski, N., and Johns, W. R., "The Aerodynamic Instability and Disintegration of Viscous Liquid Sheets," *Chemical Engineering Sciences*, Vol. 18, 1963, pp. 203–214.

[79]Weihs, D., "Stability of Thin, Radially Moving Liquid Sheets," *Journal of Fluid Mechanics*, Vol. 87, No. 2, 1978, pp. 289–298.

[80]Crapper, G. D., Dombrowski, N., and Pyott, G. A. D., "Large Amplitude Kelvin–Helmholtz Waves on Thin Liquid Sheets," *Proceedings of the Royal Society of London, Series A: Mathematical and Physical Sciences*, Vol. 342, No. 1629, 1975, pp. 209–224.

[81]Mansour, A., and Chigier, N., "Dynamic Behavior of Liquid Sheets," *Physics of Fluids A*, Vol. 3, No. 12, 1991, pp. 2971–2980.

[82]Brown, D. R., "A Study of the Behaviour of a Thin Sheet of Moving Liquid," *Journal of Fluid Mechanics*, Vol. 10, 1961, pp. 297–305.

[83]Dombrowski, N., and Hooper, P. C., "The Effect of Ambient Density on Drop Formation in Sprays," *Chemical Engineering Science.* Vol. 17, 1962, pp. 291–305.

[84]Taylor, G. I., "The Dynamics of Thin Sheets of Fluid, III. Disintegration of Fluid Sheets," *Proceedings of the Royal Society of London, Series A: Mathematical and Physical Sciences*, Vol. 253, 1959, pp. 313–321.

[85]Fullana, J. M., and Zaleski, S., "Stability of a Growing End Rim in a Liquid Sheet of Uniform Thickness," *Physics of Fluids*, Vol. 11, No. 5, 1999, pp. 952–954.

[86]Hinze, J. O., "Critical Speeds and Sizes of Liquid Globules," *Appl. Sci. Res.*, Vol. A1, p. 273–1948.

[87]Good, R. E., and Clewell, H. J., "Drop Formation and Evaporation of JP-4 Fuel Jettisoned from Aircraft," *Journal of Aircraft*, Vol. 17. 1979, p. 450.

[88]Dabora, E. K., Ragland, K. W., Ranger, A. A., and Nicholls, J. A., "Detonation in Two-Phase Media and Drop Shattering Studies," NASA CR-72421, 1968.

[89]Nicholls, J. A., and Ranger, A. A., "Droplet Shattering," Chemical Propulsion Information Agency, Pub. 183, 1968, pp. 85–90.

[90]Sirignano, W. A., Rangel, R. H., and Megaridis, C. M., "Study of Liquid Hydrogen and Liquid Oxygen Rocket Propellant Launch Hazards," Consultants' Report, 1989.

[91]Rangel, R. H., and Sirignano, W. A., "Nonlinear Growths of Kelvin–Helmholtz Instability: Effect of Surface Tension and Density Ratio," *Physics of Fluids*, Vol. 31, 1988, pp. 1845–1855.

[92]Jazayeri, S. A., and Li, X., "Nonlinear Breakup of Liquid Sheets," *Proc. 9th Ann. Conf. Liquid Atom. Spray Sys.: ILASS, North and South America*, 1996, pp. 114–119.

[93]Kim, I., and Sirignano, W. A., "Three-Dimensional Wave Distortion and Disintegration of Thin Planar Liquid Sheets," *Journal of Fluid Mechanics*, Vol. 410, 2000, pp. 147–183.

[94]Mehring, C., and Sirignano, W. A., "Nonlinear Capillary Wave Distortion and Disintegration of Thin Planar Liquid Sheets," *Journal of Fluid Mechanics*, Vol. 388, 1999, pp. 69–113.

[95]Graff, K. F., *Wave Motion in Elastic Solids*, Dover, New York, 1975.

[96]Taylor, G. I., "The Dynamics of Thin Sheets of Fluid, II., Waves on Fluid Sheets," *Proceedings of the Royal Society of London, Series A: Mathematical and Physical Sciences*, Vol. 253, 1959, pp. 253–296.

[97]Matsuuchi, K., "Modulational Instability of Nonlinear Capillary Waves on Thin Liquid Sheets," *J. Phys. Soc.*, Vol. 37, No. 6, 1974, pp. 1680–1687.

[98]Matsuuchi, K., "Instability of Thin Liquid Sheet and Its Break-Up," *J. Phys. Soc.*, Vol. 41, No. 4, 1976, pp. 1410–1416.

[99]Hashimoto, H., and Suzuki, T., "Experimental and Theoretical Study of Fine Interfacial Waves on Thin Liquid Sheet," *JSME Int. J.*, Vol. 34, No. 3, 1991, pp. 277–283.

[100]Mehring, C., and Sirignano, W. A., "Planar-Liquid-Stream Distortion from Kelvin–Helmholtz and Capillary Effects," presented at ICLASS 2000, July 16–20, Pasadena, CA.

[101]Mehring, C., and Sirignano, W. A., "Free Planar Liquid Films Impacted by Gas Jets," *ILASS, 2001*, Dearborn, MI, May 20–23, 2001.

[102]Mehring, C., and Sirignano, W. A., "Disintegration of Planar Liquid Film Impacted by Two-Dimensional Gas Jets," *Physics of Fluids*, Vol. 15, No. 5, 2003, pp. 1158–1177.

[103]Arai, T., and Hashimoto, H., "Behavior of the Gas-Liquid Interface in a Liquid Film Jet: Instability of a Liquid Film Jet in a Co-Current Gas Stream," *JSME Bull*, Vol. 28, No. 245, 1985, pp. 2652–2659.

[104]Ibrahim, E. A., and Akpan, E. R., "Three-Dimensional Instability of Viscous Liquid Sheets," *Atomization and Sprays*, Vol. 6, 1996, pp. 649–665.

[105]Ponstein, J., "Instability of Rotating Cylindrical Jets," *Applied Sci. Res. A*, Vol. 8, 1959, p. 425.

[106]Meyer, J., and Weihs, D., "Capillary Instability of an Annular Liquid Jet," *Journal of Fluid Mechanics*, Vol. 179, 1987, pp. 531–545.

[107]Lee, J. G., and Chen, L. D., "Linear Stability Analysis of Gas–Liquid Interface," *AIAA Journal*, Vol. 29, 1991, pp. 1589–1595.

[108]Shen, J., and Li, S., "Atomization of an Annular Viscous Liquid Jet," *Proc. 7th Ann. Conf. Liquid Atom Spray Sys.: ILASS*, 1994, pp. 50–54.

[109]Panchagnula, M. V., Santangelo, P. J., and Sojka, P. E., "The Instability of an Inviscid Annular Liquid Sheet Subject to Two-Dimensional Disturbances," *Proc. 8th Ann. Conf. Liquid Atom. Spray Sys.: ILASS*, 1995, pp. 54–58.

[110]Shen, J., and Li, X., "Instability of an Annular Viscous Liquid Jet," *Acta Mechanica*, Vol. 114, 1996, pp. 167–183.

[111]Mehring, C., and Sirignano, W. A., "Axisymmetric Capillary Waves on Thin Annular Liquid Sheets. Part I: Temporal Stability," *Physics of Fluids*, Vol. 12, 2000, pp. 1417–1439.

[112]Chauhan, A., Maldarelli, C., Rumschitzki, D. S., and Papageorgiou, D. T., "Temporal and Spatial Instability of an Inviscid Compound Jet," *Pheol. Acta*, Vol. 35, 1996, pp. 567–583.

[113]Lee, C. P., and Wang, T. G., "A Theoretical Model for the Annular Jet Instability-Revisted," *Physics of Fluids*, Vol. 29, 1986, pp. 2076–2085.

[114]Lee, C. P., and Wang, T. G., "Dynamics of Thin Liquid Sheets," *Proc. 3rd Int. Coll. On Drops and Bubbles*, 1988, pp. 496–504.

[115]Lee, C. P., and Wang, T. G., "The Theoretical Model for the Annular Jet Instability-Revisited," *Physics of Fluids A*, Vol. 1, 1989, pp. 967–974.

[116]Ramos, J. I., "Annular Liquid Jets: Formulation and Steady-State Analysis," *Z. Angew. Math. Mech.*, Vol. 72, 1992, pp. 565–589.

[117]Mehring, C., and Sirignano, W. A., "Nonlinear Capillary Waves on Swirling, Axisymmetric Liquid Films," *International Journal of Multiphase Flow*, Vol. 27, 2001, pp. 1707–1734.

[118]Mehring, C., and Sirignano, W. A., "Capillary Stability of Modulated Swirling Liquid Sheets," *Atomization and Sprays* (to be published).

[119]Ramos, J. I., "One-Dimensional Models of Steady Inviscid, Annular Liquid Jets," *Applied Mathematical Modelling*, Vol. 20, 1996, pp. 593–607.

[120]Panchagnula, M. V., Sojka, P. E., and Bajaj, A. K., "The Non-Linear Breakup of Annular Liquid Sheets," *Proc. 11th Ann. Conf. Liquid Atom. Spray Sys.: ILASS*, 1998, pp. 170–174.

[121]Panchagnula, M. V., Sojka, P. E., and Santangelo, P. J., "On the Three-Dimensional Instability of a Swirling Annular, Inviscid Liquid Sheet Subject to Unequal Gas Velocities," *Physics of Fluids*, Vol. 8, 1996, pp. 3300–3312.

[122]Finnicum, D. S., Weinstein, S. J., and Ruschak, K. J., "The Effect of Applied Pressure on the Shape of a Two-Dimensional Liquid Curtain Falling Under the Influence of Gravity," *Journal of Fluid Mechanics*, Vol. 255, 1993, pp. 647–665.

[123]Ramos, J. I., "Analysis of Annular Liquid Membranes and Their Singularities," *Meccanica*, Vol. 32, 1997, pp. 279–293.

[124]Yarin, A. L., *Free Liquid Jets and Films: Hydrodynamics and Rheology*, Longman Scientific and Technical, 1993.

[125]Lighthill, M. J., "Group Velocity," *J. Inst. Math. Appl.*, Vol. 1, 1965, pp. 1–28.

[126]Trefethen, L. N., "Instability of Difference Models for Hyperbolic Initial Boundary Value Problems," *Communications on Pure and Applied Mathematics*, Vol. 37, 1984, pp. 329–367.

[127]Schultz, W. W., and Davis, S. H., "One-Dimensional Liquid Fibers," *Journal of Rheology*, Vol. 26, 1981, pp. 331–345.

[128]Housiadas, K., and Tsamopoulos, J., "Unsteady Flow of an Axisymmetric Annular Film Under Gravity," *Physics of Fluids*, Vol. 10, 1998, pp. 2500–2516.

[129]Ramos, J. I., "Planar Liquid Sheets at Low Reynolds Numbers," *International Journal for Numerical Methods in Fluids*, Vol. 22, 1996, pp. 961–978.

[130]Mehring, C., and Sirignano, W. A., "Three-Dimensional Capillary Stability of Modulated Swirling Liquid Films," *Proceedings of ILASS 2001*, Dearborn, MI, 2001, May 20–23.

[131]Mehring, C., "Nonlinear Distortion of Thin Liquid Sheets," Ph.D. Dissertation, Univ. of California, Irvine, CA, 1999.

[132]Poo, J. Y., and Ashgriz, N., "Numerical Simulation of Capillary Driven Viscous Flows in Liquid Drops and Films by an Interface Reconstruction Scheme," *Proc. 3rd Int. Coll. On Drops and Bubbles*, 1998, pp. 235–245.

[133]Mehring, C., Xi, J., and Sirignano, W. A., "Dynamic Stretching of a Planar Liquid Bridge," *Physics of Fluids* (to be published).

[134]Moseler, M., and Landman, U., "Formation, Stability, and Breakups of Nanojets," *Science*, Vol. 289, 2000, pp. 1165–1169.

[135]Sirignano, W. A., *Fluid Dynamics and Transport of Droplet and Sprays*, Cambridge Univ. Press, New York, 1999.

[136]Delplanque, J.-P., and Sirignano, W. A., "Boundary Layer Stripping Effects on Droplet Transcritical Convective Vaporization," *Atomization and Sprays*, Vol. 4, 1994, pp. 325–349.

[137]Ranger, A. A., and Nicholls, J. A., "Aerodynamic Shattering of Liquid Drops," *AIAA Journal*, 1969, pp. 285–290.

[138]Ranger, A. A., and Nicholls, J. A., "Atomization of Liquid Droplets in a Convective Stream," *International Journal of Heat and Mass Transfer*, Vol. 15, 1972, pp. 1203–1211.

Chapter 6

Modeling Liquid-Propellant Spray Combustion Processes

H. H. Chiu*
National Cheng-Kung University, Taiwan, Republic of China

and

J. C. Oefelein[†]
Sandia National Laboratories, Livermore, California

I. Introduction

THIS chapter presents an overview of basic physical concepts and method-ologies for modeling aerothermochemical processes in reacting sprays. Emphasis is placed on the structures, phenomena, and mechanisms associated with gas-liquid spray combustion processes and prediction of the overall combustion and fluid dynamic characteristics under prescribed operational configurations.

Section II begins with the fundamental formulation for a two-phase flow. Aspects of the conservation equations for a multiphase turbulent reacting flow are discussed in conjunction with the physical background associated with various closure laws. The baseline formulation is cast in the Eulerian and Lagrangian frames to describe the gas and dispersed phases, respectively.

Section III describes the basic modeling approaches employed to date. Emphasis is placed on separated flow models, which adapt drops as the building blocks of homogeneous sprays, and alternatively meso-scale structures, which consist of aggregate drop clusters. Kinetic and statistical formulations for a dense dispersed spray are also presented for prediction of inhomogeneous meso-scale structures. The dynamic equation for drop motion based on the semi-empirical but generalized Basset, Bousinesq, and Ossen (BBO) equation is presented to provide the detailed dynamic response of a drop in general flow

---

*Professor Emeritus.
[†]Senior Member, Technical Staff.

environments, with the analogous equations for heat and mass transfer. The section is concluded with a discussion of morphological and kinetic models.

Section IV is devoted to the treatment of turbulence interactions, generation, and modulation. Models for predicting drop-turbulence interactions are presented, with emphasis placed on drop dispersion and the coupled processes associated with the modulation of turbulence and the turbulent energy spectrum. Issues related to the physics and modeling of dispersed phase interactions and related phenomena are discussed in detail.

Section V builds on the material presented in Section IV to describe collective phenomena in reacting sprays. Here, issues related to group combustion are presented, with emphasis placed on the mechanisms and the conditions required for the excitation of four principal group combustion modes.

The last section discusses issues related to multiphase combustion at supercritical conditions. Cryogenic liquid propellants, both at subcritical and supercritical pressures, burn at group combustion numbers on the order of 10^3. Scaling laws describing the pressure effects on external group flames at subcritical and supercritical states are developed by a canonical method. These laws account for the effects of reduced transport properties at high pressures.

The chapter concludes with a general assessment of the comprehensive versatility in performing both synthetic and reductive analyses of present models and the areas of future research.

II. Fundamental Formulation for Two-Phase Flow

Spray combustion in a liquid propellant combustor is governed by the principles of conservation of mass, momentum, and energy.[1-4] Dispersed liquid elements in the spray system interact collectively with the hot gas and ultimately combust to liberate heat. This section presents the physical concepts and mathematical formulation of two-phase flow to establish the basis for modeling key processes.

A. Local Instantaneous Equations

Conservation laws of a scalar or component of a vector variable, f_k, of phase k are derived by formulating the integral flux balance through a control volume with a distributed source term ϕ_k and the interfacial source term ϕ_I as follows:

$$\sum_{k=1}^{M} \frac{\mathrm{d}}{\mathrm{d}t} \int_{vk} \rho_k f_k \, \mathrm{d}V = \int_{SI} \rho \phi_I \, \mathrm{d}S + \sum_{k=1}^{M} \left(- \int_{Ak} \rho_k f_k u_k \cdot n_k \, \mathrm{d}S \right.$$

$$\left. + \int_{Vk} \rho_k \phi_k \, \mathrm{d}V - \int_{Sk} \xi_k \cdot n_k \, \mathrm{d}S \right) \tag{1}$$

where ρ_k is the density, ξ_k is the molecular transport flux, and n_k is the outwardly directed unit normal vector to the interface. By applying Gauss's theorem, Eq. (1) can be expressed as the sum of the volume integral over the region occupied by respective phases, and a surface integral expressing the following interfacial junction conditions:

$$\sum_{k=1}^{M} \int_{vk} \left[\frac{\partial(\rho_k f_k)}{\partial t} + \nabla \cdot (\rho_k f_k \boldsymbol{u}_k) + \nabla \cdot (\xi_k) - \rho_k \phi_k \right] \mathrm{d}V$$

$$= \int_{SI} \phi_I \, \mathrm{d}S + \sum_{k=1}^{M} \left(\int_{S_k} [\dot{m}_k f_k + \xi_k \cdot \boldsymbol{n}_k] \, \mathrm{d}S \right) \tag{2}$$

Here, $\dot{m}_k$ is the rate of mass flux transfer at the interface given by $\dot{m}_k = \rho_k f_k (\boldsymbol{u}_k - \boldsymbol{u}_s) \cdot \boldsymbol{n}_k$, where $\boldsymbol{u}_s = (\partial r / \partial t)$, is the velocity of a point on the interface s. Because the equation holds for an arbitrary volume and surface, we rewrite Eq. (2) in differential form as follows:

$$\frac{\partial(\rho_k f_k)}{\partial t} + \nabla \cdot (\rho_k f_k \boldsymbol{u}_k) + \nabla \cdot (\xi_k) - \rho_k \phi_k = 0 \tag{3}$$

The equations of continuity, momentum, and energy and the jump conditions are obtained by setting $f_k = 1$, u_k, and e_k, with $\xi_k = 0$, T_k, and $q_k - T_k \cdot u_k$, respectively. This yields a system of the form

$$\frac{\partial(\rho_k)}{\partial t} + \nabla \cdot (\rho_k \boldsymbol{u}_k) = 0 \tag{4}$$

$$\sum_{k=1}^{M} \dot{m}_k = 0 \tag{5}$$

$$\frac{\partial}{\partial t}(\rho_k \boldsymbol{u}_k) + \nabla \cdot (\rho_k \boldsymbol{u}_k \boldsymbol{u}_k) - \nabla \cdot T_k - \rho_k f_\ell = 0 \tag{6}$$

$$\sum_{k=1}^{M} (\dot{m}_k \boldsymbol{u}_k - T_k \cdot \boldsymbol{n}_k) = 0 \tag{7}$$

where T_k is the stress tensor, f is the body force, and

$$\frac{\partial}{\partial t}[\rho_k e_k] + \nabla \cdot [\rho_k \boldsymbol{u}_k e_k] + \nabla \cdot (q_k - T_k \cdot \boldsymbol{u}_k) - \rho_k f_k \cdot \boldsymbol{u}_k = 0 \tag{8}$$

$$\sum_{k=1}^{M} [\dot{m}_k e_k + (q_k - T_k \cdot \boldsymbol{u}_k) \cdot \boldsymbol{n}_k] = 0 \tag{9}$$

where e_k is the specific internal energy and q_k the energy flux due to heat conduction and mass diffusion.

B. Averaged Equations

Because each phase in the flow system may exist at a point in space r and time t, a selected averaging method is used in the representation of the conservation equations. The volume average of a flow variable f is defined around a fixed point in space and time as follows[4]:

$$\langle f \rangle_{\text{Vol}} = V^{-1} \iiint_V f \, \mathrm{d}V \tag{10}$$

The time average, on the other hand, over a time interval T is obtained as follows[4]:

$$\langle f \rangle_{\text{Time}} = T^{-1} \int_{t-T/2}^{t+T/2} f \, \mathrm{d}\tau \tag{11}$$

The ensemble average is a statistical average of the variable f at a fixed point in space and time over a large number of experiments N with the same initial and boundary conditions:

$$\langle f \rangle_{\text{Ensemble}} = \lim \left(\frac{1}{N}\right) \sum_{i=1}^{N} f_n(x, t) \tag{12}$$

Alternatively, we may use the concept of the probability of observing a process λ,

$$\langle f \rangle_{\text{Ensemble}} = \int_{\varepsilon} f \, \mathrm{d}P(\lambda) \tag{13}$$

where $P(\lambda)$ is the probability of the realization of a λ process, and ε is the set of all realizations.[5]

We introduce the phase indicator function θ_k for phase k, which is a step function defined by $\theta_k = 1$, for $r \in k$ and $\theta_k = 0$, otherwise. The average, in either volume, time, or ensemble, of the phase indicator function represents the average occurrence of phase k, i.e.,

$$\Omega_k = \langle \theta_k \rangle \tag{14}$$

where

$$\Sigma \Omega_k = 1 \tag{15}$$

The phase indicator satisfies the following relation:

$$\frac{\partial \theta_k}{\partial t} + u_i \cdot \nabla \theta_k = 0 \tag{16}$$

This equation represents the material derivative of the phase indicator.

The averaged conservation equation of a variable in phase k, denoted f_k, is given as follows:

$$\frac{\partial \langle \theta_k \rho_k f_k \rangle}{\partial t} + \nabla \cdot \langle \theta_k \rho_k f_k u_k \rangle + \nabla \cdot \langle \theta_\kappa \xi_k \rangle - \langle \theta_k \rho_k f_k \rangle = \left\langle (\dot{m}_k f_k + \xi_k \cdot n_k) \frac{\partial \theta_k}{\partial n} \right\rangle \tag{17}$$

$$\frac{\partial \theta_k}{\partial n} = -\delta_k \tag{18}$$

where δ_k is the delta function associated with phase k. The averaged mass continuity equation is obtained by setting $f_k = 1$ in the preceding equation:

$$\frac{\partial \langle \theta_k \rho_k \rangle}{\partial t} + \nabla \cdot \langle \theta_k \rho_k u_k \rangle = \langle S_{km} \rangle \tag{19}$$

$$\langle S_{km} \rangle = \left\langle \dot{m}_k \frac{\partial \theta_k}{\partial n} \right\rangle \tag{20}$$

where S_{km} is the rate of the generation of the k phase per unit volume. By the law of mass conservation, the sum of the average interfacial exchange of mass of all of the phase must vanish:

$$\sum_{k=1}^{M} \langle S_{km} \rangle = \sum_{k=1}^{M} \left\langle \dot{m}_k \frac{\partial \theta_k}{\partial n} \right\rangle = 0 \tag{21}$$

Similarly, the averaged momentum equation is given by

$$\frac{\partial \langle \theta_k \rho_k u_k \rangle}{\partial t} + \nabla \cdot \langle \theta_k \rho_k u_k u_k \rangle = \nabla \cdot \langle \theta_k T_k \rangle + \langle \theta_k \rho_k f_k \rangle + \langle S_{ku} \rangle \tag{22}$$

$$\langle S_{ku} \rangle = \left\langle (\dot{m}_k u_k - T_k) \cdot n_k \frac{\partial \theta_k}{\partial n} \right\rangle \tag{23}$$

where T_k is the stress tensor given by $T_k = -P_k I + \tau_k$, τ_k is the viscous shear stress, S_{ku} is the interfacial momentum exchange, and f_k is a body force. The averaged total interfacial exchange of momentum balances with the surface tension force as follows:

$$\sum_{k=1}^{M} \langle S_{ku} \rangle = \sum_{k=1}^{M} \left\langle (\dot{m}_k u_k - T_k \cdot n_k) \frac{\partial \theta_k}{\partial n} \right\rangle = \sigma \langle \kappa \nabla \theta_k \rangle \tag{24}$$

Here σ is the surface tension and κ is the curvature of the interface.

C.   Decomposition of Variables and Averaging Procedures

Equations (19–24) contain averages of products of the dependent variables that are generally represented in the form of products of the averaged variables. Two procedures are required: 1) transformation of variables by weighting and

2) decomposition of variables. Decomposition of variables is achieved by splitting a variable into the weighted mean value $\langle f \rangle^w$, and f', the deviation from the mean value. This yields

$$f = \langle f \rangle^w + f' \tag{25}$$

$$\langle f \rangle^w = \frac{\langle wf \rangle^w}{\langle w \rangle} \tag{26}$$

where w is a selected weighting factor. Weighting factors are selected according to specific applications. The Reynolds average is defined with the weighting factor $w = 1$. The weighted average given by the phase indicator function of the kth phase, θ_k, is defined with $w = \theta_k$, as follows:

$$\langle f \rangle^{\theta_k} = \frac{\langle \theta_k f \rangle}{\langle \theta_k \rangle} = \frac{\langle \theta_k f \rangle}{\Omega_k} \tag{27}$$

The combined phase average and mass weighted average is called a Favre average. This term is defined using the weighting factor $w = \theta_k \rho_k$ as

$$\langle f \rangle^{\theta k, \rho k} = \frac{\langle \theta_k \rho_k f \rangle}{\langle \theta_k \rho_k \rangle} \tag{28}$$

In addition to the Reynolds and Favre average, a weighted mean average can be extended to cover two types of averages. For example, the combination of a volume average in a specified region of space followed by a specific time interval centered around an arbitrary time t is given by

$$f(x, t) = \langle\langle f(x, t) G(x - \chi, t - \tau, \Delta, \delta\tau) \rangle^x \rangle^\tau \tag{29}$$

Alternatively, such averages can be expressed in the form of convolution integrals:

$$f(x, t) = \int \left[\int\int\int_{V_g} f(x, \tau) G(x - \chi, t - \tau, \Delta, \delta\tau) \, d\chi \right] d\tau \tag{30}$$

Here G represents an arbitrary filter function, V_g is the volume occupied by gas, $\delta\tau$ is the temporal interval, and Δ is the subgrid-scale field or filter width over which filtering is performed, and x and χ are position vectors, respectively. Filter functions are typically normalized such that

$$\int \left[\int\int\int_{V_g} f(x, \tau) G(x - \chi, t - \tau, \Delta, \delta\tau) \, d\chi \right] d\tau = 1 \tag{31}$$

and are infinitely differentiable.

The averaged conservation equations based on the Reynolds, Favre, and filtered-Favre procedures provide the fundamental framework to model key multiphase processes.

III. Basic Modeling Approaches

Numerous spray combustion models have been proposed over the past three decades. Reviews by Faeth[2,6] systematically organize relevant works up to 1987. With subtle differences in specific detail, current methodologies have generally been classified as either locally-homogeneous-flow (LHF) models, where the condensed phase is assumed to be in local thermodynamic equilibrium with the gas phase, or separated-flow (SF) models, where finite rate interphase transport is considered. Of these two classes, SF models have received the widest acceptance.

SF models are generally categorized within three groups: 1) continuous-particle models, 2) discrete-particle models, and 3) continuum-formulation models. Continuous-particle models employ a multidimensional drop distribution function that gives a statistically accurate field description of the spray. Discrete-particle models, in contrast, employ a finite number of computational parcels to achieve the same description. Parcels act as delta function point sources that represent classes of drops in various states. Continuum-formulation models treat the gas and dispersed particle phase as interpenetrating continua. Respective formulations result in similar governing equations for both phases.

Of the three SF models just described, the continuous-particle model proposed by Williams[7] and the discrete-particle model proposed by Crowe et al.[8] are fundamentally important. Consider a drop distribution function $f(s;x,t)$ where s represents an M dimensional vector with components that represent those properties required to adequately describe the state of the spray. This function is defined such that $f(s;x,t)\,\mathrm{d}s$ represents the probable number of drops, per unit system volume, per unit time, which exist in a given state s_1, $s_2, \ldots, s_M$. The continuous-particle model accommodates the distribution function directly using a phenomenological Liouville equation typically referred to as the spray equation.[9]

Difficulties arise with continuous-particle models because of the multidimensional character of the distribution function and the level of resolution required to minimize numerical diffusion when steep gradients with respect to f exist. Practical implementation requires one of two treatments: 1) specifying the functional dependence of the drop distribution function in terms of statistical variances or 2) implementing discrete-particle models. There are major limitations associated with the first treatment because of a general lack of knowledge regarding the correct functional form of the drop distribution function.

Discrete-particle models minimize problems associated with numerical diffusion by solving a set of Lagrangian equations of motion and transport for the life histories of a significant sample of particles. Because of the large number of drops present in an actual spray, a sampling technique is typically employed whereby characteristic groups of drops are represented by computational parcels. Travis et al.[10] have shown that continuous-particle models can be directly related to discrete-particle models by taking various moments of the spray equation.

Early works, such as the investigations conducted by Crow et al.,[8] did not account for turbulence in the calculation procedure. Subsequently, however, stochastic methods were adopted to simulate the instantaneous velocity field. Dukowicz,[11] O'Rourke,[12] and Gosman and Ioannides[13] were the first to adopt stochastic methods to study drop dispersion due to turbulence. Shuen[14] (see Faeth[6]) extended the methodology of Gosman and Ioannides to include the effects of turbulence on interphase heat and mass transfer. Here, turbulent eddy lifetimes are specified in the vicinity of each parcel using the scaling arguments associated with the turbulence model. A particle is assumed to interact with an eddy for a time defined as the smaller of the eddy lifetime or the characteristic transit time, as given by the Lagrangian equations of motion. This methodology, which is commonly referred to as the stochastic-separated-flow (SSF) model, has been evaluated in a wide variety of flows and appears to be the most efficient way to handle particulate dynamics in most spray combustion algorithms.

A.　Eulerian-Lagrangian Formulation

The governing conservation equations for multiphase systems include terms similar to those that appear in single-phase flow. Also present are source terms that account for the interphase exchange of mass, momentum, total energy, and species. Treatment of the particulate phase in a Lagrangian frame requires that three aspects be incorporated in the formulation: 1) the variation of the liquid void fraction Ω to account for cluster relaxation, vaporization, and the effects of short-range interactions on interphase exchange rates, 2) the mass transfer rate $\dot{m}$, heat transfer rate q, and drop drag C_d, and 3) the effects of drop collisions.

The conservation equations in the Eulerian frame are given as follows:

i) Continuity equations

$$\frac{\partial \Omega \bar{\rho}}{\partial t} + \nabla \cdot (\Omega \bar{\rho} \tilde{u}) = \tilde{S}_{\ell m} \tag{32}$$

ii) Void fraction equations

$$\frac{\partial \Omega}{\partial t} + \nabla \cdot (\Omega \tilde{u}_\ell) = \frac{\bar{S}_{\ell m}}{\rho_\ell} + \nabla \cdot \tilde{u}_\ell \tag{33}$$

Here $S_{\ell m}$ represents the drop gasification rate. This term is described next.

A poly-dispersed phase is frequently divided into a finite number of groups designated by initial size, position and velocity, and calculations track the histories of each group. The source terms are determined by computing the property of the dispersed phase group as it crosses a computational cell and then summing over all groups in the computational cell as just described. Ellail and Khalil,[15] for example, adapt $S_{1m} = \sum_k \sum_n (\dot{m}_{n,k} \, p_{n,k} \, \Delta t_n) \Delta n_k$, where the summations are performed over the time step n and group size k, respectively. The term $P_{n,k}$ spans the range from 1 to 0, which represents full or no interaction, respectively. The term

$\dot{m}_{n,k}$ is the gasification rate of an isolated drop in a dilute spray. The renormalized drop law[16] is used for interacting drops for nondilute sprays. With this description in place, the remaining equations are written as follows:

iii) Momentum equation

$$\frac{\partial(\Omega\bar{\rho}\tilde{u})}{\partial t} + \nabla \cdot (\Omega\bar{\rho}\tilde{u}\tilde{u}) = -\nabla p + \nabla \cdot [\Omega(\mu + \mu_T)]\nabla\tilde{u} + \bar{S}_{lU} \tag{34}$$

where $\bar{S}_{lU}$, is given by

$$\bar{S}_{lU} = \sum\sum [\dot{m}_{n,k}(\tilde{u}_{lk} - \tilde{u}) - (F_{D,n,k} + F_{\text{coll},kj})](P_{n,k}\Delta t_n)\Delta n_k \tag{35}$$

in which $F_{D,n,k}$ is the drag force,

$$F_{D,n,k} = \frac{C_{Dk}}{2}\pi\rho\, d_{lk}^2(\tilde{u} - \tilde{u}_{D,n,k})|\tilde{u} - \tilde{u}_{D,n,k}| \tag{36}$$

and the collision force between the k and j group particle is

$$F_{\text{coll},kj} = \eta_{kj}\left(1 + \frac{m_j}{m_k}\right)\left(\frac{\rho_\ell}{\rho}\right)\frac{\sigma}{d_{\ell k}}(d_{\ell k} + d_{\ell j})^2|We|_{kj}e_{kj} \tag{37}$$

where e_{kj} is the vector in the direction linking the k and j drops, σ is the surface tension constant, and We is the Weber number.

iv) Mixture fraction equation

$$\frac{\partial(\Omega\bar{\rho}\tilde{Z})}{\partial t} + \nabla \cdot (\Omega\bar{\rho}\tilde{u}\tilde{Z}) = \nabla \cdot \Omega\left(\mu + \frac{\mu_T}{\sigma_z}\right)\nabla\tilde{Z} + \bar{S}_{lZ} \tag{38}$$

where $\bar{S}_{lZ} = \bar{S}_{lm}$ for the gasification rate of the disperse phase.

v) Thermal energy equation

$$\frac{\partial}{\partial t}(\Omega\bar{\rho}\tilde{h}) + \nabla \cdot (\Omega\bar{\rho}\tilde{u}\tilde{h}) = \frac{\partial P}{\partial t} + \nabla\Omega\left(\frac{\lambda + \lambda_T}{C_P}\right) \cdot \nabla\tilde{h}$$

$$+ \nabla \cdot \tilde{q}_R + \sum\bar{\rho}\dot{w}_iQ_i + \sum_i\nabla \cdot \Omega\left(\frac{\lambda}{C_P} - \rho D\right)\tilde{h}_i\nabla Y_i + \bar{S}_{\ell h} \tag{39}$$

where $\bar{S}_{\ell h} = -\sum\sum\dot{m}_{nk}(\tilde{L}_k + \tilde{h} - \tilde{h}_{sk})(P_{n,k}\Delta t_n)\Delta n_k$. $\tilde{L}_k$ is the effective latent heat of vaporization, and q_R is the radiation flux.

vi) Turbulent kinetic energy equation

$$\frac{\partial(\Omega\bar{\rho}\tilde{\kappa})}{\partial t} + \nabla \cdot (\Omega\bar{\rho}\tilde{u}\tilde{\kappa}) = \nabla \cdot \Omega\left(\mu + \frac{\mu_T}{\sigma_k}\right)(\nabla\tilde{\kappa}) + \Omega S_\kappa + S_{l\kappa} \qquad (40)$$

where $S_\kappa = \Omega\rho\overline{u''u''}{:}(\nabla\tilde{u}) - \Omega\rho\tilde{\varepsilon}$, and $S_{\ell\kappa} = \tilde{u}\tilde{S}_{\ell u} - \tilde{u}\tilde{S}_{\ell u}$.

vii) Energy dissipation equation

$$\frac{\partial(\Omega\bar{\rho}\varepsilon)}{\partial t} + \nabla \cdot (\Omega\bar{\rho}\tilde{u}\varepsilon) = \nabla \cdot \left(\mu + \frac{\mu_T}{\sigma_\varepsilon}\right)(\nabla\tilde{\varepsilon}) + \Omega S_\varepsilon + S_{l\varepsilon} \qquad (41)$$

where $C_{\varepsilon 1} = 1.44$, $C_{\varepsilon 2} = 1.92$, $\sigma_\varepsilon = 1.22$ (Ref. 17),

$$S_\varepsilon = -\left[C_{\varepsilon 1}\bar{\rho}\left(\frac{\tilde{\varepsilon}}{\tilde{\kappa}}\right)\overline{u''u''}{:}(\nabla\tilde{u}) + C_{\varepsilon 2}\bar{\rho}\left(\frac{\tilde{\varepsilon}^2}{\tilde{\kappa}}\right)\right] \qquad (42)$$

$$S_{l\varepsilon} = 2v\overline{\nabla u''{:}\nabla S_{lu}} \qquad (43)$$

viii) Variance equation

$$\frac{\partial(\Omega\bar{\rho}g)}{\partial t} + \nabla \cdot (\Omega\bar{\rho}\tilde{u}g) = \nabla \cdot \left[\left(\mu + \frac{\mu_T}{\sigma_g}\right)\nabla g\right] + S_g + S_{lg} \qquad (44)$$

where $C_{g1} = 2.8$, $C_{g2} = 1.92$, $\sigma_g = 0.7$ (Ref. 2), and

$$S_g = C_{g1}\mu_T(\nabla\tilde{Z})^2 - C_{g2}\rho\left(\frac{\varepsilon}{\kappa}\right)\tilde{Z} \qquad (45)$$

$$S_{lg} = 2(\overline{Z S_{lm}} - \tilde{Z}\tilde{S}_m) \qquad (46)$$

Note that the deterministic-separated-flow (DSF) model accounts for the finite interphase exchange rates but neglects the interaction of the drops with turbulence, i.e., $S_{l\kappa}$, $S_{l\varepsilon}$, and S_{lg}, are zero. The stochastic-separated-flow (SSF) model adopts random walk models for the dispersed phase and all the interaction source terms, $S_{l\kappa}$, $S_{l\varepsilon}$, and S_{lg}, are included.

The rate of the change of the size, position, velocity, and energy of each drop appears in the source terms S_{lm}, S_{lu}, and S_{ln}. These quantities are obtained in the Lagrangian frame as follows:

$$\frac{dr_l}{dt} = -\frac{\dot{m}}{4\pi\rho_l r_l^2} \tag{47}$$

$$\frac{d\mathbf{x}_l}{dt} = \mathbf{u}_l \tag{48}$$

$$\frac{d\mathbf{u}_l}{dt} = -\frac{3(\tilde{F}_D + \tilde{F}_{\text{coll}})}{4\pi\rho_l r_l^3} + \frac{\rho}{\rho_l}\frac{d\mathbf{u}}{dt} \tag{49}$$

$$\tilde{F}_{\text{coll}} = \eta_{\ell j}\left(1 + \frac{m_j}{m_\ell}\right)\frac{\rho_\ell}{\rho}\frac{\sigma}{d_\ell}(d_\ell + d_j)^2 We_{\ell j} e_{\ell j} \tag{50}$$

$$\frac{de_l}{dt} = -\frac{3\dot{m}}{4\pi\rho_l r_l^3}\left(e_l - e_{ls} + \frac{\dot{q}_l}{\dot{m}}\right) + \frac{\sum \dot{Q}_k}{\rho_l(1-\Omega)} + \frac{de_{\text{coll}}}{dt} \tag{51}$$

where $\dot{m}$ is the gasification rate, r_l the drop radius, F_{coll} is the collision force, and e_l the liquid internal energy per unit mass. The term $\dot{q}_l$ is the conductive heat flux to the drop, and $\dot{Q}_k$ is the energy flux given by

$$\dot{Q}_k = \nabla \cdot \Omega_k \lambda \nabla T + \nabla \cdot \rho\Omega_k D \sum_j h_j \nabla y_j \tag{52}$$

In conventional spray models, this term represents the effect of a non-uniform field, as identified by Sirignano.[18]

B. Dynamic Transport and Heat Transfer Equations of a Drop

Drop motion in nonsteady and turbulent gaseous environments is significantly complicated and has been historically modeled using empirical equations. The Basset, Bousinesq, and Ossen (BBO) approximation is by far the most widely used for dilute spray applications. This equation was originally derived for low Reynolds number flow and has since been generalized for use over a wide range of Reynolds numbers, in transient or unsteady compressible flows.

The BBO equation has the general form

$$\frac{4\pi}{3}\rho_l r_l^3 \frac{d\mathbf{u}_l}{dt} = 6\pi\mu r_l(\mathbf{u} - \mathbf{u}_l) + \frac{2}{3}\pi\rho r^3\left(\frac{d\mathbf{u}}{dt} - \frac{d\mathbf{u}_l}{dt}\right)$$

$$+ 6r_l^2\sqrt{\pi\rho\mu}\int_0^t \frac{\left(\dfrac{d\mathbf{u}}{dt} - \dfrac{d\mathbf{u}_l}{dt}\right)}{\sqrt{t-t'}}\,dt' + \frac{4}{3}\pi\rho r_l^3\frac{D\mathbf{u}}{Dt} + \frac{4}{3}\pi r_l^3(\rho_l - \rho)g_i \tag{53}$$

The first term represents the steady-state drag force and depends on the local flowfield structure. It has been shown that the internal circulation within the drop will in general reduce the coefficient associated with this term from 6 to 4.[18] The second term represents the effect of virtual or "added" mass due to local acceleration of ambient fluid by the motion of the drop. The third term represents the Basset history. This term provides a corrective force that accounts for the accumulated effects of temporal variations in the relative velocity. The fourth term represents the effect of pressure gradients. Force corrections can also be added to account for vaporization and reacting drops. Being an empirical expression, the BBO equation has been frequently modified to account for various dynamic and kinetic factors. These modifications account for the effects of initial conditions, added mass, Stokes stresses, nonsteady flow, acceleration and deceleration, and high Reynolds numbers effects.

By adapting the analogous approach, Michaelides and Feng[19] formulated an equation of transient heat transfer from a sphere. They separated the imposed external temperature field into an unperturbed temperature, $T^0(x,t)$, and the disturbance due to the sphere, $T_1(x,t)$, and formulated the following equation:

$$m_l C \frac{dT_l}{dt} = mC_P \frac{DT^0}{Dt} - 4\pi r k \left\{ T_l(t) - T^0(x,t)\big|_{X=0} - \frac{1}{6} r_l^2 \nabla^2 T(x,t)\big|_{X=0} \right\}$$

$$- 4\pi r_l^2 k \int_0^t \frac{d}{d\tau} \frac{\left\{ T_l(\tau) - T^0(x,t)\big|_{X=0} - \frac{1}{6} r_l^2 \nabla^2 T^0(x,t)\big|_{X=0} \right\} d\tau}{\pi r_l(t-\tau)} \tag{54}$$

The first term represents the rate of the change of the temperature of the sphere. The second term represents the rate at which heat would have entered the control volume occupied by a sphere. This term is analogous to the added mass term in the BBO equation. Sirignano[18] has shown that this term appears because of variations in the freestream. The third term represents heat transfer due to conduction from the sphere into the fluid. The conduction process is a function of both the temperature difference and the curvature of the temperature field. The last term represents heat transfer due to diffusion of the temperature gradients inside the temperature field.

C. Non-Dilute Spray Models

Modeling non-dilute sprays accurately requires consideration of several additional phenomena. This includes 1) the spray morphology, structural inhomogenity, and formation of meso-scale structures in high void fraction regions, 2) direct short-range drop-drop interactions such as collisions and coalescence, 3) indirect drop-drop interactions at a distance comparable to the instantaneous drop spacing, 4) modulation effects on interphase exchange rates, 5) long-range interaction at distances greater than interdrop spacing, which leads to group ignition and group combustion modes, and 6) interaction of meso-scale structures with Richardson's cascade eddy structure. Dispersed and carrier gas interactions induced by both short-range and long-range drop-drop interactions are referred to as collective phenomena and have been the subject of broad interest in the spray modeling community.

Various classifications have been used to characterize sprays. Elghobashi,[20] for example, has suggested that there are three relevant regimes that can be identified as a function of the void fraction. These regimes are given in Fig. 1. For void fractions less than 10^{-6}, particles will most likely have a negligible effect on turbulence, and a one-way coupled dilute suspension can be assumed. When the

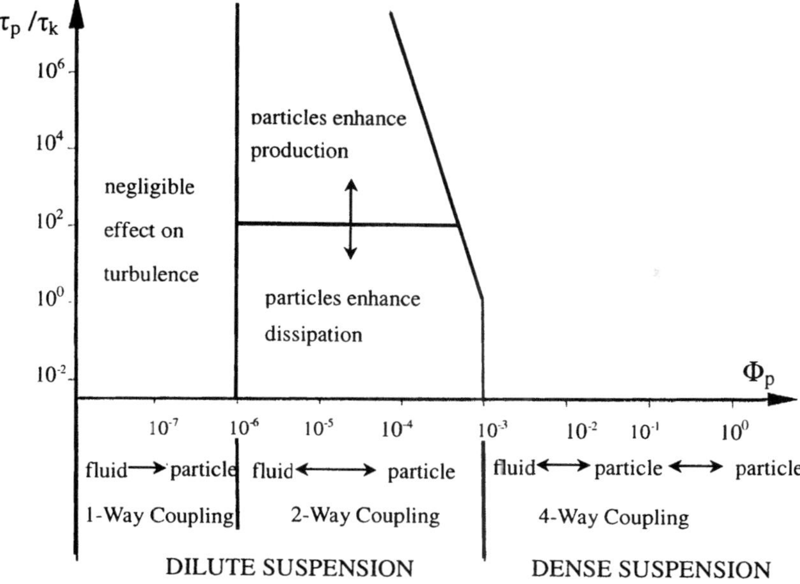

Fig. 1 Two-phase turbulent flow regimes in particle-laden flows (Sirignano,[18] redrawn from Elgohbashi[20] and contributed by S. E. Elghobashi, with permission of Cambridge University Press).

void fraction spans a range from approximately 10^{-6} to 10^{-3}, a two-way coupled dilute suspension should be assumed, and particle induced turbulence modulation must be considered. Dense suspensions occur in regions where the void fraction is greater than 10^{-3}. Here four-way coupling must be considered to account for fluid-particle and particle-particle interactions.

The renormalization number, G_{RN},[16] provides another criteria to characterize sprays. This term provides a measure of non-diluteness to allow the characterization of drag, vaporization, and combustion of drops under collective indirect interactions. It is related to the void fraction Ω_1 as follows:

$$G_{RN} = 4\pi n r_l^3 \eta_{st}^2 \frac{\mu}{(1 - \eta_{st}^{-1})} = 3\Omega_1 \eta_{st}^2 \frac{\mu}{(1 - \eta_{st}^{-1})} \tag{55}$$

where η_{st} is a nondimensional correlation length of droplet short-range interaction. The renormalization number expresses the strength of the short-range interaction between drops.

D. Morphological and Kinetic Models

Williams[7,21] proposed the following spray equation for the drop distribution function f_i:

$$\frac{\partial f_i}{\partial t} + \nabla \cdot (u f_i) + \nabla_u \cdot (F_i f_i) + \frac{\partial}{\partial r}(R_i f_i) = Q_i + \Gamma_i \tag{56}$$

Here, R_i represents the drop regression rate, $\boldsymbol{F}_i$ the force acting on the drop, Q_i the rate of drop formation or destruction, and Γ_i the collision induced variation in the distribution function. In dense, inhomogeneous spray structures, the drop distribution function becomes inadequate to describe the detailed mophography of spray. Instead, distribution functions with higher order correlations are required.

A kinetic theoretic formulation for non-dilute many-drop systems by Chiu[22] adopts the generalized Born, Bogoliubov, Green, Kirkwood, and Yovn (BBGKY) hierarchical theory by which a N-drop distribution function is used to generate a hierarchy of distribution functions of meso-scale structures. The N-drop distribution function $f^{(N)}$ for finding a system of N drops in the particular ranges of $3N$ spatial coordinates $x_1 \ldots x_N$, $3N$ velocity coordinates, $V_1 \ldots V_N$, and N drop size coordinates $r_1 \ldots r_N$, is defined by

$$f^{(N)}dV^{(N)} = f^{(N)}(x_1 \ldots x_N, v_1 \ldots v_N, r_1 \ldots r_N, t)$$
$$dx_1 \, dV_1 \, dr_1 \ldots dx_N \, dV_N \, dr_N \, dt \qquad (57)$$

The distribution function is governed by the following generalized Liouville equation[9]:

$$\frac{\partial f^{(N)}}{\partial t} + \sum_i \nabla_{x_i} \cdot (f^{(N)} v_i) + \sum_i \nabla_{v_i} \cdot (f^{(N)} \dot{v}_i) + \frac{\partial}{\partial r}(f^{(N)} r_i) = \frac{\delta f^{(N)}}{\delta t} \qquad (58)$$

where the right-hand term represents the source term that accounts for production or reduction of drops due to collision, coalescence, and breakup.

The hierarchy of reduced distribution functions $f^{(n)}$ are defined by

$$f^{(n)}(1, n, t) = G^{(n)}(N, n) \int f^{(N)}(1, \ldots, N) \, dV_{N-n} \qquad (59)$$

where $G^{(n)}$ is a normalization factor and a function of time

$$G^{(n)}(t) = [N(t)/\rho]^N / [N(t) - n]! \qquad (60)$$

wherein ρ is the mean number density of drop.

By an appropriate integration of Eq. (58), we obtain the following hierarchy of equations governing reduced distribution functions:

$$\frac{\partial}{\partial t} f^{(n)}(1,\ldots,n) + \sum_i^n \sum_{j \neq i}^n \frac{\partial}{\partial x_i} \cdot [f^{(n)}(1,\ldots,n)v_i]$$

$$+ \sum_i^n \sum_{j \neq i}^n \frac{\partial}{\partial v_i} \cdot [f^{(n)}(1,\ldots,n)\dot{v}_i] + \sum_i^n \sum_{j \neq i}^n \frac{\partial}{\partial r_i} \cdot [f^{(n)}(1,\ldots,n)\dot{r}_i]$$

$$+ \varepsilon^{(n,n+1)} \sum_{i=1}^n \sum_{\beta=n+1}^N \frac{\partial}{\partial v_i} \cdot \int f^{(n+1)}(1,\ldots,n,\beta)\dot{v}_i(i,\beta)\,\mathrm{d}\tau_\beta \quad \text{for } n = 1,\ldots,N \quad (61)$$

$$+ \varepsilon^{(n,n+1)} \sum_{i=1}^n \sum_{\beta=n+1}^N \frac{\partial}{\partial r_i} \cdot \int f^{(n+1)}(1,\ldots,n,\beta)\dot{r}_i(i,\beta)\,\mathrm{d}\tau_\beta$$

$$- \left\{ \frac{\mathrm{d}}{\mathrm{d}t} \ell n\, G^n(N,n) \right\} \cdot f^{(n)}(1,\ldots,n) = 0$$

where

$$\varepsilon^{(n,n+1)} = \frac{G^{(n)}}{G^{(n+1)}} = N - (n+1) \tag{62}$$

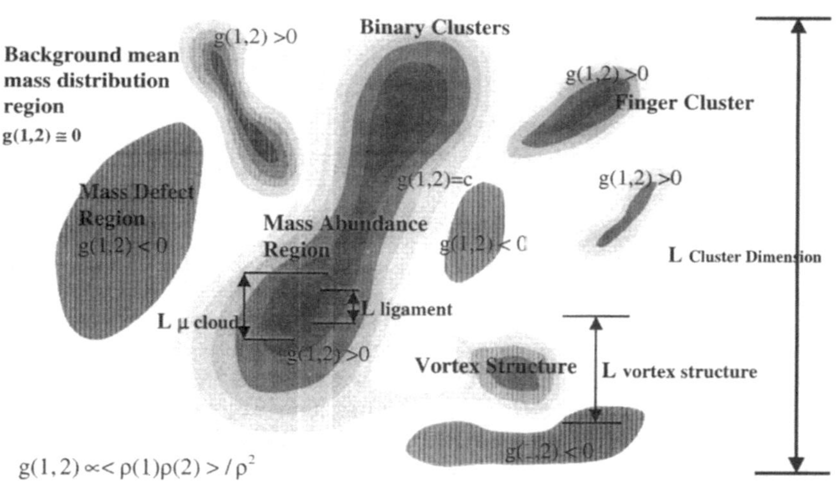

Fig. 2   Characteristics of large-scale structures by correlation functions.

We adopt the following representation for distribution functions:

$$f^{(1)} = f(1) \tag{63}$$

$$f^{(2)}(1,2) = f(1)f(2)[1 + g(1,2)] \tag{64}$$

$$f^{(3)}(1,2,3) = f(1)f(2)f(3)[1 + g(1,2) + g(2,3) + g(3,1) + h(1,2,3)] \tag{65}$$

The single drop distribution function $f(l)$ represents a locally homogeneous distribution of drop number density without drop clustering. The correlation functions, $g(i,j)$, stand for the excess joint probability of finding ith and jth drops in the phase space over a uniform random distribution, as shown in Fig. 2. This function represents the extent of clumping associated with meso-scale structures in the local drop distribution. Variations of the distribution functions in phase space are caused by the two types of source terms. The first type, termed graininess, induces short-range drop-drop interactions due to the discrete nature of the drop distribution. The second type of source term represents the effects of long-range collective interactions.

Conservation equations of non-dilute two-phase flows with locally inhomogeneous drop distribution are given by a system of equations similar to (32) and (46), with the exception that the source terms S_{lm}, S_{lu}, S_{lz}, $S_{l\kappa}$, and $S_{l\varepsilon}$ are expressed differently from those of the homogeneous sprays. The first three source terms are described in the following.

i) Source terms with single drops and drop pairs

The source term $S_{l\zeta}$ appears in mass momentum or energy and is represented by $S_{l\zeta} = \int \zeta_i(x,t,r_i)\, dr_i$, where ζ represents the gasification rate $\dot{m}_i$, drag force F_i, and heat transfer rate $\dot{q}_i$ of the ith drop in the meso-scale structure. This term includes single and drop pairs and is described by single and second-order correlation functions as follows:

$$S_{\ell s}(x,t,r_i) = \int\!\!\int f(1,r_i)\dot{\zeta}_i(1)\, dv_1\, dv_2$$
$$+ \int\!\!\int\!\!\int \{f(1,r_i)f(2,r_j)[1 + g(1,r_i,2,r_j)]\dot{\zeta}_{ij}(1)\}\, dv_1\, dv_2\, dr_j \tag{66}$$

Here $\dot{\zeta}_{ij}(1)$ is the exchange rate of drops of size i that are located at 1 and under the effects of a short-range interaction with drops of size j located at 2. The term $g(1,r_i,2,r_j)$ is the normalized pair distribution function.[16]

ii) Source terms with single, double, and triple drop configurations

The source terms for mass, momentum, or energy are represented by the general expression ζ up to the third-order distribution functions and are given by the following multi-pole source expressions:

$$S_{\ell\zeta}(x,t) = \int\int\int\int \{f(1)f(2)f(3)\zeta(1)\}\; d\mathbf{v}_1\; d\mathbf{v}_2\; d\mathbf{v}_3\; dr_i$$

$$+ \int\int\int\int\int\int \{f(1)f(2)f(3)g(1,2)\zeta(1,2)f(1)f(2)f(3)g(2,3)\zeta(1)$$

$$+ f(1)f(2)f(3)g(1,3)\zeta(1,3)\}\; d\mathbf{v}_1\; d\mathbf{v}_2\; d\mathbf{v}_3\; dr_i\; dr_j \qquad (67)$$

$$+ \int\int\int\int\int\int \{f(1)f(2)f(3)g(1,2,3)\zeta(1,2,3)\}$$

$$\times\; d\mathbf{v}_1\; d\mathbf{v}_2\; d\mathbf{v}_3\; dr_i\; dr_j\; dr_k$$

Here $\zeta(i)$ is the exchange rate of the ith isolated drop, $\zeta(i,j)$ is the exchange rate of the ith drop in a doublet source, with the second drop located at x_j and $\zeta(i,j-k)$. Observation of the source terms S_{lm}, S_{lu}, and S_{lm} reveals that the first term appears in the integrands of each equation. The single distribution function, for example, $f(1,r)\,\dot{m}$, represents the gasification of a drop without short-range interactions with its neighbors. The effects of the non-dilute characteristics are expressed by the short-range interaction terms appearing in the source terms and involve second-order distribution functions such as $f(1)f(2)g(1,2)$. The exchange rates associated with short-range interactions in higher order clusters must be predicted from the many-drop problems involving, for example, drop pairs, triplets, as well as clusters with n drops.

IV. Turbulence Generation and Modulation

Liquid spray systems contain complex turbulent interactions including two-phase transport, exchange of scalar and vector properties, drop dispersion, formation of drop clusters, heating, vaporization, and momentum exchange under the effects of turbulent interactions.[23-29] Modification by turbulence, both in the spray core and within clusters, also occurs.[30,31] Here we present some of the basic physical mechanisms and various methods for the prediction of turbulence interactions, cluster formulation, drop dispersion and the modulation of turbulent kinetic energy and the turbulent dissipation rate. Modulation of energy spectra is also discussed.

A. Drop-Turbulence Interactions

The nature of drop-turbulence interactions is frequently described in terms of drop evolution within the carrier gas. The likelihood of eddy confinement is characterized as a function of the Stokes number S_r, which compares the characteristic response time of a given drop to the residence time within an eddy.[32] The standard definition of the Stokes number is defined by an expression of the form

$$S_t = \frac{2\rho_l r_l^2 |u - u_l|}{9\mu L} \tag{68}$$

where L is a characteristic length associated with flow. When $S_t > 1$, drops follow the gas motion. When $S_t < 1$, drops deviate from the carrier gas.

Alternatively, the centrifugal Stokes number S_{tc} can be expressed by the ratio of the centrifugal force to the drag force acting on the drop.[33] This yields an expression of the form

$$S_{tc} = \rho_\ell W d_1^2 / 18 r_0 \mu \tag{69}$$

where W represents the tangential swirl velocity and r_0 is the drop radius. This quantity characterizes the extent of crossing trajectories. Sornek et al.[33] suggest that smaller drops with $S_{tc} < 1$ are confined within an eddy, whereas drops with $S_{tc} > 1$ tend to be thrown out of an eddy. For this latter situation drops form a dense region or cluster at the outer boundary of the eddy.

B. Drop Dispersion Models

Drops are dispersed because of interactions with turbulent eddies. Models to characterize drop dispersion include gradient induced transport methods and SSF methods. Jurewicz and Stock[34] and Dukowicz[35] modeled turbulent dispersion using the gradient diffusion approximation with a Lagrangian formulation for the motion and transport of the particles. Abbas et al.[36] adapted the concept of a diffusion force rather than a convective drift velocity in the formulation of drop motion. For this model, the drop acceleration is obtained using an equation of the form

$$\frac{d u_l}{dt} = f_d + f_g + f_{td} \tag{70}$$

where f_d and f_g are the drag and gravity forces and f_{td} is the effective diffusion force given by $f_{td} = V_{td}/\tau_d$. The term V_{td} is given by

$$V_{td} = -\frac{\rho}{\rho_t} D_{\text{eff},\ell} \nabla \left(\frac{\rho_\ell}{\rho} \right) \tag{71}$$

where $D_{\text{eff},l} = v_{\text{eff},l}/\sigma_l$, σ_l is an effective Schmidt number, and $\rho_l = nm$, where n is the drop number density and m is the mass of each drop.

The SSF model adapts a random walk approach that describes drops interacting with successions of integral scale eddies. Eddy velocities are randomly selected from the pdf of velocity fluctuations. The SSF approach has been used to treat dilute sprays with interaction times dependent on the eddy lifetime. Brown and Hutchinson[37] proposed that the turbulence-drop interaction time is determined by the smaller of either the eddy lifetime or the residence time required for a drop to traverse the eddy. Another criteria for drop-eddy interaction is given by the expression

$$|\Delta X_l - \mu_\ell \Delta t_l| \le l_e \qquad \Delta t_l \le t_e \tag{72}$$

where the left side represents the relative displacement of the drop with respect to the eddy, Δt_l is the interaction time, t_e is the eddy lifetime, and l_e is the characteristic length scale.[6,14,23] The characteristic eddy length and time scales are given by

$$l_e = \frac{\rho_\mu^{\frac{3}{4}} \kappa^{\frac{3}{2}}}{\varepsilon} \qquad t_e = \frac{l_e}{(2\kappa/3)^{\frac{1}{2}}} \tag{73}$$

Other methods developed to characterize drop dispersion include approximating simulated turbulence properties along a drop trajectory using statistical time series[38] and use of randomly selected Fourier modes from an isotropic turbulent environment.[39]

C. Drop Gasification and Wake-Induced Turbulence Modulation

Modulation of turbulent kinetic energy and energy dissipation by wake-induced turbulence is typically modeled by analyzing time-dependent turbulent flow over a spherical drop. By the application of the canonical interaction method to the equation for turbulent kinetic energy, we obtain the distribution of the kinetic energy κ_{sp} of a single-phase flow over an isolated drop as follows:

$$\kappa_{sp}(r, t) = \kappa_{sps}^{\gamma_{sp}} \exp\left\{ (4\pi r_s)^{-1} \mu_p \Xi_{\kappa,spr} \left(\frac{\partial \dot{m}}{\partial \omega}\right)_{sp} + \sum_{i=1}^{3} \Xi_{\kappa,spr} \Psi_{sp,i} \right\} \tag{74}$$

Here, $\Psi_{sp,i}$ is given in Tables 1 and 2, and $\kappa_{sp}(r, t) = \tilde{\kappa}_{sp}(r, t)/\tilde{\kappa}_{sp\infty}$, $\tilde{\kappa}_{sp\infty}$, is the kinetic energy of the freestream carrier gas, sp refers to single-phase flow, s stands for the drop surface, and ∞ the freestream. The terms

$$\gamma_{sp} = \frac{\left[\mu + \dfrac{\mu_T}{\sigma_\kappa}\right]_{sps}}{\left[\mu + \dfrac{\mu_T}{\sigma_\kappa}\right]_{spr}} \qquad \Xi_{\kappa,spr} = \left(\mu + \frac{\mu_T}{\sigma_\kappa}\right)_{spr}^{-1} \tag{75}$$

and κ_{sps} is the characteristic value of Dirichlet's boundary-value problem of kinetic energy given by a function of $\tilde{\kappa}_{sp\infty}$ as follows:

$$\kappa_{sps}(r_s, t) = \kappa_{sp\infty}^{\gamma_{sp\infty}^{-1}} \exp\left\{ \frac{1}{\gamma_{sp\infty}} \left[(4\pi r_s)^{-1} \mu_p \Xi_{\kappa,sp\infty} \left(\frac{\partial \dot{m}}{\partial \omega}\right)_{sp} \right.\right.$$

$$\left.\left. - \sum \Xi_{\kappa,sp\infty} \Psi_{sp,i\infty} \right]\right\} \tag{76}$$

Table 1   Source terms of distribution functions of scalar flow variables in $\kappa-\epsilon-\alpha_T$ chain

		Variable turbulent transport property potential	Distributed source potential	Dispersed source potential
$\dfrac{\alpha_T + r}{(\alpha_T + r)_{sp}}$	θ_i	$\theta_1 = -\displaystyle\int_{r_s}^{r} \ell n(\alpha_T + r)\frac{\partial}{\partial r}\left[\frac{\lambda}{C_p}\left(1 + \frac{\lambda_T}{\lambda}\right)\right]dr$	$\theta_2 = \displaystyle\int_{r_s}^{r} r'^{-2}dr' \int_{r_s}^{r'} D_{\theta_2} r^2 dr$	$\theta_3 = \displaystyle\int_{r_s}^{r'} r'^{-2}\,dr' \int_{r_s}^{r'} \frac{S_{lh}}{\alpha_T + r} r''^2\,dr''$
$\dfrac{\kappa}{(\kappa)_{sp}}$	ψ_i	$\psi_1 = -\displaystyle\int_{r_s}^{r} \ell n\left(\frac{\kappa}{k_0}\right)\frac{\partial}{\partial r}\left(\mu + \frac{\mu_T}{\sigma_\kappa}\right)dr$	$\psi_2 = \displaystyle\int_{r_s}^{r} r'^{-2}dr \int_{r_s}^{r'} D_{\psi_2} r''^2\,dr''$	$\psi_3 = \displaystyle\int_{r_s}^{r} r^{-2}\,dr \int_{r_s}^{r'} \frac{S_{l\kappa}}{\kappa} r''^2\,dr''$
$\dfrac{\varepsilon}{(\varepsilon)_{sp}}$	ϕ_i	$\psi_1 = -\displaystyle\int_{r_s}^{r} \ell n\left(\frac{\varepsilon}{\varepsilon_0}\right)\frac{\partial}{\partial r}\left(\mu + \frac{\mu_T}{\sigma_\kappa}\right)dr$	$\phi_2 = \displaystyle\int_{r_s}^{r} r'^{-2}dr' \int_{r_s}^{r'} D_{\phi_2} r''^2\,dr''$	$\phi_3 = \displaystyle\int_{r_s}^{r} r^{-2}\,dr \int_{r_s}^{r'} \frac{S_{l\varepsilon}}{\varepsilon} r''^2\,dr''$

Note: θ_{isp}, ψ_{isp}, ϕ_{isp}, for the single phase flow, are given by the same expressions as in the table with the exception that θ_3, ψ_3, and ϕ_3 are zero.

Table 2 Functions of distributed source of two phase flow: D_{θ_2}, D_{ψ_2}, D_{ϕ_2}

$$D_{\theta_2} = -J \cdot \nabla \ell n(\alpha_T + \gamma) - \rho \frac{\partial \ell n(\alpha_T + \gamma)}{\partial t} + \frac{(DP/Dt)}{(\alpha_T + \gamma)} + \frac{(\Phi + S_h)}{(\alpha_T + \gamma)} + \sum \frac{\rho h_a^0 \omega_a}{(\alpha_T + \gamma)} + \frac{\partial}{\partial \theta} \left\{ \frac{\partial}{\partial \theta} \left[\frac{\lambda}{C_P} \left(1 + \frac{\lambda_T}{\lambda} \right) \right] [\partial \ell n(\alpha_T + \gamma)(\sin \theta)] - \chi_{aT\theta} \right\}$$

$$+ \frac{\partial}{\partial \phi} \left\{ \frac{\partial}{\partial \phi} \left[\frac{\lambda}{C_P} \left(1 + \frac{\lambda_T}{\lambda} \right) \right] [\partial \ell n(\alpha_T + \gamma)] - \chi_{aT\phi} \right\}$$

$$D_{\psi_2} = -J_\kappa \cdot \nabla \ell n \left(\frac{\kappa}{\kappa_0} \right) - \rho \frac{\partial}{\partial t} \ell n \left(\frac{\kappa}{k_0} \right) + \frac{S_\kappa}{\kappa} + \left\{ \frac{\partial}{\partial \theta} \left(\mu + \frac{\mu_T}{\sigma_k} \right) \left[\frac{\partial}{\partial \theta} \left(\ell n \frac{\kappa}{k_0} \right) \sin \theta \right] + \frac{\partial}{\partial \phi} \left(\mu + \frac{\mu_T}{\sigma_\kappa} \right) \frac{\partial}{\partial \phi} \left(\ell n \frac{\kappa}{k_0} \right) \right\} \frac{1}{\sin \theta}$$

$$D_{\phi_2} = -J_\varepsilon \nabla \ell n \left(\frac{\varepsilon}{\varepsilon_0} \right) - \rho \frac{\partial}{\partial t} \ell n \frac{\varepsilon}{\varepsilon_0} + \frac{S_\varepsilon}{\kappa} + \left\{ \frac{\partial}{\partial \theta} \left[\mu + \frac{\mu_T}{\sigma_\varepsilon} \right] \frac{\partial}{\partial \theta} \left[\left(\ell n \frac{\varepsilon}{\varepsilon_0} \right) \sin \theta \right] + \frac{\partial}{\partial \phi} \left(\mu + \frac{\mu_T}{\sigma_\varepsilon} \right) \frac{\partial}{\partial \phi} \left(\ell n \frac{\varepsilon}{\varepsilon_0} \right) \right\} \frac{1}{\sin \theta}$$

$$J_h = \rho v - \left(\frac{\lambda + \lambda_T}{C_P} \right) \nabla \ell n(\alpha_T + r) \quad J_\kappa = \rho u - \left(\mu + \frac{\mu_T}{\sigma_\kappa} \right) \nabla \ell n \left(\frac{\kappa}{\kappa_0} \right)$$

$$J_\varepsilon = \rho v - \left(\mu + \frac{\mu_T}{\sigma_\varepsilon} \right) \ell n \left(\frac{\varepsilon}{\varepsilon_0} \right) \quad \mu_T = \rho C_\mu \frac{\varepsilon^2}{\kappa}$$

$$\Gamma_{\alpha_T} = - \left[\rho D(L_e - 1) \sum_{i=1}^{N} w_i v_i h_i \alpha_i \right]$$

where the differential gasification rate is given by

$$\left(\frac{\partial \dot{m}}{\partial \omega}\right)_{\mathrm{sp}} = 4\pi r_s \mu_p^{-1} S_{\mathrm{sp}} + 4\pi r_s \mu_p^{-1} \int_{r_s}^{r} L_{\mathrm{sp}}\, dr + 4\pi r_s \mu_p^{-1} \int_{r_s}^{r} r^{-2}\, dr \int_{r_s} D_{\mathrm{sp}} r^2\, dr \quad (77)$$

in which

$$S_{\mathrm{sp}} = \left[\frac{(\lambda + \tilde{\lambda}_T)}{C_P}\right] \ell n\,[\tilde{\alpha}_T(r,t) + \gamma]$$

The terms L_{sp} and D_{sp} are the laminated and distributed gasification potentials appear in the canonical drop law given by Chiu.[40] Three major mechanisms contributing to the kinetic energy distribution in Eq. (74) are 1) the gasification-induced mechanism associated with the term $(\dot{m}/\omega)_{\mathrm{sp}}$, 2) residual transport due to the spatial variation of the transport properties, $\Psi_{\mathrm{sp},1}$, and 3) the distributed potentials $D_{\theta 2}$, $D_{\Psi 2}$, and $D_{\Phi 2}$ appearing in $\Psi_{\mathrm{sp},1}$, listed in Tables 1 and 2, which include convective transfer of Spalding potentials, accumulation of kinetic energy, generation and/or dissipation of kinetic energy, and polar and azimuthal angular transport of Spalding potential of kinetic energy by molecular and turbulent diffusion.

The total kinetic energy K contained in the flow over a gasifying drop is obtained by integrating Eq. (74) over the flow region extending outside of the drop as follows:

$$K = 4\pi \int_0^{2\pi} \int_0^{\pi} \int_{r_s}^{\infty} \kappa_{\mathrm{sp}} r^2\, dr\, d\omega = 4\pi \iiint (\kappa_{\mathrm{gasif}} \kappa_\psi) r^2\, dr\, d\omega \quad (78)$$

where

$$\kappa_{\mathrm{gasif}} = \kappa_{\mathrm{sp}s}^{\mathrm{rsp}} \exp\left\{(4\pi r_s)^{-1} \mu_p \left(\Xi_{k,\mathrm{spr}} \left[\frac{\partial \dot{m}(r_s,t)}{\partial \omega}\right]_{\mathrm{sp}}\right)\right\} \quad (79)$$

$$\kappa_\psi = \exp\left(-\sum \Psi_{\mathrm{sp}}\right) \quad (80)$$

The term κ_{gasif} represents the contribution due to gasification. The term κ_ψ represents the kinetic energy due to all mechanisms other than gasification. The total kinetic energy is given by

$$K = \langle K_G \rangle \langle K_\Psi \rangle \quad (81)$$

where

$$\langle K_G \rangle = 4\pi \iiint k_{\mathrm{gasif}} r^2\, dr\, d\omega \quad (82)$$

$$\langle K_\Psi \rangle = \frac{\iiint k_{\mathrm{gasif}} k_\Psi r^2\, dr\, d\omega}{\iiint k_{\mathrm{gasif}} r^2\, dr\, d\omega} \quad (83)$$

The gasification factor $\langle K_G \rangle$ is expressed by

$$\langle K_G \rangle = 4\pi \int\!\!\int\!\!\int \kappa_{\text{gasif}}\, r^2 \, dr \, d\omega$$

$$= 4\pi \int\!\!\int\!\!\int \kappa_{\text{sps}}^{r_{\text{sp}}} \exp\left[\frac{\left(1 - \dfrac{r_s}{r}\right)\left(\dfrac{\partial \dot{m}}{\partial \omega}\right)_{\text{sp}}}{4\pi r_s \left(\mu + \dfrac{\mu_T}{\sigma_K}\right)}\right] r^2 \, dr \, d\omega \qquad (84)$$

To estimate the gasification factor $\langle K_G \rangle$, we consider the following special case. We assume that the turbulent viscosity on the drop surface, μ_{Ts}, is the maximum value and decreases at a great distance from the drop, $\mu_{Ts} > \mu_{Tr}$. If the kinetic energy on the drop surface, $\tilde{\kappa}_{\text{sps}}$, is smaller than that of the freestream $\tilde{\kappa}_{\text{sps}}$, i.e., $\kappa_{\text{sps}} < 1$, then $\kappa_{\text{sp}}^{\gamma_{\text{sp}}}$ is a monotonically decreasing function of r. Thus, we assume the following decay pattern for $\kappa_{\text{sps}}^{\gamma_{\text{sp}}}$:

$$\kappa_{\text{sps}}^{\gamma_{\text{sp}}} = \kappa_{\text{sps}}\left(\frac{r_s}{r}\right)^{\nu} \qquad (85)$$

where ν is a constant that will be specified later. The term $(\mu + \mu_T/\sigma_K)$ appearing in the denominator of the exponential function in the right side of Eq. (84) is expanded in a Taylor series in terms of $\eta = 1 - r_s/r$, i.e.,

$$\mu + \frac{\mu_T}{\sigma_K} = \left(\mu + \frac{\mu_T}{\sigma_K}\right)_{\eta=0} + \sum \frac{1}{n!}\frac{\partial^n}{\partial \eta^n}\left(\mu + \frac{\mu_T}{\sigma_K}\right)\eta^n \qquad (86)$$

where $0 < \eta < 1$ in η domain.

Substituting Eqs. (85) and (86) into (84) yields the gasification factor for turbulent kinetic energy production as follows:

$$\langle K_G^{(\nu)} \rangle = \kappa_{\text{sps}} \int\!\!\int\!\!\int (1 - \eta)^{\nu-4} \exp\left[\dot{\mu}_K \eta\left(1 + \sum_{n=1}^{\infty} \alpha_n \eta^n\right)\right] d\eta \, d\omega \qquad (87)$$

where $\dot{\mu}_K = \dot{m}/4\pi r_s(\mu + \mu_T/\sigma_K)_s$ and α_n are the constants associated with the coefficients of the Taylor series. Analysis of Eqs. (86) and (87) reveals that the largest contribution of the integration of the exponential function is made at a smaller value of η, i.e., $\eta < 1$. Thus, as a first-order approximation, we evaluate integral Eq. (87) with $\alpha_n = 0$, for $n \geq 1$, and for different values of $\nu \geq 4$ as follows:

$$\langle K_G^{(4)} \rangle = \left(\kappa_{\text{sps}} r_s^3\right)\frac{1}{\dot{\mu}_K} H_0(\dot{\mu}_K) \qquad (88)$$

$$\langle K_G^{(5)} \rangle = \left(\kappa_{\text{sps}} r_s^3\right)\left[\frac{1}{\dot{\mu}_K} H_0(\dot{\mu}_K) - \frac{1}{\dot{\mu}_K^2} H_1(\dot{\mu}_K)\right] \qquad (89)$$

$$\langle K_G^{(6)} \rangle = \left(\kappa_{\text{sps}} r_s^3\right)\left[\frac{1}{\dot{\mu}_K} H_0(\dot{\mu}_K) - \frac{2}{\dot{\mu}_K^2} H_1(\dot{\mu}_K) - \frac{1}{\dot{\mu}_K^3} H_2(\dot{\mu}_K)\right] \qquad (90)$$

where

$$H_0(\dot{\mu}_K) = e^{\dot{\mu}_K} - 1 \tag{91}$$

$$H_1(\dot{\mu}_K) = 1 + e^{\dot{\mu}_K}(\dot{\mu}_K - 1) \tag{92}$$

$$H_2(\dot{\mu}_K) = 2 + e^{\dot{\mu}_K}(\dot{\mu}_K^2 + 2\dot{\mu}_K - 2) \tag{93}$$

It is interesting to note that in the limit of a high gasification rate, i.e., $\dot{\mu}_K = R_{eK} \gg 1$, all expressions $K_{\text{gasif}}^{(v)}$ for $v \geq 4$ reduce to

$$K_G \cong \frac{3}{4\pi} \frac{K_{\text{drop}}^*}{R_{eK}} (e^{R_{eK}} - 1) \tag{94}$$

where $K_{\text{drop}}^* = 4/3 \pi r_s^3 \kappa_{\text{sps}}$ is the total turbulent kinetic energy contained in the volume of the drop with a kinetic energy density of κ_{sps}, and

$$R_{eK} = \frac{\rho_s u_s r_s}{\left(\mu + \dfrac{\mu_T}{\sigma}\right)}$$

Contributions to the modulation of turbulent kinetic energy by wake turbulence are due to both generation and dissipation mechanisms. These contributions can be expressed by

$$\kappa_{\text{wake}} = \kappa_{\text{sps}}^{\gamma_{\text{sp}}} \exp\left(\mu + \frac{\mu_T}{\sigma_k}\right)_n^{-1} \left\{ \int r^{-2}\, dr \int \left\{ \left[\frac{\bar{\rho}(\widetilde{u''u''} : \nabla\tilde{u} - \bar{\rho}\tilde{\varepsilon})}{\kappa\tilde{\kappa}_\infty} \right] \right\} r^2\, dr \right\}_{\text{wake}} \tag{95}$$

Kinetic energy increases when the first integral is greater than the second, as is physically expected. Integration is performed over the region occupied by the wake. By adapting a drop wake size r_{wake}, mean velocity U, characteristic turbulence turnover time $t_T = \kappa/\varepsilon$, residence time $t_{\text{res}} = r_{\text{wake}}/U$, and molecular and turbulent viscosities in the freestream, μ_∞ and μ_{T_∞}, respectively, as the reference quantities, the net kinetic energy is scaled by the Reynolds number of the drop wake as follows:

$$\kappa_{\text{wake}} = \kappa_{\text{sps}}^{\gamma_{\text{sp}}} \exp\left\{ Re_w \phi_{kw} - Re_w \frac{t_{\text{res}}}{t_T} \phi_{\varepsilon w} \right\}_{\text{wake}} \tag{96}$$

Here, $Re_w = (\rho_\infty U_\infty r_{\text{wake}})/[\mu_\infty + (\mu_T/\sigma_k)_\infty]$, and ϕ_{kw} and $\phi_{\varepsilon w}$ are universal functions that depend on the Reynolds numbers and turbulence properties. Note that the net rate of kinetic energy generation is anticipated to change when the Reynolds number becomes greater than the critical value for vortex shedding. This is consistent with the experimental observation reported by Wu and Faeth.[41]

By using a similar canonical integration method, the total energy dissipation rate E can be expressed by the product of the gasification factor $\langle E_G \rangle$ and mean dissipation rate $\langle E_\Phi \rangle$ as follows:

$$E = \langle E_G \rangle \langle E_\Phi \rangle \tag{97}$$

where

$$\langle E_G \rangle = 4\pi \int\int\int \varepsilon_{sps}^{\beta_{sp}} \exp\left\{ (4\pi r_s)^{-1}\mu_p \Xi_{sspr}\left(\frac{\partial \dot{m}}{\partial \omega}\right)_{sp} \right\} r^2 \, dr \tag{98}$$

$$\langle E_\Phi \rangle = \frac{\int\int\int \varepsilon_{gasif} \varepsilon_\Phi r^2 \, dr \, d\omega}{\int\int\int \varepsilon_{gasif} r^2 \, dr \, d\omega} \tag{99}$$

The terms ε_{gasif} and ε_Φ correspond to the first and second terms of the exponential function appearing in the energy dissipation rate expression.

The rate of energy dissipation of a vaporizing drop at a large gasification rate, $Re_\varepsilon > 1$, is given by

$$E_{gasif} = \frac{3}{4\pi}\frac{E_{drop}^*}{Re_\varepsilon}(e^{Re_\varepsilon} - 1) \tag{100}$$

where $Re_\varepsilon = \dot{m}/4\pi r_s(\mu + \mu_T/\sigma_\varepsilon)$ and $E_{drop}^* = 4/3\pi r_s^3 \varepsilon_{sps}$ is the total energy dissipation rate in the volume occupied by a drop with the mean dissipation rate ε_{sps}.

Turbulence modification in the energy dissipation rate of a drop due to local generation and dissipation in the wake region is given by

$$\varepsilon_{wake} = \varepsilon_{sps}^{\beta_{sp}} \exp\left(\mu + \frac{\mu_T}{\sigma_\varepsilon}\right)_r^{-1} \left\{\left[\int r^{-2} \, dr \right.\right.$$

$$\times \int \left\{ \left[(\tilde{\varepsilon})^{-1}\left[C_{\varepsilon 1}\left(\frac{\tilde{\varepsilon}}{\tilde{\kappa}}\right)\overline{\rho(u''u'')} : \nabla\tilde{u} - C_{\varepsilon 2}\bar{\rho}\left(\frac{\tilde{\varepsilon}^2}{\tilde{\kappa}}\right)\right]\right]\right\} r^2 \, dr \right\}_{wake} \tag{101}$$

Using the same non-dimensional scheme adapted for the generation of kinetic energy described previously, the net rate of dissipation of energy can be expressed as follows:

$$\varepsilon_{wake} = \varepsilon_{sps}^{\beta_{sp}} \exp\left\{ C_{\varepsilon 1} Re_w \chi_{w\kappa} - C_{\varepsilon 2} Re_w \frac{t_{res}}{t_l}\chi_{w\varepsilon} \right\}_{wake} \tag{102}$$

where $Re_w = (\rho_\infty U_\infty r_{wake})/\mu_\infty + (\mu_T/\sigma_\varepsilon)_\infty$, and $\chi_{w\kappa}$ and $\chi_{w\varepsilon}$ are universal functions of the Reynolds number and the turbulence properties.

D. Two-Way Coupling Model for Interphase Exchange of K_{sp} and E_{sp}

Based on the distribution of turbulence quantities just described, the rate of exchange of turbulence quantities, dK_{sp}/dt and dE_{sp}/dt, and with the same time wise variation in K_{sp} and E_{sp} can be modeled as follows:

$$\frac{dK_{sp}}{dt} = \left[\frac{T_L}{(T_L + t_l)}\right] T_L^{-1} \int\int\int (\kappa_{sp} - \kappa_0) \, dV \tag{103}$$

Similarly, the rate of exchange of the dissipation rate is given by

$$\frac{dE_{sp}}{dt} = \left[\frac{T_L}{(T_L + t_l)}\right] T_L^{-1} \iiint (\varepsilon_{sp} - \varepsilon_0)\, dV \tag{104}$$

where V is the effective sphere of influence of a drop environment, κ_0 and ε_0 are the kinetic energy and dissipation rate in the absence of the drop, T_L is the streamwise Lagrangian time, and T_ℓ is the drop response time. Substituting Eq. (74) and the corresponding expression for ε_{sp} into Eqs. (103) and (104), we have

$$\frac{dK_{sp}}{dt} = \left[\frac{T_L}{(T_L + t_l)}\right] T_L^{-1} \iiint \left\{ \kappa_{sps}^{\gamma_{sp}} \exp\left[(4\pi r_s)^{-1} \mu_p \Xi_{\kappa,sp}\left(\frac{\partial m}{\partial \omega}\right)_s\right.\right.$$
$$\left.\left. + \sum \Xi_{\kappa,sp}\Psi_{i,sp}\right] - \kappa_0 \right\} dV \tag{105}$$

$$\frac{dE_{sp}}{dt} = \left[\frac{T_L}{(T_L + t_l)}\right] T_L^{-1} \iiint \left\{ \varepsilon_{sps}^{\gamma_{sp}} \exp\left[(4\pi r_s)^{-1} \mu_p \Xi_{\varepsilon,sp}\left(\frac{\partial m}{\partial \omega}\right)_s\right.\right.$$
$$\left.\left. + \sum \Xi_{\varepsilon,sp}\Psi_{i,sp}\right] - \varepsilon_0 \right\} dV \tag{106}$$

The rate of the exchange of the kinetic energy between a stationary drop and an eddy is given by

$$\frac{dK_{sp}}{dt} = \frac{3}{4\pi}\left(\frac{1}{T_L + t_l}\right)\left\{ \frac{K_{drop,1}^*}{Re_{k,1}}(e^{Re_{k,1}} - 1)\langle K_{\psi_1}\rangle - \frac{K_{drop,0}^*}{Re_{k,o}}(e^{Re_{k,o}} - 1)\langle K_{\psi_0}\rangle\right\} \tag{107}$$

Similarly, the rate of the exchange in the energy dissipation is

$$\frac{dE_{sp}}{dt} = \frac{3}{4\pi}\left(\frac{1}{T_L + t_l}\right)$$
$$\times \left\{ \frac{E_{drop,1}^*}{Re_{\varepsilon,1}}(e^{Re_{\varepsilon,1}} - 1)\langle E_{\Phi_1}\rangle - \frac{E_{drop,0}^*}{Re_{\varepsilon,o}}(e^{Re_{\varepsilon,o}} - 1)\langle E_{\Phi_0}\rangle\right\} \tag{108}$$

E.   Modulation of the Turbulent Energy Spectrum

Turbulent energy spectra associated with sprays are generally modified because of both drop motion and drop vaporization. Al Taweel and Landau[42] have examined modulation in the turbulent energy spectrum due to a particle's dynamic response to carrier gas velocity in sinusoidal oscillation, and with negligible relative velocity. If particles are uniformly distributed with a weight concentration of W, the average rate of energy dissipation per unit wave number, per unit mass of fluid, resulting from the interaction is given by

$(18Wv\rho/\rho_1 d_1^2)u_{kRMS}'^2 R_k^2$, where $u_{kRMS}' R_k = u_k' - \mu_{lk}'$ is the relative fluctuation velocity and k is the wave number. Thus the source term for energy dissipation by drops, $S_{1\varepsilon}$, is given by

$$S_{1\varepsilon} = \left(\frac{18Wv\rho}{\rho_1 d_1^2}\right) \int (u_{kRMS}'^2 R_k^2)k^2 \, dk \tag{109}$$

Following a similar analysis, the kinetic energy of the carrier gas is given by

$$S_{1\kappa} = \left(\frac{1}{2}\right)\rho \int (u_{kRMS}'^2)k^2 \, dk \tag{110}$$

Results of numerical calculation have revealed a pronounced damping in the high wave number region of the spectrum where particles are unable to follow the gaseous phase oscillation.[42] This result is shown in Fig. 3.

Modulation effects are found to increase with increasing particle concentration for a fixed particle size. At low wave numbers, the smaller particles produce less damping effects than larger particles. At high wave numbers, smaller drop fluctuations result in increased damping. It is interesting to note that the spectrum modulation at a zero mean relative velocity corresponds to the complete confinement of a drop by a specific eddy. Modulation at a finite mean relative velocity corresponds to the non-confinement of a drop by turbulent eddies.

The modulation of the turbulent power spectra in a carrier gas by vaporizing clusters in turbulent premixed sprays has been studied by Sornek et al.[33] Clusters

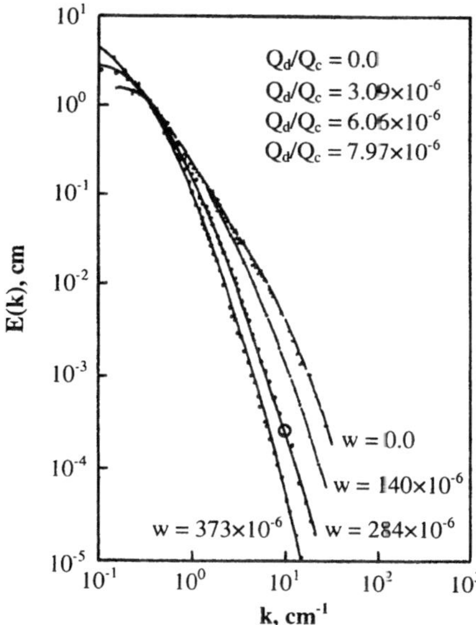

Fig. 3 **Modulation of turbulence spectra at selected particle loading W.**

of different configurations have been characterized by different values of the group combustion number G. This term is defined by

$$G = 2\pi L_e N_U (R/S)^2 (r_l/S) \tag{111}$$

where L_e is the Lewis number, N_U the drop-based Nusselt number, R the radius of a cluster, r_l the radius of a drop, and S the interdrop separation. Clusters produced by a grid of finer mesh size were found to yield larger interdrop separation S, and therefore have smaller values of G. By comparing the modulation of the spectra of high-G and low-G sprays, Sornek and his co-workers found that the reduction in the power spectra in low-G clusters is larger in the entire frequency range than that of high-G clusters, as shown in Fig. 4.

Canonical analysis of the gasification modulation of a turbulent flow gives the following asymptotic expression of the rate of energy dissipation:

$$\varepsilon = (3/4\pi)\varepsilon^* \{ [4\pi r_l(\mu + \mu_T/\sigma_\varepsilon)]/\dot{m} \} \exp \{ [4\pi r_l(\mu + \mu_T/\sigma_\varepsilon)]^{-1}\dot{m} \} \tag{112}$$

This equation leads to the conclusion that at sufficiently high gasification rates in low-G sprays the gasification modulation will yield a greater reduction in the energy spectra than drops in high-G sprays. This explains the results of Sornek

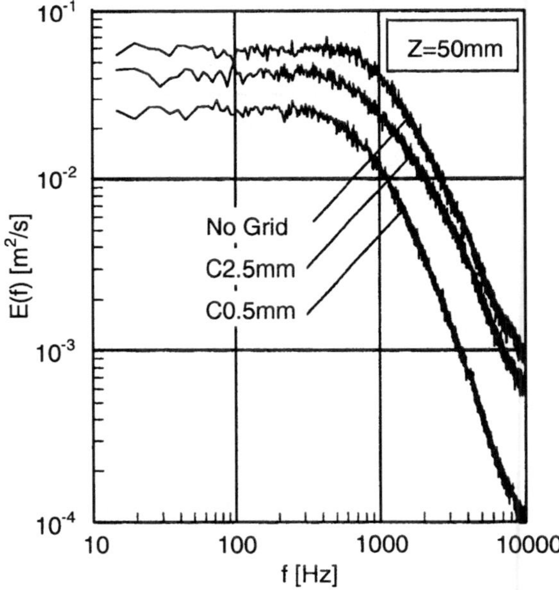

Fig. 4 Modulation of turbulence spectra for three mesh sizes at 50 mm from the nozzle exit (Sornek et al.[33] with permission of *Proceedings of the Combustion Institute*).

et al. The total power spectra reduction is the sum of the dissipation rates of drops at different values of interdrop separations, S_i, and given by

$$\varepsilon_{\text{Tot}} = \sum_{i=1} \left(\frac{3}{4\pi}\right) \varepsilon^*(i) \left\{ \frac{[4\pi r_l(\mu + \mu_T/\sigma_\varepsilon)]}{\dot{m}(S_i,d_i)} \right\} \exp\left\{ \left[4\pi r_l\left(\mu + \frac{\mu_T}{\sigma_\varepsilon}\right)\right]^{-1} \dot{m}(S_i,d_i) \right\}$$

(113)

Because the drop gasification rate $\dot{m}(S_i,d_i)$ for the different values of (S_i,d_i) exhibit different gasification characteristics, spectra modulation occurs in the wave number range corresponding to the interdrop separations.

V.   Collective Phenomena in Combusting Sprays

Many-drop systems (MDS), including sprays, exhibit group phenomena that feature excitation of a variety of combustion modes. These modes include external and internal group combustion and sheath combustion modes,[43-45] in which the totality, or part of the MDS, shares a common envelope flame surrounding a part or total aggregate of drops as shown in Fig. 5. In addition, MDS exhibit complex patterns of group vaporization[46] and group ignition,[47] caused by the retarded rates of gasification due to collective interaction.

These group phenomena in sprays are found in two different configurations. For the first configuration a core-based group combustion occurs where the vaporizing spray supports an envelope of gaseous diffusion flames that surround the core. An extreme example of this mode has been observed in a cryogenic liquid-oxygen–hydrogen propellant combustion chamber in liquid rocket engines.[48] Similar phenomena have been observed in gas turbines.[49] The second configuration is a cluster-based, or meso-scale structure, group configuration. The former configuration occurs in the upstream, whereas the latter configuration occurs in the downstream region depicted in Fig. 6. Collective phenomena occur in liquid sprays at both subcritical[30,31,44,45] and supercritical states.[50-54]

A.   Drop-Based Spray Model

To illustrate the method for prediction of group combustion phenomena, we consider a steady, axisymmetric combustion process governed by the system of conservation equations given by Eqs. (47–54). The goal for predicting group combustion modes of a spray is the determination of the state of each drop in the gaseous mixture and the ignitability of a drop at each point in a spray. A drop that lies in the lean mixture region, i.e., when the mixture fraction $Z < Z_{st}$, will combust if the ignition criterion is met. Conversely, if the local value of Z is larger than Z_{st}, a drop may be in a state of vaporization without drop-bounded combustion. This situation is described next.

The mixture fraction equation is given as follows:

$$\frac{\partial r(\rho^* u^* Z)}{\partial x} + \frac{\partial}{\partial r}\left\{ r\left[\rho^* v^* Z - \frac{\mu_T^*}{\sigma_Z^*}\frac{\partial Z}{\partial r}\right] \right\} = r S_{\ell m}^*$$

(114)

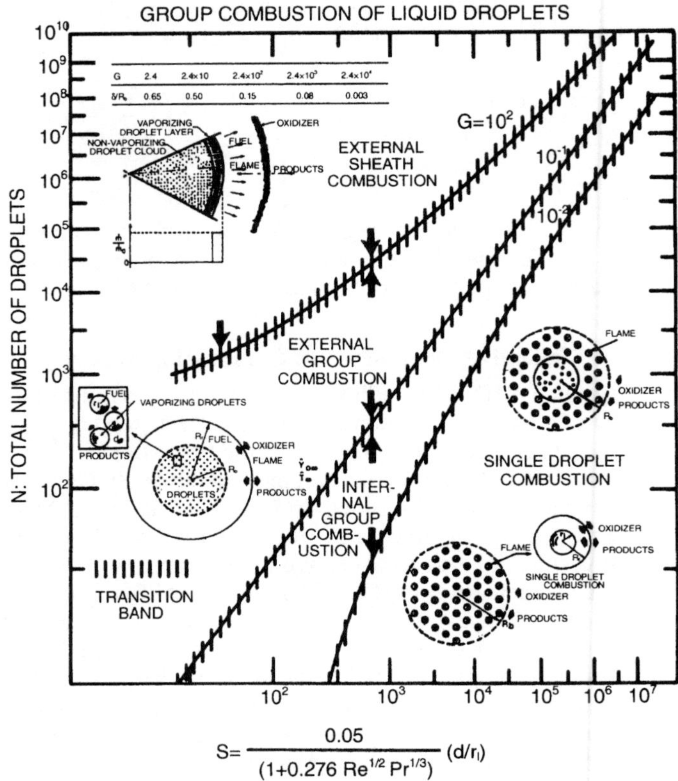

$$S = \frac{0.05}{(1+0.276\ Re^{1/2}\ Pr^{1/3})}\ (d/r_l)$$

Fig. 5 Group combustion modes for droplet clouds (Chiu et al.[44] with permission of *Proceedings of the Combustion Institute*).

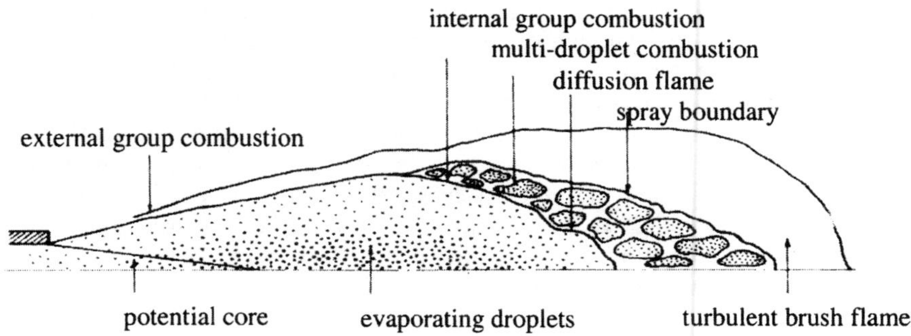

Fig. 6 Schematic diagram of core- and cluster-based group combustion of liquid spray.

where $S^*_{\ell m} = \sum \sum \dot{m}^*_{nk} P_{n,k} \Delta t_n \Delta n_k$ is the mass source[15] due to gasification and $*$ designates dimensional quantities. The preceding equation is non-dimensionalized using the reference quantities $u = u^*/u_o$, $v = v^*/u_o$, $\zeta = x/R$, $\eta = r/R$, $\eta = r/R$, $\sigma_Z = \sigma^*_Z/\sigma_{Zo}$, $\mu_T = \mu^*_T/\mu_{To}$, and $S_m = S^*_{lm}/n^*_o \dot{m}^*_r$, where R is the characteristic dimension of the spray, 0 represents the state at the center of the injector exit, and $\dot{m}_r$ is the reference gasification rate. By adapting the dimensionless mixture equation in a cylindrical coordinate system and applying canonical integration of the resulting non-dimensional equation, we obtain the axiomatic solution of the mixture fraction Z, with the appearance of a particular solution due to the source term S_{lZ}. By setting $Z = Z_{st}$, which corresponds to the flame location, we obtain

$$Z_{st} = Z_0 \exp\left\{ \sum_{i=1}^{3} A_i(\zeta_f, \eta_f) + \sum_{i=1}^{2} R_i(\zeta_f, \eta_f) \right\} \tag{115}$$

where subscript f denotes the quantities measured at the flame location. The terms A_i and R_i are given by

$$A_i(\zeta_f, \eta_f) = \left(\frac{\mu_{Tc}}{\mu_{Tf}}\right)\left(\frac{\sigma_{Zf}}{\sigma_{Zc}}\right)\left[\sum_{i=1}^{3} \Phi_{z,i}(\zeta_f, 0) \right] \tag{116}$$

$$R_i(\zeta_f, \eta_f) = \left(\frac{\mu_{Tf}}{\mu_{Zf}}\right)^{-1}\left[\sum_{i=1}^{2} \Psi_{zi}(\zeta_f, \eta_f) \right] \tag{117}$$

where

$$\Phi_{Z1} = -\int_0^\zeta (\eta \rho u Z)^{-1} \left\{ \frac{\partial}{\partial \eta}\left[\eta \rho v - (Re\sigma_{Zo})^{-1}\left(\frac{\mu_T}{\sigma_Z}\right)\left(\frac{\partial(\ell n(Z/Z_o))}{\partial \eta}\right) \right] \right\}_{\eta=0} d\zeta \tag{118}$$

$$\Phi_{Z2} = -\ell n\left(\frac{\rho_c u_c}{\rho_0 u_0}\right) \tag{119}$$

$$\Phi_{Z3} = \frac{1}{\sigma_D}\left(\frac{G}{Re}\right)\int \left[(\eta \rho u Z)^{-1} S_{lm}\right]_{\eta=0} d\zeta \tag{120}$$

$$\Psi_{Z1} = \sigma_Z Re \int_0^{\eta_f} \rho v \, d\eta + \int_0^{\eta_f} \ell n\left(\frac{Z}{Z_c}\right)\frac{\partial}{\partial \zeta}\left(\frac{\mu_T}{\sigma_z}\right) d\eta \tag{121}$$

$$\Psi_{Z2} = \sigma_Z Re \int_0^{\eta_f} \frac{d\eta}{\eta Z}\int_0^\eta \frac{\partial}{\partial \eta}(\rho u \eta Z) \, d\eta \tag{122}$$

$$\Psi_{Z3} = -\frac{\sigma_Z}{\sigma_D} G_C \int_0^{\eta_f} \frac{d\eta}{\eta Z}\int_0^\eta \eta S_{lZ} \, d\eta \tag{123}$$

Here $G = n_0 \dot{m}_0 R^2/\rho_0 D_{T0}$ is the group combustion number, and $Re = \rho_0 u_0 R/\mu_{T0}$. Equation (115) gives an implicit relation between ζ and η_f for the prescribed values of G and Re.

The necessary condition for a drop to be in a state of combustion is that it must be in the region where $Z < Z_{st}$ for all values of ζ. The sufficient condition for a drop to be in a state of combustion is that the drop satisfies the ignition criteria. These necessary and sufficient conditions for the drop will be termed criterion I.

In contrast, the necessary condition for a drop to be in a non-combusting state of vaporization, preheating, or saturation is $Z > Z_{st}$ for all values of ζ. The sufficient condition for a drop to be in the vaporization state is that it satisfies the state of non-ignition. These conditions will be termed criteria II.

By adapting these criteria throughout the spray flow domain, we predict the group combustion modes of a spray as follows:

1) External sheath and group combustion mode: When all of the drops are present in the regions where criterion II is satisfied, the spray in that region will be in external sheath, or group combustion modes (see Fig. 7).[55] The flame structure of the external group combustion is similar to that of a single-phase flow.

2) Internal group combustion mode: If part of the drop aggregate satisfies criterion I in a region and a portion of the drop aggregates in the adjacent region meet criteria II, the spray combusts in an internal group combustion mode depicted in Fig. 7. The internal group mode is complicated significantly because of the interaction of drops with the flame.

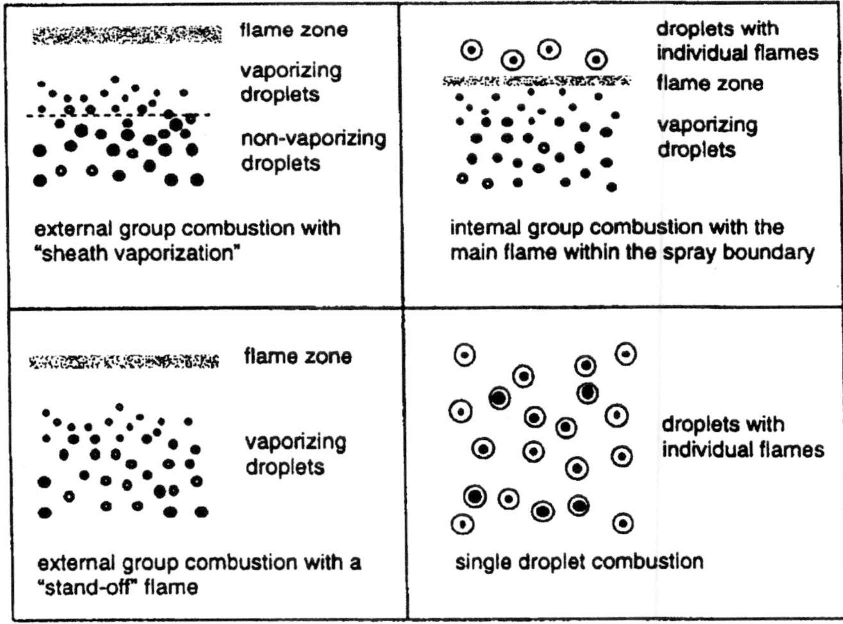

Fig. 7 Combustion models of a droplet cloud (adopted from Chiu et al.[44]; Warnatz et al.[55] with permission of Springer Verlag).

3) Isolated drop combustion mode: When all of the drops are in the region satisfying criterion I, the spray burns in the single-drop combustion mode shown in Fig. 7.

4) Spray burning rate: In addition to various group flame configurations, one of the important effects of group burning is the modification of an overall burning rate of the spray. To illustrate the effects of group phenomena on the total burning rate $\dot{M}$ of a spray, we express

$$\dot{M} = \sum_i \dot{m}_{ci} + \sum_j \dot{m}_{vj} = N_c \dot{m}_c + (N_T - N_c)\dot{m}_v \qquad (124)$$

where $\dot{m}_{ci}$ and $\dot{m}_{vj}$ are the average burning rate and the vaporization rate of a drop in each group combustion mode, respectively, whereas N_c is the total number of drops in the state of combustion. The term N_T represents the total number of drops in a spray.

B. Configuration and Structure of Group Combustion

The configuration of group flames, in particular, the external group mode and sheath mode, can be determined from Eq. (115). The analytical prediction of the flame radius R_f at an axial location of an external group and sheath mode is presented by the scaling law for a spray flame given by Eq. (127), which is presented subsequently. The flamelet equation for an external and sheath spray group flame is given by

$$-\left(\frac{\rho^* \chi^*}{2L_{ei}}\right)\left(\frac{\partial^2 \beta_i^*}{\partial Z^2}\right) = \omega_i^* + S_{lmi} \qquad (125)$$

For the external group mode, the source term $S_{bn} \cong 0$ and hence the flamelet equation is the same as that of a single-phase flow. The flame structure of an internal group combustion, on the other hand, is significantly complicated by the interaction of drops with turbulence and chemical reactions. The flame structure depends on the relative magnitudes of characteristic lengths, including the drop spacing S, drop size d, reaction zone thickness, $l_R = (\Delta Z)_R/|\nabla Z|_{st}$, and Kolmogorov length scale η_k. Events at all of these scales are contributing factors to the complexities of the flame structure. Three major families of two-phase flamelet structures have been identified as follows:

1) Two-phase flamelet structures: When the Kolmogorov eddy size η_k is greater than the other three scales (S, d, and l_R), the two-phase flame structure is embedded within the smallest eddies of turbulence. There are three distinct variations in this structure:

$$\eta_k > S > d > l_R, \qquad \eta_k > S > l_R > d, \qquad \eta_k > l_R > S > d$$

2) Interdrop flamelets or reaction zone structures: If the interdrop spacing S and size of a drop d are larger than the Kolmogorov eddies and reaction layers, flames are formed between the drops. Six variational structures described

by the following inequalities of characteristic lengths may be realized:

$$S > d > \eta_k > l_R, \qquad S > d > l_R > \eta_k, \qquad S > \eta_k > d > l_R$$
$$S > \eta_k > l_R > d, \qquad S > l_R > d > \eta_k, \qquad S > l_R > \eta_k > d$$

3) Distributed two-phase diffusion flame: In this special configuration, the smallest eddies of turbulence are embedded in the reaction layer with the following three variational structures:

$$l_R > S > d > \eta_k, \qquad l_R > S > \eta_k > d, \qquad l_R > \eta_k > S > d$$

The configuration and structure of two-phase flamelets similar to those associated with internal group combustion modes have also been studied extensively in counterflow configurations under laminar flow conditions, where the length scale of Kolmogorov eddies does not enter into consideration. The majority of these studies have focused on the effects of drop-flame interactions on laminar flame structure, distributions of chemical species, temperature, and velocities in the gas phase, and the effects of drops of different sizes on loading. Typical results are shown in Fig. 8. Gutheil and Sirignano[56] have shown that the counterflow spray at extinction under a high strain rate—say 1400/s—exhibits a structure significantly different from that of a low strain rate. The results given in Fig. 8 show that at high strain rate, both the flame and the regime of vaporization are moved toward the air side of the flame. Vaporization and combustion processes overlap, and drops exhibit a trajectory reversal and oscillate around the stagnation plane.

VI. Multiphase Combustion at Supercritical Conditions

Multiphase combustion at supercritical conditions poses a unique set of circumstances. Here we focus only on those circumstances where drops potentially exist. A series of experiments with cryogenic LOX-H_2 injected from a shear coaxial injector into a combustion chamber at transcritical and supercritical states have been conducted in recent years by Mayer and Tamaru,[48] Candel et al.,[54] Snyder et al.,[57] and most recently by Juniper et al.[50] Analogous numerical simulations of these experimental studies has been conducted by Yang,[51] and Oefelein and Yang.[52-53]

Experimental configurations exhibiting external group flames with LOX-H_2 propellants at various chamber pressures have been considered by Juniper et al.[50] The group combustion number of the spray in these experimental studies was found to be on the order of 3000. Hence, the spray burns with the external group combustion mode shown in Fig. 9. The geometrical configuration of the flame is a pseudoparabolic shape between the injector exit and the location of the maximum flame radius, beyond which the flame size decreases.

Figure 10 shows that the rate of the flame radius growth in the axial direction is reduced at a higher chamber pressure. These trends of the variations of flame geometry with respect to the chamber pressure suggest the presence of scaling laws that relate the flame topology to key system and process parameters. These

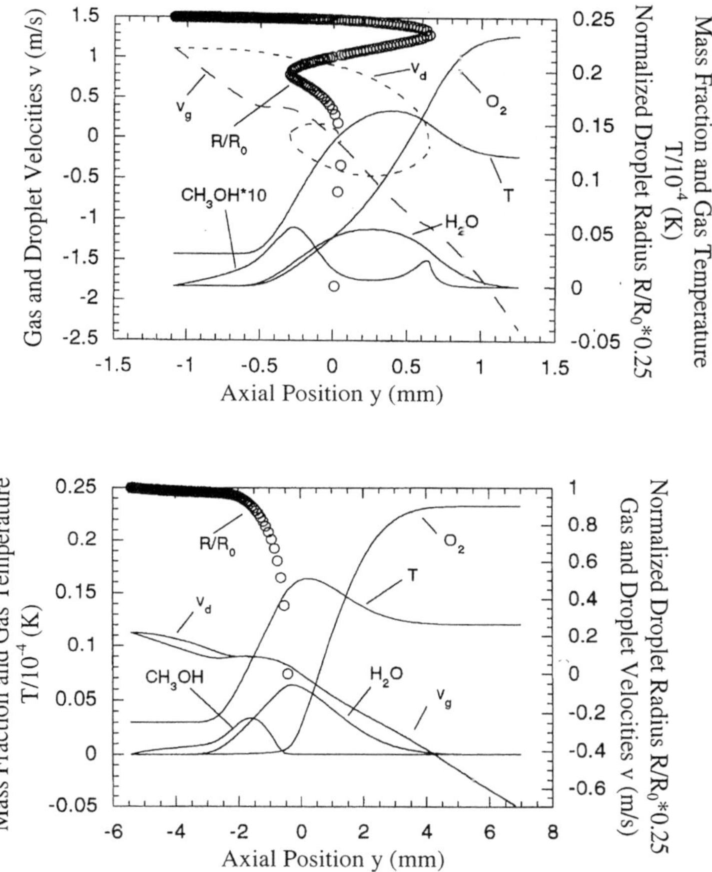

Fig. 8   Counterflow combustion structure of a monodisperse spray of methanol and air: (top) high strain and (bottom) low strain (Gutheil and Sirignano[56] with permission of *Combustion and Flame*).

parameters include the chamber pressure, group combustion number, Reynolds number, and other thermochemical parameters, as described next.

A.   General Scaling Law for Flame Radius

To formulate the scaling laws of a dense spray flame, we extend the Z_{st}/Z_o counter expression given by Eq. (115) to the corresponding form for non-dilute spray by including the void fraction and applying a scale transformation. The following transformation variables are employed:

$$\eta = \eta_f \hat{\eta}, \qquad \zeta = \hat{\zeta} \tag{126}$$

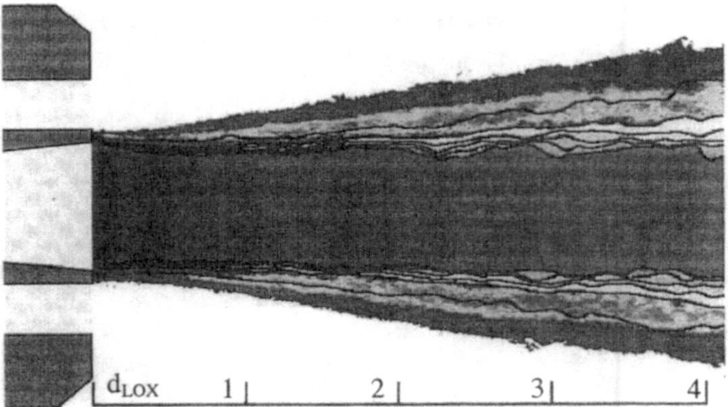

Fig. 9   Time-averaged OH* emissions and the counters of averaged jets represented by black counters, and the liquid jet in the central gray region (Juniper et al.[50] with permission of *Proceedings of the Combustion Institute*).

Here η_f is the dimensionless flame radius R_f/R, and $\hat{\zeta} = x/R$, where R is the radius of the injector. The scaling law for a non-dilute sprays is then given by

$$\ell n\left(\frac{Z_{st}}{Z_0}\right) = \left[\alpha(R_e\sigma_Z)^{-1}\hat{\phi}_{Z1,1}\right]\eta_f^{-2} + \left(\alpha\hat{\phi}_{Z1,1}\right)\eta_f^{-1} + \alpha\hat{\phi}_{Z2} + \frac{\alpha}{\sigma_D}\frac{G_C}{R_e}\hat{\phi}_{Z3} + \beta\hat{\psi}_{Z1,2}$$

$$+ \beta(\sigma_Z R_e\hat{\psi}_{Z1,1})\eta_f + \beta\left(\sigma_Z R_e\hat{\psi}_{Z2} + \frac{\sigma_Z}{\sigma_D}G_C\hat{\psi}_{Z3}\right)\eta_f^2 \qquad (127)$$

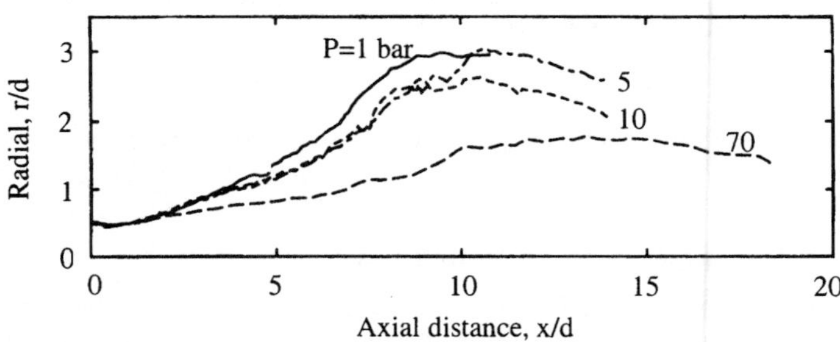

Fig. 10   Counter of the maximum intensity of the OH* emission representing the Abel transformed images of LOX-H_2 flame at different chamber pressures but similar momentum rate (Juniper et al.[50] with permission of *Proceedings of the Combustion Institute*).

where

$$\alpha = \left(\frac{\mu_{TC}/\sigma_{ZC}}{\mu_{T_j}/\sigma_{Z_f}}\right), \qquad \beta = \frac{\mu_{T_o}/\sigma_{z_o}}{\mu_{T_f}/\sigma_{z_f}}, \qquad z = \frac{Z}{Z_0}, \tag{128}$$

$$\hat{\phi}_{Z1,1} = -\int_0^{\hat{\zeta}} (\Omega\hat{\eta}\rho uz)^{-1} \frac{\partial}{\partial\hat{\eta}} (\Omega\hat{\eta}\rho v)_{\hat{\eta}=0} \, d\hat{\zeta} \tag{129}$$

$$\hat{\phi}_{Z1,2} = \int_0^{\hat{\zeta}} (\Omega\hat{\eta}\rho uz)^{-1} \frac{\partial}{\partial\hat{\eta}} \left(\Omega\frac{\mu_T}{\sigma_Z}\frac{\partial}{\partial\hat{\eta}} \ell nz\right)_{\hat{\eta}=0} \, d\hat{\zeta} \tag{130}$$

$$\hat{\psi}_{Z,2} = -\ell n\left(\frac{\rho_c u_c \Omega_c}{\rho_0 u_0 \Omega_0}\right), \qquad \hat{\phi}_{Z,3} = \int_0^{\hat{\zeta}} \frac{S_{IZ}}{\Omega\rho uz} \, d\hat{\zeta} \tag{131}$$

$$\hat{\psi}_{Z1,1} = \int_0^1 \rho v \, d\hat{\eta}, \qquad \hat{\psi}_{Z1,2} = \int_0^1 (\ell n\ z)\frac{\partial}{\partial\hat{\zeta}} \left(\frac{\mu_T}{\sigma_Z}\right) \, d\hat{\eta} \tag{132}$$

$$\psi_{Z2} = \int_0^1 \frac{d\hat{\eta}}{\hat{\eta}z\Omega} \int_0^{\hat{\eta}} \frac{\partial}{\partial\hat{\zeta}} (\Omega\rho u\hat{\eta}z) \, d\hat{\eta}, \qquad \hat{\psi}_{Z3} = -\int_0^1 \frac{d\hat{\eta}}{\hat{\eta}z\Omega} \int_0^{\hat{\eta}} \hat{\eta}S_{IZ} \, d\hat{\eta} \tag{133}$$

Equation (127) is a complex polynomial function of η_f that can be used to examine flame configurations and their dependence on various process parameters.

B. Scaling Law for the Initial Flame Expansion

Close examination of Eq. (127) reveals that there are three dominant mechanisms contributing to the initial flame development near the exit of the injector: 1) the gasification contribution in the axial direction represented by the fourth term on the right side of Eq. (127), i.e., $(\alpha/\sigma_D)(G_C/Re)\hat{\phi}_{Z3}$; 2) the rate of change of convective transport of the mixture fraction expressed by the seventh term, i.e., $\beta\sigma_Z Re\hat{\psi}_{Z2}$; and 3) the gasification contribution in the radial direction represented by the last term on the right side of Eq. (127), i.e., $\beta(\sigma_Z/\sigma_D)G_C\hat{\psi}_{Z3}$. According to these estimations, the approximate flame radius is given by

$$\eta_f \cong \left[\frac{\dfrac{\alpha}{\sigma_D}\dfrac{G_C}{R_e^2}\hat{\phi}_{Z3} + \ell n\left(\dfrac{Z_0}{Z_{s^*}}\right)}{-\beta\sigma_Z R_e\left(\hat{\psi}_{Z2} + \dfrac{G_C}{\sigma_D R_e}\hat{\psi}_{Z3}\right)}\right]^{1/2} \tag{134}$$

Using typical values of the functions ψ and ϕ, the flame configuration takes a pseudoparabolic shape with its principal axis coinciding with the $\hat{\zeta}$ axis. This observation is in qualitative agreement with the experimental data obtained by Juniper et al.[50]

The experimental results shown in Fig. 10 also reveal that the rate of expansion of the flame radius in the vicinity of the injector exit section at a chamber pressure of 70 bar is approximately 0.2–0.3 of the corresponding value at 1 bar. This latter trend can be explained by Eq. (127) by neglecting the second term appearing in the denominator. Near the injector exit at $\hat{\zeta} \sim 0$, we have $\eta_f - 1$, and $(d\phi_{Z3}^*/d\hat{\zeta}) = (S_{lZ}/\rho u z)_{\hat{\zeta}=0} = S_{lZ}(\hat{\zeta} = 0)$. By substituting $G_C = (1/3)(n_o \dot{m}_o R^2/\rho_o D_{T_o})$, $R_e = \rho_o U_o R/\mu_{T_o}$, and $\rho_o D_{T_o} = \mu_{T_o}/\sigma_D$ in the scaling law, we have

$$\frac{d\eta_f}{d\hat{\zeta}} = -\frac{\alpha}{2\beta\sigma_Z}\left(\frac{n_o \dot{m}_o v_{T_o}}{\rho_o^2 U_o^2}\right)\frac{S_{lZ}(\hat{\zeta} = 0)}{\psi_{Z2}^*(\hat{\zeta} = 0)} \tag{135}$$

If the characteristic mass flux $\rho_o U_o$ and volumetric gasification rate $n_o \dot{m}_o$ at two different chamber pressures, P_1 and P_2, remain the same, the ratio of the flame expansion rate reduces to $(d n_f/d\hat{\zeta})_{P_1}/(d\eta_f/d\hat{\zeta})_{P_2} = (v_{T_o,P_1})/(v_{T_o,P_2})$. The ratio of the flame radius expansion can be estimated by adapting the data of turbulent viscosity obtained by Oefelein and Yang,[52,53] who predicted that the ratio of the turbulent viscosity μ_T to molecular viscosity μ is 0–36 at 100 bar, and 0–0.27 at 1 bar. By taking the mean value of the viscosity ratio, i.e., $(\mu_T/\mu) = (v_T/v) = 18$ at 100 bar, and $(\mu_T/\mu) = (v_T/v) = 0.14$ at 1 bar, we obtain the ratio of the flame expansion rate by

$$\frac{(d n_f/d\hat{\zeta})_{100\text{atm}}}{(d n_f/d\hat{\zeta})_{1\text{atm}}} = \frac{v_{T,100\text{atm}}}{(v + v_T)_{1\text{atm}}} = \frac{18}{1.14}\frac{v_{100\text{atm}}}{v_{1\text{atm}}} \tag{136}$$

The ratio of the molecular kinematic viscosity at 100 atm and 1 atm at 1000 K is 10^{-2}. By substituting this value, the flame expansion at 100 atm and 1 atm is found to be 0.17, which is in qualitative agreement with the experimental data of 0.2 obtained by Juniper et al.[50] in Fig. 9.

C. Scaling Law for the Maximum Flame Radius

The maximum flame radius $\eta_{f\text{max}}$ that occurs downstream of the injector, as shown by experimental observation, can be determined by differentiating the scaling law with respect to $\hat{\zeta}$ and setting $d\eta_f/d\hat{\zeta}$ to zero. This criterion gives the following result:

$$A\eta_{f\text{max}}^2 + B\eta_{f\text{max}} + C = 0 \tag{137}$$

where

$$A = \beta\left(\sigma_z R_e \frac{\partial\hat{\psi}_{Z2}}{\partial\hat{\zeta}} + \frac{\sigma_Z}{\sigma_D}G_C\frac{\partial\hat{\psi}_{Z3}}{\partial\hat{\zeta}}\right) \tag{138}$$

$$B = \beta\left(\sigma_z R_e\frac{\partial\hat{\psi}_{Z1,1}}{\partial\hat{\zeta}}\right) \tag{139}$$

$$C = \alpha\frac{\partial\hat{\phi}_{Z2}}{\partial\hat{\zeta}} + \frac{\alpha}{\sigma_D}\frac{G_C}{R_e}\frac{\partial\hat{\phi}_{Z3}}{\partial\hat{\zeta}} + \beta\frac{\partial\hat{\psi}_{Z1,2}}{\partial\hat{\zeta}} \tag{140}$$

In the limit, when $G_C/R_e^2 < 1$, the preceding expression gives

$$\eta_{f\max} = \frac{\alpha\dfrac{\partial\hat{\phi}_{Z2}}{\partial\hat{\zeta}} + \dfrac{\alpha}{\sigma_D}\dfrac{G_C}{R_e}\dfrac{\partial\hat{\phi}_{Z3}}{\partial\hat{\zeta}} + \beta\dfrac{\partial\hat{\psi}_{Z1,2}}{\partial\hat{\zeta}}}{\beta\sigma_Z R_e\dfrac{\partial\hat{\psi}_{Z1,1}}{\partial\hat{\zeta}}} \tag{141}$$

Note that the derivatives of $\hat{\phi}$ and $\hat{\psi}$ in the preceding expression are determined at the location where $\mathrm{d}\eta_f/\mathrm{d}\hat{\zeta} = 0$. Also observe that the value of the maximum flame radius depends on the gasification factor, represented by G_C/R_e^2, and the effects of convective transport associated with the term R_e^{-1}. Since G_C/R_e^{-1} and R_e^{-1} are both proportional to the kinematic viscosity ν_T, the maximum flame radius is scaled with ν_T, i.e., $\eta_{f\max} \sim \nu_T$. Hence, the ratio of $\eta_{f\max}$ at 100 atm to the value at 1 atm is

$$\frac{\eta_{f\max,100\mathrm{atm}}}{\eta_{f\max,1\mathrm{atm}}} = \frac{\nu_{T,100\mathrm{atm}}}{(\nu_l + \nu_T)_{1\mathrm{atm}}} \simeq 0.17 \tag{142}$$

which again agrees well with the experimental value 0.17 given by Juniper and his co-workers,[50] as shown in Fig. 10.

VII. Conclusions

Fundamental physics and related mathematical methods for modeling turbulent spray combustion processes have been presented, with emphasis placed on the versatility and accuracy of modeling key processes, phenomena, and structure of spray systems. Extensive research over the last century has brought significant advances toward the understanding and prediction of the complex phenomenology associated with sprays. Major new advances in the following areas constitute the key phenomenological framework for contemporary spray system modeling.

It is well known that turbulent combustion is strongly influenced by a series of multiple-time and multiple-length processes. The principle of scale separation among the turbulence time, chemical time, and drop time scales provides a useful criteria for the categorization of various interactive process models. Various models that account for turbulence-chemistry interactions have been developed to predict the burning rates, combustion characteristics, and structures of dilute and non-dilute sprays.

Dispersed phase interactions with turbulence plays a fundamental role in the development of structures and the dynamic evolution of sprays through two- and four-way intercoupling between phases. Four principal physical problems of interest are 1) the effects of turbulence on the rates of interphase exchange, 2) the modulation of turbulent kinetic energy and dissipation rate, 3) the formation and deformation of drop clusters, and 4) turbulent dispersion of drops. Existing models and the canonical integration method provide a useful means of identifying and predicting these physical mechanisms with an acceptable level of validity and universality.

Collective phenomena due to drop-drop interactions in non-dilute systems create clusters of various length scales. These clusters are the structural factor that influence the inhomogeneous morphological characteristics practical sprays. Drops in clusters collectively affect the motion, vaporization, ignition, and combustion of the sprays. Thus, modeling spray combustion processes requires a variety of submodels essential for the determination of the rates of interphase exchange under the effects of drop-drop interaction, coupled with a comprehensive procedure for the prediction of various group combustion modes. Prediction of the flame-drop interactions that occur in an internal group combustion mode is largely unsolved.

Finally, the combustion of cryogenic propellant at transcritical and supercritical states in modern liquid rockets engines requires basic modeling of phase equilibria, thermodynamic and transport properties, along with the effects of turbulent transport properties. These collective phenomena have significant impacts on the configuration and structure of the flame.

Challenges for the development of future spray models include the development of a general methodology for optimal treatment of turbulence for specific levels of resolution. Development and implementation of the canonical integration method, renormalization group method, wavelet models, and development of efficient numerical schemes for each of these methodologies will significantly enhance the analytical versatility and predictive capability of the modeling. In addition to macrosystem models, there are also increasing needs for understanding the micro- to meso-scale and nano-scale processes. These processes play unique roles with respect to phase changes and chemical reactions at both subcritical and supercritical conditions.

Acknowledgments

Portions of the material contained in this article are the result of research projects supported by the National Science Council under contracts NSC 90-2612-E-006, 123, 178, 179, and the National Energy Commission EC 90-D0124, 91-D0124. H. H. Chiu wishes to express his appreciation to C. T. Lin, Robert H. Y. Cheng, J. C. Su, Evensa Su, and H. L. Tsai of SPRAX Advanced Science and Technology Laboratory at National Cheng-Kung University, C. F. Wu at Carnegie Mellon University, and Y. Shieh of the University of Illinois at Chicago for their contributions in preparation of this manuscript. J. C. Oefelein wishes to express his appreciation to his co-author Paul Chiu and to Vigor Yang of The Pennsylvania State University for the invitation to contribute to this manuscript. We would also like to thank Yanxing Wang for preparing the final figures for publication.

References

[1]Enwald, H., Peirano, E., and Almstedt, A. E., "Eulerian Two-Phase Flow Theory Applied to Fluidization," *International Journal of Multiphase Flow*, Vol. 22, 1996, pp. 21–66.

[2]Faeth, G. M., "Evaporation and Combustion of Sprays," *Progress in Energy and Combustion Science*, Vol. 9, 1983, pp. 1–76.

[3]Drew, D. A., "Mathematical Modeling of Two-Phase Flow," *Annual Review of Fluid Mechanics*, Vol. 15, 1983, pp. 261–291.

[4]Ishii, M., "Thermo-Fluid Dynamic Theory of Two-Phase Flow," Eyrolles, Paris, 1975.

[5]Delhaye, J. M., and Achard, J. L., "On the Use of Averaging Operations in Two Phase Flow Modeling," *Thermal and Hydraulic Aspects of Nuclear Reactor Safety, 1: Light Water Reactors*, ASME Winter Meeting, 1977.

[6]Faeth, G. M., "Mixing, Transport and Combustion in Sprays," *Progress in Energy and Combustion Science*, Vol. 13, 1987, pp. 293–345.

[7]Williams, F. A., "The Transport Equations for Multi-Phase Systems," *Physics of Fluids*, Vol. 1, No. 6, 1958, pp. 541–545.

[8]Crow, C. T., Sharma, M. P., and Stock, D. E., "The Particle-Source-In-Cell (PSI-CELL) Model for Gas-Droplet Flows," *Journal of Fluids Engineering*, Vol. 99, 1977, pp. 325–332.

[9]Williams, F. A., "Combustion Theory II The Fundamental Theory of Chemically Reacting Flow Systems," 2nd ed., Addison-Wesley, Menlo Park, CA, 1985.

[10]Travis, J. R., Harlow, F. H., and Amsden, A. A., "Numerical Calculation of Two-Phase Flows," *Nuclear Science and Engineering*, Vol. 81, 1976, pp. 1–10.

[11]Dukowicz, J. K., "A Particle-Fluid Numerical Method for Liquid Sprays," *Journal of Computational Physics*, Vol. 35, 1980, pp. 229–253.

[12]O'Rourke, P. J., "Collective Drop Effects on Vaporizing Liquid Sprays," Ph.D. Dissertation, Princeton Univ., Princeton, NJ, 1981.

[13]Gosman, A. D., and Ioannides, E., "Aspects of Computer Simulation of Liquid-Fueled Combustors," AIAA Paper 81-0328, 1981.

[14]Shuen, J. S., "A Theoretical and Experimental Investigation of Turbulent Sprays," Ph.D. Dissertation, Pennsylvania State Univ., 1984.

[15]Abou Ellail, M. M., and Khalil, E. E., "Mathematical Modelling of Turbulent Spray Flames," *Proceeding of International Power Engineering Conference*, 2, Paper 19, Cairo, 1978.

[16]Chiu, H. H., and Su, C. P., "Theory of Drops(II): States, Structures and Laws of Interacting Drops," *Atomization and Sprays*, Vol. 7, 1997, pp. 1–32.

[17]Gatski, T. B., "Turbulence Flows: Model Equations and Solution Methodology," *Hand Book of Computational Fluid Mechanics*, edited by R. Peyret, Academic Press, 1996, Chap. 6.

[18]Sirignano, W. A., *Fluid Dynamics and Transport of Drop and Sprays*, Cambridge Univ. Press Cambridge, MA, 2000.

[19]Michaelides, E. E., and Feng, Z., "Heat Transfer from a Rigid Sphere in a Nonuniform Flow and Temperature Field," *International Journal of Heat and Mass Transfer*, Vol. 37, 1994, pp. 2069–2076.

[20]Elghobashi, S., "On Predicting Particle Laden Turbulent Flows," *Applied Scientific Research*, Vol. 52, 1994, pp. 309–329.

[21]Williams, F. A., "Proceedings in Spray Combustion Analysis," *Eighth Symposium (International) on Combustion*, Combustion Inst., Pittsburgh, PA, 1962, pp. 50–69.

[22]Chiu, H. H., "Theory of Large Scale Structures and Collective Phenomena in Liquid Sprays," First Asia-Pacific Conference on Combustion, 1997, pp. 390–407.

[23]Faeth, G. M., "Current Status of Drop and Liquid Combustion," *Progress in Energy and Combustion Science*, Vol. 3, 1977, pp. 191–224.

[24]Law, C. K., "Recent Advances in Drop Vaporization and Combustion," *Progress in Energy and Combustion Science*, Vol. 8, 1982, pp. 171–1982.

[25]Bellan, J., and Harstad, K., "Unsteady Injection of Sequences of Drop Clusters in Vorticies Depicting Portions of Spray," *Atomization and Sprays*, Vol. 5, 1995, pp. 17–44.

[26]Rangel, R. H., and Continillo, G., "Theory of Vaporization and Ignition of a Drop Cloud in the Field of a Vortex," *Twenty-Fourth Symposium (International) on Combustion*, Combustion Inst., Pittsburgh, PA, 1992, pp. 1493–1501.

[27]Hinze, J. O., "Turbulent Fluid and Particle Interaction," *Progress in Heat and Mass Transfer*, Vol. 6, 1972, p. 433.

[28]Owen, P. R. J., "Saltation of Uniform Grains in Air," *Journal of Fluid Mechanics*, Vol. 20, 1964, pp. 225–272.

[29]Owen, P. R. J., "Pneumatic Transport," *Journal of Fluid Mechanics*, Vol. 39, 1969, pp. 407–432.

[30]Akamtsu, F., Mizatani, Y., Katsuki, U., Tsushima, S., Cho, Y. D., and Nakabe, K., "Group Combustion Behavior of Drops in a Premixed-Spray Flame," *Atomization and Sprays*, Vol. 7, 1997, pp. 192–218.

[31]Karpetis, A. N., and Gomez, A., "An Experimental Investigation of Non-Premixed Turbulent Spray Flames and Their Self-Similar Behavior," *Proceedings of the Combustion Institute*, Vol. 27, 1996, pp. 2001–2008.

[32]Crowe, C. T., Chung, T. N., and Troutt, T. R., "Particle Dispersion by Organized Turbulent Structures," *Particulate Two-Phase Flow*, edited by M. C. Roco, Butterworth-Heinemann, 1993, pp. 626–669.

[33]Sornek, R. J., Dobashi, R., and Hirano, T., "Effects of Turbulence on Dispersion and Vaporization of Droplets in Spray Combustion," *Proceedings of the Combustion Institute*, Vol. 25, 2000, pp. 1055–1062.

[34]Jurewicz, J. T., and Stock, D. E. A., "Numerical Model for Turbulent Diffusion in Gas-Particle Flows," American Society of Mechanical Engineers, 76-WA-FE-33, 1976.

[35]Dokowicz, J. K., "A Particle-Fluid Numerical Model for Liquid Sprays," *Journal of Computational Physics*, Vol. 35, 1980, pp. 229–240.

[36]Abbas, A. S., Koussa, S. S., and Lockwood, F. C., *Eighteenth Symposium (International) on Combustion*, Combustion Inst., Pittsburgh, PA, 1981, pp. 1427–1438.

[37]Brown, D. J., and Hutchinson P., *Journal of Fluids Engineering*, Vol. 101, 1979, pp. 265–269.

[38]Gouesbet, G., Berlemont, A., and Picart, "Dispersion of Discrete Particles by Continuous Turbulent Motion," *Physics of Fluid A*, Vol. 27, 1984, pp. 827–837.

[39]Maxey, M. R., "The Gravitational Settling of Aerosol Particles in Homogeneous Turbulent and Randon Flow Fields," *Journal of Fluid Mechanics*, Vol. 174, 1987, pp. 441–465.

[40]Chiu, H. H., "Advances and Challenges in Drop and Spray Combustion 1. Toward a Unified Theory of Drop Aerothermo Chemistry," *Progress in Energy and Combustion Science*, Vol. 26, 2000, pp. 381–416.

[41]Wu, J. S., and Faeth, G. M., "Effects of Ambient Turbulence Intensity on Sphere Wakes at Intermediate Reynolds Number," *AIAA Journal*, Vol. 33, 1995, pp. 171–173.

[42]Al Taweel, A. M., and Landau, J., "Turbulent Modulation in Two-Phase Jets," *International Journal of Multiphase Flow*, Vol. 3, 1977, p. 341.

[43]Chiu, H. H., and Liu, T. M., "Group Combustion of Liquid Drops," *Combustion Science and Technology*, Vol. 17, 1977, pp. 127–142.

[44]Chiu, H. H., Kim, H. Y., and Croke, E. J., "Internal Group Combustion of Liquid Drops," *Nineteenth Symposium (International) on Combustion*, Combustion Inst., Pittsburgh, PA, 1983, pp. 971–980.

[45]Chiu, H. H., "Mesoscale Structure of Turbulent Sprays," *Proceedings of the Combustion Institute*, Vol. 28, 2000, pp. 1095–1102.

[46]Bellan, J., and Harstad, K., "Analysis of the Convective Evaporation of Non-Dilute Clusters of Drops," *International Journal of Heat and Mass Transfer*, Vol. 30, 1987, pp. 125–136.

[47]Correa, S. M., and Sichel, M., "The Group Combustion of a Spherical Cloud of Monodisperse Fuel Drops," *Nineteenth Symposium (International) on Combustion*, Combustion Inst., Pittsburgh, PA, 1983, pp. 981–991.

[48]Mayer, W., and Tamura, H., "Propellant Injection in a Liquid Oxygen/Gaseous Hydrogen Rocket Engine," *Journal of Propulsion and Power*, Vol. 12, 1996, pp. 1137–1147.

[49]Chiu, H. H., and Zhou, X. Q., "Spray Group Combustion Process in Airbreathing Propulsion Combustors," AIAA Paper, 83-1323, 1983.

[50]Juniper, M., Tripathi, A., Scouflaire, P., Rolin, J. C., and Candel, S., "Structure of Cryogenic Flames at Elevated Pressure," *Proceedings of the Combustion Institute*, Vol. 28, 2000, pp. 1103–1109.

[51]Yang, V., "Modelling of Supercritical Vaporization, Mixing and Combustion Processes in Liquid-Fueled Propulsion Systems," *Proceedings of the Combustion Institute*, Vol. 28, 2000, pp. 925–942.

[52]Oefelein, J. C., and Yang, V., "Modeling High Pressure Mixing and Combustion Progresses in Liquid Rocket Engines," *Journal of Propulsion and Power*, Vol. 14, 1998, pp. 843–857.

[53]Oefelein, J. C., and Yang, V., "Simulation of High Pressure Spray Field Dynamics in Recent Advances in Spray Combustion: Spray Combustion Measurements and Model Simulation," Progress in Astronautics and Aeronautics, Vol. 171, AIAA, Reston, VA, 1996, pp. 263–304.

[54]Candel, S., Herding, G., Synder, R., Scouflaire, P., Rdon, C., Vingert, L., Habiballah, M., Grisch, F., Pealat, M., Bouchardy, P., Stepowski, D., Cessou, A., and Colin, P., "Experimental Investigations of Shear Coaxial Cryogenic Jet Flames," *Journal of Propulsion and Power*, Vol. 14, 1998, pp. 826–834.

[55]Warnatz, J., Maas, U., and Dibble, R. W., *Combustion*, 2nd ed., Springer Verlag, Berlin, 1999, Chap. 15.

[56]Gutheil, E., and Sirignano, W. A., "Counterflow Spray Combustion Modeling with Detailed Transport and Detailed Chemistry," *Combustion and Flame*, Vol. 113, 1998, pp. 92–105.

[57]Snyder, R., Herding, G., Rolon, J. C., and Candel, S., "Analysis of Flame Patterns in Cryogenic Propellant Combustion," *Combustion Science and Technology*, Vol. 124, 1997, pp. 371–370.

Chapter 7

Liquid-Propellant Droplet Vaporization and Combustion

Vigor Yang,[*] Patrick Lafon,[†] and George C. Hsiao[‡]
Pennsylvania State University, University Park, Pennsylvania

Mohammed Habiballah[§]
ONERA, Châtillon, France

and

Feng-Chen Zhuang[¶]
Institute of Command Technology, Beijing, China

I. Introduction

$\mathbf{T}$HE physiochemical processes in a liquid-rocket combustion chamber are extremely complicated, involving an array of intricacies such as jet atomization, spray formation, droplet transport, multiphase flow mixing, and chemical reactions. Many strong interactions among these processes exist with wide disparities in time and length scales. Because the transport characteristics of individual droplets play a decisive role in determining the local flow behavior in a spray field, the study of droplet vaporization and combustion constitutes a necessary step in approaching the complete problem. General reviews of liquid droplet transport and combustion were given by Williams,[1] Faeth,[2] Law,[3] Sirignano,[4] and Dwyer.[5] Recent advances in the modeling of supercritical droplet and mixing-layer dynamics were summarized by Bellan[6] and Yang.[7]

---

[*]Distinguished Professor, Department of Mechanical Engineering. Fellow AIAA.

[†]Post-Doctoral Research Associate, Department of Mechanical Engineering; currently in France.

[‡]Post-Doctoral Research Associate, Department of Mechanical Engineering; currently at General Electric Aircraft Engines, Cincinnati, OH.

[§]Head, Liquid Propulsion, Heat Transfer, and Advanced Propulsion System Unit.

[¶]Professor; Academician, Chinese Academy of Sciences

This chapter is concerned with the behavior of single droplets in both stagnant and forced-convective environments under conditions representative of liquid rocket thrust-chamber operations. Situations involving droplet interactions are treated in Chapter 8 of the volume. The purpose is to provide an overview of recent advances in droplet vaporization and combustion research. All three commonly used liquid propellants (i.e., hydrocarbon, cryogenic, and hypergolic propellants) are treated over a broad range of fluid thermodynamic states, with special attention given to the effects of ambient pressure and cross flow on droplet gasification and burning characteristics. Although much of the discussions given herein have direct applications to rocket engines, results can be effectively used in other liquid-fueled combustion devices such as internal combustion engines, ramjet and gas-turbine engines, and liquid-propellant guns.

The physical problem considered here is the transient vaporization and combustion of an isolated droplet when suddenly placed in a combustion chamber. As a result of heat transfer from the surrounding gases, the droplet starts to heat up and evaporation occurs due to the vapor concentration gradient near the surface. Two scenarios, subcritical and supercritical conditions, as shown in Fig. 1, must be treated to provide a complete description of various situations encountered in rocket engine combustors. If the chamber condition is in the thermodynamic subcritical regime of the injected liquid propellant, the droplet surface provides a well-defined interfacial boundary. To facilitate the analysis, physical processes in the droplet interior and ambient gases are treated separately, and then matched at the interface by requiring liquid-vapor phase equilibrium and continuities of mass and energy fluxes. The procedure eventually determines the droplet surface conditions and evaporation rate. The situation, however, becomes qualitatively different in the supercritical regime where the chamber pressure and temperature exceed the thermodynamic critical states of liquid propellants. The droplet may continuously heat up, with its surface reaching the critical mixing point prior to the end of droplet lifetime. When this occurs, the sharp distinction between the gas and liquid disappears. The enthalpy of vaporization reduces to zero, and no abrupt phase change is involved in the vaporization process. The fluid properties and their gradients vary continuously across the droplet surface. The droplet interior, however, remains at the liquid state with a subcritical temperature distribution. For convenience of discussion, the droplet regression is best characterized by the motion of the surface that attains the critical mixing temperature of the system. The process becomes totally diffusion controlled.

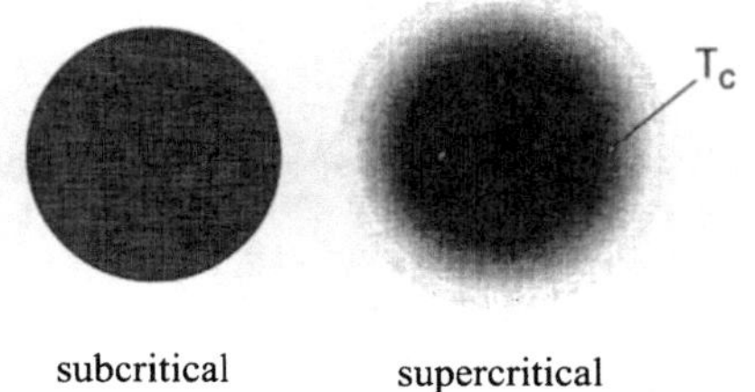

Fig. 1 Schematic diagrams of droplets vaporizing in subcritical and supercritical environments.

This chapter is organized as follows. Section II describes a unified evaluation scheme of fluid thermodynamic and transport properties. The vapor-liquid phase equilibrium is discussed in Section III. Sections IV and V deal with, respectively, droplet vaporization in quiescent and forced-convective environments under steady conditions. Various unique features of droplet transport under supercritical conditions are summarized. Section VI is concerned with droplet combustion. Finally, Section V addresses dynamic responses of droplet vaporization to ambient pressure oscillations. The resulting vaporization response function can be directly used in the analysis of combustion stability of a rocket engine.

II. Thermodynamic and Transport Properties

For subcritical droplet vaporization and combustion, the conventional approach that treats the liquid droplet and surrounding gases separately functions well, and property evaluation schemes developed for these two distinct phases[8] can be effectively used. Situations with supercritical fluids, however, become quite different. Because of the continuous variations of fluid properties in supercritical environments, classical techniques dealing with liquids and gases individually often lead to erroneous results of droplet dynamics. The problem becomes even more exacerbated when the droplet surface approaches the critical state. Fluid properties usually exhibit anomalous variations and are very sensitive to both temperature and pressure in the vicinity of the critical point, a phenomenon commonly referred to as near-critical enhancement. Thus, a prerequisite of any realistic treatment of supercritical droplet behavior lies in the establishment of a unified property evaluation scheme capable of treating thermophysical properties of the system and its constituent species over the entire fluid thermodynamic state, from compressed liquid to dilute gas.

All of the thermodynamic properties can be derived from a modified Benedict-Webb-Rubin (BWR) equation of state proposed by Jacobsen and Stewart[9] due to its superior performance over conventional cubic equations of state.[10] This equation of state has been extremely valuable in correlating both liquid and vapor thermodynamic and volumetric data; however, the temperature constants involved are available only for a limited number of pure substances.[11] To overcome this constraint, an extended corresponding-state (ECS) principle[12,13] is used. The basic idea is to assume that the properties of a single-phase fluid can be evaluated via conformal mappings of temperature and density to those of a given reference fluid. As a result, only the BWR constants for the reference fluid are needed. For a multicomponent system, accounting for changes in properties due to mixing is much more complicated. A pseudo-pure-substance model is adopted to evaluate the properties of a mixture, treating the mixture as a single-phase pure substance with its own set of properties evaluated via the ECS principle. This method improves prediction accuracy and requires only limited data (i.e., critical properties and Pitzer's acentric factor) for each constituent component.

Successful application of the corresponding-state argument for the evaluation of fluid p-V-T properties also encourages similar improvement in the prediction of thermophysical data. In the following, a brief summary of the corresponding-state method in conjunction with the mixture combining rule is given, followed by the BWR equation of state for the reference fluid. Techniques for evaluating thermodynamic and transport properties are then addressed.

A. Extended Corresponding-State Principle

The extended corresponding-state model of Ely and Hanley[12,13] is used to evaluate volumetric and transport properties of a mixture over its entire thermodynamic fluid state. The scheme assumes that the configurational properties (such as temperature, density, viscosity, thermal conductivity, etc.) of a single-phase mixture can be equated to those of a hypothetical pure fluid, which are then evaluated via corresponding-state principles with respect to a given reference fluid. For example, the viscosity of a mixture, μ_m, can be related to that of a reference fluid, μ_0, at the corresponding thermodynamic state as

$$\mu_m(\rho, T) = \mu_0(\rho_0, T_0)F_\mu \qquad (1)$$

where F_μ represents the mapping function. The correspondence of temperature and density between the mixture of interest and the reference fluid can be characterized by the following two scaling factors:

$$f_m = \frac{T}{T_0}; \quad \hbar_m = \frac{\rho_0}{\rho} \qquad (2)$$

The former represents the conformation of potential distribution of energy, while the latter characterizes the effect of mixture molecular size.

B. Equation of State

Under the assumption of the ECS principle, the density of a mixture can be evaluated by

$$\rho_m(T, p) = \frac{\rho_0(T_0, p_0)}{\hbar_m(T, p)} \qquad (3)$$

where ρ_0, T_0, and p_0 denote the corresponding density, temperature, and pressure of the reference fluid, respectively. Since the temperature at the conformal state is calculated by Eq. (2), the corresponding pressure can be derived based on the general compressibility theory,

$$p_0 = p\left(\frac{\hbar_m}{f_m}\right) \qquad (4)$$

To ensure the accuracy of the density prediction, a generalized BWR equation of state[9] is adopted for the reference fluid:

$$p_0(T, \rho) = \sum_{n=1}^{9} a_n(T)\rho^n + \sum_{n=10}^{15} a_n(T)\rho^{2n-17}e^{-\gamma\rho^2} \qquad (5)$$

where γ is 0.04, and the temperature coefficients $a_i(T)$ depend on the reference fluid used.

Although this equation of state must be solved iteratively for density at given pressure and temperature, the prediction covers a wide range of thermodynamic states, and as such promotes the establishment of a unified evaluation scheme of thermophysical properties. Figure 2 shows the comparison of oxygen density between experimental data[14] and the prediction by the BWR equation of state in conjunction with the ECS principle. The reference fluid is selected to be propane because of the availability of sufficiently reliable data correlated over a wide range of experimental conditions for this substance. The result shows excellent agreement over the entire fluid state. Figure 3 presents the relative errors of density prediction based on three commonly used equations of state, namely, the Benedict-Webb-Rubin (BWR), Peng-Robinson (PR), and Soave-Redlich-Kwong (SRK). The ECS principle is embedded into the evaluation procedure of the BWR equation of state and shows its superior performance with the maximum relative error of 1.5% for the pressure and temperature ranges under consideration. On the other hand, the SRK and PR equations of state yield maximum errors around 13% and 17%, respectively.

C.   Thermodynamic Properties

Thermodynamic properties such as enthalpy, internal energy, and specific heat can be expressed as the sum of ideal-gas properties at the same temperature and departure functions that take into account the dense-fluid correction. Thus,

$$h = h^0 + \left\{ \int_\infty^v \left[T\left(\frac{\partial p}{\partial T}\right)_v - p \right] dv + RT(Z - 1) \right\} \tag{6}$$

$$e = e^0 + \left\{ \int_\infty^v \left[T\left(\frac{\partial p}{\partial T}\right)_v - p \right] dv \right\} \tag{7}$$

$$C_p = C_p^0 + \left\{ \int_\infty^v T\left(\frac{\partial^2 p}{\partial T^2}\right)_v dv - T\left(\frac{\partial p}{\partial T}\right)_v^2 \Big/ \left(\frac{\partial p}{\partial v}\right)_T - R \right\} \tag{8}$$

where superscript 0 refers to ideal-gas properties. The second terms on the right sides of Eqs. (6−8) denote the thermodynamic departure functions, and can be obtained from the equation of state described previously.

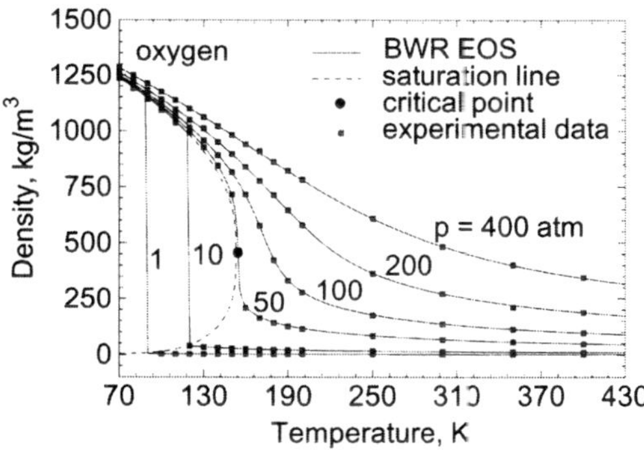

Fig. 2   Comparison of oxygen density predicted by the BWR equation of state and measured by Sychev et al.[14]

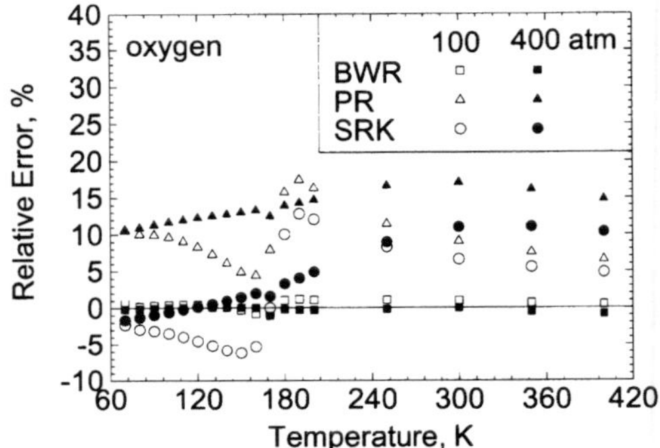

Fig. 3   Relative errors of density predictions by three different equations of state, experimental data from Sychev et al. [14]

D.   Transport Properties

Estimation of viscosity and thermal conductivity can be made by means of the ECS principle. The corresponding-state argument for the viscosity of a mixture can be written in its most general form as

$$\mu_m(\rho, T) = \mu_0(\rho_0, T_0)F_\mu\chi_\mu \tag{9}$$

where F_μ is the scaling factor. The correction factor χ_μ accounts for the effect of noncorrespondence, and has the magnitude always close to unity based on the modified Enskog theory.[15]

Because of the lack of a complete molecular theory for describing transport properties over a broad regime of fluid phases, it is generally accepted that viscosity and thermal conductivity can be divided into three contributions and correlated in terms of density and temperature.[16] For instance, the viscosity of the reference fluid is written as follows:

$$\mu_0(\rho_0, T_0) = \mu_0^0(T_0) + \Delta\mu_0^{exc}(\rho_0, T_0) + \Delta\mu_0^{crit}(\rho_0, T_0) \tag{10}$$

The first term on the right-hand side represents the value at the dilute-gas limit, which is independent of density and can be accurately predicted by kinetic-theory equations. The second term is the excess viscosity which, with the exclusion of unusual variations near the critical point, characterizes the deviation from μ_0 for a dense fluid. The third term refers to the critical enhancement that accounts for the anomalous increase above the background viscosity (i.e., the sum of μ_0^0 and $\Delta\mu_0^{exc}$) as the critical point is approached. However, the theory of nonclassical critical behavior predicts that, in general, properties that diverge strongly in pure fluids near the critical points diverge only weakly in mixtures due to the different physical criteria for criticality in a pure fluid and a mixture.[17] Because the effect of critical enhancement is not well defined for a mixture and is likely to be small, the third term $\Delta\mu_0^{crit}$ is usually not considered in the existing analyses of supercritical droplet gasification.

Evaluation of thermal conductivity must be carefully conducted for two reasons: 1) the one-fluid model must ignore the contribution of diffusion to conductivity, and 2) the effect of internal degrees of freedom on thermal conductivity cannot be correctly taken into account by the corresponding-state argument. As a result, thermal conductivity of a pure substance or mixture is generally divided into two contributions,[13]

$$\lambda_m(\rho, T) = \lambda'_m(T) + \lambda''_m(\rho, T) \tag{11}$$

The former, $\lambda'_m(T)$, arises from transfer of energy via the internal degrees of freedom, while the latter, $\lambda''_m(\rho, T)$, is due to the effects of molecular collision or translation that can be evaluated by means of the ECS method. For a mixture, $\lambda'_m(T)$ can be evaluated by a semi-empirical mixing rule.

Estimation of the binary mass diffusivity for a mixture gas at high density represents a more challenging task than evaluating the other transport properties, due to the lack of a formal theory or even a theoretically based correlation. Takahashi[18] suggested a simple scheme for predicting the binary mass diffusivity of a dense fluid by means of a corresponding-state approach. The approach appears to be the most complete to date and has demonstrated moderate success in the limited number of tests conducted. The scheme proceeds in two steps. First, the binary mass diffusivity of a dilute gas is obtained using the Chapman-Enskog theory in conjunction with the intermolecular potential function. The calculated data are then corrected in accordance with a generalized chart in terms of reduced temperature and pressure.

III. Vapor-Liquid Phase Equilibrium

The result of vapor-liquid phase equilibrium is required to specify the droplet surface behavior prior to the occurrence of the critical state. The analysis usually consists of two steps. First, an appropriate equation of state is employed to calculate fugacities of each constituent species in both gas and liquid phases. The second step lies in the determination of the phase equilibrium conditions by requiring equal fugacities for both phases of each species. Specific outputs from this analysis include 1) enthalpy of vaporization, 2) solubility of ambient gases in the liquid phase, 3) species concentrations at the droplet surface, and 4) conditions for criticality. The analysis can be further used in conjunction with the property evaluation scheme to determine other important properties such as surface tension.

As an example, the equilibrium compositions for a binary mixture of oxygen and hydrogen are shown in Fig. 4, where p_r represents the reduced pressure of oxygen. In the subcritical regime, the amount of hydrogen dissolved in the liquid oxygen is quite limited, decreasing progressively with increasing temperature and reducing to zero at the boiling point of oxygen. At supercritical pressures, however, the hydrogen gas solubility becomes substantial and increases with temperature. Because of the distinct differences in thermophysical properties between the two species, the dissolved hydrogen may appreciably modify the liquid properties and, subsequently, the vaporization behavior. The phase equilibrium results also indicate that the critical mixing temperature decreases with pressure.

The overall phase behavior in equilibrium is best summarized by the pressure-temperature diagram in Fig. 5, which shows how the phase transition

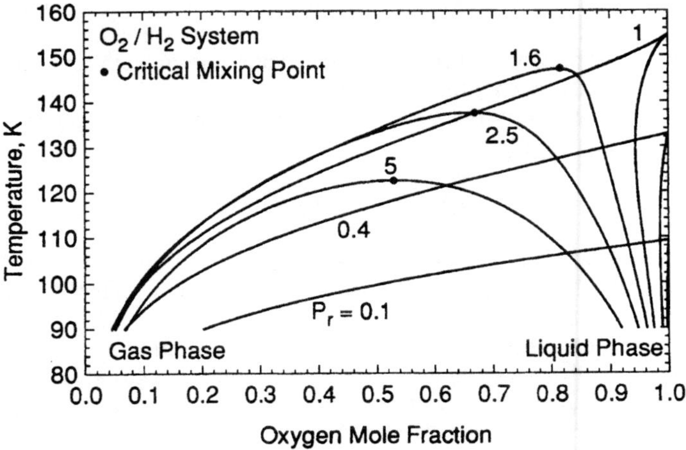

Fig. 4 Vapor-liquid phase equilibrium compositions for O_2/H_2 system at various pressures.[23]

occurs under different thermodynamic conditions. The boiling line is made up of boiling points for subcritical pressure. As the temperature increases, an equilibrium vapor-liquid mixture may transit to superheated vapor across this line. The critical mixing line registers the variation of the critical mixing temperature with pressure. It intersects the boiling line at the critical point of pure oxygen, the highest temperature at which the vapor and liquid phases of an O_2/H_2 binary system can coexist in equilibrium.

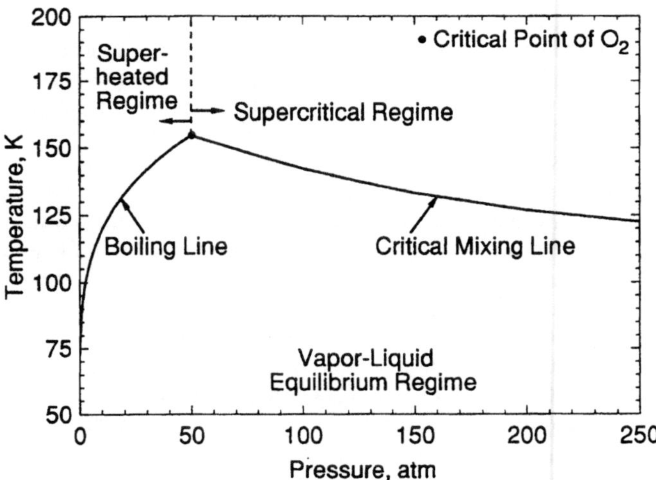

Fig. 5 Pressure-temperature diagram for phase behavior of O_2/H_2 system in equilibrium.[23]

IV. Droplet Vaporization in Quiescent Environments

Several theoretical works have recently been devoted to the understanding of droplet vaporization and combustion under high-pressure conditions. Both hydrocarbon droplets in air[19-22] and liquid oxygen (LOX) droplets in hydrogen[23-29] were treated comprehensively, with emphasis placed on the effects of transient diffusion and interfacial thermodynamics. The works of Lafon et al.,[26,30] Hsiao, Meng, and Yang,[24,31] and Harstad and Bellan[21,22,28,29] appear to be the most comprehensive to date because of the employment of a unified property evaluation scheme as outlined in Section II.

A. Cryogenic Propellants

We first consider the vaporization of LOX droplets in either pure hydrogen or mixed hydrogen/water environments[26,30] due to its broad applications in cryogenic rocket engines using hydrogen and oxygen as propellants. Figure 6 shows the time variations of droplet surface temperature at various pressures. The ambient hydrogen temperature is 1000 K. Three different scenarios are noted. First, at low pressures (i.e., $p = 10$ atm), the surface temperature rises suddenly and levels off at the pseudo-wet-bulb state, which is slightly lower than the oxygen boiling temperature because of the presence of hydrogen on the gaseous side of the interface. For higher pressures (i.e., $p = 50$ atm), the surface temperature rises continuously. The pseudo-wet-bulb state disappears, and the vaporization process becomes transient in nature during the entire droplet lifetime. For $p = 100$ atm, the droplet surface even reaches its critical state at 1 ms. Figure 7 shows the distributions of the mixture specific heat at various times. The weak divergence of the specific heat near the droplet surface (i.e., defined as the surface attaining the critical-mixing temperature) is clearly observed.

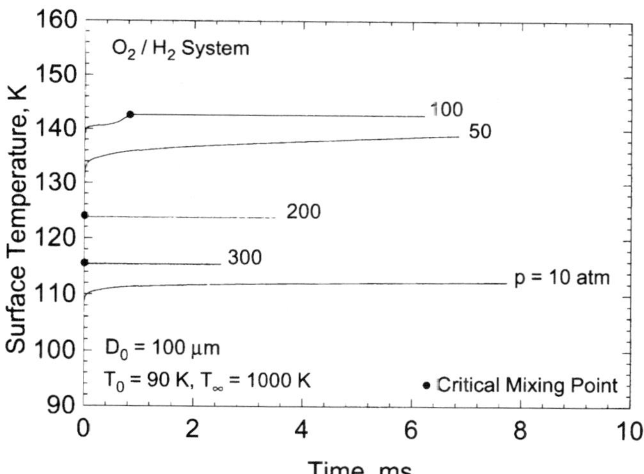

Fig. 6 Time variations of droplet surface temperature at various pressures; oxygen/hydrogen system, $T_\infty = 1000$ K, $T_0 = 90$ K, $D_0 = 100$ μm.

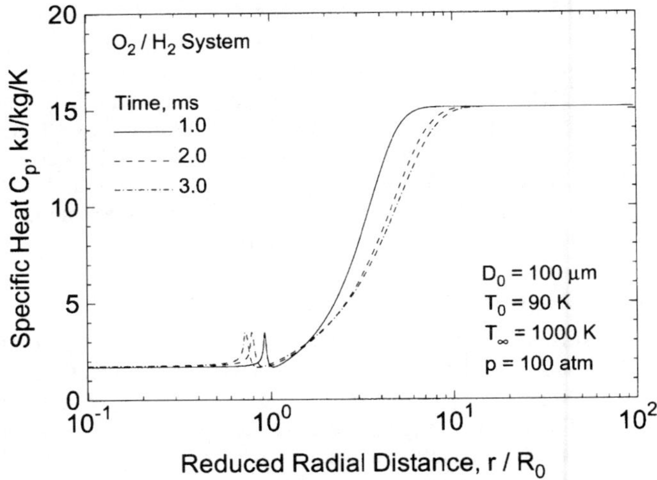

Fig. 7 Instantaneous distributions of mixture specific heat in the entire field at various times; $T_0 = 90$ K, $T_\infty = 1000$ K, $p = 100$ atm, $D_0 = 100$ μm.

An extensive series of numerical simulations were conducted for a broad range of ambient temperatures ($500 \leq T_\infty \leq 2500$ K) and pressures up to 300 atm. The calculated LOX droplet lifetime in pure hydrogen can be well correlated using an approximate analysis that takes into account the effect of transient heat diffusion in terms of the reduced critical temperature, $T_c^* = (T_\infty - T_c)/(T_\infty - T_0)$, where the subscripts ∞, c, and 0 denote the ambient, critical, and droplet initial conditions, respectively. The resultant expression of the droplet lifetime τ takes the form

$$\tau = [0.0115 + 0.542(1 - T_c^*)]\frac{R_0^2}{\alpha_0^l}f(\alpha_\infty/\alpha_0^l) \tag{12}$$

where α_∞ and α_0^l denote the thermal diffusivities of the ambient hydrogen gas and liquid oxygen droplet at its initial state, respectively. The correction factor is chosen as

$$f(\alpha_\infty/\alpha_0^l) = 1 + 3.9\lfloor 1 - \exp(-0.035(\alpha_\infty/\alpha_0^l - 1))\rfloor \tag{13}$$

The expression of T_c^* may be related to the Spalding transfer number, as follows:

$$B_T = \frac{T_\infty - T_c}{T_c - T_0} = \frac{T_c^*}{1 - T_c^*} \tag{14}$$

This correlation clearly shows that pressure affects the droplet lifetime through its influence on the mixture critical temperature and ambient thermal diffusivity.

B.   Hydrocarbon Propellants

The behavior of hydrocarbon fuel droplets basically follows the same trend as cryogenic propellants in terms of their vaporization characteristics. In general, the droplet lifetime decreases smoothly with increasing pressure, and no discernible variation occurs across the critical transition. As a specific example, the case of n-heptane droplets vaporizing in air[19,32] is considered. Only the initial diameter of 100 μm is treated, since the problem of pure vaporization involves just one length scale. The entire process is diffusion controlled, and the droplet lifetime is proportional to the square of initial droplet diameter in accordance with a dimensional analysis.[33]

Figure 8 shows the time histories of the droplet surface temperature at various pressures. The ambient and droplet initial temperatures are $T_\infty = 1000$ K and $T_0 = 300$ K, respectively. The impulsive rise of the surface temperature at $t = 0$ results from the large temperature gradient near the droplet surface (i.e., a step function was used as the initial temperature distribution). At the low pressure of $p = 5$ atm, the droplet surface temperature increases rapidly and reaches a constant value slightly lower than the n-heptane boiling temperature, a condition referred to as the pseudo-wet-bulb state.[19] At higher pressures, the surface temperature rises continuously. The pseudo-wet-bulb state disappears, and the vaporization process becomes transient in nature during the entire droplet lifetime. At $p = 150$ atm, the droplet surface even reaches its critical state just before the end of the droplet lifetime. Because three species (i.e., n-heptane, nitrogen, and oxygen) are involved in determining the thermodynamic liquid-vapor phase equilibrium at the surface, for a given pressure, interface compositions in both the liquid and gaseous phases are not unique functions of surface temperature. The thermodynamic variance is three. In contrast to conventional approaches for low-pressure applications in which the ambient gas solubility is ignored and the enthalpy of vaporization depends only on temperature, the

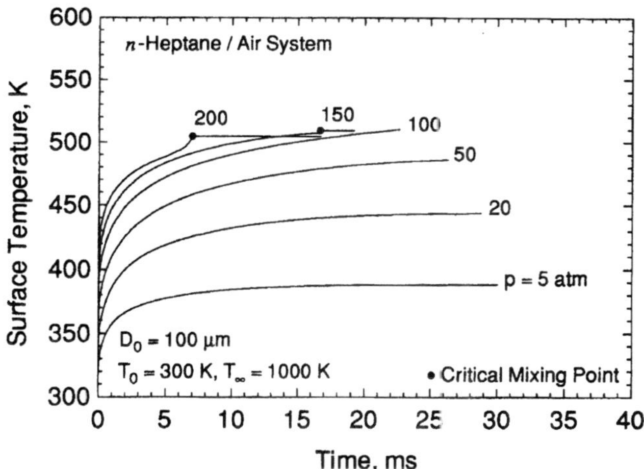

Fig. 8   Time variations of droplet surface temperature at various pressures; n-heptane/air system, $D_0 = 100$ μm, $T_0 = 300$ K, and $T_\infty = 1000$ K.

enthalpy of vaporization varies with both temperature and pressure as well as the mixture composition at the droplet surface. It decreases with increasing pressure and drops rapidly to zero at the critical mixing point.

Figure 9 presents the instantaneous distributions of temperature for a supercritical case, $p = 50$ atm. The penetration of thermal waves in regions away from the droplet surface is clearly observed. At the end of the droplet lifetime, the temperature is almost uniform within the droplet interior. For hydrocarbon droplets vaporizing in air at high pressures, the characteristics times for liquid thermal inertia and vaporization rate have the same order of magnitude, thereby resulting in the transient nature of the vaporization process.

A parametric study of the dependence of droplet lifetime on ambient temperature and pressure is conducted, giving the result shown in Fig. 10. A total of four different ambient temperatures ($T_\infty = 1000$, 1500, 2000, and 2500 K) and three different initial droplet temperatures ($T_0 = 300$, 325, and 350 K) over a pressure range of 5–200 atm are considered. The reduced lifetime, defined as the ratio of the droplet lifetime to that at the reference pressure of 5 atm, is correlated well with the ambient pressure:

$$\tau_r = \frac{\tau}{\tau_{p_{\text{ref}}}} = \exp[-a(p - p_{\text{ref}})/p_{\text{ref}}] \tag{15}$$

For the n-heptane/air system, the exponent a is equal to 0.016. The average deviation is smaller than 2%, while the maximum deviation is around 10%. The same kind of study was successfully carried out for the n-decane/air system, with the exponent a being 0.01.

It is worth noting that at high ambient temperature, droplet lifetime decreases with increasing pressure. Furthermore, the lifetime decreases with pressure

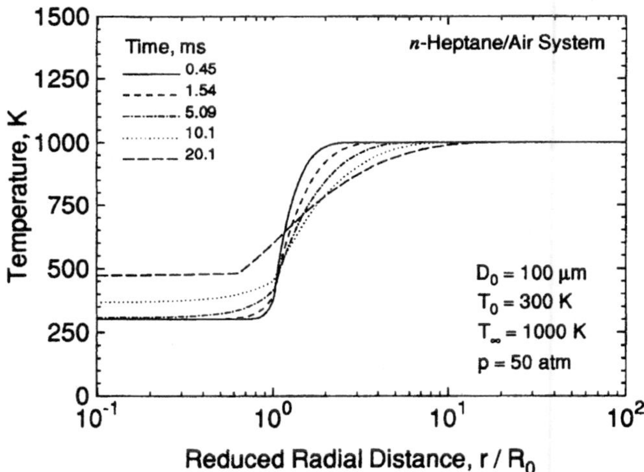

Fig. 9 Instantaneous temperature distributions in the entire field at various times; n-heptane/air system, $D_0 = 100$ μm, $p = 50$ atm, and $T_\infty = 1000$ K.

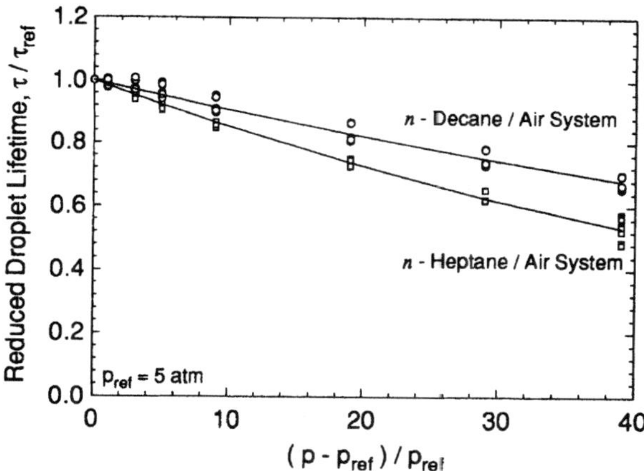

Fig. 10   Reduced droplet lifetime as a function of ambient pressure; *n*-decane/air and *n*-heptane/air systems.

smoothly, and no discernible variation occurs across the critical transition. In the subcritical regime, the decrease in enthalpy of vaporization with pressure facilitates the vaporization process and consequently leads to a decrease in droplet lifetime. The situation becomes different at supercritical conditions under which the interfacial boundary disappears with vanished enthalpy of vaporization. In this case, the decrease in droplet lifetime is mainly attributed to the increase in thermal diffusivity at the droplet surface. As the critical state is approached, the divergence of the specific heat compensates the effect of reduced enthalpy of vaporization. Consequently, no abrupt change takes place during the critical transition.

C.   Hypergolic Propellants

As a representative of hypergolic propellants, unsymmetrical dimethylhydrazine (UDMH) droplets are studied because of the least toxicity of UDMH in the hydrazine family. Zhuang and co-workers conducted a series of fundamental investigations into the vaporization and combustion characteristics of UDMH droplets at elevated pressures.[34-38] A UDMH droplet was suspended by a thermocouple wire in an electric furnace pressurized with either air or nitrogen. Ignition of the droplet was achieved using a pulse CO_2 laser. A companion theoretical analysis was conducted to provide more detailed information. Figure 11 shows the effect of pressure on the evaporation constant for three different initial droplet radii. The ambient air was at room temperature. The vaporization rate increases with increasing pressure and initial size of the droplet. The high-pressure condition does indeed facilitate the droplet gasification process as expected. The dependence of the vaporization constant on the initial droplet diameter is associated with the ignition mechanism and other length scales introduced in the problem as a result of reactive processes.

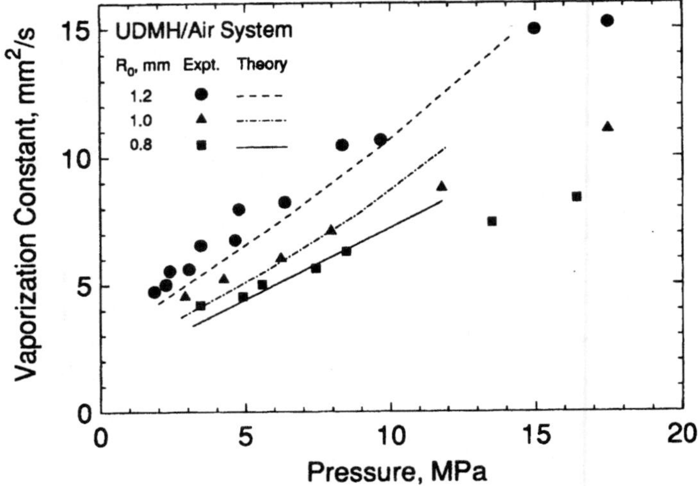

Fig. 11   Effect of pressure on evaporation constant of UDMH droplet in air at room temperature.

V.   Droplet Vaporization in Convective Environments

When a droplet is introduced into a crossflow, the forced convection results in increases of heat and mass transfer between the droplet and surrounding gases, which consequently intensifies the gasification process. Although many studies have been conducted to examine droplet vaporization in forced-convective environments, effects of pressure and freestream velocities on droplet dynamics, especially for rocket engine applications that involve supercritical conditions, have not yet been addressed in detail. Hsiao, Meng, and co-workers[24,31,39] developed a comprehensive analysis of liquid-oxygen (LOX) droplet vaporization in a supercritical hydrogen stream, covering a pressure range of 100–400 atm. The model takes into account multidimensional flow motions and enables a thorough examination of droplet behavior during its entire lifetime, including dynamic deformation, viscous stripping, and secondary breakup. Detailed flow structures and thermodynamic properties are obtained to reveal mechanisms underlying droplet gasification as well as deformation and breakup dynamics.

Figure 12 shows six frames of isotherms and isopleths of oxygen concentration at a convective velocity of 2.5 m/s and an ambient pressure of 100 atm. The freestream Reynolds number Re is 31 based on the initial droplet diameter. Soon after the introduction of the droplet into the hydrogen stream, the flow adjusts to form a boundary layer near the surface. The gasified oxygen is carried downstream through convection and mass diffusion. The evolution of the temperature field exhibits features distinct from that of the concentration field due to the disparate time scales associated with thermal and mass diffusion processes (i.e., Lewis number $\neq$ 1). The thermal wave penetrates into the droplet interior faster than the surrounding hydrogen species does. Because the liquid core possesses large momentum inertia and moves slower than the gasified

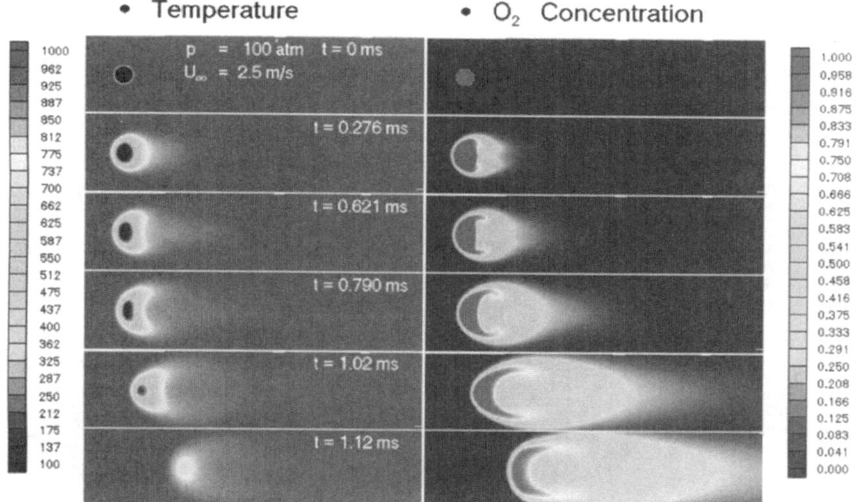

Fig. 12   LOX droplet gasification in supercritical hydrogen flow; $p = 100$ atm, $U_\infty = 2.5$ m/s.

oxygen, at $t = 0.79$ ms, the droplet (delineated by the dark region in the temperature contours) reveals an olive shape while the oxygen concentration contours deform into a crescent shape with the edge bent to the streamwise direction. At $t = 1.08$ ms, the subcritical liquid core disappears, leaving behind a puff of dense oxygen fluid that is convected further downstream with increasing velocity until it reaches the momentum equilibrium with the ambient hydrogen flow.

Figure 13 summarizes the streamline patterns and oxygen concentration contours of the four different modes commonly observed in supercritical droplet gasification. The droplet may remain in a spherical configuration, deform to an olive shape, or even break up into fragments, depending on the local flow conditions. Unlike low-pressure cases in which the large shear stress at the gas-liquid interface induces internal flow circulation in the liquid core,[40] no discernible recirculation takes place in the droplet interior, regardless of the Reynolds number and deformation mode. This may be attributed to the diminishment of surface tension at supercritical conditions. In addition, the droplet regresses so fast that a fluid element in the interphase region may not acquire the time sufficient for establishing an internal vertical flow before it gasifies. The rapid deformation of the droplet configuration further precludes the existence of stable shear stress in the liquid core and consequently restricts the development of a recirculation pattern.

The spherical mode shown in Fig. 13a typically occurs at very low Reynolds numbers. Although flow separation is encouraged by LOX gasification, no recirculating eddy is found in the wake behind the droplet. The vorticity generated is

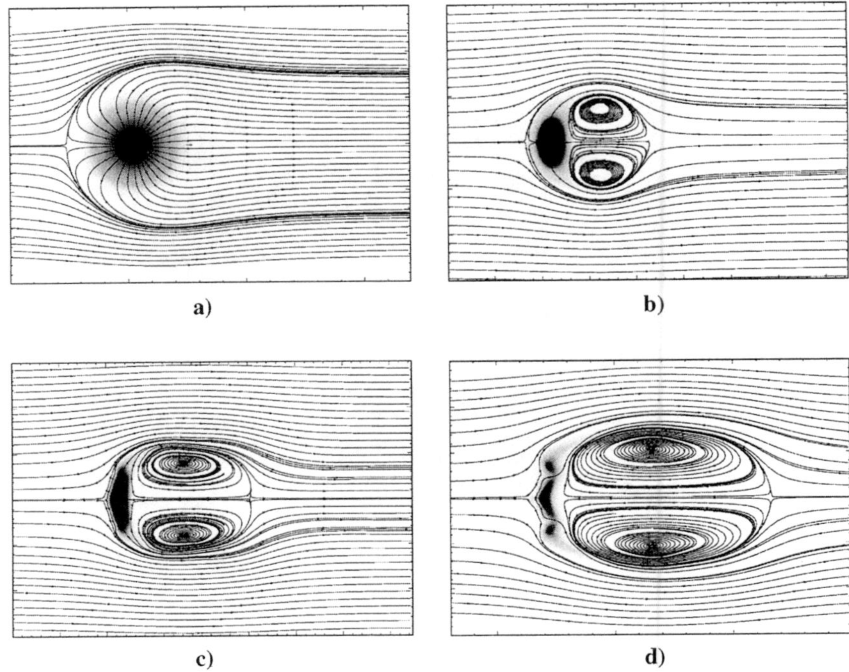

Fig. 13   LOX droplet gasification in supercritical hydrogen flow for $p = 100$ atm and $D_0 = 100$ μm: a) spherical mode, $U_\infty = 0.2$ m/s, $t = 0.61$ ms; b) deformation mode, $U_\infty = 1.5$ m/s, $t = 0.61$ ms; c) stripping mode, $U_\infty = 5$ m/s, $t = 0.17$ ms; d) breakup mode, $U_\infty = 15$ m/s, $t = 0.17$ ms.

too weak to form any confined eddy. When the ambient velocity increases to 1.5 m/s, the droplet deforms into an olive shape with a recirculating ring attached behind it, as shown in Fig. 13b. Because of droplet deformation and gasification, the threshold Reynolds number above which the recirculating eddy forms is considerably lower than that for a hard sphere. Figure 13c depicts the flow structure with viscous stripping at an ambient velocity of 5 m/s, showing an oblate droplet with a stretched vortex ring. The flattened edge of the droplet enhances the strength of the recirculating eddies and as such dramatically increases the viscous shear stress. Consequently, a thin sheet of fluid is stripped off from the edge of the droplet and swept toward the outer boundary of the recirculating eddy. At a very high ambient velocity of 15 m/s, droplet breakup takes place, as is clearly shown in Fig. 13d. The hydrogen flow penetrates through the liquid phase and divides the droplet into two parts: the core disk and surrounding ring. The vortical structure in the wake region expands substantially as a result of the strong shear stress.

The effect of ambient pressure and Reynolds number on droplet lifetime can be correlated with respect to the reference value at zero Reynolds number. The result takes the form

$$\frac{\tau_f}{\tau_{f,Re=0}} = \frac{1}{1 + 0.155 Re^{1.26}(p_{r,O_2})^{-1.58}} \quad (0 < Re < 300 \text{ and } 1 < p_{r,O_2} < 8)$$

(16)

The Reynolds number Re is based on the initial droplet diameter, and p_{r,O_2} is the reduced pressure in reference to the critical pressure of oxygen. The reference droplet lifetime, $\tau_{f,Re=0}$, is chosen to be the lifetime of an isolated droplet vaporizing in a quiescent environment, which can be obtained from Yang et al.[23] and Lafon.[25,30] This correlation bears a resemblance to the popular Ranz-and-Marshall correlation[41] for droplet heat transfer correction due to convective effect. The Ranz-and-Marshall correlation applies only to low-pressure flows and has a weaker Reynolds-number dependency.

Drag coefficient has been generally adopted as a dimensionless parameter to measure the drag force acting on a droplet. Chen and Yuen[42] found that the drag coefficient of an evaporating droplet is smaller than that of a nonvaporizing solid sphere at the same Reynolds numbers. Several researchers[43,44] numerically studied this issue by solving the Navier–Stokes equations for a spherical droplet at low to moderate pressures. No deformation was considered, in order to simplify the analysis. The results lead to the following correlation:

$$C_D = \frac{C_D^0}{(1 + B)^b}$$

(17)

where C_D^0 denotes the reference drag coefficient for a hard sphere. The exponent b has a value of 0.2 in Renksitzbulut's model[43] and 0.32 in Chiang's correlation.[44] A transfer number B is adopted to account for the effect of blowing on momentum transfer to the droplet. For droplet vaporization at low to moderate pressures ($p_r \leq 0.5$), the Spalding transfer number is widely used to characterize the vaporization rate:

$$B = \frac{C_p(T_\infty - T_s)}{\Delta h_v}$$

(18)

where T_s denotes the droplet surface temperature. The enthalpy of vaporization Δh_v becomes zero at the critical point, rendering a singular value for the transfer number. This deficiency may be remedied by introducing a transfer number suited for supercritical droplet vaporization[24,31]:

$$B_D = \left(\frac{T_\infty - T_c}{T_c - T_l}\right)$$

(19)

where T_l is the instantaneous average temperature of droplet, and T_c the critical mixing temperature. Since B_D diverges at $T_c = T_l$ at the end of droplet lifetime, the calculation of drag force was terminated when $T_c = T_l$ dropped below 1 K, at which point the droplet residual mass is usually less than one thousandth of the initial mass. The influence on the accuracy of data reduction is quite limited. Following the procedure leading to Eq. (17), a correlation for LOX droplet drag coefficient is obtained:

$$C_D = \frac{C_D^0}{(1 + aB_D)^b} \tag{20}$$

where a and b are selected to be 0.05 and $1.592\,(p_{r,O_2})^{-0.7}$, respectively. The data cluster along the classical drag curve in the low Reynolds number region, but deviate considerably at high Reynolds numbers (i.e., $Re > 10$). Although a shape factor may be employed to account for this phenomenon, which arises from the increased form drag due to droplet deformation, the difficulty of calculating this factor and conducting the associated data analysis precludes its use in correlating the drag coefficient herein. Instead, a simple correction factor $Re^{0.3}$ is incorporated into Eq. (20) to provide the compensation. The final result is

$$C_D = \frac{C_D^0 Re^{0.3}}{(1 + aB_D)^{1.592(p_{r,O_2})^{-0.7}}} \quad (0 < Re < 300 \text{ and } 1 < p_{r,O_2} < 8) \tag{21}$$

VI. Droplet Combustion

Much effort was devoted to the study of supercritical droplet combustion.[20,26,33] In spite of the presence of chemical reactions in the gas phase, the general characteristics of a burning droplet are similar to those involving only vaporization. We consider here the combustion of a hydrocarbon (e.g., *n*-pentane) fuel droplet in air.[20] The droplet initial temperature is 300 K, and the ambient air temperature is 1000 K. Figure 14 shows the time histories of the droplet surface temperature at various pressures for $D_0 = 100\ \mu\text{m}$. Once ignition is achieved in the gas phase, energy feedback from the flame causes a rapid increase in droplet surface temperature. At low pressures ($p \leq 20$ atm), the surface temperature varies very slowly following onset of flame development and almost levels off at the pseudo-wet-bulb state. As the ambient pressure increases, the high concentrations of oxygen in the gas phase and the fuel vapor issued from the droplet surface result in a high chemical reaction rate, consequently causing a progressive decrease in ignition time. Furthermore, the surface temperature jump during the flame-development stage increases with increasing pressure. Because the critical mixing temperature decreases with pressure, the droplet reaches its critical condition more easily at higher pressures, almost immediately following establishment of the diffusion flame in the gas phase for $p \geq 110$ atm.

Figure 15 presents the effect of pressure on various milestone times associated with droplet gasification and burning processes for $D_0 = 100\ \mu\text{m}$.[20] Here gasification lifetime is the time required for complete gasification; droplet burning lifetime is the gasification lifetime minus ignition time; single-phase combustion lifetime is the time duration from complete gasification to burnout of all fuel vapor; combustion lifetime is the sum of single-phase combustion lifetime and droplet burning lifetime. The gasification lifetime decreases continuously with pressure, whereas the single-phase combustion lifetime increases progressively

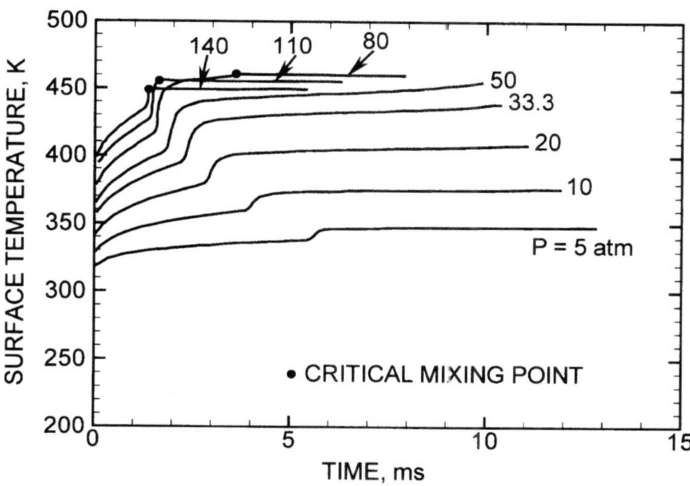

Fig. 14   Time variations of droplet surface temperature at various pressures; *n*-pentane/air system, $T_0 = 300$ K, $T_\infty = 1000$ K, $D_0 = 100$ μm.

with pressure due to its adverse effect on mass diffusion. More importantly, the pressure dependence of combustion lifetime exhibits irregular behavior. This phenomenon may be attributed to the overlapping effects of reduced enthalpy of vaporization and mass diffusion with increasing pressure.

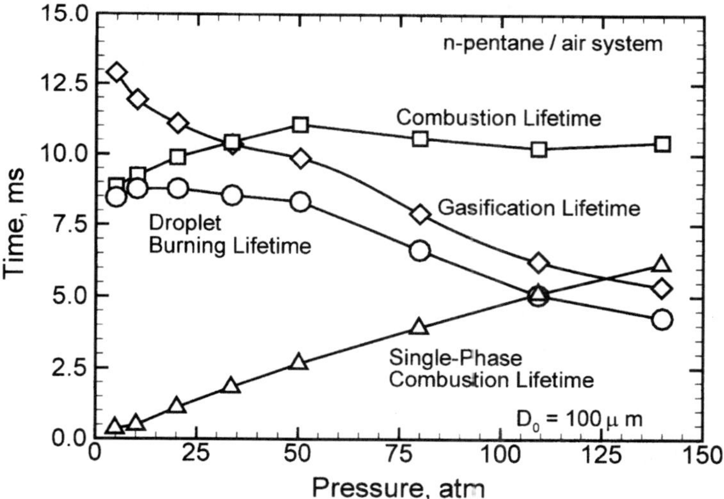

Fig. 15   Effect of pressure on milestone times associated with droplet gasification and burning processes; *n*-pentane/air system, $T_0 = 300$ K, $T_\infty = 1000$ K, $D_0 = 100$ μm.

Because the time scales for diffusion processes are inversely proportional to the droplet diameter squared, it is important to examine the effect of droplet size on the burning characteristics. In this regard, calculations were conducted for large droplets having an initial diameter of 1000 μm, which is comparable to the sizes considered in most experimental studies of supercritical droplet combustion.[45-47] Furthermore, the ignition transient occupies only a small fraction of the entire droplet lifetime for large droplets. The uncertainties associated with the ignition procedure in determining the characteristics of droplet gasification can be minimized. Consequently, a more meaningful comparison with experimental data can be made. The combustion behavior of a large droplet reveals several characteristics distinct from those of a small droplet. First, ignition for large droplets occurs in the very early stage of the entire droplet lifetime. The influence of gasification prior to ignition on the overall burning mechanisms appears to be quite limited. Second, the combustion lifetime decreases with increasing pressure, reaching a minimum near the critical pressure of the liquid fuel. As the pressure further increases, the combustion time increases due to reduced mass diffusivity at high pressures. The gasification lifetime decreases continuously with pressure, whereas the single-phase combustion lifetime increases progressively with pressure. At low pressures, the gasification of liquid fuel primarily controls the combustion process, whereas in a supercritical environment, the transient gas-phase diffusion plays a more important role.

VII.   Droplet Response to Ambient Flow Oscillation

Although unsteady droplet vaporization and combustion have long been recognized as a crucial mechanism for driving combustion instabilities in liquid-fueled propulsion systems,[48] they are extremely difficult to measure experimentally. In particular, the measurement of the effect of transverse oscillations on instantaneous evaporation and/or burning rates is formidable. The droplet volume dilatation arising from rapid temperature increase may overshadow the surface regression associated with vaporization, and thereby obscure the data analysis, especially in the early stage of the droplet lifetime. Furthermore, conventional suspended droplet experiments may not be feasible in the presence of gravity due to reduced or diminished surface tension. In view of these difficulties, it is advantageous to rely on theoretical analyses to study the responses of droplet vaporization and combustion to ambient flow oscillations. The model is based on the general approach described in Refs. 20, 23–25, and 30, but with a periodic pressure oscillation superimposed in the gas phase. Both cases involving hydrocarbon droplets in nitrogen[24] and LOX droplets in hydrogen[26] are examined. The purpose of these studies is to assess the effect of flow oscillation on vaporization process as a function of frequency and amplitude of the imposed oscillation, as well as its type.

A.   Hydrocarbon/Air System

The analysis was first carried out to study the vaporization response of n-pentane fuel droplets in a nitrogen environment. The initial droplet diameter and temperature are 100 μm and 300 K, respectively, and the ambient temperature is 1500 K. Figure 16 shows the time histories of droplet mass evaporation-rate fluctuation, $\dot{m}'$, at various mean ambient pressures, where the overbar denotes mean quantity. The oscillation frequency is 3000 Hz. The most significant result is the enhanced droplet vaporization response with increasing pressure. Among the various factors contributing to this phenomenon, the effect of pressure on enthalpy of vaporization plays a decisive role. At high pressures, the enthalpy of vaporization decreases and becomes more sensitive to variations of ambient pressure and temperature than low-pressure cases. A small fluctuation in the surrounding gases may considerably modify the interfacial thermodynamics, and consequently enhance the droplet vaporization response. The situation is most profound when the droplet surface reaches its critical condition at which a rapid amplification of vaporization response function is observed. The enthalpy of vaporization and related thermophysical properties exhibit abnormal variations in the vicinity of the critical point, thereby causing an abrupt increase in the vaporization response.

The pressure effect can be characterized and correlated by examining the vaporization response in the middle of droplet lifetime. Figures 17 and 18 show, respectively, the effect of mean ambient pressure on the magnitude and phase angle of the vaporization response function, R_p, defined as

$$R_p = \frac{\dot{m}'/\bar{\dot{m}}}{p'/\bar{p}} \tag{22}$$

The axial coordinate represents the normalized liquid thermal inertia time, where f denotes the oscillation frequency, R the droplet radius, and α_l the liquid thermal

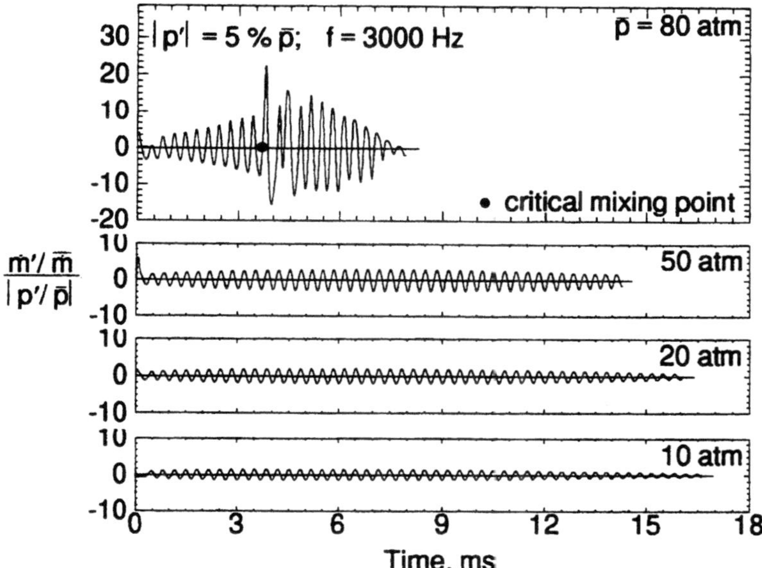

Fig. 16   Time histories of droplet evaporation rate fluctuation at various mean ambient pressures; *n*-heptane droplet in nitrogen, $D_0 = 100\ \mu m$, $T_0 = 300$ K, $T_\infty = 1500$ K, $f = 3000$ Hz.

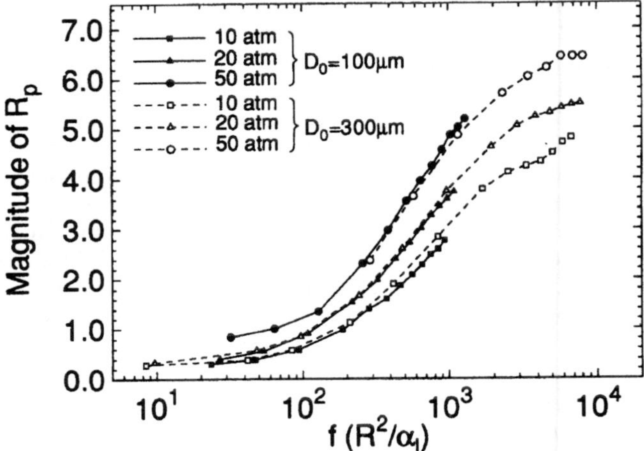

Fig. 17 Effect of mean ambient pressure on magnitude of droplet vaporization response; *n*-pentane/nitrogen system, $T_0 = 300$ K, $T_\infty = 1500$ K.

diffusivity. The response function R_p is a complex variable accounting for the phase difference between the fluctuations of ambient pressure and droplet evaporation rate. Results for the initial droplet diameters of 100 and 300 μm, shown respectively by the filled and hollow symbols, scatter along the same curve at a given pressure, regardless of the droplet size. The amplitude of the response function increases with increasing pressure due to enhanced sensitivity of enthalpy of vaporization to ambient flow oscillations at high pressures. On the other hand, the effect of mean ambient pressure on the phase angle appears limited. The phase angle decreases from zero in the low-frequency limit to -180 deg at high

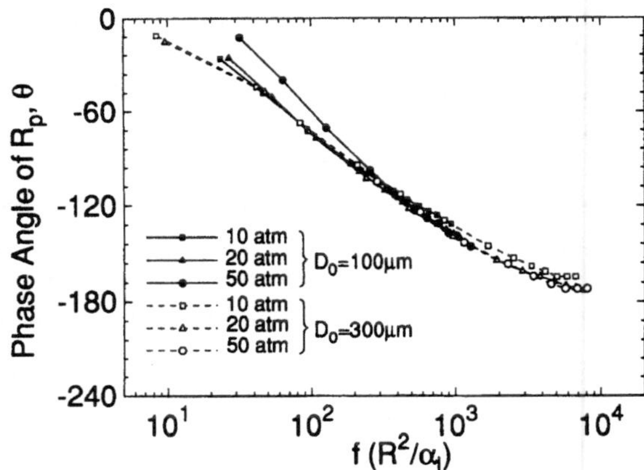

Fig. 18 Effect of mean ambient pressure on phase angle of droplet vaporization response; *n*-pentane/nitrogen system, $T_0 = 300$ K, $T_\infty = 1500$ K.

frequency, a phenomenon that can be easily explained by comparing various time scales associated with fluid transport and ambient disturbance.

B.   Oxygen/Hydrogen System

The same kind of study has been carried out for the oxygen/hydrogen system.[26] Figures 19 and 20 present the magnitudes and phase angles of the droplet vaporization response functions for different pressures over a broad range of liquid thermal inertia times. For each pressure, various values of fR_0^2 were considered. After an initial transient period, all magnitude and phase-angle data collapse on an envelope curve that depends only on the ambient pressure. The oxygen/hydrogen system exhibits an important feature distinct from the hydrocarbon/air system in that the phase angle falls between 0 and -90 deg. The real part of the response function remains always positive, even for high frequencies. For an oxygen droplet with an initial diameter of 100 μm and pressure $p = 30$ atm, the cutoff frequency exceeds 20,000 Hz. The magnitude of the droplet vaporization response is also smaller than the corresponding hydrocarbon/air system, mainly due to different liquid thermal inertia times. For oxygen droplets, vaporization proceeds faster than liquid heat-up. For hydrocarbon droplets, thermal waves penetrate faster in the liquid interior and consequently affect the vaporization response. It is worth noting that existing studies only consider droplet responses to pressure oscillations. The situation involving velocity oscillations is expected to play a substantial role in determining the combustion stability behavior of a rocket engine, and thus should be studied carefully.

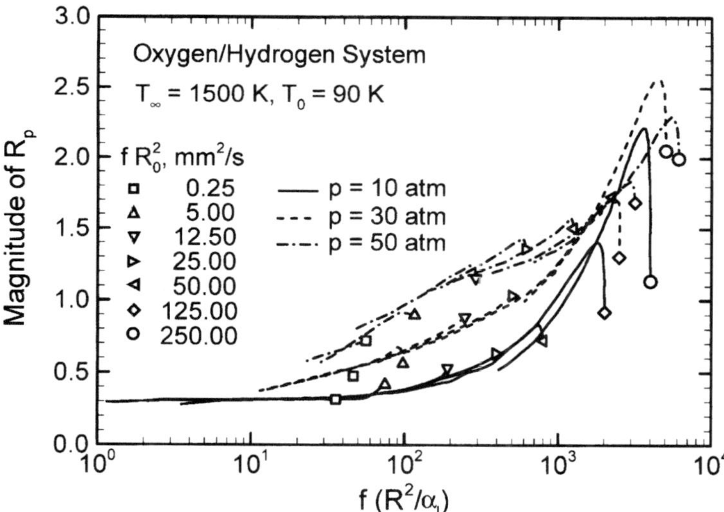

Fig. 19   Effect of mean ambient pressure on magnitude of droplet vaporization response; oxygen/hydrogen system, $T_0 = 90$ K, $T_\infty = 1500$ K.

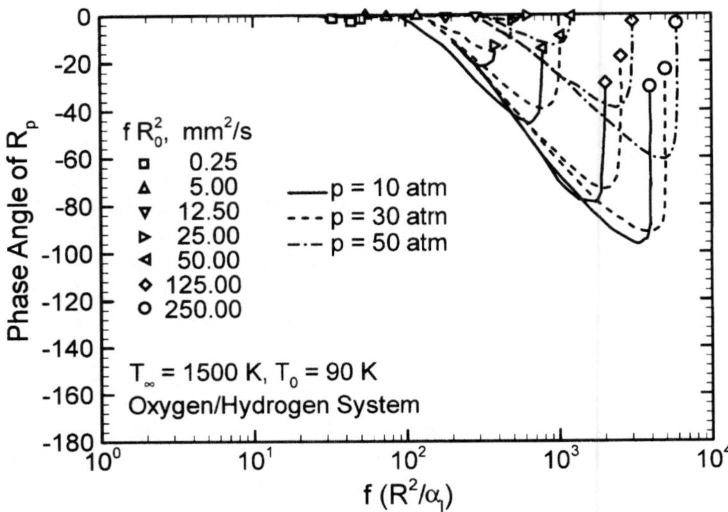

Fig. 20 Effect of mean ambient pressure on phase angle of droplet vaporization response; oxygen/hydrogen system, $T_0 = 90$ K, $T_\infty = 1500$ K.

VIII. Conclusions

An overview of recent advances in liquid-propellant droplet vaporization and combustion has been conducted. All three major types of liquid propellants (i.e., hydrocarbon, cryogenic, and hypergolic propellants) over a wide range of thermodynamic states are considered, with special attention given to situations representative of practical rocket-engine combustor conditions. Various transport characteristics of droplets are examined in both stagnant and forced-convective environments. The ambient flow conditions exert significant influences on droplet vaporization and burning processes through their effects on fluid transport, gas-liquid interfacial thermodynamics, and chemical reactions. The droplet gasification rate increases progressively with pressure and can be correlated well with dimensionless temperature and pressure parameters. Information obtained not only enhances the knowledge base of liquid droplet vaporization and combustion, but can also be used in analyses of spray combustion. In spite of the improved understanding achieved thus far through theoretical modeling, systematic verification of these models is still lacking, especially for droplets under supercritical conditions. Future work on supercritical droplet dynamics and transport should emphasize model validation against experimental data.

Acknowledgments

The research conducted at Pennsylvania State University was sponsored in part by the NASA Marshall Space Flight Center and in part by the U.S. Air Force Office of Scientific Research. The support and encouragement of Charles F. Schafer, Marvin Rocker, and Mitat A. Birkan were greatly appreciated. Patrick Lafon was supported by the European Space Agency post-doctoral fellowship. The authors wish to express their sincere thanks to Yanxing Wang and Tao Liu for their help with the illustrations.

References

[1]Williams, A., "Combustion of Droplets of Liquid Fuels: A Review," *Combustion and Flame*, Vol. 21, 1973, pp. 1–31.

[2]Faeth, G. M., "Current Status of Droplet and Liquid Combustion," *Progress in Energy and Combustion Science*, Vol. 3, 1977, pp. 191–224.

[3]Law, C. K., "Recent Advances in Droplet Vaporization and Combustion," *Progress in Energy and Combustion Science*, Vol. 8, 1982, pp. 171–201.

[4]Sirignano, W. A., "Fuel Droplet Vaporization and Spray Combustion Theory," *Progress in Energy and Combustion Science*, Vol. 9, 1983, pp. 291–322.

[5]Dwyer, H. A., "Calculations of Droplet Dynamics in High Temperature Environments," *Progress in Energy and Combustion Science*, Vol. 15, 1989, pp. 131–158.

[6]Bellan, J., "Supercritical (and Subcritical) Fluid Behavior and Modeling: Drops, Streams, Shear and Mixing Layers, Jets, and Sprays," *Progress in Energy and Combustion Science*, Vol. 26, 2000, pp. 329–366.

[7]Yang, V., "Modeling of Supercritical Vaporization, Mixing, and Combustion Processes in Liquid-Fueled Propulsion Systems," *Proceedings of the Combustion Institute*, Vol. 28, 2000, pp. 925–942.

[8]Reid, R. C., Prausnitz, J. M., and Poling, B. E., *The Properties of Gases and Liquids*, McGraw-Hill, New York, 1988.

[9]Jacobsen, R. T., and Stewart, R. B. J., "Thermodynamic Properties of Nitrogen Including Liquid and Vapor Phases from 63K to 2000K with Pressure to 10,000 Bar," *Journal of Physical and Chemical Reference Data*, Vol. 2, No. 4, 1973, pp. 757–922.

[10]Prausnitz, J. M., Lichtenthaler, R. N., and de Azevedo, E. G., *Molecular Thermodynamics of Fluid-Phase Equilibria*, Prentice-Hall, Englewood Cliffs, NJ, 1986.

[11]Perry, R. H., Gree, D. W., and Maloney, J. O., *Perry's Chemical Engineering Handbook*, McGraw-Hill, New York, 1984.

[12]Ely, J. F., and Hanley, H. J., "Prediction of Transport Properties. 1. Viscosity of Fluids and Mixtures," *Industrial and Engineering Chemistry Fundamentals*, Vol. 20, 1981, pp. 323–332.

[13]Ely, J. F., and Hanley, H. J., "Prediction of Transport Properties. 2. Thermal Conductivity of Pure Fluids and Mixtures," *Industrial and Engineering Chemistry Fundamentals*, Vol. 22, 1983, pp. 90–97.

[14]Sychev, V. V., Vasserman, A. A., Kozlov, A. D., and Spiridonov, G. A., "Liquid and Gaseous Oxygen," *Property Data Update*, Vol. 1, No. 3–4, 1987, pp. 529–536.

[15]Ely, J. F., "An Enskog Correction for Size and Mass Difference Effects in Mixture Viscosity Prediction," *Journal of Research of the National Bureau of Standards*, Vol. 86, No. 6, 1981, pp. 597–604.

[16]Vesovic, V., and Wakeham, W. A., "Transport Properties of Supercritical Fluids and Fluid Mixtures," *Supercritical Fluid Technology*, edited by T. J. Bruno and J. F. Ely, CRC Press, Boca Raton, FL, 1991, p. 253.

[17]Levelt Sengers, J. M. H., "Thermodynamics of Solutions Near the Solvent's Critical Point," *Supercritical Fluid Technology*, edited by T. J. Bruno and J. F. Ely, CRC Press, Boca Raton, FL, 1991, p. 25.

[18]Takahashi, S., "Preparation of a Generalized Chart for the Diffusion Coefficients of Gases at High Pressures," *Journal of Chemical Engineering (Japan)*, Vol. 7, 1974, pp. 417–420.

[19]Hsieh, K. C., Shuen, J. S., and Yang, V., "Droplet Vaporization in High-Pressure Environments, 1: Near-Critical Conditions," *Combustion Science and Technology*, Vol. 76, 1991, pp. 111–132.

[20]Shuen, J. S., Yang, V., and Hsiao, G. C., "Combustion of Liquid-Fuel Droplets in Supercritical Conditions," *Combustion and Flame*, Vol. 89, 1992, pp. 299–319.

[21]Harstad, K., and Bellan, J., "The Lewis Number under Supercritical Conditions," *International Journal of Heat and Mass Transfer*, Vol. 42, 1999, pp. 961–970.

[22]Harswtad, K., and Bellan, J., "An All-Pressure Fluid Drop Model Applied to a Binary Mixture: Heptane in Nitrogen," *International Journal of Multiphase Flow*, Vol. 26, No. 10, 2000, pp. 1675–1706.

[23]Yang, V., Lin, N. N., and Shuen, J. S., "Vaporization of Liquid Oxygen (LOX) Droplets in Supercritical Hydrogen Environments," *Combustion Science and Technology*, Vol. 97, 1994, pp. 247–270.

[24]Hsiao, G. C., "Supercritical Droplet Vaporization and Combustion in Quiescent and Forced-Convective Environments," Ph.D. Thesis, Pennsylvania State Univ., University Park, PA, 1995.

[25]Lafon, P., "Modélisation et Simulation Numérique de L'Evaporation et de la Combustion de Gouttes à Haute Pression," Ph.D. Thesis, L'Université D'Orléans, 1995.

[26]Lafon, P., Yang, V., and Habiballah, M., "Pressure-Coupled Vaporization and Combustion Responses of Liquid Oxygen (LOX) Droplets in Supercritical Hydrogen Environments," AIAA Paper 1995-2432, 1995.

[27]Meng, H., and Yang, V., "Interactions of Liquid Oxygen Vaporization in High Pressure Hydrogen Environment," AIAA Paper, 1998-3537, 1998.

[28]Harstad, K., and Bellan, J., "Isolated Liquid Oxygen Drop Behavior in Fluid Hydrogen at Rocket Chamber Pressures," *International Journal of Heat and Mass Transfer*, Vol. 41, 1998, pp. 3537–3550.

[29]Harstad, K., and Bellan, J., "Interactions of Liquid Oxygen Drops in Fluid Hydrogen at Rocket Chamber Pressures," *International Journal of Heat and Mass Transfer*, Vol. 41, 1998, pp. 3551–3558.

[30]Lafon, P., Yang, V., and Habiballah, M., "Supercritical Vaporization of Liquid Oxygen Droplets in Hydrogen and Water Environments," *Journal of Fluid Mechanics*. (submitted for publication).

[31]Meng, H., Hsiao, G. C., Yang, V., and Shuen, J. S., "Transport and Dynamics of Liquid Oxygen Droplets in Supercritical Hydrogen Streams," *Journal of Fluid Mechanics* (to be published).

[32]Lafon, P., and Yang, V., "On the Effect of Pressure on Hydrocarbon Fuel Droplet Evaporation," *Combustion Science and Technology* (submitted for publication).

[33]Daou, J., Haldenwang, P., and Nicoli, E., "Supercritical Burning of Liquid (LOX) Droplet with Detailed Chemistry," *Combustion and Flame*, Vol. 101, 1995, pp. 153–169.

[34]Zhuang, F.-C., and Liu, X.-D., "On Theory of Unsteady Decomposition Combustion for a Liquid Monopropellant Droplet at High Temperature and High Pressure," *Acta Aeronautica*, Vol. 6, No. 5, 1985.

[35]Zhuang, F.-C., and Liu, X.-D., "A Subcritical Unsteady Vaporization Model of a Hypergolic Propellant Species," *Journal of Chinese Society of Astronautics* (in Chinese), Vol. 2, No. 4, 1986.

[36]Hu, X.-P., Zhong, L.-S., and Zhuang, F.-C., "The Decomposition Combustion of a Monopropellant Droplet in Elevated Temperature Environment," *Proceedings of the Second Symposium on Engineering Thermophysics of Higher School* (in Chinese), 1986.

[37]Zhuang, F.-C., and Zhang, C. X., "A Combustion Model of FY-2 Engine," *Journal of National University of Defence Technology* (in Chinese), Vol. II, No. 3, 1989.

[38]Liao, H., and Zhuang, F.-C., "Combustion Modeling of Nitrogen Tetrooxide/ Hydrazin Type Fuel Propellant Rocket," *Proceedings of the First Asian-Pacific International Symposium on Combustion and Energy Utilization*, 1990.

[39]Meng, H., "Liquid-Fuel Droplet Vaporization and Cluster Behavior at Supercritical Conditions," Ph.D. Thesis, Pennsylvania State Univ., University Park, PA, 2001.

[40]Prakash, S., and Sirignano, W. A., "Liquid Fuel Droplet Heating and Internal Circulation," *International Journal of Heat and Mass Transfer*, Vol. 21, 1978, pp. 885–895.

[41]Ranz, W. E., and Marshall, W. R., "Evaporation from Drops I and II," *Chemical Engineering Progress*, Vol. 48, 1952, pp. 141–146.

[42]Chen, L. W., and Yuen, M. C., "On Drag of Evaporating Liquid Droplets," *Combustion Science and Technology*, Vol. 14, 1976, pp. 147–154.

[43]Renksizbulut, M., and Haywood, R. J., "Transient Droplet Evaporation with Variable Properties and Internal Circulation at Intermediate Reynolds Numbers," *International Journal of Multi-phase Flow*, Vol. 14, No. 2, 1988, pp. 189–202.

[44]Chiang, C. H., Raju, M. S., and Sirignano, W. A., "Numerical Analysis of Convecting, Vaporizing Fuel Droplets with Variable Properties," *International Journal of Heat and Mass Transfer*, Vol. 35, No. 5, 1992, pp. 1307–1324.

[45]Faeth, G. M., Dominics, D. P., Tulpinsky, J. F., and Olson, D. R., "Supercritical Bipropellant Droplet Combustion," *Proceedings of the Combustion Institute*, Vol. 12, 1969, pp. 9–18.

[46]Sato, J., Tsue, M., Niwa, M., and Kono, M., "Effects of Natural Convection of High-Pressure Droplet Combustion," *Combustion and Flow Flame*, Vol. 82, 1990, pp. 142–150.

[47]Sato, J., "Studies on Droplet Evaporation and Combustion in High Pressures," AIAA Paper 1993-0813, 1993.

[48]Yang, V., and Anderson, W. E., *Liquid Rocket Engine Combustion Instability*, Progress in Astronautics and Aeronautics, Vol. 169, AIAA, Washington, DC, 1995.

Chapter 8

Subcritical/Supercritical Droplet Cluster Behavior in Dense and Dilute Regions of Sprays

Josette Bellan*
*Jet Propulsion Laboratory, California Institute of Technology
Pasadena, California*

I. Introduction

CONCERN for the efficiency, stability, and safety margins of bipropellant combustion in rocket engines has prompted the investigation of many specific aspects of spray behavior previously not studied. Thus, early studies of combustion in liquid rocket engines were based on the results of the classical single-component, isolated drop combustion at atmospheric pressure.[1] Although the results from these studies provided a baseline for understanding some of the phenomena occurring in liquid rocket engines, they fail to explain important observations and facts obtained from examining rocket performance after many flights. Examples are the loss of about 3% of the liquid oxygen (LOX, one of the propellants) that exits unburned, and the existence of striations on the inner wall of engines after a flight. It then becomes apparent that many significant issues of liquid rocket spray combustion were not addressed by the early models, and that to mitigate existing problems, it is necessary to understand aspects previously unexplored.

The phenomenon of bipropellant spray behavior in a highly turbulent environment at elevated pressure is as follows: When oxygen and hydrogen sprays enter the combustion chamber, because of the higher critical temperature of oxygen with respect to that of hydrogen, the drops of hydrogen quickly become a fluid, whereas the drops of oxygen may or may not remain liquid, depending on the chamber pressure; if the pressure is subcritical with respect to O_2, the drops will remain liquid. Thus, one of the relevant phenomena to be studied is the behavior of LOX drops in surroundings initially composed of fluid H_2.

*Senior Research Scientist. Associate Fellow AIAA.

Because of the high pressure conditions, solubility of H_2 into LOX becomes important, so that although the two component fuels are initially separated, very rapidly the liquid drops become composed of a binary fuel. Because the species concentration in the drops varies in time, the mixture might be alternately at supercritical and subcritical conditions for the prevailing temperature and pressure as the composition of the drops changes.

Experimental observations of atomization of coaxial jets by Hardalupas et al.[2] and Engelbert et al.[3] reveal the initial formation of ligaments, each ligament quickly disintegrating into a cluster of drops. Similarly, visualizations of impinging liquid jets[1] have shown that under both cold and hot flow conditions the jets break up in a periodic manner into ligaments that further break up into drops thereby creating clusters of drops. Recent observations by Ryan et al.[4] have documented this typical breakup process while also yielding information regarding the breakup length and the drop size for both laminar and turbulent impinging jets. Poulikakos[5] made similar observations. These observations indicate that fluid drops created during atomization in liquid rocket chambers should not behave as isolated, and instead should have a collective behavior. This realization is very important for controlling high-frequency combustion instabilities because the response function should then incorporate this fluid-drop interaction.

This chapter reviews the evaporative/diffusive collective behavior of binary-fuel drops at subcritical/supercritical pressures and emphasizes the difference of behavior between these two regimes. In particular we show that, based on model predictions, drop interactions at supercritical pressures occur only for much closer relative drop proximity, measured by the ratio (drop interdistance)/ (drop radius), than at subcritical pressures.

II.   Clusters of Binary-Species Drops in Air (Subcritical)

The first step toward the general goal of understanding the respective role of each component of a binary-species drop during evaporation of collections of drops is to study their behavior in subcritical conditions. The studies of clusters of binary-species drops reviewed in this chapter were conducted at atmospheric pressure.

For binary-fuel isolated drops in convective flows, liquid mass diffusion has been found[6] to drive the evaporation of the solute. Indeed, examination of the characteristic time associated with liquid mass diffusion under quiescent conditions shows that it does not play an important role in the evaporation of the solute because this time is much larger than the drop lifetime. However, under convective conditions, the slip velocity at the surface of the drop induces, through shear, a circulatory motion inside the drop in the form of Hill vortices. This circulatory motion enhances liquid mass diffusion whose characteristic time becomes accordingly reduced and eventually of comparable magnitude to the drop lifetime. It is under these conditions that liquid mass diffusion becomes important in bringing solute from the drop core to the surface, and preferentially induces solute evaporation.

Consistent with the experimental observations of burning and nonburning sprays at atmospheric pressure that revealed drop clustering and interaction, the collective drop behavior was quantified by Bellan and Cuffel[7] through the concept of the "sphere of influence" around each drop in a cluster. For clusters of single-size drops, each sphere of influence has the same center as the drop and has a radius that is the half-distance between the centers of adjacent drops; the non-dimensional radius of the sphere of influence, $R_2 \equiv R_{si}/R$, is the ratio of the radius of the sphere of influence, R_{si}, to the drop radius R. Because by definition the spheres of influence are tightly packed in the cluster volume, this volume is composed of the ensemble of all of the spheres of influence and the spaces between the spheres of influence; the cluster surface is the envelope of the spheres of influence. This definition of the sphere of influence allows the accounting for the drop interaction through the magnitude of R_2. Results from simulations of single-component drop heating (the initial gas temperature exceeds that of the drops) and evaporation of drops congregated in clusters suggest the following classification: for $R_2 \leq 10$, the evaporation time is generally twice that of the isolated drop; the cluster is then called very dense. When $10 < R_2 \leq 15$, the evaporation time is approximately 50% larger than that of the isolated drop; the cluster is then denoted as dense. For $15 < R_2 \leq 30$, the evaporation time is about 10% larger than that of the isolated drop; the cluster is then labeled dilute. Finally, for $R_2 > 30$, the evaporation time tends asymptotically to the value for the isolated drop; the cluster is then called very dilute. This classification of different spray regimes, according to the value of R_2, has been adopted by experimentalists.[8]

During drop heating and evaporation, the volume of a cluster of drops will change as a result of several processes: 1) drop heating decreases the cluster volume as it represents an energy sink for the gas; 2) vapor release from the drops to the interstitial gas increases the cluster volume as it represents an energy source for the gas; and finally 3) heat transfer from the cluster surroundings to the cluster increases the cluster volume. Under adiabatic conditions, the volume of a very dilute cluster of drops will remain approximately constant because the small number of drops has little impact upon the fate of the interstitial gas. For increasingly larger drop number density, the cluster volume progressively decreases due to the larger net amount of heat transferred to the drops.

According to whether the cluster evaporates in the very dense to very dilute configurations, drop evaporation might assume different characteristics. Harstad and Bellan[9] studied a spherical cluster of binary-fuel drops assuming that the solute is much more volatile than the solvent. The gas inside the cluster was initially at rest, whereas the drops had a non-null initial velocity u_d^0; as the drops moved, they entrained the gas, which acquired a velocity of its own (only laminar situations are studied). A number $Be \equiv -[R/(D_m u_l)]^{0.5}\, dR/dt$ (D_m is the mass diffusivity, u_l is the circulatory velocity inside the drop, and t is the time) was defined representing the ratio of the drop mass regression rate to a characteristic solute diffusion rate. Thus, when $Be \ll 1$, diffusion into the boundary layer governs the rate of solute transfer from the liquid core to the drop surface and evaporation from the surface occurs at a rate defined by the Langmuir-Knudsen evaporation law (no thermodynamic equilibrium assumed). Because the processes are sequential, it is in fact the slower of these two rates that governs evaporation. In contrast, when $Be \gg 1$, the transfer of solute from the liquid core to the gas phase is governed by surface layer stripping, that is by the regression rate of the drop.

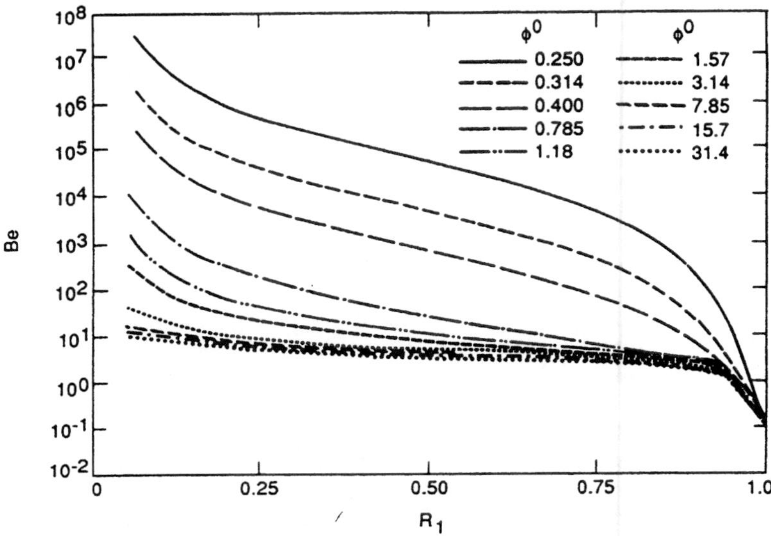

Fig. 1 *Be* **with the residual drop radius for several initial air/fuel mass ratios. The solvent is No. 2-GT (light diesel) fuel oil; the solute is n-decane. (Reprinted with permission of Begell House.)**[9]

The reader is referred to Harstad and Bellan[9] for the details of the model. Figure 1, from Harstad and Bellan,[9] shows the variation of Be with the residual drop radius, $R_1 = R/R^0$ (superscript 0 denotes initial conditions), for an extended range of initial air/fuel mass ratios Φ^0, corresponding to an extended range of values of R_2^0, as R_2^0 is a monotonically increasing function of Φ^0. The simulations were performed for a mixture of No. 2-GT (light diesel) fuel oil, as solvent, and n-decane, as solute, under the following conditions: initial far field gas temperature $T_{ga}^0 = 1000\,\text{K}$, initial far-field fuel mass fraction $Y_{Fva}^0 = 0$, initial drop surface temperature $T_{gs}^0 = 350\,\text{K}$, initial volatile fuel mass fraction in the liquid $Y_{HV,c}^0 = 0.02$, $u_d^0 = 200\,\text{cm/s}$, $R^0 = 2 \times 10^{-3}\,\text{cm}$, and initial cluster radius $R_C^0 = 3\,\text{cm}$. In the very dense and dense regimes, although Be is initially $O(1)$, it quickly becomes $\gg 1$ during the drop lifetime. In contrast, in the dilute and very dilute regimes, $Be = [O(1) - O(10)]$ during the entire drop lifetime, being much smaller than in the dense and very dense regimes. This is due to the different drop dynamics in dense and dilute clusters of drops. A denser cluster exposes more area to the flow, and thus the drag force is stronger. The result is a quicker relaxation of the slip velocity between phases, u_s; consequently, the shear at the drop surface decays very fast during the drop lifetime, and u_l becomes accordingly small. When the regression rate of the drop is smaller than the decay rate of u_l (because drop heating is hindered by the presence of other drops in close proximity), Be becomes $\gg 1$. Plots of the mass fraction of the solute in the drop core, $Y_{HV,c}$, vs R_1, shown in Fig. 2 (also reproduced from Harstad and Bellan[9]) concur with this physical picture: except for an initial

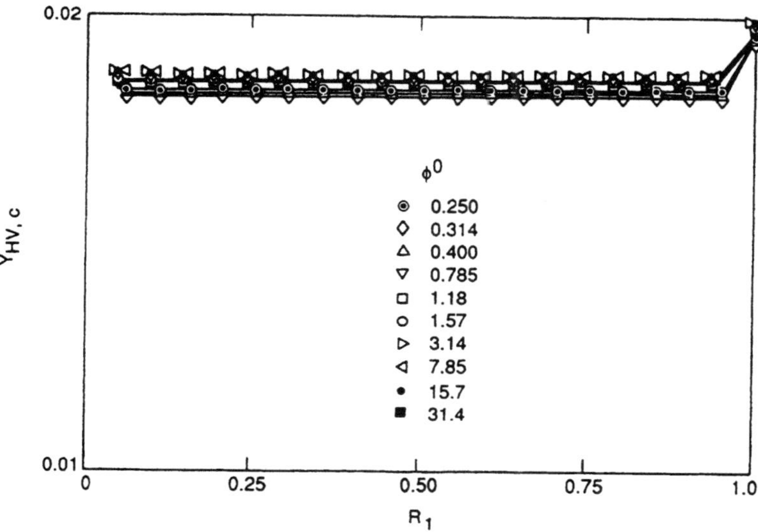

Fig. 2 Mass fraction of solute in the drop core vs the residual radius for several initial air/fuel mass ratios. Same initial conditions as Fig. 1. (Reprinted with permission of Begell House.)[9]

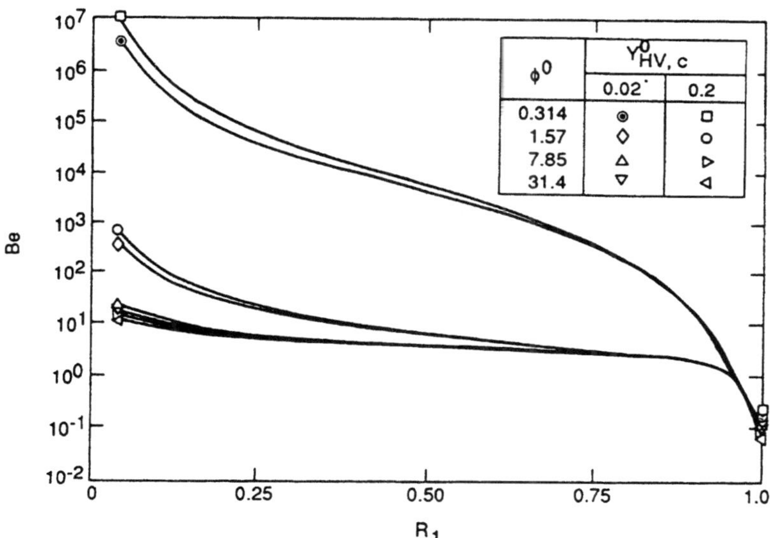

Fig. 3 *Be* with the residual drop radius for several initial air/fuel mass ratios. Same initial conditions as Fig. 1 except for $Y^0_{HV,c}$ and the fuel solute, which is n-hexane. (Reprinted with permission of Begell House.)[9]

effect of the liquid mass diffusion resulting in preferential evaporation of the solute, $Y_{HV,c}$ remains constant inside the drop, showing that it evaporates at the rate of the solvent. The conclusions do not change when the initial solute mass fraction is increased to 0.2, as seen in Fig. 3 (reproduced from the same study[9]); the calculations were performed for No. 2-GT/n-hexane. No simulations were performed with even larger $Y^0_{HV,c}$ because the validity of the model is restricted to an initial solute mass fraction smaller than that of the solvent.

Figure 4 (reproduced from Harstad and Bellan[9]) shows that when u^0_d is increased by a factor of 5 to 1000 cm/s, these results remain valid in the very dense and dense regimes; however, in the dilute and very dilute regimes the influence of liquid mass diffusion remains important during the major portion of the drop lifetime, and the solute mass fraction continues to decrease with R_1. These results support the above physical interpretation related to the variation of Be and are consistent in the very dilute regime with the isolated drop results of other investigators.[10–12]

III. Clusters of Fluid O_2 Drops in H_2 (Supercritical)

It is well known that at supercritical pressures, the drops that were liquid at subcritical pressures no longer maintain a surface because the surface tension is null. This means, in particular, that the effect of any convective motion past the drops will be negligible in terms of inducing a circulatory internal motion[13] because in the absence of surface there is no boundary layer at the drop boundary; thus, the physics of fluid drops in quiescent flow discussed here, is the simplest, yet relevant, situation to the rocket chamber environment. Under supercritical conditions, the concept of boundary is somewhat unclear.

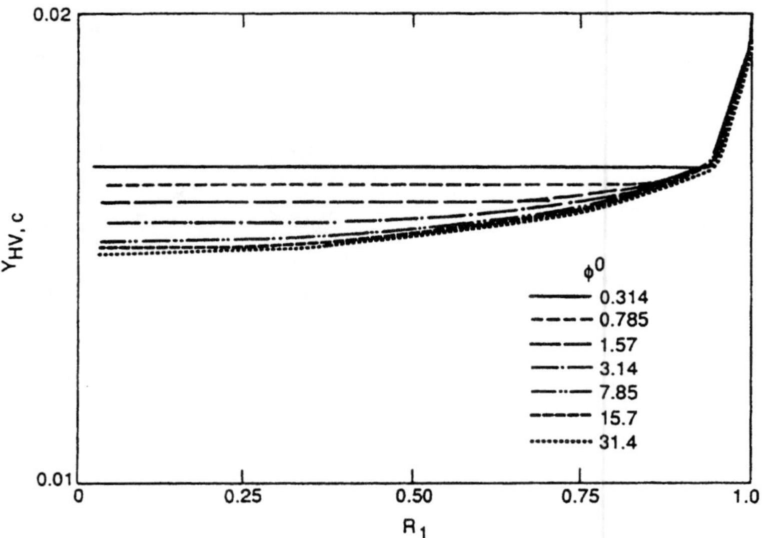

Fig. 4   Mass fraction of solute in the drop core vs the residual radius for several initial air/fuel mass ratios. Same initial conditions as Fig. 1 except for u^0_d.

Various investigators have located it at either the critical locus[13,14] or at the locus of the saturation temperature.[15] Examination of the boundary location issue based on the predictions of Harstad and Bellan[16] discussed in the following and on optical measurement techniques (which detect density variations) suggests that the point of maximum gradient in the density profile might be a more appropriate definition of the boundary location.

This discussion assumes that fluid drops are spherical. Despite the absence of material interface, they shall still be called drops. Most models of single, isolated, subcritical/supercritical O_2 drops[14,15,17] retain the subcritical classical formalism (including the assumption of thermodynamic equilibrium at the drop surface) while incorporating the solubility and binary-fuel aspects, high pressure variable properties, and real gas equations of state. In these models, it is also assumed that the drop surface suddenly vanishes when the supercritical point is attained. The model of Harstad and Bellan[16] is unique in that: 1) the difficulty of the vanishing surface tension is bypassed by formulating the conservation equations for a general binary fluid system, 2) Soret and Dufour effects are included, 3) evaporation/mass transfer is not constrained to be an equilibrium process, and 4) the phase of the system (if considered of interest) may be determined by the equation of state (the number of molar roots). In this model, the conservation equations are formulated based on the fluctuation-dissipation theory of Keizer.[18,19] One advantage of this theory is that it inherently accounts for nonequilibrium processes and naturally leads to the most general form of the fluid equations relating the partial molar fluxes and the heat flux to thermodynamic quantities. Thus, the molar fluxes J_i (i denotes the species) and the heat flux q are related to the transport matrix L through

$$J_i = L_{iq}\nabla\beta - \sum_{j}^{N} L_{ij}\nabla(\beta\mu_j), \qquad q = L_{qq}\nabla\beta - \sum_{j}^{N} L_{qj}\nabla(\beta\mu_j) \tag{1}$$

where N is the total number of species. Here L_{ij} are the Fick's diffusion elements, L_{qq} is the Fourier thermal diffusion element, L_{jq} are the Soret diffusion, and L_{qj} are the Dufour diffusion elements, μ_j is the chemical potential and $\beta \equiv 1/(R_u T)$ where R_u is the universal gas constant, T is the temperature. The Onsager relations state that $L_{ij} = L_{ji}$ and $L_{iq} = L_{qi}$. Additionally, conservation of fluxes and mass in the system imply that

$$\sum_{i}^{N} m_i J_i = 0 \quad \text{and} \quad \sum_{i}^{N} L_{ij} m_i = 0$$

for $j \in [1, N]$ and $j = q$, where m_i is the molar mass. Using the thermodynamic relationship

$$d(\beta\mu_j) = (v_j\,dp - h_j\,d\ell n\,T) + \left(\sum_{1}^{N-1} \alpha_{Dji}\,dX_i\right)\bigg/X_j \tag{2}$$

where X_i are the molar fractions, and the mass diffusion factors defined by

$$\alpha_{Dij} \equiv \beta X_i \partial \mu_i / \partial X_j = \partial X_i / \partial X_j + X_i \partial \ell n\, \gamma_i / \partial X_j \tag{3}$$

Where γ_i are the activity coefficients, one can calculate J_i from Eqs. 1 and 2. In general, this formalism proceeds with the definition of a symmetric matrix whose elements are the pair-wise mass diffusion coefficients for the mixture $D_{\mathrm{mix}}^{(ij)}$, and an antisymmetric matrix whose elements are the thermal diffusion factors $\alpha_T^{(ij)}$. For a binary-species mixture, the formalism simplifies considerably to yield

$$J_1 = -(m_2/m)(J_a + X_1 X_2 \alpha_T n D_m \nabla \ell n\, T) \tag{4}$$

where $D_m = L_{11}(m/m_2)^2 v/(X_1 X_2)$ is a mass diffusion coefficient, $\alpha_T = (L_{1q}/L_{11})\beta m_2/m$ is the ratio between the thermal and mass diffusivities, and

$$J_a = n D_m \{ \alpha_D \nabla X_1 + \beta[m_1 m_2 X_1 X_2/m][(v_1/m_1 - v_2/m_2)\nabla p$$
$$+ (h_2/m_2 - h_1/m_1)\nabla \ell n\, T]\} \tag{5}$$

with h_1 and h_2 being partial molar enthalpies; p is the pressure. According to Eq. 3, $\alpha_D = 1 + X_1(\partial \ell n\, \gamma_1/\partial X_1)_{T,p}$, and from the Gibbs-Duhem relationship $\alpha_{D11} = \alpha_{D22} = -\alpha_{D12} = -\alpha_{D21} = \alpha_D$. Similarly,

$$q = -(\alpha_T R_u T)J_a - k\nabla T \tag{6}$$

where consistent with the previous definitions $k = \beta L_{qq}/T$ is the heat conductivity of the mixture. The complete derivation of the model (equations and boundary conditions) is presented in Harstad and Bellan.[16] In this model, the transport matrix is no longer composed of terms derived uniquely from Fick's and Fourier's laws, which implies that the traditional Lewis number, $Le = \lambda/(\rho C_p D)$, is no longer an appropriate measure of the relative importance of heat diffusion to mass diffusion. The definition of an effective Lewis number Le_{eff} for application to general fluids is presented in Harstad and Bellan.[20]

Using the preceding formalism, results from simulations of isolated drops of LOX in H_2 were obtained for a far-field boundary condition taken at $R_2^0 = 20$; this value satisfies the definition of the dilute regime in Section I and the expectation that due to the essentially diffusional behavior, the very dilute regime will occur for smaller values of R_2^0. Results are illustrated in Figs. 5 and 6 (reproduced from Harstad and Bellan[16,20]) for the following initial conditions: $R_0 = 50 \times 10^{-4}$ cm, $R_{si}^0 = 0.1$ cm, $T_{d,b}^0 = 100$ K, $T_{si}^0 = 1000$ K, and $p = 20$ MPa, where subscripts si, d, and b denote, respectively, sphere of influence, drop, and boundary as opposed to surface in Section I. The illustrations clearly show that 1) as expected, the general behavior is characterized by diffusion, 2) heat diffusion is greater than mass diffusion (Figs. 5a and 5b), 3) among temperature, mass fraction, and density, it is the density that maintains its gradients for a longer time (Fig. 5), and 4) the traditional Le does not give a quantitatively or qualitatively accurate measure of the ratio of heat to mass transfer (Fig. 6). In contrast to Le, Le_{eff} gives a correct evaluation of the relative importance of heat and mass

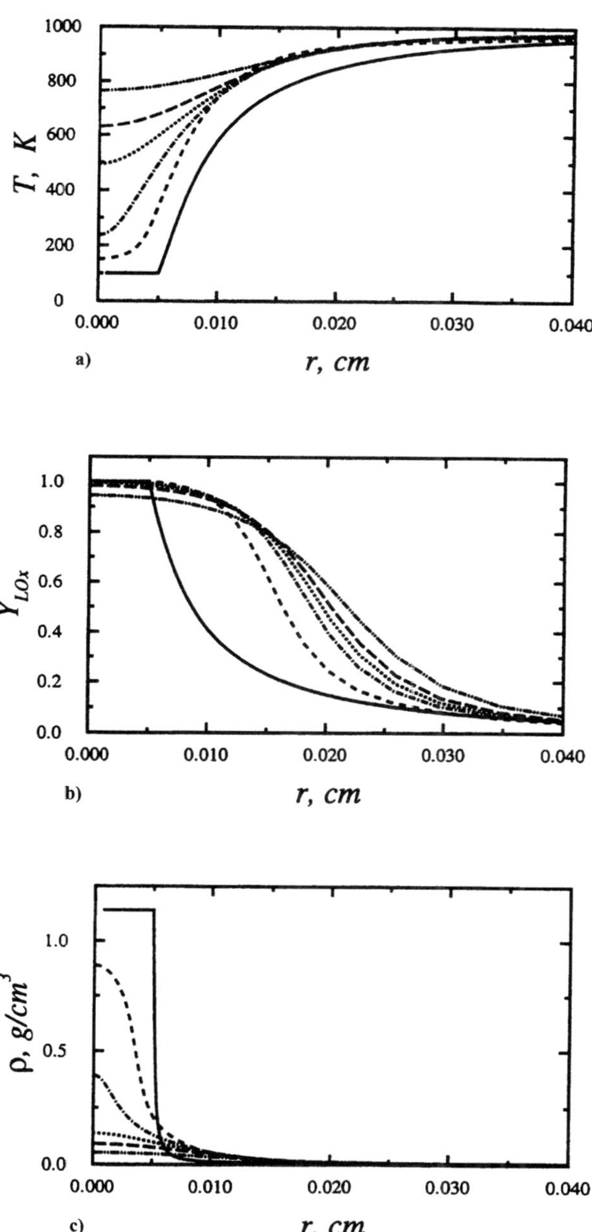

Fig. 5 Spatial variation of the a) temperature, b) oxygen mass fraction, and c) density at various times: 0.0 s (——), 7.5×10^{-3} s (- - -), 1.25×10^{-2} s (–·–·–), 1.5×10^{-2} s, (···), 1.75×10^{-2} s (———), 2.414×10^{-2} s (–··–). (Reprinted with permission of Elsevier.)[16]

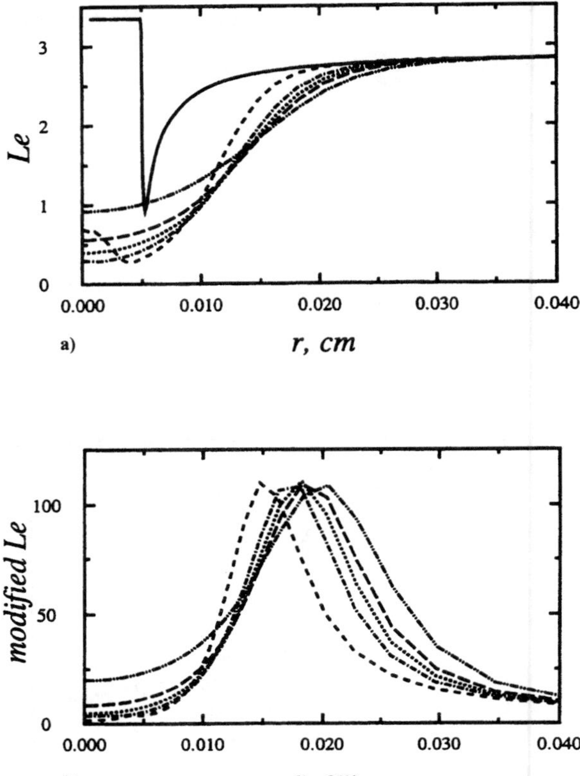

a)

b)

Fig. 6　Spatial variation of the traditional and effective Lewis number at different times. Initial conditions and legend are the same as in Fig. 5. (Reprinted with permission of Elsevier.)[20]

transfer (see Harstad and Bellan[20]). However, the difference between Le and Le_{eff} is not due to Soret and Dufour effects, which are found to be negligible in the range of α_T investigated,[20] but rather to the difference between the specific enthalpies of the two species. This difference enhances heat diffusion with respect to mass diffusion.[20]

This model of isolated fluid drops[16] constitutes the basis of the formulation describing fluid drop interactions.[21] Additional to the isolated drop model, the formulation of drop interactions includes conservation equations (based upon the concept of the sphere of influence described in Section II) for the volume of the cluster of drops; the details are described in Harstad and Bellan.[21] Transport between the cluster and its surroundings is modeled through an equivalent Nusselt number Nu defined relative to an external cluster length scale and to the cluster radius R_C. Extensive simulations with $Nu \in \lfloor 10^2,\ 10^5 \rfloor$ show that the results are very weakly dependent upon Nu.[21] Figure 7 illustrates the effect

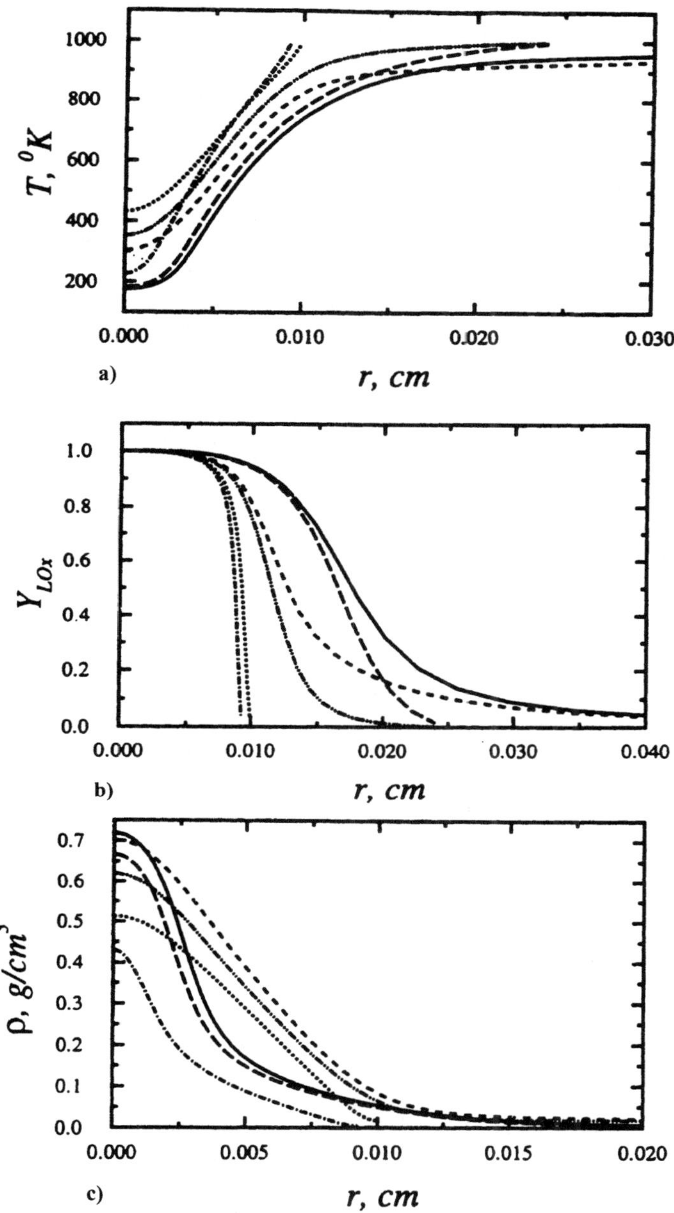

Fig. 7 Spatial variation of the a) temperature, b) oxygen mass fraction, and c) density at $t = 10^{-2}$ s for $R_{si}^0 = 10R_d^0$ [$p = 20$ MPa (——), $p = 80$ MPa (- - -)], $R_{si}^0 = 5$ [$p = 20$ MPa (——— ———), $p = 80$ MPa (– ·· –)], and $R_{si}^0 = 2R_d^0$ [$p = 20$ MPa (– · – · –), $p = 80$ MPa (····)]. (Reprinted with permission of Elsevier.)[21]

of the radius of the sphere of influence for $R_2^0 = 2, 5$, and 10 for the same initial conditions as Fig. 5 with additionally $R_C^0 = 2$ cm and $Nu = 10^2$ (a smaller value of Nu amplifies the effect of cluster denseness on the interstitial fluid). Figure 8 depicts the temporal behavior for $R_2^0 = 2$ with otherwise the same initial conditions as those of Fig. 7. Both figures were reproduced from Harstad and Bellan.[21] Comparison of the figures shows that the results for $R_2^0 = 10$ are nearly identical with those obtained for the isolated drop (Fig. 5, $R_2^0 = 20$). Thus, the very dilute fluid drop regime occurs under supercritical conditions at smaller values of R_2^0 than in subcritical situations. The explanation for this difference is in the two processes that govern the motion of the cluster boundary: one is the transfer of heat between the interstitial cluster region and the drops and the other is the heat transfer from the cluster surrounding to the cluster. For the values of Nu used in these calculations, the second process is much slower that the first one. Because under supercritical conditions there is no latent heat, which in the subcritical case is carried away by the evaporated compound to the gas phase, the transfer of species from the drops to the interstitial fluid is accomplished with diminished heat transfer. In the absence of both heat transfer from the cluster surroundings and latent heat release to the interstitial fluid, the cluster boundary motion is minimal. This results in the temperature profiles being very similar even for $R_2^0 = 5$ and 10. It is only for a very small R_2^0 ($\sim$2) that the closer drop proximity induces smaller density gradients, steeper mass fraction gradients, and larger temperatures. The most important difference between results obtained with different values of R_2^0 is that when R_2^0 is smaller ($\sim$2) there is increased accumulation of a non-negligible amount of oxygen in the interstitial region. This is the result of the cluster contraction owing to increased heat transferred to the fluid drops before additional heat from the cluster surroundings can replenish the heat lost to the drops. Furthermore, the effect of drop proximity decreases with increasing pressure (Fig. 7) and the mass fraction and temperature behavior becomes increasingly similar to that under pure diffusion.

IV. Summary and Conclusions

Studies of binary-fuel evaporation under subcritical conditions have been reviewed and compared to similar studies of oxygen fluid drops in H_2 (a binary diffusion situation) under supercritical conditions. The purpose of this comparison is the evaluation of the impact of drop interactions in these two thermodynamic regimes. Both studies were made in absence of turbulence; furthermore, the supercritical study was under quiescent conditions. Thus, these conclusions, while relevant to rocket engine studies, are not directly applicable to combustor design.

It was found that very dense clusters of drops occur at supercritical conditions for normalized (to the drop size) drop interdistances that are approximately a factor of five smaller than at subcritical conditions. Very dense clusters of drops were defined in the subcritical regime as those for which the evaporation rate is about a factor of two smaller than for isolated drops in initially identical conditions, whereas in the supercritical regime they were defined as those for which the temperature, mass fraction, or density profile differs substantially

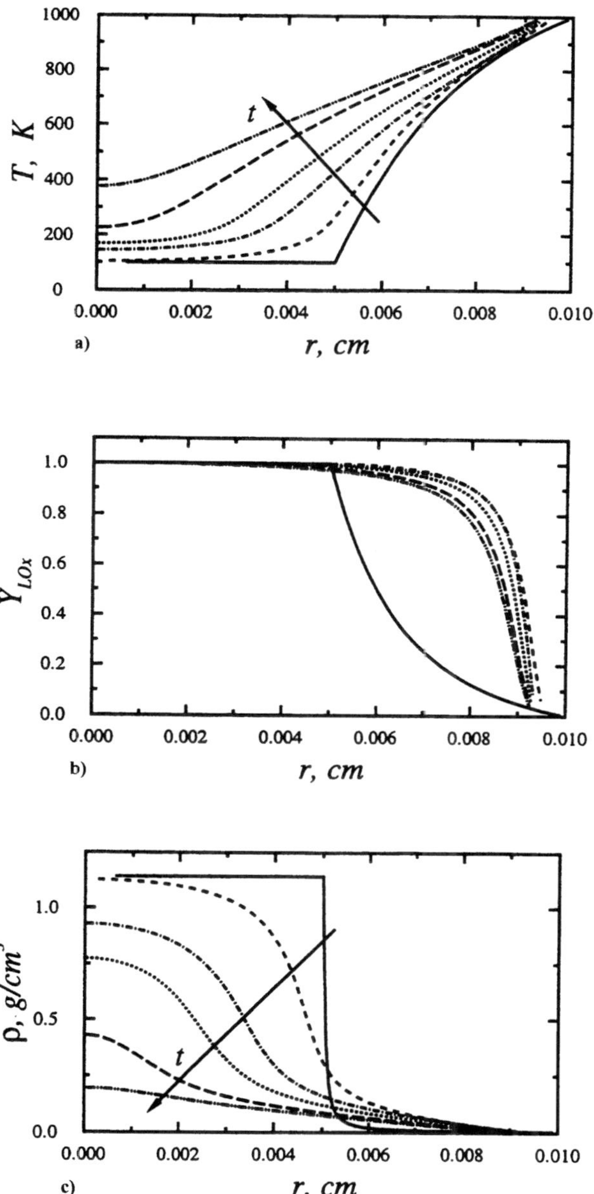

Fig. 8   Spatial variation of the a) temperature, b) oxygen mass fraction, and c) density at various times: 0.0 s (——), 2×10^{-3} s (- - -), 6×10^{-3} s (−·−·−), 8×10^{-3} s (···), 1.0×10^{-2} s (—— ——), 1.08×10^{-2} s (−··−). Same initial conditions as Fig. 7 with $p = 20$ MPa and $R_{si}^0 = 2R_d^0$. (Reprinted with permission of Elsevier.)[21]

from that of an isolated drop in initially identical conditions. For a given normalized drop interdistance, the difference in behavior between the subcritical and supercritical regimes is attributed to the absence of latent heat in the latter regime; under subcritical conditions, the latent heat carried away by the evaporated mass contributes to the energy content of the interstitial fluid in the cluster, thereby providing an additional mechanism (to drop heating) for affecting the other cluster drops. Furthermore, the results from supercritical simulations show that for a given normalized drop interdistance, the effect of drop proximity decreases with increasing pressure and the behavior of species transfer becomes increasingly similar to pure diffusion. Thus, for given initial drop proximity, under supercritical conditions, a progressive increase in pressure results in increased smearing of the gradients, a desirable aspect because it promotes mixing.

An important result from some of the reviewed studies[16,21] is that the traditional Lewis number is no longer appropriate for measuring the relative effect of heat and mass diffusion at supercritical conditions. An effective Lewis number has been defined[20] whose validity extends to a general fluid.

These results have not been validated because of lack of data. Supercritical fluid drop experiments with LOX[22] utilize suspended drops, thus proving that the drop is not uniformly at supercritical conditions (drop suspension implies finite surface tension). This prevents comparisons with the theory reviewed here. Moreover, high-pressure experimental results obtained with LOX[22] in nitrogen and in helium, and with a variety of hydrocarbons (methanol, ethanol, n-hexane, n-heptane, and n-octane) in air[23] yielded data in direct qualitative contradiction with data obtained from other hydrocarbon drop (n-octane, n-heptane, and n-hexane, all in air) experiments in supercritical conditions.[24,25] The first set of experiments showed a continuous decrease in the droplet lifetime with increasing pressure with a leveling off above the critical pressure of the fuel, whereas the second set of experiments identified a minimum in the drop lifetime at the critical pressure of the fuel. This shows that there is no consensus on the real physics of isolated fluid drops, and that clever experiments are needed to guide the further development of accurate models. Just as important, results from numerical simulations can guide in the choice of clever and uncontroversial experiments. For example, it has been found[16] that under supercritical conditions gradients increase with drop size, increasing pressure and decreasing temperature. Because accurate measurements of drop size rely upon the accurate detection of density gradients, this means that less optical fuzziness will be obtained for large drops in high pressure and low temperature surroundings.

Very recently, the model of Harstad and Bellan[16] has been validated[26] with the heptane/nitrogen experimental data of Nomura et al.,[27] but LOX/H_2 data are still lacking, presumably because of the explosive nature of the substance.

Acknowledgments

This research has been conducted at the Jet Propulsion Laboratory under sponsorship from the U.S. Department of Energy, Energy Conversion and Utilization Technologies and Advanced Industrial Concepts with M. Gunn, Jr., and G. Varga, respectively, serving as contracts monitors, under agreement with NASA; and under sponsorship from NASA Marshall Space Flight Center with K. Gross Serving as contract monitor.

References

[1]Harrje, D. T., and Reardon, F. H., *Liquid Propellant Rocket Combustion Instability*, NASA SP-194, 1972.

[2]Hardalupas, Y., Liu, C.-H., Tsai, R.-F., and Whitelaw, J. H., "Coaxial Atomization and Combustion," *Proceedings IUTAM Symposium on Mechanics and Combustion of Droplets and Sprays*, edited by N. A. Chigier and H. H. Chui, Begell House, New York, 1994, pp. 41–73.

[3]Engelbert, C., Hardalupas, Y., and Whitelaw, J. H., "Breakup Phenomena in Coaxial Airblast Atomizers," *Proceedings of the Royal Society of London*, Vol. 451, 1995, pp. 189–229.

[4]Ryan, H. M., Anderson, W. E., Pal, S., and Santoro, R. J., "Atomization Characteristics of Impinging Liquid Jets," *Journal of Propulsion and Power*, Vol. 2, No. 1, 1995, pp. 135–145.

[5]Poulikakos, D., "Determination of Structure, Temperature and Concentration of the Near Injector Region of Impinging Jets Using Holographic Techniques," *Proceedings of the Air Force Office of Scientific Research Contractors Meeting*, Rept. AD-A332874 AFOSR-97-0711TR, CASI 199980017427, 1993, pp. 275–278.

[6]Law, C. K., "Recent Advances in Droplet Vaporization and Combustion," *Progress in Energy and Combustion Science*, Vol. 8, No. 3 1982, pp. 171–201.

[7]Bellan, J., and Cuffel, R., "A Theory of Non-Dilute Spray Evaporation Based Upon Multiple Drop Interaction," *Combustion and Flame*, Vol. 51, No. 1, 1983, pp. 55–67.

[8]Mizutani, Y., Nakabe, K., Fuchihata, M., Akamatsu, F., Zaizen, M., and El-Emam, S. H., "Spark-Ignited Spherical Flames Propagating in a Suspended Droplet Cloud," *Atomization and Sprays*, Vol. 3, 1993, pp. 135–135.

[9]Harstad, K., and Bellan, J., "A Model of the Evaporation of Binary-Fuel Clusters of Drops," *Atomization and Sprays*, Vol. 1, No. 1, 1991, pp. 367–388.

[10]Law, C. K., Prakash, S., and Sirignano, W. A., "Theory of Convective, Transient, Multi-component Droplet Vaporization," *16th Symposium (International) on Combustion*, Combustion Institute, 1977, pp. 605–617.

[11]Sirignano, W. A., and Law, C. K., "Transient Heating and Liquid-Phase Mass Diffusion in Fuel Droplet Vaporization," *Advances in Chemistry Series*, Vol. 166, 1978, pp. 3–26.

[12]Lara-Rubaneja, P., and Sirignano, W. A., "Theory of Transient, Multicomponent Droplet Vaporization in a Convective Field," *18th Symposium (International) on Combustion*, Combustion Institute, 1981, pp. 1365–1374.

[13]Yang, V., Lafon, P., and Hsiao, C. C., "Liquid Propellant Droplet Vaporization and Combustion," *Liquid Rocket Combustion Devices: Aspects of Modeling, Analysis and Design*, edited by V. Yang and M. Habiballah, Progress in Astronautics and Aeronautics, AIAA, Reston, VA, 1997, Chap. 7.

[14]Yang, V., Lin, N., and Shuen, J.-S., "Vaporization of Liquid Oxygen (LOX) Droplets in Supercritical Hydrogen Environments," *Combustion Science and Technology*, Vol. 97, No. 4–6, 1994, pp. 247–270.

[15]Haldenwang, P., Nicoli, C., and Daou, J., "High Pressure Vaporization of LOX Droplet Crossing the Critical Condition," *International Journal of Heat and Mass Transfer*, Vol. 39, No. 16, 1996, pp. 3453–3464.

[16]Harstad, K., and Bellan, J., "Isolated Fluid Oxygen Drop Behavior in Fluid Hydrogen at Rocket Chamber Pressure," *International Journal of Heat Mass Transfer*, Vol. 41, No. 22, 1998, pp. 3537–3550.

[17]Delplanque, J.-P., and Sirignano, W. A., "Numerical Study of the Transient Vaporization of an Oxygen Droplet at Sub- and Super-critical Conditions," *International Journal of Heat and Mass Transfer*, Vol. 36, No. 2, 1993, pp. 303–314.

[18]Keizer, J., *Statistical Thermodynamics of Nonequilibrium Processes*, Springer-Verlag, New York, 1987.

[19]Peacock-Lopez, E., and Woodhouse, L., "Generalized Transport Theory and Its Applications in Binary Mixtures," *Fluctuation Theory of Mixtures*, edited by E. Matteoli and G. Mansoori, Taylor and Francis, New York, 1990, pp. 301–333.

[20]Harstad, K., and Bellan, J., "The Lewis Number Under Supercritical Conditions," *International Journal of Heat and Mass Transfer*, Vol. 42, No. 6, 1998, pp. 961–970.

[21]Harstad, K., and Bellan, J., "Interactions of Fluid Oxygen Drops in Fluid Hydrogen at Rocket Chamber Pressure," *International Journal of Heat and Mass Transfer*, Vol. 41, No. 22, 1998, pp. 3551–3558.

[22]Chesneau, X., Chauveau, C., and Gökalp, I., "Experiments of High Pressure Vaporization of Liquid Oxygen Droplets," AIAA Paper 94-0688, 1994.

[23]Vieille, B., Chauveau, C., Chesneau, X., Odeïde, A., and Gökalp, I., "High Pressure Droplet Burning Experiments in Microgravity," *26th Symposium (International) on Combustion*, Combustion Institute, 1996, pp. 1259–1265.

[24]Sato, J., Tsue, M., Niwa, M., and Kono, M., "Effects of Natural Convection on High-Pressure Droplet Combustion," *Combustion and Flame*, Vol. 82, No. 2, 1990, pp. 142–150.

[25]Sato, J., "Studies on Droplet Evaporation and Combustion in High Pressure," AIAA Paper 93-0813, 1993.

[26]Harstad, K., and Bellan, J., "An All-Pressure Fluid Drop Model Applied to a Binary Mixture: Heptane in Nitrogen," *International Journal of Multiphase Flow*, Vol. 26, No. 10, 2000, pp. 1675–1706.

[27]Nomura, H., Ujiie, Y., Rath, H. J., Sato, J., and Kono, M., "Experimental Study on High Pressure Droplet Evaporation Using Microgravity Conditions," *Proceedings of the Combustion Institute*, Vol. 26, 1996, pp. 1267–1273.

Chapter 9

Fundamentals of Supercritical Mixing and Combustion of Cryogenic Propellants

Wolfgang O. H. Mayer*
DLR, German Aerospace Research Center, Hardthausen, Germany

and

Joshua J. Smith[†]
University of Adelaide, Adelaide, Australia

Nomenclature

d_i = coaxial injector LOX tube inner diameter
d_o = coaxial injector LOX tube outer diameter
Da_C = chemical Damköhler number
Da_V = vaporization Damköhler number
D = coaxial injector total diameter
J = propellant momentum flux ratio
Ka = Karlovitz number
P_{ch} = combustion chamber pressure
P_{crit} = critical pressure
t_C = chemical reaction time
t_{kol} = Kolmogorov turbulent time scale
t_M = propellant mixing time
t_V = propellant vaporization time
T = temperature
T_{crit} = critical temperature
u_{LOX} = liquid oxygen injection velocity
u_{GH2} = gaseous hydrogen injection velocity
y = injector coaxial slit (H_2) thickness

---

*Head, Propellant Injection Research, Space Propulsion Institute, Lampoldshausen. Senior Member AIAA.

[†]Guest Scientist, DLR Lampoldshausen, and Ph.D. Student, Department of Mechanical Engineering. Member AIAA.

I. Introduction

CRYOGENIC rocket engines are used in a number of operational launch vehicles worldwide. The thrust chamber is the core of any liquid rocket engine and consists of injector, combustion chamber, and nozzle. High performance and reusability are the most challenging requirements for future developments requiring large technical advances for most of the engine components.[1]

Research during the last 10 years on liquid oxygen/gaseous hydrogen (LOX/ GH_2) propellant injection has provided an improved understanding of the processes in cryogenic rocket engine combustion chambers. Propellant injection and related processes have been the object of significant investigation.[2-8] Various institutions worldwide have devoted considerable resources to researching experimentally and theoretically the various processes and phenomena surrounding subcritical and supercritical injection and combustion processes. Despite this, many basic phenomena, such as turbulent mixing of propellants, cannot be predicted quantitatively.

Until recently, a majority of propellant injection and combustion research was performed at subcritical chamber pressures, that is, at pressures below the thermodynamic critical point of the injected medium. As a substance exceeds its critical point, it undergoes drastic changes physically. Large disparities in density, thermal conductivity, and mass diffusivity occur near the critical point. At the critical point, constant pressure specific heat increases rapidly and surface tension vanishes. The distinction between liquid and gas no longer exists and the substance must be described as a fluid.

As in diesel engines, the steady-state injection and combustion processes in cryogenic rocket engines occur at conditions above the critical pressure of the injectant. Upon injection the propellant temperatures are often below the critical mixing temperature, and there is a transcritical mixing behavior of the propellants.

Surface tension between the propellant components is a precondition for phase equilibrium. As shown in previous investigations, surface tension plays a dominant role in atomization.[9] A remarkable change of the atomization mechanisms from wind-induced capillary instability under subcritical conditions to turbulent mixing with expansion under supercritical conditions has been observed[10] and is dependent on local conditions at the jet surface. The phenomena are investigated and discussed in this chapter. Within this atomization regime, liquid–gas-like breakup (see Ref. 9) and gas–gas-like mixing behavior[11] can be observed. This process is visualized very well from injection tests using simulant fluids liquid nitrogen (LN_2) and liquid nitrogen/helium (LN_2/He). These cold flow experiments are discussed extensively in the first part of this chapter.

The second section of this chapter will describe the principal features of trans- and supercritical jet breakup in comparison with classic turbulent jet breakup. The discussion starts with a description of the phase equilibrium of single- and binary-component systems. The breakup phenomena are investigated using shadow photography.

The third and fourth sections of this work present an overview of recent findings in transcritical and supercritical combustion studies. The predominant area of interest is the application of findings pertaining to the LOX/GH_2 bipropellant system. This section will outline an assortment of recent findings with discussion.

Table 1 lists the critical properties of propellants and simulants outlined within this chapter.

Table 1   Critical properties of common simulants and propellants

Critical properties	N_2	He	O_2	H_2
Critical pressure P_{crit}	3.39 MPa	0.23 MPa	5.05 MPa	1.3 MPa
Critical temperature T_{crit}	126 K	5.2 K	154 K	33.2 K

II.   Cold-Flow Research

Cold-flow experimentation and research are very important in understanding the fundamental phenomena surrounding typical injection processes such as mixing, atomization, and vaporization at ambient or elevated pressure conditions. To characterize these intricate processes, a vast amount of experimental work has been performed in the past.[3,9,12]

Although lacking the realism of combustion, cold-flow studies enable one to visualize the atomization and breakup processes without the complexity of mixing combined with combustion and without the optical distortions introduced by combustion. Cold-flow studies have been performed at full scale, where injector geometries and flow rates are similar to those experienced in actual rocket engine applications.[10]

A.   Single-Component Systems

The terms sub- and supercritical conditions are generally made in reference to the quiescent chamber pressure with respect to the critical point of the injected propellant(s). Care must be taken when considering multicomponent systems, as properties such as the critical point can shift. For multicomponent mixtures, the critical mixing temperature and pressure determine the critical point. Critical mixture lines represent critical regions for mixtures; examples of these lines for common binary systems are displayed later. This critical mixture region may be dependent on relative concentrations of the injected species and is often in excess of the critical point of the individually injected components.

Woodward and Talley[13] have shown this with LN_2 jets injected into supercritical N_2 and He mixtures. By increasing the He concentration in the chamber, surface-tension-dominated jet breakup (subcritical behavior) could be observed at double the critical pressure of N_2. This fact provides an explanation for earlier findings by Newman and Brzustowski,[14] who claimed no evidence of elevated pressure effects on atomization and vaporization with CO_2 jets into an ambient CO_2/N_2 mixture. However, it is not certain that the critical mixing *line* of the CO_2/N_2 ambient combination was ever actually exceeded.

Studying pure component systems is important to provide a fundamental understanding of sub-, trans-, and supercritical injection regimes. With a defined point at which there is a transition from subcritical to supercritical behavior, one can clearly characterize and identify jet behavior below, above, and near the critical region. Nonreacting single-component investigations with liquid nitrogen have been performed to understand the effects of density, viscosity, and surface tension on atomization at high pressure. Nitrogen has been used in the study presented here to obtain a better understanding of pure substance behavior at elevated pressures.

For a single-component system, a phase equilibrium between liquid and vapor phase exists when the system pressure is equal to the component vapor pressure. At the critical point, which is characterized by the critical temperature and pressure, the density of the gas and liquid phase become equal. The critical pressure of nitrogen is 3.39 MPa. A phase diagram and properties of nitrogen are shown in Figs. 1 and 2, respectively.

One very important characteristic of exceeding the critical point is that a phase equilibrium no longer exists and the surface tension tends toward zero (see Fig. 3). Figure 2 indicates that small increases in temperature near the critical point cause large expansions of the fluid. Thus, at supercritical conditions the breakup process is influenced not only by lack of surface tension but also through enhanced heat transfer to the jet. Other properties, such as density (Fig. 4), also vary remarkably around the critical point, although, they remain between the limits of pure gas and liquid properties, except for specific heat (Fig. 2).

B. Cold-Flow Investigation Experimental Setup

Figure 5 shows the pressurized chamber and the coaxial injection element used in the pure N_2 cold-flow study. Four windows offer optical access to the chamber, which can be pressurized up to 6 MPa and is equipped with an electric heater to maintain constant ambient temperature. The temperatures of the injected fluids are measured in the injection manifold close to the injector. Temperature measurements were recorded at the injector exit directly before tests to account for heat transfer from the injector tube to the injected liquid.[15]

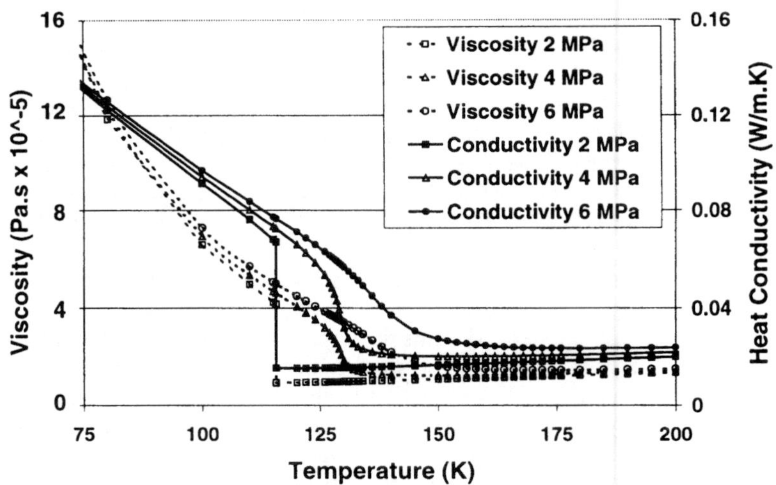

Fig. 1 Phase diagram of pure nitrogen at 2.0, 4.0, and 6.0 MPa.

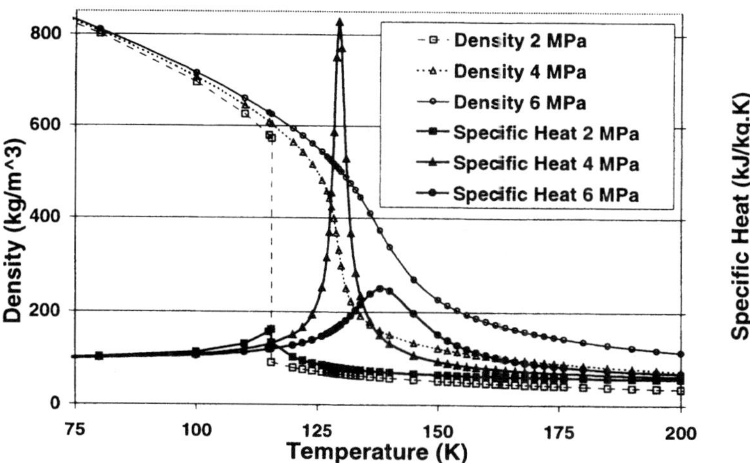

Fig. 2 Viscosity and thermal conductivity of pure nitrogen at 2.0, 4.0, and 6.0 MPa.

This cold-flow study enabled the visualization of a liquid nitrogen jet injected into gaseous nitrogen. Shadowgraphy was utilized as the diagnostic method. A simple tube injector was used with and without coaxial gas and with $d_i = 1.9$ mm. The tube diameter-to-length ratio was approximately 12. Various chamber pressures ranging from 1.0 to 6.0 MPa were investigated. For more details of the experimental setup see Fig. 5 and Refs. 10 and 15.

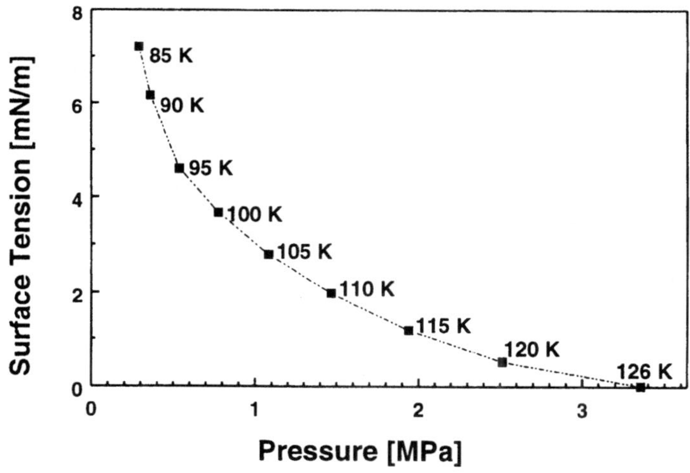

Fig. 3 Surface tension of nitrogen as a function of pressure and temperature.

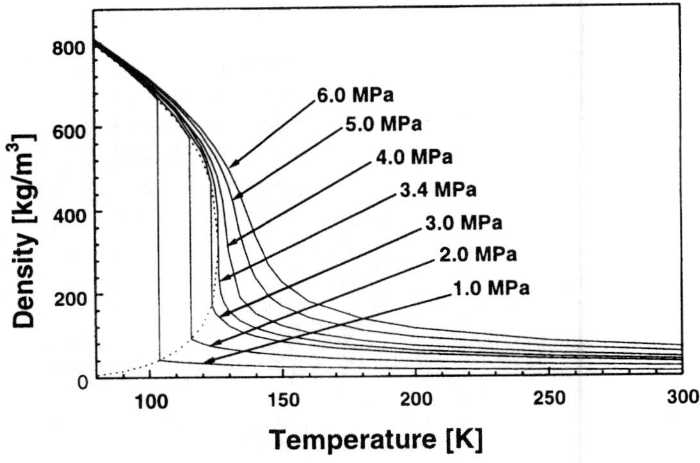

Fig. 4   Density dependence on temperature for nitrogen.

C.   Single-Component Behavior

Capillary, aerodynamic, and shear forces, as well as turbulence and expansion, all contribute to atomization, depending on injection conditions. During injection, the system is not in thermodynamic equilibrium. The property values computed for experimental analysis can be used only for forecasting trends or to accurately describe conditions downstream. Local injectant conditions can vary significantly at the point of injection.

The images in Figs. 6 and 7 show the observed atomization phenomena at different chamber pressures for a single, turbulent LN_2 jet with an injection temperature of 100 K into gaseous nitrogen at 300 K. Figure 6 shows the near-injector region with constant injection velocity. To demonstrate the breakup phenomena at the jet surface, a magnification of jet segments is presented in Fig. 7, showing also the effect of increased jet velocity when approaching and exceeding the critical pressure.

Images A, B, and C in Fig. 6 have been taken at subcritical chamber pressure, whereas images D, E, and F are at supercritical pressures. At subcritical conditions (1.0 and 2.0 MPa), the flow is controlled by aerodynamic and capillary forces, and the jet shows a wavy surface and droplet detachment. In this regime, higher pressure, higher relative velocities, and lower surface tension all cause smaller jet surface structures and droplets.

At near-critical conditions (3.0 MPa), the capillary forces are reduced considerably, but at low injection velocity ($v = 5$ m/s), droplet detachment still occurs (Fig. 7). At the higher injection velocity ($v = 10$ m/s), shear forces exceed the capillary forces and the atomization phenomenology appears to be a gas-like mixing process. This is also the case at supercritical conditions (4.0 MPa and more), where shear forces and expansion of the jet dominate. Here, heat transfer to the jet plays an important role. An increase of the observed

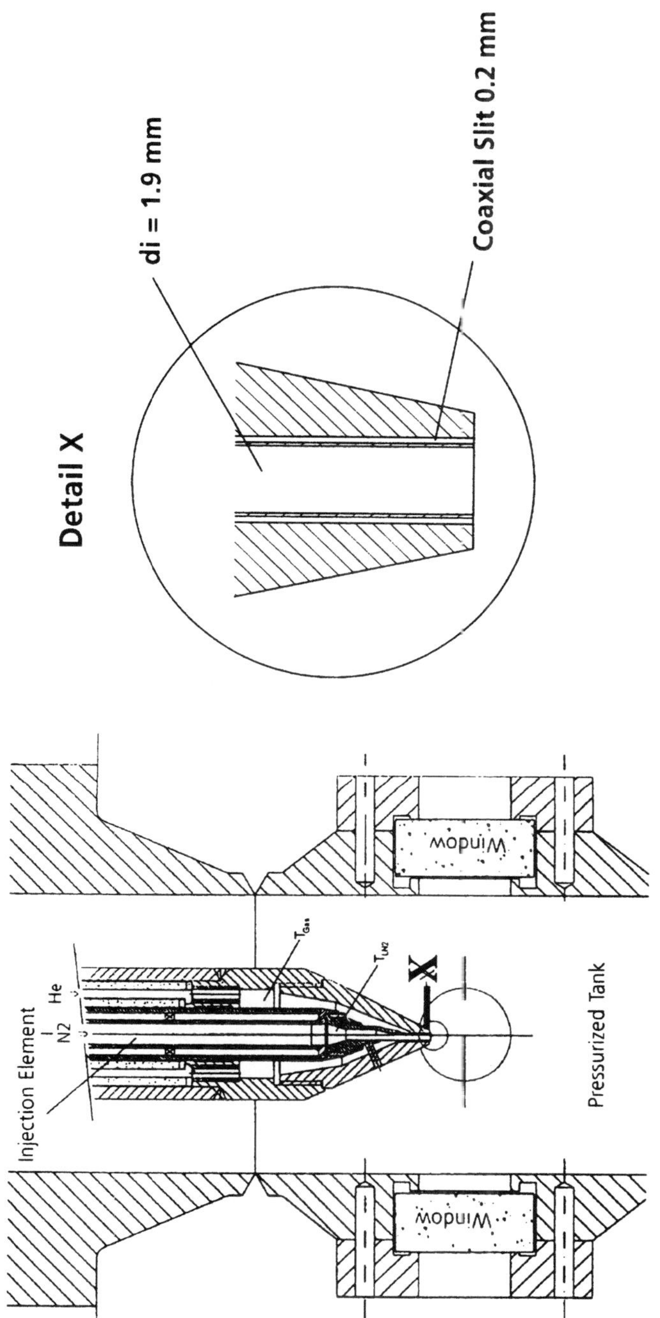

Fig. 5 Four-window cold-flow test chamber with coaxial injector.

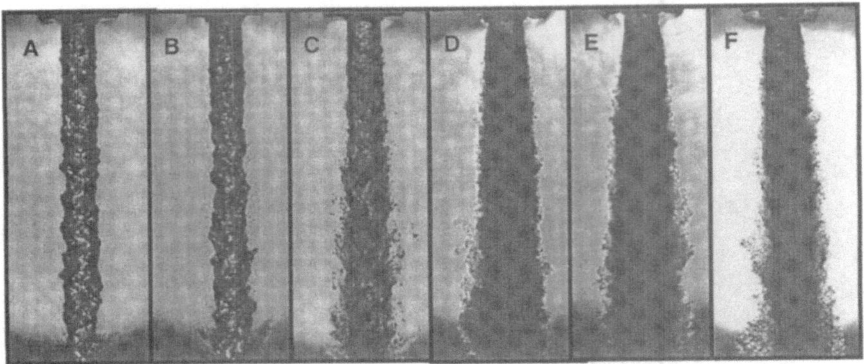

Fig. 6 LN$_2$ jets in GN$_2$: $v_{LN2} = 1$ m/s; $d_{LN2} = 1.9$ mm; $T_{LN2} = 100$ K; $Re_L \cong$ 18,000; $T_{chamber} = 300$ K. Chamber pressure from left to right: A, 1.0 MPa; B, 2.0 MPa; C, 3.0 MPa; D, 4.0 MPa; E, 5.0 MPa; and F, 6.0 MPa.

jet contour length scales results from the change of the atomization mechanism and will be discussed further later.

D.   Binary-Component Systems

For a multicomponent system a phase equilibrium can exist between a liquid and a gaseous phase[16,17] when each species is in mechanical, thermal, and chemical equilibrium. The phase equilibrium diagram for the two component system H$_2$/O$_2$ is shown in Fig. 8. For the reduced pressures $p_r = p/p_{crit,O_2}$ ($p_{crit,O_2} = 5.04$ MPa), the boundaries of the two-phase region are indicated. At a given pressure,

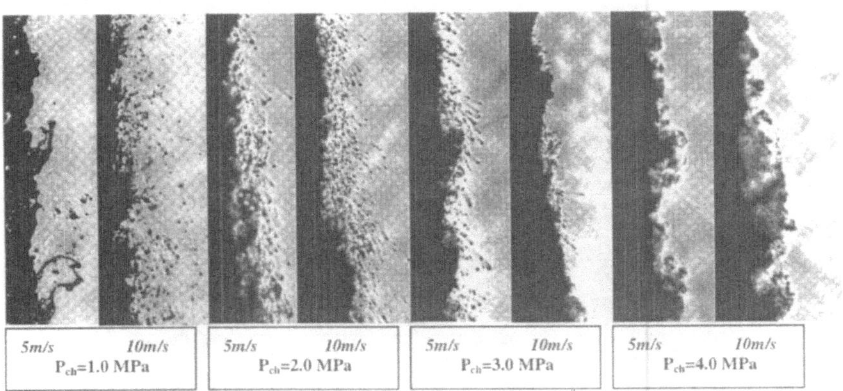

Fig. 7 Magnified shadowgraphs of LN$_2$ jet segment surfaces at various injection conditions; image size 3.1 × 7.7 mm, position 8 mm downstream from injector; $T_{LN2} = 105$ K, $T_{ch} = 300$ K, $D = 1.9$ mm, $P_{ch} = 1.0, 2.0, 3.0, 4.0$ MPa from left to right; injection velocities 5.0 and 10.0 m/s.

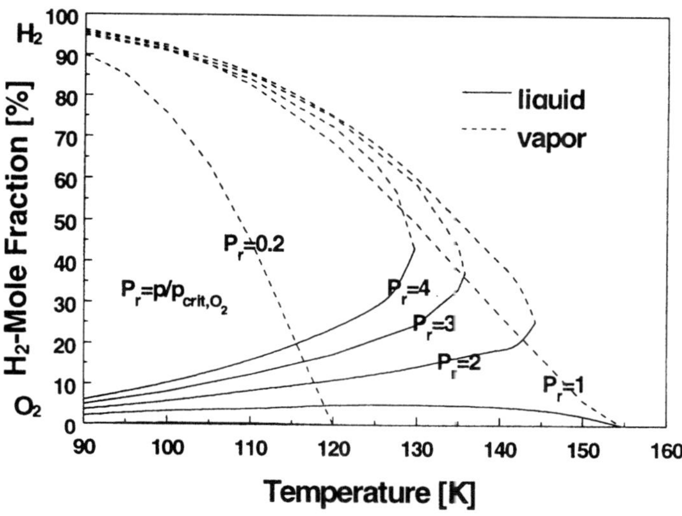

Fig. 8   Phase equilibrium of the oxygen–hydrogen (O_2/H_2) binary system.

the solubility of hydrogen in the liquid oxygen increases with increasing temperature and the amount of hydrogen in the gaseous phase decreases. This is the transcritical region. The behavior of the N_2/He system shows the same tendencies as the H_2/O_2 system but has differences in the absolute property values (e.g., critical pressures and mixing temperatures, see Fig. 9). A binary mixture at a given system pressure has a maximum boundary temperature for the existence of a

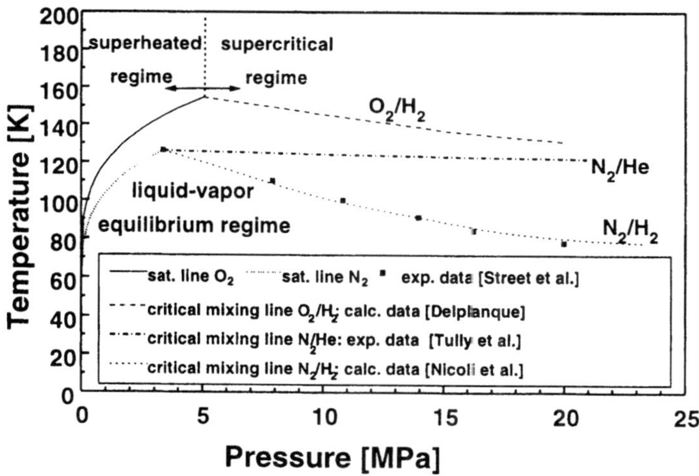

Fig. 9   Critical mixing lines of common binary systems.[17-20]

phase equilibrium. This temperature is called the critical mixing temperature. Above the critical mixing temperature, the species fractions in the gas and liquid phase are identical: the phase boundary disappears and supercritical conditions exist.

Figure 9 shows the critical mixing temperatures as a function of the system pressure for different binary systems. Below the critical mixing temperature, a phase equilibrium with a liquid and a gaseous phase exists. Above this line supercritical conditions prevail. The general trend is that the critical mixing temperatures decrease with increasing pressure.

Using the Macleod-Sugden correlation[20,21] the surface tension of the H_2/O_2 binary system is computed as shown in Fig. 10. The solid line is the saturation line for O_2. The dashed lines are constant pressures of the H_2/O_2 system. It is interesting to note that, even above the critical pressure, surface tension is present as long as the critical mixing temperature is not exceeded. In typical injection applications (rocket engines), this temperature margin can be very small.

This transcritical phenomenon is confirmed by results from liquid nitrogen (LN_2) and gaseous helium (GHe) cold-flow simulations (see Fig. 11). The cold-flow simulation tests use a coaxial injector with an inner diameter $d_i = 1.9$ mm, an outer diameter $d_o = 2.8$ mm, and a coaxial slit $y = 0.2$ mm (see Fig. 5). The change in the atomization mechanism at reduced surface tension is evident: spray formation at low pressure and dense–light fluid turbulent mixing under supercritical pressure conditions. The critical mixing temperature of the N_2/He system is 125.7 K. In the mixing layer between LN_2 and He, transcritical zones may exist. The visible surface of the LN_2 jet is assumed to be the layer that reaches the critical mixing temperature. The effect of surface tension compared to shear forces appears to be negligible in the high-pressure case.

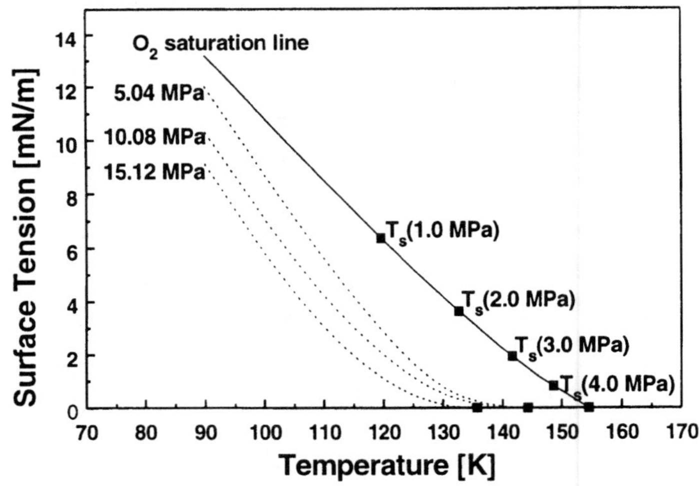

Fig. 10 Surface tension of the oxygen–hydrogen binary system.

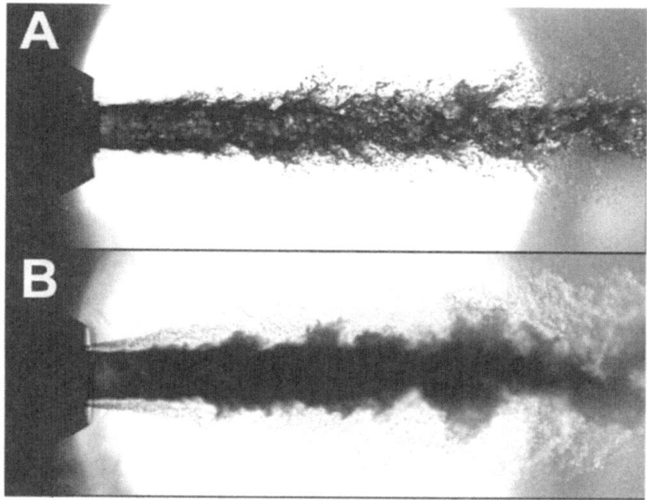

Fig. 11 Binary liquid nitrogen/gaseous helium (LN$_2$/GHe) system at A, 1.0 MPa; B, 6.0 MPa; d_{LN2} = 1.9 mm; v_{LN2} = 5 m/s; v_{He} = 100 m/s; T_{LN2} = 97 K; T_{He} = 280 K.

III. Combusting Flow Research

Combusting flow tests with liquid oxygen and hydrogen provide actual rocket engine operating conditions. Flow visualizations of injection and combustion processes under these extreme conditions demonstrate real condition flow phenomena. Mayer and Tamura[2] performed subcritical and supercritical tests with liquid oxygen and gaseous hydrogen. The primary objective of the work was to gain a realistic insight into the injection and combustion processes in LOX/GH$_2$ rocket engines by observing directly the propellant behavior at various pressures.

Furthermore, combusting flow visualization enables identification and characterization of other high-pressure combustion mechanisms that occur within rocket engine thrust chambers. Flow behavior prior to ignition and flame interaction upon ignition can be studied with high-speed visualization and diagnostic techniques. These important components of rocket engine processes are described from an experimental viewpoint in the following section.

A. Experimental Procedure

An experimental rocket motor with four flat windows in opposite walls was developed and is sketched in Fig. 12. It consists of an injector head with a single shear coaxial element, a cylindrical, water-cooled combustion chamber with a length and diameter of 430 and 50 mm, respectively, and a variable (exchangeable) nozzle. The window section is cooled by a layer of gaseous hydrogen. The cooling flow is injected parallel to the windows in the direction of the nozzle. Metal dummy windows equipped with thermocouples were used to develop the startup and shutdown sequences and a suitable window cooling technique.

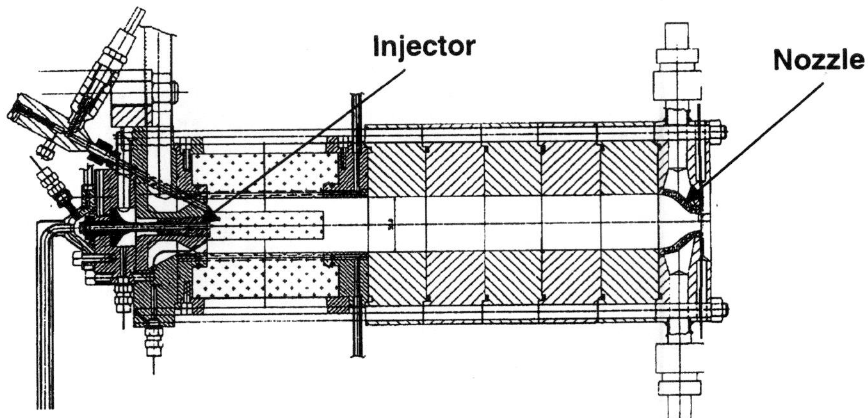

Fig. 12   Hot-fire H_2/O_2 windowed combustion chamber.

Several different visualization techniques were utilized for this study. At low chamber pressures (2.0 MPa or less), flow visualization studies were performed with a standard camera and flashlight in a shadowgraph and schlieren setup. Attempts were made to observe flow processes under severe conditions: with flow velocities up to 300 m/s, high chamber pressure with a range of fluid temperatures ranging from 100 to 3600 K. As expected, the applicability of flow visualization or measurement techniques using standard optical or laser-optical methods is limited under such extreme conditions. Thus, it was necessary to develop a unique diagnostic and optical setup to adequately visualize the injection processes. At high pressures, the flame radiation exceeds the diagnostics flashlight intensity so that a means of reducing the flame light emission is required. A detailed description of the experimental chamber and apparatus can be found in Mayer and Tamura.[2] Injection conditions for combusting test cases are outlined in Table 2.

Table 2   Injection conditions for LOX/H_2 combustion tests

Injection conditions	Case 1	Case 2	Case 3	Case 4
Core fluid	LOX	LOX	LOX	LOX
Core fluid injection temperature	100 K	100 K	100 K	100 K
Core fluid injection velocity	10 m/s	30 m/s	30 m/s	30 m/s
Coaxial fluid	GH_2	GH_2	GH_2	GH_2
Coaxial fluid injection temperature	300 K	300 K	300 K	300 K
Coaxial fluid injection velocity	300 m/s	300 m/s	300 m/s	300 m/s
Chamber pressure	1.5 MPa	4.5 MPa	6.0 MPa	10.0 MPa
$P_{ch}/P_{crit}O_{2t}$	0.30	0.89	1.19	1.98
$P_{ch}/P_{crit}H_{2t}$	1.15	3.46	4.62	7.69

The test cases were performed with a non-recessed injector at 1.5, 4.5, 6.0, and 10.0 MPa chamber pressure P_{ch}. The corresponding oxygen and hydrogen velocities (u_{LOX}, u_{GH2}) were 30 and 300 m/s, respectively, except for the 1.5-MPa case (test case 1), where LOX velocity was reduced to 10 m/s. This velocity reduction was necessary because of the reduced mixing efficiency at the 1.5-MPa chamber pressure, which results in the LOX jet and ligaments leaving the chamber unburned. Unless otherwise stated, LOX temperature T_{LOX} was 100 K, and hydrogen temperature T_{GH2} was 300 K. Injector dimensions for the referenced cases (test cases 1–4) were as follows: $d_i = 1$ mm, $d_o = 1.6$ mm, and $D = 3.9$ mm (see Fig. 13). Nozzle throat diameter was 5.8 mm. A more detailed description of the test conditions can be found in Ref. 2.

B. Subcritical Combustion

Flame visualization studies showed that the flame in the chamber is generally attached to the LOX injector post and envelops the LOX jet core. The following visualizations exhibit the flowfield with the flame light suppressed.

Figure 14 shows photographs of the 1.5-MPa combusting condition (test case 1) taken with a standard shadowgraph setup. At chamber pressures significantly lower than the critical pressure of oxygen ($\ll 5.09$ MPa), the LOX jet is atomized, forming a spray comparable to the flow pattern visualized from cold-flow studies. Ligaments are detached from the LOX jet surface, which form round droplets and finally evaporate. The 1.5-MPa chamber pressure test case exhibited secondary breakup (droplet vibrational- and bag-type break up). The droplet number density is rather low compared with the cold-flow case with similar injection conditions. This fact can be attributed to the rapid vaporization of the droplets in a burning spray. Most of the droplets are nonspherical. The very smooth surface of the LOX jet close to the injector should be noted.

Figure 15 compares the subcritical cold-flow and combusting injection conditions and demonstrates the principal effect of the flame on the LOX jet close

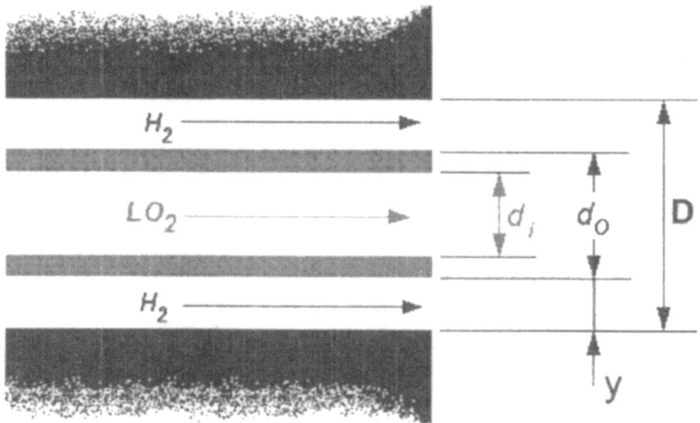

Fig. 13   Typical shear coaxial injector configuration with dimensions.

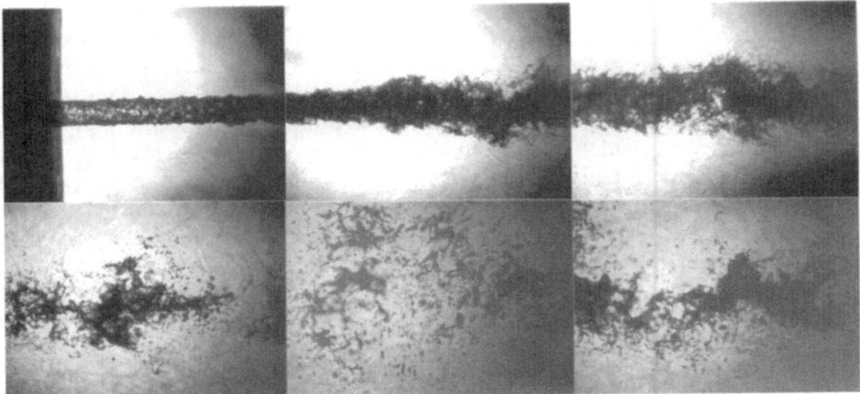

Fig. 14   Subcritical hot-fire condition at test case 1, $d_i = 1\,\text{mm}$ (Mayer and Tamura[2]).

to the injector. This cold-flow situation is similar to the conditions prior to ignition in a rocket engine combustor. Further details on ignition transients are discussed later.

In both cases the chamber pressure is 1.5 MPa (test case 1). In the cold-flow case, fine oxidizer threads and droplets are visible. In the combusting case at low chamber pressures, the fine surface structures rapidly vaporize and droplets are never formed. No droplets or detached ligaments can be observed in the vicinity of the injector. The remarkable smoothness of the LOX jet under combusting conditions is in part due to the rapid vaporization of the oxygen surface waves. The density of the hot reaction zone (3500 K) is very low, especially at low chamber pressures, causing a considerable reduction of aerodynamic interaction

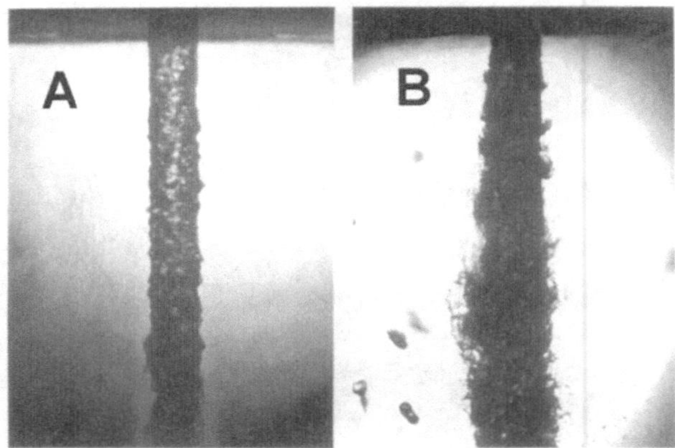

Fig. 15   LOX near-injector region, test case 1; image A: burning condition; image B, cold flow prior to ignition (Mayer and Tamura[2]).

between the LOX jet and the hydrogen. It is noteworthy that the residual surface tension forms a round liquid jet. This effect cannot be observed at chamber pressures equal to or higher than 4.5 MPa.

The combustion product water may also influence the phase equilibrium, that is, surface tension, and thereby the atomization and mixing process. It can be concluded that, with high heating rates, diffusion processes predominate prior to atomization. This conclusion is valid only for the low-pressure case chosen for this study.

IV. Supercritical Combustion

Supercritical combustion is combustion at conditions exceeding the critical properties of the injected propellants. Often the injected propellants (LOX/H_2) are injected at supercritical pressures but below the critical mixing temperature. This situation has been labeled a transcritical injection regime. Shortly after injection, however, heat transfer via convection from the combustion process reaches the injected propellant jet surface and increases the local temperatures, and the jet enters the supercritical regime.

A. Principal Flow Characteristics

The astonishing difference in jet behavior at transcritical and supercritical pressures becomes evident with visualizations such as that in Fig. 16. As the supercritical pressure is approached and exceeded, droplets no longer exist. From the LOX jet core, stringy or thread-like structures develop and grow, which do not detach, but rapidly dissolve. At elevated chamber pressures the radiance of the flame increases dramatically.

Unlike the subcritical combusting condition (Fig. 14), aerodynamic effects are diminishing and the flow is dominated by turbulent gas-like mixing. At supercritical conditions, there exists no definitive boundary separating LOX from

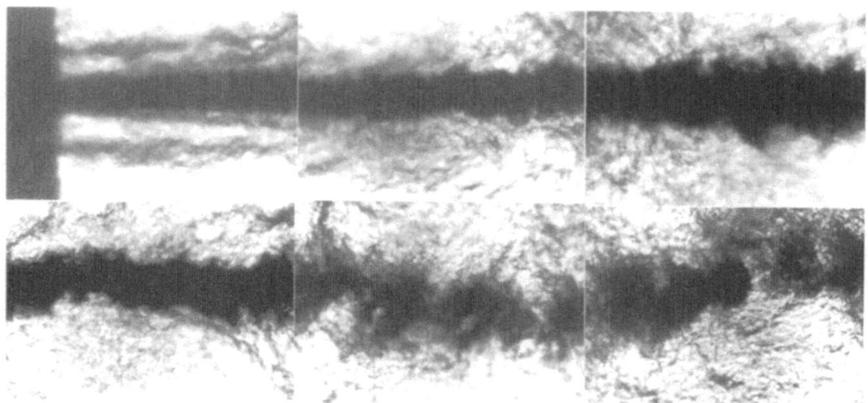

Fig. 16 Supercritical hot-fire conditions at test case 3, $d_i = 1$ mm (Mayer and Tamura[2]).

the coaxial H_2, and the binary propellant system has entered a transcritical regime in which its behavior is best described as a fluid.

Various other phenomena have been observed with visualization of the combusting flowfield. An example is the flame and flow image pair of Fig. 17. The top image is an average flame radiation photograph with an exposure duration of 16.7 ms. A bright region very near to the injector is an intense mixing and combusting region due to a recirculation of propellants near the LOX post. This fact and other observations are outlined and discussed in following sections of the chapter.

B. Oxygen Jet Breakup

Further downstream from the injector, LOX jet oscillations have been observed and reported at steady-state operation[2] and have been described as a helical instability.[22] This area may be characterized as the heavy mixing zone, which includes the maximum amplitudes of oxygen jet oscillations, widening of the oxygen jet and flame, and emission of the strongest combustion radiation. The disintegrating, helically oscillating core observation is evident from Fig. 14 and can also been found in Ref. 2, where a high-speed photo system was utilized.

Several tens of jet diameters downstream the LOX core disintegrates into large LOX lumps, which rapidly dissolve and diminish. The jet breakup length (average length of connected LOX) has been observed to decrease with increasing supercritical chamber pressure due to decreased chemical reaction time scales. This trend is consistent with various core length correlation theories developed for subcritical, noncombusting conditions. However, the correlations developed to date can be applied only within the narrow range of operating conditions for which they were developed.[23]

In the 4.5-MPa (test case 2) and 6.0-MPa cases (test case 3), mixing and combustion were incomplete at the downstream extremity of the visualized region. The 10.0-MPa case (test case 4), however, exhibited a depleted oxygen jet within the visualized region (position $x = 60-72$ mm). Several tests confirmed this with not a single oxygen lump observed beyond 70 mm downstream of the injector ($x = 70$ mm). The end of the oxygen jet coincides with the vanishing flame, i.e., ultraviolet (UV) radiation (see Fig. 17).

Figure 18 shows magnified photographs of the flowfield at a location in the chamber from 60 mm downstream of the injector faceplate for the cases with 1.5, 6.0, and 10 MPa chamber pressure, respectively. At this position the oxygen jet is disintegrating. Figure 18 demonstrates once again the remarkable difference between sub- and supercritical pressure conditions. At low chamber pressures, oxygen droplets, ligaments, and threads disperse and break up as observed in cold-flow atomization studies. Supercritical combusting condition exhibits no droplet or ligament formation whatsoever.

C. Flame Structure and Radiation

The flame of LOX/H_2 combustion radiates in a continuous range between infrared (IR) and UV. The dependency of line intensities on mixture ratio (O/F) is small, which indicates that combustion gases undergo various degrees of reaction with a wide distribution in gas temperature and composition. Radiation

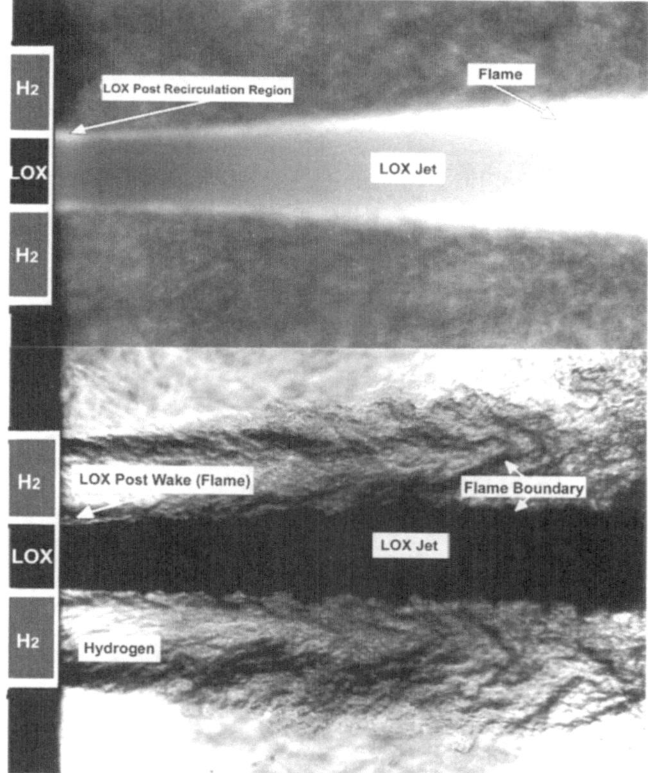

Fig. 17 Long-exposure flame radiation image (top) and flow visualization (bottom) of the near-injector region for test case 2 (Mayer and Tamura[2]).

measurements at chamber pressures between 1.0 and 10.0 MPa demonstrated that the flame spectrum (wavelength band) is similar, but the radiation intensity R increases dramatically with pressure ($R \propto P_{ch}^{2.0,...3.2}$). Another important observation is that radiation intensity increases with downstream distance. Detailed radiation measurements can be found in Mayer and Tamura.[2] Figure 19 shows

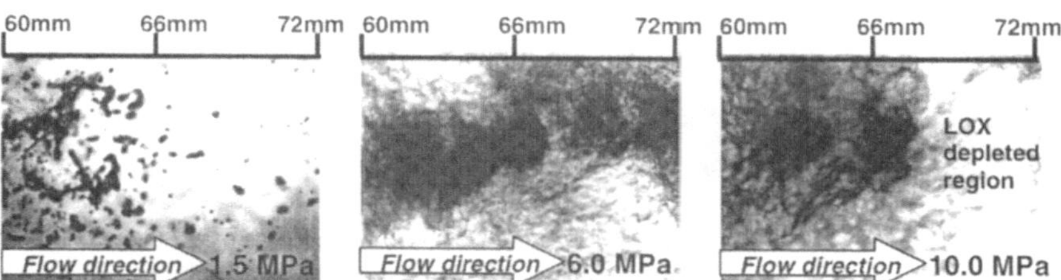

Fig. 18 Visualization of LOX jet disintegration dependence on chamber pressure.

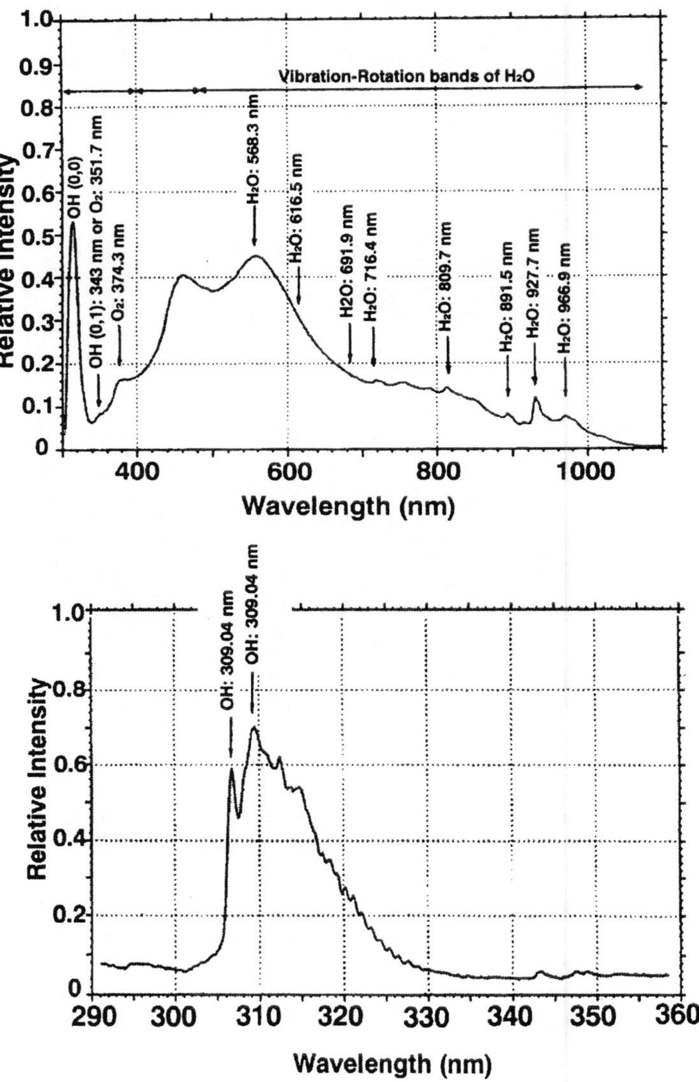

Fig. 19 LOX/H₂ flame spectra at 6.0 MPa measured 70 mm downstream (case 3).

the radiation spectrum of a LOX/H$_2$ flame at 6.0-MPa chamber pressure with a spectrometer resolution of 0.9 nm. The hydroxyl radical (OH) peaks around 306.76 and 309.04 nm can be clearly identified. The emission bands are well separated from those of O$_2$ and H$_2$O at 6.0 MPa, which simplifies the optical diagnostic process for visualizing OH and H$_2$O emission. The peak identification is consistent with the flame spectra measurements published by Gaydon.[22]

OH and H_2O are biproducts of the O_2/H_2 combustion process. The OH and H_2O emission provides an indication as to the relative intensity of the combustion processes and locations in the flame. OH is formed in the reaction zone and more specifically within the shear layer between the rate-controlling oxygen and the coannular hydrogen. This can be observed in the near-injector OH emission studies. The study of OH and H_2O emission provided single-shot images as shown in Fig. 20 with flow from left to right. Line-of-sight emission images require a mathematical deconvolution (Abel's transform) to indicate the true location of the reaction zones.

Studies were performed utilizing radical emission imaging in an attempt to characterize the near-injector region.[7,24] The primary initiative for this work is to gain a better understanding of the flame-holding effect on the LOX post of the injector. A more detailed description of this phenomenon is outlined in the upcoming section. Considerable work has been performed at subcritical pressures regarding single-injector element LOX/H_2 combustion, with Ref. 7 providing a good overview of the studies.

Ivancic and Mayer[25] have performed detailed high-pressure studies of the near injector field in an attempt to characterize the mixing and combustion processes. OH emission imaging has been performed within a 30 mm axial distance from the injector face. Figure 21 shows a deconvoluted OH emission image (lower half) overlaid upon a shadowgraph image at a chamber pressure of 6.3 MPa. It can be seen that the flow and flame behavior correspond to a situation in which the majority of combustion occurs in close proximity to the shear layer.

The lines on Fig. 21 represent OH emission boundaries and a line of maximum measured OH intensity. The outer boundaries correspond to regions of measured OH representing 3% of the maximum OH intensity recorded. The maximum intensity was found to be very close to the LOX core, which implies that the flame is also very near to the LOX core. This fact has also been mentioned by Candel et al.,[7] who used laser-induced fluorescence (LIF) diagnostics and elastic light scattering (ELS) measurements performed at pressures up to 1 MPa.

The propellant mixing, vaporization, and combustion processes can be characterized by the Damköhler numbers. The chemical reaction Damköhler number represents the mixing time relative to the chemical reaction time [Eq. (1)].

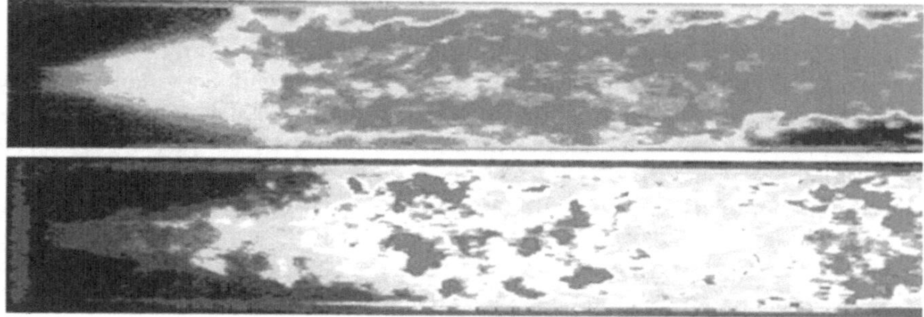

Fig. 20   OH radical emission (top) and H_2O emission (bottom) single-shot images.

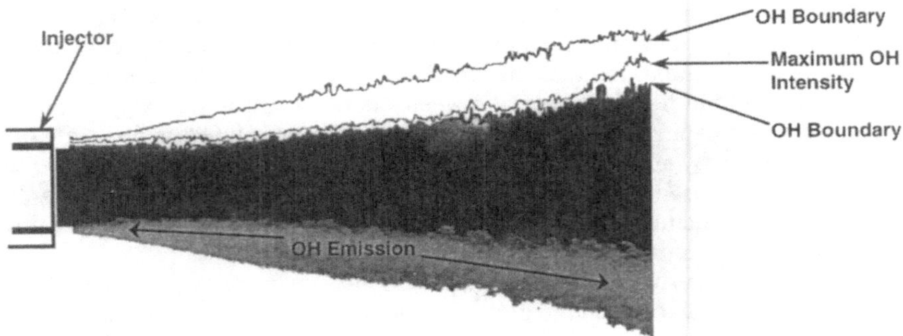

Fig. 21 Deconvoluted OH emission image overlaid upon shadowgraph with lines of OH intensity.

The vaporization Damköhler number represents the mixing time relative to the vaporization time [Eq. (2)]:

$$Da_C = \frac{t_M}{t_C} \tag{1}$$

$$Da_V = \frac{t_M}{t_V} \tag{2}$$

It has been proposed from Damköhler number estimations that liquid droplet vaporization controls the combustion process.[7] However, there exists an entire spectrum of chemical time scales that are highly dependent on local conditions in high pressure LOX/H_2 combustion. Furthermore, there are a variety of flame types that can be characterized by injection Reynolds number, Damköhler criteria, and Karlovitz number. The Karlovitz number compares the chemical time scale with the Kolmogorov turbulent time scale [Eq. (3)]:

$$Ka = \frac{t_C}{t_{kol}} \tag{3}$$

Borghi[26] has characterized flame types dependent on time and length scales and the primary thermodynamic and turbulent combustion properties outlined in Eqs. (1–3). Other variables adopted by Borghi's convention are flame velocity and turbulence intensity. Analysis of high-pressure LOX/H_2 combustion processes indicates that it is indeed difficult to characterize the flame types with the limitations summarized by the Borghi diagram. In fact, almost all of the proposed flame models appear locally at some point during cryogenic combustion, whether it be during subcritical ignition, transcritical, or supercritical, steady-state operation.[26] Wrinkled and corrugated flamelet type reaction zones become broken and distributed and coexist with a homogeneous reaction zone further downstream. This fact complicates the modeling of high-pressure combustion phenomena.

D. Flame-Holding Mechanisms

The presence of a recirculation zone at the LOX coaxial injector post has been previously identified.[2,3,25,27-29] Within this region there is an enhanced mixing of propellants causing an instantaneous, high-radiation flame anchored upon ignition[24] (see Fig. 22). This mechanism is represented in Fig. 23 and can also be distinguished in the flame radiation image of Fig. 17, which shows clearly the flame-holding mechanism introduced through LOX post recirculation.

The experimental study of Mayer and Tamura[2] demonstrated that for all steady-state combusting conditions performed, the flame was attached to the LOX post tip. In this study the LOX post tip thickness was 0.3 mm, and the LOX and GH_2 velocities were varied in the range from 10 to 30 m/s and 200 to 300 m/s, respectively. The propellant temperatures were 100 K for the LOX and varied between 100 and 300 K for the hydrogen. The effect of chamber pressure on the flame-holding mechanism was investigated in the range from 1.0 to 10.0 MPa.

Investigations by Juniper et al.[27] have suggested that the flame is always stabilized at the injector LOX post, even as H_2 temperatures descend to 90 K. Candel et al.[7] have mapped the instantaneous flame with LIF of GO_2 at pressures of 0.1 MPa and less. The result indicates that the flame is always attached at the LOX post. Preliminary results of current investigations by the authors at supercritical conditions and very low H_2 temperatures (<70 K) suggest otherwise. Further investigation into injection velocity effects and propellant injection temperatures is necessary.

To date, it can be concluded that if heating rates are high, diffusion processes can predominate prior to atomization. It seems that there exists a flame stabilization mechanism that is based on the backflow of hydrogen and precombusted

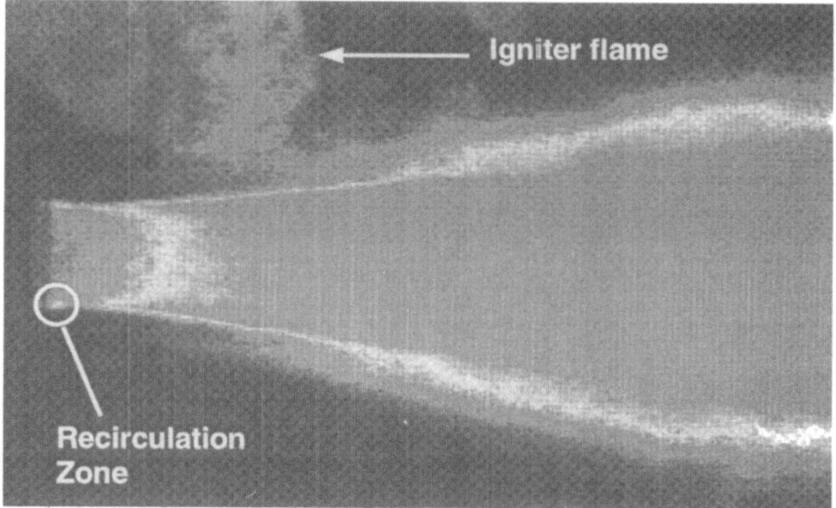

Fig. 22 OH emission image indicating intense recirculation upon ignition ($\sim + 0.44$ s).

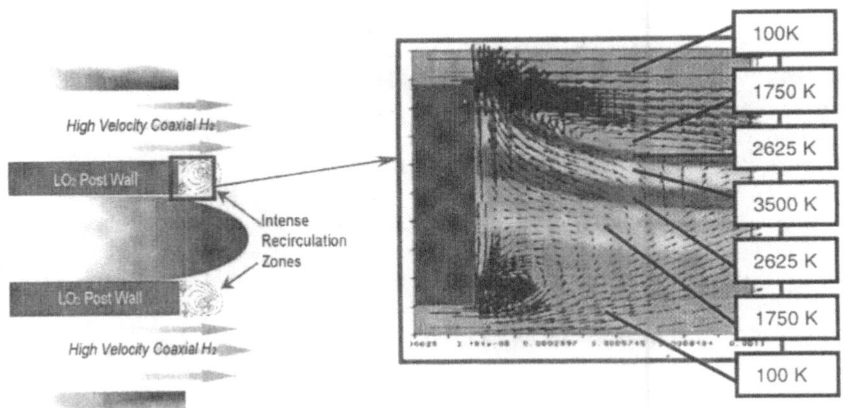

Fig. 23　LOX post recirculation zone principal sketch (left) with velocity and temperature distribution at P_{ch} = 60 bar (right).[25]

gases driven by several small recirculation eddies in the vicinity of the LOX-post lip. There, the flow velocities are partly low enough that stationary combustion is possible. Near the LOX jet, the flow is toward the faceplate (upstream with respect to the main flow direction). The downstream part of the recirculation eddy consists of partly preburned, but hydrogen rich, combustion gases. Near the LOX jet surface, the evaporating oxygen mixes into this gas and combusts. The combusted gas again mixes into the hydrogen jet and partly recirculates back.

Ivancic and Mayer[25] have performed numerical simulations of the injector near field including the flame-holding mechanism at elevated pressures (supercritical). This work has indicated a strong flame anchoring to the LOX post induced by propellant recirculation.

The flame-holding mechanism is demonstrated by Oefelein and Yang's[30] large eddy simulations (LES). The simulations showed a trend of increased unsteadiness with respect to the flame-holding mechanism at supercritical conditions.

It can be concluded that the flame-holding mechanism is not based on a simple diffusion principle, as partly wrinkled and corrugated reaction zones may appear. It can be assumed that the energy for LOX gasification mainly comes from the hot recirculating combustion gases. It is still not clear whether the energy transport is solely due to convective heat transfer, or whether flame radiation plays a role.

E.　Propellant Interface Phenomena

The influence of the hydrogen injection temperature on mixing performance in the near-injector region has been studied in the range between 150 and 300 K.[2] Two chamber pressure levels were selected for comparison (6.0 and 10.0 MPa). Other test parameters (e.g., injector geometry) were maintained constant. Oxygen and hydrogen injection velocities were kept constant at 30 and 300 m/s, respectively.

Results indicated that variations in hydrogen temperature had less effect on mixing than absolute chamber pressure. Increased chamber pressure showed stronger mixing. The mixing process in the shear layer is dependent on the velocity gradient between the injected propellants and also the density of the fluids. The density of the fluids in this layer is influenced by the combustion reaction (temperature) and is also related to the ambient chamber pressure. It appears that the flame extends axially and effectively tends to separate the LOX and annular H_2 propellants.

Juniper et al.[27] have investigated pressure effects on flame expansion angle and claim that an increase in chamber pressure decreases the expansion angle. This behavior is due to volumetric contraction of the LOX core with increasing pressure. This trend has been documented by Puissant et al.[31] for subcritical pressures with nitrogen. The authors have witnessed the opposite trend under supercritical pressure with combustion, because of the large jet expansion beyond the critical point. A secondary contributor to the flame expansion under combusting conditions is due to the heat transfer to the liquid jet from the hot combustion gases. This observation for supercritical conditions was also made in Ref. 30.

The authors have performed tests with H_2 injection temperatures as low as 65 K. The low fuel injection temperature has been found to have a significant effect on flame expansion at supercritical conditions with constant velocity ratio. At such low injection temperatures, water formation has seriously hampered flow visualization. This fact has also been previously reported with low fuel injection temperatures.[27]

At typical fuel injection temperatures around 110 K, ice has been found to build up rapidly around the injector. Figure 24 displays the phenomenon. The intense recirculation of reactant products (H_2O) onto the very cold faceplate and injector causes an ice buildup along the outer hydrogen boundary. The ice buildup has been seen to develop and decay rapidly on the injector and effectively "recess" the injector. Injector recess is detailed further in the subsequent section.

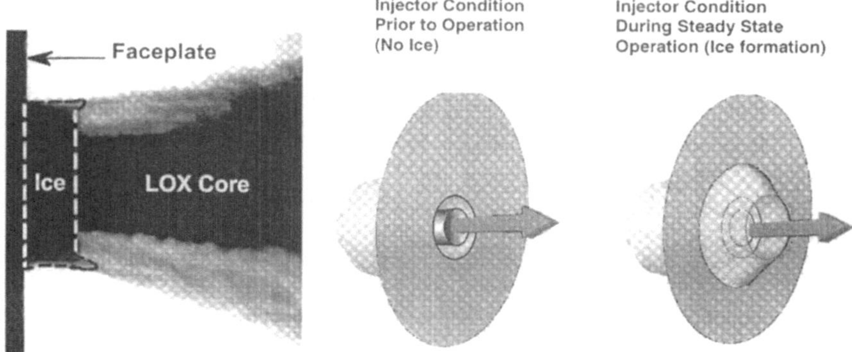

Fig. 24   Ice formation in shadowgraph (left) and phenomenological sketch (right).

F. Effects of Injector Design on Flowfield

It is widely accepted that injector geometry and design have a direct impact on injection and combustion behavior. The pressure drop across the injector element of a combustion device is of primary importance and essentially determines the low-frequency stability of the device. The LOX post tip wall thickness influences the development of the flame, as is evident from work performed by Mayer and Tamura[2] and Juniper et al.[28] Stability rating tests performed at subcritical pressures[32] have shown that injector recess improved engine performance by up to 3%. The same tests also indicated improved stability in terms of high-frequency transition temperature with injector recess.

Mayer and Tamura[2] performed tests to investigate the flow tendencies under combusting conditions when the LOX post was significantly recessed. The injector was recessed 2.9 mm, which was equivalent to 2.9 LOX post internal diameters (i.d.). The fuel annulus diameter was 3.0 mm. Without recess, the propellants enter the combustion chamber as parallel jets that diverge smoothly. The LOX jet and LOX post wake are straight (i.e., where the flame zone develops).

In the case of the recessed LOX post, the fuel enters the combustion chamber as a divergent jet. The LOX post wake is more developed, and outer geometrical boundaries of the hydrogen flow are realized sooner. The LOX jet itself seems much more disrupted in comparison with the nonrecessed injector case. Further downstream (>20 mm) the difference between a recessed and nonrecessed LOX post becomes less obvious with increasing axial distance.

The effects of the LOX post recess study have shown that the far-field region is very much the same with and without recess. Cold-flow tests have shown that injector recess has led to improved atomization.[2,28] The far field with and without recess appears to be the same. Recess has been shown to have no major effect on performance at the conditions presented in Ref. 2. In terms of improved stability, it seems that the recess region can be regarded as a small and undisturbed combustion chamber, which always keeps the propellants ignited and well mixed in an area close to the injector and is insensitive to disturbances in the chamber.

Juniper et al.[28] report that heat is transferred to the annular hydrogen from the combustion process inside the recessed region. This heat increases the gas injection velocity and leads to an increase in momentum flux ratio J. Results from Herding[33] suggest that the jet breakup is dependent on J and is enhanced above a critical value of 10. The mechanism of performance increase through injector recess was concluded to be an acceleration of H_2 by heat transfer and with a direct effect on momentum flux ratio.

G. Ignition Transients

The aim of this study was to visualize and to analyze the flow and flame transients during engine ignition. Very low hydrogen temperature ignition and combustion instability were of primary concern. Ignition poses many problems to launch vehicles. Late ignition results in a hard engine start with dangerously high chamber pressure peaks. Total failures of launch missions have been encountered due to ignition problems, as in Ariane Flight 18, 1986.[16]

A special experimental setup has been developed to study the ignition processes in a LOX/GH$_2$ rocket combustor. A combustor with 150×30 mm windows for optical access was used. Early tests failed due to condensed water vapor contaminating the optical windows in the initial phase, due to the window temperature being under the boiling temperature of the combustion product water. To counteract this, a window heating device was installed on the model combustor.

The igniter (oxygen-hydrogen pilot burner) was mounted at the center of the chamber 4 cm downstream from the injector plane. Injector baseline dimensions are a LOX post inner diameter $d_i = 1.2$ mm, outer injector diameter $D = 7$ mm, and a coaxial slit $y = 2.5$ mm.

Visualization of the ignition process at a sufficient (high) frequency proved difficult. At least 10^5 frames/s were deemed necessary for satisfactory visualiza-

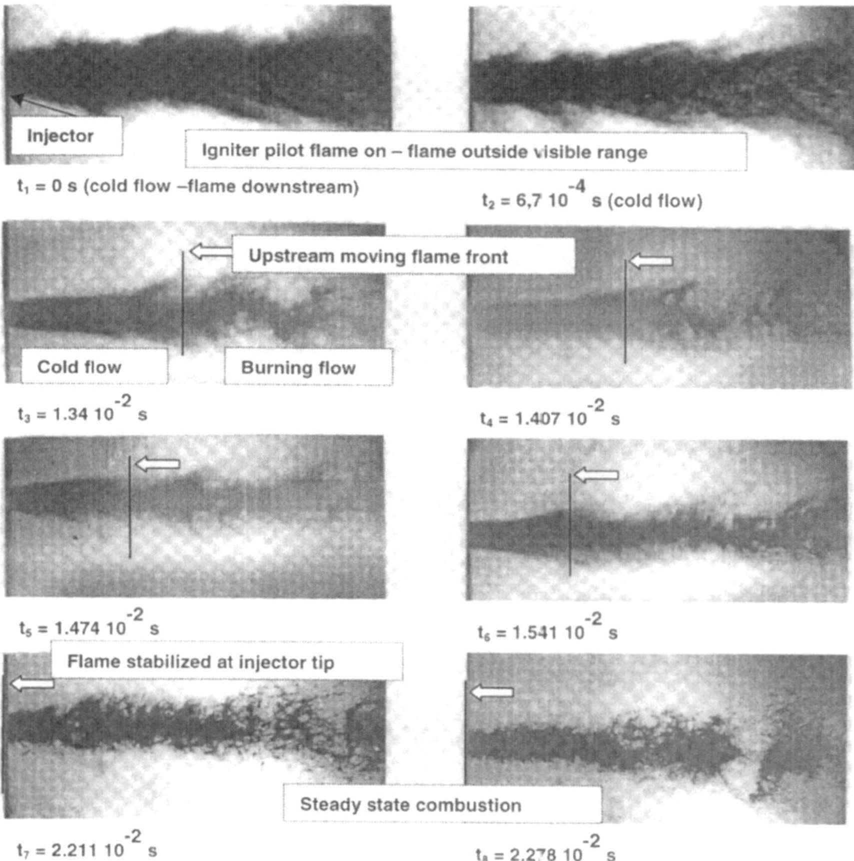

Fig. 25 LOX/GH$_2$ jet: cold flow (top row), during ignition (two middle rows), and steady-state combustion (bottom row); $P_{ch} = 0.18$ MPa, $u_{LOX} = 14$ m/s, $u_{GH2} = 340$ m/s, $T_{LOX} = 77$ K, and $T_{GH2} = 200$ K.

tions. The onset of ignition varies (in this study typically ± 10 ms) because of inconsistent valve opening times (opening time ~ 5 ms) and mixing times. The number of frames on the high-speed camera equipment was limited to around 100 per test. Therefore, the onset of ignition had to be caught exactly, and several tests were necessary to satisfactorily visualize the ignition process. The ignition tests showed that fluctuations in the igniter temperature influenced the onset of ignition considerably.

Figure 25 shows the high-speed ignition sequence and the behavior of the flame front at different times during the ignition process. The LOX jet was illumi-

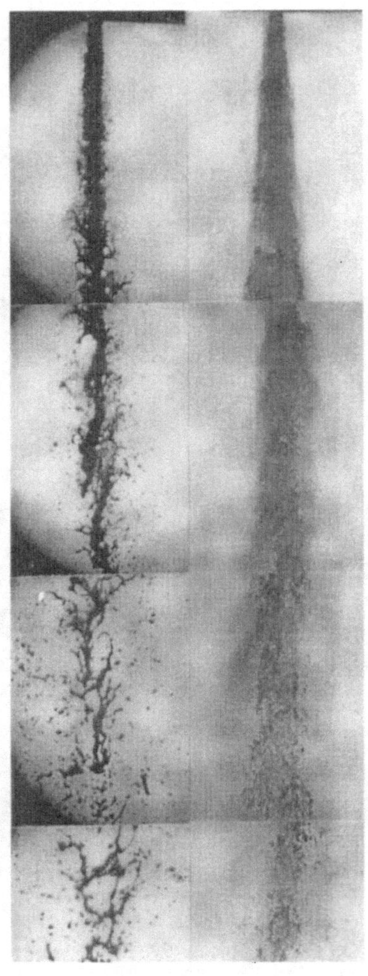

Fig. 26 LOX/GH$_2$ jet: cold flow (right) and burning (left) at $P_{ch} = 0.18$ MPa, $u_{LOX} = 14$ m/s, $u_{GH2} = 340$ m/s, $T_{LOX} = 77$ K, $T_{GH2} = 200$ K, and $d_i = 1.2$ mm.

nated with back-lighting using a short-pulse, high-repetition-rate flash lamp with a pulse duration of less than 50 ns. Further details can be found in Mayer et al.[24]

The top row of Fig. 25 shows the LOX/GH_2 injection prior to ignition. A thick spray of fine LOX droplets is visible with a large divergent spray angle. The two middle rows show the upstream movement of the flame through the diminishing spray angle and the droplet density. Small droplets cannot be found within the flame. The observed flame propagation relative to the LOX jet is on average 0.5–1.0 m/s. Therefore, the absolute propagation velocity is on the order of $10–10^2$ m/s. The bottom row of Fig. 25 already exhibits steady-state combustion conditions. Once the flame front reaches the injector (Fig. 25, $\sim 2.2 \times 10^{-2}$ s), the flame appears to anchor upon the LOX post.

Under combusting conditions, the LOX spray consists of a very small amount of ligaments and drops, which rapidly vaporize. The effect of the flame on the LOX jet near the injector can be seen in more detail in Fig. 15.

Under cold-flow conditions, very fine liquid oxygen droplets are visible, which are accelerated by aerodynamic forces from the fast-flowing hydrogen. Upon ignition, only large ligaments remain visible (Figs. 24 and 26). In the cold-flow case, the jet surface shows forward swept droplet bursts, whereas under the combustion condition, the oxygen ligaments are bowed backward, indicating that the LOX is faster than the ambient gas. In the latter case, the production of a fine spray is suppressed, and the ligaments are even larger than in the cold-flow case (Fig. 26).

V. Conclusions

Real injection conditions must be considered when characterizing propellant atomization, mixing, and combustion processes in liquid rocket engines. Typical rocket engine combustion chamber pressures range from approximately 3 MPa to typically 10–20 MPa, and thus consideration of supercritical conditions is important in all aspects of research and modeling of combustion and injection processes.

Cold-flow injection experiments have highlighted the effect of pressure on the atomization and mixing of cryogenic propellants. At low, subcritical pressures, injected fluids demonstrate a smooth, well-defined jet with sinusoidal surface instabilities that break up further downstream. An increase in chamber pressure causes a decrease in surface tension, enabling small droplets to leave the surface of the jet.

The tendency of smaller drops and ligaments at increased chamber pressure is inverted as the effect of capillary forces decreases approaching and exceeding the critical point. Because of the vanishing capillary forces in the trans- and supercritical region, a change in the atomization mechanism can be observed. Shear forces, expansion, and turbulence play a more dominant role in mixing.

In the combustion case, during ignition and combustion the interface between the propellants is always separated and affected by a layer of hot reacting gas. The flame is generally attached to the LOX post. A diffusion flame resides within the annular post wake and is anchored by the small intensive recirculation zones that exist behind these posts. Radiation intensity of the flame strongly depends on chamber pressure and increases downstream. The spectrum of the flame ranges continuously between UV and IR.

New observations of significant combustion and injection behaviors are continually being made. Injector design continues to be an important area of research in the attempt to develop optimal geometries and design features for different applications.

Further research into ignition phenomena will be fundamental in understanding transient behavior, which has an impact on engine stability. Ensuring consistent startup routines is imperative for reproducibility.

Combustion and injection research at transcritical and supercritical pressures is essential for comprehension of the processes taking place. The entire injection and combustion behavior is fundamentally different from that experienced at low, subcritical pressures.

References

[1]Mayer, W., "TEKAN-Research on Cryogenic Rocket Engines at DLR Lampoldshausen," AIAA Paper 2000-3219, July 2000.

[2]Mayer, W., and Tamura, H., "Propellant Injection in a Liquid Oxygen/Gaseous Hydrogen Rocket Engine," *Journal of Propulsion and Power*, Vol. 12, No. 6, 1996, pp. 1137–1147.

[3]Mayer, W. O. H., Schik, A. H. A., Vielle, B., Chauveau, C., Gökalp, I., and Talley., D. G., "Atomization and Breakup of Cryogenic Propellants under High Pressure Subcritical and Supercritical Conditions," *Journal of Propulsion and Power*, Vol. 14, No. 5, 1998, pp. 835–842.

[4]Chehroudi, B., Talley, D., and Coy, E., "Initial Growth Rate and Visual Characteristics of a Round Jet into a Sub- to Supercritical Environment of Relevance to Rocket, Gas Turbine and Diesel Engines," AIAA Paper 99-0206, 1999.

[5]Mayer, W., "Coaxial Atomization of a Round Liquid Jet in a High Speed Gas Stream: A Phenomenological Study," *Journal Experiments in Fluids*, Vol. 16, No. 6, 1994, pp. 401–410.

[6]Farago, Z., and Chigier, N., "Morphological Classification of Disintegration of Round Liquid Jets in a Coaxial Air Stream," *Atomization and Sprays*, Vol. 2, 1992, pp. 137–153.

[7]Candel, S., Herding, G., Snyder, Scouflaire, P., Rolon, C., Vingert, L., Habiballah, M., Grisch, F., Péalat, M., Bouchardy, P., Stepowski, D., Cessou, A., and Colin, P., "Experimental Investigation of Shear Coaxial Cryogenic Jet Flames," *Journal of Propulsion and Power*, Vol. 14, No. 5, 1998, pp. 826–834.

[8]Pal, S., Moser, M. D., Ryan, H. M., Foust, M. J., and Santoro, R. J., "Shear Coaxial Injector Atomization Phenomena for Combusting and Non-Combusting Conditions," *Atomization and Sprays*, Vol. 6, 1996, pp. 227–244.

[9]Lefebvre, A., *Atomization and Sprays*, Hemisphere, 1988, pp. 37–58.

[10]Mayer, W., Schik, A., and Schäffler, M., "Injection and Mixing Processes in High-Pressure LOX/GH2 Rocket Combustors," *Journal of Propulsion and Power*, Vol. 16, No. 5, 2000, pp. 823–828.

[11]Pitts, W., "Effects of Global Density and Reynolds Number Variations on Mixing in Turbulent, Axisymmetric Jets," National Bureau of Standards Internal Rept. 86-3340, 1986.

[12]Vingert, L., Gicquel, P., Lourme, D., and Ménoret, L., "Coaxial Injector Atomization," Progress in Astronautics and Aeronautics, *Liquid Rocket Engine Combustion Instability*, edited by V. Yang and W. E. Anderson, Vol. 169, AIAA, New York, 1994, pp. 145–189.

[13]Woodward, R. D., and Talley, D. G., "Raman Imaging of Transcritical Cryogenic Propellants," AIAA Paper 96-0468, 1996.

[14]Newman, J. A., and Brzustowski, T. A., "Behavior of a Liquid Jet Near the Critical Region," *AIAA Journal*, Vol. 9, No. 8, Aug. 1971, pp. 1595–1602.

[15]Mayer, W., Telaar, J., Branam, R., Schneider, G., and Hussong, J., "Characterization of Cryogenic Injection at Supercritical Pressure," AIAA Paper 2001-3275, July 2001.

[16]Jasper, J. J., "The Surface Tension of Pure Liquid Components," *Journal of Physical and Chemical Reference Data*, Vol. 1, 1972, pp. 841–1009.

[17]Delplanque, J.-P., and Sirignano, W. A., "Numerical Study of the Transient Vaporization of an Oxygen Droplet at Sub- and Supercritical Conditions," *International Journal of Heat and Mass Transfer*, Vol. 36, 1993, pp. 303–314.

[18]Nicoli, C., Haldenwang, P., and Daou, J., "Substitute Mixtures for LOX Droplet Vaporization Study," *Combustion Science and Technology*, Vol. 12, 1996, pp. 55–74.

[19]Street, W. B., and Calado, J. C. G., "Liquid-Vapor Equilibrium for Hydrogen and Nitrogen at Temperatures from 63 to 100 K and Pressures to 57 Mpa," *J. Chem. Therm.*, Vol. 10, 1978, pp. 1089–1100.

[20]Tully, DeVaney, and Rhodes, "Phase Equilibria of the Helium-Nitrogen System from 122 to 126 K," *Advanced Cryogenic Engineering*, Vol. 15, U.S. Bureau of Mines, Helium Research Center, Amarillo, TX, 1970, pp. 88–95.

[21]Macleod, D. B., *Transactions of the Faraday Society*, Vol. 19, No. 38, 1923.

[22]Gaydon, A. G., *The Spectroscopy of Flames*, Chapman and Hall, London, 1974.

[23]Boniface, Y., Reeb, A. B., Woodward, R. D., Pal, S., Santoro, R. J., and Mayer, W., "Hot-Fire Studies of LOX Primary Atomization from Rocket Engine Coaxial Injectors," 4th International Symposium on Liquid Space Propulsion, Heilbronn, Germany, March 13–15, 2000.

[24]Mayer, W. O. H., Ivancic, B., Schik, A., and Hornung, A., "Propellant Atomization and Ignition Phenomena in Liquid Oxygen/Gaseous Hydrogen Rocket Combustors," *Journal of Propulsion and Power*, Vol. 17, No. 4, 2001, pp. 794–799.

[25]Ivancic, B., and Mayer, W., "Time and Length Scales of Mixing and Combustion Processes in High Pressure LOX/GH2-Rocket Combustors," *Journal of Propulsion and Power*, Vol. 18, No. 2, 2002, pp. 247–253.

[26]Borghi, R., "Turbulent Combustion Modelling," *Progress in Energy and Combustion Science*, 1988, Vol. 14, 1988, pp. 245–292.

[27]Juniper, M., Tripathi, A., Scouflaire, P., Rolon, C., and Candel, S., "The Structure of Cryogenic Flames at Subcritical and Supercritical Pressures," *Combustion dans les Moteurs Fusées*, Centre Nationale d'Etudes Spatiales, 2001.

[28]Juniper, M., Tripathi, A., Leroux, B., Lacas, F., and Candel, S., "Stabilization of Cryogenic Flames and the Effect of Recess," *Combustion dans les Moteurs Fusées*, Centre Nationale d'Etudes Spatiales, 2001.

[29]Glogowski, M., Bar-Gill, M., Puissant, C., Kaltz, T., Milicic, M., and Micci, M. M., "Shear Coaxial Injector Instability Mechanisms," AIAA Paper, June 1993.

[30]Oefelein, J. C., and Yang, V., "Modeling High Pressure Mixing and Combustion Processes in Liquid Rocket Engines," *Journal of Propulsion and Power*, Vol. 14, No. 5, 1998, pp. 843–860.

[31]Puissant, C., Glogowski, M., and Micci, M. M., "Experimental Characterization of Shear Coaxial Injectors Using Liquid/Gaseous Nitrogen," ICLASS 94, pp. 672–679.

[32]Wanhainen, J. P., Parish, H. C., and Conrad, W., "Effect of Propellant Injection Velocity on Screech in 20000 lb Hydrogen-Oxygen Rocket Engines," NASA TN D-3373, 1996.

[33]Herding, G., "Analyse experimentale de la combustion d. ergols cryotechniques," Ph.D. Dissertation, École Centrale Paris, ECP 19, 1997.

Chapter 10

CARS Measurements at High Pressure in Cryogenic LOX/GH$_2$ Jet Flames

F. Grisch,* P. Bouchardy,[†] and L. Vingert[‡]
ONERA, Palaiseau Cedex, France

W. Clauss[¶] and M. Oschwald[¶]
DLR, Hardthausen, Germany

and

O. M. Stel'mack[¶] and V. V. Smirnov[§]
Academy of Sciences of Russia, Moscow, Russia

Nomenclature

C = mole of fraction
E_J = energy of the lower state of the molecules
F = error function
g_τ = nuclear statistical weight
G = relaxation matrix
I = intensity of the signal
J_m = momentum flux ratio
J, K_a, K_c = rotational quantum numbers
l = collisionally transferred angular momentum

L = length of the probe volume
M = mixture ratio
p = pressure
$Q(T)$ = partition function
Re = real part of a complex value
T = temperature
V = velocity
W_{ij} = off-diagonal elements of the G matrix
x, y = axial, radial distance
α = isotropic polarizability of the molecules
$\chi^{(3)}$ = third-order susceptibility
Δk = phase mismatch
Δ_j = pressure-induced line shift
γ_v = vibrational dephasing broadening
Γ = homogeneous linewidth of the Raman transitions
ν_i = vibrational modes of the molecules
ρ = density
ω = frequency

I. Introduction

A COOPERATIVE research program on combustion in liquid rocket engines has been conducted in Germany and France in the framework of a joint venture between industry [Société Nationale d'Etudes et de Construction de Moteurs d'Aviation (SNECMA) Moteurs-Fusées, Astrium], the French space agency [Centre National d'Etudes Spatiales (CNES)], and several research organizations [ONERA, DLR, and Centre National de la Recherche Scientifique (CNRS)]. The objective of this research is to get insight into the main physical processes to account for in designing new systems or optimizing and consolidating existing ones. In particular, an extensive experimental program has been conducted to generate a fundamental understanding of the spray characteristics and combustion mechanisms, and to develop an experimental database required for computer modeling. These experiments were run on two cryogenic test facilities. The first one, called Mascotte, was developed by ONERA. The second one, the P8 test facility, was designed for higher mass flows and pressures and was built at DLR-Lampoldhausen through a German-French cooperative agreement. At both test sites, combustion chambers with similar geometric dimensions are operated. Both chambers have optical access for application of nonintrusive optical diagnostics. In addition to investigations of phenomena controlling combustion, these experimental facilities are used to improve and assess the diagnostic techniques (especially optical diagnostics) for cryogenic high-pressure combustion. Obtaining reliable data from cryogenic combustion under such conditions still represents a big challenge because of the experimental constraints imposed by the high-pressure two-phase flow and the hostile environment at the test sites with strong sound-induced vibration.

Although sometimes affected by optical breakdown due to the presence of droplets and the problems of strong refractive index gradients, optical techniques, with their high spatial and temporal resolution, offer the greatest potential for these studies. Several optical diagnostic techniques have been applied recently, such as emission and laser-induced fluorescence (LIF) on the OH radical, to get information on the location and the structure of the flame front.[1,2] However, the quantitative analysis of the LIF data is difficult because the quenching rates are often not known to sufficient accuracy for the high-pressure environment and absorption of the excitation laser beam by the liquid phase.

Relative to these techniques, coherent anti-Stokes Raman scattering (CARS), a molecule-specific laser spectroscopic diagnostic tool, is well suited to perform quantitative temperature and concentration measurements in practical combustion systems. Because most cryogenic test facilities operate for only a short duration, a need exists for simultaneous temperature and multiple-species measurements to maximize data. This would minimize the sources of possible temperature biases, especially in diffusion flames, where the probe species can be missing in some regions. Furthermore, the density derived from the simultaneous CARS signals of multiple species can give important information about mixing processes in the flame.

Our attention was focussed on H_2 and H_2O as probe molecules because they represent the main species present into the flowfield. O_2 was considered not suitable because it is present only in vapor in thin regions close to the flame front. Although multispecies CARS approaches exist, such as multicolor CARS techniques,[3,4] these techniques require multiple dye lasers and complex optical arrangements. Moreover, such techniques complicate interpretation of the data because of the complex interactions between signals resulting from multiple-probe species and the effects of the environment within the measurement volume. The use of two synchronized multiplex CARS systems has been chosen as a better approach to overcoming the disadvantages mentioned previously and to simultaneously probing both species in time and space.

In this chapter, the CARS diagnostics for temperature and species concentration measurements under the conditions in cryogenic liquid rocket combustors are described. Results of measurements in the Mascotte facility at combustion chamber pressures up to 6.5 MPa are given and discussed.

II.   Experimental Facilities

A.   Mascotte Facility

This facility is aimed at feeding a single element combustor with actual propellants.[5] Three successive versions of this test facility were designed, each one representing a new step toward operating conditions closer to those of actual rocket engines. The first version (V01) featured a low-pressure chamber (<1 MPa) and used hydrogen at room temperature. In the second version (V02), a heat exchanger was implemented for gaseous hydrogen (GH_2) cooling to 100 K. This allowed an increase of the maximum flow rate of GH_2 at low pressure with subsonic flow at the injector exit. The last version (V03) is now operating at supercritical pressures of oxygen in the combustor with an increased maximum liquid oxygen (LOX) mass flow rate (up to 400 g/s). A specific high-pressure combustor was developed for this purpose (Fig. 1).

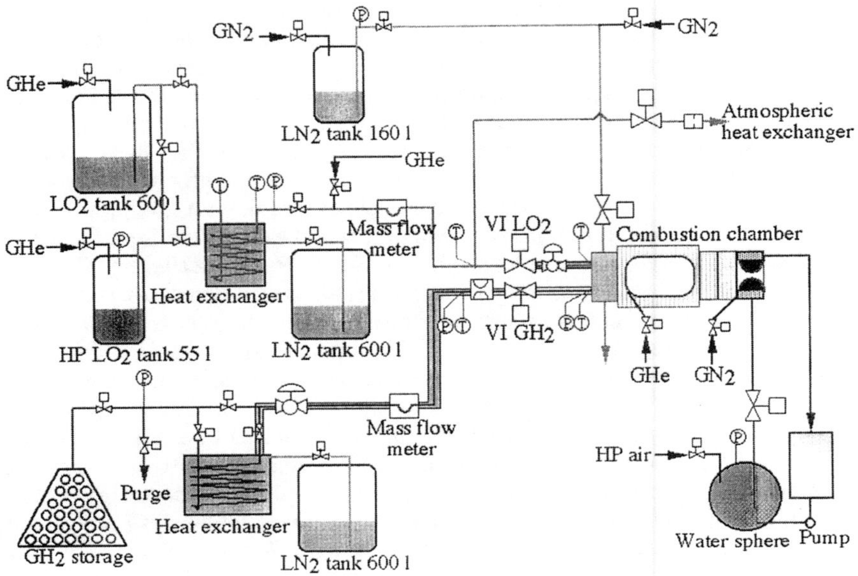

Fig. 1 Mascotte combustion facility.

The LOX tank is pressurized with helium to minimize pollution of the liquid with dissolved gas. LOX issuing from the tank passes a liquid nitrogen heat exchanger and is cooled down to 80 K to avoid vaporization and cavitation problems in the line and two-phase flow in the injector, when the combustor is operated at low pressure. The most important difference from V02 to V03 consists of replacing the 0.18 m^3 LOX tank with two tanks: a medium pressure storage of 0.70 m^3, suitable for pressure tests up to 1.5 MPa, and a high-pressure tank of 0.05 m^3 designed for 20 MPa. The maximum available pressure is 16 MPa, which is sufficient to reach 10 MPa in the combustor. The available mass flow rate is in the range 0–400 g/s.

The hydrogen storage facility comprises four containers of 28 bottles pressurized at 20 MPa, each container representing 252 m^3 of hydrogen under normal conditions. The flow rate is regulated by means of a sonic throat. V02 includes an additional heat exchanger in the high-pressure segment of the hydrogen line (on the upstream side of the sonic throat). The temperature can be decreased to 100 K. The mass flow rate specification may then reach 100 g/s. To keep the hydrogen at 100 K, the entire line downstream of the heat exchanger is immersed in a liquid nitrogen flow. The change from V02 to V03 consists of implementing a mass flow meter downstream of the heat exchanger, which allows use of a regulation valve. In this configuration, the sonic throat may be suppressed.

In addition to LOX and GH$_2$, additional fluids are available for different purposes on the bench: 1) gaseous nitrogen is used as the auxiliary fluid for operating the pneumatic and relief valves and to pressurize the liquid nitrogen tank; it may also be injected into the chamber at the nozzle entrance to drive the pressure independently from the combustion (at low pressure) or to facilitate the cooling of the nozzle; 2) gaseous helium is used to pressurize the LOX and to cool the internal face of the combustion chamber windows; 3) liquid nitrogen is used to cool the injection head and the main lines downstream of the heat exchangers as well as to feed these exchangers; and 4) high-pressure water circuit was designed to cool the nozzle of the high-pressure combustor. Water is injected into the supersonic jet for reducing the noise produced by the facility at high pressure.

The chamber pressure is limited to 1 MPa in the V01 and V02 versions. The last version is designed for chamber pressures up to 10 MPa. Experiments are generally conducted at a pressure of 7 MPa, which is sufficient to operate above the LOX critical pressure. The GH_2 mass flow rate for V01 ranges from 5 to 20 g/s of gaseous hydrogen at room temperature and at atmospheric pressure. This upper limit is related to the requirement of subsonic flow in the injector to be representative of a rocket coaxial injector. The upper limit of hydrogen flow rate that the test bench could deliver is related to the pressure of the storage (maximum 20 MPa) and to the cross section of the sonic throat in the feed line. With the current storage of 1000 m^3, the available flow rate of hydrogen would be approximately 75 g/s. The lower limit corresponds to the lowest upstream pressure for which the throat becomes sonic. On the LOX side, the inner diameter of the line is 10 mm. The maximum mass flow rate corresponding to a velocity of 5 m/s in this line would be approximately 450 g/s, while the specified ranges were 20–100 g/s of liquid oxygen for V01 and V02 and 20–400 g/s for V03.

In addition to the mixture ratio M, one of the most important similarity parameters, especially for the atomization performance of an injector, is the gas-to-liquid momentum ratio of the injected propellants, defined as

$$J_m = (\rho V^2)_{H_2}/(\rho V^2)_{LOX} \tag{1}$$

Two reference points, A and C, have been chosen for the Mascotte V01 test bench at atmospheric pressure (see Table 1). For both of them, the LOX mass flow rate is 50 g/s. Changing from A to C, the mixture ratio M is increased from 3.3 to 5, while the J ratio is approximately divided by 2. When running the bench at 1 MPa with

Table 1   Mascotte operating conditions

Point	Pressure, MPa	LOX mass flow rate, g/s	Gas mass flow rate, g/s	M	J_m
A	0.1	50	15.0	3.3	13.4
C	0.1	50	10.0	5.0	6.3
A-10	1.05	50	23.7	2.1	14.5
C-10	0.95	50	15.8	3.2	6.5
A-30	3	50	25.2	2.0	15.5
C-30	2.8	54	17.0	3.2	6.6
A-60	6.6	100	70.0	1.4	14.4
C-60	5.8	100	45.0	2.2	6.8

cold hydrogen (V02), or for nonreactive flow simulation tests, equivalent operating points have been defined by keeping the LOX mass flow rate and the J ratio roughly constant. To achieve this, the injector had to be changed by reducing the hydrogen annulus outer diameter DG. At higher pressures (around 6 MPa, for instance), new similar operating points have been defined by keeping the J_m ratio constant and further reducing DG, but the LOX mass flow rate had to be increased. Trying to reach 6 MPa in the combustor, with only 50 g/s of LOX injected, would lead to insurmountable technological problems. The needed throat would be so small that it would become impossible to cool it down and it would not survive the heat fluxes encountered in that zone.

The low-pressure combustion chamber (versions V01 and V02) is a square duct of 50-mm inner dimension, made of stainless steel and fitted with four fused-silica windows for optical access. Two lateral windows of 100 mm in length and 50 mm in height are used for visualization. Their internal face is cooled by a helium film. The upper and lower ones are also 100 mm in length, but their width is only 10 mm. They will be used for longitudinal laser sheet entrance and exit. The combustor is assembled of different interchangeable modules, which allow visualization of the entire combustion chamber by putting the so-called visualization module at different longitudinal locations. The nozzle of the chamber is usually made of graphite, but the use of a copper one, either water- or heat-sink cooled, is also possible. The combustion chamber is designed with thermomechanical models and conservative assumptions for 30 s of operation duration at atmospheric pressure, with the maximum mass flow rate of 120 g/s at a mixture ratio of 6. The test duration is reduced to 20 s at 1 MPa. This combustor may be fired 6–10 times in one day, with 5–10 min between runs. The V01 and V02 combustors are not suitable for use above 1 MPa. A new high-pressure system had to be developed based on the experience gained with the previous ones (Fig. 2). Its specifications allow a run time between 15 and 20 s at high pressure up to 10 MPa without degradation; it also offers the possibility of performing optical diagnostic experiments as well as classical measurements (pressure and temperature transducers). The combustor consists of three parts: an injection head, cooled with liquid nitrogen, similar to the V01 and V02 heads, a combustion chamber including metallic modules and visualization modules, and a water-cooled nozzle.

B. P8 Facility

The P8 facility is a test facility for research and technology studies in the field of high-pressure rocket combustion devices.[6] Operation of subscale combustors is possible using liquid oxygen and gaseous or liquid hydrogen as propellants. Thus, the installed equipment consists of tanks for liquid oxygen (LOX), gaseous and liquid hydrogen (GH$_2$, LH$_2$), and cooling water (H$_2$O). A heat exchanger, which is operated with liquid nitrogen (LN$_2$), is used to cool down gaseous hydrogen to approximately 150 K. The facility is equipped with two test cells, allowing parallel test programs with different models. Because of safety requirements the entire facility is operated from the control room, which is located about 100 m from the P8.

Fig. 2 High-pressure combustion chamber.

The run tanks for propellants and cooling water as well as the hydrogen heat exchanger are located between the test cells to minimize pressure losses. Both oxygen and cooling water are operated in a blow-down mode. Therefore, the run tanks, with maximum pressure of 50 MPa for oxygen and 32 MPa for water, respectively, are connected to a high-pressure nitrogen supply system. Tank pressurization and tank pressure control are achieved by using electrically operated control valves, which were especially designed for the P8. With additional closed-loop control valves in each of the feedlines, the mass flow rate can be controlled between 0.2 and 8 kg/s for oxygen and between 1 and 50 kg/s for water, respectively.

The supply system for hydrogen is also operated in blow-down mode, but a separate run tank is not necessary because the medium is already stored under high pressure (63 MPa). Pressure and mass flow control are again managed by closed-loop control and regulation valves. A bypass line along the GH_2 cooler is used for temperature regulation within another closed control loop. Therefore, the hydrogen temperature at the interface of the test specimen can be set to the desired level between 100 and 150 K. Most important for successful tests is an efficient control of the mass flow rate for propellants as well as cooling water. The algorithms for these control loops are implemented in routines on the main computer located in the control room.

Because of the important role of nonintrusive diagnostics in basic combustion research, the requirements of optical diagnostic setups have been taken into account in the design of the test facility. On both sides of each test cell there are rooms in which optical diagnostics and data acquisition systems are installed.

The size of the rooms and their infrastructure allow the remote control of setups for different optical diagnostics at the same time.

Additional systems, such as a LN_2-jacket cooling system and a GH_2 fuel line, can be extended to be within the scope of the partners' test requirements. The implemented LN_2-jacket cooling system allows an exact adjustment of the propellant temperatures independently of the mass flow rate and faster achievement of steady-state conditions, especially at lower propellant flow rates. In this way a more realistic ignition simulation can be obtained. LN_2 offers a maximum mass flow of 0.11 kg/s. The additional GH_2 supply allows testing of film cooling, transpiration cooling devices and engine cycle simulation for expander cycle operation or staged combustion. A maximum interface pressure of 36 MPa is available, and the maximum mass flow rate is 1 kg/s.

For safety reasons all experimental setups must be operated by remote control from a control building about 100 m away from the test facility. Control and data exchange between the test facility P8 and the control building for the application of optical diagnostics had to fulfill the following conditions: 1) independence from the measurement, control, and command computer system that operates the facility and 2) high flexibility, satisfying the needs of different diagnostic techniques and users.

For research purposes, a single coaxial injector design with a rectangular layout of 50 mm × 50 mm was designed for combustion chamber pressures up to 10 MPa and is operated at P8.[2] The combustion chamber consists of modules made of a special copper alloy (Fig. 3). Water cooling of all modules allows long run times. The window module has observation windows of 50 mm × 100 mm for imaging, spectroscopy, CARS, and LIF detection. A laser sheet can be sent into the chamber through smaller windows of 20 mm × 100 mm to perform LIF and PLIF diagnostics. The windows are cooled and protected inside by a gas flow of hydrogen or helium. For CARS measurements, the combustor was run at 6 MPa with a LOX mass flow of 240 g/s at 120 K and a GH_2 mass flow of 80 g/s at 110 K. GHe was used to cool down the windows at a flowrate of 160 g/s. These mass flows, although still higher than at Mascotte, represent the lowest operating regime of the P8.

III.   CARS Overview

CARS has received considerable attention over the last two decades for combustion diagnosis, based on the pioneering investigations of Taran.[7] Excellent review papers and textbooks are available that deal with CARS theory, derive expressions for the signal intensity observed in real experiments, and describe the numerous technical approaches in practical measurement systems.[8-10]

As shown in Fig. 4, CARS is an example of a four-wave parametric process in which three waves, two at the pump frequency ω_P and one at the Stokes frequency ω_S, are focused to the measurement point in the sample to produce a new coherent beam at the anti-Stokes frequency ($\omega_A = 2\omega_P - \omega_S$). The strength of the interaction depends on the third-order susceptibility of the medium, which is greatly enhanced when the frequency difference $\omega_P - \omega_S$ matches a Raman

Fig. 3   P8 combustion chamber during a high-speed schlieren imaging experiment at $p = 3$ MPa.

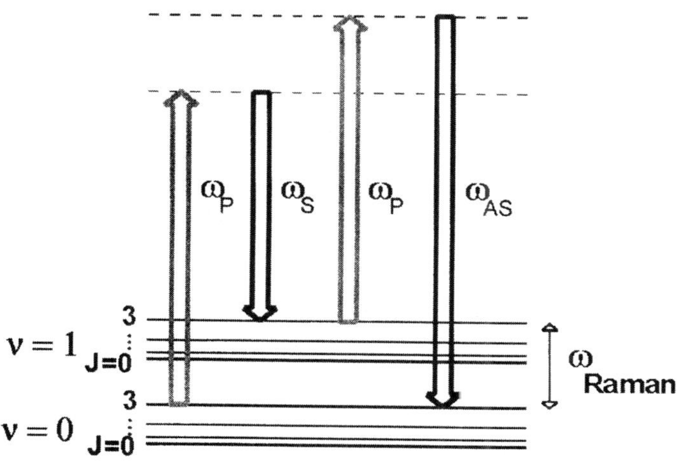

Fig. 4   Energy level diagram for CARS.

active vibrational resonance in the medium. By tuning the Stokes frequency, a spectrum of the species under study can be recorded.

The spectral intensity of the CARS signal is given by[8]

$$I_{\omega_{AS}}(\omega) = I_{\omega_P}^2 I_{\omega_S}(\omega) |\chi^{(3)}(\omega)|^2 L^2 \sin c^2\left(\frac{\Delta k L}{2}\right) \tag{2}$$

where $I_{\omega_{AS}}$, I_{ω_P}, and I_{ω_S} are the laser intensities of the three laser beams, $\sin c(x) = \sin(x)/x$; Δk is the phase mismatch of the CARS beam and the exciting laser beams ($\Delta k = k_{AS} + k_S - 2k_P$), k_i is the wave vector of each laser beam, and L is the length of the probe volume. The expression is obtained assuming that all the beams have the same diameter and that their divergence is limited by diffraction. It is independent of the focal length of the lenses and of the beam diameter. The CARS intensity scales with $I_{\omega_P}^2 I_{\omega_S}$ and has no lower threshold but is generally limited at elevated pressure by laser-induced breakdown.

The temperature is deduced from experimental CARS spectra by analyzing the spectral profile formed by the different intensities of the excited rotational-vibrational transitions of the probe molecules. Effects of temperature and concentration on the CARS spectrum are then contained in the squared modulus of $\chi^{(3)}(\omega)$ given by

$$\chi^{(3)}(\omega) = \chi_R^{(3)}(\omega) + \chi_{NR}^{(3)}(\omega) \tag{3}$$

The $\chi_{NR}^{(3)}(\omega)$ nonresonant contribution contains the effects of remote vibrational resonances and those associated with the nonlinear distortion of the electron cloud of the molecules caused by the intense laser fields. $\chi_{NR}^{(3)}(\omega)$ is a real dispersionless number predicted by a mole-fraction weighted sum of contributions from all the species present. The Raman resonant susceptibility term $\chi_R^{(3)}(\omega)$ is a complex quantity with real and imaginary parts. It describes the response of the medium when ($\omega_P - \omega_S$) is scanning across the vibrational resonance. It is a function of the number density of the probe molecule, temperature, pressure, and molecular parameters such as the rotational linewidths.

Generally, each molecule in the medium has several specific vibrational resonances by which it can be identified; each resonance is split in a fine structure of lines associated with rotational sublevels. Mathematically, the expression $\chi_R^{(3)}(\omega)$ is obtained by summing over the resonances in question. Following the work of Hall et al.,[11] this term is derived as

$$\chi_R^{(3)}(\omega) = \frac{iN}{\hbar} \sum_j \alpha_j \sum_k \alpha_k \Delta\rho_k^{(0)} G_{jk}^{-1} \tag{4}$$

where N is the number density of the probe molecule, α_j denotes the isotropic polarizability of the molecules, $\Delta\rho_k$ is the population difference between the levels involved in the transition, and G is the relaxation matrix. The elements of the G matrix are expressed by

$$G_{jk} = i(\omega_P - \omega_S - \omega_R)\delta_{jk} + \left(\frac{\Gamma_j}{2} - i\Delta_j\right)\delta_{jk} + W_{jk}(1 - \delta_{jk}) \tag{5}$$

where Γ_j is the homogeneous linewidth of the Raman transition, Δ_j is the pressure-induced line shift, and W_{jk} represent the off-diagonal elements of the G matrix. The two first terms of Eq. (5) describe the isolated-lines approximation simulating the case in which the overlap of individual rotational lines in the CARS spectra is weak, i.e., the low-pressure regime. The last term is governed by the off-diagonal elements W_{jk}, which describe the state-to-state rotational relaxation rates between the rotational states, that is, the rates of inelastic energy transfer between vibration-rotation states in the active molecule, which is mainly important in a high-pressure regime.

Certain relationships involving these transfer rates follow, one of them being the principle of detailed balance,

$$\rho_{kk} W_{jk} = \rho_{jj} W_{kj} \tag{6}$$

and the other following from conservation of probability in inelastic collisions (scattering matrix unitary),

$$\Gamma_j = -p \sum_{J \neq k} \mathrm{Re}\, W_{jk} + \gamma_v \tag{7}$$

where γ_v is the contribution arising from the vibrational dephasing collisions, and p is the pressure. If vibrationally inelastic processes are slow, there can be no coupling between different vibrational bands, and so the G matrix will be block-diagonal for the case of high-temperature gases where vibrational hot bands are important. At high temperatures, where large numbers of transitions must be included to accurately describe a CARS spectrum, the analysis of line overlap effects can lead to rather cumbersome and time-consuming calculations because, at each frequency of interest, a large, complex matrix must be inverted.

CARS spectra also offer the potential for species concentration measurements. Concentration measurements are based on the strength of the CARS signals in general. The absolute CARS intensities measured in the flame are compared with those measured in a reference flame where temperature, pressure, and species concentration are well known. Unfortunately, such measurements are complex to perform in practical situations because the presence of beam steering effects markedly alters signal levels. Fortunately, the shape of the CARS spectrum is also concentration sensitive, thus allowing fractional concentration measurements from spectral profiles. This occurs in gas mixtures in the presence of a significant, but not overwhelming, nonresonant susceptibility. This technique provides relative densities by indirectly comparing resonant to nonresonant susceptibilities, but lacks sensitivity and accuracy for low concentrations.[12]

IV. Probe Species in LOX/GH$_2$ Combustion

CARS thermometry is usually performed with majority species or species in sufficient abundance that nonresonant background effects are relatively insignificant. For LOX/GH$_2$ rocket engine applications, the possible Raman active probe species are H$_2$, H$_2$O, and O$_2$. Generally, rocket combustors are operated in fuel-rich conditions, which means that hydrogen is present downstream of the flame front and therefore is the prominent probe molecule into the flowfield. O_2 is present only in vapor in the thin shear reactive mixing layer located around the LOX jet, while the major product of combustion, H_2O, is available in abundance everywhere. We concentrated our efforts on CARS of H_2 and H_2O for the following reasons: 1) both molecules are the major species into the flowfield and 2) a simultaneous detection of H_2 and H_2O gives valuable information on the mixing processes in the turbulent reacting flow.

A. Hydrogen

H_2, when present, is ideal for thermometry because of its few isolated Q-branch lines. The rotational constant B_e and the vibration-rotation interaction constant α_e are both very large, and so the Q-branch structure of H_2 is characterized by widely spaced isolated rovibrational lines in the wave number region $4000-4155$ cm^{-1}. This spectral feature has two consequences: 1) in opposition to N_2 and O_2 CARS spectra,[13,14] collisional interferences between neighboring lines do not occur for H_2 in the investigated pressure range, and this makes the interpretation of H_2 spectra easier; and 2) the number of the rotational levels that are significantly populated and so have measurable intensities is small. We decide to derive the flame temperature from only seven rovibrational lines. CARS line intensities are strongly dependent on the Raman linewidths of the individual transitions located in that range. Temperature measurements from the H_2 Q-branch require accurate knowledge of the Raman linewidths and of their temperature dependence.

At low density (i.e., pressure below 0.1 MPa), the CARS linewidth is mainly Doppler broadened. Consequently, the Raman linewidths are identical for all the Q-lines, and the temperature can be deduced by plotting the amplitudes of the Q-lines in a Boltzmann diagram. In the case of higher density (i.e., pressure above 0.1 MPa), two other phenomena become important. The first one is pressure broadening, which is variously caused by collisional processes, and the second one is related to the Dicke narrowing.[15] This last effect is a result of a coupling of the Doppler and the collisional effects as collision frequency increases. It results from a coherent averaging of frequencies within the normal Doppler line profile, induced by velocity-changing collisions that conserve vibrational phase. If the frequency of velocity-changing collisions is greater than that of dephasing collisions, line narrowing is observed. This phenomenon is represented by a soft collision model such as the well-known Galatry model,[16] which has been suitably used for many collisional systems. The complex Galatry function describes the observed spectra whatever the density. This property is important in CARS thermometry because, with the temperature being unknown, the density is unknown in the considered probed gas volume. The complex Galatry profile is calculated with Varghese's procedure,[17] which uses the fast and simple Fortran subroutine developed by Humlicek.[18] The pressure shift of lines has been omitted because it produces merely a small overall shift of the Q-branch spectrum. The important parameter is then the collisional broadening. This coefficient depends on the density, the J quantum number, the temperature, and the composition of the mixture.

In the Q-branch, the collisional broadening can be of two types. First, a contribution is due to elastic collisions that change the phase of the vibrational motion. No energy is exchanged during this type of collision. Second, another contribution is relative to inelastic collisions that remove molecules from one molecular state to another. During this inelastic process, the energy can be transferred to the translational motion of the collider or to internal motion if the collider is a molecule. The vibrational dephasing contribution denoted as γ_v is negligible compared to inelastic broadening for some molecules of interest in combustion, such as N_2 or O_2.[13,14] It is then possible to model the behavior of the broadening coefficient using statistical fitting laws or dynamically scaling laws.[11,19,20] For the H_2 molecule, γ_v is not negligible and must be taken into account; that complicates the extrapolation of the broadening coefficients at high temperature. An alternative way to determine the broadening coefficient at high temperature is quantum or semiclassical calculations. Such calculations require an accurate knowledge of the interaction potential between the active molecule and the perturber. At the present time, quantum or semiclassical calculations have not yet been applied to complex collisional systems, such as H_2-H_2 and H_2-H_2O, because of the large number of relaxation channels that are encountered. In the present study, broadening coefficients in H_2-H_2O mixtures are obtained from experiments performed in a static cell at a density of about 10 amagat (corresponding to pressures of 22, 29, 36, and 44 bars at temperatures of 600, 800, 1000, and 1200 K, respectively).[21] Within the experimental uncertainties, the behavior of the broadening coefficient $\gamma_{H_2-H_2}$ was found to be linear with temperature, whereas the H_2-H_2O broadening coefficients of the $Q(J)$ transitions were simulated according to a polynomial temperature law.[21]

A specific analytical procedure was developed and employed to reduce CARS spectra to temperatures, despite the lack of knowledge of species composition. For each pressure condition, a library of theoretical CARS spectra was created at 50-K increments over the range 300–3000 K and at 5% increments for the H_2 and H_2O molar fractions. The experimental spectra were compared to the theoretical library spectra with a simple least-squares fitting procedure to get temperatures. The error function is then defined by

$$ F = \Sigma \omega_j [I_j^{exp} - I_j^{theo}(T, C_{H_2}, C_{H_2O})]^2 \tag{8} $$

where ω_j is a statistical weight characteristic of the line magnitude. C_{H_2} and C_{H_2O} are the mole fractions of H_2 and H_2O using the assumption $C_{H_2} = 1 - C_{H_2O}$.

B. Water Vapor

The CARS spectra of the reaction product water vapor have been processed both for temperature and for species concentration. The species concentration measurement serves as an indicator of the extent of the chemical reaction and of overall combustion and is furthermore needed to improve the temperature determination from H_2–CARS as explained in the preceding discussion. For a limited concentration range, H_2O concentration measurements were obtained from the shape of the CARS spectrum by modeling the interference of the nonresonant CARS signal with the resonant CARS signal.

H_2O is an asymmetric top molecule, characterized by three vibrational modes, a symmetric stretch v_1, a bending mode v_2, and an asymmetric stretch v_3. All modes are Raman active, but the only strong mode, sufficiently intense to be of interest for CARS diagnosis, is the symmetric stretch with a Raman shift of 3657.1 cm^{-1}. The mode is also known to be almost totally polarized, indicating that the Raman scattering is dominated by the trace or isotropic part of the polarizability tensor, so that isotropic Q-branch transitions turn out to be important. The vibrational states are denoted by v_1, v_2, v_3, where v_i represents the vibrational quantum number to the ith mode. In addition to the quantum number J giving the total rotational angular momentum, each rotational state is characterized by two other quantum numbers, K_a and K_c, representing the projection of J along the molecular axis. For convenience, the rotational states are often labeled by J and a pseudoquantum number $\tau = K_a - K_c$, with $-J \leq \tau \leq J$. Each J level has $2J+1$ magnetic sublevels or azimuthal degeneracy.

The calculation of H_2O CARS spectra is fairly straighforward and follows the treatment given by Grisch and Péalat.[22] The pure Raman part of $\chi^{(3)}$ is deduced from Eq. (4). For the present study, 475 rotational lines belonging to the (000) → (100) vibrational band and 158 rotational lines of the (010) → (110) vibrational band were used to calculate the CARS spectra in a spectral range of 155 cm^{-1} from 3501.682 cm^{-1} to 3657.053 cm^{-1}. The energy levels associated with these vibrational bands were derived from the survey of data published by Flaud et al.[23] Only the selection rules $\Delta J = \Delta \tau = 0$ associated with the Q-branch transitions were used. Furthermore, the variation of the Raman cross section with the Q transitions were ignored.

The Boltzmann population $\rho_{J\tau}$ is expressed by

$$\rho_{J\tau} = g_\tau \frac{(2J+1)}{Q(T)} \exp\left(-\frac{hcE_{J\tau}}{kT}\right) \tag{9}$$

where g_τ is the nuclear statistical weight (1 or 3 depending on whether the level is symmetric or not), $E_{J\tau}$ is the energy of the lower state, and the partition function $Q(T)$ is given by

$$Q(T) \cong 2\sqrt{\frac{\pi T^3}{ABC}} \prod_i \left(1 - \exp\left(-\frac{hv_i}{kT}\right)\right)^{-1} \tag{10}$$

The determination of the linewidths requires knowing population changing rate constants to compute the collisional narrowing observed in high-pressure conditions. Although no direct measurements of these parameters are available for water vapor, substantial progress has been possible using models based on the concept of statistical fitting laws. An excellent general overview of collisional effects on Q-branch spectra and their description, with different approaches for the modeling of relaxation rates, has been given by Robert.[24] There are exponential gap laws and inverse power laws, and combinations of these models, but they remain for the most part semi-empirical fitting laws with incompletely understood physical bases. In particular, a large variety of models based on scaling laws have been investigated for the rotational relaxation rates of H_2O.[25] The most successful of these relationships is the so-called collisional angular momentum exponent law (CAMEL); a simplified statement of this law is based on the concept of a collisionally transferred angular momentum induced by the angular momentum reorientation produced during collisions. In these investigations, CAMEL has been employed in the following form:

$$W_{J\tau,J'\tau'} = A_0 \left(\frac{T_0}{T}\right)^N (2J+1)\rho_{J\tau}\exp\left(-\beta\frac{|l|}{kT}\right) \tag{11}$$

where the collisionally transferred angular momentum l is

$$l^2 = J^2 + J - 2K_a K'_a - 2K_c K'_c + J'^2 + J'$$
$$- 2(J^2 + J - K_a^2 - K_c^2)^{1/2}(J'^2 + J' - K_a'^2 - K_c'^2)^{1/2} \tag{12}$$

The parameters A_0, N, and β are determined from the inversion of a large number of experimental Q-line broadening coefficients and using the sum rule [Eq. (7)], which ensures the conservation of the total rotational distribution. For the H_2O–H_2O self-broadening collisions, the best parameters are the following: $A_0 = 1.90 \text{ mK} \cdot \text{atm}^{-1}$, $\beta = 60 \text{ cm}^{-1}$, $N = 1.11$, $\gamma_\lambda = 0$, and $T_0 = 300 \text{ K}$. Influences of the H_2O–O_2 and H_2O–H_2 collisions were neglected in the present study because H_2 and O_2 act as less efficient collision partners for rotational energy transfer.

The code has been expanded to simulate the interference between the resonant and the nonresonant CARS signal. Figure 5 shows the sensitivity of CARS spectra calculated at 2300 K for various nonresonant susceptibility levels. The results indicate well-defined modulated spectral profiles in the concentration range of 0.5–40%. Outside this domain, this technique becomes insensitive to the H_2O concentration. At low concentrations, the modulation disappears into a baseline level derived from the nonresonant susceptibility, preventing any concentration measurements. For concentrations large enough that there is no measurable modulation in the CARS spectrum, concentration cannot be deduced automatically from the spectral profile. The H_2O concentration must be deduced from the integrated signal intensity. Unfortunately, this technique could not be used in the present study because of the harsh experimental conditions encountered in the combustion chamber.

After creating a database of $\chi^{(3)}(\omega)$, the calculated spectra were then convoluted with the apparatus function. This function was obtained by measuring the response of the detection system to the O_3-line of H_2 at 3568.24 cm^{-1}. The spectrum was calibrated with the same O_3-line by recording a sequence of single-shot CARS spectra on a known mixture of H_2 and N_2. The dispersion coefficient was deduced from the hydrogen Q-lines present on the vicinity of the H_2O CARS spectra. For each pressure condition, a library of theoretical CARS spectra was created at 50-K increments over the range 300–3300 K and at 0.01 increments for χ_{NR}. Temperatures and H_2O concentration were deduced by fitting the experimental CARS spectra to the theoretical ones. The fitting program uses a nonlinear least-squares fitting procedure based on a Marquardt algorithm.[26]

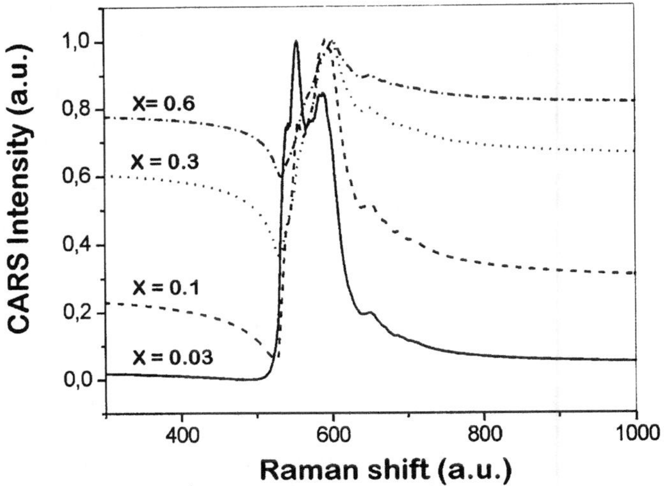

Fig. 5 Theoretical H_2O spectra at 2300 K showing the effect of varying χ_{NR} on spectral shape. X is defined as the ratio between the nonresonant susceptibility of the medium and the nitrogen nonresonant susceptibility.

Normalization for the dye laser spectral profile was also done in the fitting procedure. This profile was obtained experimentally on pure nitrogen by acquiring a sequence of 1000 single-shot nonresonant CARS signals. The single-shot measurements were averaged to remove statistical dye laser noise and to get a smooth envelope of the profile of this normalization spectra.

V. Experimental Setup

As illustrated in Fig. 6, the hardware required to implement CARS can be divided into three subsystems: the lasers, the handling optics, and the detection systems.

The laser system is composed of two separate optical benches, which produce the pump and Stokes beams required for H_2 and H_2O CARS spectroscopy. For H_2, the pump beam is the doubled-frequency output of an Nd:YAG laser chain composed of a single-mode Q-switched oscillator followed by an amplifier. The laser delivers 140 mJ in 13-ns pulses with a frequency rate of 10 Hz, thus providing single-shot measurements that can be used to study the dependence/ variability of a process. Half of the green energy is used to pump the Stokes dye laser, which emits the broadband ω_2 beam. LDS 698 dye is diluted in a mixture of methanol and ethanol to produce the Stokes beam centered at 683 nm with a $200\,cm^{-1}$ bandwidth (FWHM). At the output of the laser bench, the pump beam is split in two parallel beams, and one of them is overlapped with the ω_2 beam (planar BOXCARS arrangement).[8] The energy per pulse of the beams is 30 mJ for each of the pump beams and 4 mJ for the Stokes one. All beams are horizontally polarized.

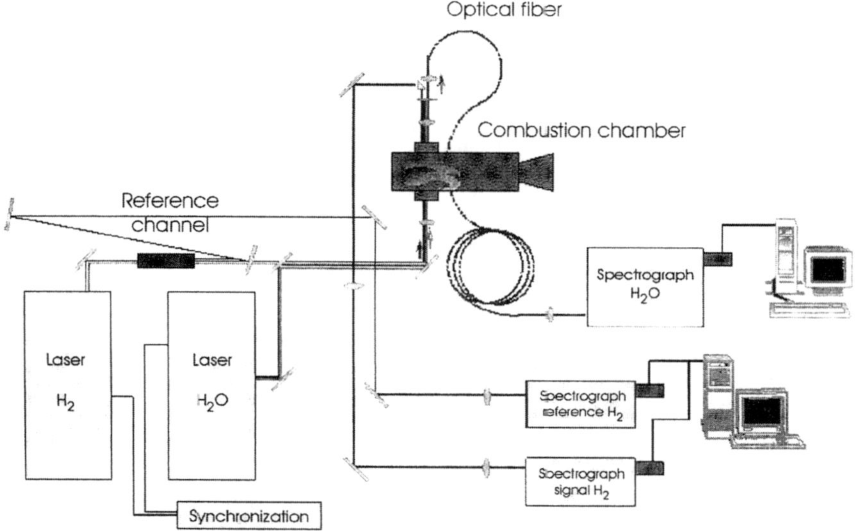

Fig. 6 Schematic diagram of CARS experimental setup.

The H_2O pump laser is a Quanta-Ray DCR 3D Nd : YAG unstable oscillator-amplifier system whose output is frequency doubled to produce a 532-nm laser pulse with a duration of 8 ns. The output is multimode and the bandwidth about $0.7\ cm^{-1}$. The green output is split by a dichroic mirror to pump a broadband dye laser and to act as the pump beam in the CARS process. DCM dye was dissolved in Dimethyl Sulfoxide (DMSO) to match the water vapor Q-branch, which is located on the anti-Stokes side from $3200\ cm^{-1}$ to $3700\ cm^{-1}$. The bandwidth of the Stokes laser was narrowed by an intracavity etalon. The narrowed excitation profile of the Stokes laser ranged from $3100\ cm^{-1}$ to $4200\ cm^{-1}$. By angle-tuning the etalon, the Stokes intensity maximum can be shifted spectrally to the maximum of the H_2O Q-branch at $3655\ cm^{-1}$. All beams have a linear and parallel polarization. The unstable-resonator spatially enhanced detection CARS (USED-CARS) beam geometry was chosen because it is easier to align and not so sensitive to loss of beam overlap caused by refractive index variations.[9]

A trigger signal issuing from the electronics of the H_2 CARS laser and delayed by a delay generator (Stanford research, DG535) was used to fire the H_2O CARS laser. Using this technique synchronous operation of both lasers at 10 Hz with a jitter of less than 200 ns was achieved.

The H_2 beams are focused first in an atmospheric pressure flux flow of argon where a nonresonant CARS signal is created to monitor the shot-to-shot fluctuations of direction and of pulse energy of the laser beams and of the spectral shape of the Stokes laser. Then, that reference signal is split off, and the H_2 laser beams are combined spatially with the H_2O laser beams onto the same path using a single mirror (i.e., the H_2 beams passed over the mirror while the

H_2O beams are reflected on the mirror). All laser beams pass to the testing facility through optical windows where they are focused in the chamber by means of a single 200-mm-focal-length achromat yielding a 1-mm-long and 50-μm-diam probe volume for H_2 and 2-mm-long and 100-μm-diam for H_2O. The adjustment of the same focus point for all beams is performed by burning a hole in a metal foil with the H_2 laser beams and adjusting the passage of the H_2O laser beams through the same hole. According to this strategy, the beam waists were not more than 50 μm apart. The focal volume is positioned axially and radially in the combustion chamber by moving the optical lenses by means of translational stages. An OG 530 Schott long-pass filter is placed along the H_2 optical path just before the first reactor window to reduce the energy of the pump beams by 2. The resulting pump and Stokes beam pulse energies, 15 and 2 mJ, respectively, prevent problems like damage to windows and laser-induced breakdown into the flowfield. For H_2O, the Nd:YAG laser is operating at full power to reach stable thermal conditions. The pump beam energy is adjusted using a rotating zero-order half-wave plate and a Glan-Thompson prism while the Stokes beam energy is fixed. Energies at the probe volume are 18 mJ for the pump beam and 4 mJ for the Stokes beam.

The H_2 CARS signal is sent to the spectrograph by a series of mirrors. Reference and sample H_2 CARS spectra are dispersed using two separate spectrometers equipped with holographic gratings (2100 grooves/cm^{-1}, radius of curvature 800 mm). The hydrogen spectrum and the broadband reference are formed in the output plane of the spectrographs and detected by means of intensified 512-photodiode arrays (OMA III 1420B/512/HQ-FP). The spectral dispersion is 0.25 cm^{-1} per diode and the resolution is 1.6 cm^{-1} (FWHM). After leaving the combustion chamber, the H_2O CARS signal, which was geometrically separated from the H_2 CARS signal, was guided off the optical path by means of a prism. The H_2O laser beams were blocked by a dichroic filter, and the CARS signal was focused into a single 550-μm-diam quartz fiber by means of a 120-mm-focal-length achromat. The fiber guided the CARS signal onto a 1-m spectrograph (Jobin Yvon HR1000) equipped with a 1200-grooves/cm^{-1} grating. The spectral dispersion per diode was 0.95 cm^{-1}. An input coupler optics matched the aperture of the signal beam to the spectrograph aperture of 1/10. The water vapor spectra have been sequentially recorded using an intensified gated diode array (Princeton Instruments IRY1024 Blue). The 10-Hz sequence of H_2O CARS spectra was stored on a personal computer and displayed in real time on the personal computer. The two data acquisition systems were synchronized using the same external trigger issued manually after combustion was stabilized in the chamber.

VI.   Results

Because this chapter focuses mainly on diagnostic techniques, only results obtained in the Mascotte combustor for condition A are presented. This allows discussion of the main features of the experiment, as well as some characteristic properties of the flame. Measurements are carried out in the combustion chamber for different locations downstream from the injector by placing the combustor modules with and without optical ports as necessary.

A. H_2 Data Reduction

Depending on the location in the flame, the H_2 CARS spectra present some distortion in shape and in intensity. Figure 7 shows some typical signals from single-shot H_2 CARS spectra with different levels of perturbations. Experimental observation indicates that the variation could be related to the presence of a background signal, which may swamp the CARS signal (Figs. 7b, 7c). This background may be attributed to the optical breakdown of the liquid phase, which becomes ionized by the intense electric fields of the laser radiation. The resulting plasma, which is composed of ions and atoms, possesses a large third-order non-resonant susceptibility, which produces a nonresonant CARS signal large enough to compete with the hydrogen CARS signal. This trend is clearly illustrated in Fig. 7c, where the spectral signature corresponds to laser-induced breakdown emitted by the neutral O transition located at 436.8 nm. The extent of this effect is related to the intensity and spatial quality of the laser beams, the droplets loading, and the molecular number density of the medium. Generally, this serious problem can be circumvented by reducing the beam intensities to the breakdown threshold of the medium. Breakdown can be totally avoided, but a considerable reduction of the signal strength is also noted. Preserving the CARS signals requires then a longer focal length lens to focus and cross the beams, which leads to an accompanying increase in the interaction length in the Boxcar phase-matching geometry. This method, well adapted in experiments in which the loss of spatial resolution is not critical, was inapplicable in our experiment because of the reduced size of the combustion chamber and the desired spatial resolution (≈ 1 mm). The method selected to minimize this effect was simply to reduce the pump beam intensities to a threshold compatible with a simultaneous decrease in the laser breakdown threshold by liquid and generation of CARS spectra of good quality.

Another source of experimental error can be related to the molecular composition of the probe volume caused by some intense turbulent mixing flow processes in the flame. It was observed that the H_2 CARS signal varies with temperature and mole fraction over the range of interest in combustion by more than several orders of magnitude, in excess of the linear dynamic range of optical multichannel detectors. The consequence of this large dynamic is that the detector diodes could be overexposed for some laser shots, leading then to measurement error. Consequently, the validation rate of the measurements, defined as the ratio between the number of spectra successfully processed and the total number of laser shots during a run, could range between 0 and 100% according to the location in the flame.

For all the H_2 CARS signals successfully processed, the temperature was obtained by two data reduction methods depending on the pressure regime. At atmospheric pressure, the temperature is deduced by plotting the amplitudes of the rotational Q-lines in a Boltzmann diagram. The rms deviation of the instantaneous temperature is defined as the uncertainty given by the least-squares fitting routine, reflecting the accuracy of the complete apparatus including the data processing. Several main sources of uncertainties can be distinguished in our experiments: amplitude fluctuations that arise from the response characteristics of the optical detection, measurement uncertainties on peak amplitude related to the

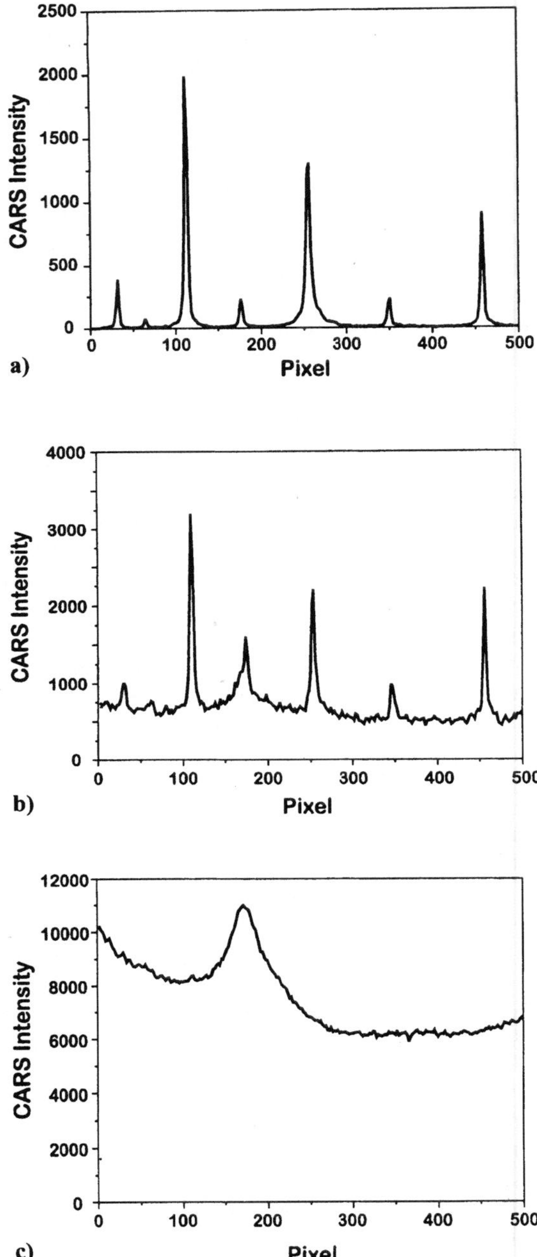

Fig. 7 Experimental H_2 CARS spectra recorded with different breakdown levels: a) no breakdown case, b) intermediate case, and c) breakdown case.

Poisson uncertainty, random dye laser intensity distribution, spectral distortions and disruption of the CARS signals, and beam steering and defocusing due to density gradients. Each of these effects is discussed separately.

First, it is well known that multichannel detectors can bias data processing with regard to their nonlinearity responses.[27] This one, checked with single-calibrated neutral density filters (NG Schott filters) and H_2 CARS spectra recorded at room temperature, was found to be linear on 90% of the dynamic range of the detector.

Second, the spectral dye intensity distribution can contribute significantly as a source of error for temperature measurement. According to nonlinear dependencies of the CARS signal upon the power density as well as the mode structure of the Stokes laser, a reference scheme must be used systematically to reproduce spatial and temporal characteristics of the sample beam focal volume to a high degree.[8] This task was accomplished by using a power reference channel mentioned previously to record for each laser shot a reference CARS spectrum displaying the intensity distribution of the laser beams. Within this arrangement, the correlation between sample and reference intensities are consistently $\approx 95\%$.

Finally, mixing with the liquid phase, leading to possible beam absorption by droplets, and large density gradients, causing more beam defocusing, are all likely sources of noise. The signal generation efficiency can obviously be reduced depending on the density of liquid present in the probe volume. As an illustration of the capabilities of the measurement technique, a large number of spectra were recorded for different thermodynamic conditions to evaluate the accuracy of the overall fitting procedure, including the background and response corrections, and the reference normalization. Figure 8a presents a typical single-shot H_2 CARS spectrum recorded at room temperature in a sample gas mixture of 10% hydrogen and 90% argon. Analyzing the $J = 1-3$ Q-line distribution by the Boltzmann diagram gives a temperature of 294 K with a resultant uncertainty of the order of 10 K. The average temperature obtained from a series of 150 consecutive single-shot measurements was 296 K with an rms error of 10 K in good agreement with thermocouple temperature measured simultaneously (295 K). The second significant result was obtained during a run at 0.1 MPa at a location where the measurements are not affected by the optical breakdown, i.e., $x = 50$ mm and $z = 20$ mm (Fig. 8b). In that case, the $J = 1-7$ Q-lines distribution is used to measure the temperature. Analyzing the rms deviation of 200 single-shot CARS spectra suggests that the accuracy of single-shot measurements is on the order of 3–4%.

At pressures higher than 0.1 MPa, the effects of collisional linewidths on the CARS spectra become significant and instantaneous temperatures are obtained using the fit procedure described previously. Note that the method provides an expression of CARS intensities as a function of two parameters, namely, the temperature and the H_2 molar fraction. Shown in Fig. 9 are typical fits of single-shot CARS spectra recorded at $x = 100$ mm downstream the injector. The spectrum plot in Fig. 6a is a comparison between the best-fit spectrum and the single-shot spectrum at $p = 3$ MPa. The theoretical spectrum fits the data very well with parameters of temperature $T = 1900$ K, and molar fractions of H_2 and H_2O, $C_{H_2} = 5\%$ and $C_{H_2O} = 95\%$. Another example is displayed in Fig. 9b, where measurements were performed under supercritical conditions ($p = 6.5$ MPa). In both cases, the overall agreement between experimental and

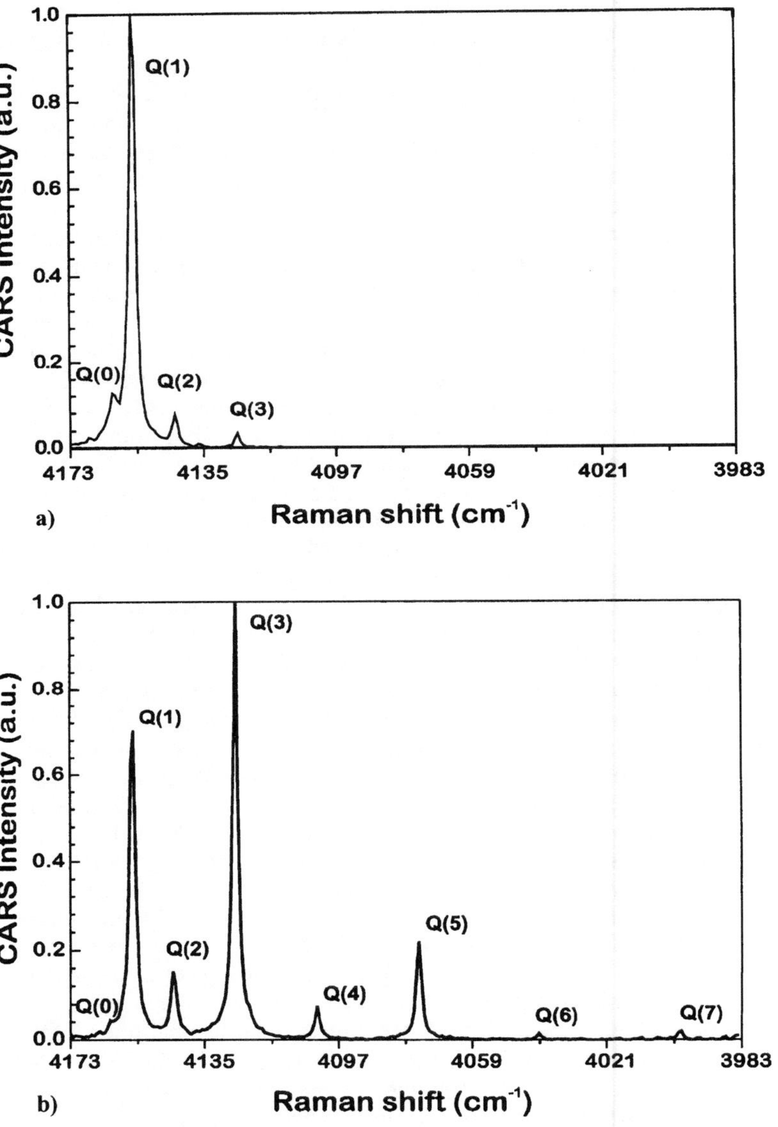

Fig. 8   Typical H_2 CARS spectra recorded at atmospheric pressure: a) inert case at room temperature and b) reactive case during a test run.

theoretical spectra is good. Detailed small discrepancies between the theoretical and experimental spectra probably arise from laser power fluctuations not totally corrected and from uncertainties relating to the imprecision of the linewidth parameters at high temperature.

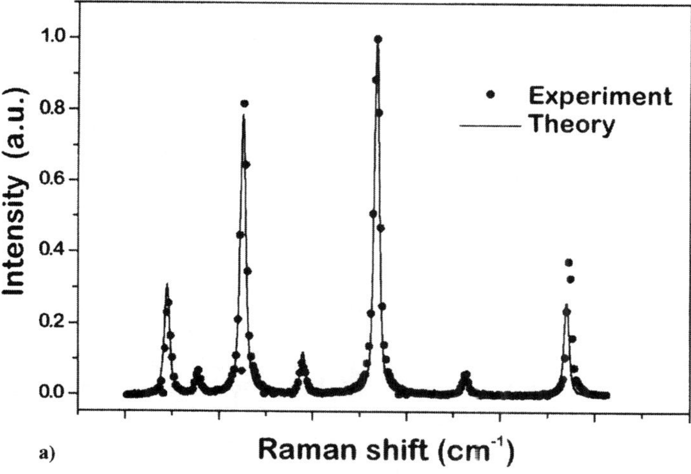

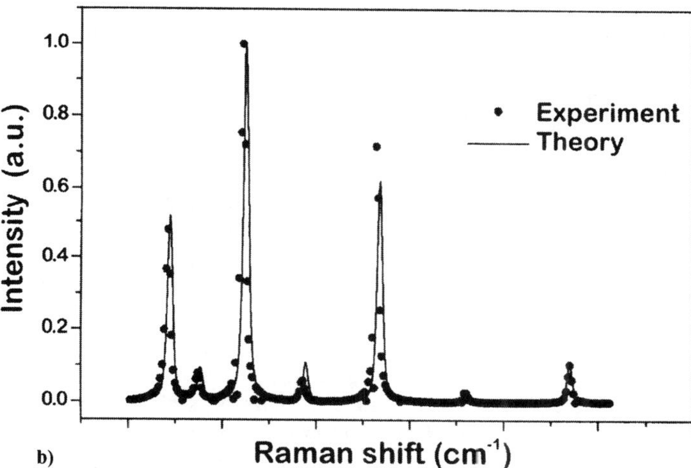

Fig. 9 **Comparison of experimental single-shot H_2 CARS spectra to calculated spectra presenting the best fit. Operating conditions: a) A-30, $p = 3$ MPa and b) A-65, $p = 6.5$ MPa.**

As stated for the atmospheric pressure case, accuracy of measurements depends on the same experimental sources of error, and also on the precise knowledge of collisional linewidths. We must remember that the collisional line parameters used in the present work are obtained only in static cells temperatures lower than 1500 K. Above this limit, the collisional parameters are only extrapolated from the linewidth model.[21] Effects of uncertainty in the collisional linewidths on the accuracy of temperatures were estimated by calculating CARS spectra with a 10% variation of the collisional linewidths. Whatever the range encompassing the expected flame temperature, the resulting maximum error in the inferred temperature never exceeds 50 K. Adding all the possible sources of errors provides an uncertainty in temperature of at most twice that of the atmospheric pressure conditions, i.e., about 6–8%.

B. H_2O Data Reduction

Similar distortions of the CARS spectra in shape and intensity were also experienced in the H_2O CARS signals. Significant beam attenuation occurred mainly in the central part of the combustion chamber and in the region of the flame fronts, where LOX is present. Additionally, a proportion of the CARS spectra depending on the flame conditions was also subject to the presence of a laser-induced breakdown induced by the liquid phase leading to a similar variation of the validation rate of the H_2O measurements such as H_2.

As discussed previously, the modulation of the CARS spectrum exhibited by interferences between the resonant and nonresonant contributions to third-order nonlinear susceptibility makes it possible in certain ranges to perform concentration measurements from the spectral shapes. An example of concentration and temperature measurements inferred from the spectral shapes of CARS measurements is shown in Fig. 10. The upper trace displays the comparison between a single-shot experimental CARS spectrum recorded into a "cold" region of the flowfield with the theoretical spectrum presenting the best fit. The lower trace shows the case at higher temperature. The experimental noise structure noted on the experimental spectra arises from the quantum statistical uncertainty in the ratio of the single-shot experimental spectra by the averaged reference signal. Although agreement between theory and experiment may be good, it is difficult to quantify the accuracy of the temperature and concentration measurements. The accuracy of the fitted temperatures and concentrations was found to be sensitive to a number of factors, including collisional linewidths, the value of the nonresonant susceptibilities, and the reference normalization. For all these factors, we have used the most up-to-date models presented previously, and we have attempted to quantify, where possible, the errors associated with remaining uncertainties in these models and parameters. The estimated total uncertainty for the measurements, corresponding to a single standard deviation for a single data point, was found to be 10%.

C. Flame Measurements

At atmospheric pressure, a sequence of 150 single-shot measurements was recorded during each run for each probe species. As mentioned previously, the validation rate of measurements during a run varies from 0 to 100%. For locations presenting a convenient validation rate, typically above 50%, the temperature evolution vs the sequential laser shot number indicates no low-frequency variation (Fig. 11). The combustion can then be considered as stationary during

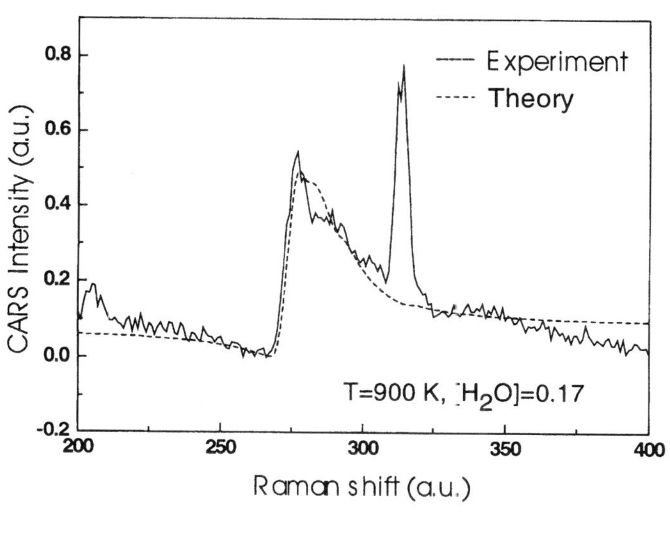

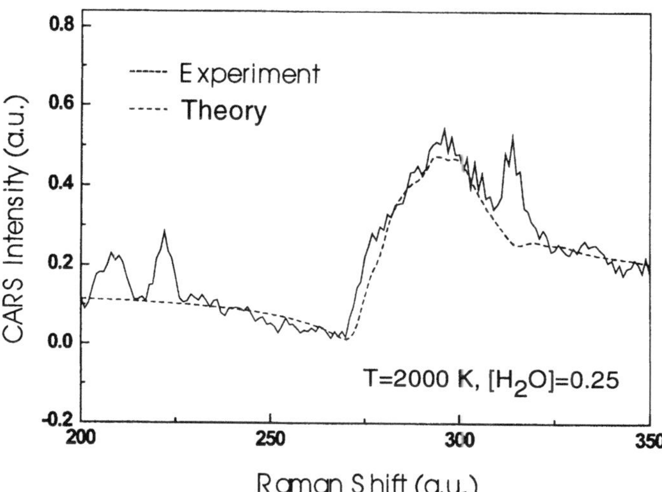

Fig. 10   Comparison of experimental single-shot H_2O CARS spectra to calculated spectra presenting the best fit. Operating conditions: A-10, $p = 1$ MPa.

the data acquisition. Histograms, mean values, and standard deviations can be deduced from the single-shot measurements. Figure 12 shows typical results for H_2 displayed with a temperature column step of 100 K matching the apparatus uncertainty. The scatter in the results, which is larger than the measurement uncertainty, gives insight into the degree of turbulence of the combustion.

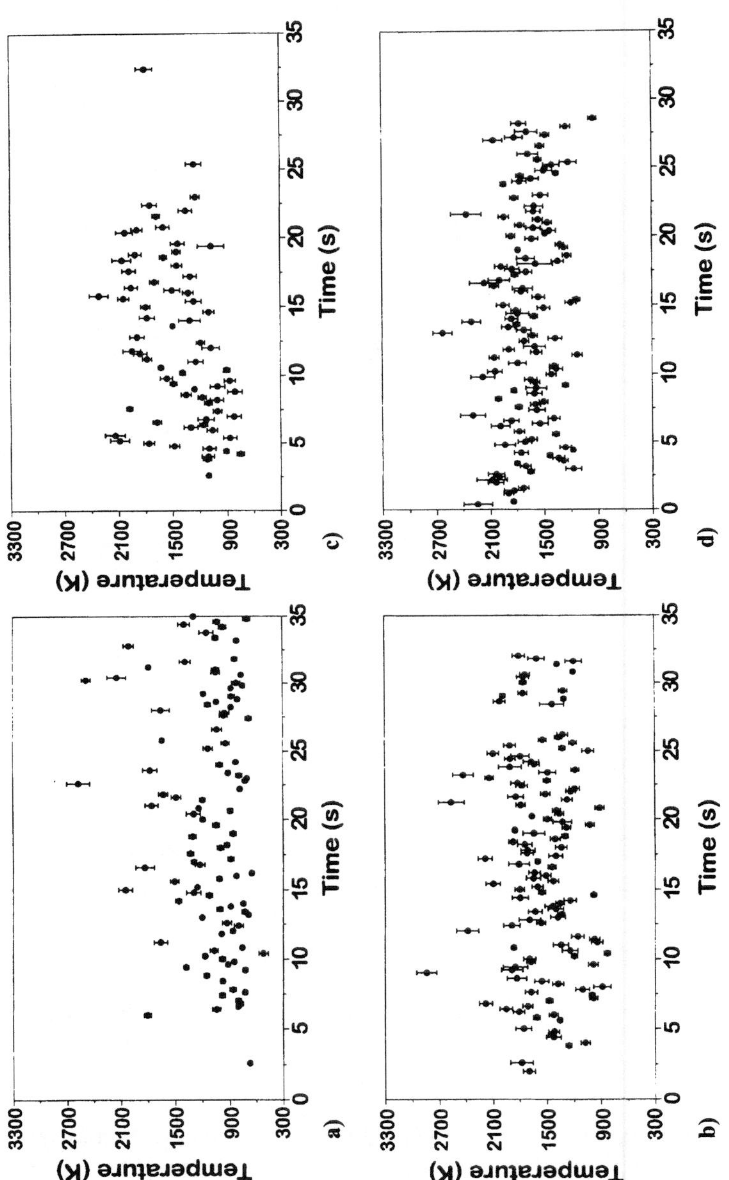

Fig. 11   Time dependence of single-shot H_2 temperature measurements recorded at several locations downstream from the injector. Operating conditions: A-1, $p = 0.1$ MPa. The data were collected at $y = 20$ mm.

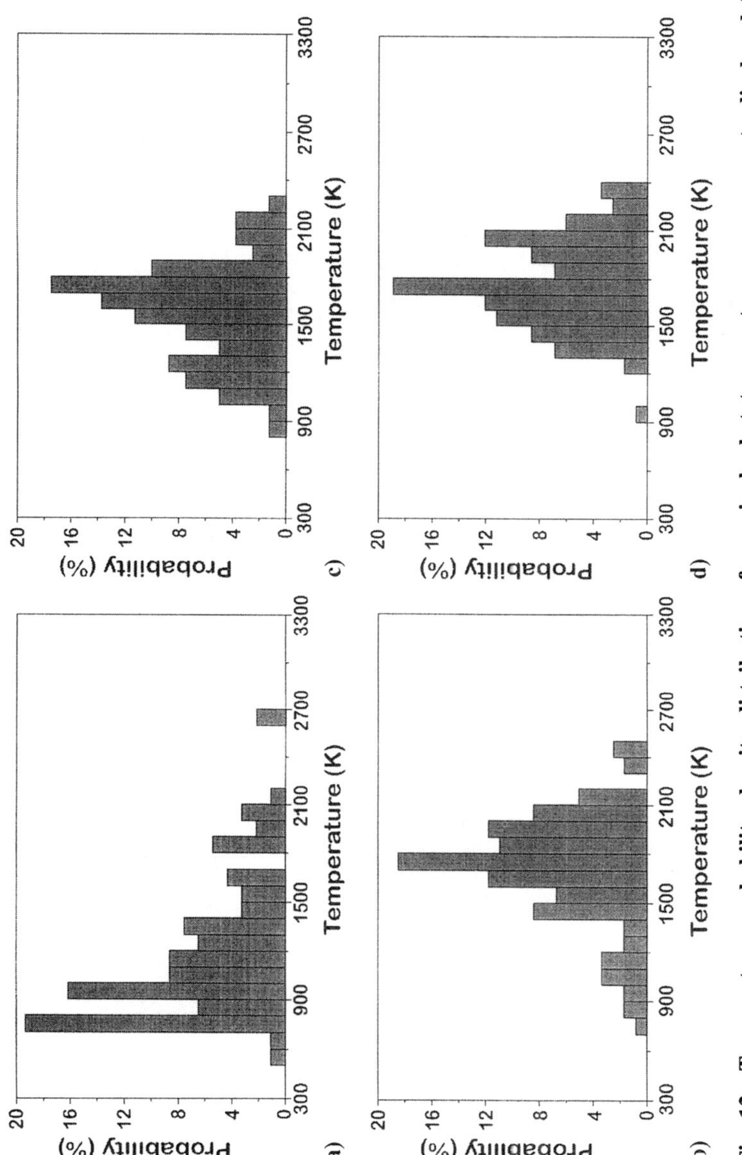

Fig. 12 Temperature probability density distributions from single-shot temperature measurements displayed in Fig. 11.

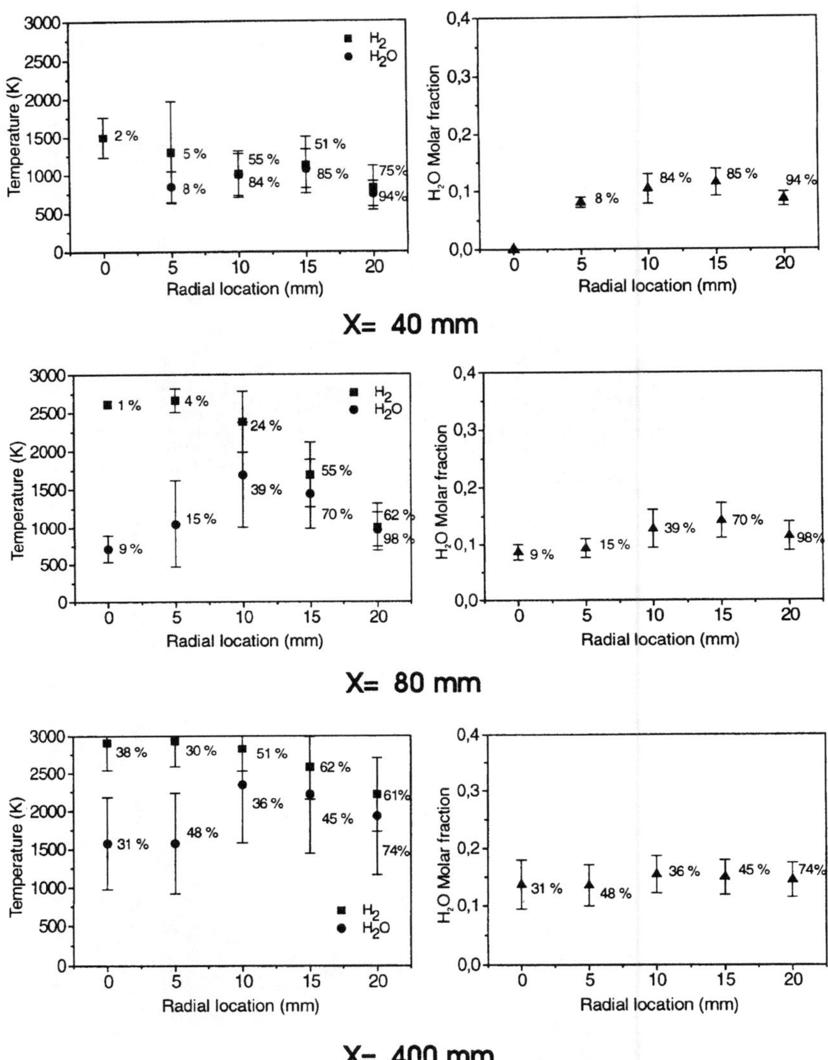

Fig. 13   Distribution of temperature and H_2O concentration recorded in several axial sections as a function of the distance from the axis. Operating conditions: A-1, p = 0.1 MPa.

Radial mean temperature and H_2O concentration profiles deduced from the data reduction of H_2 and H_2O CARS spectra are displayed in Fig. 13 for different locations downstream from the injector. The error bars represent rms fluctuations seen in measured temperatures. For each measurement, the validation percentage of the measurements is also indicated.

Near the centerline, the results show an absence of detection of H_2 CARS signals until the location $x = 200$ mm. This result indicates that most of the combustion occurs in the mixing zone between the two fluids and that in fact the flame seems to be located at the boundary between the two fluids, preventing any diffusion of H_2 in the LOX core flow. Beyond this location, the progressive H_2 diffusion occurs into the core of the flame from intense fluid mixing in the flame (increase of the validation percentage of the measurements). The mean temperature increases downstream from the injector to 2800 K at $x = 200$ mm and then remains stable. Similar tendencies are also observed in the outside periphery of the flame where the temperature profiles measured at different radial locations present similar behaviors. Finally, whatever the locations downstream of the injector, the H_2 radial temperature profiles decrease monotonically from the walls to the centerline axis, indicating thermal transfer between the wall and the gas mixture issuing from the reactive flow and the helium film.

For H_2O, the temperature and concentration profiles present some analogy with those usually encountered in gaseous diffusion flames. The H_2O temperature on centerline is cooler than at the peak temperature radial location, where the fuel and oxygen diffuse together and react. The profiles of H_2O concentration also exhibit similar shapes, with a peak coinciding approximately with the peak of the temperature. However, the behavior of the mean radial temperature profiles deduced from the H_2O CARS spectra differ from those measured in H_2. A maximal difference between temperatures of about 2000 K is obtained at $x = 100$ mm. Beyond this location, this difference decreases progressively. Note also that this difference in temperature is extreme on the flow axis and vanishes close to the walls.

These apparently different behaviors are undoubtedly due in part to possible biases of the temperature measurements by CARS and to the nature of the mixing flow processes occurring in this flame. Concerning H_2, the rejection of some distorted spectra in the data reduction can bias the mean temperature of H_2 toward higher values because of the systematic single-shot measurements of elevated temperatures. Existence of these high temperatures probably derive from the conditionally sampled data on H_2 mainly present in the region surrounding the LOX spray. For H_2O, the cooler temperature in the central flow region could be attributable to thermal effects induced by the presence of LOX. The H_2O molecules, initially produced at high temperature in the reactive zone, are transported as a relatively stable product in the core of the flow where a diphasic mixture composed of LOX (droplets, ligaments, etc.) and gaseous species takes place. During these processes, thermal transfer occurs between the two fluids leading to a decrease in the energy of the water vapor penetrating into the core flow. Consequently, the H_2O temperature level will be cooler on centerline than in the flame front. These thermal processes will probably control the combustion process on centerline; this is understandable because the thermal and convective diffusions will regulate the evaporation and the distribution of LOX at the time of combustion. Examination of Fig. 14 also reveals that the flame is characterized primarily by turbulence, driven by the shear produced between the two fluid flows. For instance, plots of the combination of the single-pulse H_2 and H_2O temperatures vs time recorded during a run on centerline at $x = 400$ mm

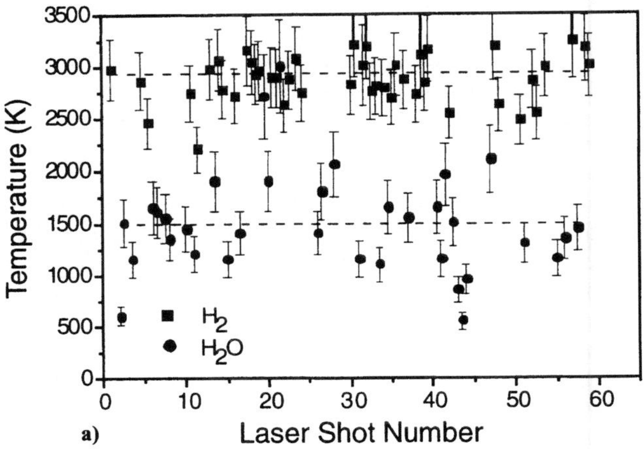

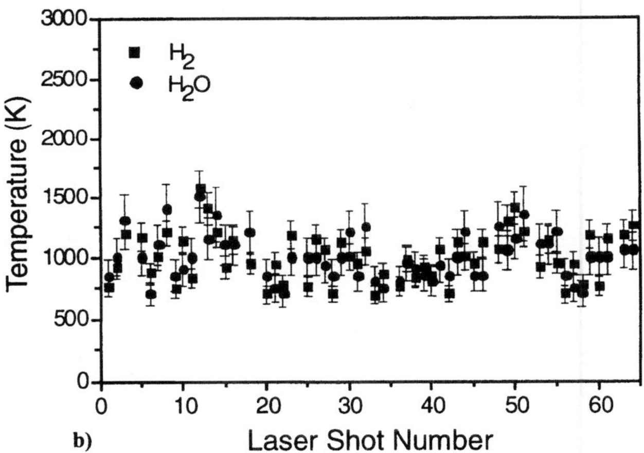

Fig. 14   Time dependence of temperature measurements at the locations a) $x = 80$ mm, $y = 20$ mm, and b) $x = 400$ mm, $y = 0$ mm.

show differences in behavior between temperature distributions coming from the mutually exclusive H_2 and H_2O signals. The fact that detection of high H_2 temperatures and cooler H_2O temperatures occurs is probably the result of the presence of a large-scale turbulent mixing reactive layer taking place around the LOX spray with the appearance of smaller packets of burning fuel separated by nonreacting regions of a mixture of combustion products and LOX. Outside the inner region, the large validation percentage of the measurements and the well-correlated temperatures for both species indicate the presence of a gaseous mixture (H_2 and H_2O) in thermodynamic equilibrium. All of these results confirm visual features of the dynamic processes responsible for the time-averaged flame shape obtained recently from OH-LIF, O_2-LIF, and emission measurements in the same flame.[1,28] Measurements indicate that the mean flame can be considered initially as a thin, cylindrical surface that expands downstream into a thick shell surrounding the gaseous oxygen and LOX spray formed after jet breakup. Further downstream, the mean flame becomes annular. The inner and outer diameters of the flame volume slowly decrease with axial distance as the oxygen convected in the central region is being consumed.

Further downstream (i.e., $x = 400$ mm), the inner jet is fully atomized due to the dominance of the outer hydrogen jet. On centerline, the validation rate of the measurements rises considerably, indicating that hydrogen starts to be entrained and mixed with reactant products within the inner jet flow. Complementary measurements devoted to probing O_2 with CARS at the exit of the combustor also show that combustion is still not complete at this location. A validation rate as high as 40% was obtained while detecting O_2 by CARS on the centerline of the combustor. All these results allow us to state that the decrease of the validation rate on H_2 near the flame axis is linked to stratification of the flowfield near the injector, which vanishes progressively further downstream when mixing is achieved.

At higher pressures (i.e., $p > 0.1$ MPa), temperature and concentration measurements were performed in sections located between 10 and 300 mm downstream of the injector. The combustor runs were 15 s long, reducing the CARS data sample to 75. Figures 15 and 16 display the radial temperature profiles recorded respectively for the conditions A-10 and A-30 at several locations downstream of the injector.

Similar trends are observed at a pressure of 1 MPa. The difference between the temperatures is still present. However, the length of the zone where the temperatures differ is reduced to about 200 mm, compared to the one observed at 0.1 MPa. The mean temperatures reach 2500 K, with a standard deviation of 600 K imputable to similar fluctuation rates induced by the turbulent flow. Between $x = 100$ mm and $x = 200$ mm and on centerline, the low validation percentage of H_2 measurements indicates that the H_2 mole fraction is quite low near the inner flow, indicating that LOX is still present. The difference between temperatures, smaller than the one measured at 0.1 MPa, indicates atomization and evaporation enhancements of LOX with pressure leading in a straightforward manner to faster burning, as is commonly thought. This explanation is also in agreement with results observed at positions farther downstream: larger validation percentage of measurements and similar temperature distributions suggesting the disappearance of droplets and the fast homogenization of the flow. Increasing the pressure to 3 MPa enhances these trends. Distributions of the mean temperature of H_2 and H_2O become similar from the location $x = 50$ mm, indicating the presence of a homogeneous mixture at this position. The mean H_2O concentration profiles present flat shapes with peak concentration of about 25% indicating that the flame produces more H_2O, which is in accordance with the lower mixture ratio level used at these pressure regimes. All of these results lead to the conclusion that an increase in pressure enhances

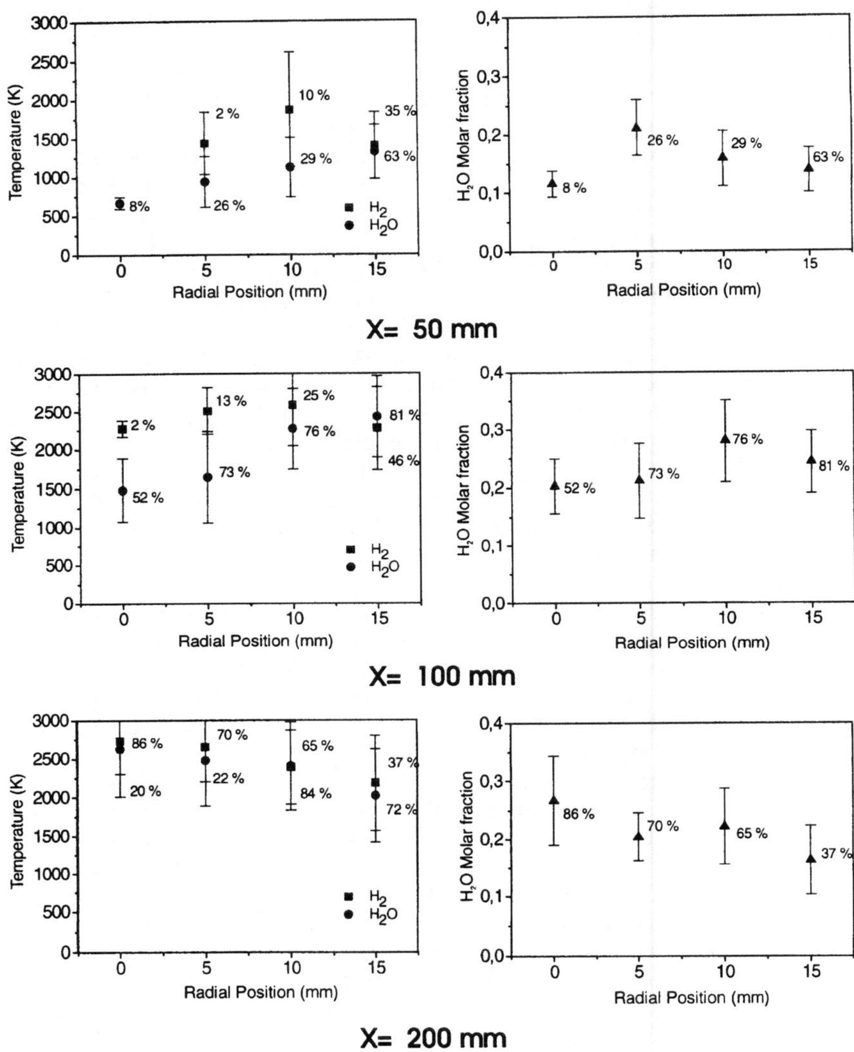

Fig. 15 Distribution of temperature and H_2O concentration recorded in several axial sections as a function of the distance from the axis. Operating conditions: A-10, $p = 1$ MPa.

the atomization and evaporation of the LOX, which in turn leads to a bigger portion of the O_2 with the fuel in the gas phase and changes in flame lengths and combustion efficiencies.

Adiabatic flame temperatures were also calculated with a simplified chemical equilibrium code that uses a reduced scheme with six species: H_2, O_2, H_2O, OH, O, and H. To that end, the theoretical temperatures were calculated for

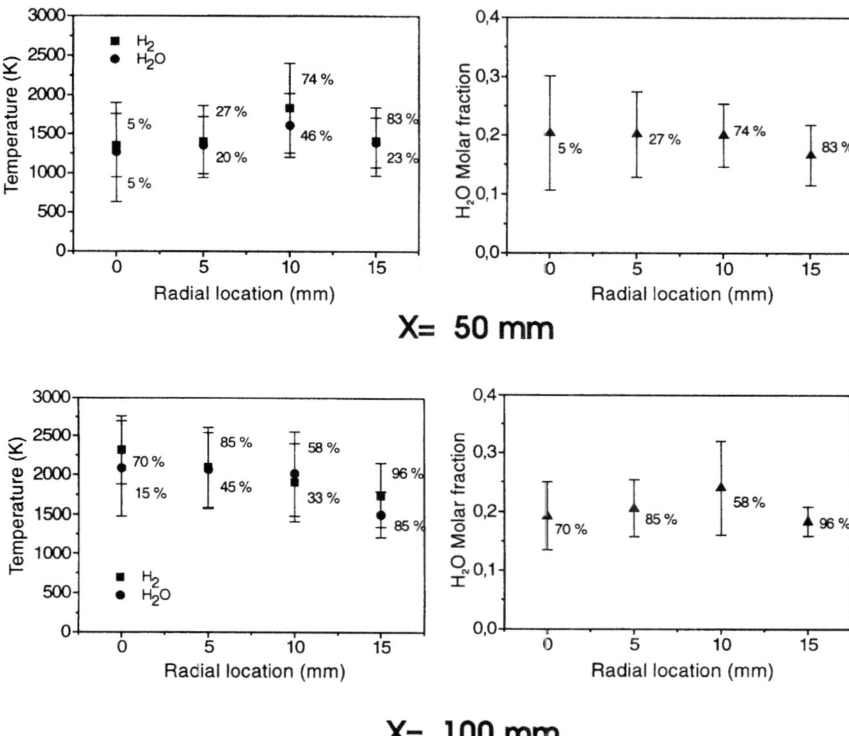

Fig. 16 Distribution of temperature and H$_2$O concentration recorded in several axial sections as a function of the distance from the axis. Operating conditions: A-30, $p = 3$ MPa.

equilibrium conditions and equivalence ratios equal to the experimental ones. Figure 17 shows the comparison between the measured temperature and calculations. The experimental temperatures are taken from results measured at $y = 10$ mm in sections close to the chamber exit, where measurements present large validation rates, assuming a quasi-achieved combustion for all the pressure conditions. As indicated in the figure, the experimental results are in good agreement with the numerical predictions. The mean temperature was slightly higher than the adiabatic flame temperature for all of the studied pressure regime, which suggests a local diffusion zone within the flame that is close to a stochiometric fuel equivalence ratio.

VII. Conclusions

The application of CARS for temperature and H$_2$O concentration measurements in high-pressure cryogenic GH$_2$/LOX combustors was successfully demonstrated. Single-shot measurements were performed in the Mascotte test facility with a spatial resolution in beam direction of $\sim$1 mm. Two CARS

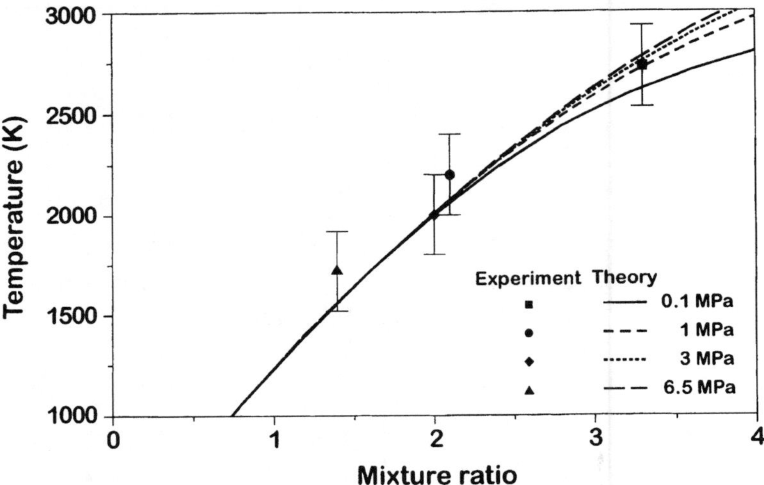

Fig. 17 Comparison of mean temperature profiles as a function of pressure. The lines correspond to equilibrium-based calculation, and the symbols refer to the experimental mean temperature recorded at the extremity of the combustion chamber and $y = 15$ mm.

setups were used to detect simultaneously H_2 and H_2O into the flowfield. Temperature was derived from the H_2 and H_2O broadband CARS spectra. H_2O concentration was deduced from the ratio between the resonant and the nonresonant background signals. The estimated accuracy of single-shot temperatures for H_2 is 3–8% depending on the pressure regime. For H_2O, the accuracy of temperature and species concentration is estimated to be about 10%, whatever the pressure range.

The experiment reveals that laser-induced droplet breakdown can cause serious problems in making temperature measurements using CARS in the core flow, where LOX is mainly present. Rejecting spectra with a low signal-to-noise ratio in the data reduction provides reasonable CARS temperature measurements that can be used to characterize the performance of the combustors, in particular the fuel/LOX mixing and the combustion efficiency. Furthermore, simultaneous measurements of both species give valuable information about the homogeneity of the combustion process about the state of the mixing and the temperature level and about the efficiency of the combustion. All of these results allow us to state that the mean flame can be considered initially as a thin, cylindrical surface that expands downstream into a thick shell surrounding the gaseous oxygen and LOX spray formed after jet breakup. This stratification of the two fluids vanishes progressively further downstream when mixing is achieved. Finally, we showed that fast evaporation and atomization of LOX by increasing the pressure is a viable means to achieve faster GH_2/LOX combustion.

Acknowledgments

This work was financially supported by SNECMA-Division Moteurs Fusées and by CNES. The authors acknowledge the technical support of J. Champy and A. Mouthon.

References

[1]Candel, S., Herding, G., Snyder, R., Scouflaire, P., Rolon, C., Vingert, L., Habiballah, M., Grisch, F., Péalat, M., Bouchardy, P., Stepowsky, D., Cessou, A., and Colin, P., "Experimental Investigation of Shear Coaxial Cryogenic Jet Flames," *Journal of Propulsion and Power*, Vol. 14, 1998, pp. 826–834.

[2]Ivancic, B., and Mayer, W., et al., "Experimental and Numerical Investigation of Time and Length Scales in LOX/GH$_2$-Rocket Combustors," AIAA Paper, June 1999.

[3]Eckbreth, A. C., Anderson, T. J., and Dobbs, B. M., "Multi-Color CARS for Hydrogen Fueled Scramjet Applications," *Applied Physics B*, Vol. 45, 1988, pp. 215–223.

[4]Anderson, T. J., and Eckbreth, A. C., "Simultaneous CARS Measurements of Temperature and H$_2$, H$_2$O Concentration in Hydrogen-Fueled Supersonic Combustion," AIAA Paper 90-0158, 1990.

[5]Vingert, L., Habiballah, M., Vuillermoz, P., and Zurbach, S., "Mascotte, a Test Facility for Cryogenic Combustion Research at High Pressure," 51st International Astronautical Congress, Rio de Janeiro, Brazil, 2000.

[6]Haberzettl, A., Gundel, D., Bahlmann, K., Thomas, J. L., Kretschmer, J., and Vuillermoz, P., "European Research and Technology Test Bench P8 for High Pressure Liquid Rocket Propellants," AIAA Paper, month, 2000.

[7]Regnier, P. R., and Taran, J. P., "On the Possibility of Measuring Gas Concentrations by Stimulated Anti Stokes Scattering," *Applied Physics Letter*, Vol. 23, 1973, pp. 240–242.

[8]Druet, S., and Taran, J. P., "CARS Spectroscopy," *Progress in Quantum Electronics*, Vol. 7, 1981, pp. 1–72.

[9]Eckbreth, A. C., *Laser Diagnostics for Combustion Temperature and Species*, edited by A. K. Gupta and D. G. Lilley, Abacus Press, Tunbridge Wells, England, UK, 1988.

[10]Greenhalgh, D. A., "Quantitative CARS Spectroscopy," *Advances in Non-Linear Spectroscopy*, edited by R. J. H. Clark and R. E. Hester, Vol. 15, Wiley, New York, 1988.

[11]Hall, R. J., Verdieck, J. F., and Eckbreth, A. C., "Pressure-Induced Narrowing of the CARS Spectrum of N$_2$," *Optics Communication*, Vol. 35, 1980, pp. 69–75.

[12]Eckbreth, A. C., and Hall, R. J., "CARS Concentration Sensitivity With and Without Nonresonant Background Suppression," *Combustion Science and Technology*, Vol. 25, 1981, pp. 175–192.

[13]Lavorel, B., Millot, G., Saint-Loup, R., Wenger, C., Berger, H., Sala, J. P., Bonamy, J., and Robert, D., "Rotational Collisional Line Broadening at High Temperatures in the N$_2$ Fundamental Q-Branch Studied with Stimulated Raman Spectroscopy," *Journal de Physique*, Vol. 47, 1986, pp. 417–425.

[14]Millot, G., Saint-Loup, R., Santos, J., Chaux, R., and Berger, H., "Collisional Effects in the Stimulated Raman Q-Branch of O$_2$ and O$_2$–N$_2$," *Journal of Chemical Physics*, Vol. 96, 1991, pp. 961–971.

[15]Dicke, R. H., "The Effects of Collisions upon the Doppler Width of Spectral Lines," *Physical Review*, Vol. 89, 1953, pp. 472–473.

[16]Galatry, L., "Simultaneous Effect of Doppler and Foreign Gas Broadening on Spectra Lines," *Physical Review*, Vol. 122, 1961, pp. 1218–1223.

[17]Varghese, P. L., and Hanson, R. K., "Collisional Narrowing Effects on Spectral Shapes Measured at High Resolution," *Applied Optics*, Vol. 23, 1984, pp. 2376–2385.

[18]Humlicek, J., "An Efficient Method for Evaluation of the Complex Probability Function: The Voigt Function and Its Derivatives," *Journal of Quantitative Spectroscopy and Radiative Transfer*, Vol. 21, 1979, pp. 309–313.

[19]Rosasco, G. J., Lempert, W., and Hurst, W. S., "Line Interference Effects in the Vibrational Q-Branch Spectra of N_2 and CO," *Chemical Physics Letters*, Vol. 97, 1983, pp. 435–440.

[20]Bonamy, L., Thuet, J. M., Bonamy, J., and Robert, D., "Local Scaling Analysis of State to State Rotational Energy-Transfer Rates in N_2 from Directs Measurements," *Journal of Chemical Physics*, Vol. 95, 1991, pp. 3361–3370.

[21]Grisch, F., Vingert, L., Bouchardy, P., Habiballah, M., Michaut, X., Berger, J. P., Saint-Loup, R., and Berger, H., "Coherent Anti-Stokes Raman Spectroscopy Measurements of Temperature in Shear-Coaxial Cryogenic Jet Flames," *Combustion and Flame* (submitted for publication).

[22]Grisch, F., and Péalat, M., "CARS Studies of H_2O Motional Narrowing in High-Pressure and High-Temperature H_2O-N_2 Mixtures," *Journal of Raman Spectroscopy*, Vol. 25, 1994, pp. 145–151.

[23]Flaud, J. M., Camy-Peyret, C., and Maillard, J. P., "Higher Ro-Vibrational Levels of H_2O Deduced from High Resolution Oxygen–Hydrogen Flame Spectra Between 2800–6200 cm^{-1}," *Molecular Physics*, Vol. 32, 1976, pp. 499–521.

[24]Robert, D., "Collisional Effects on Raman Q-Branch Spectra at High Temperature," *Vibrational Spectra and Structure*, Vol. 173, 1989, pp. 57–62.

[25]Porter, F. M., and Williams, D. R., "Quantitative CARS Spectroscopy of the v_1 Band of Water Vapour," *Applied Physics B*, Vol. 54, 1992, pp. 103–108.

[26]Press, W. H., Flannery, B. P., Teukolsky, S. A., and Vetterling, W. T., *Numerical Recipes, the Art of Scientific Computing*, Cambridge Univ. Press, 1986.

[27]Snelling, D. R., Sawchuk, R. A., and Smallwood, G. J., "Multichannel Light Detectors and Their Use for CARS Spectroscopy," *Applied Optics*, Vol. 23, 1984, pp. 4083–4089.

[28]Juniper, M., Tripathi, P., Scouflaire, P., Rolon, J. C., and Candel, S., "Structure of Cryogenic Flames at Elevated Pressures," *Twenty-Eighth International Symposium on Combustion, Edinburgh, The Combustion Institute*, Vol. 28, 2000, pp. 1103–1109.

Chapter 11

Propellant Ignition and Flame Propagation

Eric A. Hurlbert* and Robert J. Moreland*
NASA Johnson Space Center, Houston, Texas

and

Sebastien Candel[†]
École Centrale Paris and CNRS, Chatenay-Malabry, France

Nomenclature

AB = Arrhenius preexponentials
C = concentration
c_p = heat capacity
d_q = quenching distance
E = activation energy
E_{min} = minimum ignition energy
F = fuel
g = gas
h = heat
k = conductivity
l = liquid
m = reaction order with respect to fuel
N = total order of reaction
n = reaction order with respect to oxidizer
O = oxidizer
o = initial condition or total

---

*Propulsion System Engineer.

[†]Professor and Head of Mechanical and Aerospace Engineering, EM2C Laboratory. Associate Fellow AIAA.

p = pressure
Q = heat release per unit mass
R = universal gas constant
r = reactor radius
S = surface area
T = temperature
V = volume
W = molar mass
X = mole fraction
Y_F = fuel mass fraction
Y_O = oxidizer mass fraction
α = heat diffusivity of the mixture
δ = reaction vessel geometrical factor
$\dot{\omega}$ = reaction rate
ρ = volumetric mass

I. Introduction

IGNITION of propellants in rocket engines gives rise to many fundamental and practical problems. Because this process is so critical, it is important to carefully define the sequence of propellant injection and combustion initiation that will lead to controlled, reliable ignition and subsequently to flame stabilization under nominal conditions. Detailed analytical studies combined with numerical simulation and systematic testing are required to achieve this goal. The purpose of this chapter is to review some of the basic principles underlying ignition in rocket engines and to describe the technology of some ignition systems. Much information can be obtained in previous articles (see, for example, Altman and Penner[1]) and in books on rocket propulsion (for example, Barrère et al.,[2] Huzel and Huang[3]). Fundamental aspects of ignition are treated in many articles and in most textbooks (for example, Oppenheim,[4] Williams,[5] Strehlow,[6] and Glassman[7]).

First it is important to understand that the physical mechanisms that lead to combustion at the start (designated as ignition) must be distinguished from those that control flame initiation under steady-state operation. In the first case, propellants are injected in a relatively cold environment featuring ambient pressure (low pressure) and temperature, whereas in the second the reactants are delivered to a region of much higher pressure and temperature. We will consider in what follows only problems of ignition at the start.

Now, it is quite standard to describe ignition by first distinguishing two categories of propellants: 1) hypergolic propellants that ignite spontaneously after contact without any source of external energy and 2) nonhypergolic propellants that are inert under standard injection conditions and require an external source of energy to initiate the reaction. In some cases a catalyst is used to start the reaction kinetics.

For both types of propellants, the ignition delay τ_i is the fundamental parameter of the process. This delay governs the ignition overpressure (the pressure peak observed during ignition). It is important to a have a precise knowledge of this parameter under representative operating conditions because the value of τ_i

conditions the starting sequence. For hypergolic propellants the ignition delay is essentially governed by kinetics of the propellants during which exothermic liquid/vapor reactions initiate combustion throughout the chamber at the many surfaces of contact of the reactants.

For nonhypergolic propellants the delay depends on the gas-phase kinetics of the reactants, but it is also governed by the initial heat deposition, the rate of vaporization of the propellant injected as a liquid, and by the rate of turbulent mixing during the starting phase. Because the external source of energy is generally localized, ignition will involve flame propagation from the first flame kernel toward the surrounding reactants. It is then important to describe this phase to account for the ignition delay.

This review begins with some fundamental considerations and basic concepts (Section II). Specific aspects of ignition of nonhypergolic propellants are then treated in Section III, and various features of ignition of hypergolic propellants are described in Section IV.

II.   General Background and Fundamental Considerations

Among the many design factors, the ignition delay τ_i and the ignition overpressure p_{max} stand as the most significant. It is natural to compare the delay τ_i to the mean residence time in the chamber $\tau_r = M/\dot{m}_e$, where $\dot{m}_e$ is the mass flow rate through the nozzle, and M is the mass of gas inside the chamber. The ratio τ_i/τ_r essentially determines the overpressure p_{max}. To see this more clearly, we use a simple argument of Barrére et al.[2] Consider first the rocket motor operating in a steady state. The mass of propellants residing in the chamber is at any instant $\bar{\dot{m}}\tau_r$ and the pressure in the chamber is

$$p_c = \bar{\dot{m}}\tau_r \frac{R\,T_c}{W\,V} \tag{1}$$

where W designates the mean molar mass of the gases in the chamber. During the starting phase the quantity of reactants introduced in the motor is

$$\int_0^{\tau_i} \dot{m}_i(t)\,\mathrm{d}t = \bar{\dot{m}}_i\tau_i \tag{2}$$

If this quantity of propellants is burnt instantaneously at time $t = \tau_i$, the corresponding pressure will reach a value

$$p_{max} = \bar{\dot{m}}_i\tau_i \frac{R\,T_c}{W\,V} \tag{3}$$

The ratio of this ignition pressure to the nominal pressure is then given by

$$\frac{p_{max}}{p_c} = \frac{\bar{\dot{m}}_i\tau_i}{\bar{\dot{m}}\tau_r} \tag{4}$$

This expression should not be taken as exact. It provides only an estimate of the ignition overpressure and indicates that a longer delay τ_i will induce a higher pressure peak. To reduce the amplitude of this pressure, one has to decrease the delay and inject the propellants progressively into the chamber.

More precise estimates of the pressure evolution and ignition overpressure may be obtained by considering the fundamental balance equations describing the state of flow inside the chamber. It is convenient to first consider the balance of mass. Let $M(t)$ designate the mass of gas inside the chamber. The variation of this quantity as a function of time is due to the difference between the mass flow rate of burnt gases produced by the reacting propellants $\dot{m}_b(t)$ and the mass flow rate of gases exhausted through the nozzle $\dot{m}_e(t)$:

$$\frac{\mathrm{d}M}{\mathrm{d}t} = \dot{m}_b - \dot{m}_e \tag{5}$$

The mass M is related to the pressure through the perfect gas law:

$$M = \rho_g V = \frac{pW}{RT} V \tag{6}$$

For a constant molar mass W and temperature $T = T_c$, the rate of change $\mathrm{d}M/\mathrm{d}t$ is directly proportional to $\mathrm{d}p/\mathrm{d}t$, and the balance of mass governs the evolution of pressure. It is then quite easy to calculate the ignition overpressure. To account for variations of molar mass and temperature, one has to use a more complete set of equations, including the balance of species and energy. This will provide a more precise estimate of the ignition overpressure.

At this point it is worth reviewing some principles of ignition. We will successively consider the autoignition of a homogeneous volume of reactants, the minimum ignition energy, effects of turbulence and liquid droplet vaporization on ignition delay, modeling of flame propagation from an initial flame kernel, and compressibility effects on the evolution of flame kernels. This section ends with some considerations on numerical modeling of the ignition processes.

A. Autoignition of Homogeneous Volume of Reactants

Let us consider a volume V filled with premixed reactants. This mixture is brought to an initial temperature T_o. One assumes that the volume is perfectly isolated and that there is no heat exchange with the surroundings. A single-step irreversible reaction

$$v_F' F + v_O' O \longrightarrow P \tag{7}$$

may take place in the mixture with a rate given by an Arrhenius law of the form:

$$\dot{\omega} = B C_F C_O \exp\left(-\frac{E_a}{RT}\right) \tag{8}$$

where B, T_a, C_O, and C_F are, respectively, the preexponential factor, the activation energy, and molar concentrations of oxidizer and fuel. These concentrations may be expressed in terms of mass fractions Y_O, Y_F and molar masses W_O, W_F by $C_O = \rho Y_O / W_O$, $C_F = \rho Y_F / W_F$. Because there is no heat exchange with the environment, the heat release Q serves to augment the temperature of the mixture. The dynamics of the system are then easily described with asymptotic methods. One may analyze the thermal runaway by assuming that the temperature is only slightly perturbed from its initial value. It is then possible to show that temperature increment $T_1 = (T - T_o)(E_a/RT_o)$ follows a logarithmic law:

$$\frac{T_1}{T_o} \approx -\ell n\left(-\frac{t}{\tau_i}\right) \tag{9}$$

where the ignition delay τ_i is given by

$$\tau_i = \frac{RT_o}{E}\frac{c_v T_o}{QB\rho}\frac{W_F W_o}{Y_{F0}Y_{O0}}\exp\left(\frac{E}{RT_o}\right) \tag{10}$$

This analysis indicates that the ignition of a reactive mixture is determined by a characteristic time that varies exponentially with E/RT_o. This delay is diminished if the initial temperature is increased. It is interesting to compare the ignition delay to the kinetic time characterizing the chemistry of the single-step reaction τ_c and evaluated at the initial temperature T_o. This ratio is given by

$$\frac{\tau_i(T_o)}{\tau_c(T_o)} = \frac{RT_o^2}{E_a}\frac{c_v}{Q}\frac{1}{Y_{F0}Y_{O0}} \tag{11}$$

In typical situations this ratio is much smaller than unity, indicating that the thermal runaway plays a central role.

B. Minimum Ignition Energy

The minimum ignition energy is the quantity of energy that, when added to a flammable system, will lead to sustained flame propagation. This quantity is closely related to a critical dimension, which corresponds to the size at which a combustible element of reactants (kernel) brought at its adiabatic flame temperature will grow unaided. This critical size d_c is given by expressions of the form

$$\frac{d_c S_L}{\alpha} = \text{constant} \tag{12}$$

where $\alpha = k/\rho c_p$ is the heat diffusivity of the mixture, and S_L is the laminar burning velocity. The constant is about 4 in the plane case and about 12 for a spherical pocket. The critical size is thus defined by a Peclet number based on the flame speed, and it is closely connected to the quenching distance d_q. Below the critical size, combustion is extinguished because the heat removed from the kernel exceeds the energy produced by the chemical reaction. The minimum ignition energy is then given by

$$E_{\min} = \rho\pi\frac{d_c^3}{6}c_p(T_b - T_c) \tag{13}$$

To show how this quantity changes with pressure, one may write

$$E_{\min} \approx k\pi \frac{d_c^2}{S_L} c_p (T_b - T_0) \tag{14}$$

For a bimolecular reaction S_L is independent of pressure and $d_c \sim p^{-1}$ so that $E_{\min} \sim p^{-2}$. For a reaction of order N, one finds $E_{\min} \sim p^{1-3N/2}$.

C. Effects of Turbulence and Droplet Evaporation

The value of the minimum ignition energy is modified in the presence of turbulence. This point is examined, for example, by Ballal and Lefebvre,[8] who provide empirical expressions for the quenching distance d_c. For low levels of turbulence ($u' < 2S_L$), the fluctuations increase the heat losses from the kernel and consequently increase the quenching distance: $d_c = 10\alpha/(S_T - 0.16u')$. For high values of turbulence, it is argued that the wrinkling is not augmented any further but that small-scale eddies move fresh reactants into the combustion zone, increasing the reaction rate. The net effect is to increase the quenching distance according to $d_c = 10\alpha(S_T - 0.63u')$, where S_T is the turbulent flame speed. Although these expressions are probably quite specific to the reactants and geometry in which they were obtained, they also indicate that the minimum ignition energy corresponds to the maximum flame speed.

If one of the reactants is injected as a spray, evaporation effects are expected to modify the quenching distance and minimum ignition energy. Experiments of Ballal and Lefebvre[9] indicate that the ignition energy is smaller for a well-atomized spray having a low mean droplet diameter than for a spray of larger droplets. It was also found by Arai et al.[10] that formation of an envelope flame around the droplets was important for sustaining the droplet kernel.

For turbulent mixtures including one reactant in the form of a spray, one may develop an ignition correlation on the basis of three characteristic times as proposed by Peters and Mellor.[11,12] A turbulent mixing time $\tau_d = d_q/v_{\mathrm{ref}}$ based on the quenching distance d_q and typical velocity v_{ref} describes the time taken by the flow to dissipate a flame kernel having the critical size. A typical droplet evaporation time $\tau_b = d_o^2/\kappa$ evaluated from the d^2 law describes the typical time of the droplets. In this last expression d_o is the initial droplet diameter, and κ is the evaporation constant. A typical kinetic time τ_k represents the autoignition delay of a homogeneous stoichiometric mixture of propellants. The ignition boundary is then given by an expression of the form:

$$\tau_{dq} = a\tau_b + b\tau_k \tag{15}$$

If the energy deposited initially E exceeds the minimum energy $E_{\min}$, the kernel diameter $d = \{E/[(\pi/6)\rho c_p(T_b - T_0)]\}$ is greater than d_q, the kernel dissipation time τ_d exceeds the critical dissipation time τ_{dq}, and ignition may take place.

Although the ignition criterion just discussed was established for gas turbine applications, similar ideas have value in the analysis of spark ignition of propellants.

D. Propagation from a Flame Kernel

When the initial energy is deposited in a localized region in the engine, propagation of a flame from this kernel will be needed to ignite the successive rows of injectors. This propagation phase is not instantaneous, and it may significantly lengthen the ignition delay. In very large motors or when the injectors are separated in compartments, it may be necessary to form multiple initial flame kernels to accelerate the ignition of the chamber.

Propagation will take place in a highly inhomogeneous, turbulent mixture of reactants, including dispersed phases in the form of liquid droplets. The propagation process will then take place in a partially premixed mode, eventually leading to flame stabilization in a nonpremixed configuration. Some of the features of this process are illustrated by experiments carried out by McManus et al.[13] The experimental setup comprises three injectors fed with gaseous hydrogen. Air is flowing around these elements. Ignition is obtained with a spark plug located on the upper wall at a short distance from the injection plane. The combustor is two-dimensional and allows detailed optical investigation. Instantaneous maps of laser-induced fluorescence of the OH radical obtained on this device show the structure of the flame at various instants after the spark. Typical images obtained in this way are shown in Fig. 1. The following evolution is made evident.

1) In a first step, reaction begins near the spark plug where a flame kernel is formed, and while being convected downstream the reaction grows in size and reaches an apparently stable position at a distance from the first injector.

2) The rate of reaction increases rapidly in the pocket; the reaction zone reaches the injection plane and ignites two diffusion flames near the injection slits of the first injector.

3) The premixed reactants located close to this injector ignite and produce a first combustion peak. Volumetric expansion of the products pushes the flame front toward the second injector.

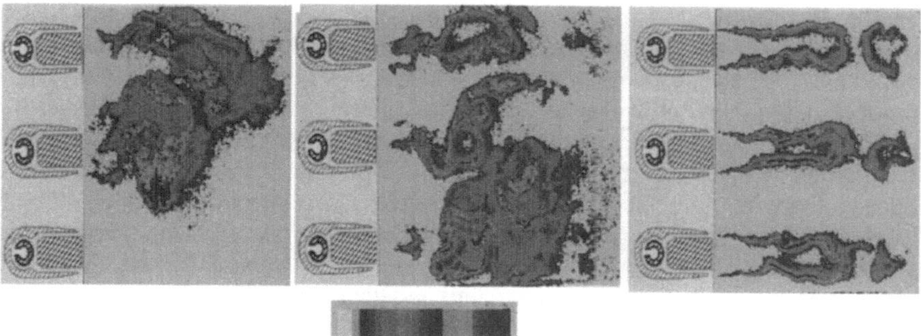

Fig. 1   Ignition and flame stabilization in a multiple injector combustor. Images obtained by laser-induced fluorescence of OH. The flame is initiated by a spark plug at the upper wall of the system.

4) Two diffusion flames form near this element while the premixed gases formed downstream react and produce a second combustion peak, pushing the flame front toward the third injector.

5) The same process is repeated once more at the third injector. After a period in which the flow is relatively perturbed, one observes the stabilization of three pairs of diffusion flames.

This well-controlled experiment indicates that the dynamics of ignition are quite complex. One distinguishes pockets of premixed propellants producing premixed reaction fronts. The propagation of these fronts through the heterogeneous turbulent medium gives rise to diffusion flames that are finally stabilized in the vicinity of the injection plane. More complex patterns are expected if one of the reactants is injected as a liquid spray, as is the case in most high-performance rocket engines (fed by liquid oxygen and gaseous hydrogen).

The transition in combustion mode during this propagation phase is a key aspect of the problem. Although the initial development of the flame is turbulent and premixed, the final outcome is turbulent nonpremixed. To describe this feature, Veynante et al.[14] developed a turbulent combustion model combining premixed and nonpremixed flame elements, allowing a progressive transition from premixed to nonpremixed flame configurations (see, for example, Fichot,[15] Candel et al.,[16] and Fichot et al.[17]). The model provides a suitable representation of the typical times and flame patterns observed in the experiment and may be extended to deal with the more complicated situation prevailing in rocket engines at the start.

E. Compressibility Effects

Although ignition is usually treated as a constant pressure process, many studies have concerned the dynamics of reactive kernels under the combined effect of pressure (see Oppenheim[4] for a review of this subject). It is, for example, possible to study the evolution of an exothermic center or "hot spot" in which the mixture, initially homogeneous, undergoes chemical reaction and induces in the surroundings a spatially inhomogeneous blast wave. It is thus possible to identify a mild ignition regime taking place at near-constant pressure in which the evolution of the system is governed by the chemical conversion of reactants, resulting in the release of heat and the losses of heat to the surroundings, and a strong ignition regime in which a compression wave is generated by the exothermic kernel and interacts with the surroundings. The energy release in the kernel causes its volumetric expansion, which produces a compression wave that in turn contributes to the temperature increase.

The significance of such effects in the case of rocket engine ignition is not fully elucidated. It is known that ignition of a flammable mixture in a tube produces a flame that may be strongly accelerated by pressure waves reflected at the tube end. The pressure waves may produce the strong ignition regime just described. In extreme cases the flame speed might approach sonic velocity, and this might lead to a detonation. This, however, is generally believed to be quite improbable as the energy required to initiate a detonation is several orders of magnitude greater than that required for mild ignition. The question of direct initiation of detonations is a matter of considerable interest, and it is studied by many different authors (a recent investigation of this subject containing many other references is given by He and Clavin[18,19]).

F. Numerical Modeling of Ignition Processes

Numerical modeling may be used at different levels to analyze the elementary mechanisms leading to ignition or the more global dynamics of the process. One may, for example, calculate in detail and with complex kinetics the evolution of homogeneous mixtures brought at elevated temperatures and thus deduce the autoignition times for specific situations. It is also possible to consider the laminar flame structure of mixtures of propellants under conditions representative of those found during ignition (see, for example, He and Clavin[18]). The ignition of droplets and droplet sprays may also be studied numerically with simple or complex chemistry. Processes involving coupled effects in turbulence like flame propagation from the initial kernel to the surrounding injectors require a considerable amount of modeling. It is, in particular, necessary to account for turbulent mixing, transport, and combustion. Such modeling is at the root of simulations carried out by Fichot et al.[17] Further modeling is needed if one wishes to describe liquid propellant injection, atomization, and combustion. This type of comprehensive simulation is well illustrated in a study of Baudart et al.[20] on the ignition of cryogenic propellants using a pyrotechnic igniter.

A three-dimensional numerical code is used to describe the transient, two-phase reactive flour in the rocket engine chamber and feedlines. An arbitrary Lagrangian-Eulerian method is used, in which the droplet trajectories are described in a Lagrangian manner and mapped onto an Eulerian grid made up of finite volume cells. Mean flow quantities are density weighted using Favre averaging. Low Reynolds number and compressibility effects are included in the k-ε model of turbulence, where k is the turbulent kinetic energy and ε is the turbulence dissipation rate. An eddy breakup model including an Arrhenius correction for thermal ignition provides the volumetric rate of reaction. Interaction between droplets and turbulence is modeled by considering the velocity fluctuation intensity, eddy life time, and droplet residence time in the eddy.

For a single coaxial element geometry, the model was able to simulate the interaction between the igniter hot gases and the oxygen and hydrogen. Because the igniter exhaust gases are a reducing agent, it is necessary for the hot gases to mix with the oxygen from the central element of the coaxial injector. However, with gaseous oxygen, ignition is difficult due to substantial mixing of the oxygen with the surrounding hydrogen stream occurring before contact with the hot gases originating from the solid propellant igniter. The model also showed that, with liquid oxygen droplets, ignition is improved due to the inertia of the droplets that cross the hydrogen stream and mix with the igniter gases.

On a global scale, the engine chamber is modeled assuming a premixed propellant flow with an oxygen droplet distribution, and not as individual coaxial elements. The model demonstrated that overpressures were produced with the igniter pointed in the axial direction because of the fact that ignition occurred downstream close to the nozzle throat. This caused a thermal blockage, and a flame propagated upstream toward the injector, which ignited the propellants that clumped in the chamber. The model showed that, with the igniter jets at an angle to the chamber flow, ignition is produced near the injector face and overpressures are reduced. The igniter angle and timing of liquid oxygen valve injections were design modifications to the HM7B engine and resulted in improved ignition characteristics.

III.　Ignition of Nonhypergolic Propellants

Ignition of nonhypergolic propellants may be achieved in various ways. This section describes various systems developed for that purpose. These systems are grouped as thermal, thermal resonant, catalytic, chemical, photochemical laser, and spark igniters. These devices may be used to directly ignite the engine chamber or to ignite a smaller torch, typically called an augmented igniter, which is then used to ignite the main chamber. In the past, most engine ignition devices have been spark igniters using a torch to light the main engine. Armstrong[21] provides an excellent review of these devices and includes information necessary for the selection of the appropriate device based on criteria, such as duty cycle, power, electromagnetic interference sensitivity, reliability, and ignition delay. The relative importance of each of the criteria depends on the application, such as a booster engine or reaction control system thruster.

A.　Thermal Ignition Devices

Thermal ignition has been demonstrated using glow plugs, electrical wires, and the laser heating of target materials. Glow plugs and electrical wires use electrical resistance to heat a conductive material to the autoignition temperature of the propellants. Typically after ignition, power is removed. For subsequent ignitions, the electrical wire needs additional power, but the glow plug may be sized to remain above the ignition temperature for some duration to allow additional ignitions without electricity. Both electric devices must be designed to balance current, wire or plug size, and heating time with the propellant flow conditions. Testing of hot wires in hydrogen-air and methane-air mixtures was reported by Stout and Jones.[22] Some select design details for these devices are given by Shorr.[23] It is also indicated that hotter ignition temperatures are required for higher propellant velocities. The range of velocities studied varied from 3.6 m/s to 152 m/s. Typical ignition delay times were reported at roughly 60 ms.

The use of lasers to heat aluminum target materials was reported by Homan and Sirignano.[24] They reported minimum ignition energies for use in hydrocarbon-air mixtures. An application to rocket engines was performed by Duncan Technologies, Inc. (DTI).[25] DTI used a low-power diode laser to heat a target above the ignition temperature of the propellant gaseous hydrogen and oxygen (GO_2/GH_2). Using 0.7 W, the diode laser heated the target to over 1000°C. Rocket engine ignition was demonstrated at sea level on a 20-N thruster. A parametric study in a static bomb test chamber indicated reliable ignition at pressures >41.3 kPa (>6 psia) for mixture ratios of >6 using hydrogen and oxygen, as shown in Fig. 2. Additional static testing with a hydrogen and oxygen mixture ratio of 10 demonstrated a cold-start ignition delay of 65 ms, and a firing test life of 4000 cycles with no degradation.

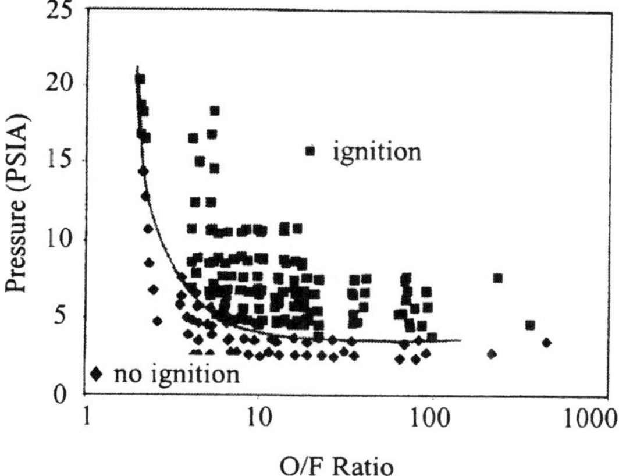

Fig. 2   Experimental laser ignition data: GO_2/GH_2 ignition region at low pressures.

B.   Resonant Ignition Devices

Resonant igniters use acoustic oscillations in an open tube with one dead end to raise the propellant gas above its autoignition temperature. A schematic of a resonant igniter is shown in Fig. 3. Ignition is achieved by flowing propellant gases through a sonic nozzle, expanding them across a small chamber, and compressing the stream into the resonant tube. A bow shock forms at the tube inlet. Additional flow from the nozzle maintains the bow shock and is forced out of the ignition chamber. The impinging flowfield causes the bow shock to become unstable and vibrate at the entrance to the resonant tube. This movement gives rise to strong longitudinal pressure oscillations inside the resonant tube at a

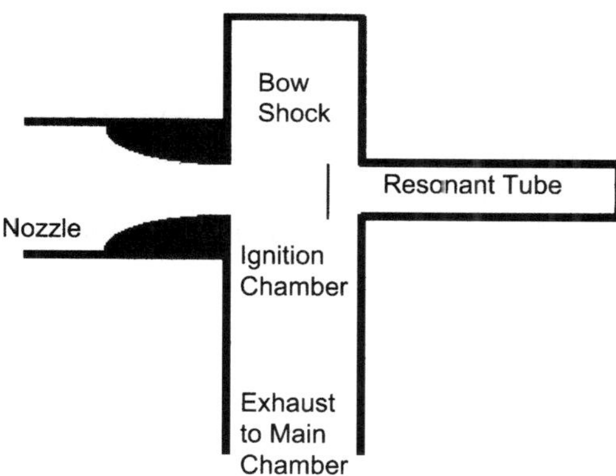

Fig. 3   Resonant igniter.

natural frequency that is a function of the speed of sound and the length of the tube. The bow shock at the tube entrance resonates with the pressure oscillations inside the tube, and temperatures increase above the autoignition temperature of the propellant combination. A flame front develops, exits the resonant tube, and moves through the ignition chamber and into the main engine chamber of the rocket engine, causing ignition.

Conrad and Pavli[26] demonstrated the resonant tube igniter for premixed GO_2/GH_2. Their investigation determined that a massive brass resonance tube conducted the heat away and prevented temperatures in the tube from exceeding the autoignition temperature. The brass was replaced with an insulated, thin wall, stainless steel tube. The maximum temperature inside the tube then exceeded 866 K and resulted in ignition. Additionally, results were presented for various resonant tube depths and gap lengths between the nozzle and the tube, but they recommend that their data was not generally applicable for designing a specific resonant tube igniter. Because the design is sensitive to system parameters such as the heat loss in the tube and the initial gas temperature, the resonant tube igniter device must be tuned for each specific application.

In a subsequent study, Phillips and Pavli[27] developed a method to render the resonant igniter insensitive to ambient pressure by enclosing the nozzle exit in a can, with choked orifices at the can exit. Testing demonstrated ignition in 130 ms (from initial temperature rise to ignition) with a 241-kPa nozzle inlet pressure and a 294-K hydrogen/oxygen gas temperature. The main parameters, such as pressure, gas supply temperature, the gap between nozzle and resonant tube inlet, nozzle diameters, resonator tube geometry, resonant tube materials, ambient pressure, and gases, were systematically varied in the tests. Some of the results included: 1) the shock structure of the underexpanded nozzle remained unchanged whether or not there is a resonance tube obstruction; 2) the temperature of the resonating gas decreased as the length of the resonant tube increased; 3) a tapered geometry for the resonant tube was superior to a cylindrical tube; 4) ignition delay time decreased with nozzle inlet pressure up to 35 psia, and then no improvement was recorded; 5) a mixture ratio around 0.6 was optimum; 6) the ignition delay was very sensitive to inlet temperature; and 7) ignition was not obtained when the gases were supplied at a temperature lower than 108 K, and at 108 K the ignition delay was 10 s. Result 2, however, conflicts with data obtained by Przirembel and Fletcher,[28] who found that the resonant temperatures increased with increasing resonant tube length, with no upper limit tested. Result 7 was improved by a subsequent study by Stabinsky,[29] when low-temperature oxygen and hydrogen were ignited by using a hydrogen lead to the resonant igniter.

A key design goal is to reach the autoignition temperature in the minimum time. Przirembel and Fletcher[28] determined that the maximum temperature in the resonant igniter occurred when the resonance tube inlet was located in the third compression cell of the underexpanded jet. Przirembel et al.[30] varied geometries to determine optimal nozzle orifice size, gap separation distance, and resonant tube length. Lauffer[31] described a resonant igniter for oxygen/ hydrogen ignition in reaction control system (RCS) application. The thermocouple response to hydrogen resonance indicated that a temperature of 811 K was reached in less than 20 ms. With a 3-ms hydrogen injection lead, delay times times of 20–30 ms were measured for the igniter chamber to reach 90% of nominal chamber pressure from start of the electrical signal.

C. Catalytic Ignition Devices

Catalytic ignition operates by tapping a small amount of the propellant flow (less than 10% of total flow for GO_2/GH_2) and directing this propellant through a catalyst bed. A standard arrangement typically consists of a cylinder filled with granules, pellets, or a monolithic structure, which are coated with the catalyst material. When the propellant flows over the catalyst, spontaneous ignition occurs, and the hot effluent gases are ejected into the main propellant chamber. Thousands of qualified flights have used such devices with monopropellant hydrazine, and by 1969 Johnson[32] demonstrated ignition for chilled and ambient GO_2/GH_2 mixtures using noble metal catalysts.

The investigation by Johnson[32] compared two catalysts (Shell-405 and Engelhard MSFA) over a range of environmental conditions. Additional tests were configured with various igniter geometries integrated into a 89-N (20-lbf) thruster. The results indicated that Shell 405 had better performance at low temperature and was otherwise comparable to the Engelhard MSFA. Neither catalysts were affected by long-term vacuum exposure. A monolithic catalyst bed was tested by Zurawski and Green[33] and consisted of a carbon sponge substrate coated with rhenium for structural strength and iridium as the catalyst. Although this attempt was unsuccessful, with design improvements the potential advantages over pellets include lower pressure drop and longer life.

Johnson[32] determined that Shell 405 was not sensitive to inlet temperatures above 172 K. At a lower temperature of 127 K, the ignition delay was double the required time for ignition at ambient temperature (275 K). With ignition delay times always exceeding 200 ms, the system was not suited for reaction control applications. Green[34] found that ignition down to a temperature of 126.2 K could be achieved with Shell 405.

Johnson[35] attempted to reduce the ignition delay time by increasing the mixture ratio of the catalyst bed, but before realizing any improvement, flame flashback occurred into the catalyst bed and injector. Zurawski and Green[33] found that flashback was potentially damaging and impaired the performance of the igniter injector and catalyst materials. Flashback did not occur at oxygen and hydrogen mixtures ratios below 1.5 : 1, or flow velocities greater than 15 m/s (50 ft/s) for all operating pressures, and these values may serve as design guidelines. Zurawski and Green[33] found that incomplete mixing at the catalyst bed inlet could cause hot zones and flashback and concluded that the injector design should maximize propellant mixing.

A steady-state firing duration of 4000 s and life-cycle capability of 5000 pulses at 3-s durations for both Shell 405 and Englehard MSFA were demonstrated by Johnson.[35] The life of the catalyst bed was found to be extended by operating at temperatures below 922 K. Low mixture ratios improved catalyst-bed life by lowering the operating temperature for a given flow rate. This life extension is easily accommodated because low mixture ratios are preferred to prevent flashback as well. Green[34] also successfully completed 4900 pulses of 2-s duration.

To reduce ignition delay by an order of magnitude compared to any previous work, Johnson[35] injected oxygen downstream of the catalyst bed. With a GO_2/GH_2 mixture ratio of approximately $1:1$ in the catalyst bed, the downstream injection of oxygen raised the overall igniter mixture ratio to $10:1$, and reduced the ignition delay time to 25 ms. This design optimized a lower catalyst bed temperature with an overall effluent temperature sufficient to provide reliable ignition. Thermal insulation of the catalyst bed and reduction of its thermal mass also reduced the ignition delay time. However, preheating the catalyst did not significantly reduce the ignition delay when used in conjunction with the downstream oxygen injection.

Ketsen[36] modeled the hydrogen-oxygen catalytic igniter and then validated the computer model with test data.[37] This design tool describes the transient response of a continuous flow GO_2/GH_2 catalyst bed in terms of temperature, pressure, and species concentration as a function of position and time. After an initial calibration with a catalyst bed of similar configuration, the model may be used to calculate excursions from the baseline.

D.   Third-Chemical or Hypergolic Ignition Devices

Third-chemical ignition occurs when a hypergolic chemical is added to one of the propellants or injected directly into the engine chamber. Fluorine has been added to liquid oxygen (FLOX) and tested by Rollbuhler and Straight[38] and by Mosier et al.[39] Fluorine is miscible in liquid oxygen, and when the FLOX contacts the fuel, ignition is hypergolic. Trioxygen difluoride was added to LOX to get hypergolic ignition with gaseous hydrogen by Dickinson et al.[40] Direct injection of a hypergolic propellant to cause ignition was tested during the development of the M-1 engine,[41] and for RCS engines in the 100-lbf-thrust class.[42]

Rollbuhler and Straight[38] tested the effect of fluorine concentration in FLOX, injector design, and propellant temperature on the ignition delay with hydrogen/oxygen on a 1112-N thruster operating at a nominal pressure of 2.1 MPa (300 psia). They found that mass fractions of fluorine of 50–60% were required to obtain reliable ignition within an ignition delay time of 1 s. The injector design that promoted rapid mixing, atomization, and vaporization lowered the mass fraction required for ignition, as did using ambient temperature hydrogen. Most testing was performed with an oxygen injection lead, and no correlation could be established for ignition delay time as a function of mixture ratio.

Additional testing was performed by Mosier et al.[39] using FLOX with hydrocarbon propellants. Ignition delay times were measured in the range of 10–100 ms, but the fluorine mass fractions were still quite high (>70%). Dickinson et al.[40] added significantly lower concentrations of trioxygen difluoride (O_3F_2) to LOX (O_3F_2 saturated in LOX at a concentration of 0.1% by weight) and observed ignition delay times in the range of 10–100 ms. Ignition depended in this case on the thermal decomposition of O_3F_2, a process that was not reliable when the initial temperature of the engine and injector was below 230 K.

The direct injection of fluorine into a LO_2/GH_2 engine chamber was tested for the M-1 program by Dankhoff et al.[41] with many design difficulties and a long ignition delay time of >2 s. Fluorine damage to a check valve and erosion of the injection tube were finally controlled by a helium purge. Testing using chlorine trifluoride produced a shorter ignition delay time that was still above 300 ms.[42]

E. Photochemical Laser Ignition

There are three basic types of laser ignition methods: 1) thermal ignition, which uses a laser to heat a target, as described earlier, 2) laser-initiated spark (LIS) ignition, which uses laser beams with a high-power density focused to create a plasma kernel, and 3) photochemical ignition, which uses a laser to excite the molecular bond of one of the propellants, thereby producing highly reactive radicals. These types of laser ignition are summarized by Ronney.[43]

Of the three types of laser ignition, LIS and photochemical ignition apply the laser directly to the propellant, rather than to a target as in thermal ignition. LIS was demonstrated by Liou[44] using 60–80 mJ with a power density at the focus on the order of 1×10^{11} W/cm^2 using gaseous oxygen and a variety of fuels. Similar power densities were used by Weinberg and Wilson[45] in another study of LIS.

Photochemical ignition requires only a fraction of the laser energy used for LIS and is strongly dependent on the wavelength of the laser. For example, Forch and Miziolek,[46] demonstrated laser ignition energies of less than 1 mJ at a wavelength of 286 nm with a minimum ignition energy occurring in the fuel-lean region of 0.6 for H_2/O_2.

F. Spark Ignition Devices

Spark ignition uses an electrical discharge between two electrodes positioned in a propellant stream to ionize, dissociate, and heat the propellant such that ignition results. The spark igniter may be mounted either directly in the engine chamber[23,47,48] (direct spark ignition) or in a prechamber that exhausts into the main chamber[31,49] (augmented spark ignition). Maly and Vogel[50] provide a description of spark ignition. Initially the spark ionizes and dissociates the fluid into a plasma. This breakdown phase lasts 1–10 ns, has characteristic temperatures of 60,000 K, and provides a plasma current path for the arc and glow discharges. The arc and glow phases have characteristic temperatures that are an order of magnitude lower than those prevailing during the breakdown phase, and they typically last 10 μs to 100 ms. The breakdown phase completely dissociates and ionizes the fluid, while the arc phase primarily dissociates and the glow phase provides only a thermal contribution to ignition.

Ignition is achieved with a combination of the breakdown, arc, and glow phases, of which the breakdown discharge is most efficient.[51] The breakdown discharge efficiency was measured to have only 6% losses compared to 50 and 70% for the arc and glow discharge losses, respectively, primarily due to heat conduction via the electrodes. Lauffer[31] described important design considerations for selecting an inductively powered spark plug to optimize for efficiency, which included low breakdown voltage (<10 kV) and high voltage during the arc and glow phases. A recessed-surface dual-gap spark plug was selected to meet these design criteria. The recessed surface was selected to protect the electrodes from the hot combustion gases, and the dual-gap spark plug reduced the breakdown voltage.

The effects of propellant inlet temperature were studied by Lawver et al.[49] for GOX/ethanol, and ignition was sensitive only to the fuel temperature. As a worst case, an abnormal ignition delay of approximately 1 s was observed for cold fuel inlet temperature of 171 K, a low chamber pressure of 0.55 MPa (80 psia), and a low mixture ratio of 0.8. This delay was attributed to limited fuel vaporization.

Spark frequency is an important design parameter for reliable ignition. Kono et al.[52] demonstrated that increasing the spark frequency augmented the probability of ignition, which asymptotically approaches a constant value. Kono also reports the optimum spark interval is 30 $\mu\sigma$ larger than the spark duration.

Minimum ignition energies are reported by many different authors.[8,53] Frendi and Sibulkin found that to make a meaningful comparison of the minimum ignition energy data, one needs to know the ignition parameters used.[54] They observe, however, that a large number of published works fail to provide important parameters such as spark gap, electrode geometry, spark duration, spark phase, etc. An earlier criticism of published data was provided by Ballal and Lefebvre,[55] wherein they identified the importance of spark gap, spark duration, spark phase, and degree of turbulence. With that in mind, the following data are included as a representative example of the energies previously measured by test. Spark levels down to 10 mJ have been demonstrated when igniting a LOX/RP-1, 2000-psia, 20,000-lbf thrust chamber.[56] Another spark igniter[57,58] for a 2.2-N (0.5-lbf) GO_2/GH_2 thruster was sized with the following characteristics: output voltage of 4.0–6.0 kV, pulse width 2–3 μs, pulse frequency 200–300 Hz, and spark energy of 0.2–0.3 mJ per spark. Lawver et al.[49] demonstrated ignition with 10 mJ for GO_2/ethanol in a torch-type igniter for a 620-lbf engine.

The effect of spark plug gap and electrode diameter on the minimum ignition energy was tested by Kono et al.[52] The results for a flowing air-propane mixture indicate that, for a particular spark gap size (1.5 mm), the minimum ignition energy drops by a factor of 3 when the electrode diameter is reduced from 0.8 to 0.05 mm. This effect is attributed to a quenching of the flame kernel by the electrodes, and this effect is not observed when the electrode distance is increased to 3.0 mm. A previous study by Kono et al.[59] also documented a dependence of the minimum ignition energy on the electrode diameter. Ballal and Lefebvre[8] measured the minimum spark gap corresponding to the minimum ignition energy for a variety of fluid stream velocities, pressures, equivalence ratios, and turbulence intensity level. For every set of conditions, an optimum spark gap was easily identified.

Parametric studies on the importance of fluid flow velocity, mixture ratio, pressure, and turbulence on the minimum ignition energy were carried out by Ballal and Lefebvre.[8] They determined that increasing fluid flow velocity increases the minimum ignition energy. For example, the ignition energy for a fluid velocity of 15 m/s was more than double the minimum energy for stagnant ignition, when using methane and diluted oxygen at a pressure of 17.2 kPa (2.5 psia). The energy required for ignition is increased because the flame kernel is strained aerodynamically. Increasing flow velocity also decreases the optimum spark interval.[52] A substantial reduction in the ignition energy was observed for increasing mixture ratios. By doubling the percentage of oxygen in the flowing stream of premixed gases, the minimum ignition energy was reduced by an order of magnitude. Tests generally confirmed that higher pressures promote ignition. By increasing the pressure from 8.1 kPa (1.2 psia) to 35.5 kPa (5.1 psia), the minimum ignition energies generally dropped by a factor of 3–6. Experiments also indicated that low levels of turbulence tended to have little impact or slightly reduce the minimum ignition energy, but at a high level of turbulence, the minimum ignition energy increases with increasing turbulence.

IV. Hypergolic Propellant Ignition

Hypergolic ignition relies on exothermic low-temperature liquid-vapor chemical reactions to initiate combustion throughout the chamber. The vaporization of propellants in the confined space of the chamber and the initial rise in chamber pressure is critical to ignition.[60] The ensuing exothermic chemical reactions cause the pressure and temperature to rise in the chamber, which further accelerates the reaction rates, leading to thermal runaway or ignition. The ignition of a hypergolic propellant engine involves many physical processes: 1) rapid vaporization of liquid propellant as it enters the chamber at vacuum, 2) freezing of propellant droplets by evaporative cooling, 3) condensation of preignition reaction products on the chamber walls, and 4) heterogeneous liquid-vapor reaction of the propellants. This section discusses the importance of these and other engine design factors that directly affect the ignition process and the associated delay based on experimental observations and models.

A. Design Considerations for Hypergolic Engine Ignition

Hypergolic engines are widely used for orbital maneuvering and attitude control in many satellites and spacecraft. In addition, hypergolic engines are employed in booster stages and upper stages for some launch vehicles. Typically these engines are required to perform multiple pulses or restarts. Ignition can be a significant contributor to the performance of the engine, the potential for spacecraft contamination, and the potential for damage to the thruster itself through overpressure or triggering of instability.

There are several propellant combinations that are hypergolic, of which some typical propellants are listed in Table 1. Nitrogen tetroxide (NTO), N_2O_4, is a reddish-brown oxidizer that is highly volatile and boils at 294 K (70°F). Monomethylhydrazine (MMH), $CH_3N_2H_3$, is a clear liquid fuel that is not as volatile as NTO, as indicated by the boiling point of 387 K (190°F).[61] Also commonly used are propellants such as hydrazine (N_2H_4) or aerozine 50 (A-50). [A-50 is a 50–50 blend by weight of N_2H_4 and unsymmetrical dimethylhydrazine (UDMH)]. It is important to consider possible freezing of propellants in a space vacuum engine start situation. More significantly from a vehicle design consideration, the freezing point of MMH is around 221 K (-63°F), as compared with hydrazine, which has a higher freezing point of 274 K (34°F).[61] The freezing point of NTO is 262 K (11°F).[62] The critical pressures of these propellants are higher than most storable-engine chamber pressures, and so subcritical liquid behavior is typical. In military specifications, there are many forms of NTO. This is due to the addition of NO, which lowers the freezing point and which also changes the

Table 1   Hypergolic propellant properties at 298 K (Refs. 61, 62)

Density	kg/m^3	Freezing point, K	Boiling point, K	Critical temperature, K	Critical pressure, MPa	Specific heat, J/gK	Viscosity, cp, g/m-s	Thermal conductivity, cal/cm-s-K	Surface tension, dynes/cm
N_2O_4	1454	262	294	431	10.1	1.57	0.396	3.13e-4	25
N_2H_4	1003	275	387	653	14.7	3.09	0.913	7.86e-5	66
MMH	870	221	361	593	8.24	2.93	0.680	5.92e-4	34
UDMH	786	216	336	523	5.42	2.73	0.492	3.76e-4	24
A50	897	265	343	608	11.7	2.91	0.817	6.83e-4	29

color to green. For example, the propellants used in the space shuttle reaction control system are monomethylhydrazine and nitrogen tetroxide. The space shuttle uses 1.5–3.0% NO content NTO, called MON-3.[62] Typical ignition delay times for the propellants in Table 1 range from 2 to 20 ms; however, this is also a function of other parameters, as further explained.

The ignition transient also depends on other parameters, such as the opening time of the valve and the acoustics of the feedlines and combustion chamber. Many of the parameters that can affect the ignition transient are driven by steady-state combustion and envelope considerations. For small attitude control thrusters, the valves are designed to fully open rapidly (or close), as fast as 3 ms in order to provide the minimum impulse. The manifolds evenly distribute propellant to each injector element. The injector mixes the propellants efficiently and provides a thermal environment compatible with the chamber walls. The chamber provides the required combustion efficiency, envelope, stability, chamber pressure, and throat area. Although these are somewhat fixed parameters necessary for steady-state performance, the ignition transient is typically adjusted by the relative valve timing between oxidizer and fuel and by the injector manifold volume.

Interaction of the propulsion feed system with the engine affects the combustion transient. Upon opening of the thruster valve, the feed system responds to the flow demand, but it cannot maintain pressure fully at the valve during the transient. For NTO, the line pressure can drop down to the vapor pressure. This pressure disturbance causes a ringing or water-hammer in the lines, resulting in low frequency oscillations in the chamber pressure.

The duty cycle and external environment determine the conditions under which the combustion transient occurs. A thruster duty cycle is the duration and frequency of pulsing required by the spacecraft. For example, the space shuttle thrusters are designed to be fired in durations from 80 ms to 800 s. For short-pulse operation, the thruster may never reach a steady-state condition. There may also be a long period of time between pulses, during which the engine chamber walls may cool down from continuous radiation into space. In most cases, small heaters mounted on the engines supply some heat to keep the valves and injector warm.

B. Physical Processes Occurring During Ignition Transient

The ignition transient can be divided into three phases: preignition, thermal ignition, and transient combustion, which will be described in the following sections. Many of these processes have been observed and photographed through scaled models or quartz windows in rocket engine chambers.[63–65] The following discussion focuses on vacuum ignition, rather than ambient starts.

1. Preignition and Preignition Products

The engine chamber and manifolds are initially at atmospheric or at vacuum pressure for altitude engines. The valves are opened and the propellants vaporize and expand rapidly in the manifold and chamber. This evaporation cools the propellant and the engine. As the vapor and liquid enter the chamber, the chamber temperature drops down to the triple point of NTO, and some liquid droplets may actually freeze in the case of A-50 and hydrazine. More significantly, nonvolatile preignition products form and accumulate on the walls during this phase. The issue of preignition reaction products is important for two reasons. First, these products may be ejected out of the thruster and contaminate spacecraft surfaces[66]; and second, these products may accumulate and react explosively after a certain delay, producing an overpressure or hardstart[63,67,68] and resulting in damage to the engine.

Several studies have been made of the preignition products. Mayer et al.[69] experimentally studied the preignition of NTO and MMH at low pressures from 10 mm Hg (0.2 psia) to 110 mm Hg (2.1 psia). Mayer noted that NTO vaporizes immediately and dissociates readily into NO_2, as shown in Eq. (16):

$$N_2O_4 \longleftrightarrow NO_2 + NO_2 \tag{16}$$

A flameless reaction between the fuel and NO_2 produces a yellow fog in the chamber. It was also found that this fog consists of monomethyl nitroamide ($CH_3N_2HO_2$). Residues of this fog were analyzed to contain MMH nitrate salts ($CH_3N_2H_3 \cdot HNO_3$). The mechanism proposed for this reaction is

$$CH_3N_2H_3 + 2NO_2 \longrightarrow 3H_2O + 2N_2 + CO \tag{17}$$

$$H_2O + 3NO_2 \longrightarrow 2HNO_3 + NO \tag{18}$$

$$HNO_3 + H_3CNHNH_2 \longrightarrow CH_3N_2H_3 \cdot HNO_3 \tag{19}$$

The reaction is heterogeneous and forms HNO_3 in the fog. The rate constant for Eq. (18) increases as temperature decreases below 30°C, such that HNO_3 is favored at low temperatures. The amount of MMH-nitrate residue formed from this fog is inversely proportional to the molar ratio of NO_2 to fuel. MMH nitrate has a low vapor pressure and can accumulate over successive pulses when the chamber wall temperature is cold. Saad et al.[70] examined the mass spectra and infrared spectra of residues from the reaction products and found indications of methylnitrosamine (CH_3N_2HO) and MMH nitrate in the residues. However, the methylnitrosamine is unstable and decomposes readily into methanol and N_2, as shown in Eq. (20):

$$CH_3N_2HO \longrightarrow CH_3N = NOH \longrightarrow CH_3OH + N_2 \tag{20}$$

During the Apollo program, Christos et al.[71] at the U.S. Bureau of Mines examined the explosivity of residues of MMH and NTO using ballistic mortar and other tests. It was found that the mixtures had an explosive equivalence to Trinitrotoluene thermal equivalent TNT (TE) of 139%. Takimoto and Denault[72] also examined residues from thrusters and found MMH nitrate and ammonium nitrate ($NH_3 \cdot HNO_3$). They proposed that the mixture that contains MMH nitrate also includes heat-sensitive compounds, methyl azides and diazomethane, that can trigger its decomposition. They also suggested that MMH nitrate forms and then partially decomposes to CH_3NH_2 nitrate:

$$CH_3N_2H_3 \cdot HNO_3 \longrightarrow CH_3NH_2 \cdot HNO_3 \tag{21}$$

Takimoto and Denault[72] found that three engine operation factors affect the formation of MMH nitrate: 1) multiple short-duration pulses, 2) low temperature, and 3) lead/lag of oxidizer vs fuel entering the chamber. Interestingly, ammonium nitrate formed only during steady-state operation. Thus, it appears that many different intermediates can be generated depending on the engine conditions. Mills et al.[68] reported that, due to the nature of hypergolic propellants, low activation energies are involved, which leads to the formation of many complex intermediates. Mills et al. also suggested that the ignition reaction is a thermal mechanism, which is exponentially dependent on temperature, as opposed to a chain branching mechanism, which is dependent on the formation of radicals.

2. Thermal Ignition

The spontaneous ignition of the propellants is indicated by a sharp rise in pressure that can be seen in a chamber pressure trace. Spalding defines spontaneous ignition as "when a reactive mixture is formed, raised to a definite temperature and pressure, and then left alone, it may burst into flame after a certain time."[73] Kuo[74] further defines hypergolic ignition as "a heterogeneous reaction that is initiated as a surface reaction following the introduction of reactive fluids and does not require an external heat source." In one particular engine, ignition is observed to occur at approximately 5 psia, which is about 9 ms after the chamber pressure starts to rise.

Gray and Sherrington[75] studied the ignition of MMH and O_2 in a spherical reaction vessel and concluded that the mechanism is primarily thermal, with self-heating more significant than chain branching. Perlee et al.[76] conducted experiments to determine the flammability of hydrazine fuels in NO_2 (dissociated NTO), primarily from a safety standpoint. These experiments determined the regions of spontaneous ignition or spontaneous ignition temperature (SIT). Simmons[60] studied the expansion and ignition of A-50 and NTO in a vessel.

During the ignition transient, large overpressures can occur due to the explosion of residues leftover from the preignition process. This is significant because of the potential for damage to the engine or as a trigger for combustion instability. Juran and Stechman[67] indicated that large ignition overpressures were related to the accumulation of residues on the chamber walls and that, as thruster temperature increases, the pressure spike occurred less frequently. The issue of overpressures also applies to larger engines, as shown for example by Kerkam and Kahl.[63] Tests were carried out on the 44.5 kN (10,000 lbf) thrust A-50/NTO lunar module descent engine. High-speed photography showed brown, viscous, low-vapor-pressure deposits on the chamber wall just prior to ignition and the pressure spike. It is important to note that injected propellants will naturally clump during the transient. This is due to initial low-velocity propellant followed by higher velocity propellant, as the result of manifold and injector dynamics. This clump may then react explosively, causing an overpressure.

3. Combustion Transient

After spontaneous ignition, the propellants will undergo changes in injected phase from vapor to liquid because of the increase in pressure. The eventual formation of liquid propellant streams produces a more uniform spray. The chamber pressure reaches a steady-state limit. Steady state is essentially achieved when a balance in pressure exists that limits the flow of propellant into the chamber and the flow out of the chamber and when significant oscillations are damped out.

Zung and White[65] performed tests at pressures of 14.7 psia and still found significant NTO boiling upon injection. As pressure increased or as temperature decreased, the boiling would stop. It appears reasonable to assume that atomization, vaporization, mixing, and combustion undergo a fundamental change when the injected phase of the NTO evolves from vapor to liquid.

Allison and Faeth[77] studied the burning of suspended MMH, hydrazine, and UDMH droplets at atmospheric pressure in oxygen vapor. The burning rates were determined from the d^2-evaporation law, which states that the square of the droplet diameter d decreases linearly with time with slope equal to an evaporation constant. They found that the droplet burns in two distinct regions, the inner region as a monopropellant, and the outer region as a bipropellant. This is especially true for hydrazine, and less so for MMH and UDMH. Eberstein and Glassman[78] experimentally determined the monopropellant decomposition rates for gas phases of MMH, hydrazine, and UDMH in an adiabatic flow reactor. They proposed that the two methyl radicals in UDMH increase the chain branching reactions, which explains why UDMH decomposition has the lowest activation energy. With MMH this chain branching does not occur, and MMH has the highest activation energy.

Eventually, liquid NTO and fuel will impinge upon each other. Daimon et al.[79] experimentally studied the contact of hypergolic liquids to measure the time lag of ignition or explosion. The tests were performed by impacting droplets of fuel, at various velocities and drop size, onto a pool of NTO. A fuel and NTO vapor layer forms because of the gasification of the lower boiling NTO, and a turbulent combustion reaction occurs. They found that ignition time lag decreases with increasing droplet diameter.

Lawver[64] photographed a single unlike-doublet injector element under various operating conditions, as shown in Fig. 4. The dark or red stream is NTO. At a nominal chamber pressure of 152 psia, the NTO forms a coherent stream of liquid when it impinges the fuel. The initial diameter of the oxidizer stream is 0.030 in., giving an indication of scale. Lawver explained that a yellow region near the impingement exists because of NH_2 radical emission, or monopropellant combustion, and that further downstream a blue region exists because of methyl radical emission, or bipropellant combustion. The emission of blue light by the methyl radical indicates that there is some delay in the combustion as a bipropellant. Lawver also noted that hydrazine burns with a yellow flame over the entire spray because of the NH_2 radical emission.

C.   Modeling of Hypergolic Ignition Transient

For the most part quasi-steady models have been used to represent the ignition transient. These quasi-steady models were initially developed from steady-state descriptions with some appropriate changes to include effects such as droplet evaporation and freezing, and to describe the ignition time delay. This quantity

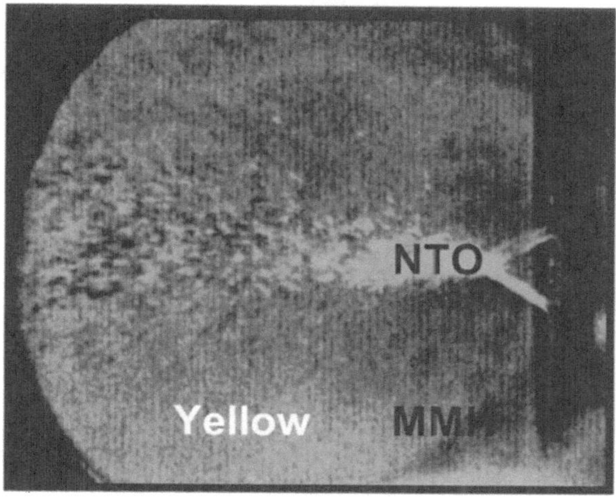

Fig. 4 Photograph of a space shuttle 870-lbf thruster injector unlike-doublet at $p_c = 124$ psia, $v_{\text{fuel}} = 73$ ft/s, and $T_{\text{fuel}} = 82°F$ (Ref. 64).

is generally deduced from a thermal ignition model as summarized in the following.

1. Thermal Ignition Model

The ignition of the NTO and MMH mixture can be computed using the Semenov and Frank-Kamenetskii method, as described by Glassman[7] or Seamans et al.,[80] which assumes that the reaction is governed by thermal effects rather than chain branching. This appears to be a good assumption, at least for hydrazine and MMH. The Semenov method considers the balance between heat conducted to the walls of the chamber and that generated by the reaction. If the mixture temperature is such that heat production rate exceeds the heat conduction rate to the walls, then the mixture will self-heat and spontaneously ignite. Applied to a control volume of gas with a wall temperature of T_o, the energy balance is given by the following expressions:

$$c_v V \frac{dT}{dt} = \dot{q}_r - \dot{q}_l \qquad (22)$$

where the heat losses may be represented in terms of an effective heat transfer coefficient

$$\dot{q}_l = hS(T - T_o) \qquad (23)$$

and the heat release

$$\dot{q}_r = V\dot{\omega}Q = VQC^N A \exp(-E/RT) \qquad (24)$$

This set of equations may be used to find the critical condition for thermal ignition, as given by Eq. (25). To determine the preexponential A and the heat of reaction Q, experimental data from a 2.35-cm-radius flow reactor was used along with Eq. (25):

$$\delta = \left(\frac{QE}{k_g RT_o}\right) r^2 X_{Fo}^m X_{Oo}^n \left(\frac{p}{RT_o}\right)^N A \exp\left(-\frac{E}{RT_o}\right) \tag{25}$$

For an infinite cylinder, δ is 2. The overall order N of the reaction is assumed to be 2, with $m = 1$ and $n = 1$. For MMH and NTO, the activation energy and conductivity are 5.2 kcal/mol and 7.5×10^{-5} cal/(s-cm-K). The test determined that the explosion limit is 5.5 mm Hg at $T_o = 298$ K for mole fractions of fuel and oxidizer, $X_{Fo} = 0.285$ and $X_{Oo} = 0.715$, respectively. From this experiment, the value of AQ is found to be 3.4×10^{14} cal-cm^3/(mol^2 s). Then, assuming a well-mixed vapor and neglecting heat losses, one may derive an ignition delay from a perturbation analysis of the energy equation. The value for AQ is then used to find the ignition time from the following expression:

$$\tau_i = \frac{R^2 T_o^3}{pEAQ}\left(1 + \frac{1}{\beta}\right)(C_{pF} + \beta C_{pO}) \exp\left(\frac{E}{RT_o}\right) \tag{26}$$

where

$$p = p_F + p_O \tag{27}$$

$$\beta = p_O/p_F \tag{28}$$

and C_{pF} and C_{pO} are specific heats per unit mole.

It should be noted that Eq. (26) applies only to a system that is at constant pressure and temperature. Although an actual engine ignition delay involves the time required to pressurize the chamber, this model provides results comparable to the ignition delay for the space shuttle 870-lbf thruster, where ignition occurred after 9 ms at a pressure of 0.37 atm (5 psia). The model assumed $T_o = 298$ K and $\beta = 2.5$. As shown in Fig. 5a, the effect of $\beta = 5$ and $\beta = 0.1$ is to delay ignition. These two values are higher and lower than the stoichiometric mixture ratio of $\beta = 2.5$ for MMH/NTO. Seamans found that optimum ignition is obtained for $\beta = 1$. As shown in Fig. 5b, colder temperatures increase the ignition delay. Under these conditions the mass of unreacted propellants in the engine will increase, and its delayed ignition will produce the overpressures as observed.

2. Engine Ignition Transient Model

Steady-state models used in the quasi-steady mode often constitute starting points for a transient description of this kind. Vaporization-controlled combustion models generally assume that the fuel is in the liquid phase and the oxidizer is in the gas phase. The fuel droplets enter the one-dimensional constant area chamber uniformly in size and velocity. The droplets, at their boiling

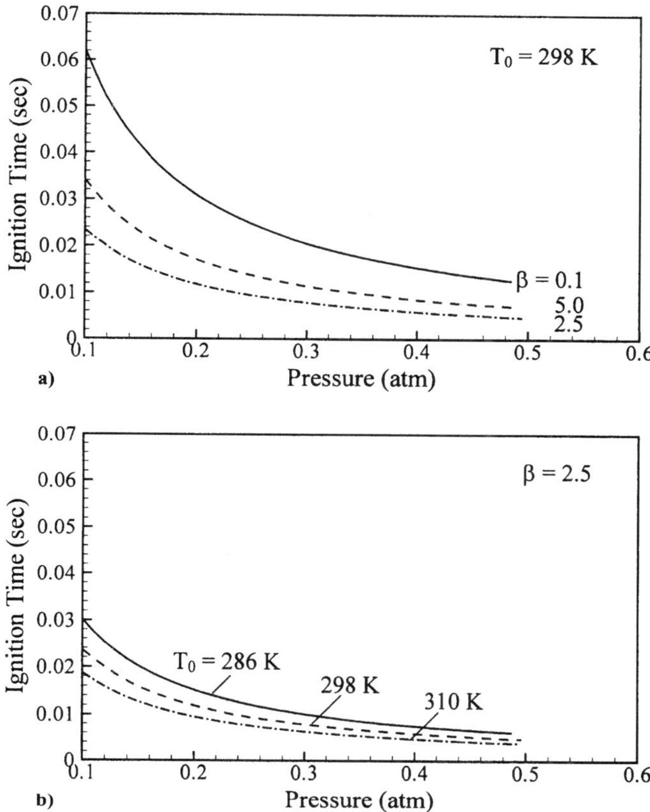

Fig. 5   Theoretical ignition delay-pressure curves for NTO/MMH.[81] a) Effect of $\beta = p_O/p_F$ at $T = 298$ K and b) Effect of gas temperature, $\beta = 2.5$.

temperature, vaporize according to the d^2 law. The fuel vapor is instantly burned, with the product species determined by equilibrium. Axial diffusion, radiation, and friction are neglected. The inputs are the mass flow rate, initial velocity, number, and droplet size.

The quasi-steady approach is used, for example, by Seamans et al.[80] to analyze the combustion transient in hypergolic engines based on a one-dimensional vaporization model. At each time step, the one-dimensional model essentially is at equilibrium and new drops are injected into the chamber. The drop sizes and number densities are specified to obtain a mass distribution. The droplets undergo convective heating and evaporative cooling. If the freezing point is reached, the temperature is kept constant as the quality of the droplet changes. Condensation of propellants on the chamber walls is also modeled. The hypergolic ignition process is described by considering the intersection of the vapor pressure vs time curve with the vapor pressure vs ignition delay time τ_{ign}, curve.

Quasi-steady assumptions in the model treat the chamber pressure increase at each time step as a result of the combustion of the fuel and oxidizer, which assumes an ideal gas for the combustion products, and includes the mass flow out of the nozzle. Calculations agree with test data except that the predicted ignition delay time is generally shorter because the pressurization of the propellant manifolds and the line dynamics are not included in the model.

An overview of spray combustion is provided by Faeth.[81] Statistical descriptions of the spray droplets generally rely on a discrete droplet model. Considering the spray as shown in Fig. 4, it would appear difficult to determine a suitable model. Both oxidizer and fuel appear to exist as droplets in the spray. The NTO droplets are evaporating, along with the MMH. The burning appears to occur in a cloud that surrounds the droplets. The droplets may be undergoing substantial secondary atomization and collisions. In addition, significant mixing will be caused by turbulence. The process then is further complicated by consideration of unsteady flow associated with the feed system dynamics. It is also reasonable to assume that, because of oscillations in the flow from the line dynamics, the droplet sizes may significantly vary with time.

Webber[82] and Hoffman et al.[66] developed a one-dimensional model that is able to predict the amplitude, frequency, and transient phenomena in a combustor. This model considers the propellant flow in the manifolds, atomization, droplet drag, and evaporation in a convective environment. Radiation and heat transfer into the droplet are neglected, so that all heat goes into evaporating the droplet. The droplet impingement distance and breakup distance were based on correlations by Priem et al.[83]

Webber[82] used experimental burning rate data based on the d^2-evaporation law to determine the vapor thermal conductivity because this was not a known property. The combustion chemistry is based on equilibrium (assuming fast chemistry), which allows the product gas temperature and properties to be a function of the mixture fraction. The resulting product gases are assumed to be ideal and at equal pressure from which exit mass flow rate is calculated. The major disadvantages to this model are that it does not compute the ignition delay, which is an input to the model, and that it does not accurately model the line dynamics.

Schuman et al.[84] combined three existing codes to create a transient performance prediction model. The one-dimensional equilibrium (ODE) code is used for the thermochemistry. The generalized propulsion system model (GPSM) is used to simulate the propellant feed system using the one-dimensional hydrodynamic equations. This analysis tool substantially improved the modeling of the transient in the feed lines by allowing for two-phase flow, frequency dependent damping, and complex manifolding. The transient combustion chamber (TCC) code is used for the thrust chamber. The ignition model uses a form similar to that developed by Seamans et al.,[80] with a delay, given by Eq. (22).

V.   Discussion and Conclusion

For hypergolic ignition, none of the models discussed fully captures all of the processes that govern the ignition transient. Future models could include the pressure- and temperature-dependent reaction rates, line dynamics, and chamber dynamics, along with a suitable spray model. The one-dimensional assumption may be adequate for initial predictions. The details for such a model would include: 1) multistep reaction kinetics, heat transfer, and spray models to predict ignition delay, 2) one-dimensional axial chamber dynamics to simulate longitudinal acoustic modes, and 3) line dynamics, including two-phase compressibility or bulk modulus. With properly coupled equations, the pressure response of the manifolds, chamber, and combustion rate could be studied. The rapid combustion of a clump of propellants at a given axial location in the chamber would create an acoustic wave that could propagate up the manifold and down the chamber. The multistep reaction kinetics would cover preignition products, ignition, and steady-state reactions. The goals of the analysis are the accurate prediction of destructive overpressures, the potential for contamination in the exhaust, and the thruster performance.

For ignition-assisted propellants, the concepts of resonance ignition and laser igniters, as an alternative to the augmented spark igniter, are attractive prospects for engines that can be used in multistart applications, such as reaction control systems where hypergolic propellants are currently used. There appears to be a substantial amount of data on ignition; a standardized presentation of ignition energy would be useful and would allow comparison of the data in terms of details such as electrode diameters, spark durations, etc. Also, in some descriptions it is not clear how ignition delay time is defined, i.e., from the valve opening or from the time the igniter is fired. With ignition-assisted propellants, the effects of turbulence and heterogeneity are important, so that emphasis should be placed on these in experimental studies.

Acknowledgments

Ignition studies carried out at the EM2C laboratory were supported by SNECMA and Centre National d'Etudes Spatiales, CNES. Fluorescence images were obtained by K. McManus and F. Aguerre. S. Candel wishes to acknowledge the friendly advice of the late M. Barrère.

References

[1]Altman, D., and Penner, S., "Ignition Phenomena in Bipropellant and Monopropellant Systems," *Combustion Processes*, edited by Princeton Univ. Press, Princeton, NJ, 1956, pp. 470–489.

[2]Barrère, M., Jaumotte, A., de Veubeke, B. F., and Vandenkerckhove, J., *Rocket Propulsion*, Elsevier, New York, 1960.

[3]Huzel, D. K., and Huang, D. H. (eds.), *Modern Engineering for Design of Liquid Propellant Rocket Engines*, Progress in Astronautics and Aeronautics, Vol. 147, AIAA, Washington, DC, 1992.

[4]Oppenheim, A. K., "Dynamic Features of Combustion," *Philosophical Transactions of the Royal Society of London*, Vol. A315, 1985, pp. 471–508.

[5]Williams, F. A., *Combustion Theory*, 2nd ed., Addison Wesley Longman, Reading, MA, 1993.

[6]Strehlow, R. A., *Combustion Fundamentals*, McGraw–Hill, New York, 1984.

[7]Glassman, I., *Combustion*, Academic International Press, New York, 1987.

[8]Ballal, D. R., and Lefebvre, A. H., "Ignition and Flame Quenching in Flowing Gaseous Mixtures," *Proceedings of the Royal Society of London*, Vol. A315, 1977, pp. 163–181.

[9]Ballal, D. R., and Lefebvre, A. H., "Ignition of Liquid Fuel Sprays at Subatmospheric Pressures," *Combustion and Flame*, Vol. 31, 1978, pp. 115–126.

[10]Arai, M., Yoshida, H., and Hiroyasu, H., "Ignition Process of Compound Spray Combustible Mixtures," *Dynamics of Heterogeneous Combustion and Reacting Systems*, edited by Progress in Astronautics and Aeronautics, AIAA, Washington, DC, 1993, pp.

[11]Peters, J. E., and Mellor, A. M., "An Ignition Model for Quiescent Fuel Sprays," *Combustion and Flame*, Vol. 38, 1980, pp. 65–74.

[12]Peters, J. E., and Mellor, A. M., "A Spark Ignition Model for Liquid Fuel Sprays Applied to Gas Turbine Engines," *Journal of Energy*, Vol. 6, 1982, pp. 272–274.

[13]McManus, K., Aguerre, F., Yip, B., and Candel, S., "Analysis of the Ignition Sequence of a Nonpremixed Combustor Using PLIF Imaging," *Nonintrusive Combustion Diagnostic*, edited by K. K. Kuo and T. P. Parr, Begell House, New York, 1993, pp. 714–725.

[14]Veynante, D., Lacas, F., and Candel, S., "A New Flamelet Model Combining Premixed and Non-Premixed Flames," AIAA Paper A19, 1989, pp. 89–487.

[15]Fichot, F., *Modeling of Ignition of a Turbulent Diffusion Flame. Application to Cryogenic Rocket Engines*, Ph.D. Dissertation, Ecole Centrale Paris, 1994.

[16]Candel, S., Veynante, D., Fichot, F., and Thevenin, D., "Modélisation de Problémes d'Allumage," *Modélisation de la Combustion*, edited by Centre National de la Recherche Scientifique, Paris, 1996, pp. 109–117.

[17]Fichot, F., Schreiber, D., Lacas, F., Veynante, D., and Yip, B., "New Flamelet Approach to Model the Transient Phenomena Following Ignition in a Turbulent Diffusion Flame," *Dynamics of Gaseous Combustion*, edited by A. L. Kuhl, J. C. Leyer, A. A. Borisov, and W. Sirignano, AIAA, Washington, DC, 1993, pp. 331–343.

[18]He, L., and Clavin, P., "Premixed Hydrogen-Oxygen Flames. Part II: Quasi-Isobaric Ignition near Flammability Limits," *Combustion and Flame*, Vol. 93, 1993, pp. 408–420.

[19]He, L., and Clavin, P., "On the Direct Initiation of Detonations by an Energy Source," *Journal of Fluid Mechanics*, Vol. 277, 1994, pp. 227–248.

[20]Baudart, P. A., Duthoit, V., and Harlay, J. C., "Numerical Simulation of Cryogenic Rocket Engine Ignition," AIAA Paper 91-2290, 1991.

[21]Armstrong, E., "Ignition Systems for Liquid Oxygen (LOX)/Hydrocarbon Booster Engines," NASA TM102033, 1989.

[22]Stout, H. P., and Jones, E., "Ignition of Gaseous Explosive Media by Hot Wires," 3rd Symposium on Combustion and Flame and Explosive Phenomena, Williams and Wilkens Co., Baltimore, MD, 1948, pp. 329–336.

[23]Shorr, M., "Ignition Techniques for Gaseous Oxygen-Hydrogen Mixtures," 11th JANNAF Combustion Meeting, Vol. 2, 1974, pp. 219–254.

[24]Homan, H. S., and Sirignano, W. A., "Minimum Mass of Burning Aluminum Particles for Ignition of Methane/Air and Propane/Air Mixtures," *18th Symposium (International) on Combustion*, Combustion Inst., Pittsburgh, PA, 1981, pp. 1709–1717.

[25]Duncan, D. B., et al., "Smart Laser Igniter for RCS Thrusters," Duncan Technologies, Inc., Phase I Final Rept., SBIR Contract NAS9-19036,

[26]Conrad, E. W., and Pavli, A. J., "A Resonance-Tube Igniter for Hydrogen-Oxygen Rocket Engines," NASA TM X-1460, 1967.

[27]Phillips, B. R., and Pavli, A. J., "Resonance Tube Ignition of Hydrogen-Oxygen Mixtures," NASA TN D-6354, 1971.

[28]Przirembel, C. E. G., and Fletcher, L. S., "The Aerothermodynamics of a Simple Resonance Tube," AIAA Paper 1975-687, 1975.

[29]Stabinsky, L., "Analytical and Experimental Study of Resonance Ignition Tubes," NASA CR-136934, 1974.

[30]Przirembel, C. E. G., Wolf, D. E., and Fletcher, L. S., "Thermodynamic Characteristics of a Blunt, Two-Dimensional Resonance Tube," AIAA Paper 1976-145, 1976.

[31]Lauffer, J. R., "Spark and Auto Ignition Devices for ACPS," Space Transportation System Propulsion Technology Conference, Vol. 2, Auxiliary Propulsion System, 1971, pp. 665–758; also NASA TM X-67246.

[32]Johnson, R. J., "Investigation of Thrusters for Cryogenic Reaction Control Systems," NASA Contract NAS 3-11227, Final Rept., June 1970–Dec. 1971, p. 214.

[33]Zurawski, R. L., and Green, J. M., "Catalytic Ignition of Hydrogen and Oxygen Propellants," AIAA Paper 1988-3300, 1988.

[34]Green, J. M., "A Premixed Hydrogen/Oxygen Catalytic Igniter," AIAA Paper 1989-2302, 1989.

[35]Johnson, R. J., "Hydrogen-Oxygen Catalytic Ignition and Thruster Investigation Volume 1: Catalytic Ignition and Low Pressure Thruster Evaluations," NASA-CR-120869, Final Rept., June 1970–Dec. 1971.

[36]Ketsen, A. S., "Study of Catalytic Reactors for Hydrogen-Oxygen Ignition," NASA CR-72567, UARL H910721, 1969.

[37]Ketsen, A. S., "Transient Model of Hydrogen/Oxygen Reactor," NASA CR-120799, UARL K910962-12, 1971.

[38]Rollbuhler, R. J., and Straight, D. M., "Ignition of Hydrogen-Oxygen Rocket Engine by Addition of Fluorine to the Oxidant," NASA TN-D-1309, 1962.

[39]Mosier, S. A., Dotson, R. E., and Moehrbach, O. K., "Hypergolic Ignition of Light Hydrocarbon Fuels with Fluorine-Oxygen (FLOX) Mixtures," Presented at the 1965 Fall Meeting of the Combustion Institute, Western States Section, Paper WSCI 65-23, 1965.

[40]Dickinson, L. A., Amster, A. B., and Capener, E. L., "Application of Trioxygen Difluoride in Liquid-Rocket Propellant Technology," *Journal of Spacecraft and Rockets*, Vol. 5, No. 11, 1968, pp. 1329–1334.

[41]Dankhoff, W. F., et al., "M-1 Injector Development—Philosophy and Implementation," NASA TN D-4730, 1968.

[42]Bellingham, R., and Sandri, R., "Ignition of the Hydrogen-Oxygen Propellant Combination by Chlorine Trifluoride," *AIAA Journal*, Vol. 5, No. 4, 1967, pp. 770–773.

[43]Ronney, P. D., "Laser Versus Conventional Ignition of Flames," *Optical Engineering*, Vol. 33, No. 2, 1994, pp. 510–521.

[44]Liou, L. C., "Laser Ignition in Liquid Rocket Engines," Internal NASA Paper.

[45]Weinberg, F. J., and Wilson, J. R., "A Preliminary Investigation of the Use of Focused Laser Beams for Minimum Ignition Energy Studies," *Proceedings of the Royal Society of London, Series A: Mathematical and Physical Sciences*, Vol. 321, No. 1544, 1971, pp. 41–52.

[46]Forch, B. E., and Miziolek, A. W., "Oxygen-Atom Two-Photon Resonance Effects in Multiphoton Photochemical Ignition of Premixed H_2/O_2 Flows," *Optics Letters*, Vol. 11, No. 3, 1986, pp. 129–131.

[47]Gregory, J. W., and Herr, P. N., "Shuttle ACPS Thruster Technology Review," NASA TM X-68146, 1972.

[48]Senneff, J. M., "High Pressure Reverse Flow APS Engine," NASA CR-120881, 1973.

[49]Lawver, B. R., Rousar, D. C., and Boyd, W. C., "Ignition Characterization of the GOX/Ethanol Propellant Combination," AIAA Paper 1984-1467, 1984.

[50]Maly, R., and Vogel, M., "Initiation and Propagation of Flame Fronts in Lean CH_4-Air Mixtures by the Three Modes of the Ignition Spark," *Proceedings of the Combustion Institute*, Vol. 17, Combustion Inst., Pittsburgh, PA, 1978, pp. 821–831.

[51]Ziegler, G. F. W., Wagner, E. P., and Maly, R., "Ignition of Lean Methane-Air Mixtures by High Pressure Glow and Arc Discharges," *Proceedings of the Combustion Institute*, Vol. 20, Combustion Inst., Pittsburgh, PA, 1984, pp. 1817–1824.

[52]Kono, M., Hatori, K., and IInuma, K., "Investigation on Ignition Ability of Composite Sparks in Flowing Mixtures," *Proceedings of the Combustion Institute*, Vol. 20, Combustion Inst., Pittsburgh, PA, 1984, pp. 133–140.

[53]Swett, C. C., "Spark Ignition of Flowing Gases," NACA Rept. 1287, 1956.

[54]Frendi, A., and Sibulkin, M., "Dependence of Minimum Ignition Energy on Ignition Parameters," *Combustion Science and Technology*, Vol. 73, 1990, pp. 395–413.

[55]Ballal, D. R., and Lefebvre, A. H. "The Influence of Flow Parameters on Minimum Ignition Energy and Quenching Distance," *Proceedings of the Combustion Institute*, Vol. 15, Combustion Inst., Pittsburgh, PA, 1974, pp. 1473–1480.

[56]Labotz, R. J., Rousar, D. C., and Valler, H. W., "High Density Fuel Combustion and Cooling Investigation," NASA CR 165117, 1980.

[57]Bjorklund, R. A., "Very Low Thrust Gaseous Oxygen-Hydrogen Rocket Engine Ignition Technology," *20th JANNAF Combustion Meeting*, edited by D. S. Eggleston, Vol. I, Chemical Propulsion Information Agency, Pub. 383, Laurel, MD, 1983, pp. 699–711.

[58]Bjorklund, R. A., and Apel, M. A., "Very Low Thrust and Low Chamber Pressure GO_2/GH_2 Thruster Technology," NASA Contract NAS7-918.

[59]Kono, M., Kumagai, S., and Sakai, T., "Ignition of Gases by Two Successive Sparks with Reference to Frequency Effect of Capacitance Spark," *Combustion and Flame*, Vol. 27, 1976, pp. 85–98.

[60]Simmons, J. A., Gift, R. D., Spurlock, J. M., and Fletcher, R. F., "Reactions and Expansion of Hypergolic Propellants in a Vacuum," *AIAA Journal*, Vol. 6, No. 5, 1968, pp. 887–893.

[61]Schmidt, and Eckardt, W., *Hydrazine and Its Derivatives: Preparation, Properties, and Applications*, Wiley, New York, 1984.

[62]Wright, A. C., "USAF Propellant Handbooks Nitric Acid/Nitrogen Tetroxide Oxidizers," Vol. II, Martin Marietta Corp., AFRPL-TR-76-76, 1977.

[63]Kerkam, B. J., and Kahl, R. C., "Hypergol Engine Restart Characteristics," The Boeing Co., NASA Johnson Space Center, 1972.

[64]Lawver, B. R., "High Performance N_2O_4/Amine Elements Blowapart," NASA Johnson Space Center, Aerojet Liquid Rocket Co., Contract NASA 9-14186, 1979.

[65]Zung, L. B., and White, J. R., "Combustion Process of Impinging Hypergolic Propellants," NASA CR-1704, 1971.

[66]Hoffman, R. J., Webber, W. T., Oeding, R. G., Nunn, J. R., "An Analytical Model for the Prediction of Plume Contamination Effects on Sensitive Surfaces," AIAA Paper 72-1172, 1972.

[67]Juran, W., and Stechman, R. C., "Ignition Transients in Small Hypergolic Rockets," *Journal of Spacecraft and Rockets*, Vol. 5, No. 3, 1968, p. 288.

[68]Mills, T. R., Breen, B. P., Tkachenko, E. A., and Lawver, B. R., "Transients Influencing Rocket Engine Ignition and Popping," NASA 7-467, Interim 14 Month Rept., 1969.

[69]Mayer, S. W., Taylor, D., and Schieler, L., "Preignition Products from Storable Propellants at Simulated High Altitude Conditions," Aerospace Corp., Air Force Contract F04695-67-C-0158.

[70]Saad, M. A., Detweiler, M., and Sweeney, M., "Analysis of Reaction Products of Nitrogen Tetroxide with Hydrazine Under Nonigintion Conditions," *AIAA Journal*, Vol. 10, No. 8, Aug. 1972, pp. 1073–1078.

[71]Christos, T., Miron, Y., James, H., and Perlee, H., "Combustion Characteristics of Condensed Phase Hydrazine Type-Fuels with Nitrogen Tetroxide," *Journal of Spacecraft and Rockets*, Vol. 4, No. 9, 1967.

[72]Takimoto, H. H., and Denault, G. C., "Combustion Residues from N_2O_4-MMH Motors," The Aerospace Corp., TR-0066(5210–10)-1, 1969.

[73]Spalding, D. B., *Combustion and Mass Transfer*, Pergamon, 1979, p. 272.

[74]Kuo, K. K., *Principles of Combustion*, Wiley-Interscience, New York, 1986, Chap. 10.

[75]Gray, P., and Sherrington, M. E., "Self-Heating and Spontaneous Ignition in the Combustion of Gaseous Methylhydrazine," *Journal of the Chemical Society, Faraday*, Vol. 70, No. 4, 1974, pp. 740–75.

[76]Perlee, H. E., Imhof, A. C., and Zabetakis, M. G., "Flammability Characteristics of Hydrazine Fuels in Nitrogen Tetroxide Atmospheres," *Journal of Chemical and Engineering Data*, Vol. 7, No. 3, 1962, pp. 377–379.

[77]Allison, C. B., and Faeth, G. M., "Decomposition and Hybrid Combustion of Hydrazine, MMH, and UDMH as Droplets in a Combustion Gas Environment," *Combustion and Flame*, Vol. 19, 1972, pp. 213–226.

[78]Eberstein, I. J., and Glassman, I., "The Gas Phase Decomposition of Hydrazine and Its Methyl Derivatives," *Proceedings of the Combustion Institute*, Vol. 10, 1965, pp. 365–374.

[79]Daimon, W., Gotoh, Y., and Kimura, I., "Mechanism of Explosion Induced by Contact of Hypergolic Liquids," *Journal of Propulsion and Power*, Vol. 7, No. 6, 1991, pp. 946–952.

[80]Seamans, T. F., Vanpee, M., and Agosta, V. D., "Development of a Fundamental Model of Ignition in Space-Ambient Engines," *AIAA Journal*, Vol. 5, No. 9, 1967, pp. 1616–1624.

[81]Faeth, G. M., "Evaporation and Combustion of Sprays," *Progress in Energy and Combustion Science*, Vol. 9, 1983, pp. 1–76.

[82]Webber, W. T., "Calculation of Low-Frequency Unsteady Behavior of Liquid Rockets from Droplet Combustion Parameters," *Journal of Spacecraft and Rockets*, Vol. 9, No. 4, 1972, pp. 231–237.

[83]Harrje, D. T. and Reardon, F. H. (eds.), *Liquid Propellant Combustion Instability*, NASA SP-194, 1972, pp. 195–205.

[84]Schuman, M. D., Ervin, J., and Taniguchi, M., "Transient Performance Program," Air Force Rocket Propulsion Lab., AFRPL-TR-80-22, 1981.

Chapter 12

Rocket Engine Nozzle Concepts

Gerald Hagemann* and Hans Immich[†]
EADS Space Transportation, Munich, Germany

Thong Nguyen[‡]
GenCorp Aerojet, Sacramento, California

and

Gennady E. Dumnov[§]
NIKA Software, Moscow, Russia

Nomenclature

$$
\begin{aligned}
A &= \text{area} \\
\text{amb} &= \text{ambient} \\
c &= \text{combustion chamber} \\
\text{cr} &= \text{critical} \\
c^* &= \text{characteristic velocity} \\
c_F &= \text{thrust coefficient} \\
e &= \text{exit plane} \\
F &= \text{thrust} \\
\text{geom} &= \text{geometrical} \\
h &= \text{flight altitude} \\
I &= \text{impulse} \\
l &= \text{length} \\
M_{\text{in}} &= \text{inlet Mach number} \\
M_{\text{id}} &= \text{internal design Mach number}
\end{aligned}
$$

*Project Manager, Research Engineer, Space Infrastructure Propulsion. Member AIAA.
[†]Project Manager, Space Infrastructure Propulsion. Senior Member AIAA.
[‡]Technical Principal. Senior Member AIAA.
[§]Division Head.

M_d = design Mach number
m = mass
p = pressure
p_c/p_{amb} = pressure ratio
r = mass ratio oxidizer/fuel
ref = reference
sp = specific
t = throat
t = time
vac = vacuum
w = wall
x, y = coordinate
ε = nozzle area ratio
ϕ_{int} = internal flow turning angle
ϕ_{ext} = external flow turning angle

I. Introduction

THE key demand on future space transportation systems is the concurrent reduction of earth-to-orbit launch costs and the increase in launcher reliability and operational efficiency. Meeting this demand strongly depends on engines that deliver high performance with low system complexity. The engine performance is characterized by the specific impulse that follows: $I_{sp} = F/(\partial m/\partial t) = c^* \cdot c_F$. The characteristic velocity c^* mainly depends on the propellant combination, mixture ratio, and combustion efficiency, and relates to the efficiency of the combustion process. The thrust coefficient c_F is mainly a function of the gas composition and pressure ratio across the thrust chamber (that is, the combustion chamber and nozzle extension), and thus of the nozzle area ratio. The thrust coefficient can be interpreted as the amplification of thrust due to the supersonic expansion process in the nozzle compared to the thrust contribution of the combustion chamber acting over the throat area only. The efficiency of the nozzle expansion process can be described with the ratio of measured thrust coefficient to its theoretically attainable value.[1]

Performance of a rocket engine is always lower than the theoretically attainable value because of imperfections in the mixing and combustion process, and also because of the further expansion of the propellants, as summarized schematically in Fig. 1. These loss effects have been the subjects of investigation for many years. Table 1 summarizes performance losses in the combustion chamber and nozzle of typical high-performance rocket engines, such as the space shuttle main engine (SSME) and Vulcain engine (central stage engine of Ariane 5).[2] The combustion process in the combustion chamber is characterized by very low losses, and measured characteristic velocities c^* reach almost their theoretically possible values with today's rocket engines. Viscous effects due to turbulent boundary layers and the nonuniform flow profile of the flow in the exit area are among the important loss sources in the further expansion process in the nozzles. Furthermore, the nonadaptation of the exhaust flow to the

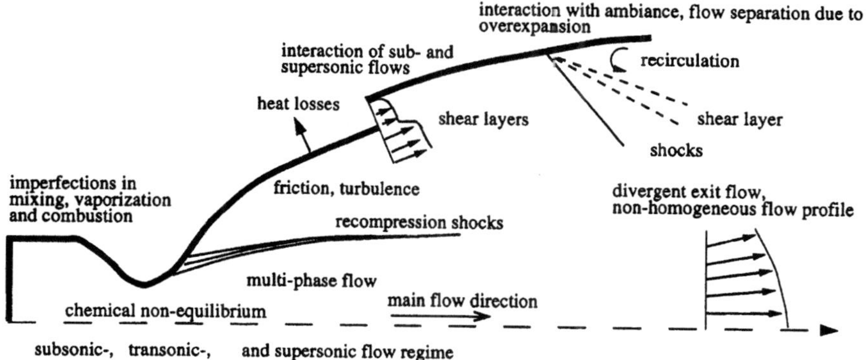

Fig. 1 Flow phenomena and loss sources in rocket nozzles.

varying ambient pressures induces a significant thrust loss compared to an ideally adapted nozzle. Biggest growth potential in performance is obviously reachable with the thrust coefficient, especially by a nozzle design allowing for altitude adaptation, which is the adaptation of the nozzle exit pressure to the varying ambient pressure.

This chapter addresses different nozzle concepts, including conventional nozzles, as well as several nozzle concepts with the capability for altitude adaptation (see Refs. 3–6).

II. Conventional Nozzles

Different design approaches have been proposed for the conventional bell-type rocket nozzles used in nearly all of today's rocket engines, such as the design approaches for 1) the conical nozzle, 2) the truncated ideal nozzle, and 3) the thrust-optimized nozzle. The conical nozzle has the simplest design, with typical divergence angles between 15 and 25 deg, but consequently it also has high divergence and profile losses. This approach is mainly used for solid rocket boosters.

The ideal nozzle establishes a one-dimensional exhaust flow profile, but it has a huge length, especially because its last contour part has only minor inclination needed to obtain uniform flow. Thrust contribution of this end portion is negligible

Table 1 Performance losses in conventional rocket nozzles

Losses	Mainly affecting	Vulcain, %	SSME, %
Imperfection in mixing and combustion	c^*	1.0	0.5
Chemical nonequilibrium	c^*, c_F	0.2	0.1
Friction	c_F	1.1	0.6
Divergence, nonuniformity of exit flow	c_F	1.2	1.0
Nonadapted nozzle flow	c_F	0–15	0–15

because of the minor wall slopes; therefore, truncation of the last nozzle portion makes this approach feasible for rocket motors without introducing significant losses in performance due to the nonuniformity of the flow. For example, the LR-115, RD-0120, and LE-7 are designed as truncated ideal nozzles. Further shortening of truncated ideal nozzles may be achieved by applying the design approach proposed by Ahlberg[7]: the compressed truncated ideal nozzle design.

For thrust-optimized nozzles, Rao[8] has proposed a variational optimization method based on Lagrange multipliers that gives the nozzle design for maximum performance at a given length. As a rough approach, the typical length of a Rao-type nozzle is 75–85% of the length of a 15-deg conical nozzle having the same area ratio. Later it was also shown by Rao[9] that the contour designed with this variational optimization method can be approximated with a skewed parabola without introducing a significant performance loss. This approach is frequently used for the nozzle design of modern rocket nozzles, for example, SSME, RS-68, Vulcain, or Vulcain 2.

A typical design approach for a conventional bell-type rocket nozzle is described in the next chapter for the European Vulcain 2 engine nozzle for the Ariane 5 launcher.

Conventional nozzles limit overall engine performance during the ascent of the launcher because of their fixed geometry. Significant performance losses are induced during the off-design operation of the nozzles, when the flow is overexpanded during low-altitude operation with ambient pressures higher than the nozzle exit pressure, or underexpanded during high-altitude operation with ambient pressures lower than the nozzle exit pressure. Figure 2 shows photographs of nozzle exhaust flows during both off-design operation modes.

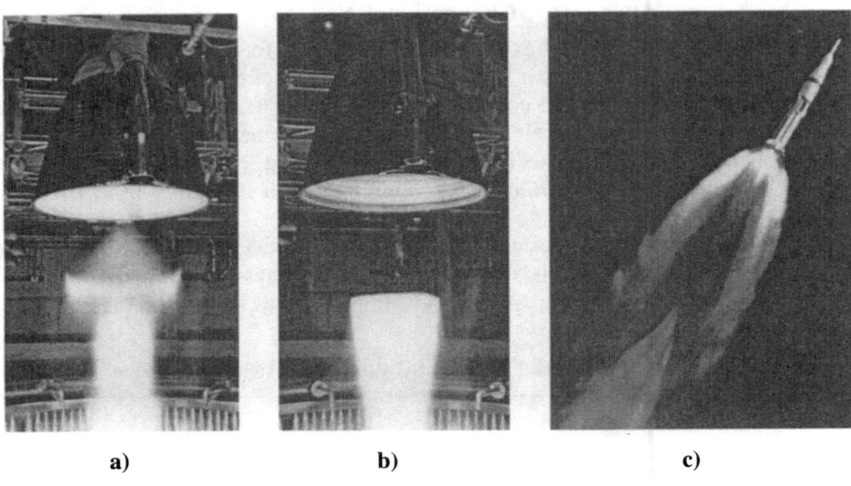

Fig. 2   Rocket nozzle flowfield during off-design operation: a) overexpanded flow with cap-shock pattern Vulcain engine, b) overexpanded flow with Mach disk Vulcain engine, and c) underexpanded flow Saturn-1B, Apollo-7. (See also the color section of figures following page 620.) (Photographs courtesy of Dasa, SNECMA, NASA.)

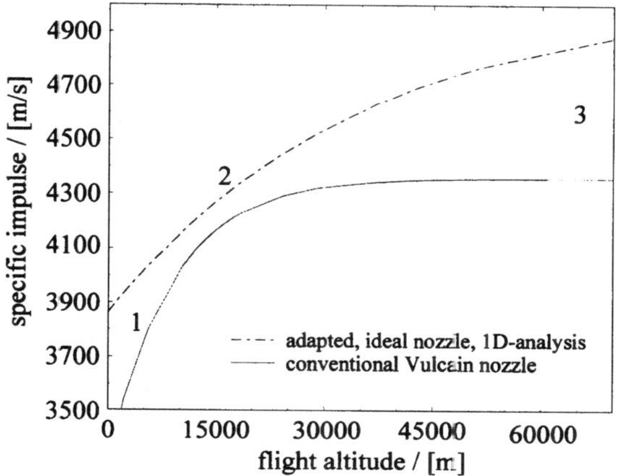

Fig. 3 **Performance data for a conventional rocket nozzle. Data estimated for Vulcain engine ($\varepsilon = 45$, $r = 5.89$, $p_c = 100$ bar).**

For overexpanded flow, oblique shocks emanating into the flowfield adapt the exhaust flow to the ambient pressure. Further downstream, a system of shocks and expansion waves leads to the characteristic barrel-like form of the exhaust flow. Different shock patterns in the plume of overexpanded rocket nozzles have been observed, including the classical Mach disk or regular shock reflection at the centerline. (For axisymmetric flow, a pure regular reflection at the centerline is not possible, and it thereby features a small normal shock at the centerline.) In addition to these two plume patterns, a third pattern is observed in the plume of the thrust-optimized or parabolic nozzles featuring an internal shock that limits the high Mach number field at the centerline (the latter is also commonly referred to as kernel). This plume pattern, called cap-shock pattern,[10-12] follows from the interaction of an inverse Mach reflection of the internal shock at the centerline with the recompression shock induced at the wall.[11] In contrast, the underexpansion of the flow results in further expansion of the exhaust gases behind the rocket.

Off-design operations with either overexpanded or underexpanded exhaust flow induce performance losses. Figure 3 includes calculated performance data for the Vulcain nozzle as a function of flight altitude, together with performance data for an ideally adapted nozzle. Flow phenomena at different pressure ratios, p_c/p_{amb}, are included in Fig. 4.* The Vulcain nozzle is designed in such a manner that no uncontrolled flow separation should occur during steady-state operation on ground or at low altitude, resulting in a wall exit pressure of $p_{w,e} \sim 0.4$ bar,

---

*The sketch with flow phenomena for the lower pressure ratio, p_c/p_{amb} (see Fig. 4), shows a normal shock (Mach disk). Depending on nozzle design and pressure ratio, p_c/p_{amb}, a cap-shock pattern, Mach disk, or regular reflection may appear (see Fig. 2).

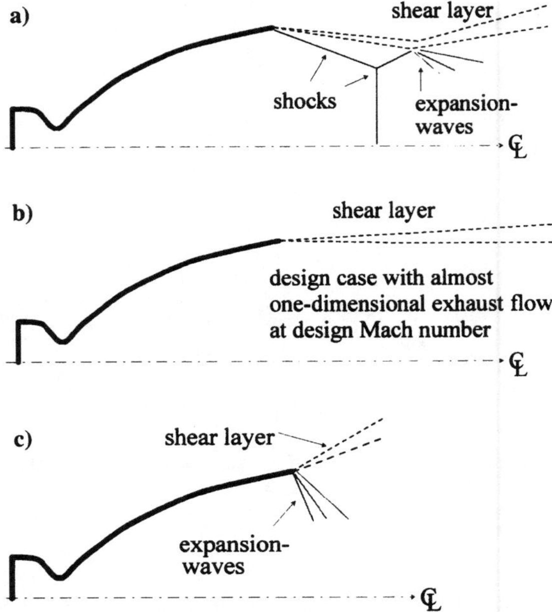

Fig. 4   Flow phenomena for a conventional rocket nozzle for a) sea-level operation, b) adapted nozzle flow, and c) high-altitude operation.

which is in accordance with the Summerfield criterion.[13,14] The nozzle flow is adapted at an ambient pressure of $p_{amb} \sim 0.18$ bar (corresponding flight altitude: $h = 15.000$ m), and performance losses observed at this ambient pressure are caused by internal loss effects (friction, divergence, mixing), as summarized in Table 1. For other flight altitudes, losses in performance during off-design operations with over- or underexpansion of the exhaust flow rise up to 15%. In principle, the nozzle could be designed for a much higher area ratio to achieve better vacuum performance, but the flow would then separate inside the nozzle during low-altitude operation with an undesired generation of sideloads.

A.   Flow Separation and Sideloads

Flow separation in overexpanding nozzles and its theoretical prediction have been the subject of investigation in the past decades,[5,6,14–17] and different physical models and hypotheses for the prediction of flow separation have been developed. In strongly overexpanding nozzles, the flow separates from the wall at a certain pressure ratio of wall pressure to ambient pressure p_w/p_{amb}. The typical structure of the flowfield near the separation point for the free shock separation is shown in Fig. 5a, together with wall pressure data.

A second flow separation pattern is the restricted shock separation, where the flow reattaches after initial separation. This structure of the flowfield is illustrated

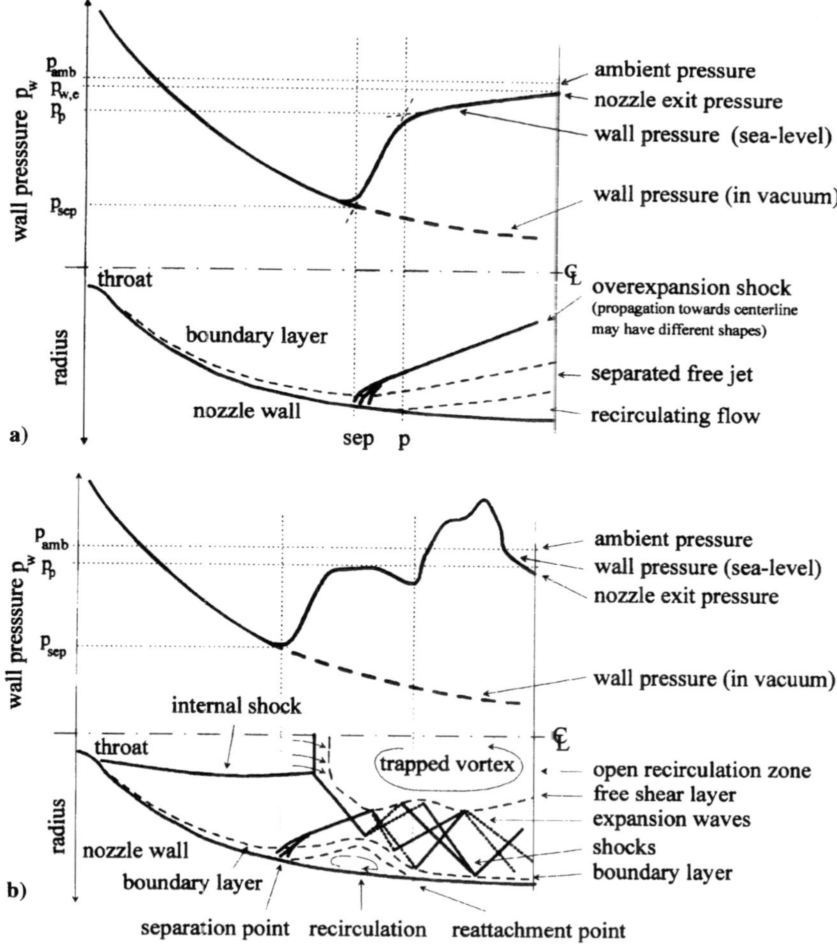

Fig. 5 Flow separation in overexpanding rocket nozzles, wall pressure profile, and phenomenon: a) free shock separation and b) restricted shock separation.

in Fig. 5b. Experimental investigation and numerical analyses have proven that the cap-shock pattern in the plume of nozzles featuring an internal shock, for example, parabolic rocket nozzles, drives this reattachment of the flow.[11−13,18] The typical core vortex downstream of the cap-shock pattern has also been proven by theoretical and experimental studies.[19,20]

The two flow patterns, free- and restricted shock separation, are visualized in Fig. 6. Both photos were exposed during a hot-firing test campaign with a GOX/ GH$_2$ demonstrator engine designed for 40-kN vacuum thrust level. For both flow conditions the thrust chamber was throttled down to 40% power level. Reference 21 includes further details on the test campaign and results.

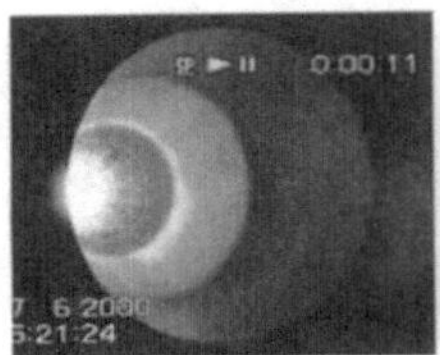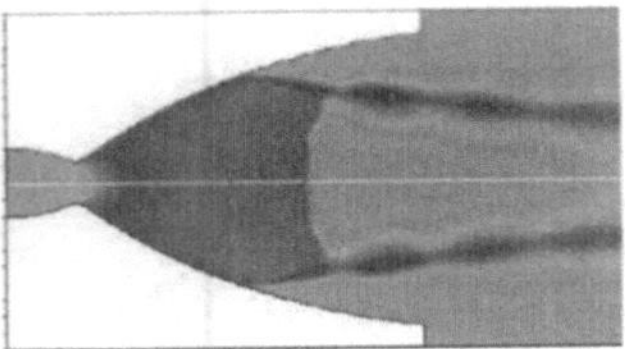

GOX/GH$_2$ thrust chamber operating at 40% power level featuring free shock separation

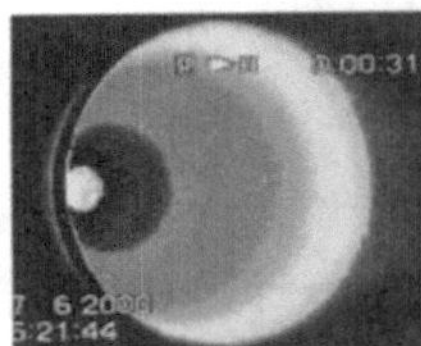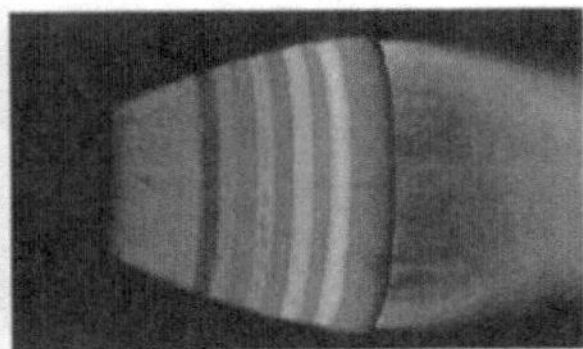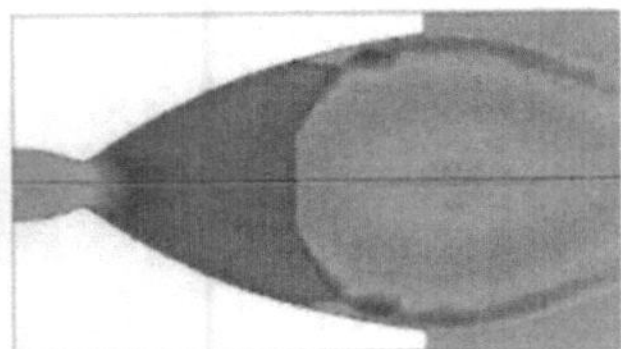

GOX /GH$_2$ thrust chamber operating at 40% power level featuring restricted shock separation

Fig. 6 Hot-firing test with GOX/GH$_2$ thrust chamber designed for 40-kN vacuum thrust level at 40% power level featuring free shock separation (top figures) and restricted shock separation (bottom figures). CFD results for flow visualization (Mach number distribution, from blue to red for increasing Mach number). (See also the color section of figures following page 620.)

Various approaches to the prediction of flow separation are used in industry and research institutes for analyses, and as design tools for the definition of new nozzle contours (see Chapter 13). Flow separation inside a rocket nozzle at nominal operation condition is undesired, as the separation front is naturally three-dimensional and features a random movement, resulting in possible generation of side-loads.[11,16,17,22,23] Sideloads in rocket nozzles may have different origins, and hence different models have been developed for their prediction. Potential origins for aerodynamic sideloads are asymmetric separation line, pressure pulsation at the separation location and in the separated flow region, aeroelastic coupling, transition of separation pattern in thrust-optimized or parabolic nozzles, and external flow instabilities and buffeting.

The assumption of a macroscopically tilted separation line is the basis of several sideload models (see, for example, Refs. 6 and 16). However, experiments have shown that even sideloads may be measured[18,22] for symmetric separation conditions. For example, random pressure pulsation near the separation line and in the recirculation region downstream of the separation line is the basic idea of the model presented in Ref. 22. The model application to Russian rocket nozzles, like RD-0120, gave reasonable agreement between measured and predicted sideloads.

The hypothesis of an aeroelastic coupling of the separated nozzle flow with the thin nozzle shell is formulated in Ref. 24.

A further sideload origin is the transition from free to restricted shock separation and vice versa. It has been shown that this transition is triggered by the cap-shock pattern.[11,12,18] For example, maximum sideloads measured in parabolic nozzles, like Vulcain or SSME, are caused by this origin.[13]

Sideloads are undesired phenomena that may result in the destruction of the rocket nozzle. In Ref. 16, the destruction of a J-2D engine as a result of sideloads is reported. Therefore, maximum area ratios of all first-stage nozzles or booster nozzles are chosen to avoid flow separation at the nominal chamber pressure operation at sea level. As a result, the vacuum performance of rocket engines that operate during the entire launcher trajectory, such as the SSME or Vulcain engines, is limited.

B. Potential Performance Improvements

Compared to existing rocket engines, a gain in performance is achieved with advanced engines such as mixed mode propulsion systems, dual mixture ratio engines, or dual-expander engines. Nevertheless, the upgrade of existing engines with better performing subsystems, such as turbines and pumps, also leads to a gain in overall performance data. This is discussed in more detail in Ref. 25. Nozzle performance of conventional rocket engines is already very high with regard to internal loss effects (friction, nonuniformity). However, for nozzles of gas-generator open-cycle engines like the Vulcain engine, a slight performance improvement can be achieved with the turbine exhaust gas (TEG-) injection into the main nozzle. This injection is realized, for example, with the F-1, J-2S, and Vulcain 2 engine (see Chapter 13), and numerical simulations[24,26,27] and experimental results[5] confirm the performance gain. Despite the slight performance gain by turbine exhaust gas injection, the low-pressure near-wall stream of the injected gas favors a reduction of the critical pressure ratio at which flow separation occurs, and therefore an earlier nozzle flow separation.[5,23]

III. Altitude Adaptive Nozzles

A critical comparison of performance losses shown in Table 1 reveals that most significant improvements in nozzle performance can be achieved through the adaptation of nozzle exit pressures to the variations in ambient pressure during the launcher's ascent through the atmosphere. This can be achieved with nozzle concepts that offer either a discrete stepwise or a continuous altitude adaptation.

A. Nozzles with Devices for Controlled Flow Separation

Several nozzle concepts with devices for controlled flow separation have been proposed in the literature, with primary emphasis on the reduction of sideloads during sea-level or low-altitude operation; however, the application of these concepts also results in an improved performance through the avoidance of significant overexpansion of the exhaust flow.

1. Dual-Bell Nozzle

This nozzle concept was first studied at the Jet Propulsion Laboratory in 1949.[15] In the late 1960s, Rocketdyne patented this nozzle concept. It has again received attention in the last decade of the 20th century in the United States,[28] Japan,[29] and Europe.[30–32] Figure 7 illustrates the design of this nozzle concept with its typical inner base nozzle, the wall inflection, and the outer

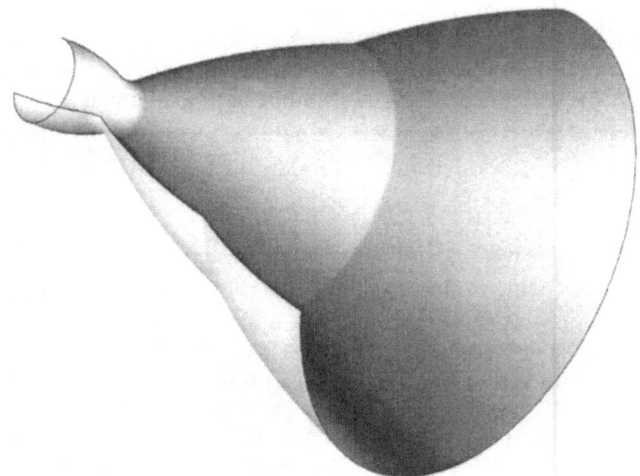

Fig. 7   Sketch of a dual-bell nozzle.

nozzle extension. Figure 8 emphasizes the essential flow pattern. In low altitudes, controlled and symmetrical flow separation occurs at the wall inflection. For higher altitudes, the nozzle flow is attached to the wall until the exit plane, and the full geometrical area ratio is used. Because of the higher area ratio, a better vacuum performance is achieved; however, additional performance losses are induced in dual-bell nozzles, as compared with two baseline nozzles having the same area ratio as the dual-bell nozzle at its wall inflection and in its exit plane. Figure 9 illustrates the dual-bell performance characteristic as a function of flight altitude. The pressure within the separated flow region of the dual-bell nozzle extension at sea-level operation is slightly below the ambient pressure, inducing a thrust loss referred to as aspiration drag. In addition, flow transition occurs before the optimum crossover point that leads to a further thrust loss as compared to an ideal switchover. The nonoptimum contour of the full flowing dual-bell nozzle results in further losses at high altitudes. Extensive computational fluid dynamics (CFD) analyses have shown that additional losses due to contour imperfections are in the order of divergence losses for conventional nozzles.[31]

Analytical considerations and experiments revealed that flow transition behavior in dual-bell nozzles strongly depends on the contour type of the nozzle extension.[5,28,31,32] A desirable sudden transition can be achieved with a "constant pressure" extension (zero pressure gradient), or an "overturned pressure" extension (positive pressure gradient). Tests have further shown that the pressure ratio at which transition occurs features a significant hysteresis between up-ramping and down-ramping.[32]

A design philosophy for dual-bell nozzles is described in Ref. 32. The base nozzle is classically designed as a parabolic or truncated ideal nozzle for high sea-level performance. The nozzle extension is designed with an inverse method of characteristics. The wall pressure profile is prescribed, and the

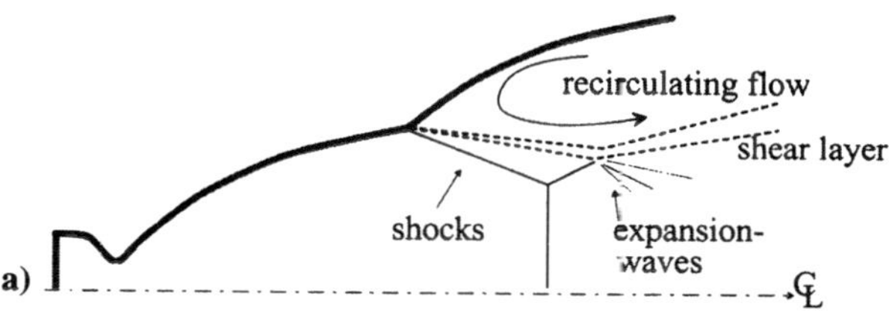

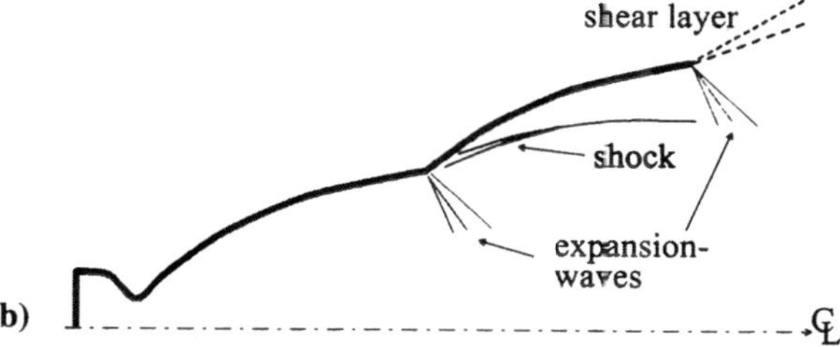

Fig. 8   Flow phenomena for a dual-bell nozzle for a) sea-level operation and b) high-altitude operation.

contour follows as a freejet streamline. The feasibility of this approach is proven by cold gas tests.

2.   Nozzles with Fixed Inserts

A trip ring attached to the inside of a conventional nozzle disturbs the turbulent boundary layer and causes flow separation at higher ambient pressures. At

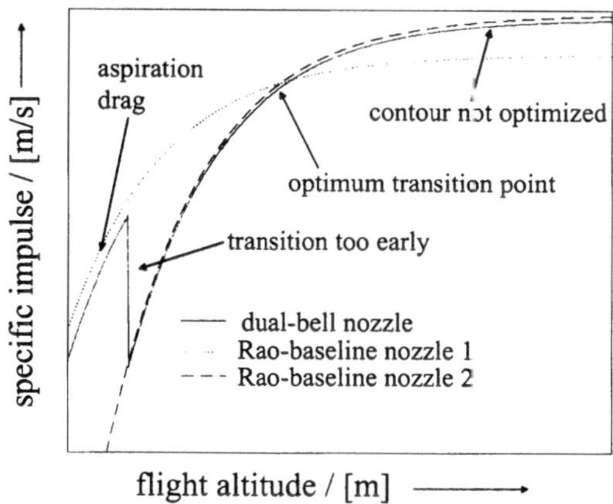

Fig. 9   Performance characteristic of a dual-bell nozzle. Performance is compared with two baseline bell-type nozzles (baseline nozzle 1: same area ratio as dual-bell base nozzle; baseline nozzle 2: same area ratio as dual-bell nozzle extension).

higher altitudes with lower ambient pressures, the flow reattaches to the wall downstream of the trip ring, and full flow of the nozzle is achieved. The transition from sea-level to vacuum mode depends on the wall pressure near the trip ring location and on the disturbance induced by the trip ring. The size of the trip ring is a compromise between stable flow separation during sea-level operation and the induced performance loss during vacuum operation. In Ref. 16, it is reported that a trip ring size of 10% of the local boundary-layer thickness is sufficient to ensure stable flow separation.

In principle, this concept is similar to the dual-bell nozzle concept with regard to performance characteristics; however, at sea level the bell nozzle with trip rings has higher divergence losses than a comparable dual-bell nozzle. The reason for this is that the nozzle contour upstream of the obstacle differs from the optimal contour for this low area ratio, because of the bell nozzle design for best vacuum performance. The additional losses induced during vacuum operation are about 1%, compared with the performance of the clean bell nozzle.[16]

The trip rings can be attached into existing nozzles and therefore represent a low-cost concept, at least for test purposes, with low technological risk. Trip rings have been demonstrated to be effective for sideload reduction during transient startup of rocket engines. The main problems with trip ring nozzles are not only performance losses but also ring resistance in high temperature boundary layers, the exact circumferential fixing, and the uncertainties in the transition behavior. These uncertainties might be the reason why active interest in this nozzle concept in the 1970s, which is documented in various publications,[16,33–35] has disappeared in recent years.

3. Nozzles with Temporary Inserts

Nozzle concepts with fixed wall discontinuities have the disadvantage of lower vacuum performance compared with a conventional bell nozzle with equal design and operation data. A promising concept for controlled flow separation is therefore temporary inserts that are removed for vacuum operation. These inserts can be either ablative or ejectible. The inserts may have the form of a complete secondary nozzle[36] (see Fig. 10), or the shape of small steps attached inside the nozzle wall. In the case of ejectible inserts, a reliable mechanism is needed to provide a sudden and symmetrical detachment. This minimizes the risk of high aerodynamic and mechanical loads due to shocks or a downstream collision with the nozzle wall.

Hot-firing tests were performed in Russia with a modified RD-0120 engine with such a secondary nozzle insert. Test analyses revealed a significant performance gain during sea-level operation of 12% at 100% chamber pressure, compared to the original RD-0120 performance.[36] These tests demonstrated the durability of nozzle material and sealing, and the release mechanism, and thus the feasibility of this concept. The principal performance characteristics of this RD-0120 nozzle with ejectible insert are included in Fig. 11. The nozzle operation with insert results in a slight performance loss as compared with an ideal bell nozzle with the same reduced area ratio due to aspiration drag and the presumably nonoptimized insert contour. The performance degradation is

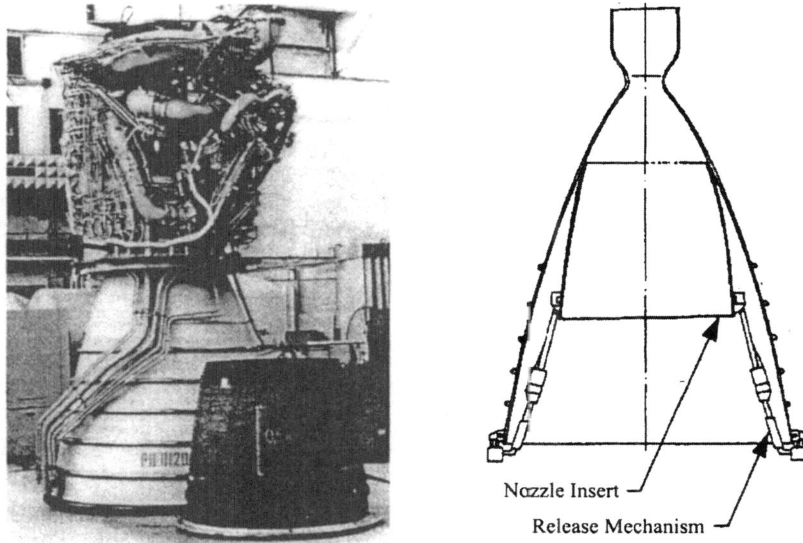

Fig. 10   RD-0120 nozzle hardware with removed nozzle insert (left), and sketch of secondary nozzle mounted inside of the RD-0120 nozzle (right; photographs from Ref. 36).

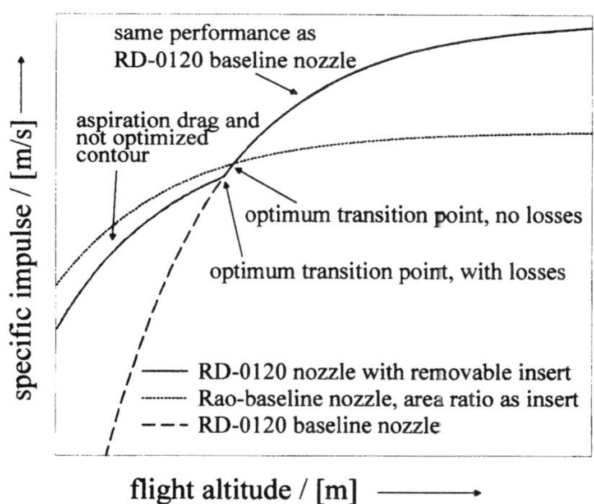

Fig. 11   Performance characteristic of a nozzle with insert.

comparable to the one induced in dual-bell nozzles during sea-level operation, as shown in Fig. 9. In vacuum or high-altitude operations, the higher performance of the baseline nozzle is achieved.

Another method for insert removal is the use of combustible or ablative elements.[16,37] During the ascent of the launcher, the size of the insert is continuously reduced until it is completely consumed, resulting finally in a full-flowing bell nozzle with a clean contour with high vacuum performance. The principal uncertainties of this nozzle concept lay within the stability and surface regression rates of the inserts. Furthermore, a homogeneous, symmetrical, and temporally defined consumption must be guaranteed despite possible local pressure and temperature fluctuations near the nozzle walls.

4.   *Nozzles with Active or Passive Secondary Gas Injection*

With this nozzle concept the flow in an overexpanding nozzle is forced to separate at a desired location by injecting a second fluid into the gas stream in the wall normal direction. The injection could be either active, such as through gas expansion from a higher pressure reservoir, or passive, using holes in the wall through which ambient gas is sucked in (vented nozzle concept). The latter concept can work only in the case of gas pressures near the wall inside the nozzle, which are lower than the ambient pressure. Experience on forced secondary gas injection gained at Aerojet shows that a large amount of injected fluid is required to induce a significant flow separation. Furthermore, no net specific impulse gain is realized when considering the additional gas flow rate.

Experiments with a modified RL10A-3 engine with the vented nozzle concept were performed at Pratt & Whitney.[38] Performance results showed that only over a small range of low-pressure ratios the perforated nozzle performed as well as a nozzle with its area ratio truncated immediately upstream of the vented area; however, at above some intermediate pressure ratios, the thrust efficiency suddenly dropped and approached that of the full flowing nozzle.

5.   *Two-Position or Extendible Nozzles*

Nozzles of this type with extendible exit cones are currently used only for rocket motors of upper stages to reduce the package volume for the nozzle, for example, at present for solid rocket engines such as the intertial upper stage (IUS) or for the liquid rocket engine RL10. The main idea of the extendible extension is to use a truncated nozzle with low expansion in low flight altitudes and have a higher nozzle extension at high altitudes. Figure 12 illustrates this nozzle concept. Its capability for altitude compensation is indisputable, and the nozzle performance is easily predictable. A minor performance loss is incorporated during low altitude operations because of the truncated inner nozzle, which has a nonoptimal contour for this interim exit area ratio. The performance characteristics as a function of flight altitude are similar to those of the nozzle with the ejectible insert, as shown in Fig. 11.

The main drawback of the extendible nozzle concept is the additional mechanical device for the deployment of the extension that reduces engine reliability and increases total engine mass. The necessity for active cooling of the extendible extension requires flexible or movable elements in the cooling system, which also

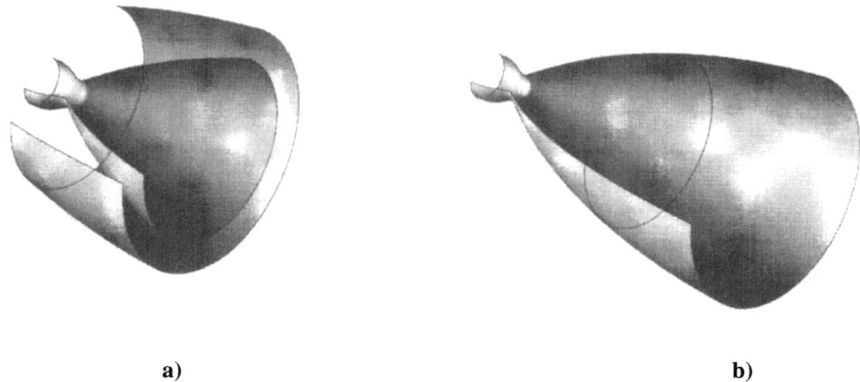

a)   b)

Fig. 12 Sketch of a two-position nozzle during a) sea-level and b) vacuum operation.

reduces system reliability. Former investigations have shown that the external flow causes both steady and unsteady pressure loads on the retracted nozzle extension, whereas the engine jet noise causes strong vibrations of the nozzle extension.

B. Plug Nozzles

Experimental, analytical, and numerical research on plug nozzles has been performed since the 1950s in the United States,[6,39–46] Europe,[30,47–52] Russia,[5,50] and Japan.[53–55] In contrast with the previously discussed nozzle concepts, plug nozzles provide at least theoretically a continuous altitude adaptation up to their geometric area ratio. Figure 13 shows a typical application of a linear plug nozzle to a winged launch vehicle, and a sketch of a linear plug engine segment with primary internal expansion investigated within a research program in Europe.[50] Because of the characteristic form of the central plug body, these nozzle types are also called "aerospike" nozzles.

Different design approaches, either for linear or axisymmetric plug nozzle, have been published in the literature. For an ideal plug contour, two methods may be applied. The first is purely based on Prandtl–Meyer expansion (Ref. 39), and the second is based on the method of characteristics (Ref. 40). Figure 14 highlights the principle of both design approaches. For an axisymmetric plug configuration, the second approach must be used to achieve uniform exit flow, as the Prandtl-Meyer equations are only valid for planar flows.

For high Mach number plug nozzles with fully external expansion, large turning angles of the flow, and thus for the throat inclination, are required. A primary internal expansion may be foreseen to avoid this large turning angle, as illustrated in Fig. 13b. This can be achieved by a symmetric or nonsymmetric primary contour design. Figure 15 gives details of both design approaches. The exhaust flow initially expands in the internal nozzle from the throat, AB, to the Mach number M_{id}. For both configurations, the plug contour, DF, must be designed with method of characteristics to achieve a uniform internal exit flow through EF.

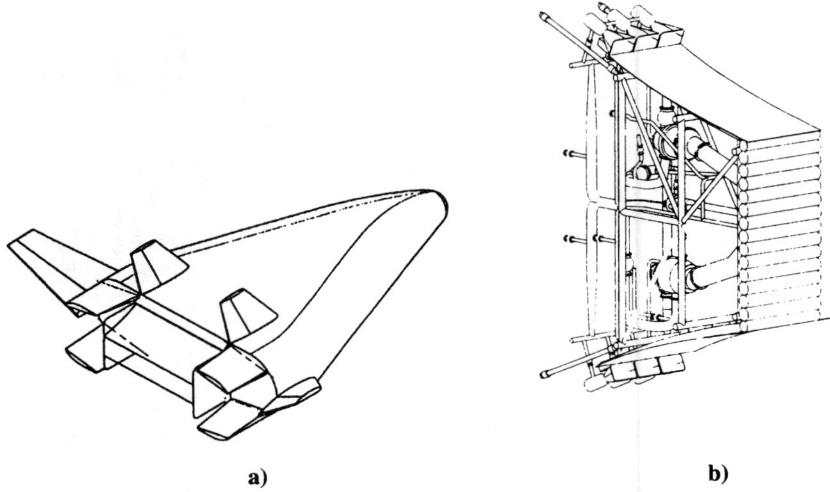

Fig. 13   Principle design of a launch vehicle with a) integrated linear plug nozzle and b) sketch of linear plug engine segment (full-scale, investigated in European research program).

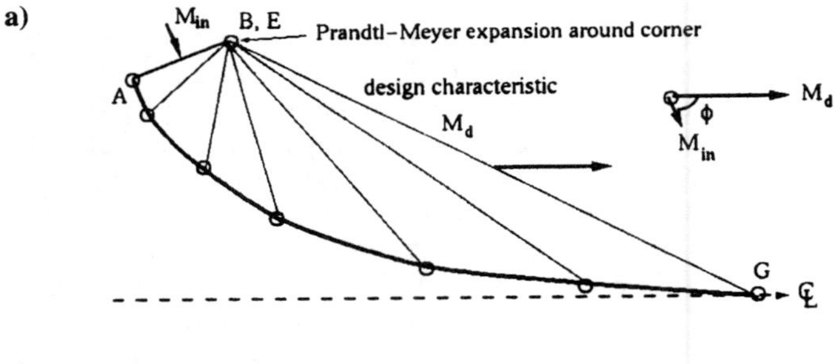

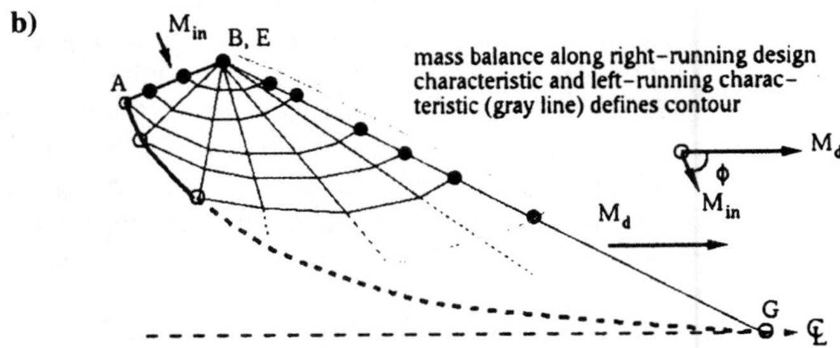

Fig. 14   Plug contour definition a) fully based on Prandtl–Meyer expansion and b) based on method of characteristics.

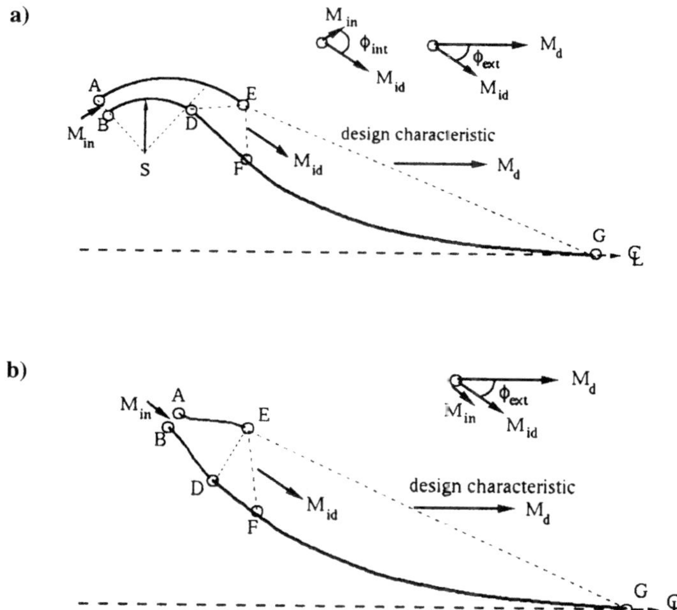

Fig. 15 Plug contour definition with primary internal expansion. Internal expansion in a) nonsymmetric nozzle and b) symmetric nozzle.

The external expansion is established along the contour, FG, to the chosen exit Mach number M_d. Again, the contour DFG is defined based on the method of characteristics, with the principles described in Fig. 14b. Ref. 50 gives further details on cold-gas tests performed for nozzle design validation purposes.

To complete the discussion on plug nozzle design approaches, it has to be mentioned, especially for truncated plug nozzles, that Rao[41] developed an additional design approach, taking into account that a simple truncation of a plug nozzle with a full-length central body designed for maximum performance does not automatically result in the best-performing truncated plug nozzle. Rao therefore proposed a design method for truncated plug nozzles based on a

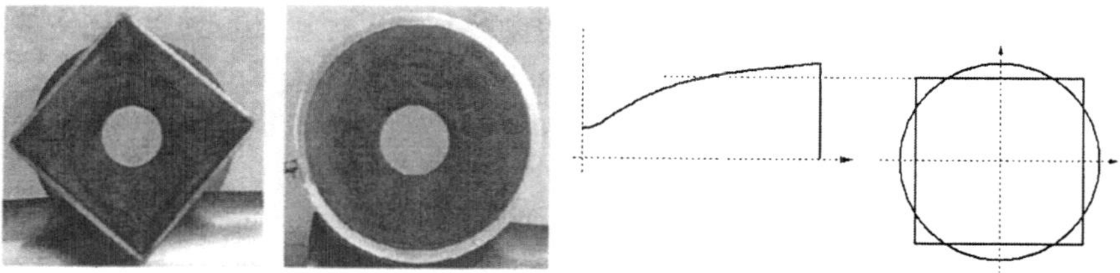

Fig. 16   Design principle of a round-to-square module nozzle.

variational method, similar to his well-known method for shortened, thrust-optimized bell-type nozzles.

To avoid any base flow for truncated plug nozzles, nonoptimized contours with respect to performance could also be applied, for example, conical plug bodies. Because a plug engine is an integral part of a launcher, even a benefit with regard to payload might be achieved because of a potential decrease of engine mass with alternative contours, despite their performance loss. In these approaches, the potential occurrence of flow separation due to the imperfect contours must be carefully studied.

A plug design with a single internal cell nozzle design is not feasible with regard to manufacturing and operations. Furthermore, thrust vector control requirements especially strengthen the need for individual internal expansion module thrusters. When a plug nozzle is clustered, the following requirements are specified for the primary expansion modules: 1) a fully attached module nozzle flow on ground conditions, 2) the absence of substantial disturbances (shock waves) in the module outflow, 3) optimum integrated thrust efficiency, 4) efficient module cooling, and 5) easy manufacture from a technological viewpoint.

Axisymmetrical bell-type module nozzles satisfy many of these requirements; however, these module nozzles are not optimal from the viewpoint of the overall plug design, because the spatial gaps between the module nozzles disturb the near-plug flow. Experiments performed with round primary module nozzles, either for axisymmetrical or linear plug configurations, have shown that the jet interactions from the individual modules produce a significant performance loss.[51] As a result, module nozzles with rectangular exit are optimal from the layout point of view, and it has been shown both numerically[42] and experimentally[50] that transition from a round-to-square primary nozzle results only in very small additional performance losses. Figure 16 illustrates the round-to-square nozzle tested in Ref. 50 for performance assessment, together with the classical reference nozzle.

Figure 17 summarizes the flow phenomena of axisymmetric or linear plug nozzles with full-length and truncated central bodies at different off-design (Figs. 17a and 17c) and design (Fig. 17b). pressure ratios. For pressure ratios lower than the design pressure ratio of a plug nozzle with well contoured central body, the flow expands near the central plug body without separation, and a system of recompression shocks and expansion waves adapt the exhaust flow to the ambient pressure. The characteristic barrel-like form with several inflections of the shear layer results from various interactions of compression and expansion waves with the shear layer. At the design pressure ratio (see Fig. 17b), the characteristic with the design Mach number should be a straight line emanating to the tip of the central plug body, and the shear layer is parallel to the centerline. The wall pressure distribution remains the same at pressure ratios above the design pressure ratio, that is, the plug nozzle behaves like a conventional nozzle, the loss of its capability of further altitude adaptation included. Figure 17c illustrates the flowfield at higher pressure ratios.

The truncation of the central plug body, which is of advantage due to the huge length and high structural mass of the well-contoured central body, results in a different flow and performance behavior. At lower pressure ratios an open wake flow establishes in the base, with a pressure level nearly equal to the

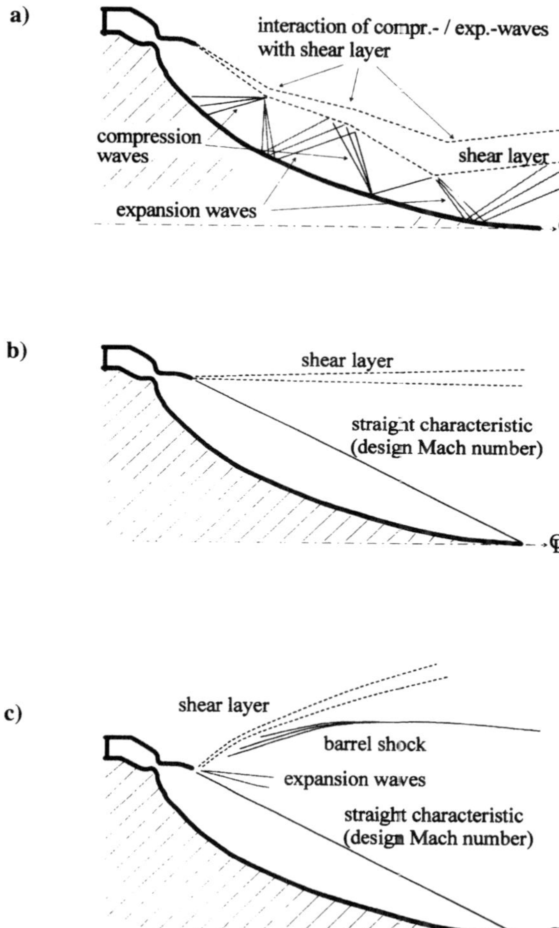

Fig. 17   Flow phenomena of a plug nozzle with full-length central body at a) sea-level condition, b) design condition, and c) high-altitude condition.

ambient pressure (Fig. 18a). At a certain pressure ratio close to the design pressure ratio of the full-length plug nozzle, the base flow suddenly changes its character and turns over to the closed form, characterized by a constant base pressure that is no longer influenced by ambient pressure. Analyses indicate that shorter plug bodies with higher truncations trigger an earlier change in wake flow. At the transition point the pressure within the wake approaches a value that is below ambient pressure, and the full base area induces a negative thrust (Fig. 18b). This thrust loss depends on the percentage of truncation and the total size of the base area. Beyond the transition point, the pressure within the

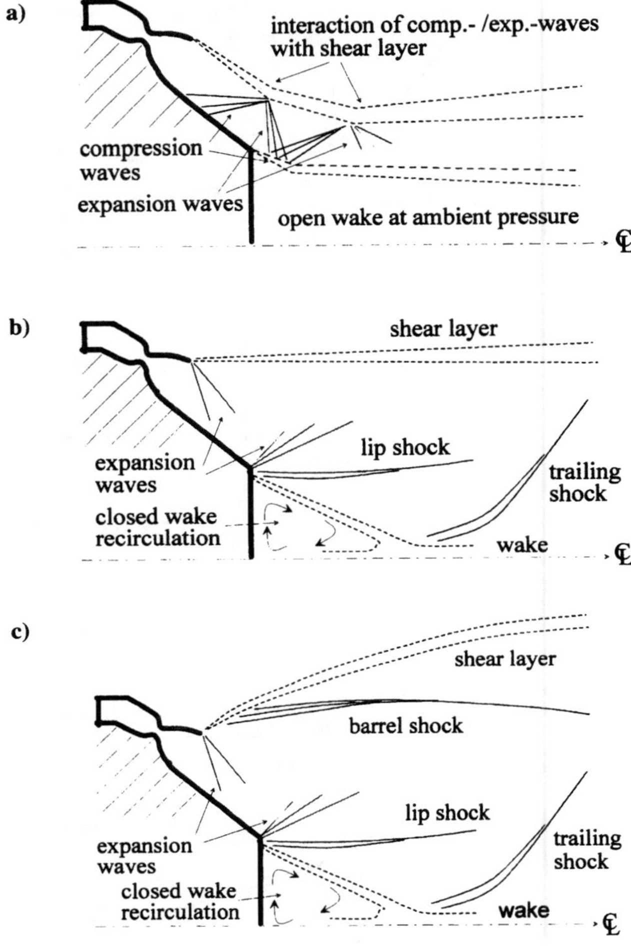

Fig. 18   Flow phenomena of a plug nozzle with truncated central body at a) sea-level condition, b) design condition, and c) high-altitude condition.

closed wake remains constant. At these lower ambient pressures, the base pressure is then higher than the ambient pressure, resulting in a positive thrust contribution of the total base area. Detailed discussion of plug nozzle flow features is included in Ref. 51.

For linear plug nozzles, special attention must be paid to the influence of both end sides, where the surrounding flow disturbs the expanding flowfield, resulting in an expansion of the flow normal to the main flow direction and therefore in an effective performance loss. For truncated plug nozzles especially, the change of wake flow behavior may be strongly influenced by the penetration of ambient pressures through both end sides. End plates could be used to avoid this ambient pressure penetration.[50] To illustrate the efficiency of side fences, Fig. 19 visualizes the expansion characteristic along a linear plug ramp with and without side fences by means of numerical Mach number distribution and also experimental pressure distribution. With side fences, a nearly two-dimensional expansion characteristic is achieved. Note also the aforementioned influence of the side fences on the base flow development.

Typical performance data of a plug nozzle are included in Fig. 20 and compared with a conventional bell nozzle with equal area ratio. The altitude compensation capability of plug nozzles is one of the main advantages compared with conventional bell nozzles. Principally, plug nozzles have a slightly lower vacuum specific impulse than conventional bell nozzles with the same area ratio because of truncation and clustering losses; however, this can be compensated by a plug design making maximum use of the vehicle base area available for integration of the engine. In this manner a very high plug nozzle area ratio can be achieved with a better vacuum performance than a bell nozzle. Because of the altitude compensation capability, a plug nozzle with a high area ratio can be operated at sea-level conditions, whereas the area ratio of bell nozzles is limited by the flow separation criterion.

C. Expansion-Deflection Nozzles

An expansion-deflection (E-D) nozzle is shown in Fig. 21. E-D nozzles were at one time thought to have capabilities for altitude compensation because the gas expansion takes place with a "constant pressure" free boundary. Thus, the aerodynamic behavior of E-D nozzles as a function of altitude is in principle quite similar to plug nozzles because the ambient pressure and hence altitude control the expansion process. In contrast to plug nozzles, however, the expansion process for E-D nozzles is controlled from inside the nozzle. At low altitude, the higher ambient pressure limits the gas expansion, resulting in a low effective expansion area ratio. The exhaust gas is adapted to the ambient pressure level by systems of recompression and expansion waves, as shown in Fig. 22. At higher altitude, the lower ambient pressure allows more gas expansion within the nozzle, resulting in a higher effective expansion area ratio; however, in contrast to plug nozzles, the pressure in the wake of the center plug is always less than the ambient pressure because of the aspiration. This occurs at low-pressure ratios when the wake is opened, and results in an aspiration loss. Furthermore, because the exhaust flow expands to this base pressure rather than to the ambient pressure level, wall pressures downstream are overexpanded. This results in an additional overexpansion loss. As the pressure ratio increases, the wake region closes and is thus totally isolated from the ambient environment (see also Fig. 22). The behavior during transition from open wake to closed wake is again equal to the plug nozzles, and the base pressure in the closed wake region is essentially independent from the ambient pressure.

The E-D nozzle concept has also been the subject of numerous analytical and experimental studies. Results from these studies have confirmed that E-D nozzle capabilities for altitude compensation are poor[46] because of aspiration

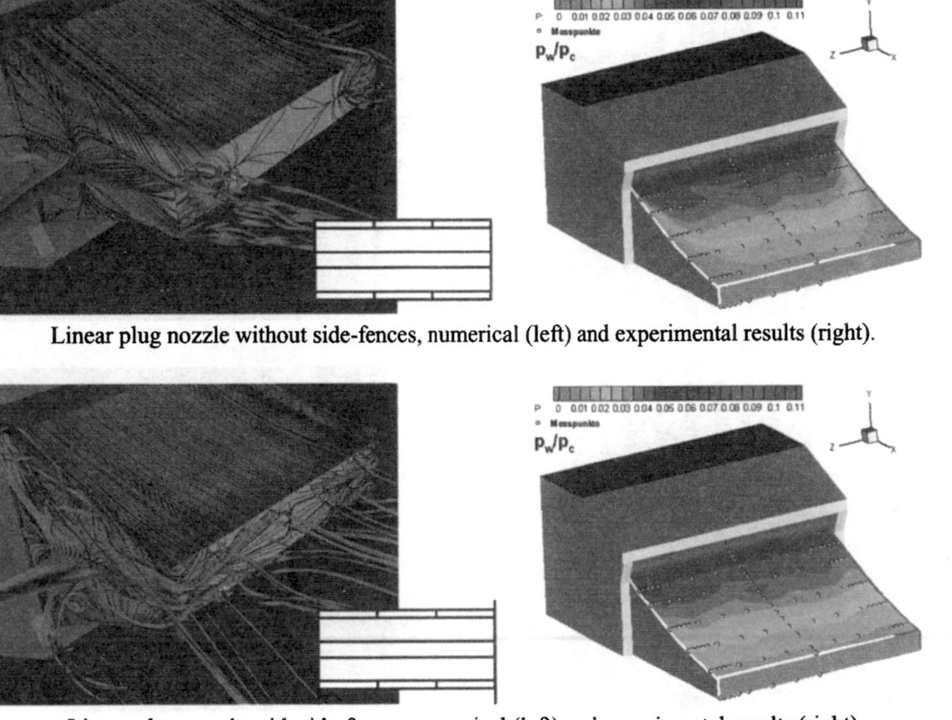

Linear plug nozzle without side-fences, numerical (left) and experimental results (right).

Linear plug nozzle with side-fences, numerical (left) and experimental results (right).

Fig. 19   Flow phenomena for a linear plug nozzle without side fences (top figure) and with side fences (bottom figure): numerical results, Mach number distribution and streamlines (left, from red to blue for increasing Mach number), and experimental results, pressure ratio p_w/p_c distribution (right, from red to blue for increasing pressure ratio). (See also the color section of figures following page 620.) (CFD images courtesy of NASA Marshall Space Flight Center; see also Ref. 56.)

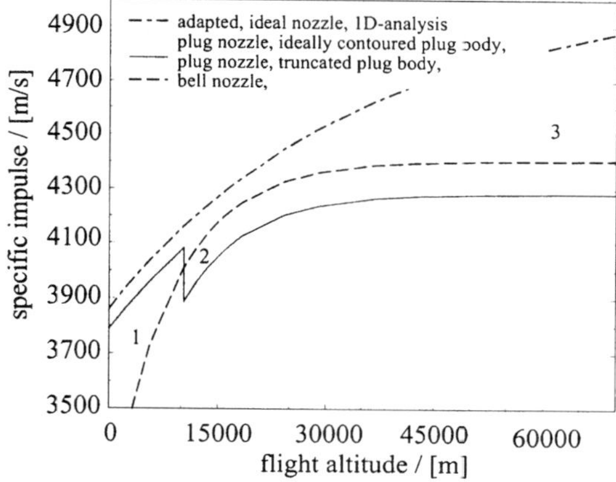

Fig. 20   Performance of a numerically simulated plug nozzle with full-length and truncated central plug body. Performance data based on hydrogen-oxygen, $r = 6.0$, $p_c = 100$ bar.

and overexpansion losses. Despite poor altitude compensation, the E-D nozzle has a potential for upper stage application, for a high area ratio nozzle with smallest engine envelope and no moving parts.

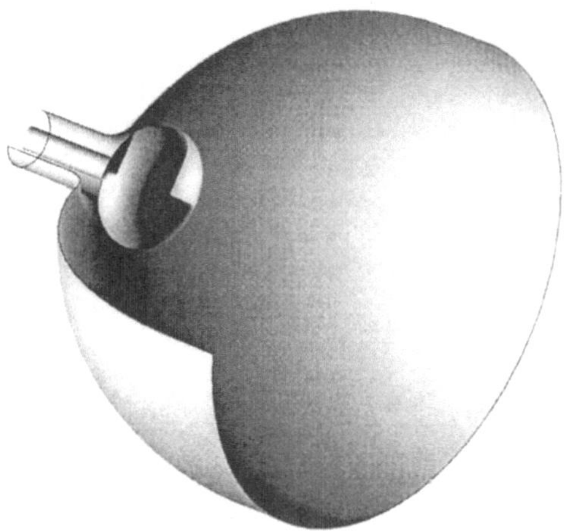

Fig. 21   Sketch of an expansion-deflection nozzle.

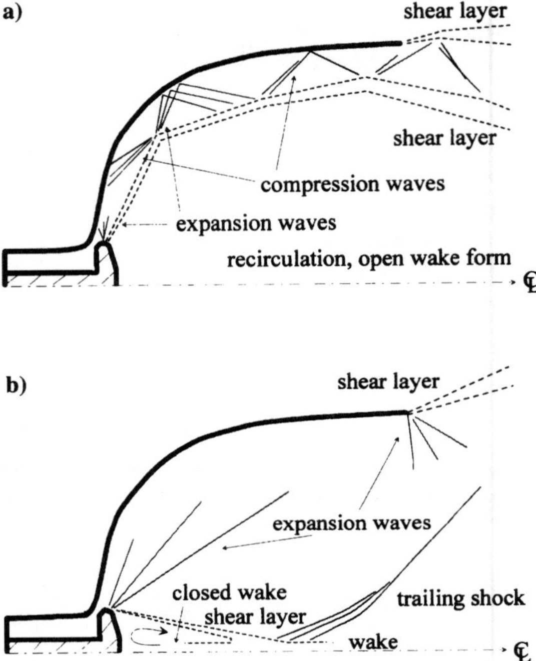

Fig. 22 Flow phenomena for an expansion-deflection nozzle with a) open and b) closed wake.

D.   Nozzles with Throat Area Varied by a Mechanical Pintle

This nozzle concept uses a conventional bell nozzle with a fixed exit area and a mechanical pintle in the combustion chamber and throat region to vary the throat area and hence the expansion area ratio.[57] The area of the nozzle throat, an annulus between the pintle and the shroud, is varied by moving the pintle axially.

The pintle concept has been used in solid rocket motors as a means to provide variable thrust. The concept, in principle, allows a continuous variation of the throat area and thus optimum expansion area ratios throughout a mission; however, it requires an actuator and a sophisticated control system. The concept raises issues of engine weight, design complexity, cooling of pintle and nozzle throat, and reliability.

E.   Dual-Mode Nozzles

Dual-mode rocket engines using one or two fuels offer a trajectory-adapted dual-mode operation during the ascent of a launcher, which may be of significant advantage especially for single-stage earth-to-orbit (SSTO) vehicles. This engine concept involves the use of a dense propellant combination with moderate performance during liftoff to provide high thrust during the initial flight phase, and a better performing propellant combination in vacuum, which results in higher specific impulse. The fuels are burned in two different combustion chambers, with one located completely inside the other in the case of engines with dual-throat nozzles, or with a conventional bell thrust chamber surrounded by an annular thrust chamber in the case of dual-expander engines. This type of engine has a built-in throttling capability, achieved by shutting down one of two thrust chambers. Apart from the indicated benefits of dual-mode engines, which will be discussed in more detail, their development and construction require considerable technological effort.

1. Dual-Throat Nozzles

A dual-throat nozzle configuration is shown in Fig. 23. At low altitude, the outer thrust chamber operates with the inner thrust chamber running in parallel. In this operation mode, the engine has a larger throat providing a moderate expansion area ratio. At a certain point during the mission, the outer thrust chamber is shut off and operation continues with only the inner engine. In this configuration, flow from the inner engine expands and attaches supersonically to the outer engine, resulting in a higher expansion area ratio for the rest of the mission. Flow phenomena in both operation modes are included in Fig. 24.

Hot-firing tests were conducted to provide heat transfer data that were very useful for the thermal analysis and design of the dual-throat nozzle configuration.[58] These tests showed that flow separation occurred in the inner engine nozzle at higher ratios of outer-to-inner chamber pressures during the first operation mode with both chambers burning in parallel. The flow separation

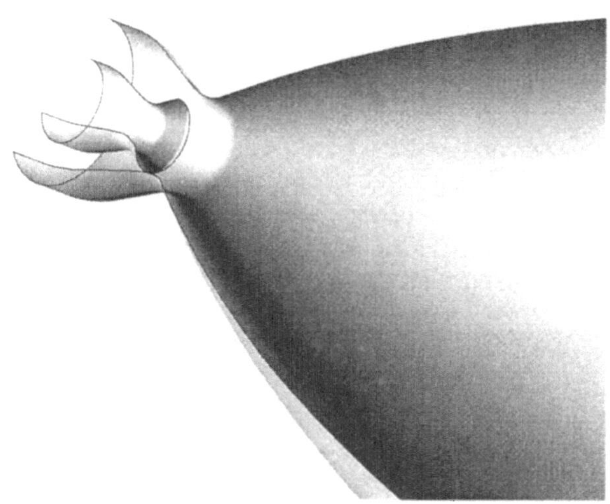

Fig. 23 Sketch of a dual-throat nozzle, view of combustion chamber and throat region.

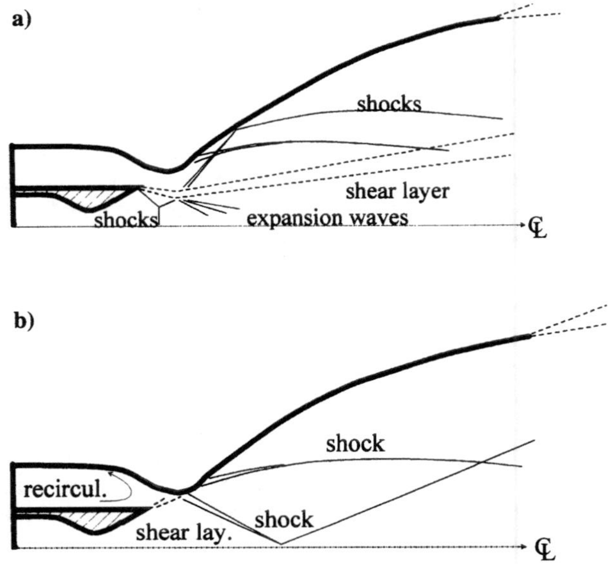

Fig. 24   Flow phenomena for a dual-throat nozzle, during a) sea-level and b) high-altitude operation.

resulted in a higher heat load to the inner nozzle. Subscale tests have shown that the additional loss caused by the nozzle contour discontinuity during vacuum operation with active inner chamber is in the range of 0.8–4% (see Ref. 5). This performance loss, which is quite high, results from the interaction of the inner chamber jet with the outer chamber nozzle wall (see also Fig. 24).

2.   Dual-Expander Nozzles

Figure 25 shows a typical dual-expander nozzle configuration. At low altitude, both thrust chambers operate sharing the same exit area, which results in a moderate expansion area ratio. One thrust chamber is shut off at a certain point during the mission, allowing the other nozzle to use the whole exit area, creating a high expansion area ratio for the rest of the burn. In principle, the two operation modes are comparable to those of dual-throat nozzles.

Numerical simulations of the flowfield in dual-expander nozzles during all of the operation modes are published in the literature.[59,60] These analyses have shown that dual-expander nozzles produce high performance in both operation modes. Figure 26 emphasizes the flow pattern for mode 1 operation with both thrust chambers burning, and for mode 2 operation with the outer thrust chamber burning.

Several analytical works on SSTO and TSTO (two-stage-to-orbit) vehicles using hydrogen/propane or hydrogen/methane as fuels revealed the lowest

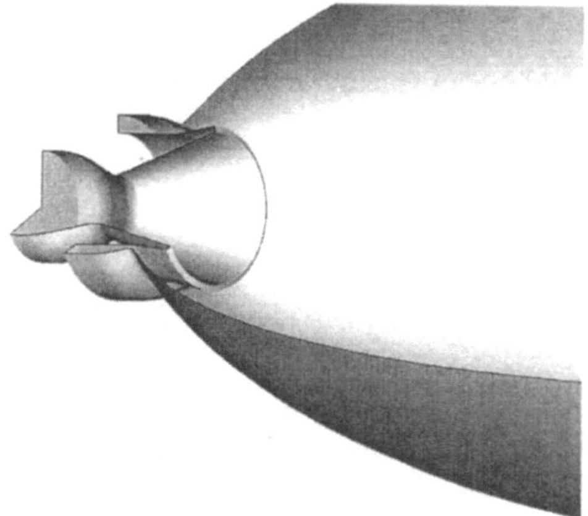

Fig. 25 Sketch of a dual-expander nozzle, view of combustion chamber and throat region.

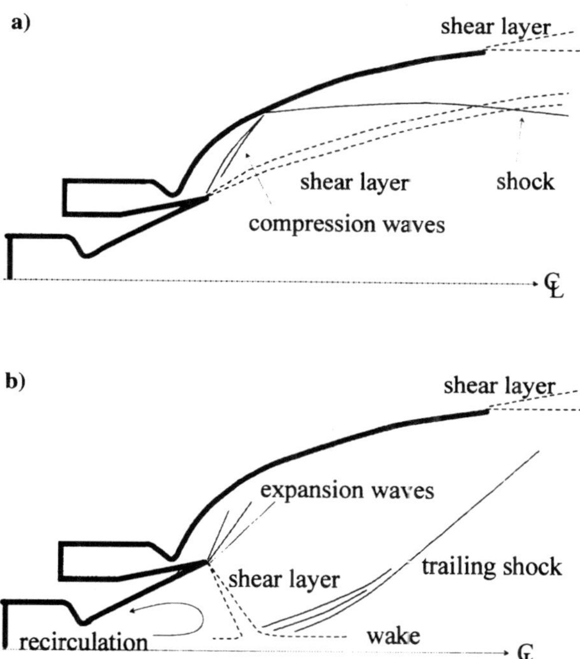

Fig. 26 Flow phenomena for a dual-expander nozzle during a) sea-level and b) high-altitude operation.

vehicle dry masses for dual-mode engines in comparison with other engines (see, for example, Refs. 25 and 61 for a literature overview). Other dual-mode engines using hydrogen as the single fuel but using two mixture ratios also revealed some benefits over conventional engines for SSTO and TSTO applications. Even a single-fuel operation with constant mixture ratios in both combustion chambers indicated a gain in launcher performance.[61]

IV. Conclusion

Several nozzle concepts that promise gains in performance over conventional nozzles have been discussed in this chapter, including performance enhancements achieved by slight modifications of existing nozzles, for example, through cool gas injection into the supersonic nozzle part. It is shown that significant performance gains result from the adaptation of the exhaust flow to the ambient pressure, and special emphasis has been given to altitude adaptive nozzle concepts.

Several nozzle concepts with altitude compensating capability have been identified and described. The performance of the nozzles must be characterized to assist the selection of the best nozzle concept for launch vehicle applications. This can be done using computational fluid dynamics and/or cold-flow tests. Existing computational fluid dynamics methods that are in use in the aerospace industry and at research institutes have been verified for a wide number of subscale and full-scale experiments, and these methods provide sufficiently reliable performance determination for the different nozzles types.

Theoretical evaluations, numerical simulations, and test results showed that the different concepts have real altitude compensating capabilities; however, the compensation capabilities are limited, and there are some drawbacks associated with each concept. Additional performance losses are induced in practically all of these nozzle concepts when compared with an ideal expansion, mainly because of non-isentropic effects like shock waves and pressure losses in recirculation zones. However, these additional performance losses are less than 1–3%, depending on the different nozzle concepts.

In addition to aerodynamic performance, other technical issues (weight, cost, design, thermal management, manufacturing, system performance, and reliability) must especially be addressed. Furthermore, before a final decision can be made as to which nozzle concept offers the greatest benefits with regard to an effective payload mass injection, combined launcher and trajectory calculations must be performed and compared to a reference launcher concept with conventional nozzles. Different nozzle efficiencies, which account for the additional losses of advanced rocket nozzles and which are extracted from numerical simulations and experiments, must be taken into account.

References

[1]Sutton, G., and Biblarz, O., *Rocket Propulsion Elements*, 7th ed., Wiley, New York, 2001.

[2]Manski, D., and Hagemann, G., "Influence of Rocket Design Parameters on Engine Nozzle Efficiencies," *Journal of Propulsion and Power*, Vol. 12, No. 1, 1996, pp. 41–47.

[3]N. N., "Liquid Rocket Engine Nozzles," NASA SP-8120, 1976.

[4]Hagemann, G., Immich, H., Nguyen, T. V., and Dumnov, G., "Advanced Rocket Nozzle Concepts," *Journal of Propulsion and Power*, Vol. 14, No. 5, 1998, pp. 620–634.

[5]Dumnov, G. E., Nikulin, G. Z., and Ponomaryov, N. B., "Investigation of Advanced Nozzles for Rocket Engines" (published in Russian), *Space Rocket Engines and Power Plants*, Vol. 4 (142), NIITP, 1993.

[6]Nguyen, T. V., and Pieper, J. L., "Nozzle Flow Separation," *Proceedings of the 5th International Symposium of Propulsion in Space Transportation*, Paris, France, May 22–24, 1996.

[7]Ahlberg, J., Hamilton, S., Migdal, D., and Nilson, E., "Truncated Perfect Nozzles in Optimum Nozzle Design," *ARS Journal*, Vol. 31, No. 5, 1961, pp. 614–620.

[8]Rao, G. V. R., "Exhaust Nozzle Contours for Optimum Thrust," *Jet Propulsion*, June 1958, pp. 377–382.

[9]Rao, G. V. R., "Approximation of Optimum Thrust Nozzle Contours," *ARS Journal*, June 1960, p. 561.

[10]Frey, M., "Shock Pattern in the Exhaust Plume of Rocket Nozzles," *Proceedings of the 3rd European Symposium on Aerothermodynamic of Space Vehicles*, ESA SP-426, ESA ESTEC, Noordwijk, the Netherlands, 1998, pp. 395–403.

[11]Hagemann, G., Frey, M., and Koschel, W., "Appearance of Restricted Shock Separation in Rocket Nozzles," *Journal of Propulsion and Power*, Vol. 18, No. 3, pp. 577–584.

[12]Frey, M., and Hagemann, G., "Restricted Shock Separation in Rocket Nozzles," *Journal of Propulsion and Power*, Vol. 16, No. 3, May–June 2000, pp. 478–484.

[13]Terhardt, M., Hagemann, G., and Frey, M., "Flow Separation and Side-Load Behaviour of the Vulcain Engine," AIAA Paper 99-2762, 1999.

[14]Summerfield, M., Foster, C., and Swan, W., "Flow Separation in Overexpanded Supersonic Exhaust Nozzles," *Jet Propulsion*, Sept.–Oct. 1954, pp. 319–321.

[15]Foster, C., and Cowles, F., "Experimental Study of Gas Flow Separation in Overexpanded Exhaust Nozzles for Rocket Motors," JPL Progress Rept. 4–103, May 1949.

[16]Schmucker, R., "Flow Processes in Overexpanding Nozzles of Chemical Rocket Engines" (published in German), Technical University of Munich, Munich, Germany, Rept. TB-7, -10, -14, 1973.

[17]Hagemann, G., Terhardt, M., Frey, M., Reijasse, P., Onofri, M., Nasuti, F., and Oestlund, J., "Flow Separation and Side-Loads in Rocket Nozzles," *Proceedings of the 4th International Symposium on Liquid Space Propulsion*, DLR Lampoldshausen, Germany, March 13–15, 2000.

[18]Mattsson, J., Högman, U., and Torngren, L., "A Sub-Scale Test Programme on Investigation of Flow Separation and Side-Loads in Rocket Nozzles," *Proceedings of the 3rd European Symposium on Aerothermodynamics of Space Vehicles*, ESA-ESTEC, ESA SP-426, Netherlands, November 24–26, 1998.

[19]Stark, R., Kwan, W., Quessard, F., Hagemann, G., and Terhardt, M., "Rocket Nozzle Cold-Gas Test Campaigns for Plume Investigations," *Proceedings of the 4th European Symposium on Aerothermodynamics of Space Vehicles*, ESA SP-487, Dec. 2001.

[20]Reijasse, P., Morzenski, L., Blacodon, D., and Birkemeyer, J., "Flow Separation Experimental Analysis in Overexpanded Subscale Rocket Nozzles," AIAA Paper 2001-3556, July 2001.

[21]Hagemann, G., Alting, J., and Preclik, D., "Scalability Discussion for Rocket Nozzle Flows Based on Subscale and Full-Scale Testing," *Journal of Propulsion and Power*, Vol. 19, No. 3, pp. 321–331.

[22]Dumnov, G. E., "Unsteady Side-Loads Acting on the Nozzle with Developed Separation Zone," AIAA Paper 96-3220, 1996.

[23]Nave, L. H., and Coffey, G. A., "Sea-Level Side-Loads in High Area Ratio Rocket Engines," AIAA Paper 73-1284, 1973.

[24]Pekkari, L. O., "Advanced Nozzles," *Proceedings of the 5th International Symposium of Propulsion in Space Transportation*, Paris, France, May 22–24, 1996.

[25]Manski, D., "Overview of Cycles for Earth-to-Orbit Propulsion," *Journal of Propulsion and Power*, Vol. 14, No. 5, 1998.

[26]Hagemann, G., Krülle, G., and Hannemann, K., "Numerical Flowfield Analysis of the Next Generation Vulcain Nozzle," *Journal of Propulsion and Power*, Vol. 12, No. 4, 1996, pp. 655–661.

[27]Voinow, A. L., and Melnikov, D. A., "Performance of Rocket Engine Nozzles with Slot Injection," AIAA Paper 96-3218, 1996.

[28]Horn, M., and Fisher, S., "Dual-Bell Altitude Compensating Nozzles," NASA CR-194719, 1994.

[29]Kumakawa, A., Tamura, H., Niino, M., Nosaka, M., Yamada, H., Konno, A., and Atsumi, M., "Propulsion Research for Rocket SSTOs at NAL/KRC," AIAA Paper 99-2337, June 1999.

[30]Immich, H., and Caporicci, M., "Status of the FESTIP Rocket Propulsion Technology Program," AIAA Paper 97-3311, 1997.

[31]Hagemann, G., and Frey, M., "A Critical Assessment of Dual-Bell Nozzles," *Journal of Propulsion and Power*, Vol. 15, No. 1, 1999, pp. 137–143.

[32]Hagemann, G., Terhardt, M., Haeseler, D., and Frey, M., "Experimental and Analytical Design Verification of the Dual-Bell Concept," *Journal of Propulsion and Power*, Vol. 18, No. 1, 2002, pp. 116–122.

[33]Luke, G., "Use of Nozzle Trip Rings to Reduce Nozzle Separation Side Force during Staging," AIAA Paper 92-3617, 1992.

[34]Chiou, J., and Hung, R., "A Study of Forced Flow Separation in Rocket Nozzle," Univ. of Alabama, Final Rept., Huntsville, AL, 1974.

[35]Schmucker, R., "A Procedure for Calculation of Boundary Layer Trip Protuberances in Overexpanded Rocket Nozzles," NASA TM X-64843, 1973.

[36]Goncharov, N., Orlov, V., Rachuk, V., Shostak, A., and Starke, R., "Reusable Launch Vehicle Propulsion Based on the RD-0120 Engine," AIAA Paper 95-3003, 1995.

[37]Clayton, R., and Back, L., "Thrust Improvement with Ablative Insert Nozzle Extension," *Jet Propulsion*, Vol. 2, No. 1, 1986, pp. 91–93.

[38]Parsley, R. C., and van Stelle, K. J., "Altitude Compensating Nozzle Evaluation," AIAA Paper 92-3456, 1992.

[39]Angelino, G., "Theoretical and Experimental Investigations of the Design and Performance of a Plug Type Nozzle," NASA TN-12, 1963.

[40]Lee, C. C., "Fortran Programs for Plug Nozzle Design," NASA TN R-41, 1963.

[41]Rao, G. V. R., "Spike Nozzle Contours for Optimum Thrust," *Ballistic Missiles and Space Technology*, Pergamon, New York, 1961.

[42]Nguyen, T. V., Spencer, R. G., and Siebenhaar, A., "Aerodynamic Performance of a Round-to-Square Nozzle," *Proceedings of the 35th Heat Transfer and Fluid Mechanics Institute*, May 1997.

[43]Beheim, M. A., and Boksenbom, A. S., "Variable Geometry Requirements in Inlets and Exhaust Nozzles for High Mach Number Applications," NASA TM X-52447, 1968.

[44]Valerino, A. S., Zappa, R. F., and Abdalla, K. L., "Effects of External Stream on the Performance of Isentropic Plug-Type Nozzles at Mach Numbers of 2.0, 1.8 and 1.5," NASA 2-17-59E, 1969.

[45]Mercer, C. E., and Salters, L. E., Jr., "Performance of a Plug Nozzle Having a Concave Central Base With and Without Terminal Fairings at Transonic Speeds," NASA TN D-1804, 1963.

[46]Wasko, R. A., "Performance of Annular Plug and Expansion-Deflection Nozzles Including External Flow Effects at Transonic Mach Numbers," NASA TN D-4462, 1968.

[47]Manski, D., "Clustered Plug Nozzles for Future European Reusable Rocket Launchers" (published in German), DLR-IB 643-81/7, Lampoldshausen, 1981.

[48]Hagemann, G., Schley, C.-A., Odintsov, E., and Sobatchkine, A., "Nozzle Flowfield Analysis with Particular Regard to 3D-Plug-Cluster Configurations," AIAA Paper 96-2954, 1996.

[49]Rommel, T., Hagemann, G., Schley, C.-A., Manski, D., and Krülle, G., "Plug Nozzle Flowfield Calculations for SSTO Applications," *Journal of Propulsion and Power*, Vol. 13, No. 6, 1997, pp. 629–634.

[50]Hagemann, G., Immich, H., and Dumnov, G., "Critical Assessment of the Linear Plug Nozzle Concept," AIAA Paper 2001-3683, July 2001.

[51]Onofri, M., "Plug Nozzles: Summary of Flow Features and Engine Performance—Overview of RTO/AVT WG 10 Subgroup 1," AIAA Paper 2002-0584, Jan. 2002.

[52]Schwane, R., Hagemann, G., and Reijasse, P., "Plug Nozzles: Assessment of Prediction Methods for Flow Features and Engine Performance," AIAA Paper 2002-0585, Jan. 2002.

[53]Tomita, T., Tamura, H., and Takahashi, M., "An Experimental Evaluation of Plug Nozzle Flow Field," AIAA Paper 96-2632, 1996.

[54]Tomita, T., Takahashi, M., Onodera, T., and Tamura, H., "Effects of Base Bleed on Thrust Performance of a Linear Aerospike Nozzle," AIAA Paper 99-2586, July 1999.

[55]Ito, T., and Fujii, K., "Flow Field Analysis of the Base Region of Axisymmetric Aerospike Nozzles," AIAA Paper 2001-1051, Jan. 2001.

[56]Ruf, J., Hagemann, G., and Immich, H., "Comparison of Experimental and Computational Fluid Dynamics Analysis for a Three-Dimensional Linear Plug Nozzle," AIAA Paper 2003-4909, July 2003.

[57]Smith-Kent, R., Loh, H., and Chwalowski, P., "Analytical Contouring of Pintle Nozzle Exit Cone Using Computational Fluid Dynamics," AIAA Paper 95-2877, 1995.

[58]Ewen, R. L., and O'Brian, C. J., "Dual-Throat Thruster Results," AIAA Paper 86-1518, 1986.

[59]Nguyen, T. V., Hyde, J. C., and Ostrander, M. J., "Aerodynamic Performance Analysis of Dual-Fuel/Dual-Expander Nozzles," AIAA Paper 88-2818, 1988.

[60]Hagemann, G., Krülle, G., and Manski, D., "Dual-Expander Engine Flowfield Simulations," AIAA Paper 95-3135, 1995.

[61]Manski, D., Hagemann, G., and Sassnick, H. D., "Optimisation of Dual-Expander Rocket Engines in Single-Stage-to-Orbit Vehicles," *Acta Astronautica*, Vol. 40, No. 2-8, 1997, pp. 151–163.

Chapter 13

Nozzle Design and Optimization

Patrick Vuillermoz*
Centre National d'Etudes Spatiales, Evry, France

Claus Weiland[†] and Gerald Hagemann[†]
EADS Space Transportation, Ottobrünn, Germany

Bertrand Aupoix[‡]
Office National d'Etudes et de Recherches Aérospatiales, Toulouse, France

Hervé Grosdemange[§]
SNECMA Moteurs, Vernon, France

and

Mikael Bigert[§]
Volvo Aero Corporation, Trollhattan, Sweden

I. Introduction

OPTIMIZATION of nozzle extension is one of the most critical issues in the design of a rocket engine. The quality of the expansion of the gases produced in the combustion chamber has a direct impact on the propulsion performance of the system. The gain or loss of specific impulse because of the nozzle design will translate directly to the gain or loss of several hundreds of kilograms

---

*Launcher Directorate; currently, Head, Liquid Propulsion Systems Section, EADS Space Transportation.
†Project Manager.
‡Senior Scientist.
§Future Projects Engineer.

of payload. This chapter describes how these issues have been considered in the design of the Vulcain 2 engine for the Ariane 5 launch vehicle.

Since the early works of Rao,[1] considerable effort has been focused on the prediction of nozzle performance and design optimization. Advanced methods in the field of numerical fluid dynamics have been developed, including the method of characteristics coupled with boundary-layer solvers,[2] Navier–Stokes solvers,[3,4] turbulence models,[5–15] and kinetic models.[16,17]

Two different situations should be considered. The first situation involves upper-stage nozzles. In the design of rocket nozzles for operation in vacuum only, the optimization of performance depends mainly on two factors: size limitations and heat transfer. The optimization of such nozzles requires very careful treatment of all wall turbulence and kinetic effects.

The second situation involves sea-level operating nozzles. The major difficulty in the design of a nozzle for sea-level operation is that optimization requires a compromise between sea-level conditions and vacuum conditions. This is the case for first- or second-stage booster engines as well as for single-stage-to-orbit (SSTO) engines. For such rocket engines, the majority of the flight will take place under vacuum conditions, so that engine performance in vacuum will be the determining factor in flight performance. However, the optimization of the design cannot be based only on vacuum conditions, as it is for upper-stage nozzles. Two specific constraints due to the operation at sea level during the early portion of the flight must be taken into account: 1) flow separation in the nozzle due to high ambient pressure and 2) sideloads during startup due to nonaxisymmetric flow separation.

The main physical issues to be considered are 1) high area ratio expansion, 2) flow separation and sideloads, and 3) heat transfer. Nozzles with high area ratio are particulary attractive for high-altitude applications. However, when the area ratio increases to very high values, the performance gain evolves very slowly. Figure 1 shows the thrust coefficient c_F as a function of the nozzle area ratio for various pressure ratios. In addition, gas expansion becomes very sensitive to local phenomena such as kinetics recombination and boundary-layer evolution. These two aspects influencing the nozzle performance have been analyzed theoretically[16] and experimentally.[17]

Flow separation in a nozzle for sea-level operation is a severe design constraint because it may lead to heavy dynamic loads on the structure that could be unacceptable for the hardware. These effects have been discussed in Chapter 12. They have been considered in the design of the Vulcain 2 engine nozzle.

Most large liquid rocket engines use regenerative cooling of the nozzle with coolant circulation in tubes or channels. The coolant is generally tapped off the fuel (e.g., hydrogen, kerosene, monomethyl hydrazine, etc.). The cooling must prevent the wall materials from exceeding the maximum operating temperatures.

For high-pressure engines or expander-cycle engines, the best technology is regenerative cooling because it does not affect the general performance of the engine. It has been used, for instance, on the space shuttle main engine and RL10 engine. This technology is, however, very demanding in turbopump power and is less applicable for moderate-pressure engines like gas-generator cycle engines or for coolant-bleed cycle engines. In these cases, the cooling

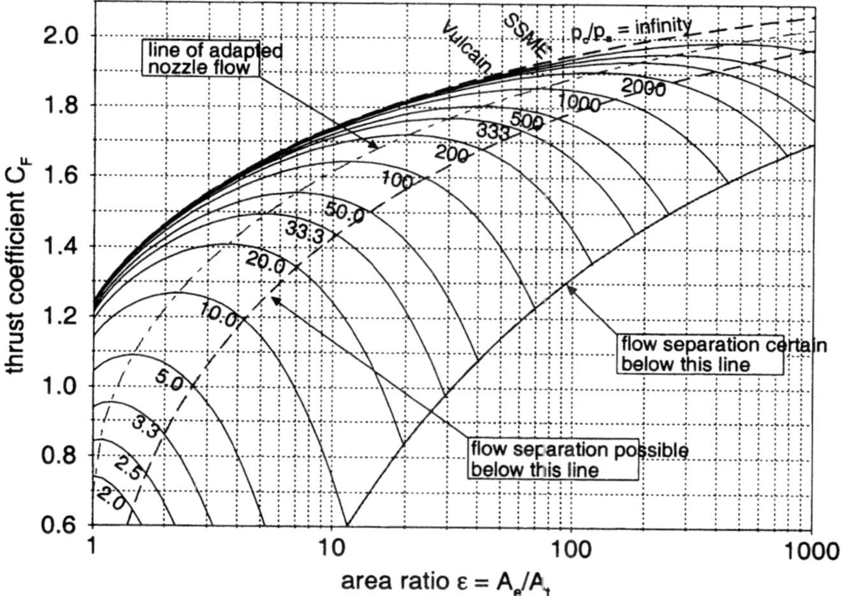

Fig. 1   Thrust coefficient C_F as a function of nozzle area and pressure ratios.

may be achieved by dump cooling. This technology is easier to master than regenerative cooling but leads to performance losses for the engine. This technology is used for the Vulcain nozzle extension of the Ariane 5 main cryogenic stage.

Another way to cool the nozzle wall is to use radiative cooling. This technology is less efficient than the active cooling methods and requires the use of high-temperature materials such as thermostructural composites or columbium alloy that can withstand the heat fluxes released by the main jet coming from the combustion chamber. For instance, the Viking engines of the first and second stages of the Ariane 4 launcher have nozzle extensions made from single-sheet cobalt base skirts. Radiative cooling has also been demonstrated on the HM7 engine, the Ariane 4 third-stage engine, with the successful testing of a carbon/silicon carbide skirt.[18] More recently, the Ariane 5 third-stage engine, the Aestus engine,[19] has been developed with this technology with a cobalt base skirt.

Radiative cooling can be used in combination with film cooling. This achieves better cooling efficiency by protecting the nozzle wall with a coolant stream. For gas-generator cycle engines, the turbine exhaust gases (TEG) can be used as coolant to gain engine performance by using lower dump mass flow. Generally, only the lower part of the nozzle can be cooled in this way. Upstream of the injection of the coolant film, the wall may be cooled either regeneratively or by dump.

Several existing open-cycle rocket engines have TEG delivered to the nozzle, including the F1 engine[20] and J2 engine[21] in the United States and the upgraded

LE5 engine[16] in Japan. The main objectives of TEG reinjection are 1) to avoid the risks of rich fuel post burning and recirculation at the base of the launcher and 2) to simplify the layout of the engine on the launcher. Recently, this technology was planned for use on the STME engine.[22]

It can be shown how optimization techniques and advanced technologies for cooling can be applied in the development of a high area ratio rocket nozzle, with the Vulcain 2 nozzle as an example. The Vulcain 2 is the second generation of the Ariane 5 Vulcain engine.[23,24] It is the engine of the main cryogenic stage of the Ariane 5 evolution launcher. The engine was built on the same principles as Vulcain, i.e., a gas-generator cycle with separated ejection of turbine exhaust gases.[25] The thrust increased from 1145 kN in Vulcain to 1350 kN in Vulcain 2. The chamber pressure and propellant mass flow are higher, and the engine mixture ratio will go up from 5.35 to 6.1. This last change has a negative impact on specific impulse.

To achieve a performance gain of at least 2 s of specific impulse, compared with the Vulcain engine, it was necessary to develop a new nozzle extension based on a modified technological concept using film cooling. The general features of this nozzle are as follows (see Fig. 2): 1) area ratio $\varepsilon = 60$ (45 for Vulcain), 2) dump cooling of the upper part of the nozzle, and 3) film cooling of the lower part of the nozzle with turbine exhaust gas reinjection. These choices achieve performance gains by 1) reducing the dump mass flow, 2) increasing the TEG expansion efficiency as compared with separated ejection, and 3) decreasing the friction losses along the lower part of the nozzle. The design optimization of the nozzle contour was performed by means of the computational codes discussed in Section II. The main conclusions of the research carried out to support nozzle development are presented in Section III. Finally, a summary of a technological program at engine level is presented in Section IV.

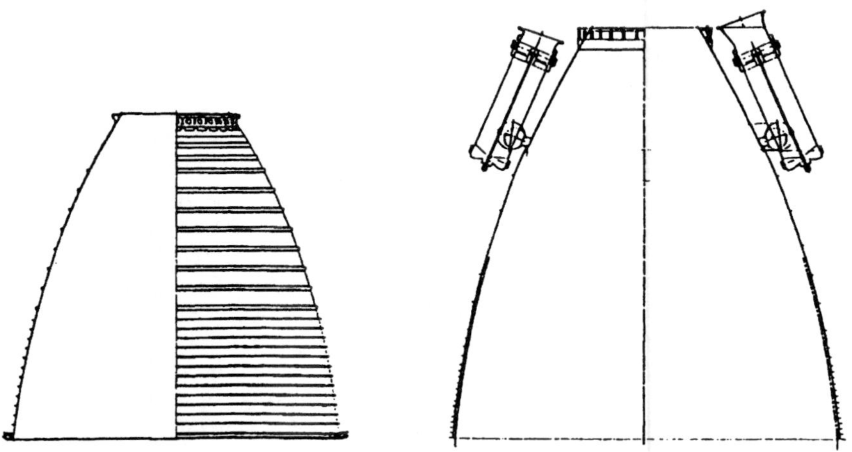

Fig. 2   Vulcain and Vulcain 2 nozzles.

II. Nozzle Contour Optimization

The design of an advanced rocket nozzle requires the use of highly accurate aerodynamic prediction tools based on Euler and Navier–Stokes codes in the design and validation process. The following sections describe a particular method that includes state-of-the-art aerodynamic prediction tools in a complex optimization process. Special care has been taken to keep design and validation time as low as possible by using low-order methods.

A. Numerical Methods

1. Inviscid-Flow Approach

A major part of the design is the optimization of the nozzle contour with respect to overall dimensions, mass, specific impulse, flow separation, and sideloads. The main goal of this optimization process is to determine a contour with maximum specific impulse at a given exit pressure and a given limit for the maximum sideloads on the engine actuators.

The exit pressure of the nozzle, as well as the maximum sideloads, are nearly independent from viscous effects. Therefore, it is possible to use Euler simulations to large inviscid-flow database for different nozzle contours with various parameters.

For the generation of this basic data, an Euler code with inclusion of real gas effects was used.[26] Based on the inviscid pressure distribution at the wall, the aerodynamic sideloads of the nozzle were calculated following Schmucker's method.[27] In this study, the coefficients of the Schmucker formula were chosen to be $k_{\mathrm{fl}} = 0.05$ and $k_g = 1.044$. The formula for separation sideloads was modified to

$$F_{\mathrm{sl}} = 2k_g R(p_a - p_w)\min\left(\Delta l_{\mathrm{fl}}, x_{\mathrm{exit}} - x_w\right) \tag{1}$$

where R is the nozzle exit radius, p_a is the atmospheric operating pressure, p_w is the exit wall pressure at average separation, $x_{\mathrm{exit}} - x_w$ is the distance between average flow separation and nozzle exit, and Δl_{fl} is the range between minimum and maximum separation location.

With a variation of the chamber pressure for one contour, the maximum sideloads were then determined.

2. Viscous-Flow Approach

To account for viscous effects on the specific impulse, calculations with a state-of-the-art Navier–Stokes code[28] were conducted. As in the case of the Euler code, chemical nonequilibrium was considered. The algebraic turbulence model of Baldwin and Lomax[6] was used to account for the turbulent motion of the fluid. Thus, the influence of wall friction on specific impulse was calculated for some selected configurations of the inviscid database.

B. Application to Vulcain 2 Design

The approach just described was used for the design and the optimization of the Vulcain 2 nozzle.

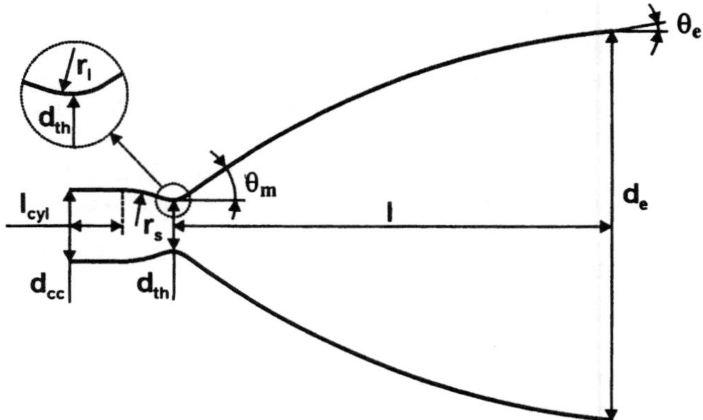

Fig. 3 Clean contour Navier–Stokes, turbulent calculation with chemical equilibrium ($\theta_e = 5.5$ deg, $\theta_m = 37$ deg, $\varepsilon = 59$, length = 2550.6 mm).

1. Inviscid-Flow Database

The supersonic part of the nozzle can be approximated using a parabola and then described completely using the following parameters (see Fig. 3): throat diameter d_{th}, radius of the throat r_l, starting angle of the parabola θ_m, end angle of the parabola θ_e, length of the supersonic part l, expansion ratio (area ratio) $\varepsilon = d_e^2/d_{th}^2$.

The parameters $d_{th} = 274.18$ mm and $r_l = 68.545$ mm were used for all the design calculations. The other parameters were varied independently in the following ranges: $-\theta_m = 33$–39 deg with each increment of one degree; $-\theta_e = 4.5$–8.5 deg with each increment of one degree; $-l = 2.3$–2.8 m with each increment of of 0.1 m, and $-\varepsilon = 55$–75 with each increment of 5.

A total of 700 different nozzle contours have been studied at the reference nozzle operating point. Additionally, calculations have been conducted at a second operating point corresponding to lower chamber pressure conditions. This was necessary to get the separation pressure according to Schmucker,[27] which is the minimum exit pressure level where separation is likely to occur. Calculations showed that an exit pressure level of 0.33 bar at the design point 5′ will fulfill this restriction for all nozzle contours under consideration. This pressure level corresponds to an exit pressure of about 0.35 bar at the engine design point S_2 ($p_c = 115.5$ bar, $O/F = 7.19$).

Based on this data, for the exit pressure levels 0.33, 0.35, and 0.38 bar at the design point S_2, the maximum nozzle length was determined for each combination of θ_m, θ_e, and ε (see Fig. 4). These 300 nozzle contours were studied at the nominal conditions of the S_2 point and at 12 low-pressure points to determine the maximum sideloads during engine startup. Thus, including some additional necessary calculations, more than 5300 inviscid-flow calculations were conducted in total. These calculations form the basic database.

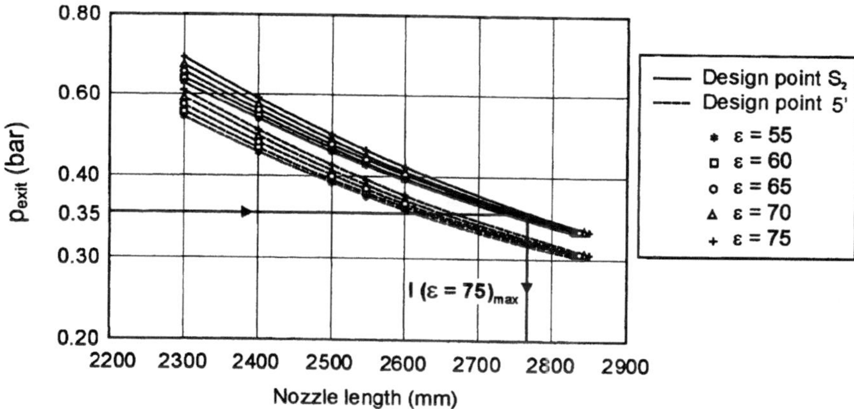

Fig. 4   Influence of exit pressure on nozzle length and area ratio ε (inviscid flow, $\theta_e = 4.5$ deg, $\theta_m = 37$ deg).

2.   *Optimization Results*

For the optimization process, three basic restrictions were set: 1) full flowing nozzle (no separation) for the low point 5' of the operating envelope ($p_c = 107$ bar, $O/F = 6.7$); 2) no adverse wall pressure gradient at the nozzle exit, and 3) quasi-static sideloads lower than 50 kN.

The first of these three restrictions was fulfilled by setting a minimum exit pressure (maximum length, Fig. 4). Figure 5 shows that the second restriction is met by most of the nozzles, especially those with a length greater than 2.6 m. This has to be checked for each nozzle under consideration. The

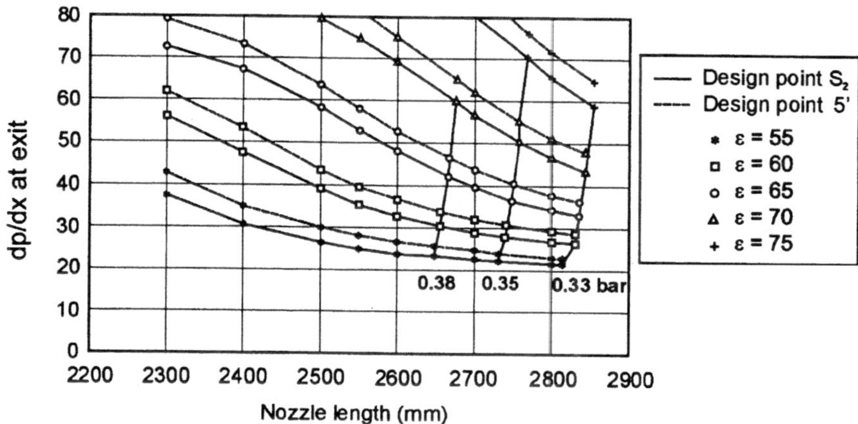

Fig. 5   Influence of exit pressure gradient on nozzle length and area ratio ε (inviscid flow, $\theta_e = 4.5$ deg, $\theta_m = 37$ deg).

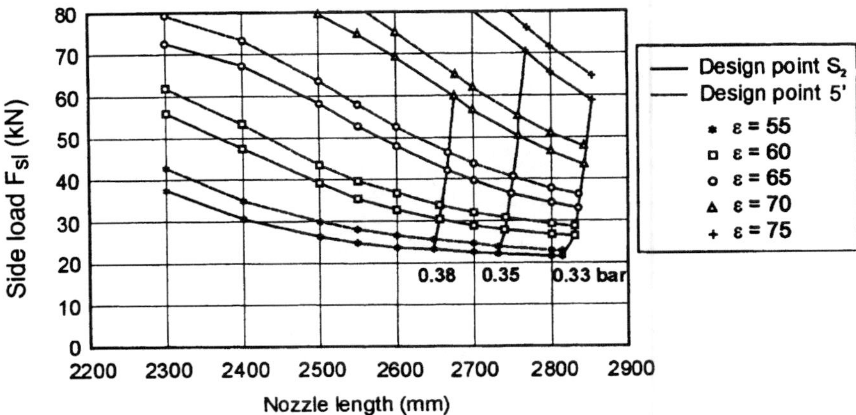

Fig. 6 Influence of sideload force F_{sl} on nozzle length L and area ratio ε (inviscid flow, $\theta_e = 4.5$ deg, $\theta_m = 37$ deg).

aerodynamic sideloads decrease with increasing length, but increase with area ratio ε (see Fig. 6). The quasi-static sideloads can be calculated from the aerodynamic sideloads by applying the amplification factor of 1.49. This factor was deduced from comparison with measurements on the HM60-Vulcain engine nozzle, where the original factor of 1.74 was reduced by 86%. With this result, the third restriction can be transformed to a limitation of maximum of aerodynamic sideloads of about 35 kN.

3. Influence of Friction

Because calculations with viscous flows based on the Navier–Stokes equations require significantly longer computing time than those for inviscid flows, it was necessary to limit the total number of computations. Thus, a preliminary optimization was achieved to select eight nozzle contours close to the optimum based on specific impulse. From these calculations, it was noted that a change in the contour has only a small effect on wall values (i.e., pressure coefficient, wall temperature, skin friction coefficient), whereas the pressure fields differ significantly for the contours under consideration.

Several effects were observed. The loss in specific impulse ΔI_{sp} increases linearly by a factor of 2.0 s/m with increasing nozzle length. This linear behavior is due to an almost constant friction coefficient near the wall exit. The variation in ΔI_{sp} at almost the same length is an effect of the variation in θ_m, θ_e, and ε. In addition, it can be seen that the loss is dependent on combustion chamber conditions and on the exit pressure level. Summarizing these observations, it is possible to deduce a correction of the inviscid specific impulse prediction at vacuum conditions for nozzles operating at identical chamber pressure and exit pressure conditions. For operation at S_2 and with exit pressure of 0.35 bar, this correction factor is

$$\Delta I_{sp} = I_{sp}(\text{viscous}) - I_{sp}(\text{inviscid}) = -4.4\,\text{s} - 2.0\,\text{s/m}\,(l - 2.55\,\text{m}) \qquad (2)$$

This formula permits deduction of optimized nozzle contours directly from the inviscid database and without further Navier–Stokes calculations.

4. TEG Reinjection Modeling

In a nozzle with TEG reinjection, such as the Vulcain 2, one can expect the presence of the film to have an effect on thrust and specific impulse. This question has been investigated by means of two-dimensional axisymetric simulations.[26,28] Comparison of Euler simulations with and without TEG injection have shown that the main flow is virtually unaffected by the film. It shows that using an inviscid-flow database to select a nozzle configuration is valid for nozzles with TEG injection. However, the final prediction of the overall nozzle performance requires an accurate description of the mixing cf the TEG stream with the main jet.

For this purpose, Navier–Stokes computations were performed to test various turbulence models, including algebraic and two-equation models (see Fig. 7). The k-ω model[13] predicts a significantly longer mixing length as compared with the Baldwin-Lomax model.[6] Moreover, this two-equation model leads to a larger estimate of specific impulse. The difference between these two results may be more than 1 s, depending on the nozzle configuration.

C. Summary

The methodology described in this section demonstrates the possibility of deducing optimized nozzle contours from inviscid-flow computations. It is

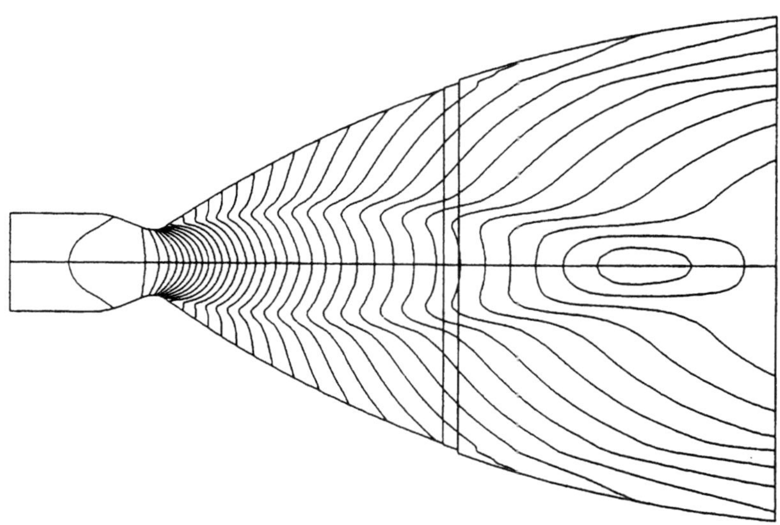

Fig. 7 Pressure contours (viscous flow computation with k-ω model).

shown that for special classes of nozzle shapes, a simple formula can take into account the difference in specific impulse between Euler predictions and Navier–Stokes predictions. This formula depends only on the combustion chamber conditions and the exit pressure at this point. The study of the Vulcain 2 nozzle reveals that slight modifications of contour angles θ_m and θ_e may have a significant effect on the level of sideloads during engine transients, even though specific impulse is not much affected. In general, aerodynamic sideloads and specific impulse behave in opposite directions. The maximum specific impulse available is closely determined by the maximum sideloads acceptable by the engine.

III.   Nozzle Film Cooling

Film injection optimization must take into account various aspects of the flow physics, such as thermal protection provided by the film, contribution of the film to the engine thrust, and impact of the film on the sideloads during engine startup or shutdown. Film cooling efficiency can be predicted in various ways. The simplest method is the use of correlations of experimental data computational fluid dynamics (CFD) can be used to solve boundary-layer or Navier–Stokes equations. A first comparison between different approaches for Vulcain 2 conditions has shown that the predicted distance downstream of film injection that is fully protected varies widely. Predictions of temperature at the nozzle exit also vary.

Therefore, it was decided to focus on basic issues of film cooling mechanisms by means of experimental techniques and to assess the validity of the various physical models according to these data.

A.   Experimental Study

In the Vulcain engine, the film is produced using turbine exhaust gases. The exhaust gases are first collected in a torus and then re-injected into the engine nozzle. The mixing of the film with the main flow first occurs through a mixing layer, which eventually merges with the wall boundary layer. It was thus important to try to reproduce the key parameters of the mixing layer. These key parameters are the Mach numbers of the film and the main flow, the Reynolds number based on the injection height, the ratio $\lambda = (\rho U)_f/(\rho U)_m$ of the film and main flow, the convective Mach number, and the temperature or density ratio of the two flows. The latter two parameters have a strong influence on the mixing layer spreading rate.

The main flow is mainly water vapor, and the film is a hydrogen/water vapor mixture, both at high temperatures. To investigate the flow structure, a conventional wind tunnel with standard gases at moderate temperatures was used. A cryogenic technology was chosen to obtain a cold film, made up of a mixture of air and vaporized liquid nitrogen. Therefore, it was impossible to reproduce all the similarity parameters. The main flow Mach number and the convective Mach number were duplicated, and the lower temperature ratio was investigated, but the temperature and density ratios, the isentropic exponent of the gases, the film Mach number, and the ratio λ were not duplicated.

1. Experimental Setup

The experiments were conducted in the ONERA S5 wind tunnel in Chalais-Meudon.[29,30] This is a continuous-flow facility; the experiment time was limited only by the liquid nitrogen storage for the cold film. The supersonic nozzle was set in a half-nozzle configuration, with the measurements performed on the symmetry axis of the complete nozzle configuration. A sketch of the model is shown in Fig. 8. The measurement plate is located in the constant Mach number region of the wind tunnel, the Mach number being about 2.78. The film exit Mach number is about 2.

Two different experiments were performed. In the first one, the film injector height was 10 mm, to duplicate the engine Reynolds number (and, fortuitously, the engine injector height). The mixing layer development was well documented, but the film did not seem to break up in the test section. Therefore, a second experiment was performed with a half-nozzle configuration, i.e., an injector height of about 5 mm was used to investigate the film breakup.

Although the focus was primarily on adapted flow conditions, i.e., when the static pressure of the two flows is equal, the model was also designed to investigate the influence of pressure mismatch. Both film pressure and temperature were automatically monitored so that pressure mismatches of about $\pm 50\%$ were achieved to reproduce flow conditions that could be encountered in the engine.

The experiment was designed to study a cold film, but a film at "ambient" temperature was also investigated to consider the influence of the change of the film temperature, which strongly affects the mixing layer spreading rate, as both the

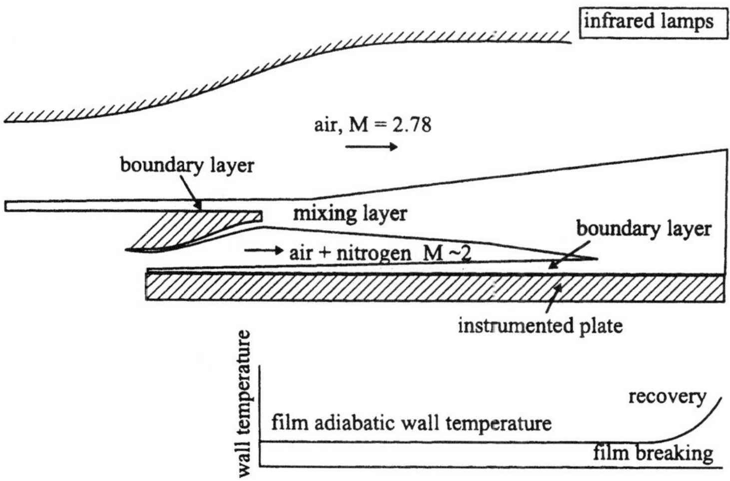

Fig. 8   Sketch of ONERA S5 experimental setup.

density ratio and the convective Mach numbers are modified. The convective Mach number is 0.73 for the cold film and 0.24 for the ambient temperature film.

2. Measurements

Schlieren photography was used to monitor the flow and capture the variations in flow structure when the film pressure or temperature is changed. The plate is equipped with 18 pressure taps and 18 thermocouples in two lines close to the symmetry axis. Two extra transverse rows of pressure taps and thermocouples are used to measure the two-dimensional character of the flow. Wall pressure, adiabatic wall temperature, and exchange coefficients were measured. To determine the exchange coefficients, a set of infrared lamps was installed in the facility in front of the plate. A thermal pulse was obtained on the plate, and the exchange coefficient was deduced from the relaxation of the plate surface temperature.[31]

Flow measurements were performed with a three-probe rake. Static and stagnation pressures, together with stagnation temperature, were recorded simultaneously at exactly the same altitude. All data were later interpolated at the same altitude. Probing of supersonic boundary layers is often performed with only a stagnation pressure probe, but this gives only the Mach number profile, provided the static pressure is constant. Here, information about variations in static pressure as well as on the thermal mixing of the two flows was sought.

B. Computational Approach

1. Boundary-Layer Approach

A boundary-layer approach was used both to compute the film development on the engine and to compare model predictions with experimental data. For the engine, a chemically reacting boundary-layer code[32] with few turbulence models was used. To simulate the experiment, an ideal gas code[33] with a large variety of turbulence models was preferred. Of course, a boundary-layer approach is restricted to matched conditions. Special attention must be paid to the singularity that is present in the boundary-layer equations at the injector lip. The numerical code has a self-adaptive grid, both along the wall normal and streamwise directions. Very small steps had to be enforced at the lip, and a fine grid resolution was required. Nevertheless, the cost of a boundary-layer computation was still negligible compared to that of a full solution of the Navier–Stokes equations.

The computational procedure, both for the experiment and for the real engine, is as follows. First, inviscid solutions for the main flow and for the injector are obtained using either an Euler solver or a method-of-characteristics approach. Then, the boundary layers that develop in the main flow and in the injector are computed in a standard way, using these inviscid solutions. At the injector lip, all of these solutions have to be piled up to generate a new initial profile.

2. Turbulence Models

A large variety of eddy viscosity turbulence models was tested. The Cebeci-Smith model[5] and the Baldwin-Lomax algebraic model[6] were investigated, as they are very simple and inexpensive in computational time, but they are a priori not well suited for this kind of flow.

Two-equation models, of the k-ε and the k-ω types, have also been used, as they are more relevant to this kind of flow situation. Five k-ε models, those of Jones and Launder,[7] Launder and Sharma,[8] Chien,[9] Nagano and Tagawa,[10] and finally So et al.[11] were tested. The first two are very similar. Chien's model mainly differs in the use of the wall distance instead of the turbulent Reynolds number in the damping terms. These three models solve an equation for a pseudo-dissipation, which is zero at the wall. The last two solve a transport equation for the real dissipation. The So et al. model differs in the use of unusual diffusion coefficients in the transport equations. The chosen k-ω models are the standard Wilcox model[12] and its extension to account for low Reynolds number flows and transition.[13]

Finally, a four-equation model by Sommer et al.[14] which is an extension of the So et al. k-ε model with two extra equations for the temperature fluctuation variance and its dissipation rate, was tested. The goal was to eliminate the assumption of a constant turbulent Prandtl number in computing the turbulent heat fluxes, as experiments tend to show that the turbulent Prandtl number is roughly constant in boundary layers and in mixing layers, but with a different level.

Moreover, it is well known that most of the turbulence models are unable to correctly predict the reduction of the mixing-layer spreading rate as the convective Mach number increases. This is usually blamed on the compressible character of the turbulent motion. Various modifications have been proposed in the literature. We tested several exploratory models but will discuss here only the compressibility correction of Sarkar[15] for dilatational dissipation.

3. Model Validation

Four cases, corresponding to two injector heights and two film temperatures, have been investigated. The higher injector gives more information on the development of the mixing layer, whereas the smaller one provides data on the film breaking and the relaxation toward an equilibrium boundary layer.

The algebraic model predictions were found to be very poor, as expected. These models deal with only global information on the turbulence length scale deduced from the flow geometry. They always assume that the wall boundary layer and the mixing layer form only one wall layer, even before they have merged, and that the thickness of this layer gives the turbulence length scale. Therefore, they overestimate the turbulence length scale and the turbulent diffusion so that they predict very rapid mixing and very quick relaxation toward an equilibrium boundary-layer solution, as shown in Fig. 9.

Predictions of the Jones and Launder k-ε model are given in Fig. 10 for the higher injector, cold film configuration. It can first be noted that the agreement in the Mach number profiles is fair for the last stations, whereas there is a visible shift between experimental and computational data in the velocity and temperature profiles (not shown). This demonstrates the importance of measuring not only the stagnation pressure but also the stagnation temperature to have access to more information about the flowfield. Here, errors in the velocity and

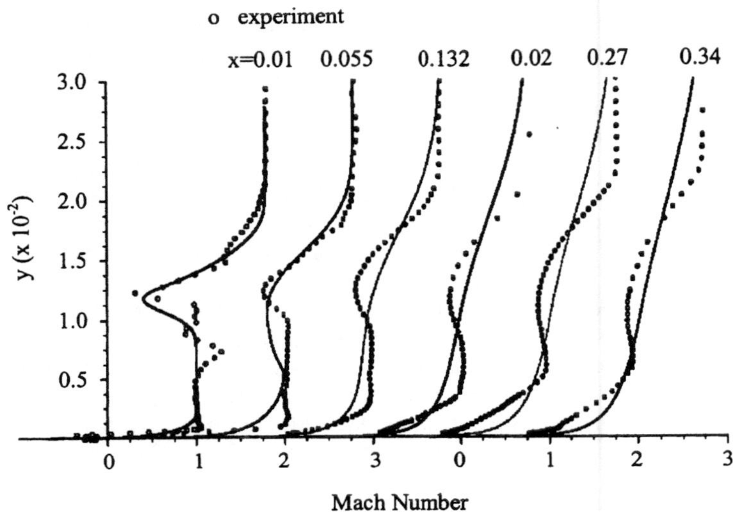

Fig. 9   Mach number profiles (cold film; injector 10 mm; Baldwin–Lomax model).

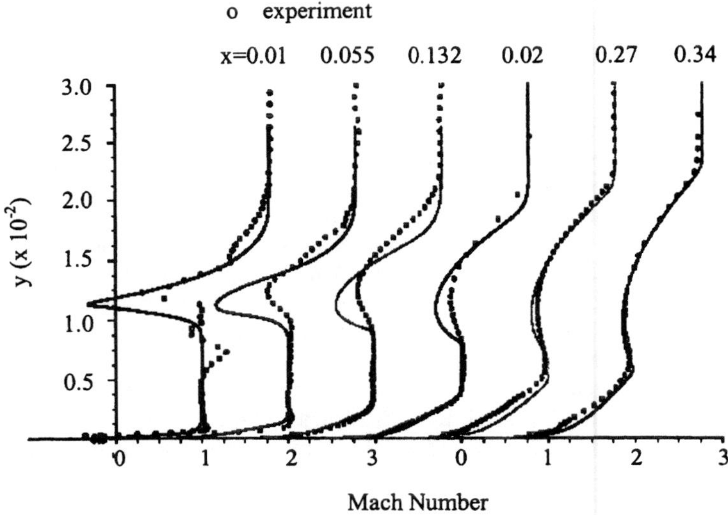

Fig. 10   Mach number profiles (cold film; injector 10 mm; Jones–Launder model).

temperature profiles roughly compensate to give a good prediction of the Mach number profile. The shift in the profiles can be explained by the use of the boundary-layer equations. Streamwise diffusion is neglected, so that streamwise gradients can be overestimated. Through the continuity equation, an error on the streamwise gradient of the longitudinal velocity component yields an overestimation of the wall normal velocity component and a downwash of the computed profiles. This may also be the cause of a poor prediction of the wake of the injector lip. Finally, it can be noted that the mixing-layer expansion rate is overestimated by this model, as expected.

The Chien k-ε model, with and without compressibility correction, was also used. Some results are presented in Fig. 11. This model also overestimates the mixing-layer spreading rate. Sarkar's compressibility correction reduces the mixing-layer spreading rate and thus improves the prediction.

The best predictions were achieved with the So et al. model, as shown in Fig. 12. This model reproduces the mixing-layer spreading rate quite well. Because there is no compressible turbulence correction in this model, its accuracy seems to be due to the use of unusual diffusion coefficients in the turbulence transport equation, which allows a better prediction of the outer region of the mixing layer. The use of extra equations to compute the turbulent Prandtl number did not improve the prediction.

Analogous conclusions can be drawn for the other test cases. The lower convective Mach number in the ambient temperature case reduces the differences between models. The smaller injector case shows that a good prediction of the mixing layer before it merges with the wall boundary layer and diffuses down

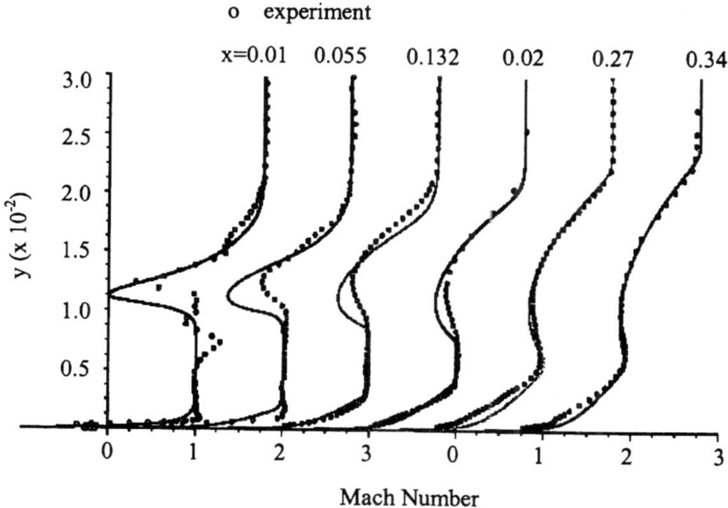

Fig. 11   Mach number profiles (cold film; Chien model with compressibility corrections).

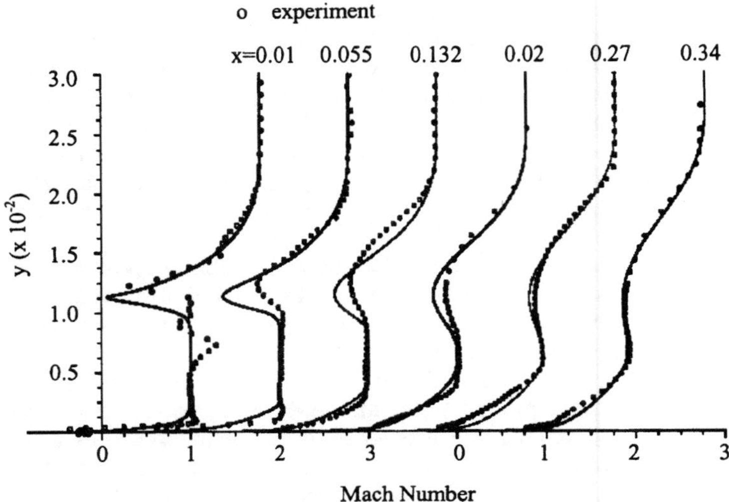

Fig. 12   Mach number profiles (cold film; injector 10 mm; So et al. model).

to the wall is required to correctly predict the relaxation of the boundary layer toward an equilibrium state downstream of film breakup. Here again, the best results are achieved with the So et al. model, as shown in Fig. 13.

The prediction of the adiabatic wall temperature shows that transport equation models can reproduce fairly well the film breakup location and the temperature increase downstream, whereas algebraic models predict a too rapid film breakup, as shown in Fig. 14. It must be noticed that small thermal leaks slightly alter the experimental data.

C.   Summary

Film cooling has been investigated for two slot dimensions and two film temperatures. This creates a large database for validation of numerical approaches and turbulence models. Although a boundary-layer approach is not correct near the injector lip and thus predicts a small downwash of the mixing layer, it is a very efficient and inexpensive way to investigate film cooling. Algebraic models are not well suited to predicting such flows, whereas two-equation models correctly reproduce the key features of the flow.

A good prediction of the mixing layer is required to reproduce all of the details of the flow; the So et al. model seems to perform the best among the tested models.

IV.   Vulcain 2 Demonstration Program

To demonstrate the film cooling concept before full-scale development of the Vulcain 2 nozzle extension, a demonstration program was started. In Vulcain 2 the lower part of the nozzle is film cooled by injection of turbine exhaust

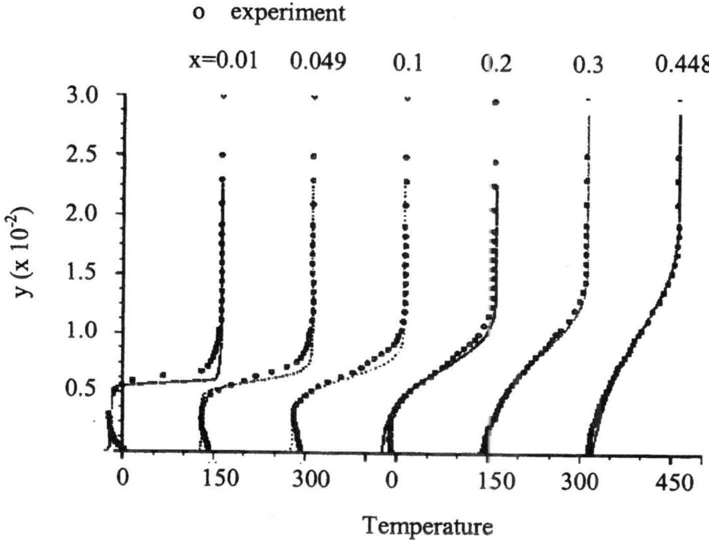

Fig. 13   Temperature profiles (cold film; injector 5 mm; So et al. model).

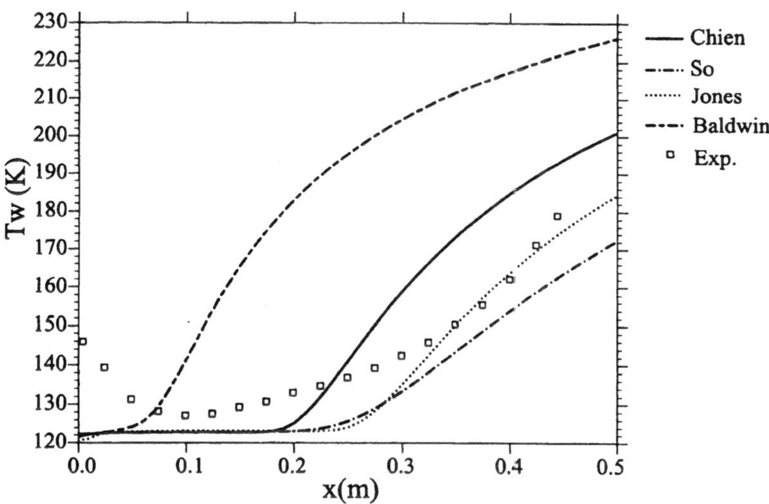

Fig. 14   Wall temperature (cold film; injector 5 mm): comparison between experimental data and computational predictions.

gases and dump coolant gas. The purpose of the demonstrator was to verify the film cooling efficiency and validate design tools. The film of combined turbine exhaust gases and dump flow should be able to cool the nozzle skirt both on and off design. To achieve a high confidence level, a full-scale demonstration nozzle was needed. The Vulcain engine is 30% smaller than the Vulcain 2 and fits well for demonstration purposes. The main objectives of the testing were 1) to prove the feasibility of the film-cooling concept, 2) to demonstrate good film cooling efficiency throughout the envelope, 3) to calibrate the design tools for film cooling, and 4) to study the impact of film cooling on separation and sideload.

A. Design of Demonstrator

The design was made based on the following principles: 1) maximum use of existing Vulcain hardware, 2) minimum scaling with respect to Vulcain 2, 3) representative length of skirt to study variations in film cooling efficiency, 4) high design margins to avoid stopping the test because of test bench safety procedures, 5) maximum instrumentation life, and 6) adequate mechanical margin, acheived by making a stiff design at the expense of weight.

The film cooling system proposed for the Vulcain 2 engine uses a supersonic film. Design and development of such technology based on injection of supersonic flow interacting with a main supersonic flow are not common design practices. However, the available analysis codes did meet the standards needed to start development, although demonstration of the technique and further development of analysis codes were necessary. The available engineering correlation codes and CFD codes were sufficient for design of the demonstrator. Test facilities for wind tunnel and engine tests were also available.

Scaling test results for film efficiency for geometrical dimensions and gas properties is a sensitive process. To obtain a relevant demonstration, the scaling effects must be calculated and kept small. One typical parameter is the convective Mach number. For comparison, in the Vulcain 2 this critical physical parameter has a 5–10% larger value than in the demonstrator, which makes the demonstration conservative. Other investigated parameters also gave conservative or neutral comparisons, except for the main jet boundary-layer thickness, which is thicker for the Vulcain 2. This could result in increased mixing.

To benefit from the cooling advantages, a film may yield the flow also to stay attached to the wall. The total expansion ratio for the nozzle must be in level to convectively cooled nozzles. The sideload characteristics may change as the film is introduced. The flow separation margin may change, and the injector step may influence the way the separation front travels. However, because the total area ratio of Vulcain 2 was smaller than that of Vulcain, no new problems were expected.

The quality of the film as it leaves the injector in part determines its downstream development. Another design criterion requires that pressure at injection must match the main jet. Separation should not occur in the injector. A good distribution of flow with minimal influence from vane is essential to keep the mixing low. The injector is formed by a lip. The control of the distorsion of the injector lip geometry during operation is essential. At the point of injection, the film may be influenced by the wake behind the injector lip or by another cooling film injected upstream.

The demonstrator main data were 1) total area ratio, 40.5, 2) dump injection area ratio, 17, 3) TEG injection area ratio, 19, and 4) total length from combustion chamber to nozzle exit, 1470 mm.

B. Test Results

The results can be summarized as follows.

1) The demonstrator was tested in three consecutive engine tests without any shutoff or other major problems. The total test time was about 1600 s.

2) The main objective of successfully demonstrating the film injection concept validity was met. The film efficiency in terms of measured wall temperatures was higher than anticipated in the preceding analysis.

3) The empirical models used for film cooling prediction have been correlated and verified by the tests.

4) Secondary test objectives, such as investigation of pressure loads and tube wall cooling, have also been met.

The covered operation points in the tests are in the range of pressure between 85 and 120 bar, $(O/F)_c$ between 4.5 and 7.2 (Fig. 15). The skirt temperature results for the first test are shown in Fig. 16. The results at points Q7, D9, and Ref-72 will be discussed. The tests have shown that the wall temperature rise can vary slightly depending on the length of the two TEG feed lines from the hydrogen and oxygen turbopumps. The corresponding pressure measurements are shown in Fig. 17.

C. Test Analysis

The film cooling models used to analyze the test results have the following main features: 1) flow calculation with the JANNAF standard code using the method of characteristics for axisymmetric flow; and 2) calculation of boundary layer with one injected flow; the turbulence model is the algebraic type. When making these calculations, the standard method was to 1) include the dump mass flow in the injected mass flow, 2) set a uniform temperature equal to the

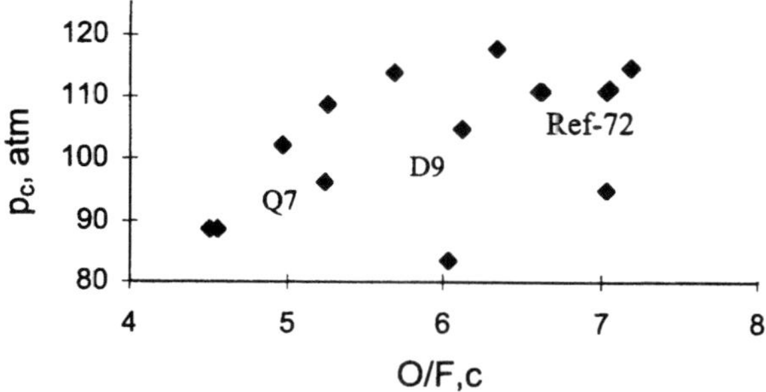

Fig. 15 Operation points during test program.

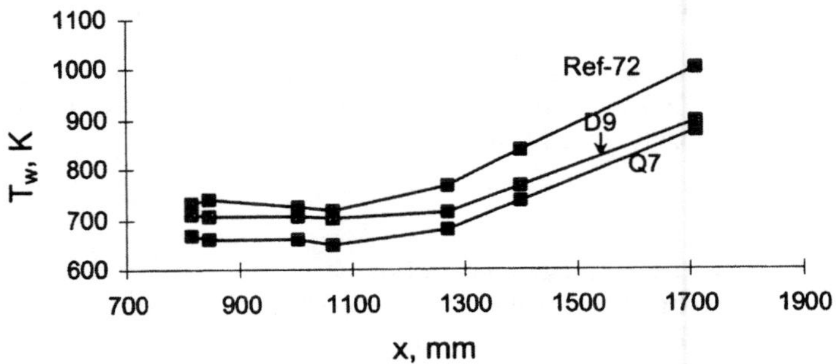

Fig. 16   Temperature distribution along the wall as a function of distance from throat *x*.

TEG manifold temperature for the respective side, the same applies for the pressure, and 3) adjust the mixing length empirically, which comprises the actual tuning.

The results of the calibration were fairly successful for most of the operating points. There are only two points where the error is considerable, namely for the D7 points with very low pressure and TEG mass flow. As an example of the complete temperature, the Ref-72 point from the first test is shown in Fig. 18.

The film cooling model yields good correspondence between measured and calculated values, taking into account all of the major design parameters.

Another type of film cooling is the cooling of the injector lip by the dump film. It was analyzed and found to be working properly for all of the points. There was found to be direct influence from both the dump film temperature and the TEG manifold temperature.

The separation and sideload characteristics were also investigated. Here, it was important to compare not only with calculations for the nominal contour

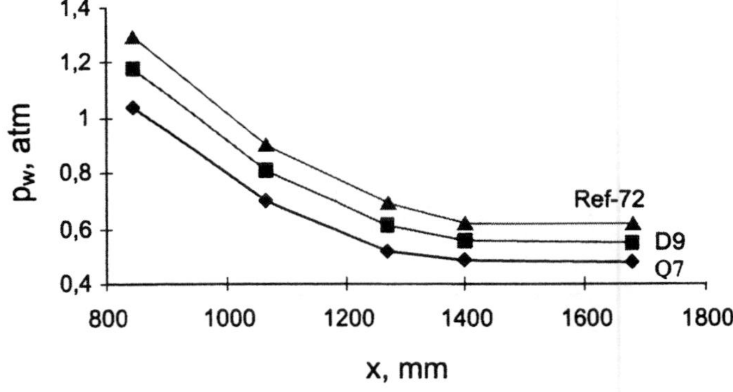

Fig. 17   Pressure distribution along the wall as a function of distance from throat *x*.

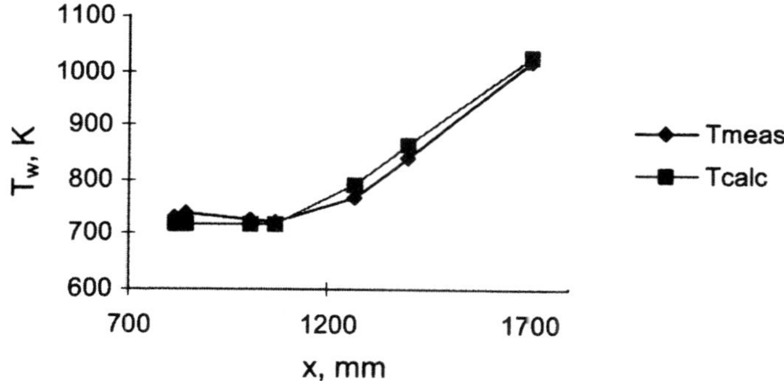

Fig. 18   Temperature distribution along the wall as a function of distance from throat x**.**

but also with the actual contour as manufactured and measured at inspections. The results on separation, based on pressure measurements on the skirt, were quite close to what could be predicted using the standard Schmucker separation formula.[10] The sideloads encountered during the tests also behaved in accordance with predictions.

D. Summary

The following main conclusions can be drawn from the demonstrator programs thus far.

1) The manufacturing experience of this program has proven that the concept is feasible from a production point of view.

2) The tests have made a considerable contribution to the understanding of film cooling. The film cooling performance was better than most predictions.

3) No great drawbacks in the areas of separation, sideload, or mechanical integrity were encountered.

V. Conclusions

The development of space systems demands the design of hardware able to match performance goals and reliability requirements. This challenge is particularly important for the rocket nozzle, as engine performance is very sensitive to the optimization of the design of the nozzle. However, this optimization must take into account a variety of different constraints, including external loads, heat transfer, transients, and the fluid dynamics of expanded gases.

Design optimization is especially difficult for sea-level operating nozzles dedicated to first- or second-stage boosters or a single-stage-to-orbit vehicle. In these cases, the design has to deal with variations in external pressure during flight. This can be handled either by adapting engine operating conditions or, in a simpler way, by controlling flow separation.

In this chapter, we have presented in detail the various components of a typical development program for an advanced cryogenic engine nozzle. This program was focused on controlling heat transfer by using an advanced nozzle cooling technique. The present chapter discusses, in the frame of the development of the Vulcain 2 nozzle, three types of complementary activities that have been conducted: 1) computational fluid dynamics activities to classify various configurations, 2) scientific experimental and theoretical approaches to improve the understanding of basic phenomena, and 3) technological development to verify the feasibility of a new concept at a representative scale.

This program has shown that the understanding of full-scale nozzle behavior requires a detailed investigation of basic phenomena. However, testing of representative models may enable the designer to calibrate the engineering codes for use in real nozzle design.

Beyond the achievement of the development of Vulcain 2, this type of technological program could be continued in the direction of flow separation control. As mentioned earlier, the optimization of performance for such a nozzle requires control of flow conditions at sea level by using either passive or active devices.[34–36] The selection and validation of such a device will demand as much effort as advanced cooling technology.

References

[1]Rao, G. V., "Exhaust Nozzle Contour for Optimum Thrust," *Jet Propulsion*, Vol. 28, No. 6, 1958, pp. 377–382.

[2]Kacynski, K. J., and Hoffman, J. D., "The Prediction of Nozzle Performance and Heat Transfer in H_2/O_2 Rocket Engines with Transpiration Cooling, Film Cooling, and High Area Ratio," AIAA Paper 94-2757, 1994.

[3]Manski, D., and Hagemann, G., "Influence of Rocket Design Parameters on Engine Nozzle Efficiencies," AIAA Paper 1994-2756, 1994.

[4]Weiland, C., Hartmann, G., and Menne, S., "Aerodynamics of Nozzle Flows for Advanced Hypersonic Systems," *CFD Techniques for Propulsion Applications*, AGARD Rept. CP-510, 1991.

[5]Cebeci, T., and Smith, A. M. O., "Analysis of Turbulent Boundary Layers," *Applied Mathematics and Mechanics*, Vol. 15, Pergamon, 1974.

[6]Baldwin, B. S., and Lomax, H., "Thin Layer Approximation and Algebraic Model for Separated Turbulent Flows," AIAA Paper 1978-256, 1978.

[7]Jones, W. P., and Launder, B. E., "The Prediction of Laminarization with a Two-Equation Model of Turbulence," *International Journal of Heat and Mass Transfer*, Vol. 15, 1972, pp. 301–314.

[8]Launder, B. E., and Sharma, B. I., "Application of the Energy-Dissipation Model of Turbulence to the Calculation of Flow near a Spinning Disc," *Letters in Heat and Mass Transfer*, Vol. 1, 1974, pp. 131–138.

[9]Chien, K. Y., "Predictions of Channel and Boundary-Layer Flows with a Low-Reynolds-Number Model of Turbulence," *AIAA Journal*, Vol. 20, No. 1, 1982, pp. 32–38.

[10]Nagano, Y., and Tagawara, M., "An Improved k-ε Model for Boundary Layer Flows," *Journal of Fluids Engineering*, Vol. 112, 1990, pp. 33–39.

[11]So, R. M. C., Zhang, H. S., and Speziale, C. G., "Near-Wall Modeling of the Dissipation-Rate Equation," AIAA Paper 1992-0441, 1992.

[12]Wilcox, D. C., "Reassessment of the Scale-Determining Equation for Advanced Turbulence Models," *AIAA Journal*, Vol. 26, No. 1, 1988, pp. 1299–1310.

[13]Wilcox, D. C., "Progress in Hypersonic Turbulence Modeling," AIAA Paper 1991-1785, 1991.

[14]Sommer, T. P., So, R. M. C., and Zhang, H. S., "Near-Wall Variable Prandtl-Number Turbulence Model for Compressible Flows," *AIAA Journal*, Vol. 31, No. 1, 1993, pp. 27–35.

[15]Sarkar, S., Erlebacher, G., Hussaini, M. Y., and Kreiss, H. O., "The Analysis and Modeling of Dilatational Terms in Compressible Turbulence," *Journal of Fluid Mechanics*, Vol. 227, 1991, pp. 473–493.

[16]Kim, S., "Numerical Study of High Area Ratio H_2/O_2 Rocket Nozzles," AIAA Paper 1991-2434, 1991.

[17]Miyajima, H., Nakahashi, K., Hirakoso, H., and Sogame, E., "Low Thrust LH_2/LO_2 Engine Performance with a 300 : 1 Nozzle," *Journal of Spacecraft*, Vol. 22, No. 2, 1985, pp. 188–194.

[18]Saucereau, D., Beaurain, A., and Soler, E., "Demonstration of Carbon-Silicon Carbide Novoltex Reinforced Composite Nozzle for LH_2/LO_2 Engine," AIAA Paper 1990-2180, 1990.

[19]Schmidt, G., Langel, G., and Zewen, H., "Development Status of the ARIANE 5 Upper-Stage Aestus Engine," AIAA Paper 1993-2131, 1993.

[20]Warren, D., and Langer, C., "History in the Making—the Mighty F-1 Rocket Engine," AIAA Paper 1989-2387, 1989.

[21]Vilja, J. O., Briley, G. L., and Murphy, T. H., "J-2S Rocket Engine," AIAA Paper 1993-2129, 1993.

[22]Tucker, P. K., and Croteau-Gillespie, M., "Combustion Devices Technology Team: An Overview and Status of STME-Related Activities," AIAA Paper 1992-3224, 1992.

[23]Jorant, P., "ARIANE 5 Family," AIAA Paper 1993-4131, 1993.

[24]Bonniot, C., "The Vulcain MK2 Engine for Ariane Performance Improvement," AIAA Paper 1992-3454, 1992.

[25]Nyden, O. B., and Rosendahl, B. G., "Development of the HM60-Vulcain Nozzle Extension," AIAA Paper 1991-2566, 1991.

[26]Hartmann, G., Menne, S., Schroder, W., and Weiland, C., "Strömungsfeldberechnungen und Leistungs Vorhersage von fortgeschrittenen Raketendüsen," *DLGR Jahrbuch*, Paper 92-03-047, DGLR-Jahrstgung, Bremen, Germany, 29.09-02.10., 1992.

[27]Schmucker, R. H., "Side Loads and Their Reduction in Liquid Rocket Engines," Technical Univ. of Munich, Rept. TB-14, Munich, Oct. 1973.

[28]Caporicci, M., Eriksson, L. E., Pekkari, L. O., Onofri, M., Popp, M., and Weiland, C., "Advanced Nozzle Technologies for the ARIANE 5 Vulcain Engine," AIAA Paper 1994-3263, 1994.

[29]Bouvier, F., Gaillard, R., Rigoulet, C., and Séverac, N., "Essai de refroidissement par film sur une plaque plane," ONERA, Rept. June 1994.

[30]Bouvier, F., Séverac, N., and Rigoulet, C., "Essai de refroidissement par film sur une plaque plane," ONERA, Rept. Feb. 1995.

[31]Balageas, D. L., Boscher, D. M., Deom, A. A., Fournier, J., and Gardette, G., "Application de la thermographie passive et stimulée à la mesure des flux thermiques en soufflerie," *La Recherche Aérospatiale*, Vol. 4, 1991, pp. 51–72.

[32]Leclère, F., and Aupoix, B., "Hypersonic Turbulent Non-Equilibrium Reactive Nozzle Flow Calculations," 19th Congress of the International Council of the Aeronautical Sciences, Anaheim, CA, Sept. 18–23, 1994.

[33]Aupoix, B., "Calcul des couches limites compressibles bidimensionnelles: Programmes CLIC et EQUI," CERT/ONERA, Rept., July 1991.

[34]O'Brien, C. J., "Dual Nozzle Design Update," AIAA Paper 1982-1154, 1982.

[35]Oechslein, W., "Extendable Nozzle for the HM60-Vulcain Rocket Engine," *Fluid Dynamics and Space*, Belgium, June 1986.

[36]Horn, M., and Fisher, S., "Dual-Bell Altitude Compensating Nozzles," NASA N94-23057, 1994.

Simulation and Analysis of Thrust Chamber Flowfields: Storable Propellant Rockets

Dieter Preclik,[*] Oliver Knab,[†] and Denis Estublier[‡]
EADS Space Transportation, Munich, Germany

and

Dag Wennerberg[§]
Tecosim, Rüsselsheim, Germany

I. Introduction

FOR over four decades, analytical modeling of liquid propellant rocket combustion processes has been pursued by the space propulsion community, with the goal of improving the general understanding of such systems and providing guidelines for their design and optimization. In the early 1960s, for example, a widely accepted attempt was made by Priem and Heidmann[1] suggesting propellant vaporization as a design criterion for rocket engine combustion chambers. Though limited to a simplified, one-dimensional description of reactive flows in liquid/gas bipropellant rocket motors, their model could show and explain fundamental effects of propellant properties, spray conditions, chamber geometry, and operating parameters on combustor performance and efficiency whenever propellant vaporization could be considered as the rate-controlling step. Several years later Breen et al.[2] investigated injection and combustion of

---

*Manager, Systems Technology and R & D Programs, Propulsion and Equipment. Senior Member AIAA.

†Manager, Combustion Devices Engineering. Senior Member AIAA.

‡Propulsion Engineer.

§Senior Engineer, Computational Fluid Dynamics.

hypergolic propellants and developed a combustion model for reactive stream impingement. A mixing vs blowapart relationship was formulated for various types of impingement elements and correlated with propellant injection parameters. Likewise, Breen and his co-authors wanted to reduce the degree of intuition and trial-and-error testing inherent until then in the design process of storable bipropellant injectors.

During the 1970s and 1980s, attempts at modeling rocket combustion processes became more and more sophisticated, employing proven finite difference schemes for solving the fully coupled fluid dynamic equations. A typical example—even it deals with liquid oxygen and gaseous hydrogen combustion—was the Advanced Rocket Injector Combustor (ARICC) code developed by Liang.[3] In trying to resolve the complex mechanisms of propellant injection, atomization, mixing and burning, however, most of the modeling efforts at that time were too ambitious and actually not suited to provide practical support for the design engineers in overcoming their technical problems.

In the recent past, increasing competition and cost reduction in the space propulsion business has led industry to reconsider the conventional trial-and-error approach based primarily on exhaustive experimental testing as sole engineering methodology in rocket engine development. Although commercial computational fluid dynamics (CFD) codes simulating turbulent spray combustion and flow evolution had already been integrated into the design process of aircraft propulsion systems and components, they were not yet ready to be used in the design process of rocket engines. As most available spray combustion codes were able to treat evaporation and burning of a single liquid fuel component within a gaseous oxidizer environment, they were not applicable to most rocket launcher and spacecraft propulsion systems, where two liquid propellants are injected and burned. Among such rocket propellant systems, one will typically find hypergolic storable propellants using hydrazine or its derivatives, monomethyl hydrazine and unsymmetrical dimethyl hydrazine (MMH, UDMH), as fuel and dinitrogen-tetroxide (NTO) as oxidizer. Examples are the Viking engine of the first and second stage of the Ariane 4 launcher and its liquid strap-on boosters (NTO/UDMH), the upper stage Aestus engine (NTO/ MMH) for the Ariane 5 launcher, or the different kinds of orbit and attitude control engines operating in the low-thrust regime below 500 N (see Fig. 1).

The analytical treatment of spray combustion comprises many physical disciplines, such as aerothermodynamics, multiphase flows, turbulence, and multispecies chemistry, as well as numerical solution techniques. In most of these disciplines, simplified standard models have evolved over the past that are applicable to a wide range of conventional flow situations and constitute the current state of the art. Unfortunately, these models and their complex interaction have not yet been sufficiently validated in the combustion environment of a rocket motor. With regard to turbulent spray combustion of liquid bipropellants, there is only a small amount of work available worldwide, including work by Chiu et al.,[4] Larosiliere et al.,[5] Tang and Schuman,[6] and Jiang et al.[7] More recently, liquid bipropellant rocket combustion modeling has been undertaken by the authors within a European technology research program funded by the European Space Research and Technology Center, ESA/ESTEC, which led to the development of the liquid bipropellant code Rocket Combustion Flow Analysis Module

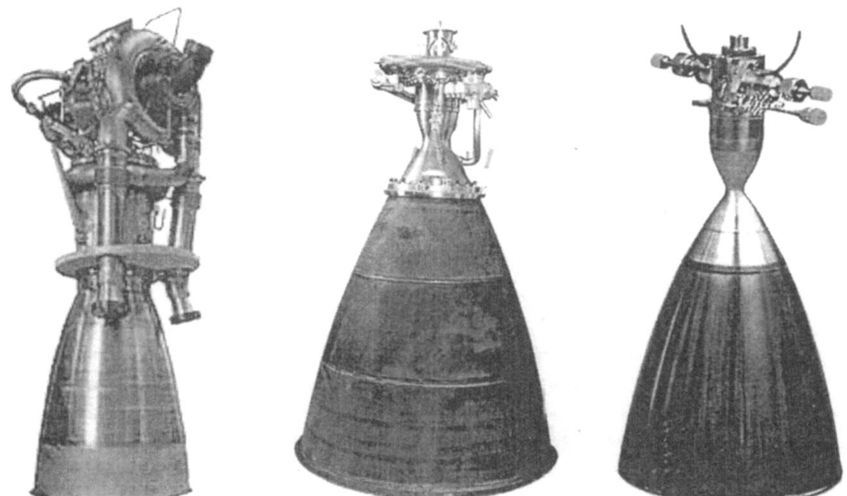

Fig. 1 Typical representatives of storable bipropellant rocket engines (scales not the same): the Ariane 4 Viking engine (left), the Ariane 5 upper stage Aestus engine (middle), and a 400-N apogee thruster (right).

(ROCFLAM).[8,9,10] In contrast to most of the previous work to be found in the literature, the authors aimed at applying their spray combustion code to a wide range of storable bipropellant rocket engines, thus exploring the capabilities of such a tool and demonstrating its usefulness in supporting the design and optimization process of modern rocket engines.

II.   General Aspects of Modeling Storable Propellant Combustion
A.   Propellant Properties and Chemistry

In almost every discussion of spray combustion modeling, an understanding of the thermodynamic and transport properties of propellants, which is the most fundamental prerequisite for such activities, is presupposed and the issue is quickly passed. Reality, however, shows that concerning hypergolic propellants there is only a very limited database available today, particularly with respect to the liquid regime. References 11 and 12 show that many properties, such as liquid density, heat capacity, vapor pressure, and latent heat of vaporization, have not been thoroughly studied at elevated temperatures, that is, beyond about 100°C (212°F). In this context, it is important to note that many property correlations for liquid hypergolic propellants completely fail if they are applied to higher temperatures, as needed, for instance, when dealing with the process of droplet evaporation in a hot gas environment. Figure 2 illustrates how fast a 50 μm hydrazine and NTO droplet is heated to temperatures beyond 100°C when exposed to a combustion chamber environment characterized by 2500 K gas temperature and 10 bar system pressure (no chemical reaction assumed). The figure also

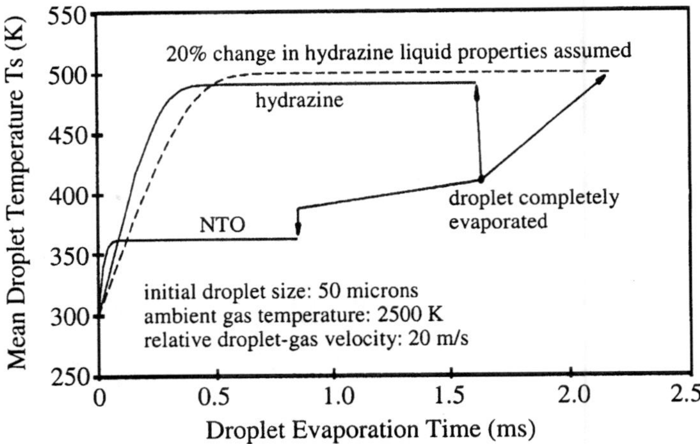

Fig. 2   Mean droplet temperature for hydrazine and NTO vs droplet evaporation time (infinite conductivity model).

contrasts the influence of a 20% inaccuracy in hydrazine liquid properties (here, increase in heat capacity and latent heat of vaporization) with respect to the overall droplet evaporation time, providing for a 35% increase in total lifetime.

Whereas modeling of most processes and mechanisms associated with tracking of the dispersed droplet phase in a rocket combustor, such as droplet evaporation, droplet breakup and coalescence, droplet-to-wall interaction, and liquid film buildup at the combustor walls, depends primarily on the quality of the liquid property database, the evolution of the reactive core flow is primarily governed by the gaseous phase properties. These properties are evaluated from the local species composition of the flow and hence depend on the local presence of evaporated oxidizer and fuel as well as on the reaction rate model and the associated chemistry scheme. The chemistry scheme in turn defines the individual species to be treated in the combustion process and prescribes the local energy release available to heat the various reaction products. Actually, a comprehensive modeling of hypergolic propellant combustion would involve a multitude of elementary reactions and require detailed information about the various mechanisms of reaction initiation, propagation, branching, and termination, which is neither feasible today nor practical for most engineering purposes. As a result, most analysts still pursue a simplified global chemistry representation by prescribing only a few governing reaction steps that approximate the real chemistry mechanisms in a combustor. This procedure calls for a careful selection of the governing reactions with their associated species for the situation of interest.

B.   Injection and Atomization

The injection principles employed in today's storable propellant combustors range from splash plate and pintle elements to various forms of impingement and coaxial-swirl-type configurations. These injectors are usually designed for injection velocities on the order of 10–20 m/s and utilize mechanical forces to disintegrate the liquid jets with the aim of providing a well-premixed spray pattern of reasonably small droplets. Priem and Heidmann's vaporization design criterion[1] suggested that the smaller the droplet sizes are, the higher the total vaporized mass of the propellant component and consequently the performance efficiency for a given geometrical combustor length. In their model, however, they had to assume that the vaporization process of one propellant component is controlling the reaction while the other component is already completely vaporized. Their model could not therefore be directly applied to combustors, which are controlled by incomplete vaporization of both fuel and oxidizer, as is the case in typical hypergolic liquid bipropellant rocket engines. In a recent investigation, Jiang et al.[7] showed that the conventional design concept linked to the common principle "the finer the spray, the better the efficiency" needs to be reassessed for liquid bipropellant systems. They simulated the combustion performance of the variable thrust engine (VTE) for different polydispersed MMH/NTO spray patterns and found highest chamber pressures when the characteristic mean droplet size (here, D_{30}) for the NTO spray pattern was around 40 μm, as compared with 20 μm for MMH. In addition to this, they predicted a drop in VTE performance of about 32% when the NTO spray pattern was shifted toward smaller droplets ($D_{30} = 10$ μm). They concluded that the primary atomization conditions for liquid bipropellant engines are most likely the dominant parameters that affect combustion efficiency.

Jiang's results clearly demonstrate that modern spray combustion codes need to treat fuel and oxidizer injection properties, i.e., droplet size classes, droplet injection velocities and orientation, separately and in as much detail and as accurately as possible. Moreover, collective bipropellant droplet interaction mechanisms are important phenomena that influence the local distribution of the dispersed phase within the combustor as well as the evolution of the reactive core flow. Unfortunately, the process of liquid jet disintegration and atomization is still incompletely understood today, and no generally applicable model is yet available for implementation in engineering CFD tools. Hence, in the near future typical models for injection and atomization will still have to be based on tentative approaches featuring various degrees of empiricism. Viable support, however, may come from today's advanced optical measurement techniques that may be employed to better characterize a specific injector configuration vs its operational envelope, thereby filling the present modeling gap between propellant injection and droplet formation.

C. Cooling Principles

Usually, hypergolic launcher propulsion systems do employ either a combination of film and radiation cooling, sometimes film and ablative cooling, or just regenerative cooling for their main combustion chamber. In most small hypergolic propellant spacecraft engines, however, cooling is typically supported only by a liquid film in the cylindrical part of the combustor and by radiation in the convergent and divergent part of the nozzle. In practice, the principle of internal film cooling requires the maintenance of a buffer layer of liquid coolant between the hot combustion gas and the chamber wall. From a modeling point of view, this kind of cooling is quite sophisticated to predict because three physical phases have to be considered simultaneously: two dispersed droplet phases moving through the reactive gaseous phase and impinging onto the liquid cooling film, as well as one annular film phase describing the interactions of the liquid film with both the combustion gas and the chamber wall.

About a decade ago wall cooling performance analysis was largely based on empirical correlations commonly derived from dimensionless numbers. The underlying prerequisite of flow similarity, however, usually caused the failure of predictions of the wall heat transfer characteristics, particularly in those parts of a combustor where the injector and its related propellant spray pattern govern the evolution of the wall boundary-layer flow. As a result, correction factors sometimes twice as large as the predicted values had to be applied to scale the theoretical heat fluxes onto the experimentally observed ones. For some cooling principles, such as film cooling, the lack of adequate physical models even forced engineers to evaluate wall heat transfer characteristics almost entirely from subscale or even full-scale chamber testing.

III. Liquid Bipropellant Spray Combustion Modeling

In this chapter the authors' approach to storable liquid bipropellant spray combustion modeling is briefly outlined. The main goal of this approach was to compile state-of-the-art models, each of which were deduced with different motivations, integrate them into a comprehensive code, and then employ the final tool to different, real thrust chamber situations, verifying the analytical results with corresponding experimental data from in-house research or from the literature. The goal was not therefore the accurate resolution of each single subprocess, starting with liquid jet breakup, followed by droplet deformation, breakup, and vaporization, and ending with gaseous species mixing and turbulent combustion, but the ability to reflect the global operational behavior of storable liquid bipropellant propulsion systems with reasonable confidence. Because the program objective focused on an engineering tool rather than a research code per se, the selected models had to achieve a balance between the details and accuracy they promised and the computational effort they required.

A. Gas-Phase Flow Modeling

The present approach treats the gaseous phase in a two-dimensional or axisymmetric fashion by solving the Favre-averaged Navier–Stokes equations extended by the species continuity and k-ε turbulence equations. The latter include appropriate modifications accounting for compressibility effects and handle the near-wall region either by a logarithmic wall function approximation or by a two-layer approach. The energy equation employs the sensible static enthalpy and contains source terms for pressure, heat transfer between gas and dispersed droplet phases, film evaporation, and gas-phase reaction temperatures. Mass conservation is enforced with the SIMPLE (Semi-Implicit Method for Pressure Linked Equations) procedure adapted to compressible flows. The set of equations is discretized according to the finite volume methodology for nonorthogonal, boundary-fitted, multiblock grids and solved by an implicit algorithm. Both central and upwind differencing schemes are applied. To allow for independent grid refinement in both grid coordinate directions, a multiblock procedure is included, which is a nonoverlapping method with matching interfaces.

1. Hypergolic Chemistry and Gas-Phase Combustion

Unfortunately, the literature gives only a little information about the reaction chemistry of hypergolic propellants. For the hydrazine/NTO system, Sawyer and Glassman,[13] for instance, proposed a two-step global reaction scheme without, however, considering the process of propellant decomposition. The latter is actually very important for combustion systems that operate under fuel-rich conditions, as is usually the case in rocket engines. Eberstein and Glassman[14] studied especially the hydrazine decomposition mechanism for low reactant concentrations at temperatures around 1000 K, identifying mainly ammonia, hydrogen, and nitrogen as decomposition products. Gaseous NTO decomposition was described in more detail by Svehla and Brokaw.[15] Based on these findings, the following simplified global reaction scheme was chosen for the hydrazine/NTO system (see also Ref. 5):

$$\text{Hydrazine Decomposition:} \quad N_2H_4 \longrightarrow NH_3 + 0.5H_2 + 0.5N_2 + 4.43 \text{ MJ/kg}_{N_2H_4}$$

$$\text{NTO Decomposition:} \quad N_2O_4 \longrightarrow 2NO_2 - 0.55 \text{ MJ/kg}_{N_2O_4}$$

$$\text{One-Step Oxidation:} \quad NH_3 + 0.5H_2 + NO_2 \longrightarrow 2H_2O + N_2 + 27.2 \text{ MJ/kg}_{NH_3}$$

$$H_2O\text{-Dissociation:} \quad H_2O \longrightarrow H^+ + OH^- - 27.7 \text{ MJ/kg}_{H_2O}$$

For reasons of simplicity, it is assumed that both propellants actually decompose instantaneously upon evaporation. Although this approximation is quite accurate for NTO, it may be somewhat too straightforward for hydrazine, leading to a stimulation of the global combustion process by shifting the onset of reaction closer to the injector.

For the MMH/NTO system, the literature provides hardly any feasible global chemistry model. Hence, a simplified reaction scheme was developed and verified by an extensive sensitivity study covering different propellant injection and chamber operational conditions. In principal, the global MMH/NTO combustion model was based on four individual reaction steps, of which three run sequentially. In the first step, liquid MMH and NTO is assumed to instantaneously decompose upon evaporation, similar to the model of the hydrazine/NTO system. However, for MMH two different decomposition reactions have been employed depending on the local temperature of the flow, i.e., decomposition 1 for the hot core flow condition and decomposition 2 for the relatively cold condition in the boundary layer along the wall (see also Ref. 14). Then, a two-step reaction mechanism was imposed for the gaseous methane, comprised of a turbulence-controlled exothermic and a kinetically controlled endothermic reaction, in such a way as to approximate equilibrium chemistry conditions under complete combustion:

MMH Decomposition 1: $MMH \rightarrow CH_4 + H_2 + N_2 + 3.67 \ MJ/kg_{MMH}$

MMH Decomposition 2: $MMH \rightarrow HCN + 2.5H_2 + 0.5N_2$
$- 0.85 \ MJ/kg_{MMH}$

NTO Decomposition: $N_2O_4 \rightarrow 2NO_2 - 0.55 \ MJ/kg_{N_2O_4}$

CH_4 Reaction Step 1: $CH_4 + 2.3NO_2 + H_2 \rightarrow 3H_2O + 1.15N_2$
$+ 0.4CO + 0.6CO_2$
$+ 62.71 \ MJ/kg_{CH_4}$

CH_4 Reaction Step 2: $CH_4 + 0.5CO_2 + 0.5H_2O \rightarrow 1.5CO + 2.5H_2$
$- 14.16 \ MJ/kg_{CH_4}$

H_2O-Dissociation: $H_2O \rightarrow H^+ + OH^- - 27.7 \ MJ/kg_{H_2O}$

Whereas fast turbulent reaction rates are calculated by means of Magnussen's eddy dissipation concept, an Arrhenius type of reaction model is utilized whenever local combustion is controlled by kinetic processes rather than turbulent ones. This applies primarily to the highly endothermic dissociation of H_2O, which is known to become important in combustion systems at temperatures beyond 3000 K. A kinetically controlled mechanism was likewise chosen for the methane reaction step 2. With these assumptions, the global MMH/NTO chemistry model was found to provide very good values for rocket engine exhaust velocities and chamber pressures.

2. Wall Boundary Layer

In turbulent reactive flows, extremely large velocity gradients exist near solid walls because of the nonslip condition at the wall and the high velocities in the fully turbulent free stream flow slightly away from the wall. In this thin viscous layer, the flow also undergoes a transition from fully turbulent to laminar conditions in the sublayer closest to the wall. An accurate numerical treatment of the entire viscous flow structure within a boundary layer has to appropriately resolve such phenomena. For practical applications, three different approaches have principally evolved comprising wall functions, low Reynolds number models, and two-layer models.

The wall function approach is based on so-called bridging functions (logarithmic-wall law) and is actually designed to avoid a detailed numerical resolution of the flow structure in the laminar sublayer. Though these models are most widely used in today's engineering tools, as they absorb a minimum of additional numerical resources, their prediction capability, in particular for wall heat transfer problems, is often unsatisfactory.

Contrary to the wall function approach, low Reynolds number models aim at solving the boundary-layer flow from the turbulent freestream edge down to the laminar sublayer. In general, these models require a very fine grid resolution as the wall is approached. Quite often, more than 100 grid points are utilized just to represent a small fraction of the total flow area in an adequate manner. As a result, three major difficulties arise in practice: first, the responsiveness of the code drops because of the enhanced requirements in computational resources limiting the potential of extending other physical submodels that have an important effect on the flowfield evolution; second, the thickness of the near-wall control volumes gets very small, thus violating a major prerequisite for two-phase, dispersed droplet flows, i.e., the volume fraction of the droplet phase has to be small in each computational cell; third, the aspect ratio of the near-wall control volumes becomes very large, thereby impairing the accuracy of the numerical scheme as well as perturbing the code convergence behavior. Therefore, low Reynolds number models do not appear to be perfectly suited for spray combustion simulations as pursued herein.

The two-layer approach tries to make a reasonable compromise between the physical description of the boundary-layer flow and the corresponding demands on computational resources. The model treats the boundary layer as consisting of two regions, the viscous sublayer close to the wall with $y^+ < 30$ and the turbulent boundary layer ranging from $y^+ > 30$ to about $y^+ \approx 300$. Whereas the standard k-ε model describes the outer region of the boundary layer closer to the wall, the dissipation equation is replaced by an algebraic relation and a specified length scale prescription. The length scale relation is employed to determine the eddy viscosity taking into account the dampening effect of the wall. A more detailed description of the two-layer methodology is given by Rodi.[16] Approximately 10 nodes are typically placed within a wall distance of $y^+ \approx 100$ as recommended in the literature, saving many numerical resources in comparison to low Reynolds number models.

To enable a more accurate simulation of wall heat transfer processes, a two-layer model as just described was added to the standard wall function approach and validated in a first step using Stoll and Straub's calorimeter experiments.[17] There, a two-dimensional convergent-divergent nozzle was employed to determine local wall heat fluxes of a single-phase, heated airflow under isothermal wall conditions. It was found that the two-layer model provides good agreement with the experimental measurements whereas the logarithmic-wall law underestimated these measurements with an increasing tendency toward the supersonic part of the nozzle. It was interesting to note that in the case of the logarithmic-wall law, the wall heat flux evolution could be satisfactorily improved when using the total enthalpy of the wall adjacent cell in the heat flux relation instead of the static one. This correction emphasized that the dynamic contribution of the enthalpy by the velocity of the flow in the first cell outside the sublayer becomes nonnegligible, especially as the trans- and supersonic sections of the nozzle are reached. This, however, demonstrates as well that the standard logarithmic-wall law cannot be used for heat flux calculations in nozzles without proper modifications. In practice this means that this type of boundary-layer model is a priori applicable only to flows where the static-to-dynamic enthalpy ratio is high, as is the case in low Mach number combustion chamber barrel sections.

B. Dispersed Phase Modeling

1. General Description

The liquid-phase solver, i.e., the droplet tracking module, uses a sequential Lagrangian approach integrating the force balance on the droplets. The liquid phase is composed of a discrete number of droplets of different sizes that are traced individually in the flow over their complete lifetime. This modeling approach is based on the assumption that the particle phases are sufficiently dilute so that particle-particle interactions and the effects of the particle volume fraction on the gas phase are negligible. In practice this means that the dispersed phases must be present at fairly low volume fractions (less than $\approx 10\%$). Consequently, any treatment of liquid jet disintegration or primary atomization is beyond the scope of this approach.

The effect of flow turbulence on droplet motion is simulated through a sto-chastic separated flow (SSF) model. It assumes that the droplets interact with a random distribution of eddies along their trajectories. Whereas the mean flow velocity is obtained by solving the Favre-averaged momentum transport equations, the fluctuating term is sampled from a Gaussian velocity distribution function characterized by the fluid root-mean-square value, which in turn is computed from the turbulent kinetic energy. To solve the droplet equations of motion, an explicit first-order integration scheme is used with an adaptive time step computed from the minimum of the eddy lifetime, the droplet motion characteristic time, the droplet characteristic vaporization time, the time for a droplet to cross an eddy, and finally the time for a droplet to cross a control volume. For this time step a minimum value is specified to prevent stationary droplet trajectories for too-small turbulence scales. It can be shown that the flow fluctuations affect primarily the smaller droplets. The trajectories of the large droplets are basically unaffected because their characteristic response times are much larger than the Lagrangian time scale.

To keep the computational time within reasonable limits, several droplets are supposed to have exactly the same behavior in the flow. Such a parcel of physical droplets is treated as a fictitious or computational droplet. Because a stochastic model is employed to simulate the effect of velocity flowfield fluctuations on droplet motion, a certain number of identical parcels has to be specified. Experience reveals that tracking 10 times the same parcel gives a reasonable estimate of the droplet turbulent diffusion process in most cases, without being too large a burden on the computational time. In contrast, however, a reasonably accurate prediction of the buildup of a liquid cooling film along the wall requires two orders of magnitude more computational droplets with the same properties. Resolving a propellant spray pattern by, for instance, 40 different droplet size classes will require some 80,000 parcels (2 propellants $\times$ 40 classes $\times$ 1000 parcels) to adequately handle a liquid film cooling situation.

2. Injection and Primary Atomization

The process of propellant injection and primary atomization is known to govern the flowfield development and combustion phenomena in a rocket motor. Although much effort has been put into studying these processes in the past, a satisfactory understanding of the physical mechanisms taking place during liquid jet breakup and droplet formation has not yet been achieved. Hence, practical models of primary atomization to be utilized within today's spray combustion codes are limited to experimental and analytical data on droplet size and velocity distributions characterizing a specific type of injector (coaxial, impinging, swirl, etc.). Such models assume that the injection mechanism creates an instantaneous atomization pattern of sparse spherical liquid droplets. However, this simplification can be valid only at some downstream distance from an injector, where the liquid core, surrounded by a dense spray, is completely disintegrated. Depending on the obtained mass mean diameters as well as the spread in size for the distributions, the combustion efficiency can vary considerably; incomplete combustion may be present and injector plate burnout may occur due to strong flow recirculation. Undesirable situations like these have been observed during many rocket engine development programs.

To avoid such problems during the design of an injector, there are more parameters to monitor than the atomization mass mean diameter per se. A very important parameter, for example, is the quality of the oxidizer spray relative to the fuel spray. For the case of a liquid hydrazine/NTO bipropellant thruster, it is particularly desirable to generate superimposed vaporization patterns to enhance chemical reaction. Because NTO evaporates much faster than hydrazine, the fineness of the corresponding sprays had to counterbalance potential differences in propellant vaporization rates, the latter being a function of the individual vapor pressures and surface tensions.

Unfortunately, representative data on droplet size and velocity distribution are scarce, and information corresponding to the conditions typically prevailing in rocket engines is difficult to find. At the current level of knowledge, one usually has to extrapolate cold-flow spray measurements made with simulation fluids instead of real propellants to obtain the required input for an analytical simulation. With regard to the velocity distributions of droplets within a spray, no suitable information at all is available at the present time.

3. Droplet Vaporization

Droplet vaporization has been described by an infinite conductivity model following Abramzon and Sirignano's theory.[18] Main assumptions are that droplets need to be spherical and to have a uniform but not constant temperature. Consequently, the model does not account for heat circulation inside a droplet due to temperature gradients, but the heat-up phase, which can take up to 25% of the total droplet lifetime, is considered, as well as its influence on the liquid properties, especially the liquid density, which creates a slight increase in diameter during the heat-up phase. During this phase, the vaporization rate is limited by the gaseous film resistance to mass and heat transfer. When the wet-bulb temperature is reached, the vaporization rate is controlled by the heat transfer capabilities of the film surrounding the droplet. In practice, the model accounts for the resistance to heat and mass exchange between the droplet surface and the hot gas flow by incorporating a quasi-steady vapor film of a certain thickness around the droplet. This thickness is modified to account for the blowing effect on the droplet film due to the high relative velocity of the droplet with respect to the gaseous flow. The highest relative velocities are obtained for the largest droplets, which are more inert and hence adapt more slowly to the gaseous flow velocity. For reasons of simplicity, it has also been assumed that the fuel droplets evaporate in a pure oxidizer environment and the oxidizer droplets evaporate in a pure fuel environment. The gaseous film around the droplets is then uniquely composed of either oxidizer or fuel.

4. Secondary Atomization and Droplet Breakup

As discussed in Section II.B, the choice of the initial droplet size distribution has a strong effect on engine performance. To partially overcome this undesired sensitivity to input data, a secondary atomization model has been developed with the objective of including as much physics as possible without increasing the computational cost of the droplet tracking procedure. The method used is based on a Taylor analogy of the evolution of a damped spring-mass system with external excitation. A second-order nonlinear differential equation for the nondimensional deformation of an oblate droplet is derived. Testing of this model in a uniform flow has yielded important results on droplet behavior and on the influence of the droplet liquid properties. Because of the relative motion of the droplet in the flow, shear pressure forces tend to deform the liquid droplet in either an oscillatory damped regime or a breakup regime, depending on the droplet Weber number. Consequently, the droplet breakup mechanism cannot be characterized simply by means of a Weber number, which tells whether or not breakup will occur. It has been observed that the breakup regime has a very sharp asymptotic behavior in time for the droplet deformation ultimately defining the breakup time.

After extensive numerical sensitivity studies on the breakup time for different Weber numbers, droplet properties, and flow conditions, a universal droplet breakup time description could be obtained being consistent with other experimental and analytical expressions available in the literature for the linear term. However, a nonlinear correction was found to be important for regimes close to the critical Weber number ($We_{cr} = 17.6$), which are present in liquid rocket engines. In general, the breakup model has shown three important characteristics of the breakup process. A first result is the existence of a critical droplet Weber number above which no steady-state deformation can be sustained and breakup occurs. Then, if a droplet is above the critical number, breakup will not occur instantaneously but will take place if the droplet is maintained above this value for a certain period of time, defined as the breakup time. Finally, the breakup time can reach a nonnegligible fraction of the overall vaporization time. These aspects were considered within a so-called delayed droplet breakup model, which avoided solution of the droplet deformation equation but still satisfied the main physics behind a secondary atomization process.[9]

5. Droplet-to-Wall/Film Interaction

To complete the description of the dispersed phase modeling, a few key notions will finally be given on the droplet-to-wall and droplet-to-film interaction modeling. Concerning droplet-to-wall interaction, two models for dry and wet droplet-to-wall impingement including droplet splashing have been considered. The so-called dry collision regime is actually defined only for low-energy impacts where the vapor layer around the droplet can exert sufficient counteracting pressure over the drop bottom surface as to permanently prevent direct liquid-solid contact. For higher energy collisions, the vapor layer is squeezed strongly between the droplet and the wall until it is broken up by wall surface proturberances. After a certain interaction time, the impingement condition switches from a dry type to a wet type, and the heat transfer rate increases abruptly due to direct

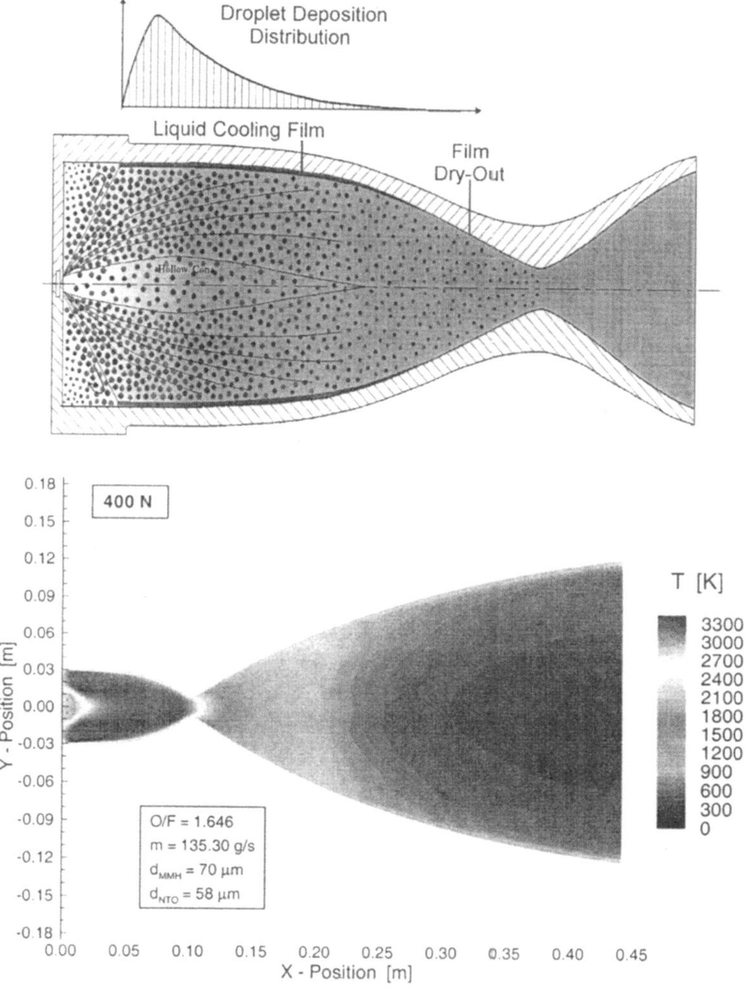

Fig. 3 Sketch of a liquid film cooled 400-N apogee thruster illustrating the droplet flowfield and annular cooling film evolution (top) and temperature contours of a corresponding CFD simulation (bottom). (See also the color section of figures following page 620.)

conduction. In the case of a liquid cooling film generated at the wall, droplet-to-film interaction distinguishes between crown-splashing, absorption, and reflection phenomena, depending on the impacting droplet's size and Weber number.

C. Liquid Film Modeling

In general, a multiphase spray combustion simulation is initiated by the prescription of a certain propellant spray pattern somewhere in the combustion chamber. As a consequence, some of the droplets may reach the combustor wall. If this droplet flow rate is large enough, a liquid film will be built up at the wall (see schematic in Fig. 3). Because of the shear forces between the gaseous core flow and the annular film, such a film moves with the gaseous flow downstream toward the throat. Additionally, the gas mixture starts to heat up and vaporize the liquid film. Both evaporation and acceleration of the film lead to a reduction of the film thickness, which will finally cause the film to disappear at the so-called film dry-out point. As long as a liquid film exists at the wall, it is assumed that the film temperature remains close to the fluid's saturation temperature.

To treat liquid film buildup, transport, and evaporation, the following processes and mechanisms need to be considered: droplet-to-film deposition, film evaporation, gas-to-film and film-to-wall heat transfer, shear forces between the film and the gas phase, as well as hypergolic liquid phase reactions. Because of the lack of modeling information on droplet entrainment, this process has been neglected so far. Hence, momentum exchange between the droplets and the film takes place solely in one direction, i.e., toward the liquid film. The coupling with the gas phase, however, is bidirectional. Heat and momentum are transferred from the gas phase to the liquid film. The transferred energy causes the film to evaporate and transfer mass back to the gas phase.

With regard to the liquid film, mass and heat conservation equations are solved in a one-dimensional form. The velocity profile within the film is assumed to follow the von Karman universal velocity profile. With this assumption the local film thickness can be determined in an iterative way in line with the solution of the mass and heat conservation equations.

D. Computational Efficiency and Flow Convergence Monitoring

The compressible Navier–Stokes flow solver has to handle a set of transport equations for fluid momentum, sensible enthalpy, turbulent kinetic energy and dissipation, and species mass fractions. For the flow with a given distribution of liquid source terms originating from the droplet and liquid film phases, convergence is obtained by an iterative approach whereby all conservation and transport equations are solved sequentially and independently by Stone's implicit procedure (SIP), which is a very fast implicit solution algorithm. To minimize the required CPU time to achieve convergence, monitoring of the process and step-wise adjustments of the convergence parameters are necessary. In conventional single-phase flows, the convergence monitoring uses residual values for mass and momentum normalized by the inlet variables. For rocket engine combustion modeling, it is necessary to compute a multiphase flow problem with or without gaseous inflow. Moreover, because the liquid mass loading is high, especially in the injector plate region, a strong coupling of the liquid droplet and the gaseous phases as well as the liquid film and the gaseous phase is present for the mass, momentum, enthalpy, and species mass fraction conservation equations. These strong couplings require a careful choice of the convergence parameters, such as underrelaxation factors, calling frequency for the recalculation of the liquid droplet and film phases and the number of droplet parcels tracked. A case-independent technique for optimizing these parameters has been developed for the code to produce satisfactory converged solutions with reasonable computational effort. Because the cross-section mass flow rate is not constant, because of evaporation of the liquid phase, the residual for each flow variable is the absolute sum of residuals along a user-defined cross section normalized by the cross-sectional flow rate of the same variable. The final residual value then is the average over all cross sections. Convergence of the overall flow problem is assumed when the various residuals are smaller than a certain threshold, which has been set to 0.001.

IV. Applied Simulations on Liquid Bipropellant Rocket Combustion

This chapter will outline some applied simulations of different storable liquid bipropellant combustion chamber and nozzle flow situations. The selected examples may highlight the general methodology pursued by the authors and show the capability of such a tool in supporting the engineering process. First, a very small hydrazine/NTO rocket combustor will be presented with the aim of generally assessing some polydispersed liquid bipropellant spray combustion performance trends and phenomena. In particular, the evolution of chamber pressure and characteristic exhaust velocity, depending on global mixture ratio and propellant mass flow rates, will be analyzed for this small-scale combustor. The second example will discuss the simulation results obtained for the upper stage MMH/NTO Aestus engine (see Fig. 1), a comparatively large propulsion system. Calculated thrust and heat load values will be compared with available experimental data, demonstrating the representativity of the simulation results over a wide operational range. Finally, a highly complex three-phase flow simulation will be presented for a liquid film cooled apogee thruster in the 400-N thrust range (see Fig. 1). This case highlights the possibilities in the analysis of hypergolic liquid bipropellant spray combustion modeling today. At this point, however, it must be recalled that the objective of all of these simulations has not been the detailed resolution of certain subprocesses—the verification of

Table 1 Operational parameters for selected hydrazine/NTO liquid bipropellant combustor tests (Ref. 19)

Test parameters	Case 1	Case 2	Case 3
Mixture ratio O/F	0.829	0.848	1.048
Total mass flow rate m_t, g/s	213.2	181.0	218.2
Chamber pressure $P_{c.\mathrm{exp}}$, bar	17.9	14.6	18.0
Exhaust velocity C^*_{exp}, m/s	1501	1441	1477
C^*_{exp}-efficiency, %	85.2	81.7	83.3

which is out of reach or impossible anyway—but the reflection of the main flow characteristics of storable liquid bipropellant rocket thrust chambers.

A.   Hydrazine/NTO Small Rocket Combustor Simulations

1.   General Test Case Introduction

The integrated code was subjected first to the hydrazine/NTO liquid bipropellant system employing a very small rocket combustor, extensively tested by Wasserbauer and Tabat[19] in 1961. Actually, three test cases were selected for this purpose; operational parameters are listed in Table 1. While test case 1 was used as reference case, case 2 was chosen to simulate primarily a change in total propellant mass flow rate, and case 3 to simulate primarily a change in mixture ratio. In general, these simulations were intended to demonstrate the combustor's behavior under various operating conditions through comparing experimental and computational global performance parameters, such as chamber pressure and C^*-exhaust velocities.

Originally, the combustor experiments[19] were performed with a capacitively cooled chamber and a multiple concentric tube injector (without recess). The injection elements (21 in number) were uniformly spread over the injector's center area (radius: 6.3 mm). The diameter of the combustor was 26.2 mm (1.03 in.), and the injector-to-nozzle-throat length was 57.8 mm (2.28 in.). The combustor's contraction ratio was 3.

For the numerical simulation, the spray pattern produced by this type of concentric tube injection was approximated by a modified Buschulte[20] correlation for mechanical atomization:

$$D_m/D_{\text{noz}} = 0.618 \, We^{0.333}(\rho_l/\rho_g)^{0.167}$$

Herein, the droplets' mass mean diameter D_m related to the hydraulic diameter of injection nozzle D_{noz} is determined by the Weber number and the ratio of liquid and gaseous density. Moreover, the droplets were assumed to follow a logarithmic-normal type of distribution function (distribution parameter $\delta = 1.171$) with a constant ratio between the maximum and mean droplet diameter in the spray ($D_{\text{max}}/D_m = 3.5$). According to Simmons,[21] these assumptions are suggested for many types of different pressure and air atomizers with reasonably good accuracy.

2.   Combustor Flowfield Development and Performance Efficiency

The Buschulte correlation was utilized to evaluate the mass mean diameters for the three test cases from the injected fuel and oxidizer mass flow rates, the injection temperature of the propellants, and an estimation of the gas condition in the combustor close to the injection plane. The droplets' injection velocity was calculated from the propellant mass flow rate per element and the cross-sectional area of the injection nozzle. For the sake of simplicity, it was assumed that all droplets, belonging to either hydrazine or NTO, have the same mean initial velocity. Table 2 shows the injection conditions as used herein for the three test cases.

Table 2 Injection parameters used for simulation of hydrazine/NTO combustor tests (Ref. 19)

Injection parameters	Case 1	Case 2	Case 3
Mass mean diameter (N_2O_4), μm	76.4	87.2	68.8
Mass mean diameter (N_2H_4), μm	61.7	71.4	64.5
Mean injection velocity (N_2O_4), m/s	13.1	11.3	15.1
Mean injection velocity (N_2H_4), m/s	14.5	12.2	13.3
Axial droplet velocity fluctuation, rms	2.5	2.5	2.5
Radial droplet velocity fluctuation, rms	5.0	5.0	5.0

The logarithmic-normal distribution for both propellants was represented by 10 discrete size classes with droplet diameters of 2, 5, 10, 20, 30, 40, 60, 85, 110, and 140 μm. The maximum size class of 140 μm was selected in such a way that the calculated pressure level for the reference case matches that observed in the corresponding experiment. Because secondary droplet breakup processes were not taken into account in these simulations, the very large droplets ranging from 140 μm up to about 3.5 times the mass mean diameter ($3.5 \cdot D_m$) could not actually be considered.

Initially, a numerical mesh consisting of 15 radial and 50 axial control volumes was set up. Propellant injection was simulated in each of the first seven radial control volumes (radial coordinates 0 to 6.3 mm) by means of a full-size spectrum ranging from 2 to 140 μm per control volume and a constant radial mass flow rate distribution for both hydrazine and NTO. The combustor walls were assumed to be adiabatic in these calculations.

Figure 4 depicts hydrazine and NTO evaporation rates for test case 1. It can be seen that the centers of high evaporation fall almost onto each other, thus furthering the initial gas-phase reaction process. In former calculations with the two-step chemistry model of Sawyer and Glassman[13] lacking the fast hydrazine exothermic decomposition step, fuel evaporation was shifted more downstream of the injector. As a result, initial gas-phase combustion was limited, causing a major recirculation zone to develop in the upper left corner of the combustor. The fast exothermic hydrazine decomposition process as taken into account in the current chemistry model, however, eliminated the problem of low initial reaction rates and near injector recirculation zones.

The gas temperature distribution of test case 1 is plotted in Fig. 5. At the point of injection the gas temperature is about 1500 K. Further downstream the temperature reaches some 3400 K in the core of the combustor, suggesting that the H_2O-dissociation step should be considered in the current chemistry model.

As mentioned earlier, quantitative code validation was expected to be based only on global combustor performance data. In this context, the prime interest was to see whether relative changes in total propellant flow rate and mixture ratio could be properly handled by the code. Table 3 includes the combustor performance data showing that computational chamber pressures as well as C^*-efficiencies are in reasonably good agreement with the values from the original experiments depicted in Table 1. The upper plot in Fig. 6 depicts the computational C^*-efficiencies of cases 1–3 in line with those experienced in the

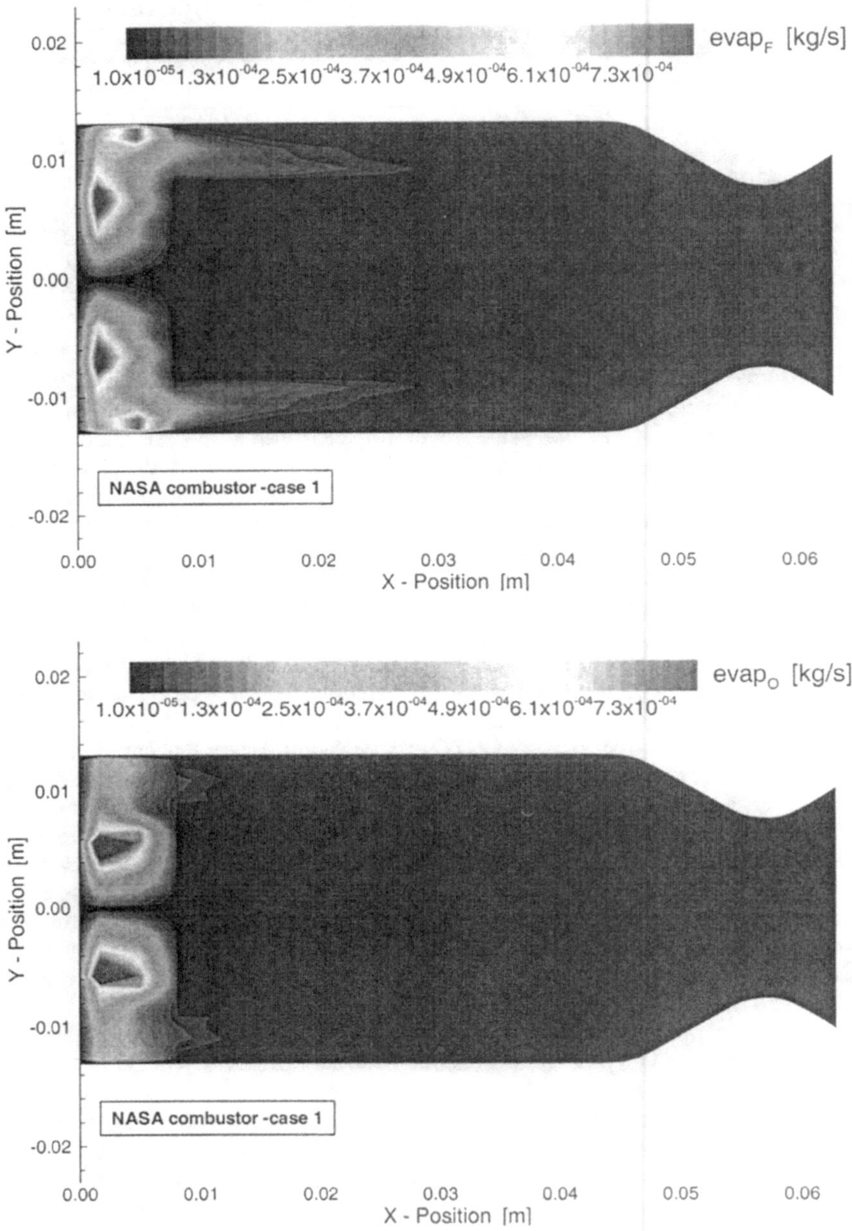

Fig. 4 Calculated hydrazine (top) and NTO (bottom) evaporation rates for combustor test case 1 (Ref. 19). (See also the color section of figures following page 620.)

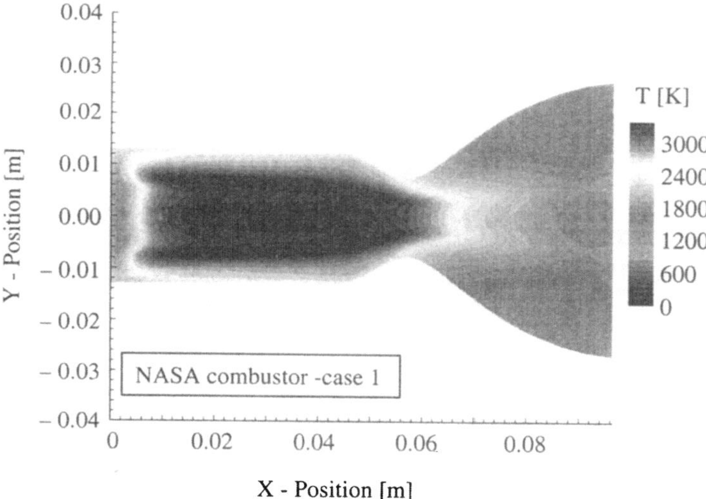

Fig. 5 Calculated gas temperature contours for combustor and nozzle test case 1 (Ref. 19). (See also the color section of figures following page 620.)

original test programs. The simulation results can be seen to lie well within the bandwidth of scattering given by the original combustor test map.[19]

The combustor simulations also revealed that the smaller the hydrazine mass mean diameter (all other parameters being the same), the better the overall combustion performance will be. Actually, this is not a new finding, because it is well known that smaller droplets evaporate faster and consequently reduce any drop in performance due to incomplete propellant vaporization. However, whether this "common engineering rule" held true for the oxidizer component in a liquid bipropellant combustor remained to be investigated. Therefore, a set of simulations was run for test case 1 by reducing the NTO mass mean diameter in a stepwise manner. The second plot in Fig. 6 shows how combustion efficiency develops under these circumstances. One can see that this engineering rule is apparently not directly applicable to NTO. A very fine oxidizer droplet size distribution ($10 < D_m < 30~\mu$m) leads to a major drop in combustion efficiency.

Table 3 Combustor performance results calculated for the three load points under investigation

Simulation	Case 1	Case 2	Case 3
Mixture ratio O/F	0.829	0.848	1.048
Total mass flow rate m_t, g/s	213.2	181.0	218.2
Chamber pressure $P_{c,\text{sim}}$, bar	17.79	14.60	17.72
Exhaust velocity C^*_{sim}, m/s	1503	1453	1459
C^*_{sim}-efficiency, %	85.3	82.4	82.3
$C^*_{\text{sim}}/C^*_{\text{exp}}$	1.001	1.008	0.988

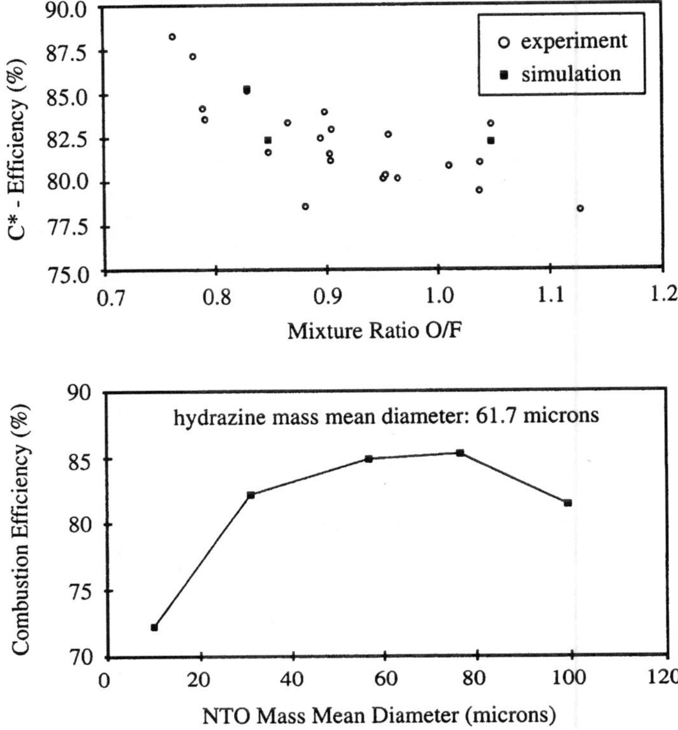

Fig. 6 Comparison of performance predictions with experimental test map (top) and computational combustion efficiency vs mass mean diameter of NTO ($D_{m,\text{hydrazine}}$ = const) (bottom).

This behavior may be explained by recalling the propellant evaporation characteristics depicted in Fig. 2. There, one can see that the process of NTO evaporation is much faster than that of hydrazine because of the higher vapor pressure of NTO. Keeping in mind that NTO evaporation is accompanied by a fast endothermic dissociation into $2 \cdot NO_2$, a too fine NTO spray is likely to remove too much energy from the gaseous flow in the injector vicinity, thereby blocking the hydrazine evaporation process in line with the chemical reaction between the two propellants. Hence, finer NTO spray patterns will shift hydrazine evaporation more downstream of a combustor, eventually leading to incomplete fuel evaporation and thus performance degradation. As mentioned in Section II.B, similar observations were made by Jiang et al.[7] in the context of numerical VTE simulations.

B. MMH/NTO Aestus Engine Simulations with Regenerative Cooling

The AESTUS engine depicted in Fig. 1 is the key element of the upper stage EPS (Etage à Propergols Stockables) of the Ariane 5 launcher. This pressure-fed engine uses the MMH/NTO propellant system and operates at a nominal chamber pressure of about 11 bar (154 psia) and a mixture ratio of 2.05. The Aestus injector consists of 132 coaxial/slot injection elements, arranged in six rows around a centerline element. The main combustion chamber is regeneratively cooled; its diameter is 210 mm (8.27 in.) at the injector and 136 mm (5.35 in.) at the throat; the total chamber length is 591.5 mm (23.3 in.). The metallic nozzle extension is radiatively cooled; its area ratio is 84.

For simulation purposes, three load points were selected from the operational matrix of the Aestus engine (mass flow rate vs mixture ratio): the reference load point R with a mass flow rate of $m = 8.77$ kg/s (19.33 lb/s) at a mixture ratio of $O/F = 2.05$, the envisaged operational load point OP with $m = 9.37$ kg/s (20.66 lb/s) at $O/F = 2.04$, and the extreme load point $Q5'$ with $m = 10.19$ kg/s (22.47 lb/s) at $O/F = 2.20$. In all simulations, the complete nozzle extension was taken into account. Figure 7 depicts the spatially discretized engine grid. A three-block mesh consisting of 73.356 computational cells was generated to realize adequate grid resolution with least computational effort. In particular, the boundary layer in the combustion chamber and the supersonic part of the thruster were better resolved to accurately predict wall heat transfer and performance.

A few specific features were employed when setting up the Aestus case: first, the boundary layer along the thrust chamber contour was resolved by the two-layer model as described in Section III.A.2 involving a computational mesh with a grid fineness characterized by dimensionless wall distances of $0.1 < y^+ < 5$; second, at the face plate the grid has intentionally been made coarser for application of the logarithmic wall function approximation; third, both fuel and oxidizer propellant sprays have been assumed to follow the Rosin-Rammler droplet size distribution, being resolved by 40 discrete droplet size classes ranging from 5 to 200 μm in diameter.

Following the crucial choice of a reasonably representative characteristic mean droplet diameter for the NTO propellant spray, a sensitivity study was conducted on how this parameter affects the near injector temperature stratification. Figure 8 shows the results for two different oxidizer spray patterns. As can be seen, the seven injector rows are distinctively resolved. Enlarging the mass mean diameter of NTO from $D_{NTO} = 110$ μm to $D_{NTO} = 150$ μm deferred the evaporation and combustion process and moved the flame front further downstream. However, because the Aestus engine has some margin with respect to the characteristic chamber length, this variation had no significant effect on the computed engine performance. Figure 8 also shows a view inside the Aestus chamber giving a qualitative idea on this stratification phenomenon.

Because the initial propellant spray characteristics described by droplet size distributions, droplet injection velocities and orientations, and injection locations are known to crucially affect the computational engine behavior, a so-called injector anchoring is among the crucial steps in defining proper initial computational setups. This injector anchoring must cover the envisaged operational range of an engine in such a way that good agreement with experimental results is achieved. In an early stage, injector anchoring may be performed using data from subscale chamber experiments, which later can be refined by more representative full-scale prototype test results. Notice that the obtained

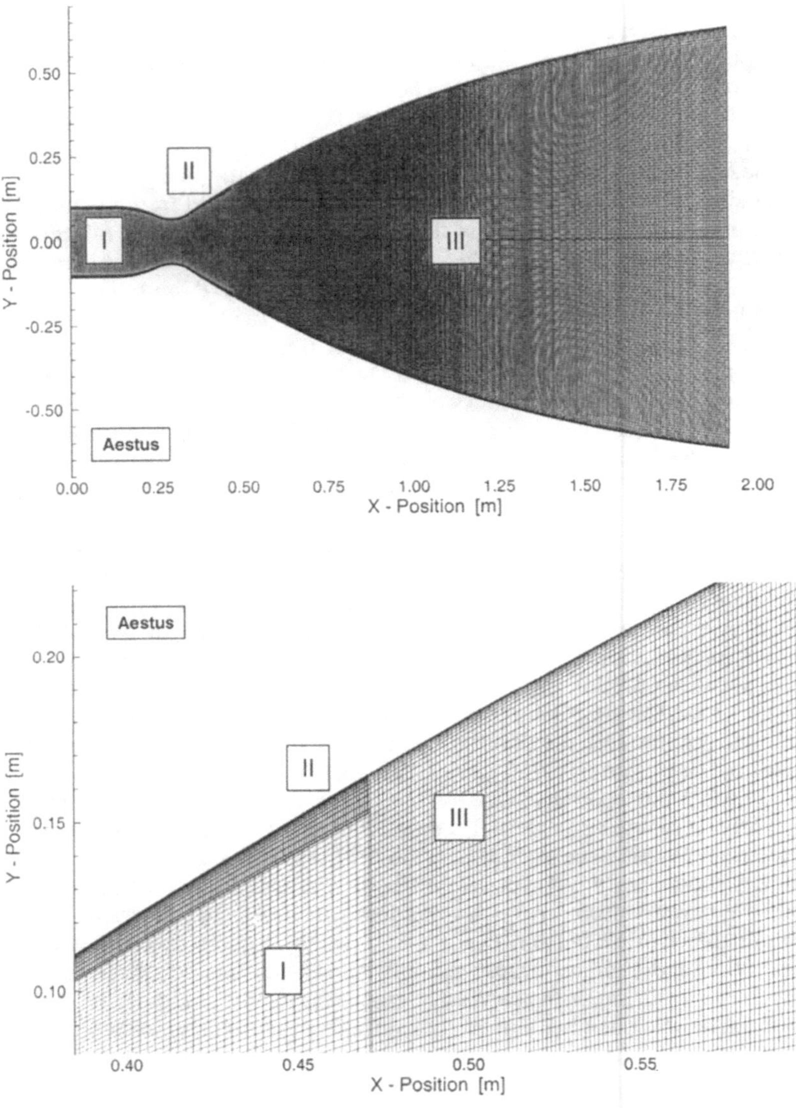

Fig. 7 Computational three-block grid for Aestus thruster chamber simulations with cell aspect ratios below 30 and refined axial resolution in the supersonic part of the thrust chamber.

data are not only representative for the injector or type of injector element under investigation, but also depend on the computational tool and its underlying modeling approach. There is no unique way of doing this anchoring, but once it has been accomplished, simulations can be undertaken to optimize the design of a

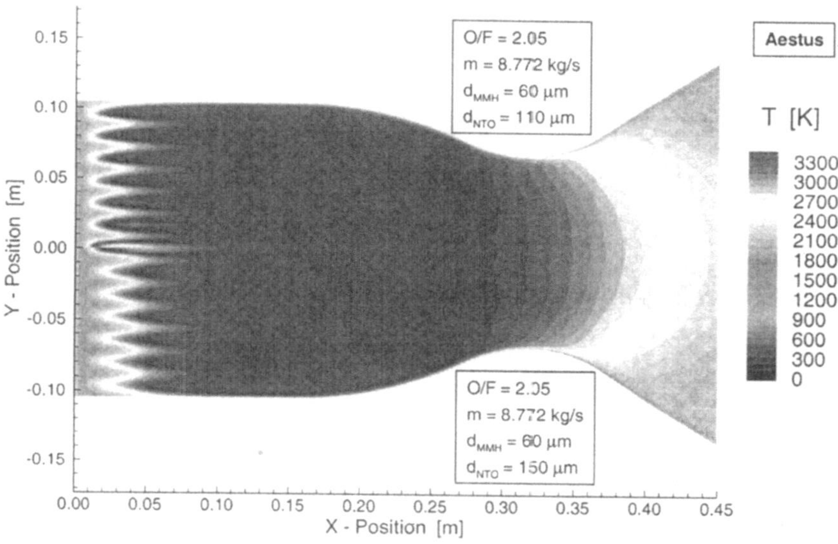

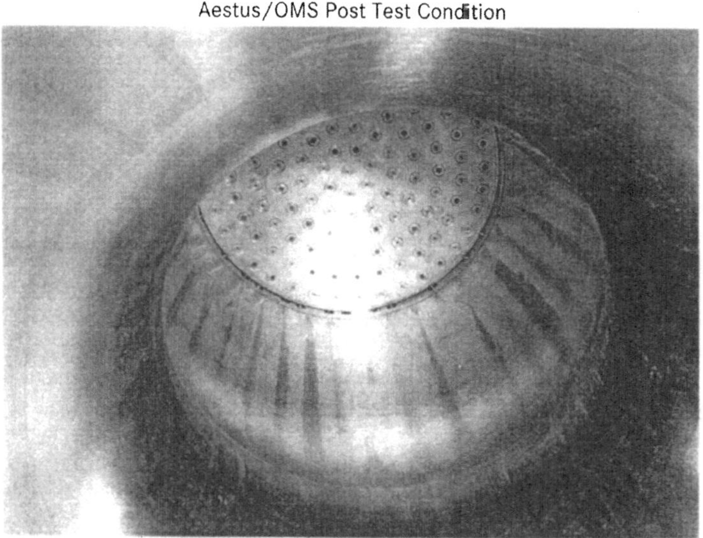

Fig. 8 Comparison of computational temperature stratifications close to the injector plate resulting from two different oxidizer spray patterns: upper half $D_{NTO} = 100\ \mu m$, lower half $D_{NTO} = 150\ \mu m$ (top), and photography taken from tested Aestus hardware illustrating the effect of temperature stratification (bottom). (See also the color section of figures following page 620.)

Table 4   Comparison of predicted and measured performance data for different load points of the Aestus engine

Load point	*R*		*OP*		*Q5′*	
Mixture ratio	2.05		2.04		2.20	
Mass flow rate, kg/s	8.77		9.37		10.19	
	Predicted	*Measured*	*Predicted*	*Measured*	*Predicted*	*Measured*
Chamber pressure, bar	10.3	10.2	11.0	10.9	11.9	11.9
Thrust, kN	27.7	27.8	29.6	29.8	31.9	32.3
Vacuum specific impulse, s	322.1	322.8	322.4	324.2	319.4	323.7

thrust chamber, such as its diameter, length, and material. Limited information can also be gained with regard to the effect of the number of injection element rows and the mass flux per injection element row, that is, the number of elements per row. Such design simulations can contribute to substantially reducing expensive and time-consuming hardware testing.

Table 4 compares predicted and measured performance data for the three Aestus engine load points considered here. In contrast to the previously discussed combustor simulations, constant mean mass diameters ($D_{\mathrm{MMH}} = 60\ \mu$m, $D_{\mathrm{NTO}} = 110\ \mu$m) have been utilized here for all simulations. It is emphasized that no pressure boundary conditions were prescribed for these simulations. The pressure distribution is a direct result of the complete chamber flowfield calculation and hence verifies to a certain extent the applied evaporation and combustion models. Clearly, the analytical predictions agree fairly well with the experimental data. Discrepancies of less than 0.9% for the chamber pressure and less than 1.3% for the thrust and vacuum specific impulse are well within the range of measurement uncertainties. Because the employed MMH/NTO chemistry, as introduced in Section III.A.1, was designed for a global chamber mixture ratio around 1.6, it is not surprising that the extreme load point $Q5'$ with $O/F = 2.2$ resulted in highest deviations here. Moreover, the computational performance figures also indicate that the assumption of keeping the characteristic droplet diameters D_{MMH} and D_{NTO} constant within an operational range defined by $R-OP-Q5'$ may eventually be acceptable for certain design investigations.

In addition to engine performance, local and total wall heat loads are likewise critical engine design parameters. An accurate prediction of the wall heat loading, for instance, is indispensable for the layout of the cooling circuit, which of course depends on the wall temperature and heat transfer properties of the boundary-layer flowfield along the chamber and nozzle contour. In practice, a coupled hot gas to coolant side heat transfer model needs to be established to analyze a regeneratively cooled design task, as in the case of the Aestus engine. Such an analysis delivers the chamber's temperature and heat flux distribution along

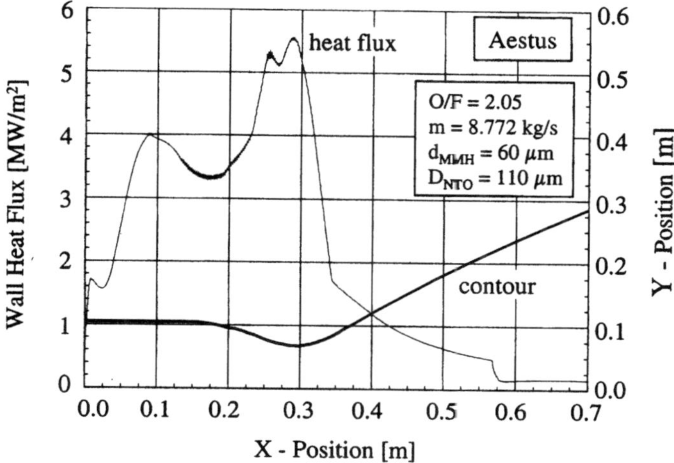

Fig. 9 Calculated heat flux distribution along the regeneratively cooled Aestus combustion chamber (reference load point R).

the wall. An example is given in Fig. 9 for the reference load point R. Two major heat flux maxima can be noted, the first at the location where the flame reaches the wall ($x \approx 0.09$ m, see Figs. 8 and 10) and the second close to the throat ($x \approx 0.29$ m). The low heat fluxes close to the injector head is typical for spray combustion modeling where the buildup of the wall heat load actually corresponds with the onset of combustion. The integration of the wall heat flux profile over the hot-gas side surface results in the total wall heat load, which, in the case of the Aestus simulation, was found to be 940 kW. In comparison with the heat load measurement of about 900 kW during testing, a deviation of only 4.4% is observed. This level of agreement is remarkable given the uncertainties in the characteristic droplet diameters representing the injector spray patterns, and usually far beyond the capabilities of Nusselt-equation based models. These simulations were thus considered to be an important milestone in verifying that hypergolic liquid bipropellant spray combustion modeling is able to describe the global flowfield inside a rocket engine with reasonably good confidence.

Figure 10 presents the calculated external (solid) and internal (dashed) wall temperature distributions for the R and $Q5'$ load points, respectively. The conspicuous temperature rise at the area ratio $\varepsilon = 10$ ($x \approx 0.6$ m) marks the chamber to nozzle interface, i.e., the switch from regenerative to radiation cooling. Because of the small wall thickness and good heat conductivity of the metallic nozzle skirt, the temperature differences between the inner and outer sides are less than 10 K. Especially at the critical interface, good agreement with available temperature measurements on the outer nozzle surface is apparent. At this point it must be emphasized that the energy recovery at the wall depends on the adiabatic temperature of the core flow. The latter, however, is affected by

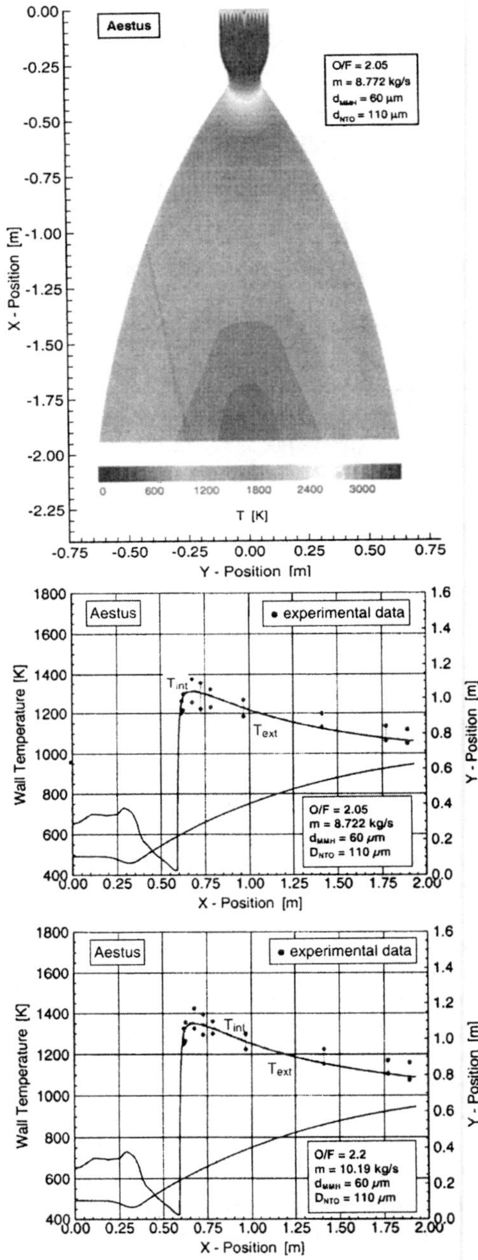

Fig. 10   Temperature contours predicted for the reference load point R (top) and external and internal wall temperature distribution for load point R (middle) and $Q5'$ (bottom). (See also the color section of figures following page 620.)

the imperfect combustion process in the combustion chamber. Hence, ordinary single-phase simulations starting with equilibrium conditions at the face plate are not able to account for combustion efficiency aspects (which are usually estimated for performance predictions,[22] or for accurate wall temperatures along the nozzle skirt. In general, such computations overestimate the nozzle wall temperatures.

Finally, the simulation of the $Q5'$ load point was repeated assuming a ceramic nozzle extension instead of a metallic one. By merely changing the nozzle wall properties in the computational input data set, the results shown in Fig. 11 were obtained. Because of the thicker wall and the reduced heat conductivity of the ceramic material, the temperature difference between the inner and outer side increased substantially ($\Delta T \approx 100$ K). Compared with the metallic nozzle skirt, an internal wall temperature increase of about 40 K is to be expected for the peak point. This may be taken as an excellent example of the applicability of such an engineering tool, once its validation and anchoring have been achieved by means of available test data.

C. MMH/NTO 400-N Engine Simulations with Liquid Film Cooling

In contrast to the costly regenerative cooling employed on most medium and heavy thrust engines for launcher stages, small orbit and attitude control satellite thrusters usually rely on the relatively simple internal wall film cooling. The 400-N engine shown in Fig. 1 is a typical example. This thruster has a chamber diameter of 60 mm (2.35 in.), a throat diameter of 16.5 mm (0.65 in.), and a nozzle expansion ratio of 220. It operates at a chamber pressure of 10 bar (145 psia) with a nominal mixture ratio of 1.65.[23]

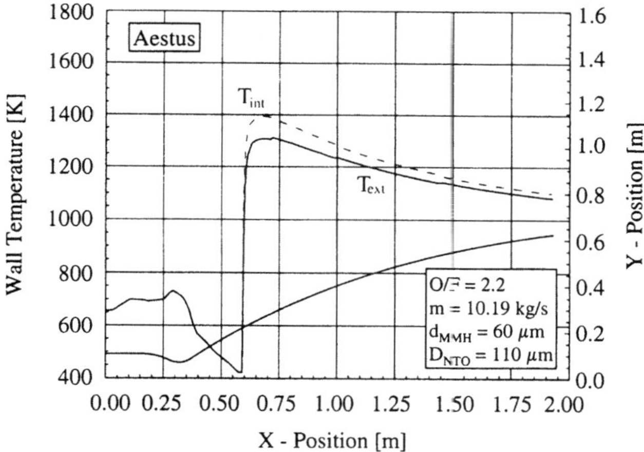

Fig. 11   External (solid) and internal (dashed) wall temperature distribution for the extreme load point $Q5'$ of Aestus utilizing a ceramic nozzle extension.

A coaxial double swirl injector is used to provide a sufficiently high deposition rate of MMH droplets on the combustor wall; the droplets coalesce to form a liquid annular cooling film (see schematic in Fig. 3). As this film moves downstream along the chamber wall, the hot combustion gases cause it to start to evaporate. The film dry-out point is usually reached in the convergent nozzle section a few centimeters upstream of the nozzle throat, so that any performance losses due to nonvaporized fuel remain acceptably low. Externally mounted wall thermocouples can detect this dry-out point quite accurately, because the wall temperature starts to rise sharply immediately after this location.

Three-phase flow simulation results have already been presented by the authors in Refs. 9 and 10. Here, however, the applicability of the tool is demonstrated for the first time over an extended operational range employing the previously mentioned 400-N liquid film cooled bipropellant thruster. Three load points were considered: the reference load point R with a mass flow rate of 135.3 g/s (0.298 lb/s) and a mixture ratio of $O/F = 1.646$, the load point $R+$ with a similar mass flow rate of 132.3 g/s (0.292 lb/s) and a slightly higher mixture ratio of $O/F = 1.746$, and the load point $R+L$ with a distinctly reduced mass flow rate of 120.1 g/s (0.265 lb/s) and an almost constant mixture ratio of $O/F = 1.744$ as compared to $R+$. All simulations were run with a three-block mesh (33,670 computational cells) similar to that presented in Fig. 7 for the Aestus engine.

As in the Aestus engine simulations, a Rosin-Rammler droplet size distribution was assumed and 40 discrete droplet size classes ranging from 5 to 200 μm were considered. The injection velocities were deduced from the prescribed injection areas, which have to be determined from the two conical propellant film sheets left behind the coaxial double swirl injector. For the sake of simplicity, these injection areas were kept constant within the considered operational envelope, so that the injection velocities were merely proportional to the propellant mass flow rates. The mass mean diameters were scaled inversely with the mass flow rate ($D_m \propto m^{-3/4}$), a relation which is often used for double swirl injectors. For the diverse load points, Table 5 presents the scaled mass mean diameters.

In general, engine wall temperature and thrust measurements were utilized herein to characterize the operational behavior of the 400-N thruster. Whereas the outer wall temperature measurements were employed to determine the length of the liquid cooling film, the thrust measurement was used to evaluate the overall performance of the engine. In Fig. 12, the external and internal wall temperature distribution, calculated for the 400-N thruster's reference load point R, is presented. The position of the liquid cooling film can be clearly related to the low wall temperature plateau. Film on-set and dry-out are associated with a conspicuous change in temperature slope. Because heat conduction inside the combustion chamber walls was considered in both radial and longitudinal direction, the temperature increase at the film dry-out position could be adequately resolved. Neglecting longitudinal heat conduction in the wall, for instance, would lead to an almost infinite temperature rise and to an unrealistic high temperature peak. The thin wall, together with the high thermal conductivity of the platinum-iridium material, causes the temperature difference between the inner and outer wall surface to be very small, i.e., only

Table 5 Comparison of predicted and measured performance data for different load points of the high performance 400-N apogee thruster; the mass mean diameters of the propellant sprays were scaled with the relation $D_{m,i} \propto m_i^{-3/4}$

	R		R+		R+L	
Mixture ratio	1.646		1.746		1.744	
Mass flow rate, g/s	135.3		132.3		120.1	
Mass mean diameter MMH, μm	70		73		79	
Mass mean diameter NTO, μm	58		58		62	
	Predicted	*Measured*	*Predicted*	*Measured*	*Predicted*	*Measured*
Thrust, N	421.8	419.5	411.8	409.5	369.7	366.3

a few Kelvin at nozzle throat. The temperature measurements plotted in Fig. 13 illustrate the excellent agreement between test and simulation.

Figure 12 also depicts the MMH film deposition and evaporation rates encountered in each computational wall cell. Notice that such smooth profiles can be obtained only if sufficiently numerous particles are traced (here 80,000). Close to the face plate where the film is built up, the deposition rate dominates. From about $x = 0.035$ m, the film evaporation rate exceeds the deposition rate. This is the location where the film thickness begins to decrease. The computed dry-out point corresponds to the location at which the deposition is equal to the evaporation rate. Downstream of this point the local MMH deposition rate is too low to maintain a film. As a result, only a gaseous film is present there.

The two plots in Fig. 13 illustrate the computed and measured cooling behavior of the thruster for the load points $R+$ and $R+L$, respectively. As can be seen, the film length increases for the $R + L$ load point, although the total propellant mass and hence the injection velocities are considerably reduced. The simple scaling relation that was applied for the mass mean diameters provided increased droplet diameters for both MMH and NTO, which overcompensated the reduction in injection velocities. On the other hand, the increase in mixture ratio, when switching over from load point R to $R+$, should lead to a certain shrinkage of the cooling film. Although the test data seem to indicate a small, almost unnoticeable, reduction of the cooling film length, the calculation did not resolve this trend, delivering nearly the same results for R and $R+$. Because the MMH mass flow rate of load point $R+$ was roughly 6% lower than that of R at almost identical NTO mass flow rates, the initial reaction in the case the of $R+$ became somewhat delayed in the simulation, thus allowing a similar MMH deposition rate at the wall and consequently a similar evolution of the film as compared to R.

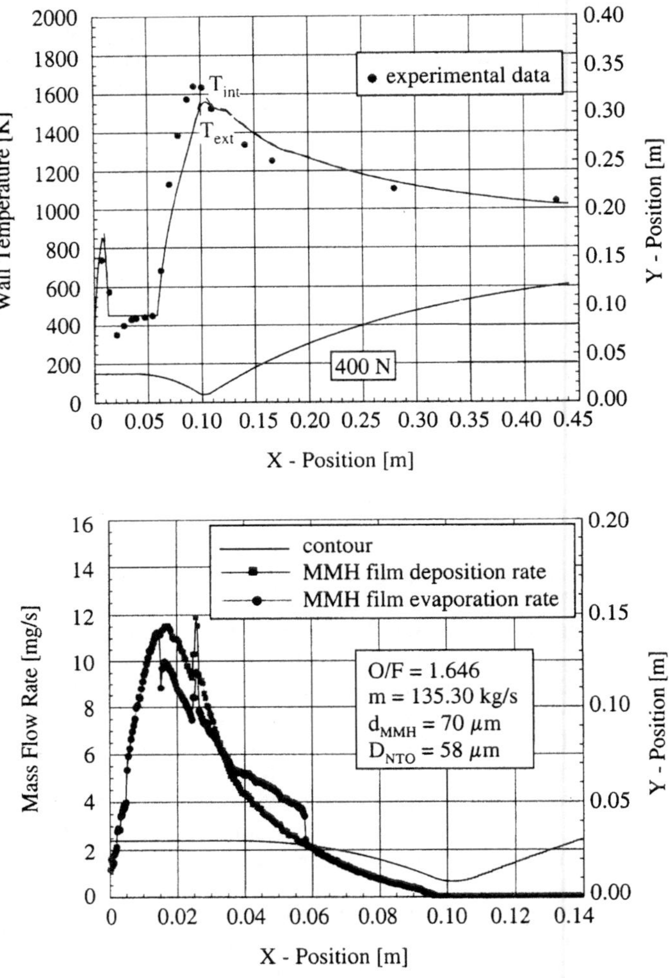

Fig. 12 Internal and external wall temperature distribution (top) and corresponding MMH film deposition and evaporation rates (bottom) calculated for the 400-N thruster reference load point *R*.

These simulations demonstrated that within an extended operational range a relatively simple scaling relation for the mass mean diameters can be employed to achieve a satisfactory agreement with the test data. This is further confirmed by comparing the calculated and measured thrust values as shown in Table 5. Deviations of less than 1% impressively outline the computational capabilities of the applied spray combustion tool.

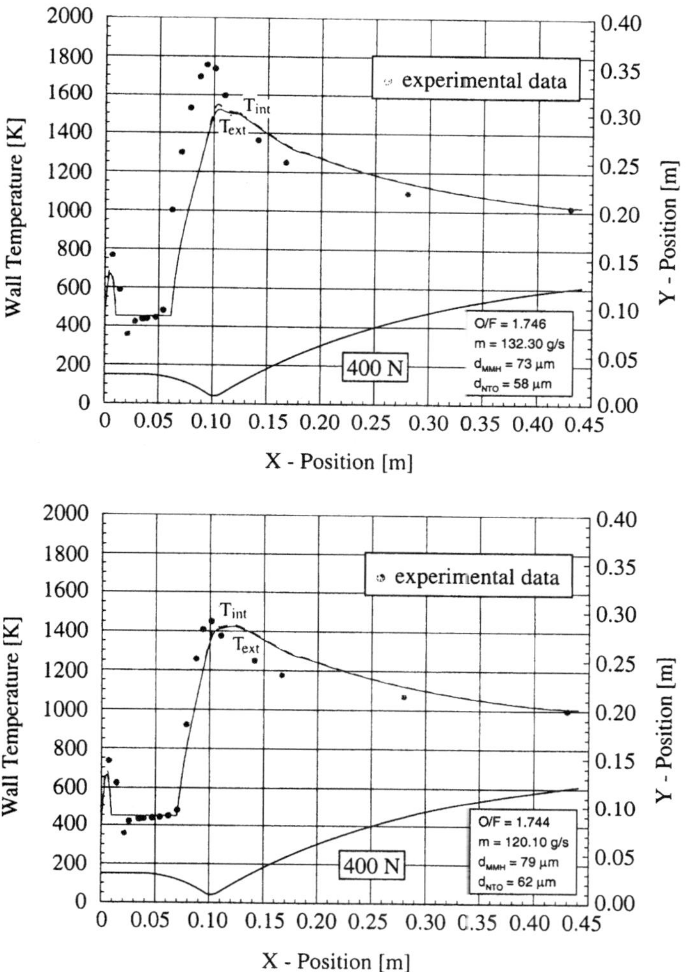

Fig. 13 Internal and external wall temperature distributions computed for the load point $R+$ (top) and $R + L$ (bottom).

V. Conclusions

This chapter has presented an axisymmetrical CFD approach to turbulent spray combustion and flow evolution in liquid bipropellant rocket motors operating under well-premixed conditions. State-of-the-art physical models have been employed for droplet tracking, droplet breakup and evaporation, droplet to gas-phase turbulent interaction, and gas-phase turbulent combustion. A prime development interest was to achieve the best possible compromise between computational responsiveness and physical model complexity. To the knowledge of the authors, this development and its way of treating such propellant systems may be considered as quite novel in Europe.

A large number of computations was carried out in the course of the code development, simulating a small hydrazine/NTO combustor under different mixture ratios and total propellant mass flow rates. Global performance figures such as chamber pressure and theoretical exhaust velocity were used to assess the computational results. Specific emphasis was put on how primary and secondary atomization parameters, as well as propellant chemistry aspects, affect the evolution of the reactive flowfield. In contrast to many other typical spray combustion approaches reported in the literature, the primary atomization pattern of both fuel and oxidizer was represented by means of polydispersed sprays with droplets ranging in size from very small to very large to obtain more realistic spray pattern resolutions. The effect of the initial droplet diameters characterizing the sprays was investigated showing that too fine an NTO spray relative to that of hydrazine (say, $D_m[\text{NTO}] < 0.5 \cdot D_m[\text{hydrazine}]$) can actually retard hydrazine evaporation and hence cause the overall combustion efficiency to drop. Therefore, an optimum liquid bipropellant injector design has to adequately balance the spray patterns for both fuel and oxidizer.

Finally, the direct simulation of the wall heat transfer process in rocket combustion chambers was introduced as a typical industrial application of such a tool. The authors believe that only a solution of the entire flow structure, including the dispersed droplet phase as well as chemical reaction effects, is capable of predicting combustion wall heat transfer rates in a sufficiently reliable and accurate manner, especially under circumstances in which the injector governs the interaction of the dispersed flow with the combustor walls. This is particularly the case for liquid bipropellant film cooled rocket engines, where the heat transfer from the combustion gas strongly affects the film evolution, but also for conventionally regeneratively cooled engines, where the knowledge of accurate heat flux profiles along the entire combustion chamber is indispensable for an efficient cooling circuit design. For both the Aestus engine and the 400-N thruster, it was demonstrated that often relatively simple correlations describing the injector characteristics and substituting extensive primary atomization models are already capable of providing satisfactory models of global rocket engine behavior.

References

[1]Priem, J. R., and Heidmann, M. F., "Propellant Vaporization as a Design Criterion for Rocket-Engine Combustion Chambers," NASA TR-67, 1961.

[2]Breen, B. P., Zung, L. B., and Lawyer, B. R., "Injection and Combustion of Hypergolic Propellants," AFRPL-TR-69-48, 1969.

[3]Liang, P. Y., "Liquid Rocket Combustor Computer Code Development," Advanced High Pressure Oxygen/Hydrogen Conference, Huntsville, AL, 1984.

[4]Chiu, H. H., Jiang, T. L., Krebsbach, A. N., and Gross, K. W., "Numerical Analysis of Bipropellant Combustion in Orbit Maneuvering Vehicle Thrust Chamber," Final Report, NASA Huntsville, 1989.

[5]Larosiliere, L., Litchford, R., and Jeng, S. M., "Hypergolic Bipropellant Spray Combustion and Flow Modelling in Rocket Engines," AIAA Paper 90-2238, 1990.

[6]Tang, Y. L., and Schuman, M. D., "Numerical Modeling of Liquid–Liquid Bi-Propellant Droplet/Gas Reacting Flows," AIAA Paper 92-0232, 1992.

[7]Jiang, T. L., Chiang, W. T., and Jang, S. D., "Numerical Simulation of Variable Thrust Engine Combustion Chamber," AIAA Paper 92-3769, 1992.

[8]Preclik, D., Estublier, D., and Wennerberg, D., "An Eulerian-Lagrangian Approach to Spray Combustion Modelling in Liquid Bi-Propellant Rocket Motors," AIAA Paper 95-2779, 1995.

[9]Knab, O., Preclik, D., and Estublier, D., "Flow Field Prediction Within Liquid Film Cooled Combustion Chambers of Storable Bi-Propellant Rocket Engines," AIAA Paper 98-3370, 1998.

[10]Knab, O., Fröhlich, A., and Wennerberg, D., "Design Support for Advanced Storable Propellant Engines by ROCFLAM Analyses," AIAA Paper 99-2459, 1999.

[11]Knox, B. P., and Marsh, W. R., "USAF Propellant Handbooks Vol. I—Hydrazine Fuels," AFRPL-TR-69-149, 1970.

[12]Wright, A. C., "USAF Propellant Handbooks Vol. II—Nitric Acid/Nitrogen Tetroxide Oxidizers," AFRPL-TR-76-76, 1977.

[13]Sawyer, R. F., and Glassman, I., "Gas-Phase Reactions of Hydrazine with Nitrogen Dioxide, Nitric Oxide, and Oxygen," *Proceedings of the Combustion Institute*, Vol. 11, 1966, pp. 861–869.

[14]Eberstein, I. J., and Glassman, I., "The Gas-Phase Decomposition of Hydrazine and Its Methyl Derivatives," *Proceedings of the Combustion Institute*, Vol. 10, 1965, pp. 365–374.

[15]Svehla, R. A., and Brokaw, R. S., "Thermodynamic and Transport Properties for the $N_2O_4 \rightarrow 2NO_2 \rightarrow 2NO + O_2$ System," NASA TN D-3327, 1966.

[16]Rodi, W., "Experience with Two-Layer Models Combining the k-ε Model with a One-Equation Model Near the Wall," AIAA Paper 91-0216, 1991.

[17]Stoll, J., and Straub, J., "Film Cooling and Heat Transfer in Nozzles," *Journal of Turbomachinery, Transactions ASME*, Vol. 110, Jan. 1988, pp. 57–65.

[18]Abramzon, B., and Sirignano, W. A., "Droplet Vaporisation Model for Spray Combustion Calculations," AIAA Paper 88-0636, 1988.

[19]Wasserbauer, J. F., and Tabat, W., "Performance of Small Rocket Motors Using Coaxial Injection of Hydrazine and Nitrogen Tetroxide," NASA TN-D 1162, 1961.

[20]Buschulte, W., "Liquid Atomization by Injector Nozzles," *Proceedings of International Society of Air-Breathing Engines*, 1989.

[21]Simmons, H. C., "The Correlation of Drop Size Distributions in Fuel Nozzle Sprays," *Journal of Engineering for Power*, 1977, pp. 309–319.

[22]Nickerson, G. R., Coats, D. E., and Dunn, S. S., "Two-Dimensional Kinetics (TDK) Nozzle Performance Computer Program," NASA NAS8-36863, 1989.

[23]Schulte, G., "High Performance 400 N MMH/NTO Bipropellant Engine for Apogee Boost Maneuvers," AIAA Paper 99-2466, 1999.

Simulation and Analysis of Thrust Chamber Flowfields: Cryogenic Propellant Rockets

Dieter Preclik,[*] Oliver Knab,[†] Josef Görgen,[‡] and Gerald Hagemann[§]
EADS Space Transportation, Munich, Germany

I. Introduction

$\mathbf{T}$HE simulation of reactive flowfields in cryogenic liquid oxygen/hydrogen rocket thrust chambers has advanced significantly over the past decade, especially in light of industrial applications. Driven by the ever growing need to economize on the development of rocket propulsion systems, the capabilities of modern spray combustion codes are being exploited in many new ways. Among these is the extended analysis of combustion performance and local wall heat transfer evolution. Currently available codes are capable of treating injection element patternation effects, hot gas temperature stratification characteristics, and secondary hot gas wall cooling by liquid or gaseous propellant films or injection element mixture ratio biasing. In this context, axisymmetric codes are currently a good compromise between the range of applicability to practical issues and the computational effort required, both factors that are of major importance in today's engineering world.

In the mid-1990s, the authors started to develop a cryogenic liquid oxygen (LOX)/gaseous hydrogen (H_2) spray combustion code as a derivative of the Rocket Combustion Flow Analysis Module (ROCFLAM) code described in more detail in Chapter 14 of this volume. This LOX/H_2 code derivative

*Manager, Systems Technology and R & D Programs, Propulsion and Equipment. Senior Member AIAA.

†Manager, Combustion Devices Engineering. Senior Member AIAA.

‡Propulsion Engineer.

§Project Manager. Member AIAA.

was largely based on the physical models and principles employed for the multi-phase bipropellant version[1-3] with proper modifications for sub- and supercritical LOX-gasification, propellant properties, and reaction chemistry and species, as well as for the wall boundary-layer treatment. The main interest of this modeling initiative was first to create—from the physical point of view—a well-balanced code that treats the various physical processes and disciplines, such as aerothermodynamics, multiphase flow, propellant evaporation, turbulence, and combustion with a similar degree of complexity and accuracy, thus avoiding a too detailed treatment of any one phenomenon. The second object of the effort was to apply this code to all existing, in-house LOX/H_2 thrust chamber types, including the HM7, Vulcain, Vulcain 2, and Vinci, as shown in Fig. 1. The methodology used was to conduct detailed comparisons and assessments of the simulation results with given engine data, be it from full- or subscale testing, in order to systematically explore the capabilities of the code to better support future engineering work. This chapter will highlight some typical examples and demonstrate the ways in which a spray combustion code may be employed as an engineering tool to reflect on and/or provide answers to typical questions arising in the course of rocket engine development.

II. General Aspects of Modeling LOX/H_2 Propellant Combustion

In modern regeneratively cooled high-pressure rocket engines like the space shuttle main engine (SSME) or Vulcain, liquid cryogenic hydrogen is usually fed through a large number of structural cooling channels to absorb the enormous heat that emanates from the combustion process using liquid oxygen at gas temperatures around 3500 K and above (see Fig. 2). Depending on the engine's pressure level, these thermal loads can typically reach heat fluxes up to $160\ MW/m^2$ (SSME) in the throat section of a combustion chamber. To guarantee the thermomechanical integrity of the chamber structure, the liner's gas side temperature has to remain below a certain threshold. Above this threshold, the mechanical properties of the liner material degrades rapidly. For the copper alloys usually employed as liner material on such engines, this maximum wall temperature is around 800–850 K.

A crucial prerequisite for optimizing the regenerative cooling circuit design of a rocket engine is therefore the accurate prediction of the axial hot gas wall heat transfer evolution. This is important for the following reasons: first, and this is particularly true for expander cycle engines, to guarantee that sufficient energy is transferred to the coolant to drive the turbopumps without making the chamber too long; second, to ensure that the wall temperature criterion just mentioned is respected everywhere in the combustion chamber; and third, to ensure that the available coolant pressure budget is most efficiently used along the cooling fluid flowpath. Any overcooling or undercooling of specific chamber areas is to be avoided, because this would constitute either a loss in overall system performance through a penalty on the hydrogen turbopump, or a loss in chamber structural lifetime. Therefore, a major goal of optimizing the hydrogen cooling circuit is to achieve an axial and circumferential hot gas wall temperature distribution that is as low and homogeneous as possible for a given chamber pressure budget and lifetime.

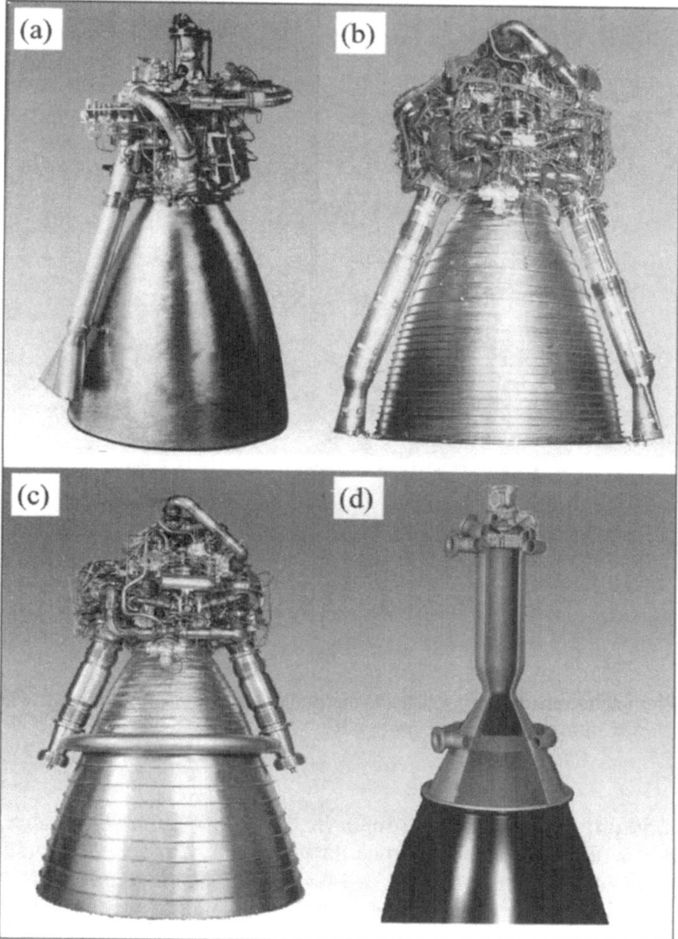

Fig. 1 Family of European LOX/H₂ rocket engines employed on the Ariane launcher (different scales). From left to right: the Ariane 4 upper stage HM7 engine, the Ariane 5 main stage Vulcain and Vulcain 2 engines, and the new upper stage Vinci engine, currently under development.

A typical LOX/H₂ rocket engine injector consists of a large number of coaxial elements, each of which under stationary operational conditions injects a central liquid oxygen jet at supercritical pressures but subcritical temperatures ($T_{LOX} \approx 100$ K) and with a velocity of about 15 m/s surrounded by a fast annular gaseous hydrogen flow with a velocity around 300 m/s. Downstream of the injection plane, the liquid oxygen jet is disintegrated, heated up, mixed, and finally burned with hydrogen.

Because the rocket injector controls the process of propellant atomization and mixing, it governs the overall combustion process and its efficiency to a large

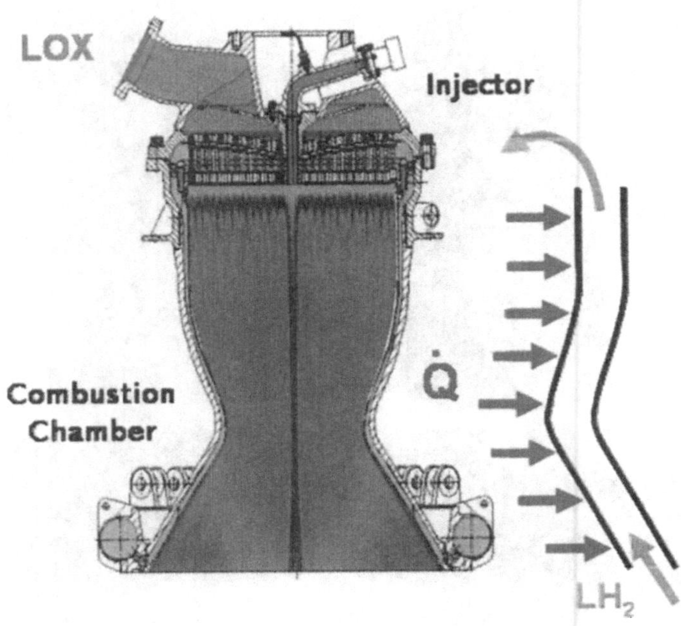

Fig. 2   Flow schematic for regeneratively cooled combustion chamber. (See also the color section of figures following page 620.)

extent. In addition, the near-wall injection elements' outlet flow characteristics impact the evolution of the boundary layer and thus affect the development of the hot gas heat flux along the wall. It is commonly found that different injection element configurations and arrangements, especially in the vicinity of the chamber wall, can cause rather different heat flux distributions inside a rocket thrust chamber. In this context, the level of stratification of the hot gas flow downstream of the injection elements is considered to be of major importance, as non-homogeneous gas conditions may limit complete combustion and hence reduce performance.

A.   Liquid Oxygen/Gaseous Hydrogen Injection

Modeling liquid oxygen/gaseous hydrogen injection has been an issue for many years. Liang et al.[4,5] was among the first to undertake a more realistic description of coaxial injection, considering a numerical, axisymmetric, three-phase approach that treats a liquid core jet, a dispersed droplet phase, and a gaseous phase. To the authors' knowledge, most of Liang's work was dedicated to a combustor with a single element involving considerable empiricism in the underlying physical models. It is not known whether a multi-element injector and chamber arrangement was ever addressed. Because even today the liquid jet disintegration process is still not completely understood, its explicit treatment in a spray combustion code appears of little help. Moreover, current LOX/H_2 main chamber injectors do typically consist of more than 500 individual injection elements (see Fig. 3), too large a number for individual consideration, which means that alternative simulation methods are necessary.

For these reasons, a three-dimensional treatment of the injection and combustion process of a rocket thrust chamber is still beyond the scope of any feasible industrial application. Nevertheless, it shall be shown herein that geometrical complexity and a willingness to compromise modeling does not generally rule out successful use of spray combustion codes in advanced rocket combustion problems.

The approximation of a real three-dimensional injector within an axisymmetric numerical scheme may be pursued in two different ways, each focusing on the radial injection element rows rather than on the individual elements. In doing so, a dispersed, non-premixed (I) or premixed (II) approach may be applied for each individual row as shown in Fig. 4. The resolution of a single element would in fact suggest methodology I, taking into account a liquid core if desired. However, for the multi-element system of a full-scale injector, methodology I would call for considerable grid refinement to adequately resolve the very small hydrogen annuli and the interfacial zones between gaseous hydrogen and liquid oxygen. In practice, this approach is difficult to adapt to the geometry of a real injector. In contrast, methodology II can largely overcome this problem in creating a well-premixed condition for the two propellants. In addition to this, methodology II foresees that in the case of hydrogen, the total propellant inlet mass, its momentum, and energy are immediately transferred within the first computational cells to the gas phase and accounted for in the corresponding source terms of the Eulerian conservation equation system.

The initial representation of hydrogen as a discrete phase in methodology II provides for two major simulation advantages:

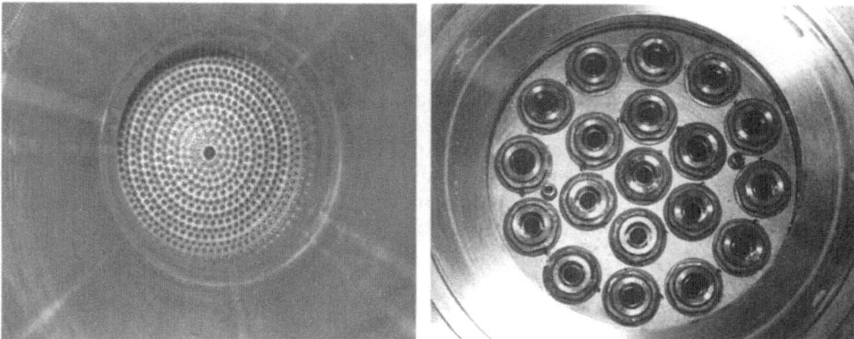

Fig. 3 View of a full-scale Vulcain injector with a diameter of 415 mm and 566 coaxial injection elements arranged in 13 injection element rows (left), and a typical subscale injector with a diameter of 80 mm and 19 coaxial elements of identical geometrical size arranged in two rows around a centerline element (right).

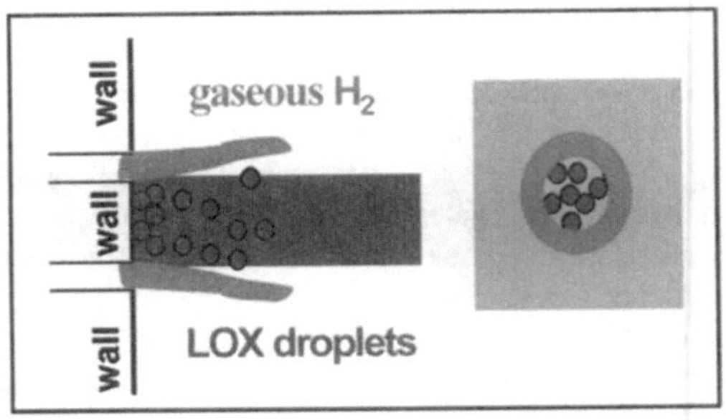

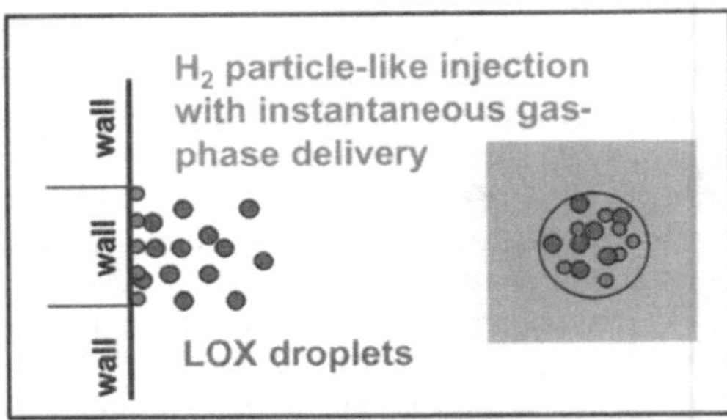

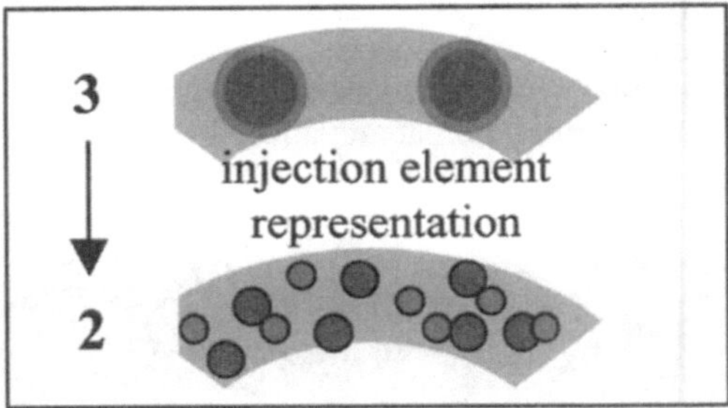

Fig. 4   Schematic of non-premixed, coaxial LOX/H_2 injection employing a liquid jet cone (top), well-premixed injection (center), and the principal representation of an injection element row through an axisymmetrical treatment (bottom). (See also the color section of figures following page 620.)

1) The injection plane can be treated as a rigid wall without true inlet areas and the corresponding grid refinements. This simplifies the overall computational grid setup and positively affects numerical convergence and thus computational time.

2) The hydrogen particle-like injection allows the hydrogen injection velocity and momentum to be treated in an axisymmetric computational frame without

artificial manipulation of the inlet flow area, as in the case of a continuum inlet flow, with its reduced cross-sectional areas and its associated problems of actual resolution and representativity.

To assess the computational quality of methodology II, a test case employing a single-element configuration (see Ref. 6) was calculated for two different grid resolutions, a standard resolution applicable to a multi-element injection system, and a strongly refined one, not actually feasible for a typical multi-element configuration. The corresponding temperature distributions are shown in Fig. 5. The result for the standard grid resolution shows that flame anchoring at the injector tip, as observed in the original experiment, could not be reproduced in the simulation. This, however, was achieved with the strongly refined grid. Moreover, the

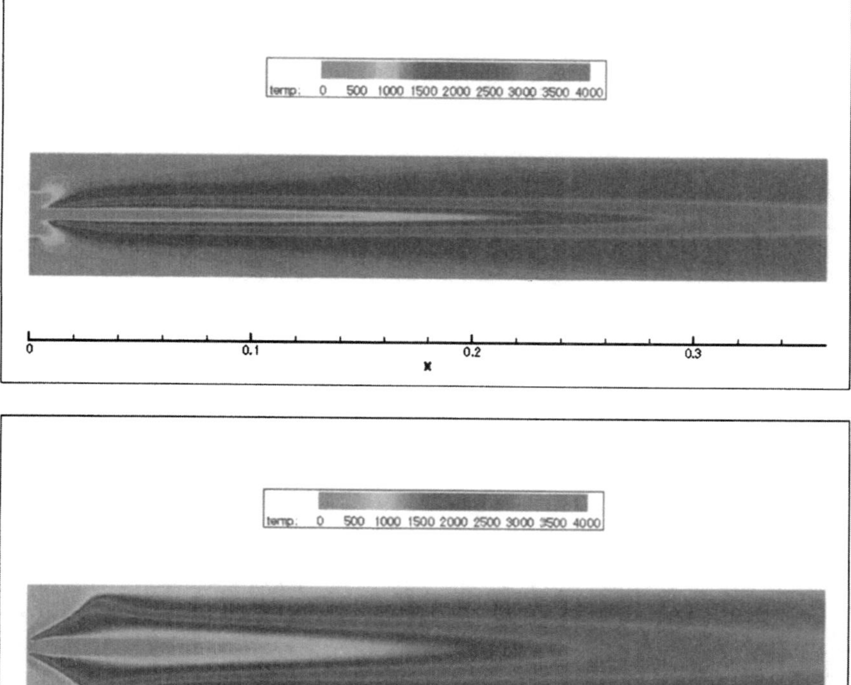

Fig. 5   Temperature distribution for single-element combustor test case ($P_c = 10$ bar) employing a standard grid resolution applicable to a multi-element injection system (top) and a strongly refined grid resolution (bottom). (See also the color section of figures following page 620.)

refined simulation generated a much broader flame shape close to the injector, because of improved recirculation zones in the corner areas of the combustor. The simulations demonstrated that, from a numerical point of view, the simplified spray initialization methodology II is definitely able to resolve the structural details of flame evolution as long as the grid is sufficiently refined. On the other hand, this spray initialization method is still able to resolve global features of the flame structure in an acceptable manner using a much coarser grid resolution. The latter is a distinct advantage when applying this spray initialization methodology to a multi-element injection system, as in a real rocket chamber configuration.

B.   Gas-Phase Combustion and Chemistry

As with all combustion codes, the liquid propellant component is typically initialized as a dispersed phase at the upstream injection plane, consisting of a number of different droplet classes with individual properties, i.e., droplet sizes, injection velocities, and flow directions. During simulation, each droplet class is traced sequentially through the flowfield until its end of life. The effect of flow turbulence on droplet motion is usually accounted for by a stochastic separated flow (SSF) model, assuming that the droplets interact with a random distribution of eddies along their trajectories. More information on the code's capabilities are given in Ref. 7.

More recently there have been controversial discussions on the representativity of discrete particle spray initialization for cryogenic oxygen injected at supercritical pressures and subcritical temperatures, suggesting a dense gas approach[8-10] rather than a dispersed droplet approach. Though the authors acknowledge that experimental evidence may suggest a change of the propellant gasification mechanism under supercritical pressure injection conditions toward a so-called dense gas, there is no readily available physical model for implementation in today's combustion codes. For this reason, a conventional multiclass spray initialization procedure has been employed as a baseline for all simulations provided hereafter. The oxygen droplet evaporation model was modified, however, to take into account transient heat-up to supercritical droplet surface conditions, variable oxygen density, and convective blowing effects as well as nonequilibrium between heat and mass transfer during the entire lifetime.

Finite rate chemistry simulations with a large number of reaction mechanisms are very time consuming, as characteristic time scales for chemistry terms are orders of magnitude lower than for the convective and viscous flowfield terms. Moreover, for the numerical integration of the stiff system of chemistry equations, time-implicit solvers are used. To avoid these difficulties, a reduced reaction mechanism may be employed as an alternative, together with a turbulence eddy dissipation concept (EDC) and/or kinetically controlled (Arrhenius) combustion model.

For the investigations herein, a turbulence controlled combustion model is used with a single-step, global O_2/H_2-reaction chemistry scheme:

$$H_2 + xO_2 \Longrightarrow aH_2O + bH + cOH + \Delta E_{reaction}$$

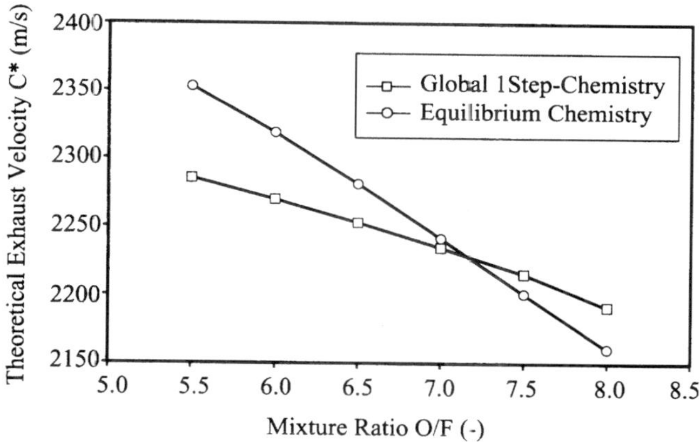

Fig. 6   LOX/H$_2$ theoretical exhaust velocity C^* vs chamber mixture ratio O/F for global one-step chemistry and equilibrium chemistry models; reaction coefficients optimized for $O/F \approx 7$.

The reaction coefficients a, b, c, and x are evaluated for the overall injection propellant mixture ratio of interest by utilizing the CET-equilibrium chemistry code,[11] such that the theoretical exhaust velocity C^* arrived at through complete combustion using the simplified global chemistry scheme is in best agreement with that produced through equilibrium chemistry. Figure 6 shows a comparison of C^* for a global, one-step chemistry with reaction coefficients a, b, c, and x optimized for $O/F \approx 7$ and for equilibrium chemistry. Because the injection methodology II employed herein does not require prescription of any pressure boundary condition along the subsonic flow in the chamber, a major mismatch of C^* provided by the simplified global chemistry model would cause the chamber pressure to develop either too low or too high with corresponding drawbacks in the heat flux calculation, the latter being dependent on the chamber pressure with the power of roughly 0.8. Therefore, the reaction coefficients of a global chemistry model have to be carefully evaluated.

C.   Wall Heat Transfer

A more detailed and reliable analysis of rocket chamber wall heat transfer characteristics has to consider representative hot-firing combustor tests. The objective of such tests is the measurement of the local wall heat flux distribution from the injector faceplate down to the nozzle throat and beyond. In practice, such tests are conducted with a subscale calorimeter chamber that consists of several independently cooled circuits. Figure 7 shows a schematic of a calorimeter chamber with 11 individually cooled segments for the barrel section and 9 segments for the convergent-divergent throat section.

For hot firing, an injector with a reduced number of original, full-scale injection elements is usually employed, as shown in Fig. 3 (right). The element pattern is usually arranged in such a way that characteristic features to be later realized in

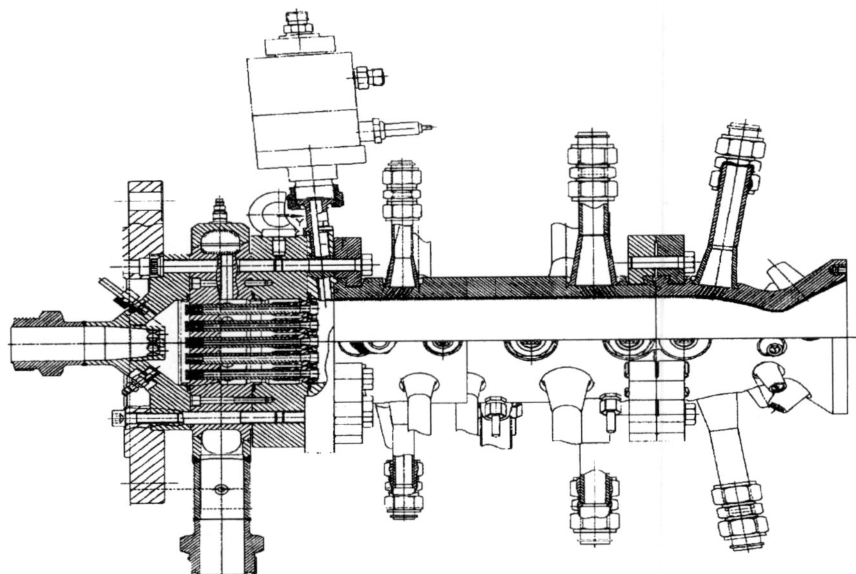

Fig. 7   Subscale calorimeter chamber with 20 individually cooled segments (see Ref. 12).

a full-scale design, such as the element spacing along the wall or the element-to-wall distance, are represented in a $1:1$ fashion. Figure 8 shows some typical experimental test data for the barrel section of this chamber, for two chamber pressures and three mixture ratios (see also Ref. 12).

The data using coaxial injection show that the influence of the mixture ratio is relatively small as compared with that of the chamber pressure. Assuming that local wall heat loads q_w depend on the chamber pressure P_c in a $q_w \sim P_c^a$ like fashion, a constant exponent a was found here only for the downstream segments 8–11, with a being 0.83. For the upstream segments 1–7, the exponent a ranged between 0.22 (segment 1) and 0.73 (segment 7).

Though such experimental measurements are the baseline for an optimum design of a rocket chamber regenerative coolant circuit, they are of minor help as long as there is no reliable procedure for scaling this data from subscale size up to the original, full-scale size of a thrust chamber, the latter often having a rather different geometrical shape. In the past, dimensionless Nusselt-number correlations were employed as a standard means to conduct this upscaling. The authors, however, have found that such dimensionless numbers do provide inherent scaling uncertainties that are not acceptable for optimization. The following example may illustrate this in more detail.

Assuming the hot gas side heat transfer coefficient h_g follows the Nusselt correlation[13]:

$$Nu = CRe^{0.8}Pr^{0.34} \quad \text{with} \quad Nu = h_g D_c/k \quad \text{and} \quad q_w = h_g(T_g - T_w)$$

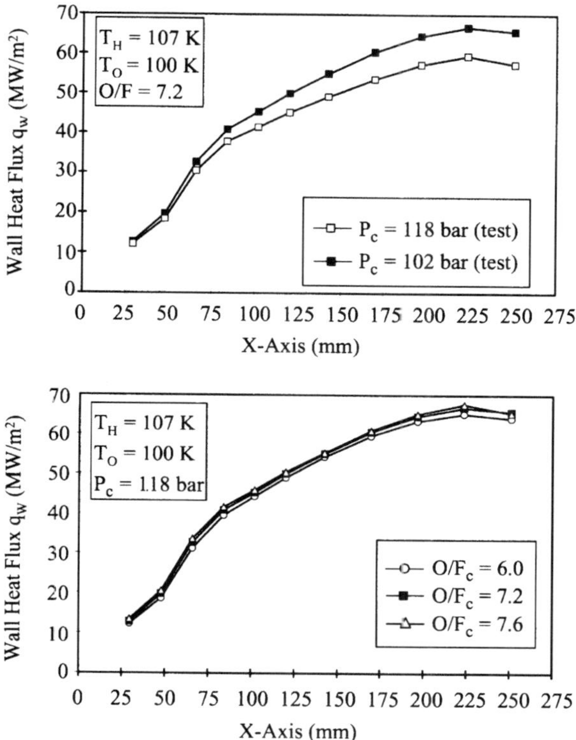

Fig. 8 Calorimeter chamber wall heat flux evolution based on pressure $P_c = 102$, **118 bar (top) and mixture ratio** $O/F = 6.0, 7.2$, **and 7.6 (bottom).**

where Re is the Reynolds number; Pr is the Prandtl number; k is the thermal conductivity; $T_g - T_w$, is the gas-to-wall temperature gradient; and C is either constant or the scale-dependent factor that adjusts for individual injector characteristics. One may transform this correlation to obtain:

$$q_w \sim \text{const}(T_g - T_w)D_c^{-0.2}$$

which means that, all other parameters being roughly the same, in particular the chamber operational load point (chamber pressure, mixture ratio, hot gas properties, hot gas-to-wall temperature gradient), the wall heat load will scale approximately with the chamber diameter D_c to the power of -0.2. In practice, this leads to a reduction of the wall heat load on any full-scale design utilizing subscale calorimeter data. Engine tests, however, demonstrated that this is actually not the case. To verify this, two simulations were conducted for both subscale and full-scale geometry: an O_2/H_2 gas/gas simulation with combustion and plug-profile, gaseous inflow conditions; and a LOX/H_2 liquid/gas simulation with

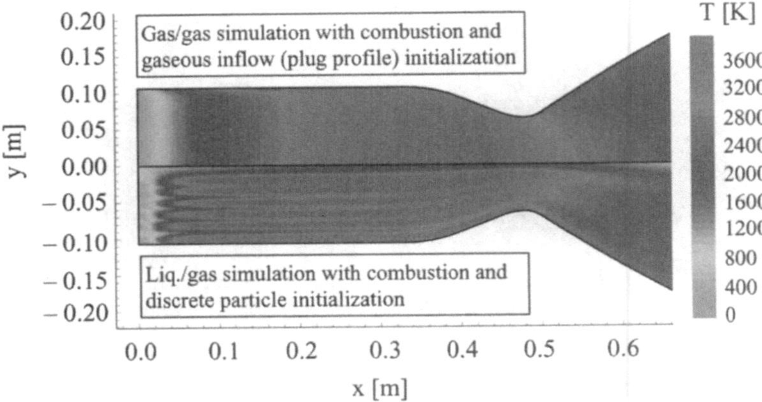

Fig. 9   Hot gas temperature distribution inside a combustion chamber obtained for a gas/gas simulation with plug-inflow profile (upper half) and liquid/gas discrete particle (lower half) initialization, both with turbulent combustion and prescribed wall temperature boundary conditions (boundary-layer model: logarithmic wall functions). (See also the color section of figures following page 620.)

discrete particle injection. A well-prescribed wall temperature boundary condition was employed in both cases, i.e., the gas/gas and the liquid/gas simulation. Figure 9 shows the temperature profiles inside the combustion chamber obtained for the two cases—here on full-scale level—disclosing a considerable gas temperature stratification for the liquid/gas discrete particle injection method.

The corresponding wall heat flux evolutions for the two cases and the two chamber sizes, i.e., subscale and full-scale, are depicted in Fig. 10. One can see that in the case of the gas/gas simulation, upscaling reduces the wall heat loads in a manner comparable to that suggested by the Nusselt correlation. In contrast, the second case, with the liquid/gas simulation and the discrete particle initialization, shows an overall increase in the wall heat load due to upscaling. The latter result is in good agreement with the authors' experience, disclosing an obvious weakness in the Nusselt correlation approach.

Because the liquid/gas simulation was conducted on both the subscale and the full-scale levels with discrete particle injection at identical element-to-wall distances for the near-wall injection row, much like the design of the actual hardware, the explanation for the heat flux scaling shown in Fig. 10 is quite straightforward: as long as subscale and full-scale injectors employ the same element-to-wall distances in their hardware, Nusselt correlations must underestimate the heat flux evolution during upscaling, because they imply a certain increase in the element-to-wall distance that is not represented by the actual design. The liquid/gas discrete particle injection simulation, however, is able to overcome this weakness by providing for a much more realistic upscaling approach.

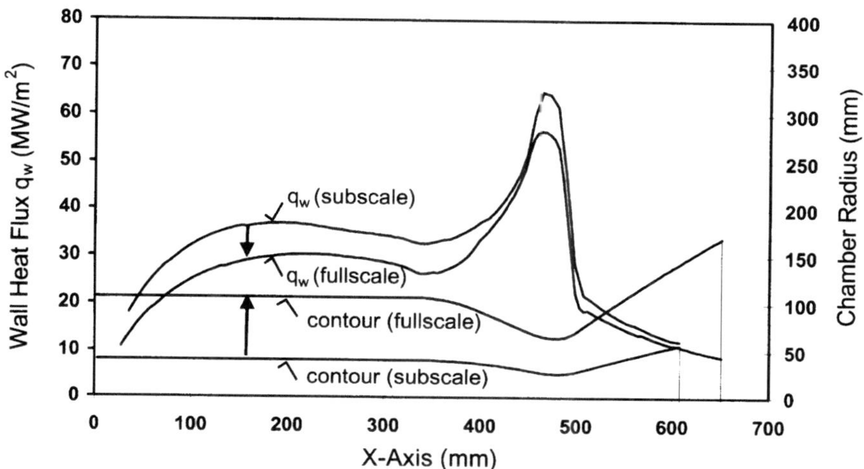

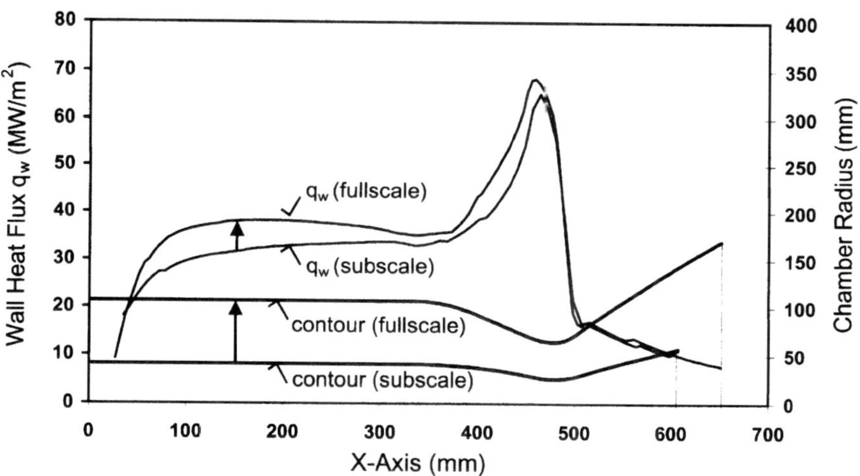

Fig. 10 Comparison of subscale and full-scale chamber wall heat flux evolutions employing gas/gas simulation with plug-inflow profile (top) and liquid/gas simulation with discrete particle initialization (bottom).

III. Applied Cryogenic LOX/H$_2$ Rocket Combustion Simulations

We will now discuss some applications of advanced LOX/H$_2$ spray combustion simulations to typical rocket thrust chamber issues. In doing so, the computational results will, whenever possible, be validated by means of experimental test data. As already mentioned, wall heat transfer and performance characteristics will be focused on because these aspects are of key importance for optimizing rocket propulsion systems.

A. Chamber Wall Heat Load and Effect of Injection Element-to-Wall Distance

The authors have recently conducted a significant number of calorimeter subscale chamber (see Fig. 7) LOX/H$_2$ hot-fire tests employing different chamber operational conditions (propellant temperatures, chamber pressures, mixture ratios, etc.), coaxial injection element designs, and element-to-wall distances, as well as secondary wall film cooling principles utilizing a slot- or jet-like design. A major goal of all of these tests was to validate the in-house spray combustion code in an extended way. Some of the results are reported in more depth in Ref. 12.

Figure 11 depicts a subscale chamber liquid/gas spray combustion simulation (left) with discrete particle initialization for typical gas generator engine operational conditions, i.e., propellant injection temperatures around 100 K for both liquid oxygen and gaseous hydrogen, a chamber pressure of 100 bar, and a mixture ratio of 6. The simulated temperature distribution inside the chamber and nozzle throat section shows a considerably stratified gaseous combustion flowfield correlating with the centerline element and the two injection rows. At first sight, this level of stratification may look surprising, but it can actually be confirmed by the photograph in Fig. 11 (right) showing a view on the wall of the divergent part of the nozzle throat during firing. One can clearly see the individual "footprints" of each of the 12 injection elements around the chamber wall. It should be mentioned that this level of stratification is dominant even at high combustion efficiencies around 99%. Figure 12 shows the corresponding evolution of the wall heat flux in line with the mean average, cross-sectional hot gas temperature inside the chamber, as simulated for the previously mentioned operational conditions. One can see that the wall heat flux evolution agrees well with the experimental test data (for these tests, only the calorimeter barrel section was employed). The evolution of the mean average, cross-sectional hot gas temperature highlights that a nozzle length of about 200 mm is needed to achieve complete combustion with a mean gas temperature around 3500 K. This temperature level also correlates well with equilibrium chemistry calculations. The similarity in the slope of the wall heat flux and the mean average cross-sectional gas temperature along the chamber barrel section suggests that the flame stratification of the injection elements, especially of those near the wall, governs the heat load buildup.

A verification of the quantitative effect of the injection element-to-wall distance (EWD) on the wall heat flux evolution was conducted by the authors employing two different 19-element injectors, one with EWD = 8 mm (0.315 in.) and one with EWD = 10 mm (0.394 in.). Both injectors were hot fired employing

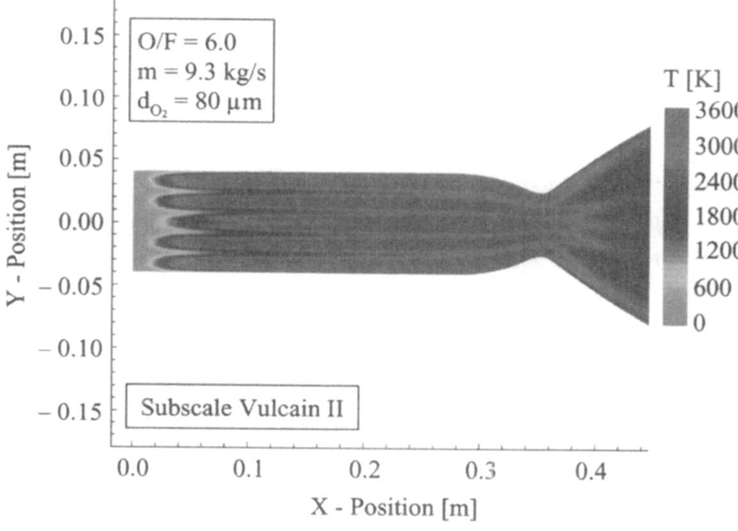

Fig. 11   LOX/H₂ temperature distribution and flowfield stratification inside subscale chamber employing the discrete particle initialization method (top); "footprints" of near-wall injection elements observed on divergent part of nozzle throat wall (bottom) during hot firing; operational load point: P_c = 100 bar; O/F = 6. (See also the color section of figures following page 620.)

chamber pressures P_c from 35 to 70 bar and mixture ratios O/F from 5 to 7. The propellant temperatures were set at around 100 K for liquid oxygen and around 225 K for gaseous hydrogen, which is representative for expander-cycle engine operational conditions. Figure 13 compares the wall heat flux evolution in the combustion chamber barrel and throat sections for the two element-to-wall

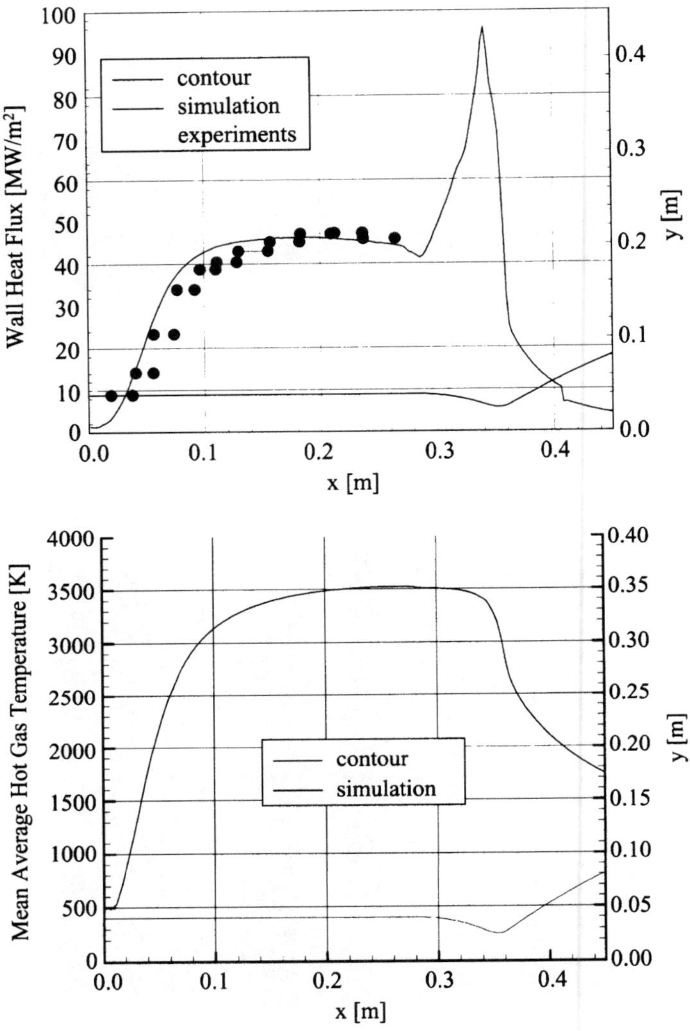

Fig. 12　Evolution of subscale combustion chamber wall heat flux (top) and mean average, cross-sectional hot gas temperature (bottom) for typical gas generator engine operational conditions; operational load point: $P_c = 100$ bar; $O/F = 6$.

distances at $P_c = 60$ bar and $O/F = 6$ in line with the experimental measurements (left figure). One can clearly see that a decrease in the element-to-wall distance of 2 mm (0.079 in.) considerably reduces the local wall heat load along the entire length of the combustion chamber and nozzle, a confirmation for high flowfield stratification inside the combustion chamber. The radial temperature

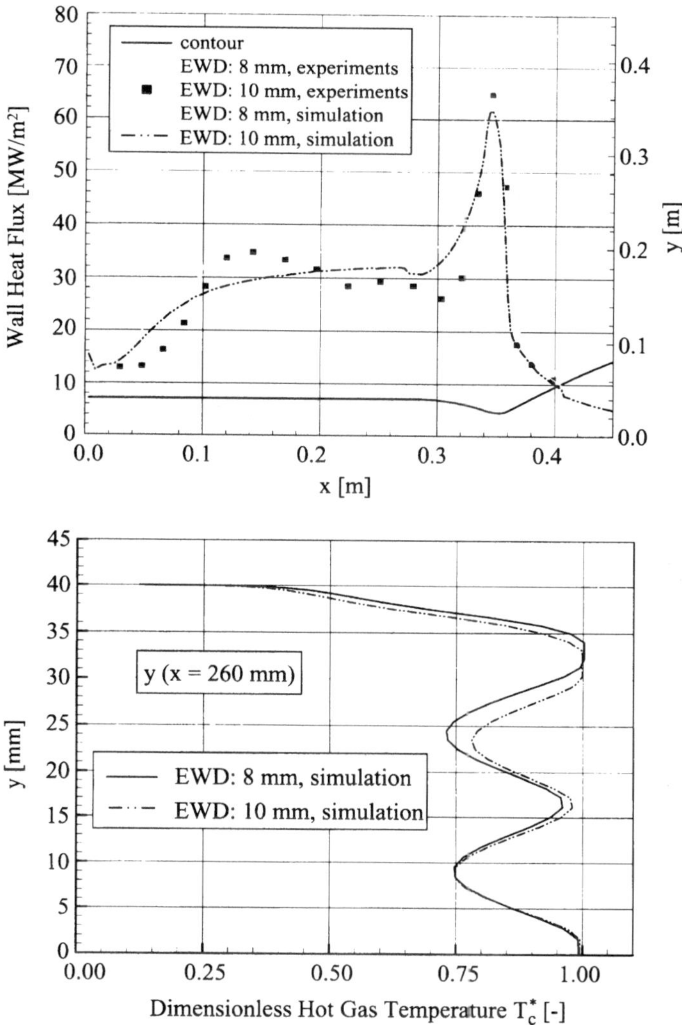

Fig. 13 Comparison of experimental and computational wall heat flux evolutions for two different injection element-to-wall distances (EWD = 8 mm and EWD = 10 mm) (top); nondimensional hot gas temperature vs chamber radius for EWD = 89 mm and EWD = 10 mm at $x = 260$ mm downstream of the injection faceplate (bottom); operational load point: $P_c = 60$ bar; $O/F = 6$.

distribution at $x = 260$ mm (right figure) shows that a smaller element-to-wall distance in fact leads to a steeper temperature gradient at the wall, with consequently higher heat fluxes. Moreover, the results prove that conventional Nusselt correlations are actually not able to adequately support the scaling of heat loads from small size chambers to full-size chambers, as Nusselt numbers cannot account for the injection element position relative to the chamber wall.

B.   Chamber Liner Crack Evolution and Performance

Extensive testing of an expendable cryogenic LOX/H_2 engine will cause cracks to develop along the liner of the combustion chamber. Usually, a small number of cracks, typically fewer than 20–40, does not really affect the performance of the engine in an adverse manner. Though a small amount of hydrogen is lost through these cracks (note that in regeneratively cooled systems the pressure along the cooling circuit is always higher than that inside the combustion chamber), leakage film cooling downstream of the cracks, mostly toward higher heat loads, will heal the cooling problem to some extent. Upstream of the crack, however, there is a cooling deficiency in counterflow cooled engines, which causes the crack to continuously grow in this direction. Under a more extreme local crack situation, additional coolant fluid can even be supplied toward the crack location from the upstream near-injector manifold, taking away some of the coolant hydrogen from other noncracked channels to be fed through the injector into the combustion chamber. As the number of cracks becomes too high, engine performance will start to drop due to incomplete combustion and unburned hydrogen lost through crack leakages.

It is of practical interest to estimate the amount of hydrogen leakage flow that starts to cause the combustion performance to drop unacceptably. Though this is a local, three-dimensional problem, one may try to estimate it using an axisymmetric scheme by assuming the crack to be a circumferential slot with an equivalent area and mass influx. In this way, the local conditions around the crack will not be completely representative, but the effect of different crack sizes on combustion performance evolution can still be estimated. In this process, the hydrogen mass flow rate injected at the upstream faceplate must be corrected by the amount lost through crack leakage. This will lead to higher mixture ratios in the vicinity of the injector.

Four different test cases (relating to the Vulcain-like engine liner; see Fig. 1) were defined and simulated, including one case without cracks for reference and

Table 1   Crack parameters for the four test cases

Test case no.	Length of crack, mm	Leakage mass flow rate, kg/s	Mean average number of cracks
1	——	——	0
2	0.34	3.0	150
3	0.57	5.0	250
4	0.80	7.0	350

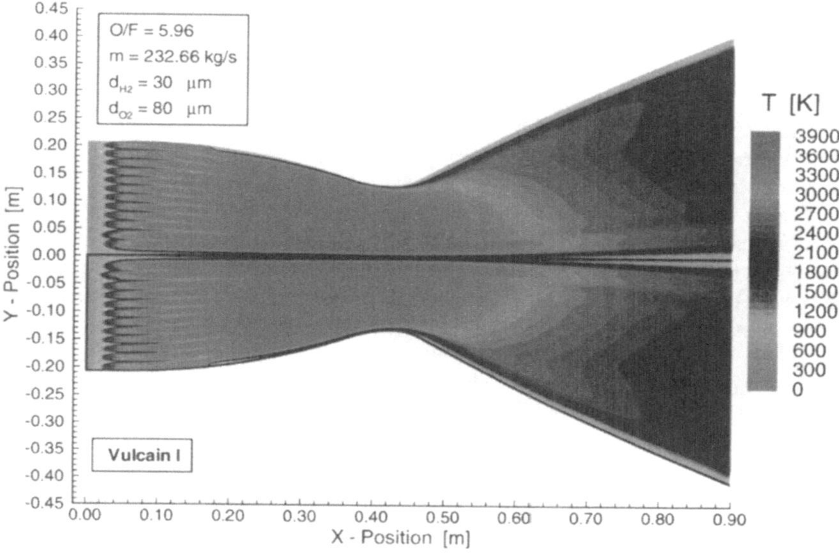

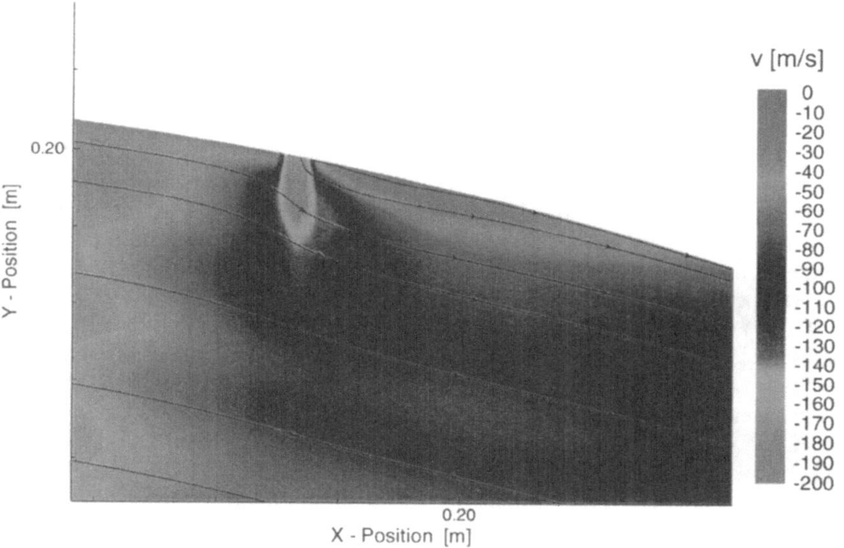

Fig. 14 Temperature distribution inside a rocket combustion chamber with an artificial crack 170 mm downstream of the injector (top), and local velocity profile with streamlines around the crack (bottom). (See also the color section of figures following page 620.)

three cases with different infiltrated mass flow rates, as shown in Table 1. Figure 14 shows the temperature distribution and local velocity profile developing inside the combustion chamber around the crack location, 170 mm downstream of the injector. Although the temperature distribution shows the film cooling effect of the hydrogen leakage flow along the downstream wall quite nicely, the radial velocity profile illustrates the hydrogen inflow at crack position in more detail.

Figure 15 depicts the influence of the leakage flow rate through the crack on combustion performance (C^*/C_{ref}^*, top plot) and vacuum specific impulse ($I_{\text{sp}}/I_{\text{sp,ref}}$, bottom plot) in relation to a noncracked condition. One can see that C^* drops more severely than I_{sp}. This is due to the fact that the hydrogen that is leaked into the chamber cannot be burned with oxygen as it flows along the chamber wall toward the throat and beyond. In the divergent part of the nozzle, however, hydrogen is more and more mixed into the main stream, heated up, and expanded in the nozzle, thus recovering some of the combustion performance losses. A comparison between the simulated performance drop and the engine test data in Fig. 15 discloses reasonably good agreement of the slopes of the two curves with slight differences in the absolute values of the loss. Unfortunately, no thrust measurements are available at engine level to compare the results for I_{sp}.

C. Supersonic Nozzle Wall Heat Transfer for High-Temperature Ceramic Materials

In a hot-fire test program performed with a 40-kN LOX/H_2 thrust chamber, valuable insight into flowfield development was gained.[14,15] The high quality of the test results allow their use for computational fluid dynamics (CFD) validation purposes. The geometry of the nozzle extension is of the Vulcain type (see Fig. 1) with a 1:5 scale. The divergent part of the nozzle extension consists of a water-cooled throat section up to the point where the area ratio is $\varepsilon = 5$, a graphite interface section in between $5 < \varepsilon < 9$, and a ceramic section up to the maximum area ratio of $\varepsilon = 45$.

Different measurement techniques were used, including high-frequency wall pressure measurements, OH spectroscopy, and wall temperature measurements by means of thermocouples and infrared thermography. Details on the test campaign, measurements, and test results are summarized in Ref. 14. Figure 16 illustrates the high chamber pressure test ($P_c = 80$ bar) at 10 s after ignition. The exhaust plume with the typical cap-shock pattern is clearly visible. Additionally, the thermographical imaging lens system is shown nearly perpendicular to the outer surface of the nozzle. The measurement data indicate that the temperature distribution reached steady-state conditions at 24 s after ignition. For this steady-state condition, the wall temperature was predicted with the current code. This simulation includes prediction of the flowfield and modeling of the hot-gas side wall heat transfer, the heat conduction across the wall, and the external heat transfer by radiation. The result for steady-state conditions at this extreme operational point is presented in Fig. 17. The two solid lines represent the nozzle temperature prediction: the upperline (T_i) shows the inner wall temperatures, while the lower one (T_o) shows the outer wall temperature. The symbol-

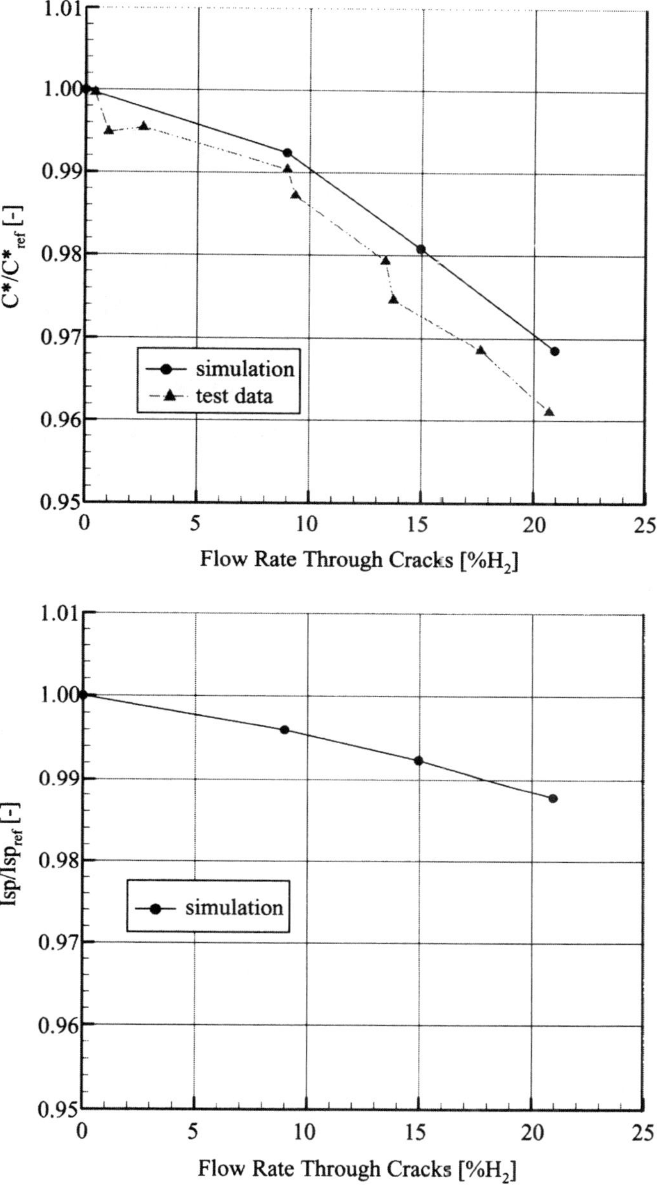

Fig. 15 Evolution of C^* combustion performance (top) and specific impulse (bottom) depending on H_2 leakage flow rate.

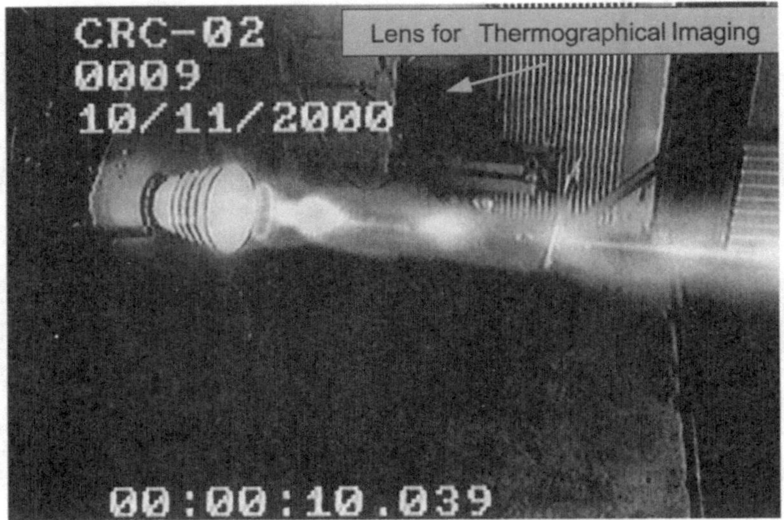

Fig. 16 Ceramic matrix composite subscale nozzle extension at $P_c = 80$ bar exhaust plume with cap-shock pattern and multiple shock reflections in exhaust plume.

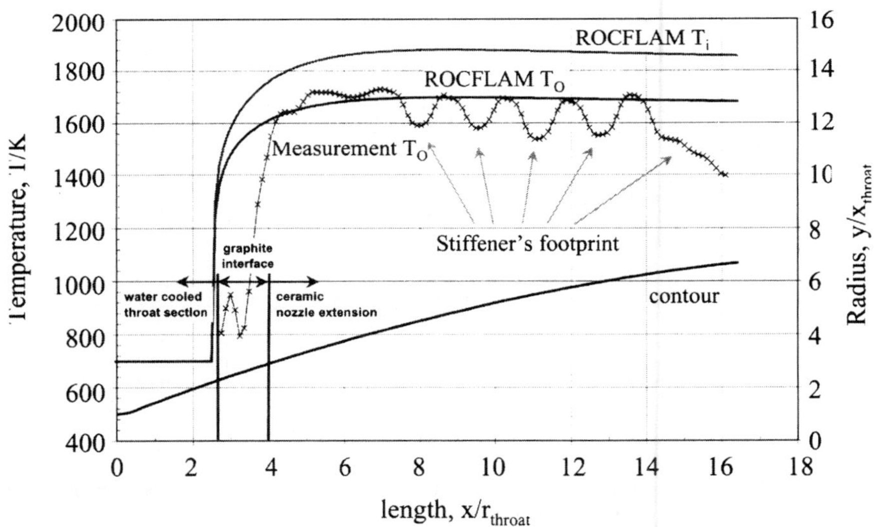

Fig. 17 Comparison of predicted and measured nozzle surface temperatures for Vulcain-type ceramic matrix composite subscale nozzle; T_o outer and T_i inner wall temperature.

marked line is the nozzle's outside temperature measured by the thermographic imaging method. Apart from the fact that the values differ in the first graphite interface region $2.5 < x/r_{throat} < 4$ where the measurement signal is disturbed due to mirror effects, the predicted temperature values agree very well with the measured ones. The shape of the measurement curve is caused by the stiffeners installed on the nozzle. These stiffeners nearly double the wall thickness and thus change the heat transfer profile. For simplicity reasons, the stiffeners were not modeled here.

A comparison between full-scale test visualization and computation of the flow pattern of the subscale nozzle is given in Fig. 18. The numerical flowfield analysis is based on the simplified assumption of an ideal gas and constant specific heats. Thermodynamic gas properties were chosen to represent the combustion gas composition as closely as possible, i.e., using a specific heat ratio of $\kappa = 1.2$ and a molar mass of $M = 13.5$ g/mol. Details on the numerical method used for this simulation are included in Refs. 16 and 17. The characteristic plume pattern of a parabolic nozzle extension, the cap-shock pattern,[17] is clearly visible in all three pictures and shows nearly identical shape. This plume pattern results from the interaction of the recompression shock and the inverse Mach reflection of the internal shock (see Ref. 17 for further details). This shock interaction is influenced very little by viscous effects (except for the inner core recirculation) and is therefore as stated by Euler's theory, largely independent of the geometric scale employed, as is clearly demonstrated by the present sub- and full-scale tests.

IV. Summary and Conclusions

This chapter has illustrated recent advancements in spray combustion simulations of cryogenic LOX/H$_2$ rocket thrust chambers employing a state-of-the-art axisymmetric, Eulerian–Lagrangian Navier–Stokes code. A major design

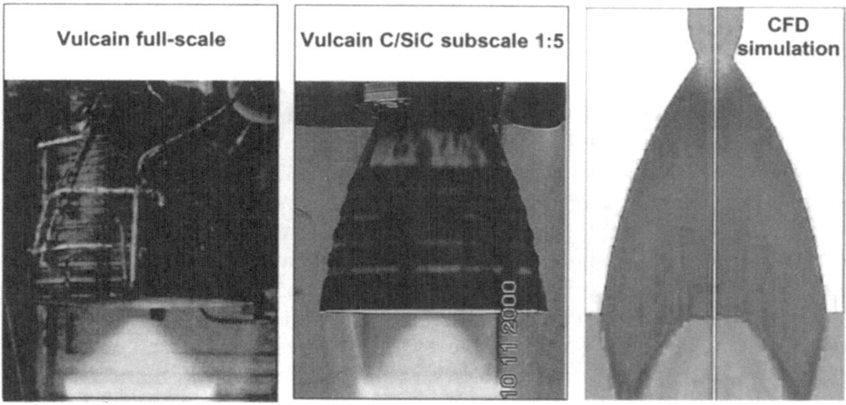

Fig. 18 Comparison of Vulcain-type full-flowing nozzles at nearly identical operational conditions: full-scale (left), long C/SiC subscale (middle), and numerical simulation for full-scale (right). (See also the color section of figures following page 620.)

objective for this engineering tool was to treat the various physical processes and disciplines, such as aerothermodynamics, multiphase flow, propellant evaporation, turbulence, and combustion, with a similar degree of complexity and accuracy, avoiding a too detailed description of any one phenomenon. Some typical code applications focusing especially on design issues related to wall heat transfer and combustion performance were outlined and discussed in light of the capabilities and drawbacks inherent in an axisymmetric flowfield approach, which in the authors' opinion is still the only feasible approach available for effectively solving industry-driven, multi-element rocket engine injection and combustion flow problems.

The discrete particle injection initialization methodology used in this study, which completely eliminates specific continuum inflow boundary conditions, was described in more detail. The advantage of this kind of simulation approach, which can approximately represent multirow coaxial injectors, was highlighted in comparison with typical gas/gassimulations. It was demonstrated that the scaling of wall heat fluxes from small-size injector and chamber testing onto full-size engine conditions may introduce nonnegligible inaccuracies when conventional gas/gas simulations or alternatively empirical Nusselt correlations are used. Because neither of these can take the injection element-to-wall distance problem adequately into account, upscaling is erroneous as long as subscale and full-scale injectors are based on identical EWDs, as is usually the case. Dedicated calorimeter chamber experiments employing injectors with two different EWDs (8 vs 10 mm) confirmed this hypothesis, which is governed by flowfield stratification emanating from the combustion process downstream of the injection elements and persisting throughout the entire chamber length toward the nozzle throat and beyond.

A second example was dedicated to the assessment of C^* and I_{sp} performance degradation due to the evolution of chamber liner cracks and associated coolant leakage rates. The simulations showed that C^* performance is much more sensitive to a liner crack leakage of coolant than I_{sp}-performance, the latter recovering some of the combustion performance loss as the coolant fluid is more and more mixed with and thus heated by the core flow downstream of the throat toward the nozzle exit.

Finally, supersonic flowfield characteristics were investigated on a $1:5$ subscale ceramic nozzle and compared with full-scale engine tests. It was found that principle flow phenomena visualized during full-scale engine testing, e.g., the cap-shock pattern in the plume of such parabolic nozzles, or free and restricted shock separation during startup and shutdown, can likewise be observed in subscale tests under representative combustion chamber operational conditions, i.e., equal propellants, chamber pressures, and mixture ratios. Subscale and full-scale flow patterns were found to compare well for this Vulcain-type parabolic nozzle problem. Moreover, the excellent agreement of the computational results with the visualized shock patterns demonstrates the important role of advanced engineering tools for a better understanding of highly complex, supersonic flow structures.

References

[1]Preclik, D., Estublier, D., and Wennerberg, D., "An Eulerian-Lagrangian Approach to Spray Combustion Modelling in Liquid Bi-Propellant Rocket Motors," AIAA Paper 95-2779, 1995.

[2]Knab, O., Preclik, D., and Estublier, D., "Flow Field Prediction Within Liquid Film Cooled Combustion Chambers of Storable Bi-Propellant Rocket Engines," AIAA Paper 98-3370, 1998.

[3]Knab, O., Fröhlich, A., and Wennerberg, D., "Design Support for Advanced Storable Propellant Engines by ROCFLAM Analyses," AIAA Paper 99-2459, 1999.

[4]Liang, P.-Y., Fisher, S., and Chang, Y. M., "Comprehensive Modeling of a Liquid Rocket Combustion Chamber," *Journal of Propulsion and Power*, Vol. 2, No. 2, 1986, pp. 97–104.

[5]Liang, P.-Y., "Liquid Rocket Combustor Computer Code Development," NASA CP-2372, *Proceedings of the High-Pressure Oxygen/Hydrogen Technology Conference*, June 1984, pp. 696–716.

[6]Görgen, J., and Knab, O., "Application of Astrium's CryoROC Code to a Single-Injector Problem—a Contribution to the RCM-3 Mascotte Test Case," *Proceedings of the 2nd International Workshop on Rocket Combustion Modeling*, 2001.

[7]Görgen, J., and Knab, O., "CryoROC—a Multi-Phase Navier–Stokes Solver for Advanced Rocket Thrust Chamber Design," *Proceedings of the 4th European Symposium on Aerothermodynamics for Space Vehicles*, 2001.

[8]Schley, C.-A., Hagemann, G., and Golovitchev, V., "Comparison of High Pressure H_2/O_2 Rocket Model Engine Reference Simulations," AIAA Paper 95-2429, 1995.

[9]Schley, C.-A., Hagemann, G., Tucker, P. K., Venkateswaran, S., and Merkle, C. L., "Comparison of Computational Codes for Modeling Hydrogen-Oxygen Injectors," AIAA Paper 97-3302, 1997.

[10]Mayer, W., and Tamura, H., "Flow Visualization of Supercritical Propellant Injection in a Firing LOX/GH_2 Rocket Engine," AIAA Paper 95-2433, 1995.

[11]McBride, B., and Gordon, S., "CET93—An Interim Updated Version of the NASA Lewis Computer Program for Calculating Complex Chemical Equilibrium with Applications." NASA TM-4557, NASA Lewis Research Center, Cleveland, OH, 1994.

[12]Preclik, D., Wiedmann, D., Oechslein, W., and Kretschmer, J., "Cryogenic Rocket Calorimeter Chamber Experiments and Heat Transfer Simulations," AIAA Paper 98-3440, 1998.

[13]Hutzel, D. K., and Huang, D. H., *Modern Engineering for Design of Liquid-Propellant Rocket Engines*, Vol. 147, Progress in Astronautics and Aeronautics, AIAA, Washington, DC, 1992.

[14]Hagemann, G., Alting, J., and Preclik, D., "Scalability Discussion for Rocket Nozzle Flows Based on Subscale and Full-Scale Testing," *Proceedings of the 4th European Symposium on Aerothermodynamics of Space Vehicles*, ESA SP-487, 2001.

[15]Alting, J., Grauer, F., Hagemann, G., and Kretschmer, J., "Hot-Firing of an Advanced 40 kN Thrust Chamber," AIAA Paper 01-3260, 2001.

[16]Frey, M., and Hagemann, G., "Status of Flow Separation Prediction in Rocket Nozzles," AIAA Paper 98-3619, 1998.

[17]Frey, M., and Hagemann, G., "Flow Separation and Side-Loads in Rocket Nozzles," AIAA Paper 99-2815, 1999.

Chapter 16

Scaling Techniques for Design, Development, and Test

Carol E. Dexter* and Mark F. Fisher*
NASA Marshall Space Flight Center, Huntsville, Alabama

James R. Hulka[†]
Aerojet General Corporation, Sacramento, California

and

Konstantin P. Denisov,[‡] Alexander A. Shibanov,[§] and
Anatoliy F. Agarkov[¶]
*Scientific Research Institute of Chemical Machine Building,
NIICHIMMASH, Sergiev Posad, Russia*

Nomenclature

A = area, m^2
A_s = effective heat transfer area of the coolant channel, m^2
c = velocity of sound, m/s
c_p = specific heat at constant pressure, $kJ/kg \cdot K$
c_v = specific heat at constant volume, $kJ/kg \cdot K$
c^* = characteristic exhaust velocity, m/s
D = diffusion coefficient, m^2/s^2
Da, i = first Damköhler number
Da, iii = third Damköhler number

---

*Aerospace Engineer.
[†]Engineering Specialist; currently with Jacobs/Sverdrup Technology. Senior Member AIAA.
[‡]Scientific Director, Vice-General Director.
[§]Head, Research Division.
[¶]Senior Engineer, Research Division.

d = diameter, m
f = frequency, Hz
F = function (functional relationship)
Fr = Froude number
g_a = gravitational acceleration, m/s^2
h = enthalpy per unit mass, kJ/kg
h_g = hot gas heat transfer coefficient, $kW/m^2 \cdot K$
h' = acoustic particle displacement, m
k = thermal conductivity, $J/m^2 \cdot s \cdot K$
K_{dil} = dilution ratio
L = length, m
L' = geometrical chamber length (injector face to geometric throat), m
L^* = characteristic chamber length (ratio of chamber volume
         to throat area), m
$\dot{m}$ = mass flow rate, kg/s
M = Mach number
n = interaction index
N = number of injection elements
N_A = amplitude criterion
p = pressure, N/m^2
P_c = thrust chamber pressure, N/m^2
p' = acoustic pressure, N/m^2
Pr = Prandtl Number
q = volume flow rate, m^3/s
q' = heat addition per unit volume, kW/m^3
Q = heat load, kW
Q/A = heat flux, kW/m^2
r = oxidizer-to-fuel mixture ratio
Rc = recovery factor
Re = Reynolds number
Sc = Schmidt number
t = thickness, spray cone sheet film thickness, m
T = temperature, K
T_p = oscillation period, s
T_{aw} = adiabatic wall temperature, K
T_{wh} = hot gas wall temperature, K
T_c = combustion chamber gas temperature, K
v = velocity, m/s
v' = acoustic velocity, m/s
We = Weber number

Greek

α = reciprocal of the equivalence ratio
γ = specific heat ratio
δ_{ac} = acoustic loss coefficient
δ_{gen} = generation coefficient module
δ_t = oscillation decrement

Δ = finite difference
ζ = mass fraction
η_{c*} = characteristic velocity efficiency
μ = absolute viscosity, kg/m·s
ν = kinematic viscosity, m^2/s
ρ = density, kg/m^3
τ = characteristic delay time, s, sensitive timelag, s
τ_{ch} = overall chemical conversion time, s
τ_i = characteristic conversion time of chemical species i, s
τ_r = relaxation time, s
ϕ = equivalence ratio
Φ = dimensionless parameter
Ω = phase criterion

Subscripts

c = chamber
comb = combustion
cool = coolant
cr = critical
d = discharge
enc = enclosure
f = fuel
fs = full scale
in = inlet
inj = injector
m = model
meas = measured
out = outlet
o = oxidizer
pred = predicted
s = scale
sat = saturated
ss = subscale
st = stoichiometric
t = throat
total = sum, complete
Σ = total

I. Introduction

$\mathbf{S}$ CALING of liquid propellant rocket engine combustion devices has been a subject of considerable difficulty and controversy since the beginning of rocketry. The phenomena are complex, the relationships unclear, and the test results often contradictory. Nevertheless, the use of scaling techniques can be invaluable, especially when there is significant emphasis on development cost, because development of rocket engine combustion devices can be prohibitively expensive. In the past, scaling directly from successful preexisting combustion devices was often the best design approach. Subsequently, using test results of specifically scaled hardware in the design phase lessened full-scale design risk and lowered development cost because of the use of smaller hardware, and smaller, lower flow rate, and possibly lower pressure test facilities. The latter approach is the only way to inexpensively evaluate novel and innovative designs by hot-fire test.

Scaling has been defined as the ability to design new combustion devices with predictable performance on the basis of test experience with old devices.[1] Historically, this meant changing—usually increasing—the thrust level of an existing combustor to meet current needs. Usually, thrust was increased by increasing combustor size and mass flow rate, rather than pressure. Often, nearly identical injection elements were packaged in a larger chamber.

An updated definition of scaling would include using design not only from old devices but also from specialized test hardware, and using not only test experience but also analysis. Specialized test hardware, which can be larger or smaller, single- or multi-element, reacting or nonreacting, at different pressure or temperature, or something unique, can improve the means to successfully design new, full-scale hardware. In Russia, for example, making the model or subscale hardware much simpler than the actual object is emphasized to isolate the phenomenon under examination. Scaling techniques are required to make use of test results of this specialized hardware for the design of the new hardware. Analysis can connect these test results to the new design, or even substitute for the testing itself. Thus, scaling techniques can be integrated throughout the design process, rather than used just as a point of departure.

Scaling techniques for liquid rocket combustion devices have evolved from the purely empirical methods of the very earliest period of rocket development, to an analytical treatment of dimensional analyses and similarities in the 1950s, to the current practice of selecting suitable experimental and analytical methods that are strongly driven by cost. This evolution was first categorized by Weller,[2] who grouped the techniques used in similarity analyses into five classes listed in increasing order of a priori knowledge required for application, and decreasing ambiguity of results: 1) empirical scaling or modeling, 2) dimensional analysis and rational scaling, 3) similarities, 4) simplified solutions, and 5) complete solutions.

Empirical scaling uses trial-and-error methods to determine rules for modeling or scaling a system.[2] Much of the development of early liquid propellant rocket combustion devices falls into this category. Often, old devices were tested after the fact in new regimes to show in hindsight how potential problems could have been predicted in new devices. In dimensional analysis, variables are listed that may influence the system, them combined into nondimensional groups. In the method of similarities, formalized equations are written and then nondimensionalized, producing nondimensional groups that avoid the potential for including unimportant variables or groups, as may happen in dimensional analysis.[2] To scale with these two methods, chosen groups are simply kept constant between the old device and the new. Simplified and complete solutions do not rely on the identification of nondimensional groups but on solutions of the formalized equations themselves, with various levels of simplifications and assumptions. The impetus to apply computational fluid dynamics (CFD) to rational design of liquid propellant combustion devices is an example of these categories. For these analyses, a preexisting device is no longer required and the design can proceed directly to full scale. More typically, however, the old device is still required to calibrate assumptions, unknowns, and inlet and boundary conditions and increase confidence in the new hardware design.

Because almost all major design requirements in combustion devices—including performance, stability, and compatibility—originate with the injector, scaling is significantly dependent on injection processes. Consequently, injection processes have received a considerable amount of study. Injector development has been conducted with both nonreacting (cold) and reacting (hot) flows, and with four sizes dependent on the number of injection elements: uni-element, multi-element, subscale, and full scale. Uni-element hardware includes only one injector element, usually without any other means for mass injection, resulting in hot-fire chambers with very large contraction ratios. Element hydraulics can be well characterized, and windowed chambers provide detailed views of injection, atomization, and mixing processes. The Russians have developed a unique methodology using single-element hardware to characterize combustion stability, which is discussed in this chapter.

Multi-element and subscale reacting flow tests have been used on later generation engine development programs, mainly as a means to generate test data inexpensively. They differ only in hardware size and purpose. Both types of tests provide information to characterize injector core performance, nonacoustic and longitudinal combustion stability, and some chamber wall compatibility. The thrust level (or flow rate) of multi-element chambers typically is 10% or less of full scale. Subscale chambers are sized to provide specific information on combustion stability at higher acoustic modes. Typically, the flow rate is about 20% of full scale, although this can be reduced with the use of innovative geometries to generate specific acoustic frequencies.

This chapter discusses the use of scaling in combustion devices for the evaluation of combustion processes, combustion chamber performance, combustion stability, heat transfer, and cycle life. Scaling techniques are derived from the previous results of scaled hardware used in various liquid rocket engine development programs, or developed from basic principles.

II.   Combustion and Performance

The combustion process is the source for understanding how to scale performance, stability, and compatibility in liquid rocket combustion chambers. Combustion and injector performance scaling is discussed in this section in three parts. First, the idea of combustion similarity is reviewed, with brief descriptions of the analytical similitude techniques that were developed in the 1950s.[1-10] Next, the extensive use of nonreacting (or cold) flow techniques in liquid propellant rocket injector development and experimental research is discussed. Last, the use of reduced-size hardware in reacting (or hot-fire) flow for development and research is discussed, along with a perspective of how the cold-flow and hot-fire techniques are related and used for scaling the combustion processes and injector performance.

A. Combustion Similarity

Exact combustion similarity between two combustion flows in chambers of different sizes is a very rigorous requirement, implying that all component processes of combustion, although occurring at different scales, occur in identical fashion.[3] Thus, the flow paths, flame patterns, locations and time histories of species generation and heat release, and contours of temperature, pressure, and velocity, are geometrically similar even though the actual scales may be different.[3]

A set of similarity parameters for liquid propellant rocket combustion flows may be obtained by writing the conservation equations for mass, momentum, and energy in nondimensional form, and identifying the nondimensional groups of parameters that multiply the dimensionless differential equations.[1] A complete set of such groups for exact combustion similarity was defined by Penner[4] for reacting multicomponent gas mixtures neglecting radiant heat transfer and thermal diffusion effects:

$$Re = \frac{\rho v L}{\mu} \tag{1}$$

$$Sc = \frac{\mu}{\rho D} \tag{2}$$

$$Pr = \frac{c_p \mu}{k} \tag{3}$$

$$M = \left(\frac{\rho v^2}{\gamma p}\right)^{1/2} \tag{4}$$

$$Fr = \frac{v^2}{g_a L} \tag{5}$$

$$\Phi = \frac{1/2 v^2}{(c_p/\gamma)T} \tag{6}$$

$$\gamma = \frac{c_p}{c_v} \tag{7}$$

$$\text{First Damköhler Group} = Da,i = \frac{L}{v\tau_i} \tag{8}$$

$$\text{Third Damköhler Group} = Da,iii = \frac{q'L}{vc_pT\tau_i} \tag{9}$$

The first seven groups are familiar from nonreacting flow processes and can be maintained constant even without chemical reactions in the system. Re is the ratio of inertial forces to viscous forces in the unit volume, Sc is the ratio of kinetic viscosity to molecular diffusivity, Pr is the ratio of momentum diffusivity to thermal diffusivity, M is the ratio of kinetic energy of the flow to internal energy (or linear velocity to sonic velocity), and Fr is the ratio of inertial forces to gravitational forces.

Chemical changes in the flow processes are introduced by the two Damköhler groups. Da,i is the ratio of the rate of convection time L/v to chemical time τ_i, or the inverse ratio of specie generation by chemical reaction and the rate of removal by convection. The third Damköhler group Da,iii is the ratio of the rate of heat addition per unit volume by chemical reaction, q'/τ_i, and the rate of removal of heat by convection of enthalpy, vc_pT/L.

Constancy of all nine dimensionless groups for all of the processes between different sized combustion chambers assures that the processes will be similar, because the different combustion flows would then be described by identical nondimensional differential equations. However, because the number of processes occurring in liquid propellant thrust chambers is so large, scaling of reacting flows with complete similarity is found to be practically impossible, because perfect similarity of all processes is impossible to achieve.[3,4] Indeed, all of the processes may not yet have even been identified. Therefore, Penner[4] and Crocco[5] concluded that reasonable conjectures about scaling procedures would be possible only by classifying the physicochemical processes of the combustion into rate-controlling chemical reaction steps, and including only the dominant processes, disregarding the others for engineering purposes.

By assuming homogeneous, low-velocity flow systems without significant external forces, which are reasonable assumptions for the head end of a combustion chamber, Penner reduced the required set of similarity parameters for assuring similar combustion processes to five groups: Re, Sc, Pr, Da,i, and Da,iii.[4] These five groups are equivalent to Damköhler's original five criteria for assuring dynamic and reaction-kinetic similarity in low-velocity flows without external forces and without heat loss to the chamber walls.[4] These five criteria can be satisfied in the scaling of liquid fuel rocket engines only if the reaction rate as a function of engine size can be varied independently (e.g., by modification of injector design),[6] which is a formidable requirement when little is understood about the functional form of the reaction rate. Note that for fixed values of Re and Pr, the Nusselt heat transfer number is expected to be constant, so that the boundary conditions corresponding to heat transfer to chamber walls introduces no new similarity parameter.[4] For a given propellant system with fixed injector temperature, the important similarity groups for stable combustion reduce to Re and Da,i,[7] although M may become important for high-velocity flow processes involving oscillations.[7]

For practical scaling laws, the critical variable remaining in these nondimensional groups is found to be the chemical conversion time τ_{ch}, the completion of all τ_i. Based on the functional form assumed for τ_{ch} a variety of scaling rules for liquid rocket engine combustion chamber geometries can be devised.[1,4,5,7,8] Crocco assumed τ_{ch} was inversely proportional to some power of chamber pressure, generating a scaling rule that preserves combustion similarity and Re, but not M, causing dimensions and thrust to scale with chamber pressure as a function of that power.[5] Penner assumed that chamber pressure be maintained constant, so that τ_{ch} increases with the square of the engine thrust or the square of the dimensions.[8]

Rocket development has provided no conclusions to these assumptions. Because early development of liquid rocket engine thrust chamber hardware was well funded, rapid, and intensely empirical, stable and efficient designs were created in a relatively short time and at great expense, using full-scale hardware almost exclusively. Even later, less well-funded engine development programs—to regulate costly test programs—had little incentive to change these developed and successful designs, even though applications were often considerably different. Thus, the analytical scaling rules had minimal impact during the formative years. Bastress et al.[10] suggested related reasons: 1) the achievement of high combustor efficiency, eliminating the necessity to develop performance scaling rules, 2) the persistence of combustion instability mostly independent of performance considerations, and 3) the increasing complexity of computer-intensive analyses, allowing for more complete analytical solutions upwards on Weller's classifications.

Nevertheless, some of the scaling rules were examined in academic and industrial research programs, as stability reviews of oxygen/hydrogen and oxygen/RP-1 propellants showed.[11] In academia, lower cost methods for rational design and scaling were developed, such as use of nonreacting flows.

B. Nonreacting Flow Testing

To understand how to model the chemical conversion time τ_{ch}, or even to determine what would be the rate-controlling steps, requires a thorough examination of the physical and chemical processes of combustion in liquid propellant rocket engine chambers. For this reason, simplified models and experiments were devised, which naturally required additional examination of how to scale the results to the real combustion process.

Because it is simple, inexpensive, and generally nonhazardous, nonreacting (or cold) flow testing is widely used in the liquid rocket industry for injector development and characterization. Cold-flow testing is almost always used in engine development programs to measure hydraulic resistance of candidate element and injector design changes before hot-fire testing. It is commonly used in research programs for simple experiments on atomization and mixing. Single-element cold-flow testing is used often for the former, and almost exclusively for the latter.

1. Single-Element Nonreacting Flows

To be most useful for scaling, cold-flow testing must simulate some process, for which the data collected can then be used directly to define that process if reactive phenomena do not significantly alter it. Naturally, cold flow does not scale with the Da,i, or Da,iii, and so the processes that can be simulated must occur before reaction has taken place, or in a region where there is less or no influence of the reaction. In a liquid propellant rocket combustion chamber, these can be the injection, atomization, and mixing processes.

Quantitative measurements of injection, atomization, and intraelement mixing processes with single-element cold-flow testing have been conducted to develop correlations to define inputs for combustion analyses. With the atomization and mixing processes defined by cold flow, the combustion analyses could then be used for scaling. An extensive program in the late 1960s and early 1970s[12-21] was a major attempt to use cold-flow testing to characterize the basic processes of atomization and mixing, and ultimately predict injector performance from mixing and vaporization losses. Atomization parameters were measured with molten wax simulants, and mixing was measured with patternators (for liquid/liquid propellants) or an intrusive probe that measured local total gas pressure and liquid mass flux (for liquid/gas propellants). With the cold-flow data input directly into steady-state combustion models (correcting atomization data for different propellant properties), reacting flow injector performance was calculated. Comparisons of these predictions with results from single-element and multi-element hot-fire testing of the same designs were found to be reasonable ($\pm 5\%$), with some exceptions.

To achieve a closer agreement between hot-fire test results and combustion model predictions with inputs defined by cold flow, modifications of the cold-flow inputs were allowed. Thus, an empirical transformation from cold flow to hot-fire was generated: a set of multipliers on intact core length, mean droplet size and size distribution, mean droplet velocity and velocity distribution, and mixing efficiency. These multipliers were model dependent, used to force the model to best match hot-fire testing. However, for a priori prediction of injector performance, some prior knowledge of the input modifications, which are not known for variant designs without the hot-fire data, would be required. This dependency limits the practical usefulness of such models to previously tested or functionally similar injection designs, and inhibits application to new or novel injection designs.

The lack of more generally applicable scaling or transformation laws between the cold and hot-fire flows is due to the lack of basic understanding of the atomization and mixing characteristics. Nearly every injector element design has its own unique injection, atomization, and mixing characteristics, and a generic model describing these processes has not been developed. Much of the cold-flow testing previously cited was conducted to address this limitation. The impact of reacting flow on these processes can be significant and unexpected, however. Scaling the Re, Sc, and Pr may be possible (but not simple) with proper selection of cold-flow simulants, but scaling the Damköhler groups is not possible because there is no heat release or chemical conversion of species. For reactive flows with longer atomization, mixing, and vaporization time scales (those processes prior to reaction), the effects of these processes on the primary atomization and mixing processes are less important, and cold-flow simulations may be useful. Cold-flow simulation would then be required only to match fluid dynamic conditions (including the effect of chamber gas density, for example). For some reacting flows, however, reactive phenomena such as blowapart (an explosive demixing common to hypergolic propellants), carbon formation (particulate condensation in kinetically limited combustion), and precombustion (shortening of the process time scales to reaction, such as in recessed coaxial elements) completely alter the processes that can be measured in cold-flow testing.

Consequently, single-element cold flow has been used predominantly for measurement of hydraulic resistance or qualitative assessment of atomization and mixing potential, or used only indirectly by the application of previously established correlations. Hydraulic testing with Space Shuttle main engine

(SSME) elements, for example, was used to tailor element inlet orifice diameters at different locations on the injector face,[22] whereas cold-flow mixing tests were used only to corroborate what had already been measured in hot-fire tests.[23] Combustion model predictions using data from these cold-flow tests underestimated the injector performance measured in hot-fire tests. On injector development for the orbital maneuvering subsystem (OMS) rocket engine for the space shuttle, single-element cold flows were examined visually to qualitatively assess injection and atomization characteristics.[24,25] The characteristics of designs that were difficult to analyze—such as the fan structure and atomization quality of platelet manifolded impinging elements, for example—were quickly assessed with these tests.

2. Multi-Element Nonreacting Flows

Cold-flow testing of multi-element hardware has been limited primarily to qualitative checkout of injector spray formation and measurement of the injector hydraulic admittance. Occasionally, the mass flow distribution across the injector face is measured, usually with small intrusive probes. Maldistributions on the SSME injector have been measured in ongoing efforts to determine the sources of performance losses and compatibility problems limiting chamber life.[26] Full-scale cold-flow patternations of OMS injectors were used to estimate the effects of injector face maldistribution on injector performance and combustion stability.[24,27,28] In research programs, Dickerson et al.,[12] Falk et al.,[13] and Falk[15] conducted multi-element injector cold-flow mixing experiments and compared these results successfully to hot-fire results.

C. Reacting Flow Testing

Cold-flow testing is simple and inexpensive, but often its usefulness is diminished by not properly assessing the influence of the reaction. Including the reaction, but reducing the combustor size, pressure, or flow rate, is another means to study the processes required to define the chemical conversion time τ_{ch}. Hot-fire testing eliminates the uncertainties and limitations in cold-flow testing because of the lack of reactive phenomena. However, additional limitations arise because of the practical impossibility of perfectly simulating all of the processes in liquid propellant rocket combustor flows. In the following section, hot-fire testing of single element,[14–18,29–38] multi-element,[24,25,27] or subscale injectors[24,25,27,29,39] conducted to investigate combustor phenomena and to reduce development costs due to full-scale hardware testing is discussed.

1. Single-Element Reacting Flows

Single-element hot-fire testing has been used primarily to measure global, externally sensed parameters (such as characteristic velocity, heat load, combustion chamber static pressure profile, injector pressure drop variations, and oscillatory pressures),[14–18,29,30,38] or to qualitatively observe flame structure or characteristics.[29–35] Single-element hot-fire testing was used on J-2 engine injector development to resolve design features being applied to full-scale injectors at that time, especially regarding the effects of coaxial element oxidizer post recess and fuel sleeve design on injector pressure losses.[29] Some single-element hot-fire testing was used to measure local, internally sensed parameters such as gas composition,[31,32] and droplet size,[36] and the use of laser diagnostics has provided a significant advance in internal capability to make these measurements.[40] Single-element hot-fire testing has also been used to attempt to develop correlations to measurements made in cold-flow testing.[31,32,34–36]

In nonwindowed combustion chambers, hot-fire measurements provided evaluation of steady-state combustion models that used atomization and mixing measurements from cold-flow experiments,[12–21] as discussed in Section II.B.1. Comparisons of mixing-limited (completely vaporized) experimental and analytical efficiencies were reasonable ($\pm 5\%$), but discrepancies were larger when vaporization losses were included, indicating that the atomization measurements were less reasonable. Addition of the effects of secondary atomization,[18–20] simulating essentially the effect of the accelerating combustion gas on the atomization, improved the prediction.

Some hot-fire test programs demonstrated conclusively that cold-flow measurements were not applicable for certain reacting flows, including where reactive stream separation occurred with hypergolic propellants[30] and carbon formation occurred with fuel-rich oxygen/hydrocarbon propellants.[33] The hot-fire tests also identified the injection geometries and operating conditions in which these conditions occurred.

In other flows, the lack of applicability of cold-flow measurements was a surprise. In gas/gas combustion,[31,32] gas samples of the hot flow were obtained with a cooled intrusive probe and analyzed with a mass spectrometer. Relating the

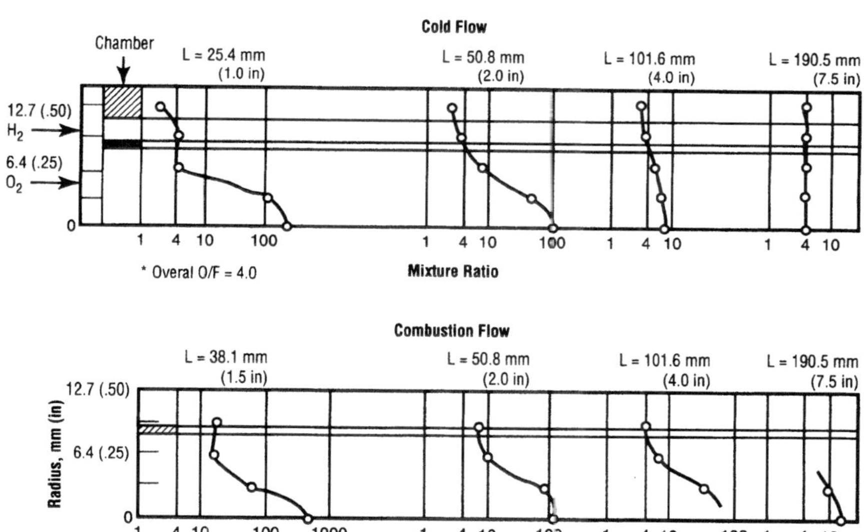

Fig. 1 Comparison of empirical mixture ratio profiles from gas/gas testing (modified from Ref. 32).

spectrometer data to the partial pressure of the individual species, mixture ratio was determined. Comparisons with similar measurements in cold but dynamically similar flows showed wide variations in mixture ratio evolution, as shown in Fig. 1.[32] The rapid chemical reaction in gas/gas combustion seemed to interrupt the mixing process prior to its completion, unlike what was observed in the cold-flow mixing experiments. In liquid/liquid and liquid/gas propellant combustion, there may be combustion delays sufficient to allow the propellants to mix before significant reaction has occurred, so that cold-flow mixing experiments may be realistic and capable of characterizing hot-fire conditions.

Spatial liquid droplet distribution measurements in cold-flow and hot-fire conditions measured with a laser holographic technique were compared by George.[36] For a like-on-like doublet element, the size intervals in which the greatest number of droplets occurred, and the range over which the droplets appeared, were reduced in hot-fire compared to cold-flow conditions. The spray field in hot-fire conditions exhibited significantly less ligamentation and discernible waves of spray, although the primary fan formation and overall fan shape appeared the same. The droplet sizes displayed little correlation between cold-flow and hot-fire conditions; the range of droplet sizes in hot fire was half that observed in cold flow for all conditions.

Recent comparisons of reacting and nonreacting shear coaxial jets showed large differences in atomization structures at subcritical conditions, but similarities at supercritical conditions.[34,35] Both subcritical hot fire and supercritical cold flow and hot fire exhibited less ligament structure than the subcritical cold flow. In the former, no droplets were observed at all.[35]

In the past decade, the increasing use of laser-based nonintrusive diagnostics has provided the beginnings of a more quantitative direct comparison between reacting and nonreacting flows. The amount of information that can be extracted with these local, internal measurements from the difficult-to-examine reacting flows has increased significantly,[40] along with the quality of the information available from nonreacting flows. One of the initial comparisons of temporal droplet distribution measurements between cold-flow and hot-fire experiments yielded larger mean drop sizes for hot-fire conditions at similar downstream locations, although large differences in Re and We were still present.[37] This unobvious measurement is probably due to the rapid vaporization of the small drops in the hot fire.

The use of reacting flows greatly improves the simulation and scaling applicability of single-element testing, because it includes the effects of the reactions identified by the Damköhler groups. However, other factors arise that limit the applicability, mainly a result of the limitations from placing a single element into an acceptable chamber. The lack of adjoining element flows results in unrealistic recirculation zones near the injection region, and in very low chamber Mach numbers. For typical injectors with tightly packed elements, interelement effects can be substantial. For this reason, tests conducted in multi-element hardware are more realistic.

2. *Multi-Element Reacting Flows*

Multi-element and subscale hot-fire testing have been used in engine development programs for injector performance characterization. In OMS injector development, injector performance was measured on multi-element injectors $10\%^{27}$ and $17\%^{41}$ of full-scale thrust (or 33% and 46%, respectively, of full-scale chamber diameter). Predicted injector energy release efficiencies on full-scale injectors with identical elements in combustion chambers of equal length agreed within $0.5\%^{41}$ and 0.5–1.5%, respectively.[24] In a development program with hydrocarbon propellants,[39] full-scale injector energy release efficiencies agreed within 0.5–1.5% of data from a subscale injector 44% of full-scale diameter with identical O-F-O triplet elements.

These variations are not surprising, in reality, because with geometrically identical injection elements spaced identically, with the same propellants and operating conditions, the combustion should be exactly the same, and hence injector performance should be similar. The reasons it is not similar, besides instrumentation accuracy, are often subtle, such as changes in chamber characteristic volume, chamber wall surface area, chamber contraction ratio, and injector manifolding, resulting in local variations and different losses. Differences with chamber proportions when scaling conventional combustion chambers were described by Ross,[9] as shown in Fig. 2. In Fig. 2a, both the injected weight flow per unit area $(\dot{m}_s/A_c)$ and characteristic length (L^*) are constant when scaling up in thrust, but the barrel section of the combustion chamber is shortened. If the processes leading up to complete reaction for a particular injector are sufficiently long, some processes may occur in the convergence portion of the nozzle, where M would influence the overall performance. The variable barrel length is exchanged for a variable nozzle convergence length in Fig. 2b,

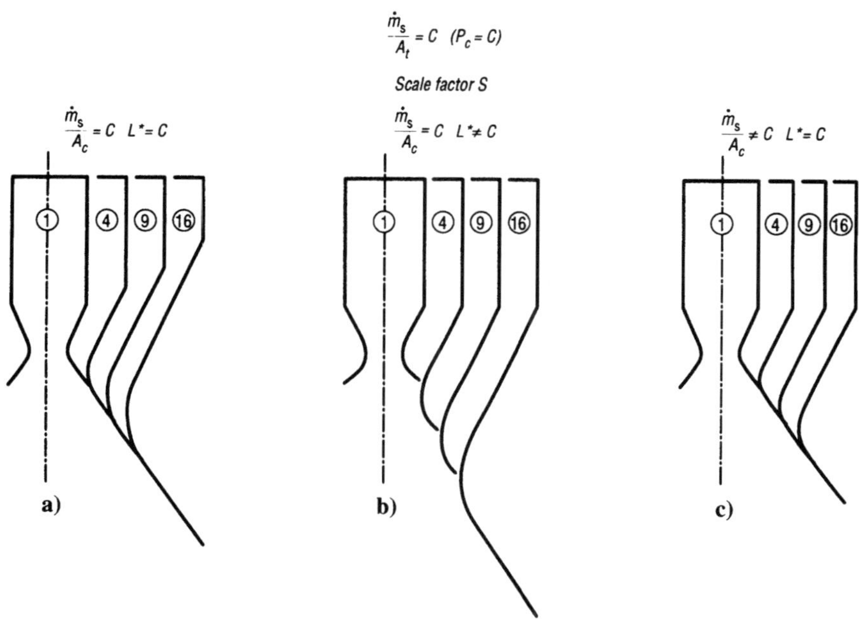

Fig. 2   Geometric scaling methods for combustion chambers.[9]

where L^* is no longer kept constant. This results in a longer mean chamber length (L') and different residence time for combustion, which will influence the performance for vaporization-limited combustion. Figure 2c shows the compromise in which $\dot{m}_s/A_c$ is no longer constant, which now introduces a different injector pattern in subscale and the resulting influence of interelement effects on performance.

In addition to changes in the combustion due to scaling the combustion chamber, subtle differences may occur when scaling the manifolding. The physical realities when scaling injectors of manifolding the propellants to achieve constant $\dot{m}_s/A_c$, even when identical elements are located the same distance apart, are not simple. Supplying propellant from a single inlet into a larger manifold may not provide the same distribution of flow across the injector face, because the injector face area has increased. Adding additional inlets presents the difficulties of creating regions of overfeeding and stagnation that were not initially present, creating regions of local higher and lower flow rate at the injector face. With these difficulties, correlating injector performance between scales within 1% is acceptable.

III. Combustion Stability

The objective of evaluating combustion stability with scaled hardware is to understand the stability characteristics sufficiently to minimize or eliminate instability in the full scale. Testing with specialized scaled hardware should be simple and quick, providing direction for effective solutions such as identifying more suitable injector elements and determining whether stability will improve or worsen with design changes. The testing should determine the boundaries of oscillation, excitation, and disappearance (including any hysteresis), along with the parametric criteria for predicting full-scale chamber operating parameters and stability based on the operating parameters of the model or subscale chamber.

In this section, two methods of stability scaling techniques are presented, one based on uni-element hardware and the other on multi-element hardware. In Russia, a unique methodology using a low-pressure, single-element test article was developed. The methodology and examples of its use are described in Section III.A.

In the United States, stability-scaling techniques have progressed using multi-element hardware that recreates the frequency ranges of interest. Generally, the injection elements and operating conditions are equivalent to full scale, and the combustion chamber is specially designed to generate a range of specific acoustic frequencies. These techniques are described in Section III.B.

A. High-Frequency Instability Modeling in a Low-Pressure Single-Element Setup

1. Concept of Low-Pressure Combustion Models

Modeling of high-frequency instability using a single-element, low-pressure hot-fire experiment is based on the following approaches and assumptions:

1) General principles of approximate partial modeling or simulation of a complex process are employed. This approach permits selecting one or several physical phenomena from the great number that constitutes full-scale combustion processes, which represent the most typical features of the aspect under study.

With this approach, satisfying all of the similarity conditions is not mandatory, and success depends on correctly detecting the governing parameters and reproducing them in modeled conditions. In addition, the physical model should not only represent the full-scale processes correctly, but also be much simpler. The comparison of model and full-scale test results is the most comprehensive means of verifying the assumptions made in the development of the approximate partial modeling method.

2) Injector elements with full-scale geometry are used in the model. This approach provides for the most convenient comparison between the model and the full scale.

3) The influence of neighboring element sprays on the combustion process in the initial section of the spray (which is mostly sensitive to disturbances) can be neglected, because bipropellant injector elements are assumed to operate, to a substantial extent, independently.

4) Selections of boundary conditions and governing criteria are based on the physical concepts of the process being studied.

5) The boundaries of regions of spontaneous excitation and damping in the model tests are determined by propellant mass flow rate variation. Changes of the boundary positions are indicative of the increased or decreased combustion stability to soft (spontaneous) or hard (dynamic) excitation.

6) Combustion chamber mean pressure p exerts no principal influence on chamber acoustic field spatial parameters, which are defined by the relative value of acoustic pressure oscillations p'/p.

7) The phases of the combustion chamber acoustics and the combustion processes in the model should be identical to that of the full-scale chamber, i.e.,

$$\Omega = (\tau f)_m^{-1} = (\tau f)_{\text{fs}}^{-1} \tag{10}$$

Here $f = 1/T_p$ is the acoustic oscillation frequency specified by oscillation period T_p and τ is the characteristic delay time, i.e., the duration of the propellant conversion to combustion gases. A set of design and operating parameters, proportional to τ, is determined by analyses of the physical features of the processes and the known analytical or experimental relations for typical atomization and mixing patterns. The processes, as far as the amplitude criterion $N_A = |\delta_{\text{gen}}|/\delta_{\text{ac}}$ is concerned, are assumed to be self-similar in most cases; therefore, the dimensionless phase criterion Ω is the only stability parametric criterion to be determined.

8) The phases of the injector manifolding acoustics and the injection processes in the model should be identical to that of the full-scale injector, i.e.,

$$\Omega_{\text{inj}} = (\tau_{\text{inj}} f)_m^{-1} = (\tau_{\text{inj}} f)_{\text{fs}}^{-1} \tag{11}$$

This identity represents the time delay τ_{inj} of propagation of acoustic disturbances along an injector passage of length L_{inj} at a sound velocity c_{inj}, where $\tau_{\text{inj}} \sim L_{\text{inj}}/c_{\text{inj}} \sim L/f_{\text{inj}}$, and f_{inj} is the natural frequency of the injector being investigated. This identity is satisfied by selecting proper geometry of the model feed manifolds and by setting such gaseous propellant temperature at which the sound velocity in model and full-scale tests would be the same, $c_{inj,m} = c_{inj,fs}$.

9) The combustion chamber transverse oscillation frequencies in the model should be the same as in the full-scale chamber,

$$f_{c,m} = f_{c,fs} \tag{12}$$

This condition is satisfied by a proper selection of model chamber diameter d_m, accounting for effective chamber combustion sound velocity c_m:

$$d_m = \frac{d_{fs} c_m}{c_{fs}} \tag{13}$$

10) Injector elements to be investigated in a single-element setup should be placed close to the combustion wall, where tangential mode oscillation amplitudes are highest. High-frequency combustion instability is most often encountered during the development of large combustion chambers and gas generators for liquid rocket engines, and instabilities in transverse tangential modes are most likely to occur.

11) Mixing is the governing factor in the whole complex of physical and chemical processes in combustion chambers and gas generators. This is because in high pressure liquid rocket engines, especially with staged combustion cycles, atomization and vaporization are not rate-limiting factors in the entire complex of physical and chemical processes involved in combustion, as they are either completed very quickly or are actually absent. Also, at these high pressures and high temperatures, chemical kinetic processes proceed very quickly and have little part in the total combustion duration in the combustion zone. Thus, because atomization, vaporization, and kinetics process times are relatively unimportant under actual conditions of rocket engine combustion, mixing should be the rate controlling stage of the entire combustion process. The response times of the combustion zone processes should be mainly defined by the mixing time τ_{mix}.

12) Propellants, actual or simulated, are in the gas form. In high-pressure liquid rocket engines, combustion occurs at pressures above the critical pressures of the propellants used, and propellant temperatures at the injector inlet are close to the critical temperature. Under these conditions, the physical properties of the oxidizer and the fuel being injected approach the properties of the dense gas. Thus, when the conditions leading to high-frequency instability in full-scale engines are modeled at low pressure, the modeled conditions will be closer to the actual ones if gaseous propellants are used instead of liquid. For closed cycle (staged combustion) engines, in which one or both propellants are fed to injectors in the gas phase, this approach using gaseous propellants seems even more justified.

13) A special expedient for simulating mixing is assumed: reactive propellants (oxidizer and fuel) are diluted with inert gases (such as nitrogen and helium). This technique permits the volume flow rates and thus the discharge velocities of fuel or oxidizer to change without changing reactive component mass flow

rate. Consequently, various ratios of the propellant velocities and densities could be provided at the injector element exit with constant values of the reciprocal of the equivalence ratio $\phi = (\dot{m}_0/\dot{m}_f)(1/r_{st})$.

14) Reversing the propellant feeds, i.e., feeding the fuel through the oxidizer passage and feeding the oxidizer through the fuel passage, is allowed. Under some conditions, this may provide a better approximation of the model conditions to the full scale.

In a low (atmospheric) pressure model chamber with the oscillation frequency $f_m = f_{fs}$, full-scale amplitudes of the acoustic velocity, $v'_m = v'_{fs}$, and of the particle acoustic displacement

$$h' = \left(\frac{1}{2\pi}\right)\frac{v'}{f}\bigg|_m = \left(\frac{1}{2\pi}\right)\frac{v'}{f}\bigg|_{fs}$$

may be achieved; these amplitudes define the acoustic field geometric similarity. These similarities are possible because v' and h' are proportional to the oscillation relative amplitude $\bar{p}' = p'/p$ and not to absolute value p'. The low oscillation amplitude p' in combination with low mean pressure, $p \approx 0.1$ MPa, simplifies the procedure of model tests for determining stability boundaries. Long-duration tests can be run in spontaneous instability regimes without damaging the hardware, and special instruments for measuring static and oscillating parameters can be used. The densities ρ of gaseous propellants in a model combustion chamber at the atmospheric pressure are smaller than full-scale liquid propellant densities by a factor of some hundreds. Therefore, with the same volume flow rates

$$q = \frac{\dot{m}}{\rho}\bigg|_m = \frac{\dot{m}}{\rho}\bigg|_{fs}$$

and the same injector discharge velocities

$$v_d = \frac{q}{A_{inj}}\bigg|_m = \frac{q}{A_{inj}}\bigg|_{fs}$$

where A_{inj} is the injector outlet area, the propellant mass flow rates $\dot{m}_m$ of injector elements in the model chamber will be hundreds of times smaller than in the full-scale chamber, $\dot{m}_{fs}$.

Note that the use of a single bipropellant element for model hot-fire tests is not a specific requirement. The best approximation of the model conditions to the full scale is provided by low-pressure, hot-fire testing of a full-scale gaseous propellant injector; however, full-scale injector testing is more laborious and expensive, and less flexible, than model testing. If the full-scale injector includes different types of bipropellant injector elements, model tests should include simulation of the combined effect of the dissimilar elements. The single-element model should be modified into a two- or three-element model (depending on the number of types), with the arrangement of the elements providing an adequate contribution of each element type into the acoustic oscillation energy.

The model chamber is driven unstable by varying the oxidizer and fuel volume flow rates q_o and q_f, which vary the propellant discharge velocities from the element. At the model chamber pressure $p \approx 0.1$ MPa, gaseous propellant densities do not change ($\rho_o = $ constant; $\rho_f = $ constant), and so volume flow rates vary with mass flow rates $\dot{m}_o$ and $\dot{m}_f$. Altering the propellant discharge velocities varies the characteristic delay time τ, which defines the phase conditions of oscillation excitation at a certain natural frequency of the combustion chamber in which the combustion takes place.

The application of the procedure just described should not be extended to the use of monopropellant injectors, or to the operating regimes in which atomization and vaporization are the governing processes. Such conditions may take place, for example, during start transients when propellant pressure and temperature are much lower than the critical values.

2. Methods for Estimating Agreement Between Model and Full-Scale System Properties

The validity of the model design and the selection of operating parameters are verified by comparing model and full-scale test results. The consistency of the results for two or three selected versions of injector elements can be considered a foundation for assuming the same consistency for the rest of the versions. The comparison may be qualitative or quantitative, but aims primarily to verify the validity of the predicted variation of the full-scale engine combustion stability. Thus, for example, the variation in full-scale stability due to changes in operating parameters or injector design can be traced by the direction of the change in the oscillation decrement δ_t and oscillation relative amplitude $\bar{p}' = p'/p$. These are determined from the spectral characteristics of combustion chamber oscillations, usually for the first tangential and the first longitudinal modes, which, depending on the combustion chamber geometry, are the most sensitive or "dangerous" natural acoustic oscillations because the oscillation amplitudes are usually the largest for these modes.

The change in combustion chamber dynamic stability resulting from artificial disturbances of finite amplitude can be determined by the change in the damping oscillation relaxation time τ_r on applying calibrated pulses to the chamber. The increase in δ_t and decrease in $\bar{p}'$ and τ_r suggest improvement in combustion process stability.

A change in spontaneous combustion stability margin of the model chamber can be estimated by the change in the boundaries of the regions of spontaneous excitation and any hysteresis, if present. Combustion stability margin can be estimated by a dimensionless quantity, $\bar{R}$, which designates how far the full-scale operating regime being modeled is from the predicted boundaries of the regions of instability and hysteresis, as shown in Fig. 3. The quantity $\bar{R}$ is determined by various operating parameters X and Y that are used as coordinates for the presentation of

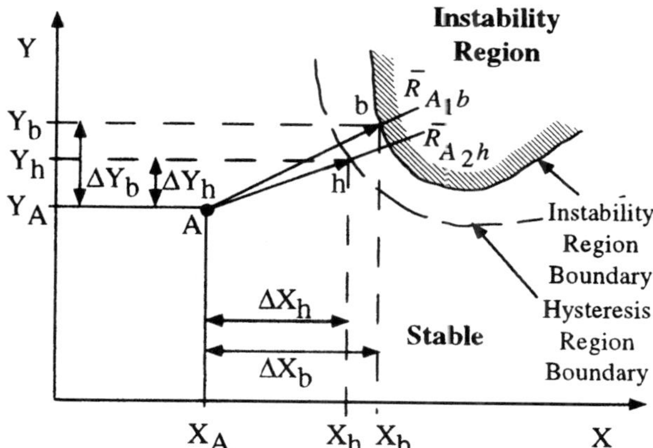

Fig. 3 The method for determining the stability margin for any characteristic parameter (A is the nominal operating point).

experimental data on stability and hysteresis region boundaries:

$$\bar{R} = \sqrt{\bar{R}_X^2 + \bar{R}_Y^2} \tag{14}$$

where

$$\bar{R}_X = \Delta X/X_A \qquad \bar{R}_Y = \Delta Y/Y_A$$
$$\Delta X_b = X_b - X_A \qquad \Delta X_h = X_h - X_A$$
$$\Delta Y_b = Y_b - Y_A \qquad \Delta Y_h = Y_h - Y_A$$

Here, X_A and Y_A are the coordinates of some point A that belongs to the operating region of the full-scale chamber and is situated close to the instability region boundaries; X_b and Y_b are the coordinates of a point on the stability region boundary; and X_h and Y_h are the coordinates of a point on the hysteresis region boundary that is the closest to the operating point A.

The proximity of the full-scale chamber combustion process to the region of hysteresis indicates the possibility of high-frequency oscillation excitation by random pulse disturbances. In rating versions of injectors or elements, those with higher $\bar{R}_b$ and $\bar{R}_h$ are considered more stable. A narrowing of the instability region, or the reduction of self-oscillation relative amplitude $\bar{p}_m = p'_m/p_m$, even without changes in the position of stability boundaries, also indicates the usefulness of the action that caused such an effect.

A similar direction of the change in full-scale chamber parameters $(\delta_t)_{fs}$, $\bar{p}'_{fs}$, and τ_r, on the one hand, and model chamber parameters $\bar{R}_b$, $\bar{R}_h$, and $\bar{p}'_m$, on the other hand, indicates a qualitative similarity of phenomena governing combustion instability by soft (i.e., spontaneous) and hard (dynamic) pressure oscillation excitation under actual and simulated conditions.

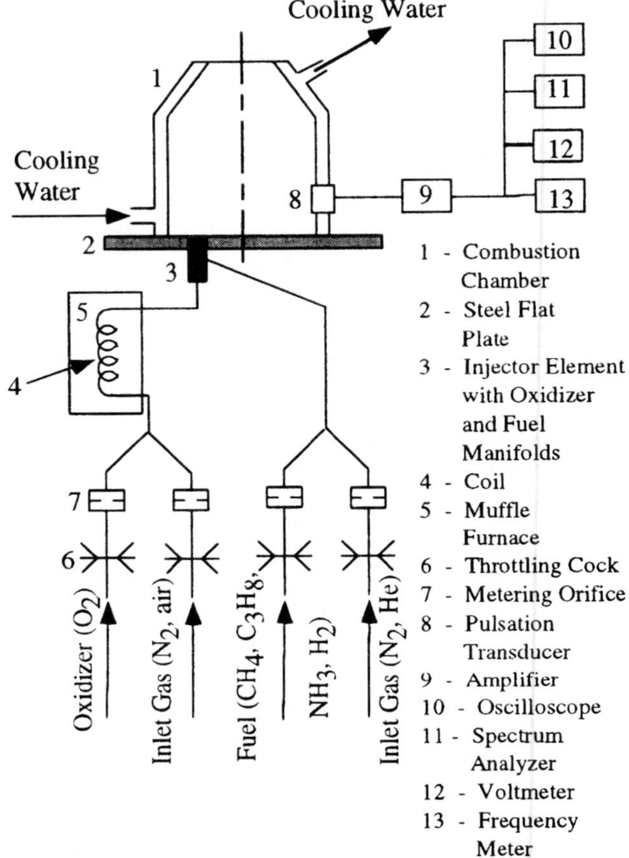

Fig. 4   Schematic of single-element model setup and instrumentation.

The quantitative comparison of model and full-scale chamber operating parameters on the stability region boundaries should be based on the parametric stability criteria that are to be determined, i.e., the homochronicity criteria and other governing criteria discussed in Section III.A.1. All these criteria are required not only for predicting the position of stability region boundaries for actual operating conditions but also for selecting model operating conditions in the course of model setup development.

3.   *Example of a Single-Element Model Experiment and Test Conditions*

A schematic of a model combustion stability experiment is shown in Fig. 4. The combustion chamber-resonator (1) is a water-cooled cylinder, the inner diameter of which is selected so that the condition $f_m = f_{fs}$ is met. Chamber pressure is atmospheric. The chamber may include a transparent section for visual observation of the combustion processes. The chamber is positioned vertically onto a steel flat plate (2), which acts as an injector faceplate simulator. The exit section of the element under study (3) is mounted flush with the plate surface and the element itself is positioned close to the chamber wall for providing excitation of the tangential mode. Ignition is provided by the flame of an external igniter.

The experiment is designed to use gaseous oxygen, which may be diluted with inert gases such as N_2 or air. The oxidizing gas is heated in a coil (4) placed in a furnace (5). CH_4, C_3H_8, NH_3, and H_2 are used as fuels, which may also be diluted with inert gases such as N_2 or He. Oxidizer and fuel flow rates are controlled with throttling valves (6) or special gas-distributing devices, and measured with supercritical pressure orifices (7).

The instability region boundaries are determined by continuously increasing or decreasing propellant flow rates until either periodic oscillations occur, establishing the location of the stability boundary, or disappear, establishing the hysteresis boundary. An example of variations in operating regimes and spontaneous and hysteresis stability boundaries is shown in Fig. 5.

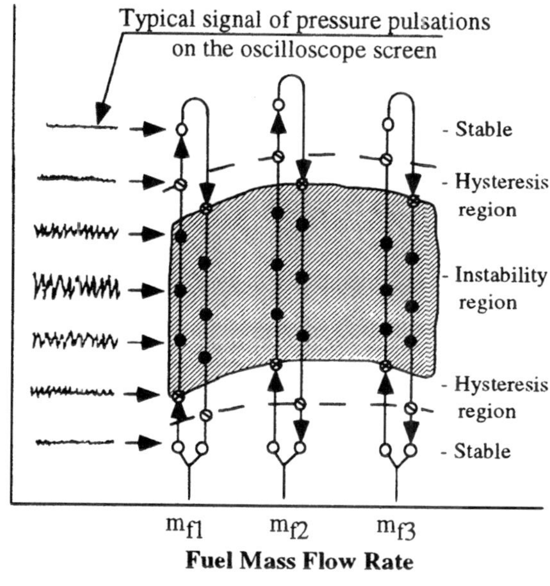

Fig. 5 Experimental method for determining instability region boundaries in a model setup.

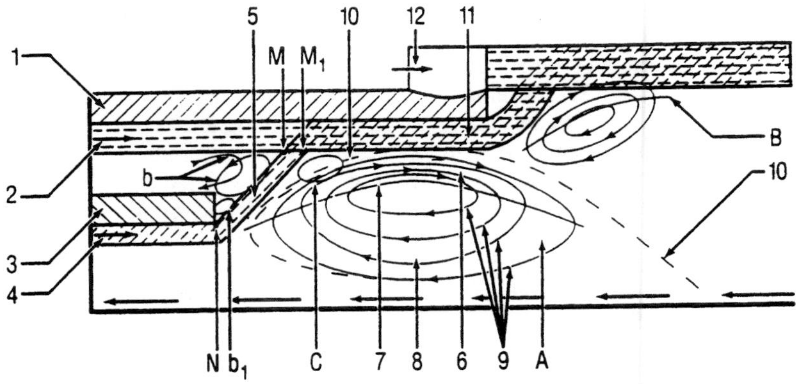

1 — oxidizer injector nozzle
2 — oxidizer sheet
3 — fuel injector nozzle
4 — fuel sheet
5 — fuel sheet cone
6 — recirculation zone direct flow
7 — boundary of reverse velocities zone
8 — recirculation zone reverse flow
   (flow of reverse velocities zone)
9 — streamlines (lines of equal velocities)
10 — recirculation zone boundary
11 — bipropellant sheet of combusting mixture
12 — oxidizer jet injection
A — main post developed recirculation zone
b — recirculation zone in the inter-sheet space
b_1 — local recirculation zone
B — exit section recirculation zone
C — local vortex
N, M, M_1 — flame stabilization characteristic points

Fig. 6 Schematic of flow structure in a swirl injection element.

4. *Example of Combustion Modeling*

In this section, a model of a full-scale bipropellant element for an oxidizer-rich preburner of an engine for the Energia launcher is presented. Hot-fire tests of the full-scale preburner element in the actual regime were performed in a self-contained liquid oxygen (LOX) and kerosene power unit. Preburner chamber pressure was about 50 MPa, i.e., much higher than the critical pressures of oxygen ($p_{cr} = 5$ MPa) or kerosene ($p_{cr} = 1.71–2.31$ MPa). Oxygen, after passing through a cooling jacket, entered the element at a temperature close to critical ($T_{cr} = 154$ K). Thus, the physical properties of the oxidizer were close to the properties of the dense gas.

The baseline version of the bipropellant injector element is illustrated in Fig. 6. Fuel was injected with swirl from a post inside a coaxial oxidizer post. The end of the fuel post was recessed 15–27 mm from the end of the oxidizer post. The oxidizer was also injected with swirl. The fuel swirl and post recess initiated the primary zone of high-temperature combustion products with $r_{inj} = \dot{m}_{o,inj}/\dot{m}_f = 8\text{--}12$. Outside the outer diameter of the oxidizer post, the remaining oxygen was injected through rectangular channels. This excess oxygen surrounded the high-temperature combustion gases, diluting them to an overall mixture ratio $r_{\Sigma} = \dot{m}_o/\dot{m}_f = 33\text{--}75$, with corresponding cooler temperatures.

During the initial modeling, the physical nature of the preburner process was analyzed, and the following items were considered:

1) A thin fuel sheet (film thickness $t_f < 0.5$ mm), propagating through the long ($L/d = 13$) swirling channel, is heated by hot combustion gases that are ingested by the gas vortex into the fuel element. Therefore, the use of gaseous propellants for simulating hydrodynamic effects of the fuel seems justified.

2) Because of the thinness of the fuel and oxidizer sheets ($t_f < 0.5$ mm and $t_o < 2.0$ mm), upon impingement the sheets are assumed to penetrate each other and mix intensively. Thus, the mechanisms that govern the burning of the bipropellant gas mixture formed inside the element approach the mechanisms of a completely homogeneous gas mixture with a fuel mass fraction $\zeta_f = 1/(1 + r_{inj})$.

3) The coaxial position of the fuel swirl element inside the oxidizer swirl element provides conditions for changing the direction of the fuel sheet because of the gas ejecting from the space between the fuel sheet cone and the oxidizer channel. The rarefaction in this space, ΔP_{enc}, may be two orders of magnitude higher than the rarefaction in the vortex zone of the central (fuel) element, and thus may become one of the governing parameters in the interaction process of the sheets. ΔP_{enc} is determined by the total surface area of the spray cones that eject gas from this space and the oxidizer and fuel velocities in the sprays, v_o and v_f. With sheets of fuel and oxidizer impinging within the bipropellant element, ΔP_{enc} does not depend on preburner chamber pressure and is governed only by v_o and v_f. The effects of discharging sheet densities to preburner combustion gas density may be neglected. As the oxidizer sheet surface area, to the point of impingement, is approximately 20 times as large as that of the fuel, ΔP_{enc} is defined mostly by v_o, whereas the fuel sheet defines the volume of the interspray zone and the position where the interaction of the sheets starts.

Oxygen/hydrocarbon propellants were chosen for the element model hot-fire testing, with gaseous oxygen as the oxidizer and gaseous methane as the fuel. For convenience, the temperatures of the propellants were made close to room temperature ($\sim 20°C$). With equal volume flow rates per one element of the liquid oxygen (full-scale conditions) and of the gaseous oxygen (model conditions), i.e., $q_{o,m} = q_{o,fs}$, the oxidizer swirl element in the model chamber will operate with Reynolds number at the element inlet, $Re_{o,in}$, close to the full scale. With equal volume flow rates per one element of the liquid fuel (full-scale conditions) and of the gaseous fuel (model conditions), i.e., $q_{f,m} = q_{f,fs}$, the Reynolds number from the selected gaseous fuel (CH_4) is reduced by more than an order of magnitude from the full scale, far beyond the region of self-similarity. This may result in the reduction of the fuel sheet cone angle and of the gas vortex radius. Therefore, Reynolds number should be considered the governing criterion of the hydrodynamic similarity for the fuel element, and the model should retain the full-scale value for self-similarity. For an element of full-scale geometry, the condition $Re_{f,m} = Re_{f,\text{fs}}$ is equivalent to the condition

$$\left.\frac{q_f}{v_f}\right|_m = \left.\frac{q_f}{v_f}\right|_{\text{fs}}$$

where v_f is the fuel kinematic viscosity. Similarly, the condition $v_{o,m} = v_{o,\text{fs}}$, required for $\Delta P_{\text{enc},m} = \Delta P_{\text{enc,fs}}$ to be satisfied, is equivalent to $q_{o,m} = q_{o,\text{fs}}$ for the full-scale element.

From this modeling of fuel and oxidizer sheet aerodynamics and geometry, element dynamic processes, and flame stabilization, three governing criteria are derived for acceptable simulation of preburner combustion instability:

1) process homochronicity criterion,

$$\Omega = \left.\frac{v_o}{L_f}\right|_m = \left.\frac{v_o}{L_f}\right|_{\text{fs}} \tag{15}$$

2) fuel injector Reynolds number criterion,

$$Re_{f,m} = Re_{f,\text{fs}} \tag{16}$$

3) fuel mass fraction in the propellant mixture within the element reaction zone,

$$\zeta_f = \left.\frac{1}{1 + r_{\text{inj}}}\right|_m = \left.\frac{1}{1 + r_{\text{inj}}}\right|_{\text{fs}} \tag{17}$$

For an element of actual geometry ($L_m = L_{\text{fs}}$), exciting model chamber oscillations of natural frequency ($f_m = f_{\text{fs}}$), the requirement of Ω identity is reduced to v_o identity or q_o identity.

If from these three criteria only Eq. (15) is defined, experimental data on stability should be presented as the functional relation $\Omega = F(Re_f, \zeta_f)$. If Eq. (16) is satisfied, then the functional relation can be simplified to $\Omega = F(\zeta_f)$, and plots on the coordinates q_o and ζ_f can be constructed.

In the example, diluting gaseous CH_4 with N_2 increased the model fuel volume flow rate to satisfy Eq. (16). The dilution ratio $K_{\text{dil}} = \dot{m}_{N_2}/\dot{m}_f$ was determined by the requirement to preserve the identity of the three criteria.

For simulating actual full-scale mixture ratio range, $r_\Sigma = 33.5-75$, the dilution ratio was varied in the range $K_{\text{dil}} = 4.72-7.16$. For the sake of model testing simplicity, the mean value of $K_{\text{dil}} = 6.0$ was adopted, which corresponded to $r_\Sigma = 51.6$.

The validity of these physical concepts and the sufficiency of the selected similarity criteria were experimentally verified during the first stage of the model operation. Various means of verification included 1) special studies of the effect of various combinations of separate parameters comprising the criteria, 2) using the results of burnout curves, optical experiments, and high-speed filming for determination of unstable combustion, and 3) direct quantitative comparison of model and full-scale tests. In the direct quantitative comparison, the coordinates q_o and ζ_f were revised to the coordinates of total flow rate $\dot{m}_\Sigma$ and total mixture ratio r_Σ, because the results of full-scale preburner hot-fire tests are usually presented using these coordinates.

The procedure of scaling the model operation to the full-scale operation was as follows:

1) The actual full-scale preburner oxidizer mass flow rate was determined on the basis of the identity of oxidizer volume flow rate per element ($q_{o,m} = q_{o,\text{fs}}$):

$$\dot{m}_{o,\text{fs}} = q_{o,m}\rho_{o,\text{fs}}N \tag{18}$$

where N is for the full-scale preburner.

2) The full-scale preburner mixture ratio was determined on the basis of the identity of fuel mass concentration inside the element ($\zeta_{f,m} = \zeta_{f,\text{fs}}$),

$$r_{\Sigma_{\text{fs}}} = \alpha_m r_m + \frac{K_{\text{dil}}}{\zeta_o} \tag{19}$$

where α_m is the reciprocal of the model chamber total equivalence ratio, $r_m = 4$ is the stoichiometric mixture ratio of the model propellants ($O_2 + CH_4$), and $\zeta_o = \dot{m}_{o,\text{inj}}/\dot{m}_o$ is the oxidizer fraction entering the oxidizer swirl channel.

3) The propellants total mass flow rate in the full-scale preburner was determined by

$$\dot{m}_{\Sigma,\text{fs}} = \dot{m}_{o,\text{fs}}\frac{r_{\Sigma,\text{fs}} + 1}{r_{\Sigma,\text{fs}}} \tag{20}$$

4) The plot of instability region boundaries was constructed on the coordinates $\dot{m}_{\Sigma,\text{fs}}$ and $r_{\Sigma,\text{fs}}$.

Five preburner injector element configurations were selected for a quantitative comparison of model and full-scale test results. The results from two of these elements are shown in Fig. 7. Very satisfactory quantitative consistencies of the stability region boundaries for the model operation and full-scale operation were found, observed by comparing the region described by the model testing (circles) and that tested by the full scale (other symbols). The testing confirmed the validity of the selected procedure so that the next stage of testing could begin, which was estimating the effects of variations of the injection configuration on combustion stability. The estimation was performed by calculating the distance of the instability region boundary to the actual operating regimes being modeled, using Eq. (14).

Forty-two versions of injector elements were tested in the model. Extensive information on the influence of injector design parameters and injector faceplate baffles on stability was obtained and used to direct combustion stability improvements during the development of the full-scale preburner. The combustion stability results obtained were interpreted on the basis of an instability mechanism because of the combined influence of injector dynamic characteristics and factors leading to the change of the flame base stabilization position inside the injector.

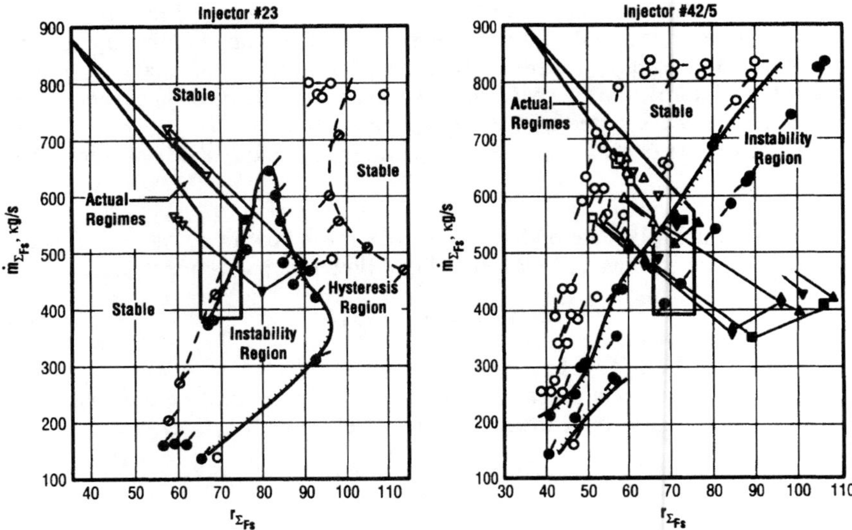

Fig. 7 Comparison of model and full-scale gas generator experimental stability results (O—model results; △, ▽, ◇, □,—full-scale results).

B.　Stability Scaling with Multi-Element Hardware

The influence of combustor size on combustion stability became evident during the Hermes, Vanguard, and Vega rocket development programs in the United States in the 1940s.[42] In successive designs, thrust was increased rapidly by increasing combustor size and pressure, often using the same injector pattern and element geometry of smaller hardware. Combustion instability occurred more frequently and became more damaging as the combustors became larger.

This development was demonstrated clearly on the LOX/kerosene Hermes project, which began in 1944 with a 156 N (35 lb$_f$) thrust chamber with the intent to scale up to 80 kN (18,000 lb$_f$) thrust. Tests were conducted in chamber diameters of 76.2, 114.3, and 215.9 mm and the full scale, 254 mm. Unstable and noisy behavior in the 76.2 and 114.3 mm diam chambers at 2.4 MPa was so prevalent it was considered normal, although it was rarely damaging. When these injector patterns were used directly in a 152.4 mm diam chamber, however, the nozzle was destroyed by high heat transfer rates due to instabilities. Subsequently, many 152.4 mm as well as 215.9 and 254 mm diam chambers were tested at 3.45 MPa, with approximately two-thirds exhibiting unstable behavior. Eventually, through extensive testing and cut-and-try methods, a stable design was produced in a 254 mm diam full-scale chamber. The Vanguard made use of this extensive subscale testing, exhibiting stable behavior in the Hermes 215.9 mm diam chamber at a chamber pressure of 4.14 MPa. The Vega— essentially the same engine with a higher expansion ratio nozzle, new coolant jacket design, and longer required duration—was also stable.

These trials in scaling from a small to a large size and thrust were later repeated during the development of the F-1 engine. The design heritage of the F-1 engine can be traced from the E-1 and H-1 engines back to the S-3D engine, used to power the Thor-Jupiter vehicles, and finally to the Navaho engine developed in the 1940s. In 1958, the 667 kN thrust LOX/RP-1 S-3D engine was upgraded to 836 kN and designated the H-1. The first designs of this new engine were operated at 734 kN and, like the S-3D, exhibited stable behavior.[43-45]

Concurrently, a larger experimental engine of 1779 kN thrust, designated the E-1, was developed for application to a larger thrust LOX/RP-1 engine. The early F-1 injectors were scaled directly from the E-1 injectors, and the design methodology was based on the E-1 program. Of 44 injector tests conducted on full-scale 4448 kN workhorse hardware between January 1959 and May 1960, 20 resulted in spontaneous instabilities with amplitudes in excess of 100% of mean chamber pressure.[46]

Injector designs based on the 734 kN thrust H-1, which had proven to be stable, were tested at 6673 kN thrust and exhibited somewhat better stability.[46] Still, in June 1962 combustion instabilities caused total loss of an F-1 engine.[47]

In early 1963, the H-1 engine was upgraded to 836 kN thrust and was bomb tested to verify its capability to recover from a substantial perturbation. The engine failed to recover within 100 ms in 6 out of 16 stability tests.[46]

During the Project First Program,[46] which was tasked with developing a flight-qualified F-1 injector, extensive testing with the H-1 engine exhibited stable behavior up to 912 kN thrust with the addition of baffles and rearrangement of orifices.[47] The essentially stable preflight rated test injectors for F-1 evolved from these tests.[47]

This 10-fold increase in size and thrust from the S-3D engine to the F-1 engine exhibits the potential problems that can occur when using small hardware with uncertain characteristics to design larger hardware. Because of funding constraints, however, a development program like the F-1 is unlikely to be repeated in the near future. Therefore, the use of smaller or subscale hardware for evaluation of combustion stability will be required in its absence. In the following section, various subscale devices are described that have been used or suggested to investigate combustor combustion stability scaling techniques. In one type, particular rules for successful scaling are discussed that have become apparent by examining historical data.

1. Chamber Designs for Stability Testing

Although the primary objective of using scaled multi-element hardware remains to minimize or eliminate combustion instability in the full scale, an additional objective is to do so at lower development cost. Reduced-size hardware can lower the development cost in two ways: 1) because thrust and flow rates are less than the full scale, the construction and operation costs of test facilities are lower; and 2) because the hardware is smaller, the hardware development and fabrication costs are lower, a potentially important feature for combustion stability testing in which destruction of hardware and numerous iterations may occur. However, these cost savings are reduced as pressure and combustor size are increased.

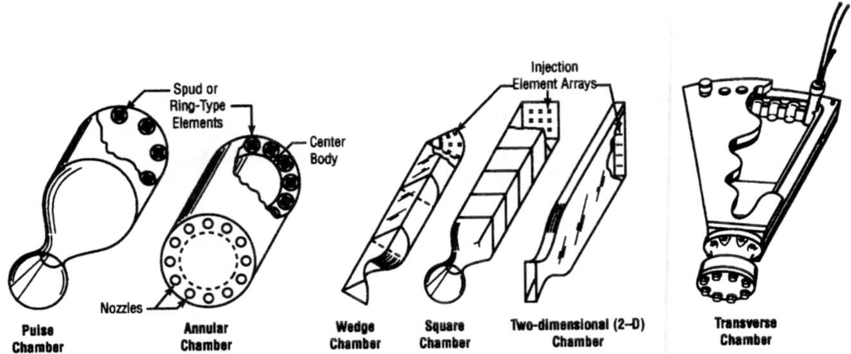

Fig. 8   Multi-element subscale chambers for stability testing.[54,57]

As in the low-pressure single-element methodology discussed in Section III.A, multi-element hardware must recreate the important stability characteristics of the full-scale engine. One feature that does the most to maintain this similarity is to use identical element geometry, while making the element pattern and manifolding as similar as possible. The other key feature is to select the frequency range to examine. Research and development activities have generated eight important multi-element subscale chamber shapes for recreating various frequency ranges in subscale hardware[39,48–57]: 1) pulse chamber, 2) annular chamber, 3) wedge chamber, 4) square chamber, 5) two-dimensional chamber, 6) transverse excitation chamber, 7) variable length combustor, and 8) circular reduced size chamber.

The first six chambers are illustrated in Fig. 8, and the variable length combustor is illustrated in Fig. 9. Other types of devices have been developed and tested

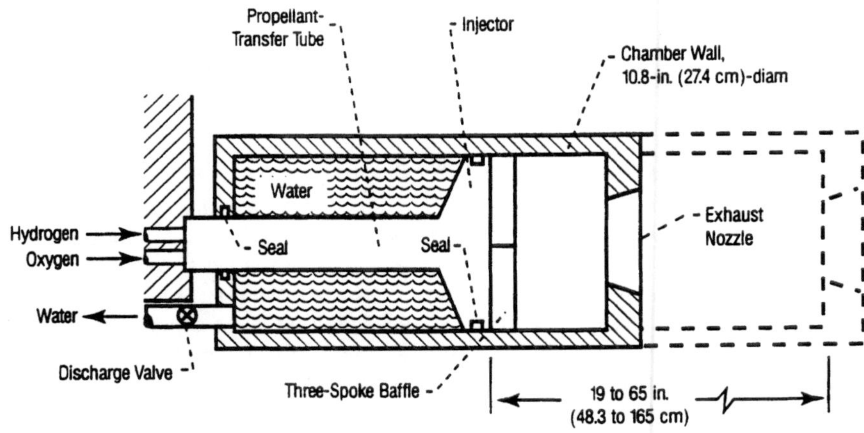

Fig. 9   Schematic of the variable length chamber.[56]

in addition to these, but are less typical for testing multi-element liquid propellant combustion devices. The classic T-burner, for example, has been tested with single-element injectors.[57]

a. Pulse and annular chambers. The pulse chamber was developed to test injection designs in full-scale combustion chambers, simulating full-scale transverse mode acoustics but with lower flow rates.[57] In some versions, replaceable injector sections, known as spuds, were located at the periphery of the combustion chamber head end, as illustrated in Fig. 8, while no mass was injected in the center of the chamber. In other versions, a fixed injector ring pattern was located at the periphery, sacrificing the modularity allowed by the spuds. Explosive charges, typically tangentially mounted pulse guns, were used to perturb the combustion processes to assess dynamic stability (hence the name pulse chamber).

With reduced injector flow rates in a full-scale chamber diameter, however, the contraction ratio of the combustion chamber is necessarily increased, the injector pattern is altered or made discontinuous, and a large recirculation zone exists in the center of the chamber. These issues create uncertainties in the use of the results for full-scale development. Some moderate success was demonstrated using the sensitive time-lag theory as the correlating tool.[48-50]

The annular chamber, illustrated in Fig. 8, was also designed to maintain full-scale combustion chamber dimensions, but a centerbody was used to eliminate the open area in the center of the chamber that occurred with the pulse motor, and hence eliminate the large recirculation zone.[57] However, the annular chamber transverse acoustics would be different from the full-scale chamber. The acoustic mode shape would be altered, and the frequency depressed for the same outer diameter, while wall effects would be accentuated.[57]

b. Square and rectangular chambers. The square chamber, illustrated in Fig. 8, has been used to examine longitudinal wave interactions with combustion processes, especially in research programs.[51] It also has been used to examine the interactions between injector elements, providing data on the influence of spacing between elements and the spray fan orientations. The chamber can be varied in length to observe characteristic velocity variations and thereby measure axial combustion distribution.

The rectangular, or two-dimensional (2-D) chamber, illustrated in Fig. 8, consists of a radial slice of the full-scale cylindrical combustion chamber. Generally, the long width of the chamber is much longer than the short width, so that long width mode frequencies are much lower and more likely to generate instabilities. The disadvantage of the two-dimensional chamber is that the width mode only simulates the acoustic field distribution of certain regions of the tangential or radial modes of the full-scale cylindrical chamber.

As with other chambers with flat walls, the two-dimensional chamber provides a simple means to introduce optical techniques to examine combustion phenomena under stable and unstable conditions, including droplet dynamics, triggering mechanisms that result in combustion instability, and instability oscillations.[57] During the Project First Program, gas and droplet velocities were measured from streak photographs during hot-fire tests through special two-dimensional chambers with Plexiglas® walls.[52] Also during the Project First Program, a two-dimensional chamber was used to research the effects of baffle parameters (length, number, and size of compartments), along with enlarged fuel orifices and new injector concepts, on combustion stability for the F-1.[52,53]

Two-dimensional combustors have also been used as part of a stability methodology in which the full-scale chamber first tangential (1T) acoustic mode was matched in frequency to the first width (1W) mode of the two-dimensional combustor to examine the effects of this mode on the combustion response.[39] Full-scale testing in this program did not proceed sufficiently to successfully validate the use of the two-dimensional chamber in the methodology.

c. Wedge and transverse excitation chambers. The wedge chamber, illustrated in Fig. 8, is a segment slice of a circular full-scale injector, providing full-scale injection density and contraction ratio but with reduced flow rate. Longitudinal, radial, and higher order tangential (i.e., those associated with the wedge angle) standing modes can be simulated.[57]

A variation of the wedge chamber was created as a research device to measure the frequency sensitivity of injection element designs.[54,55] The transverse excitation chamber (TEC), illustrated in Fig. 8, was developed to provide simple modularity in testing transverse mode frequencies by changing the wedge angle with inserts.[54] Wedge angles from 9 to 36 deg provided a range of fundamental tangential mode frequencies from 7000 to 1800 Hz. Longitudinal modes (i.e., the mode in the direction of flow) were highly damped, while transverse frequencies in the width mode were typically greater than 13,000 Hz. The injector elements, located on the wall of the outer circular segment, were made as removable inserts to enhance modularity. Chamber pressure was varied by altering the throat area at constant injection density, or by throttling the injector with the same throat area.

Hot-fire tests were conducted to demonstrate the capability of the TEC to characterize the frequency response of various injector elements and rate the stability of a particular injector design.[54,55] Spontaneous instabilities with a particular injection element provided data that identified a peak response, which compared well with previous correlations for this injector.[54] These tests indicated that the TEC would be useful as an inexpensive rating tool to evaluate many single-parameter characteristics of an injection pattern.

d. Variable length chamber. The variable length combustion chamber was developed to examine the response of injector concepts to continuously varying longitudinal modes and frequencies as functions of hydrogen injection temperatures.[56] The chamber, illustrated in Fig. 9, could vary chamber length continuously during hot-fire operation, changing the longitudinal acoustic frequencies that could couple with the combustion processes, resulting in change from stable to unstable at some critical point. Further increases in chamber length could usually stabilize operation.

Hot-fire tests with one version showed that above a hydrogen injection temperature of 45.6 K, no instability was observed at any chamber length.[56] Below

45.6 K, the coaxial injector was unstable in the first longitudinal (1L) acoustic mode at a 762 mm chamber length, and unstable in the second longitudinal (2L) mode at 1524 mm. The instability regions widened as the hydrogen injection temperature decreased.[56]

e. Circular reduced-size chamber. Among the simplest hardware with which to obtain stability data for axisymmetric chambers is an identical but smaller version. Reduced-size circular chambers provide the means to investigate combustion stability at the higher order acoustic modes of the full scale. Too small a subscale leaves too much frequency range untested, whereas too large a subscale reduces the advantages of testing using subscale hardware. A suitable size can be recommended by the results of a number of programs, which are discussed in the following paragraphs.

A program in the 1980s to develop a methodology for stability scaling included both circular reduced-size (three-dimensional) and rectangular (two-dimensional) combustion chambers.[39] The three-dimensional chamber was sized so that the frequency of its 1T acoustic mode was equal to the frequency of the third tangential (3T) acoustic mode of the full-scale chamber (i.e., $f_{1T,ss} = f_{3T,fs}$), resulting in a chamber diameter 43% of the full scale. The two-dimensional chamber was sized so that the frequency of its 1W acoustic mode was equal to the frequency of its 1T acoustic mode of the full-scale chamber (i.e., $f_{1W,ss} = f_{1T,fs}$). The injection elements in these subscale injectors were made identical to those in the full scale. Thus, the combination of two-dimensional and three-dimensional subscale hardware presented the proposed injection element and portions of the injection pattern to the full range of frequencies found in the full-scale chamber. One injector pattern, spontaneously unstable in both the two-dimensional and three-dimensional chambers, was tested in the full scale and found to be unstable at the 1T mode with first radial (1R) tuned cavities and relatively short (6 and 17% of chamber diameter) baffles.[39]

The genesis of the $f_{1T,ss} = f_{3T,fs}$ size came from development of the OMS rocket engine for the space shuttle. In a direct application of stability scaling in a development program, various sizes of subscale hardware were used to assess the stability characteristics of the full-scale design. However, the lack of a well-developed methodology at the time created a number of surprises when some designs were taken to full scale.

Subscale stability tests were conducted with chambers 33% of full-scale chamber diameter[24] and 46% of full-scale chamber diameter.[41] The undamped 1T acoustic mode frequency of the 33% subscale chamber corresponded to a frequency between the fourth tangential (4T) and fifth tangential (5T) modes of the full-scale chamber, and the 46% subscale chamber to a frequency between the 1R and 3T modes. Many of the subscale tests were conducted with head-end acoustic resonator cavities in place, resulting in a highly damped and depressed fundamental mode in the subscale chamber, which raised the sensitive acoustic frequencies of the combustion chamber to even higher values.

In tests with the 33% subscale chamber, one injector pattern exhibited stable operation and recovered from artificial perturbations, with and without cavities.[24]

However, the injector pattern was found to be unstable in the full-scale chamber at the 1T and 3T modes.[24] These modes were stabilized with cavities, and operation was satisfactory until a number of changes were made for flight weight hardware, when instabilities reappeared, and persisted despite numerous variations of cavity geometry.[27,28,41,58] The injector pattern had to be abandoned. Another pattern, higher performing but unstable in the subscale without cavities, was found to be unstable in the full-scale chamber at the 1T, second tangential (2T), and 3T modes. These modes were also stabilized with cavities, but instabilities reappeared with small changes in chamber geometries, and this injector pattern was also abandoned.[24]

These tests showed that this size subscale ($f_{1T,ss} = f_{4T,fs}$ or $f_{1T,ss} = f_{5T,fs}$) was too small to successfully characterize full-scale stability. If instabilities occur at this scale, stabilization at full scale will be too difficult. If stable, the full scale may still be unstable over a range of modes that may be too difficult to be successfully damped.

In tests with the 46% subscale chamber, one injector pattern that was bombed unstable without cavities but was stable with cavities, was found to be unstable in the full scale at the 1T, 3T, 4T, and higher modes.[41] No combination of resonator cavity tuning and chamber length was found that could completely stabilize the system and eliminate persistent 1L and chugging problems.

These tests showed that instabilities at this size subscale ($f_{1T,ss} = f_{1R,fs}$ or $f_{1T,ss} = f_{3T,fs}$) are still too difficult to damp in full scale, at least with cavities alone. Damping the subscale tests was also found to have presented a false sense of security regarding full-scale stability.

These results suggest that a suitable subscale size for circular reduced-size testing may be the $f_{1T,ss} = f_{3T,fs}$ scale. This relationship is shown in Fig. 10. An

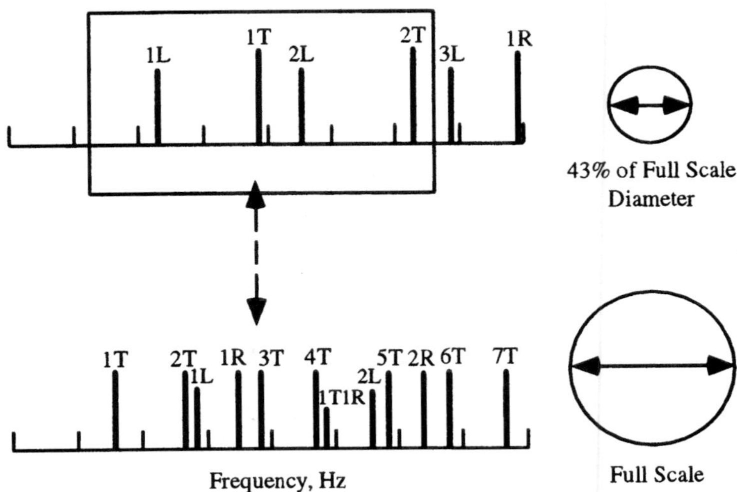

Fig. 10   **Chamber size and acoustic mode relationships in $f_{1T,ss} = f_{3T,fs}$ scaling.**

injector pattern that is stable in an undamped chamber of this size demonstrates a minimum level of stability that may be damped successfully in full scale. The previous discussion clearly shows that an injector pattern unstable in this size chamber (or smaller) will present inordinate stability problems at the full scale.

2. Data Analysis and Scaling Techniques

Various methods for generating subscale multielement stability data were described in Section III.B.1. Four approaches to coupling the subscale data to full-scale stability are now discussed, including: 1) no correlation (hardware or model independent)—simple experimental evaluation of stability, without relation to full-scale acoustics; 2) hardware-coupled or hardware-dependent correlation—experimental evaluation of full-scale stability from subscale tests; 3) model-coupled or model-dependent correlation—evaluation of full-scale stability by application of analytical or empirical combustion stability models calibrated by subscale experimental data; 4) model correlation—evaluation of full-scale stability by application of analytical combustion stability models.

In the first approach (no correlation) the relative stability of various designs is assessed or compared without relating the stability characteristics to the full scale. Unstable or less stable designs can be discarded. Relative stability can be assessed from either spontaneous or dynamic features. Spontaneous features of the combustion can include oscillation frequencies, amplitudes, or levels of combustion noise. Dynamic features include effects of an artificial perturbation, such as the pulse charge size used to trigger instability, the initial and maximum perturbation magnitudes required for instability to occur, and the damping time after the perturbation (the time required for amplitudes of chamber pressure oscillations to decrease from initial overpressure to $\pm 5\%$ of nominal chamber pressure).

However, the ultimate stability of the design will still be in question because the stability margin is not evaluated relative to the full-scale combustion chamber acoustics. Therefore, in the second approach, a hardware size is applied that is relevant to the full-scale stability. As presented in Section III.B.1.e, a minimum level of stability can be specified that has been shown to relate to full-scale stability, by requiring stable operation in the undamped subscale chamber sized so that $f_{1T,ss} = f_{3T,fs}$.

The third approach is an indirect evaluation: applying data gathered from subscale testing to models that would then be used to predict the full-scale stability behavior. The models can be analytically or empirically based, or a combination of the two, but must rely on some experimental measurements to complete a full-scale stability prediction. The most well-known example of such a model is the sensitive time-lag theory,[59,60] in which the combustion process is found to be sensitive to changes in the local conditions only during a specific period of the total time required for combustion. A later version is a response factor model, in which the ability of the combustion chamber to dissipate acoustic energy (the chamber admittance) is compared in the complex plane to the oscillatory combustion driving mechanisms (the combustion response).[61]

Experimental stability and combustion data from subscale chambers are used in these models to complete the full-scale stability prediction. Combustion data, gathered from high-speed photography or advanced laser diagnostics through windowed or transparent sections of the chamber walls, are used to describe combustion phenomena otherwise difficult to calculate, or to calibrate portions of the model. Subscale stability data are used to calibrate the stability predictions at the subscale, so that predictions can be made more reliable at the full scale.

The fourth method (an uncoupled or uncorrelated model) is purely analytical and does not rely on subscale data or empirical correlations to complete a full-scale stability prediction. Subscale data can be used to show whether the model can predict the correct results, but have no place in the model for calibration. Many first-principles models have been developed,[11,62] but few are in actual use because of the lack of detailed understanding of all of the relevant steady and unsteady combustion processes.

IV. Heat Transfer

Techniques to scale heat transfer are illustrated in the following discussion using the oxygen/hydrogen SSME main combustion chamber (MCC) liner development program as an example. Topics include generation of a hot gas heat transfer coefficient h_g profile from heat flux measurements in a subscale chamber, scaling of the subscale h_g profile to different chambers at different chamber pressures, and comparison of heat load predicted from these measurements and measured in a full-scale chamber.

Subscale development tests for SSME were conducted at 14.7 kN (3.3 Klb$_f$) and 178 kN (40 Klb$_f$) thrust levels during the 1970s.[63-68] The 40 Klb$_f$ thrust, or 40 K, program was planned to provide cycle life demonstrations and design data for thermal analyses of the SSME NARloy-Z combustion chamber liner MCC, along with the main injector liquid oxygen posts. The 40 K hardware was designed to have the same thermal strains as the full-scale SSME combustion chamber and injector elements.

Heat flux was measured circumferentially at axial locations in a 40 K water-cooled calorimeter chamber to confirm hydrogen coolant flow rates and operating parameters required for regeneratively cooled chamber testing. The throat region coolant channel wall temperatures and total cyclic strains were required to match those predicted for full-scale SSME MCC. Twenty-five tests were conducted during calorimeter chamber testing, 12 of which were of sufficient duration to provide the required data.[64] Two 40 K hydrogen regeneratively cooled chambers were tested following completion of calorimeter chamber testing.[65,66]

A. Hardware Description

40 K hardware was approximately 1/10 scale, with respect to thrust, of the full-scale SSME. The injector/combustion chamber wall heat transfer relationships of the full-scale SSME were reproduced by including a number of features: 1) warm hydrogen fuel provided by a preburner staged combustion configuration,

Table 1 Comparison of subscale to full-scale SSME design parameters

Design parameter	Full scale	40 K water cooled	40 K regeneratively cooled
Total no. of injection elements	600	61	61
No. of baffle injection elements	75	3	3
Chamber diameter, mm	450.6	143.8	43.8
Throat diameter, mm	261.75	84.1	84.1
Contraction ratio	2.96	2.92	2.92
Chamber length (L'), mm	355.6	355.6	355.6
Nozzle expansion ratio	5:1	7:1	5:1
No. of chamber coolant channels	390	116	128
Chamber coolant channel orientation	Axial	Circumferential	Axial

2) injection elements identical in design and dimension to the full scale, and 3) a transpiration-cooled faceplate. However, the hot gas manifold did not duplicate full-scale design, and there was no mixture ratio biasing of the outer element row. Also, hydrogen coolant from the regeneratively cooled chamber was not injected through the face, as on the full scale, but discharged overboard to a test facility burn stack. A comparison of 40 K hardware parameters to the full-scale SSME is provided in Table 1.

The water-cooled calorimeter chamber had 116 circumferential coolant channels manifolded into 58 independent coolant circuits, each with separate temperature and pressure measurements. The hydrogen regeneratively cooled chamber had 128 longitudinal coolant channels machined into a NARloy-Z liner. The coolant channels were closed out with electrodeposited copper/nickel in a manner similar to the full-scale MCC. The structural jacket was not identical to the MCC. Hydrogen coolant entered the chamber at the nozzle end and exited at the injector end, but unlike in the full-scale hardware, this hydrogen coolant was discharged overboard. Both subscale chambers had the same combustion chamber length (injector face to throat), throat convergence ramp angle, throat contour radius of curvature, and combustor contraction ratio as the full-scale MCC.

B. Subscale Chamber Test Programs

1. 40 K Calorimeter Chamber Heat Flux and Heat Transfer Coefficient

Heat flux profiles for chamber pressures (p_c) between 8.6 and 11.4 MPa and mixture ratios (r) between 5.5 and 6.5 were measured in the 40 K calorimeter chamber, within the mixture ratio range but less than the SSME nominal rated power level (RPL) of $p_c = 20.47$ MPa and $r = 6.0$ (corresponding to thrust = 2091 kN). The maximum allowable chamber pressure for the calorimeter chamber was limited by the throat region burnout heat flux of 10.63 kW/cm^2. The calorimeter data were used for both prediction and evaluation of the MCC design. Heat transfer rates (i.e., the heat flux profile) were calculated along the axial length of the calorimeter chamber from the water flow rate and temperature rise within each of the 58 circumferential coolant circuits.

Heat flux Q/A for each of the 58 coolant circuits was calculated by

$$Q/A = \left(\frac{\dot{m}c_p \Delta T}{A_S}\right) \tag{21}$$

The hot gas heat transfer coefficient h_g for each coolant circuit was then calculated by

$$h_g = \frac{Q/A}{T_{\mathrm{aw}} - T_{\mathrm{wh}}} \tag{22}$$

T_{aw} was calculated from

$$T_{\mathrm{aw}} = R_c T_c \tag{23}$$

where

$$R_c = \text{recovery factor} = \frac{1 + Pr^{1/3}\left(\dfrac{\gamma - 1}{2}\right)M^2}{1 + \left(\dfrac{\gamma - 1}{2}\right)M^2} \tag{24}$$

and Pr, γ, M, and T_c were calculated using one-dimensional chemical equilibrium at the appropriate test chamber pressure and mixture ratio.[69]

T_{wh} was estimated using two-dimensional finite difference algorithms with forced convection and nucleate boiling. T_{wh} was held constant when it reached $T_{\mathrm{sat}} + 283$ K.

The Q/A profile for a 40 K calorimeter chamber test with $p_c = 10.87$ MPa and $r = 6.0$ is shown in Fig. 11. The h_g profile, generated from this Q/A profile using Eq. (22), is shown in Fig. 12. The total heat load measured during this test was 9079 kW, and the injector characteristic velocity efficiency η_{C*} was calculated to be 100%.

2.   40 K Regeneratively Cooled Chamber Heat Flux and Heat Transfer Coefficient

The heat flux measurements from the 40 K calorimeter chamber, shown in Fig. 11, were used to predict the heat flux profile in the regeneratively cooled chamber at the maximum SSME chamber pressure condition. For each axial location, h_g was scaled from $p_{c1} = 10.87$ MPa to $p_{c2} = 20.47$ MPa by

$$h_g\big|_{p_{c2}} = h_g\big|_{p_{c1}} \left(\frac{p_{c2}}{p_{c1}}\right)^{0.8} \tag{25}$$

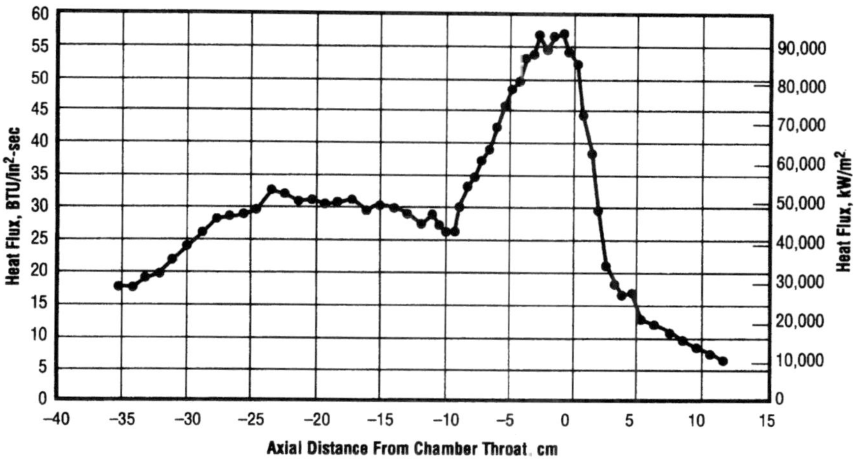

Fig. 11   SSME 40 K calorimeter chamber test 024 heat flux profile.

The predicted h_g profile for the 40 K regeneratively cooled chamber at the higher chamber pressure is plotted in Fig. 12. This profile was not confirmed experimentally because the regeneratively cooled chamber measured only total heat load and not heat transfer at discrete axial locations. The predicted regeneratively cooled chamber profile was used to calculate a heat flux profile and then a

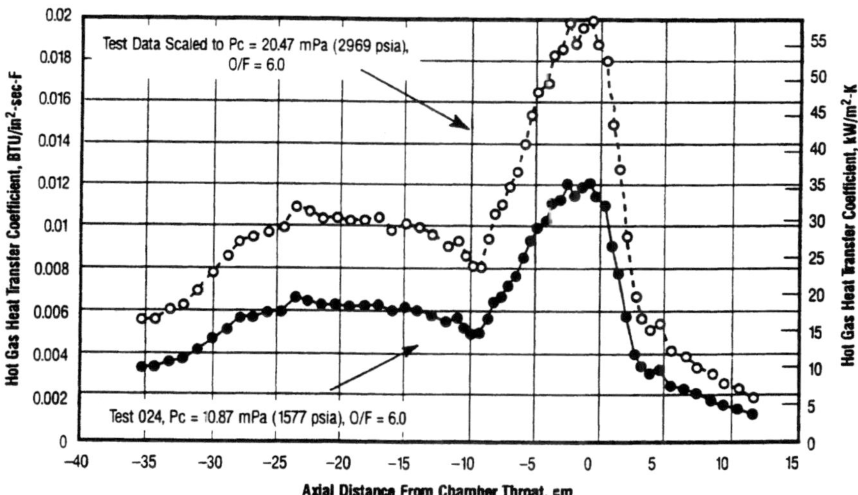

Fig. 12   SSME 40 K calorimeter chamber heat transfer coefficient profile.

total heat load, which could be compared to test data from the regeneratively cooled chamber.

The heat flux profile was calculated at each axial location by

$$Q/A = h_g(T_{\text{aw}} - T_{\text{wh}}) \tag{26}$$

The predicted heat flux profile for a test with conditions $p_c = 20.47$ MPa and $r = 6.0$ is shown in Fig. 13.

The total chamber heat load was calculated by summing heat flux at each axial location,

$$Q_{\text{total,pred}} = \sum [(Q/A)A_S] \tag{27}$$

The total heat load predicted for the 40 K regeneratively cooled chamber from the heat flux profile in Fig. 13 was 15061 kW. The total heat load for RPL chamber pressure and mixture ratio was originally predicted to be 14813 kW.[64]

The total heat load measured in the 40 K regeneratively cooled chamber was calculated using

$$Q_{\text{total,meas}} = \dot{m}_{\text{cool}}(h_{\text{out}} - h_{\text{in}}) = \dot{m}_{\text{cool}}(c_p T|_{\text{out}} - c_p T|_{\text{in}}) \tag{28}$$

from measurements of the total hydrogen coolant flow. The measured heat load was corrected for each data slice to a reference condition of chamber pressure, injector characteristic velocity, coolant orifice discharge coefficient, coolant inlet temperature, and coolant flow rate. The heat load for RPL conditions was initially measured to be about 13927 kW, which increased during the test to

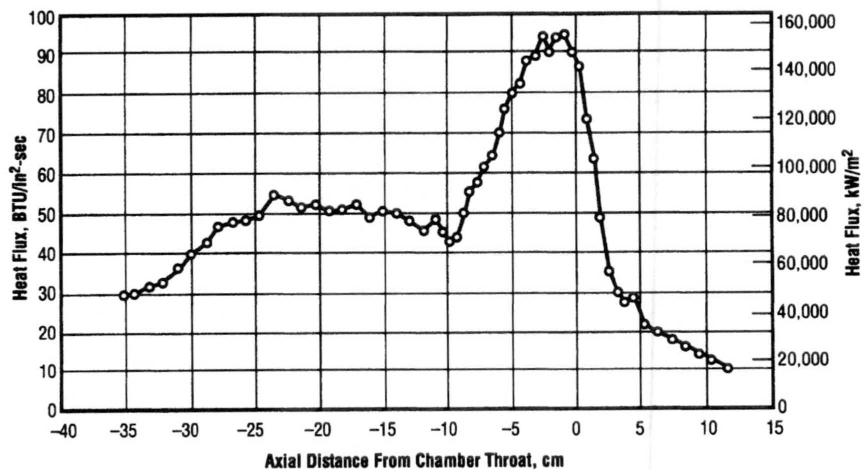

Fig. 13   Predicted SSME 40 K regenerative chamber heat flux profile.

14296 kW,[65] about 5% lower than predicted by the calorimeter data. The rise in heat load during the test has not been completely explained.[65]

C. Full-Scale SSME Heat Flux

The 40 K calorimeter chamber data were used to predict full-scale SSME MCC total heat flux. First, the full-scale h_g profile was generated from the 40 K h_g profile, shown in Fig. 11, at equivalent Mach numbers. From the injector face to approximately 127 mm downstream, the heat flux was the same as in the 40 K calorimeter chamber, because the injector elements were the same and heat transfer rates near the injector were then primarily influenced by the distance from the injector.[67] Further downstream, where heat transfer rates are primarily influenced by velocity,[67] the full-scale heat flux corresponded to axial locations where the hot gas Mach number was the same as in the 40 K chamber. The h_g profile for the SSME MCC at RPL conditions is presented in Fig. 14.

Next, with this h_g profile, the full-scale SSME MCC Q/A profile was calculated using Eq. (26), and the total chamber heat load was calculated using Eq. (27). The total heat load, normalized to an SSME MCC test at $p_c = 20.99$ MPa and $r = 5.96$, was predicted to be 58150 kW.

This prediction was compared to heat load calculated with coolant pressure and temperature measured in two places, the MCC coolant outlet manifold and the outlet line. Heat load from manifold measurements was 55697 kW, 4.2% lower than predicted, whereas the heat load from outlet line measurements was 57211 kW,[67] 1.6% lower than predicted. Manifold measurements include circumferential Q/A variations and may be more accurate, whereas the outlet line measurement represents total heat load with all variability mixed away.

Scaling 40 K data to SSME MCC conditions predicted total heat loads that were 2–5% higher than the measured heat loads. Although the heat transfer

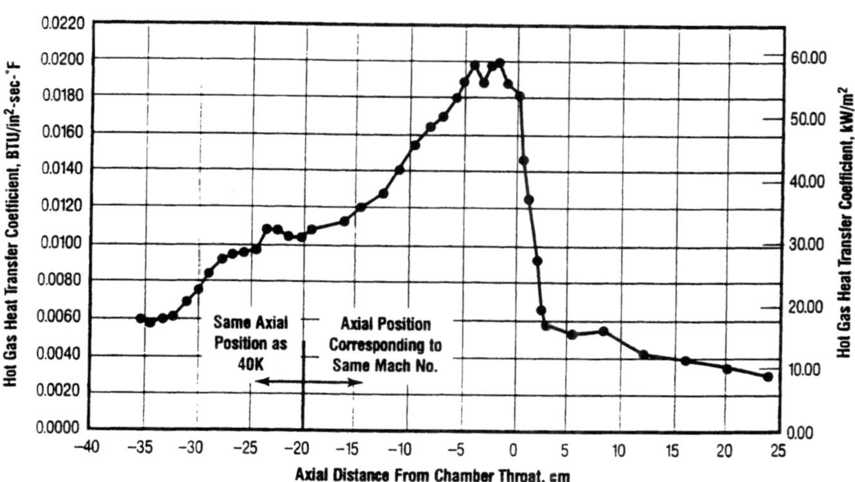

Fig. 14 Predicted SSME MCC hot gas heat transfer coefficient profile for RPL conditions.

data between subscale and full-scale hardware agreed well, there were and still are chamber life issues caused by localized overheating of the chamber wall, which are discussed in Section V.

V. Life Cycle Assessment

Hardware durability must be evaluated during engine development, especially for reusable systems. Because of the harshness and complexity of liquid rocket combustion, durability is difficult to predict and requirements are often difficult to establish. The use of subscale hardware in life cycle evaluation can identify design and manufacturing problems early in the development stages, and minimize or prevent costly program delays. However, subscale hardware and experiments for durability evaluations must be examined carefully, as results may not be conservative and can be misleading. The following discussion of durability in combustion devices is illustrated using the SSME MCC liner development program as an example. The combustion chamber is typically the limiting component for durability in a combustion device.

A. Subscale Combustion Chamber Liner Test Programs

Design life requirements for the SSME MCC liner were to survive 100 missions and five acceptance tests for 105 thermal cycles, approximately 8 h of run time. It was designed for a throat heat flux of $163479 \, kW/m^2$. A safety factor of 4 was used for the life cycle analysis; therefore, the liner was required to survive 420 cycles or 32 h.[70] Reference 70 describes the procedure used to estimate the MCC liner life and the resulting design impacts.

Two levels of subscale testing, 14.7 kN (3.3 Klb$_f$) and 178 kN (40 Klb$_f$) thrust, were conducted with SSME MCC liner hardware. Evaluation of the 3.3 Klb$_f$ thrust, or 3.3 K, hardware provided sufficient confidence in the design and manufacturing processes to scale the hardware to the 40 Klb$_f$, or 40 K, size. The 40 K testing provided data to confirm design details and proceed with full-scale component manufacture and test.

1. 3.3 K Program

The 3.3 K thrust chamber assemblies were used in two test programs to investigate preliminary durability of a cast NARloy-Z combustion chamber liner.[63] Both programs used identical injectors, and measured the heat-flux profile with water-cooled calorimeter chambers prior to testing with the NARloy-Z regeneratively cooled chambers. The calorimeter and regeneratively cooled chambers had identical hot gas sidewall contours. Testing was conducted at $p_c = 5.2 \, MPa$ and $r = 5.5$ during both test programs, using test durations of 2.2 and 15 s.

During the earlier program, the first through crack developed after 410 short-duration thermal cycles. The chamber developed premature cracking because of relatively large zirconium oxide inclusions in the material.[63]

To reduce the zirconium oxide inclusions, the NARloy-Z chamber liner in the subsequent program was manufactured from a centrifugal casting. This chamber accumulated 1013 thermal cycles before the first through crack developed, demonstrating that the centrifugal casting process improved the chamber liner material properties to allow proceeding to the 40 K level.[63]

2. 40 K Program

One 40 K regeneratively cooled combustion chamber with a NARloy-Z combustion chamber liner manufactured from a centrifugal casting was tested to provide the life cycle verification of the chamber liner design. Tests were conducted at $p_c = 20.68$ MPa and $r = 6.0$ with varying test durations. Initially the heat load to the chamber was approximately 10% lower than predicted from the 40 K calorimeter chamber testing. Following thermal cycle 29, a 6.35 mm longitudinal surface crack was observed 279.4 mm upstream of the throat, which developed into a through crack after an abnormal shutdown on cycle 47, but did not propagate. Because it did not propagate, this crack was considered anomalous and not a low cycle fatigue failure. By cycle 70, the heat load had grown to several percent above the calorimeter values. The liner surface was polished with a crocus cloth to reduce the surface roughness, resulting in 10% less heat load on cycle 71. During continued testing the heat load increased progressively with each test. The chamber was polished after 90 cycles and again after 100 cycles to reduce surface roughness. Following cycle 109, a through crack was observed 228.6 mm upstream of the throat, which was considered the first low cycle fatigue failure. The chamber was polished and testing continued. Two through cracks were observed in the throat following cycle 112. The final test of this chamber unit was cycle 118.

The longitudinal cracks were located in the middle of the channels, in local areas of high hot gas temperature. The local high temperature zones were due to local and secondary injector effects that also roughened the hot gas wall surface. Following completion of testing, the chamber was sectioned for metallurgical analyses. Surface roughening was found to have increased, which would have increased the wall temperatures further, causing local channel deformation. Midchannel pressure stresses would be redistributed because of this geometry deformation, forcing material to creep into the land region and the hot gas wall above the channel to thin.[66] The low cycle fatigue crack propagation appeared oriented to this severe channel deformation.[66] Because of the demonstrated 109 thermal cycles of this chamber prior to attaining a low cycle fatigue crack and other low cycle fatigue analyses, the service life of the SSME MCC was predicted to be 103 thermal cycles.

B. Full-Scale Chamber Liner Testing

Full-scale testing with an autonomous thrust chamber did not demonstrate the life of 103 starts predicted from 40 K testing. In initial testing, high temperatures in the acoustic cavity were observed. To reduce the heat load and minimize chamber damage, the main injector was redesigned to reduce the mixture ratio at the wall by including larger fuel face nuts, fuel film cooling, and canting of the LOX posts. In addition, the chamber liner was, and continues to be, polished regularly to minimize roughened areas.

Subsequent engine system testing also did not demonstrate the life predicted during the 40 K testing.[71] Chamber U/N 0008, tested on SSME 0009, was tested for a total of 9000 s (21 tests) prior to the first observed crack, located 25.4 mm upstream of the throat perpendicular to the hot gas flow. Discoloration and blanching were noted in this region prior to crack formation, and chamber wall roughness had increased throughout testing. Longitudinal cracks were observed at 10,040 s (23 tests), and there were five cracks in the chamber liner after 10,900 s (25 tests) had been accumulated on this chamber.

Chamber U/N 0007, tested on SSME 0008, developed a crack after accumulating 1143 s (5 tests).[71] Roughening was observed in the area of crack formation in prior tests. No additional cracks were observed after accumulating 6750 s (21 tests).

Chamber U/N 2007, tested on SSME 2008, developed a crack after accumulating 1020 s (7 tests).[71] Two areas of increased surface roughness were observed after the following two tests. Four fuel film cooling holes were enlarged to minimize further damage. No additional problems were observed during testing with this engine. When engine 2008 was rebuilt as 2018, surface roughening was observed again and fuel film cooling was increased again. However, fine surface cracking developed.

Although these three examples of crack formation occurred early in the SSME program, cracks are still observed early in testing. In one chamber, cracks were observed after only three tests.

During 40 K testing, roughening of the chamber wall was regularly noted, which was attributed to local high temperature in the combustion flow. The total heat load increased with roughening of the wall, requiring chamber walls to be regularly polished. Total heat loads also increased during the duration of single tests, particularly during longer duration tests.

Roughening of the chamber wall was also observed in full-scale chamber testing, which was also attributed to local high temperature in the combustion flow. There is significant circumferential and engine-to-engine variation in the outer element row mixture ratio. Any given injector is consistent in creating hot spots from one test to another, but there is no injector-to-injector consistency. The maximum energy release of the injector was predicted to be approximately 3 in. downstream of the injector face, which corresponds to the locations with the most frequent localized overheating in both the 40 K and full-scale chambers. However, correlation of the hot spots on the chamber walls to any particular injector element design feature has not been successful. Because of this heat-flux variability, predicted life of the SSME MCC was reduced.[68]

In this example, the subscale durability testing did not predict full-scale durability closely. One reason may be producibility methods; increased chamber damage observed between 40 K and full-scale SSME testing has been partially attributed to the 40 K hardware being built to more stringent tolerances, as is typical of development hardware. Full-scale hardware is built to standard production manufacturing procedures; hence, there is typically more allowable variation except for critical features.

VI. Conclusions

The use of scaled hardware can provide valuable guidance in the design, development, evaluation, and validation of liquid propellant rocket combustion devices. Used during design, properly scaled hardware can reduce the risks of critical or uncertain features of the design. Scaled hardware can be used to evaluate combustion processes, and predict full-scale performance, stability, and heat transfer. During development, subscale or reduced-size testing can evaluate design features prior to committing to full-scale fabrication and development, often at substantially lower overall costs to the program. Although still controversial, component durability may be evaluated and validated during testing with scaled hardware, as long as manufacturing differences are understood and included.

The difficulty in using scaled hardware successfully is that successful simulation depends on understanding the critical full-scale characteristics, and using that understanding to design the scaled hardware. This is circular logic. To be most useful, the scaled hardware must provide the critical desired information for design and evaluation of the full scale, as well as be influenced and designed itself by it. To be most cost effective, scaled hardware testing should be performed prior to final full-scale component design, fabrication, and testing, so that the scaled hardware can be used to drive and define the full-scale design as much as possible. Generally, the scaled hardware must duplicate the full-scale design and environment as accurately as possible. Unfortunately, the full-scale environment is rarely understood as completely as desired, so that design information for scaled hardware is often limited.

This chapter presented the use of scaling techniques from two perspectives. The first was a historical perspective, in which the varied successes of previous scaled hardware were discussed and examined. The second used this historical information—often in hindsight—to define scaling techniques for performance, combustion stability, chamber heat transfer, and cycle life to be used during design, development, and test.

Because combustion is the central process that defines the purpose of combustion devices, scaling of combustion processes has received a significant amount of previous consideration. This chapter summarized this previous work, and considered the improvement in understanding of combustion processes due to uni-element and multi-element tests with both reacting and nonreacting flows.

Scaled hardware can be used to understand combustion stability characteristics sufficiently to eliminate or minimize instability from occurring in the full-scale engine. The scaled hardware testing is required to provide simple, quick, and effective solutions for problems of instability, and at less cost than that of developing a full-scale engine that exhibits unstable characteristics and must undergo design modifications. There are several methods to scaling hardware for evaluation of combustion stability, including single- and multi-element designs with a wide variety of flow conditions, chamber pressures, and geometries.

A novel low-pressure, single-element procedure for evaluation of combustion stability that was developed in Russia was described. The procedure involves the determination of instability and hysteresis region boundaries for actual full-scale frequencies of transverse mode oscillations, for combustion chambers operating at atmospheric pressure. For simulation of a typical high-pressure combustion chamber typical of staged combustion cycle engines, propellants or their simulants are fed into the test chamber in gaseous form. The procedure is intended for prompt estimation of the effects of design variations on stability, and for preliminary selection of injector element configurations with the best stability characteristics.

Scaling multi-element hardware for evaluating combustion stability characteristics was described. There are a multitude of geometries that may be applied to generate frequencies of interest to the combustion stability problem. One that has been shown to have historical merit is a circular subscale sized so that the third tangential acoustic frequency of the full-scale chamber is made to match the first tangential acoustic frequency of the subscale chamber. An injector pattern that is stable in an undamped chamber of this size demonstrates a minimum level of stability that may be damped successfully in full scale. The previous discussion clearly shows that an injector pattern unstable in this size chamber (or smaller) will present inordinate stability problems at the full scale.

The use of subscale hardware is particularly well suited to scaling heat flux. Subscale calorimeter and regeneratively cooled chambers were used during SSME development to provide both heat flux and total heat load data. The data were scaled to full-scale conditions and were used to predict total heat loads. Measured total heat load data correlated very well with predictions from the scaled data.

However, using subscale hardware to scale durability or cycle life did not correlate well when considering the SSME. Element-to-element variation in the injector outer row and differences in manufacturing tolerances are the most likely causes of the discrepancy between durability predicted from subscale testing and observed in full scale.

References

[1]Penner, S. S., *Chemical Problems in Jet Propulsion*, Pergamon, London, 1957, pp. 345–347, 376–388.

[2]Weller, A. E., "Similarities in Combustion, A Review," *Selected Combustion Problems, II*, AGARD Combustion Colloquium, Butterworths, London, 1956, pp. 371–383.

[3]Stewart, D. G., "Scaling of Gas Turbine Combustion Systems," *Selected Combustion Problems, II*, AGARD Combustion Colloquium, Butterworths, London, 1956, pp. 384–413.

[4]Penner, S. S., "Similarity Analysis for Chemical Reactors and the Scaling of Liquid Fuel Rocket Engines," *Combustion Research and Reviews*, AGARD, Butterworths, London, 1955, pp. 140–162.

[5]Crocco, L., "Considerations on the Problem of Scaling Rocket Engines," *Selected Combustion Problems, II*, AGARD Combustion Colloquium, Butterworths, London, 1956, Chap. 12, pp. 457–468.

[6]Penner, S. S., and Datner, P. P., "Combustion Problems in Liquid-Fuel Rocket Engines," *Fifth Symposium (International) on Combustion*, Reinhold, New York, 1955, pp. 11–28.

[7]Penner, S. S., and Fuhs, A. E., "On Generalized Scaling Procedures for Liquid-Fuel Rocket Engines," *Combustion and Flame*, Vol. 1, 1957, pp. 229–240.

[8]Penner, S. S., "On the Development of Rational Scaling Procedures for Liquid-Fuel Rocket Engines," *Jet Propulsion*, Sept. 1957, pp. 156–161.

[9]Ross, C. C., "Scaling of Liquid Fuel Rocket Combustion Chambers," *Selected Combustion Problems*, II, AGARD Combustion Colloquium, Butterworths, London, 1956, pp. 444–456.

[10]Harrje, D. T., and Reardon, F. H. (eds.), *Liquid Propellant Rocket Combustion Instability*, NASA SP-194, 1972, pp. 221–226.

[11]Yang, V., and Anderson, W. E. (eds.), *Liquid Rocket Engine Combustion Instability*, Progress in Astronautics and Aeronautics, Vol. 169, AIAA, Washington, DC, 1995.

[12]Dickerson, R., Tate, K., and Barsic, N., "Correlation of Spray Injector Parameters with Rocket Engine Performance," TR. AFRPL-TR-68-147, June 1968.

[13]Falk, A. Y., Clapp, S. D., and Nagai, C. K., "Space Storable Propellant Performance Study, Final Report," NASA CR-72487, Nov. 1968.

[14]Mehegan, P. F., Campbell, D. T., and Scheuerman, C. H., "Investigation of Gas-Augmented Injectors, Final Report," NASA CR-72703, Sept. 1970.

[15]Falk, A. Y., "Space Storable Propellant Performance Gas/Liquid Like-Doublet Injector Characterization," NASA CR-120935, Oct. 1972.

[16]Burick, R. J., "Space Storable Propellant Performance Program, Coaxial Injector Characterization," NASA CR-120936, Oct. 1972.

[17]McHale, R. M., and Nurick, W. H., "Noncircular Orifice Holes and Advanced Fabrication Techniques for Liquid Rocket Injectors (Phases I, II, III, and IV), Comprehensive Program Summary Report," Rocketdyne Rept. R-9271, Contract NAS9-9528, Canoga Park, CA, May 1974.

[18]Nurick, W. H., "Study of Spray Disintegration in Accelerating Flow Fields," NASA CR-114479, June 1972.

[19]Zajac, L. J., "Droplet Breakup in Accelerating Gas Flows, Part I: Primary Atomization," NASA CR-134478, Oct. 1973.

[20]Zajac, L. J., "Droplet Breakup in Accelerating Gas Flows, Part II: Secondary Atomization," NASA CR-134479, Oct. 1973.

[21]Nagai, C. K., Gurnitz, R. N., and Clapp, S. D., "Cold-flow Optimization of Gaseous Oxygen/Gaseous Hydrogen Injectors for the Space Shuttle APS Thruster," AIAA Paper 71-673, June 1971.

[22]"Verification Complete Package, Injector Element Cold Flow Testing—Pressure Drop," Rocketdyne Rept. DVS-SSME-303, Contract NAS8-27980, Canoga Park, CA, 21 May 1974.

[23]"Verification Complete Package, Injector Element Cold Flow Performance Characterization," Rocketdyne Rept. DVS-SSME-303, Contract NAS8-27980, Canoga Park, CA, 21 May 1974.

[24]"Space Shuttle Orbital Maneuvering Engine Platelet Injector Program," Aerojet Liquid Rocket Co., Final Rept. 13133-F-1, Contract NAS9-13133, Sacramento, CA, Dec. 1975.

[25]Ito, J. I., "Development Test Report, OMS Injector Subscale Pattern Evaluation," Aerojet Liquid Rocket Co., Rept. PDRD TMO5-25, Contract M4J7XMA-483030H, Sacramento, CA, March 1976.

[26]Mahorter, L., Chik, J., McDaniels, D., and Dill, C., "Airflow Model Testing to Determine the Distribution of Hot Gas Flow and O/F Ratio Across the Space Shuttle Main Engine Injector Assembly," Chemical Propulsion Information Agency, Pub. 550, Vol. II, Oct. 1990, pp. 117–126.

[27]Blubaugh, A. L., "Development Test Report, OMS Injector, Early Injector Program," Aerojet Liquid Rocket Co., Rept. PDRD TMO5-18 Rev. A Contract M4J7XMA-483030H, July 1975.

[28]Pieper, J. L., "Performance, Injector Pressure Drop, and Cold Flow Test Correlations," Aerojet Liquid Rocket Co., Rept. PDRD TMO5-22, Contract M4J7XMA-483030H, June 1975.

[29]"J-2 Engine Quarterly Progress Reports," Rocketdyne Engineering, Repts. R-2600-10 to R-2600-15, Contract No. NAS8-19, Canoga Park, CA, June 1963 to August 1964.

[30]Lawver, B. R., "High-Performance N_2O_4/Amine Elements—Blowapart, Final Report," Rept. 14186-DRL-5, Contract NAS9-114186, Sacramento, CA, March 1979.

[31]Calhoon, D. F., Ito, J. I., and Kors, D. L., "Investigation of Gaseous Propellant Combustion and Associated Injector/Chamber Design Guidelines," NASA CR-121234, July 1973.

[32]Kors, D. L., and D. F. Calhoon, "Gaseous Oxygen/Gaseous Hydrogen Injector Element Modeling," AIAA Paper 71-674, June 1971.

[33]Judd, D. C., "Photographic Combustion Characterization of LOX/Hydrocarbon-Type Propellants, Final Report," Rept. MA-129T, Contract NAS9-15724, Aug. 1980.

[34]Meyer, W., and Tamura, H., "Propellant Injection in a Liquid Oxygen/Gaseous Hydrogen Rocket Engine," *Journal of Propulsion and Power*, Vol. 12, No. 6, 1996, pp. 1137–1147.

[35]Meyer, W., Ivancic, B., Schik, A., and Hornung, U., "Propellant Atomization in LOX/GH_2 Rocket Combustors," AIAA Paper 98-3685, July 1998.

[36]George, D. J., "Rocket Injector Hot Firing and Cold Flow Spray Fields," AIAA Paper 73-1192, Nov. 1973.

[37]Pal, S., Moser, M. D., Ryan, H. M., Foust, M. J., and Santoro, R. J., "Flowfield Characteristics in a Liquid Propellant Rocket," AIAA Paper 93-1882, June 1993.

[38]Denis, L., and Georges, P., "An Experimental Study of LOX/LH_2 Coaxial Injection Elements for the Vulcain Gas Generator," AIAA Paper 87-2113, June 1987.

[39]Pieper, J. L., "Oxygen/Hydrocarbon Injector Characterization," Rept. PL-TR-81-3029, Contract FO4611-85-C-0100, Sacramento, CA, June 1991.

[40]Santoro, R. J., "A Summary of the JANNAF Workshop on Diagnostics," Chemical Propulsion Information Agency, Pub. 573, Nov. 1991.

[41]Schindler, R. C., and Mercer, S. D., "FY 1973 Annual Report, Space "ΔV" Engine, ELOX Injector, and Nontubular Thrust Chamber," Aerojet Liquid Rocket Co., Rept. 73-F, Dec. 1973.

[42]"Combustion Instability Experience, Final Report," General Electric Aerospace and Defense Service Engineering, Contract NAS8-5293, July 1963.

[43]Bilstein, R. E., *Stages to Saturn: A Technological History of the Apollo/Saturn Launch Vehicles*, NASA SP-4206, 1980.

[44]Bostwick, L. C., "Development of LOX/RP-1 Engines for Saturn/Apollo Launch Vehicles," AIAA Paper 68-569, June 1968.

[45]McCool, A. F., and McKay, G. H., Jr., "Propulsion Development Problems Associated with Large Liquid Rockets," NASA-TM-X-53075, Aug. 1964.

[46]"History of Project First, the F-1 Combustion Stability Program," Vol. 1, Book 3, Rocketdyne Engineering, Canoga Park, CA, Oct. 1962.

[47]Oefelein, J. C., and Yang, V., "Comprehensive Review of Liquid-Propellant Combustion Instabilities in F-1 Engines," *Journal of Propulsion and Power*, Vol. 9, No. 5, 1993, pp. 657–677.

[48]"An Experimental Investigation of Combustion Stability Characteristics at High Chamber Pressure," Interim Rept., Aerojet General Corp., Rept. 11741-I, Contract NAS8-11741, Sacramento, CA, Sept. 1965.

[49]"An Experimental Investigation of Combustion Stability Characteristics at High Chamber Pressure," Final Rept., Phase II, Aerojet General Corp., Rept. 11741/SA6-F, Contract NAS8-11741, Sacramento, CA, Aug. 1966.

[50]"High Chamber Pressure Operation for Launch Vehicle Engine Program," Final Rept., Aerojet General Corp., Rept. 4008-SA4-F, Contract NAS8-4008, Supplemental Agreement No. 4, Sacramento, CA, June 1964.

[51]Crocco, L., and Harrje, D. T., "Combustion Instability in Liquid Propellant Rocket Motors," Chemical Propulsion Information Agency, Vol. 1, Pub. 68, Sacramento, CA, 1964.

[52]Arbit, H. A., "Design, Assembly, and Operation of a High-Pressure Two-Dimensional Research Motor," Rocketdyne Research Rept. 64-16, Canoga Park, CA, May 1964.

[53]Coultas, T. A., and Levine, R. S., "Baffles and Chemical Additives as Acoustic Combustion Stability Suppressers in a Two-Dimensional Thrust Chamber," Chemical Propulsion Information Agency, Pub. 105, 1965.

[54]"Stability Characterization of Advanced Injectors," Final Rept. Phase I, Aerojet General Corp., Rept. 20672-P1, Contract NAS8-20672, Sacramento, CA, Oct. 1968.

[55]"Design Guide for Stable H_2/O_2 Combustors, Vol. 1: Design Application," Aerojet Liquid Rocket Co., Rept. 20672-P2D, Contract NAS8-20672, Sacramento, CA, May 1970.

[56]Morgan, J. C., and Sokolowski, D. E., "Longitudinal Instability Limits with a Variable-Length Hydrogen-Oxygen Combustor," NASA TN-D-6328, April 1971.

[57]Harrje, D. T. and Reardon, F. H. (eds.), *Liquid Propellant Rocket Combustion Instability*, NASA SP-194, 1972, pp. 451–459.

[58]Blubaugh, A. L., "Development Test Report, OMS Injector Stability/Stability Screening Evaluation," Aerojet Liquid Rocket Co., Rept. PDRD TMO5-24, Contract M4J7XMA-483030H, Sacramento, CA, Feb. 1976.

[59]Crocco, L., and Cheng, S., *Theory of Combustion Stability in Liquid Propellant Rocket Motors*, AGARD No. 8, Butterworths, London, 1956.

[60]Smith, A. J., Jr., and Reardon, F., "The Sensitive Time Lag Theory and Its Application to Liquid Rocket Combustion Instability Problems," Rept. AFRPL-TR-67-314, March 1968.

[61]Muss, J. A., Nguyen, T. V., and Johnson, C. J., "User's Manual for Rocket Combustor Interactive Design (ROCCID) and Analysis Program," NASA CR-187109, May 1991.

[62]Priem, R. J., and Morrell, G., "Application of Similarity Parameters for Correlating High Frequency Instability Behavior of Liquid Propellant Combustors," *Detonation and Two-Phase Flow*, edited by S. S. Penner and F. A. Williams, Progress in Astronautics and Rocketry, Vol. 6, Academic Press, 1962, pp. 305–320.

[63]Farley, B. B., "NARloy-Z 3.3K Thrust Chamber Cycle Life Test Program," NASA MSFC Memo ET14 (75-18), 1975.

[64]Romine, W. D., "Thermal Analysis of the Data from the 40 K Calorimeter Thrust Chamber Tests," Rocketdyne SSME Paper 75-2736, Canoga Park, CA, Nov. 1975.

[65]Romine, W. D., "Thermal Analysis of the Data from the 40 K Subscale Regeneratively Cooled Thrust Chamber Cyclic Life Tests," Rocketdyne SSME Paper 76-2523, Canoga Park, CA, Sept. 1976.

[66]Cook, R. T., "40 K Subscale Main Injector and Regen-Chamber Post Test Analysis," Rocketdyne SSME Paper 76-2709, Canoga Park, CA, Sept. 1976.

[67]Romine, W. D., "SSME Main Combustion Chamber Heat Load Summary for Coca 4B Tests 028 through 045," Rocketdyne SSME Paper 76-2867, Canoga Park, CA, Oct. 1976.

[68]Cook, R. T., "Verification of Main Combustion Chamber and Nozzle Cooling (DVS-SSME-303 Para. 4.5.1.6.2)," Contract NAS8-27980, Canoga Park, CA, 8 Dec. 1977.

[69]Gordon, S., and McBride, B. J., "Computer Program for Calculation of Complex Chemical Equilibrium Compositions, Rocket Performance, Incident and Reflected Shocks, and Chapman-Jouguet Detonations," NASA SP-273, March 1976.

[70]Cook, R. T., and Coffey, G. A., "Space Shuttle Orbiter Engine Main Combustion Chamber Cooling and Life," AIAA Paper 73-1310, Nov. 1973.

[71]Cook, R. T., Fryk, E. E., and Newell, J. F., "SSME Main Combustion Chamber Life Prediction," NASA CR-168215, May 1983.

Chapter 17

Assessment of Thrust Chamber Performance

Douglas E. Coats*
Software and Engineering Associates, Inc., Carson City, Nevada

Nomenclature

A = cross-sectional area
C^* = characteristic velocity, $p_c A^*/\dot{m}$
Isp = specific impulse, $F/\dot{m}$
F = thrust
M = Mach number
$\dot{m}$ = mass flow rate
O/F = propellant mixture ratio
p = pressure
r = nozzle radius
Re = Reynolds number
y = distance from the nozzle wall

Greek

α = nozzle wall angle
δ = boundary-layer thickness
δ^* = boundary-layer displacement thickness
γ = ratio of specific heats, Cp/Cv
ε = expansion ratio (A_e/A^*)
η = efficiency
θ = boundary-layer momentum thickness, also nozzle half-angle

Subscripts

a = ambient
c = chamber

*President. Associate Fellow AIAA.

          div = divergence
            e = nozzle exit
          eq = chemical equilibrium
            f = finite contraction ratio
        froz = chemically frozen
            i = ith zone or striation
          kin = finite rate chemistry
          th = theoretical or ideal
          xz = interzonal
          ∞ = infinite contraction ratio
            0 = stagnation, Eq. (1)

Superscripts

          * = nozzle throat plane

I. Introduction

THE assessment of thrust chamber performance is not a new topic, and there are many excellent sources of information on it. In the United States the reports of Pieper[1] and Evens[2] address all of the issues discussed in this chapter. The works of Sutton,[3] and Zucrow and Hoffman[4] cover many of the topics in detail.

Liquid propellant rocket engines (LREs) are devices that convert the latent energy of the propellants into sensible heat in the combustion chamber and then convert it again into kinetic energy in the nozzle. To make comparisons between different engine and system designs, we must be able to assess the performance of the LRE. The thrust chamber of an LRE produces the measurable output of a liquid propellant powered rocket, i.e., thrust, and hence the assessment of its performance is of great importance in evaluating the overall performance of the entire system. Because of the tremendous energy flow in LREs, these engines are characterized by small performance losses because of heat loss, friction, vaporization, and mixing inefficiencies. Even small losses, however, have a large impact on delivered payload or on range of the system and are therefore important.

To assess the performance of a system, one must establish a figure of merit that characterizes the system. The figure of merit must also be a measurable quantity. The most used such quantity for LRE thrust chambers is the specific impulse *Isp*. The specific impulse is defined as the engine thrust divided by the mass flow rate of the propellants, and thus it tells us how effectively the thrust chamber converts propellant into thrust. The specific impulses delivered to vacuum and to ambient pressure conditions are both commonly used.

Although performance as measured by specific impulse is important, the design of a thrust chamber is quite often driven by other factors, including weight, power extraction, physical packaging constraints, materials and life cycle, heat transfer/cooling, combustion stability considerations, environmental and safety concerns, and propellant shelf life. To describe methods of assessing thrust chamber performance, we shall first discuss how to estimate the maximum possible performance obtainable, and then the losses experienced by real engines and the elements required to model these losses. Specific requirements for modeling these losses are then addressed. Also discussed are those phenomena that are not well characterized and hence lack adequate analytical models to describe them. Finally, descriptions of specific codes and procedures in use in the United States are presented.

II.   Definition of Ideal or Theoretical Performance

Maximum performance of a thrust chamber, sometimes called ideal or theoretical performance, is achieved if the propellants entering the thrust chamber react completely and chemical equilibrium is maintained throughout the expansion process. Additionally, the flow should be isentropic and one-dimensional. Under these ideal conditions, the thrust chamber performance is dependent on the physical, chemical, and thermodynamic properties of the propellants and their combustion products, and on the operating conditions of the engine, that is, propellant mixture ratio O/F, chamber pressure p_c, expansion ratio ε, and ambient pressure p_a. For the ideal thrust chamber, we neglect real-world design parameters, such as nozzle geometry, size, injector element design, engine coolant configuration, and baffles.

Even with such a seemingly simple definition of ideal performance, a certain amount of disagreement can occur. For example, the value of the total pressure to be used for the expansion process depends on the assumption of either an infinite or a finite contraction ratio for the combustion chamber. From simple one-dimensional relationships, the ratio of these two total pressures can be approximated as

$$p_{0f}/p_{0\infty} = \frac{\left[1 + \frac{(\gamma - 1)}{2}M_f^2\right]^{\gamma/(\gamma-1)}}{1 + \gamma M_f^2} \approx 1 - \frac{\gamma}{2}M_f^2 \qquad (1)$$

The decrease in total pressure due to accelerating the flow affects primarily the mass flow rate of the nozzle. However, when computing the performance delivered to ambient pressure, there is a small decrease in performance due to the $p_a/\dot{m}$ effect.

Thus, our theoretical maximum performance is defined as an isentropic one-dimensional flow in chemical equilibrium (often called shifting equilibrium) at the thrust chamber O/F and chamber pressure (infinite or finite contraction ratio). More information on the chemical equilibrium solutions to this problem is given by Gordon and McBride.[5,6]

III.   Real Engine Losses

We have broken down the deviations from ideal performance into two classes of real engine losses: those that are well characterized and those that are not. Such a breakdown is very subjective and really just rates our confidence level in our understanding of the phenomena. Included in the well-characterized losses are the boundary-layer and heat transfer losses, finite rate chemical kinetics losses, and divergence or two-dimensional losses. The poorly characterized losses include the injection, atomization, vaporization, and mixing processes in the combustion chamber.

A.   Well-Characterized Losses

1.   Boundary-Layer Loss

Propulsive LREs are generally characterized by high Reynolds number flow. Table 1 lists the Reynolds numbers based on throat conditions for a variety of engines. Because the mass flux is highest in the throat region, the throat Reynolds number is almost always the largest encountered during the nozzle expansion. Other characteristics of importance in LREs that affect boundary layers are the methods of wall cooling. Because the enthalpy of the combustion products is very high, the chamber and nozzle walls need to be protected. Some standard ways of protecting the walls include regenerative cooling, barrier or film cooling, radiation cooling, ablative walls, and slot injection or transpiration cooling. A high Reynolds number means that the viscous layer next to the wall is very thin, which in turn indicates that the classical thin shear or boundary-layer assumptions are valid. Hence, except for the smallest engines, the core flow in the engine can be treated as inviscid, and the solution of the wall shear layer can be uncoupled from the core flow. The true singular perturbation nature of the boundary-layer equations becomes quite apparent in rocket engine flows, because the outer or core flow is not uniform and can vary significantly in the radial direction over the distance of a boundary-layer thickness. In addition, when film cooling is used in the engine, there is a significant total enthalpy gradient near the wall, and hence the outer flow can be highly rotational. The standard simplistic ways of looking at the boundary-layer thicknesses can be very misleading, and questions about the quantity of mass flow in the boundary layer have limited meaning.

Because solution procedures for the boundary-layer equations are well established, the only real questions are what physical phenomena are important and how best to model them. Smaller engines tend to have laminar boundary layers, whereas the larger engines are almost always turbulent. One rule is that engines with less than 45,000 N (10,000 lbf) thrust are laminar. A slightly

Table 1   Nozzle characteristics for various engines

Engine	Thrust, 10^3 N	p_c, bars	r^*, mm	ε	Re_{r^*}
Hughes 5 lbf	0.111	1.72	2.37	296.6	1.10×10^4
NASA/LeRC Hi-E	2.40	24.82	12.7	1025	1.73×10^5
XLR-134	2.28	35.16	10.06	767.9	1.80×10^5
STS/RCS	3.84	10.34	25.93	28.46	1.75×10^5
Advanced Space Engine	100.67	157.68	31.85	400.7	2.20×10^6
RL 10	60.05	27.19	65.28	205.03	1.29×10^6
RD-170	7915.73	244.65	117.75	36.9	1.62×10^7
SSME	2062.45	226.49	130.88	77.5	1.18×10^7
F1	7786.55	68.4	444.5	15.76	1.81×10^7

more appealing transition criterion is that transition occurs when the Reynolds number, based on the boundary-layer momentum thickness Re_θ, exceeds 360. Because most engines are high-thrust engines, one of the first choices to be made is the selection of the turbulence model. Coats et al.[7] have estimated that the maximum calculated variation in boundary-layer loss results is approximately 25% when a κ-ε turbulence model is compared to an algebraic eddy viscosity (e.g., Cebeci and Smith[8]) model. Because the boundary loss is usually on the order of 1–2% and almost never more than 4% of the total performance, the variation of calculated loss with turbulence model will be in the range of 0.25–1% of the total performance. Without high quality experimental data to validate turbulence models for rocket engine flows, there is no way of knowing which of the available turbulence models should be used.

Other questions arise as to which chemistry model should be used in the boundary-layer calculation. For most simple flows, i.e., single O/F core flows, almost any chemistry model will give results within the known accuracy range of the boundary-layer equations. Once higher fidelity core flow models are used, however, at least a variable gas properties treatment should be used in the boundary-layer model. If heat transfer results are required in addition to performance losses, then the choice of chemistry model can be quite important. For example, the adiabatic wall temperature at the nozzle exit plane for the Vulcain engine is 350 K greater for the finite rate calculation than for a chemical equilibrium calculation, an important difference if you are determining the cooling requirements of the engine! Another consideration in selecting the boundary-layer chemistry model is the need to predict what happens to turbine exhaust gases that are injected into the engine downstream of the throat. These injected gases have a pronounced effect on the boundary-layer profiles, as shown in Fig. 1, and can lead to either endothermic or exothermic reactions. Transpiration cooling modeling requirements will also have an impact on the chemistry model selection.

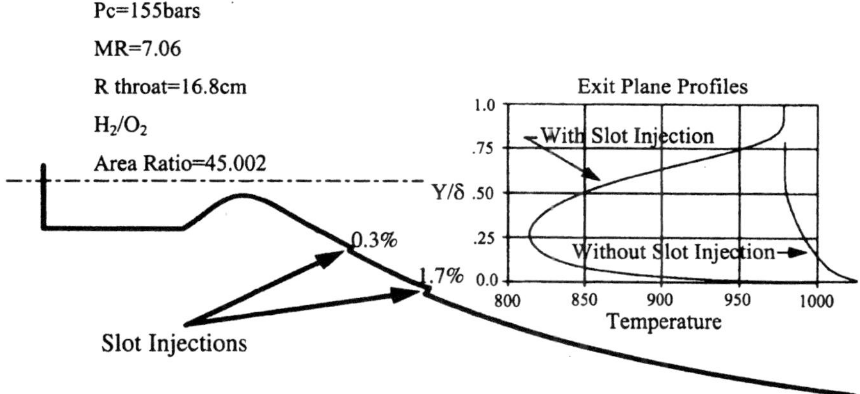

Fig. 1 **Boundary-layer profile with tangential slot injection.**

2. Divergence

The divergence or two-dimensional loss is conceptually the simplest of all of the losses. The velocity vectors of the gases exiting the nozzle are not necessarily aligned with the axis of the nozzle or vehicle. The result of the misalignment is that not all of the kinetic energy of the flow results in axial thrust. This loss is mainly a function of the nozzle geometry, and to a lesser extent of the ratio of specific heats γ.

It is generally agreed that there is little coupling between the finite rate chemical kinetics and the two-dimensional flowfield. Furthermore, except for the smallest engines or very large area ratio nozzles, the wall boundary layer is very thin and interacts with the core flow only to second order. Other performance loss considerations are due to the formation of shock waves, large gradients in the flow due to O/F striations, two-phase flow due to poor vaporization of droplets, and large sources of rotationality due to geometric or engine/injector phenomena.

For most well-designed engines these considerations are not applicable. Hence, the divergence loss can be adequately modeled by a variable gas properties solution to the full Euler equations. However, solutions to these equations, although isoenergetic along streamlines, do allow for shock waves and variations in the total enthalpy of the flow, and thus can be used for off-design analysis.

One last consideration is that the sonic line or surface in the nozzle throat region is not flat but curved. The result of this curvature is that there is less mass flow through the nozzle than would be predicted by one-dimensional flow analysis. The series solution by Hall,[9] modified by Kliegel and Levine,[10] gives a good estimate of the nozzle discharge coefficient.

3. Finite Rate Kinetics

When highly energetic propellants are burned in the combustion chamber, the resulting high temperatures cause many of the normally stable molecules to dissociate. During the subsequent expansion in the nozzle, the kinetic rate process tends to recombine these molecules, making sensible heat available to further drive the expansion. Most notably, it is the recombination of hydrogen molecules to form H_2, and the formation of CO_2 from CO and O, which release the bulk of the energy. The short residence time in the nozzle, coupled with rapidly decreasing pressure and temperature, do not allow the flow to stay in chemical equilibrium (maximum heat release).

For most propellant systems, the reaction rate mechanisms and their associated rate data are reasonably well understood. The rate data are usually known within an order of magnitude, which is adequate for determining the finite rate kinetics loss. There are several ways in which this loss is expressed. The two most common are the ratio of specific impulse with and without the loss, η_{kin}, and the fraction of the difference between equilibrium and frozen flow, η'_{kin}:

$$\eta_{kin} = Isp_{kin}/Isp_{eq} \tag{2a}$$

$$\eta'_{kin} = (Isp_{kin} - Isp_{froz})/(Isp_{eq} - Isp_{froz}) \tag{2b}$$

The loss is mainly a function of the propellant system, chamber pressure, and residence time in the nozzle. High-pressure systems tend to have smaller losses, because of the large number of molecular collisions. The nozzle length scale is also an important parameter in that it sets the residence time in the nozzle. Although high-area-ratio nozzles tend to have larger losses, this effect is less important than the other factors discussed. Table 2 shows some typical values for the kinetic loss efficiencies. As can be seen from table, amine fueled (e.g., hydrazine, UDMH, MMH) engines usually exhibit low kinetic efficiencies.

Another consideration is the starting point for the expansion. If the species in the combustion chamber do not start out in a state near chemical equilibrium, then there is the potential that they will not approach equilibrium within the nozzle. The nonequilibrium starting condition problem is especially important at off-optimum mixture ratios.

4. Radiation

Energy can escape the thrust chamber in the form of radiation. The foremost methods are from the combustion chamber walls and the nozzle, and from the hot gases. The first two losses are usually coupled with the boundary-layer loss, while the second is generally small. The majority of the energy radiated from the hot gases is in the form of infrared radiation. The largest common emitters of infrared radiation are CO, CO_2, and H_2O. These molecules emit at discrete spectral lines, and black or gray body radiation treatments are inappropriate. The majority of the radiation emitted by these species is transferred to the walls or is reabsorbed by the gases. Only a small fraction of the energy escapes out of the nozzle. In terms of the overall energy flow rate, the loss out of the nozzle is insignificant. However, for radiation-cooled nozzles the performance loss should be treated as a boundary condition for the boundary-layer calculation. Non-performance-related applications in which radiation can be the dominant effect include base heating and exhaust plume signatures.

B. Poorly Characterized Losses, the Energy Release Efficiency

The purpose of an injector element in the thrust chamber is to introduce the fuel or oxidizer in such a way that both components mix and combust completely. There are many different injector types, and the fuel and oxidizers may be

Table 2   Kinetic loss ratios

Engine	Propellants	η'_{kin}
F-1	LOX/RP-1	0.98
Atlas Booster	LOX/RP-1	0.90
Atlas Sustainer	LOX/RP-1	0.90
TR201	NTO/A50	0.50
R-4D	NTO/MMH	0.30
Titan III (Stage I)	NTO/A50	0.75

injected as gas-gas, liquid-gas, gas-liquid, and liquid-liquid. Each method of injection has its own advantages and associated problems. To illustrate the sources of potential losses, we will concentrate on liquid-liquid injection.

The process of injection, atomization, vaporization, combustion, and mixing is reminiscent of turbulence in fluid flow. That is, there are a great many length scales that must be resolved to characterize the problem. First, the liquid stream of fuel or oxidizer experiences an instability that causes it to start to break up. Prior to, during, or after breakup, the stream can collide with another stream. The original breakup is called primary atomization, and breakup of the larger drops is referred to as secondary atomization. The processes associated with secondary atomization include the dynamic stability of the droplets and collisions among droplets. Finally, the droplets vaporize in the presence of the hot combustion gases. In the case of supercritical combustion, the problems are less well understood.

Liquid rocket engines do not always vaporize all of the propellants within the combustion chamber. In many engines using hydrocarbon fuels, the fuel tends to vaporize much more slowly than the oxidizer (fuel-controlled burning). This slow vaporization can cause a large shift between the injected O/F and the effective gas-phase O/F at the exit of the combustion chamber. Engine designers often trade combustion chamber length and ease of injector fabrication for vaporization efficiency.

One of the well-known characteristics of dense sprays is that they do not behave like collections of individual droplets. Incomplete vaporization and poor mixing of the vaporized propellants can cause significant losses in performance. Many people think that of the two, the mixing loss is much larger than the vaporization loss. This type of mixing loss is sometimes referred to as cluster effect, fine mixing, intrazonal mixing, or micromixing, as compared with coarse mixing, which is discussed next.

The thrust chamber walls are sometimes cooled by injecting a fuel (or oxidizer) film spray on the wall. The lower (or higher) O/F in this region reduces the flame temperature, and thus the heat transfer rate. Because these propellants do not combust in a way that releases the maximum amount of heat, there is a loss associated with this process, as compared with our theoretical performance at the overall engine mixture ratio. This loss is referred to as coarse mixing, interzonal mixing, or macromixing loss.

Of the two mixing losses, the interzonal is the easiest to conceptualize because it is part of the engine design and the percentage of fuel near the wall is known. Figure 2 illustrates the amount of noncombusted fuel required to produce the 2.5% loss that is typical of many engines. These results, which are from equilibrium calculations for an RP-1/LOX system and an NTO/UDMH system, show that approximately 25% of the RP-1 and 35% of the UDMH would have to be unburned in order to reduce the Isp by 2.5% from its maximum value.

Sometimes, all of the above losses are lumped together and referred to as combustion efficiency or energy release losses. The most direct measurement that we have of these losses is the measured C^* efficiency of the engine. If the C^* efficiency is used to back out this loss, then the effect of the nozzle discharge

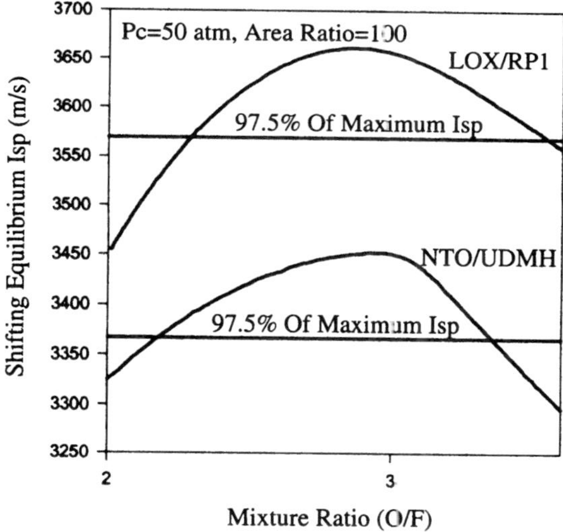

Fig. 2   Vacuum *Isp* variation with mixture ratio.

coefficient C_D must be subtracted out. In the absence of any losses, the C^* efficiency η_{C^*} is

$$\eta_{C^*} = 1/C_D \tag{3}$$

Note that the C^* efficiency is not the same as combustion efficiency, although they are very close for most engines and propellant systems.

IV.   Modeling

Analytical modeling of thrust chamber performance encompasses a wide range of approaches. It can apply to anything from using the isentropic perfect gas relations to a full Navier–Stokes (FNS) simulation. The information to be derived from the modeling drives the complexity of the selected model. For example, a value of 91% of the theoretical *Isp* provides a reasonable prediction of performance. It does not, however, provide any information on how the losses occurred or how they could be reduced. Another important point is that the model output is no better than the quality of the input data.

In this section we will explore a variety of modeling approaches, try to assess the accuracy of the approaches, and make some recommendations. Starting first with the calculation of ideal performance, we will cover the well-characterized losses and then cover the approaches to modeling the poorly characterized losses.

Three types of modeling will be discussed: simple modeling, moderately complex modeling, and complex modeling. In the first category, very simple relations will be used to estimate the magnitude of each loss, i.e., Eq. (2). For the moderate category, fairly sophisticated models are used to calculate each loss. In the final category, all (or at least most) of the losses are modeled in a coupled manner.

One point should be made with respect to the calculation of the performance of a thrust chamber; this is simply that *Isp* is a wonderful quantity in its insensitivity to uncertainty. If, when modeling an engine, the energy of the propellants is correctly specified and the conservation of mass, energy, and momentum is achieved to reasonable accuracy, then the answer will be almost independent of the small variations in modeling assumptions. On the other hand, there are those who seek accuracies of within a few tenths of a percent.

A.   Ideal or Theoretical Performance

The calculation of thrust chamber performance, assuming chemical equilibrium with complex products of combustion, is now easily accomplished.[6] Recent controversies aside, there are a number of chemical equilibrium codes available that give the same answers to the known accuracy of the thermodynamic data. Only two thermodynamic states and the elemental composition are required to fully determine the state of chemical equilibrium. The usual assumptions made are that the composition and enthalpy (heat of formation plus sensible heat) of the propellants are known. The enthalpy of the propellants in the tank is often used if known; otherwise the enthalpy at the normal boiling point (NBP) is a good choice. The chamber pressure is assumed and an enthalpy-pressure (HP) solution is found. The products of combustion are then expanded to different pressures using the entropy of the products (PS) in the chamber to close the set. From the conservation of mass and energy, the area ratios and velocities can be found at each solution point. As stated earlier, the differences between finite area and infinite area combustion are minimal, and Eq. (1) can be used to apply the correction to the chamber pressure.

One-dimensional chemical equilibrium solutions are among the most useful modeling tools available. Not only do they give us the theoretical maximum performance, but they also allow us to explore the effects of modifying propellant formulations and mixture ratios, with a minimal expenditure of time and effort.

B.   Well-Characterized Losses

Earlier in this chapter we discussed the well-characterized losses and indicated some of the more important features of these losses. Now we will concentrate on methods of solutions, the required input for the solutions, and the potential accuracy of the modeling approach. It should always be remembered that the accuracy of a solution is dependent not only on the correctness of the model but also on the accuracy of the input data. There is no advantage in using complicated models if the input required is not known to reasonable accuracy. Such approaches mask the uncertainty of the answers and are often misleading.

To this point we have not formally addressed the interactions of the various loss mechanisms. Obviously, almost every mechanism in the combustion and flow process affects every other mechanism to some extent. As we pointed out, however, for high Reynolds number flows, the boundary layer and core flow can be decoupled. These are not the only losses that can be uncoupled. Table 3,

Table 3 Interaction of physical phenomena with performance loss calculations (from Ref. 1)

Phenomena	Divergence loss	Boundary-layer loss	Finite rate kinetics loss	O/F Maldistribution loss	Energy release loss
Non-one-dimensional flow	——	1st order,[a] $>0.2\%$	2nd order,[b] $<0.2\%$	Not Imp.[c]	Not Imp.
Viscous and heat transfer	Not imp.	——	Not imp.	Not imp.	Not imp.
Finite rate chemistry	Not imp.	2nd order, $<0.2\%$	——	Not imp.	Not imp.
Nonuniform mixture ratio	Not imp.	1st order, $>0.2\%$	1st order, $>0.2\%$	——	Not imp.
Incomplete energy release	Not imp.	2nd order, $<0.2\%$	1st order, $>0.2\%$	Not imp.	——

[a]1st order, $>0.2\%$ = primary importance (could be $>0.2\%$ on Isp).
[b]2nd order, $<0.2\%$ = secondary importance (probably $<0.2\%$ on Isp).
[c]Not imp. = generally not important.

which was taken from Ref. 1, shows the estimated coupling between loss mechanisms in LREs. This uncoupling of losses is very important for simple and intermediately complex modeling.

For all but the most complex of modeling approaches, the most common method is to separate the individual losses and treat them as efficiencies η or decrements Δ to the theoretical Isp. That is,

$$Isp_D = Isp_{th} \cdot \prod_i \eta_i \tag{4a}$$

or

$$Isp_D = Isp_{th} - \sum_i \Delta Isp_i \tag{4b}$$

Although both approaches have their ardent supporters, there are really no significant differences between the two. In addition to its simplicity, the separation approach has the extra advantage that it identifies both the sources and magnitudes of the losses.

1. Boundary-Layer Loss

Before proceeding into actual methods of modeling the boundary-layer loss, the coupling of regenerative heat transfer with the theoretical specific impulse must be established. For performance prediction in most regeneratively cooled engines, it is quite acceptable to model the nozzle wall as adiabatic and the propellants at their tank enthalpies, because the energy transferred to the wall is put back into the fuel. If the nozzle wall is modeled as a cold wall, however, then the enthalpy of the incoming fuel must be adjusted upward by the amount of heat extracted. This observation is particularly important when the modeler is using some measured data. For example, if measured fuel and oxidizer inlet temperatures are used to establish the enthalpies of the propellants, the cold wall heat transfer model must be used. Otherwise, the predicted thrust chamber performance will be much too high.

Simple equations for predicting the nozzle boundary-layer loss have for the most part not been very successful. Our recommendation is that simple empirical equations not be used unless the user has a great deal of experience with the specific engine. Although engine-specific empiricism is acceptable, simplified general empiricism does not work well for the boundary-layer loss.

Part of our reluctance to endorse any general empirical relation is that boundary-layer codes have become so easy and inexpensive to use. Integral method codes, such as the TBL[11] code, have been available for decades and have a significant amount of rocket engine empiricism built into them. Modern boundary-layer codes using finite difference are now robust and easy to use. Computer codes based on parabolized and full Navier–Stokes equations are also available to compute the wall shear layer loss.

Considering the state of our knowledge in turbulence modeling for rocket engine flows, variable gas properties boundary-layer codes are adequate for performance prediction loss modeling. However, if the boundary-layer results will also be used for wall heat transfer calculations, then chemistry modeling should also be included.

2. Divergence

For a simplified approach to divergence loss, the loss expression given in Eq. (5) can be used:

$$\eta_{\text{div}} = (1 + \cos \theta)/2 \tag{5}$$

where θ is the cone half-angle (or average nozzle expansion angle). This expression was derived for conical nozzles and its development is presented in Zucrow and Hoffman.[4] Figure 3 illustrates the nomenclature.

Euler solvers are recommended for the moderately complex modeling approach because the divergence loss is essentially an inviscid phenomenon. There are many perfect gas and variable properties programs to choose from. Both the method of characteristics (MOC) and fixed grid finite difference or finite volume solvers are entirely adequate for determining the two-dimensional or divergence loss in most propulsive nozzles of interest. Although steady-state MOC solvers are fast and very accurate, they require a supersonic start line to begin their solution. This requirement means that some other method is needed to generate the transonic solution. Also, most MOC solvers can handle only weak shock waves. The latter condition is usually met in most propulsive nozzles of interest.

However, for very small nozzles or systems with large mixture ratio distributions, more complex codes are recommended. When the boundary layer starts to engulf a significant fraction of the nozzle flow, the use of parabolized or full Navier–Stokes (PNS or FNS) solvers is indicated. These types of codes should be used at least once in the analysis process to ensure that the simpler solvers are doing an adequate job. For systems with large O/F gradients, chemistry should be incorporated into the models. At least an equilibrium chemistry

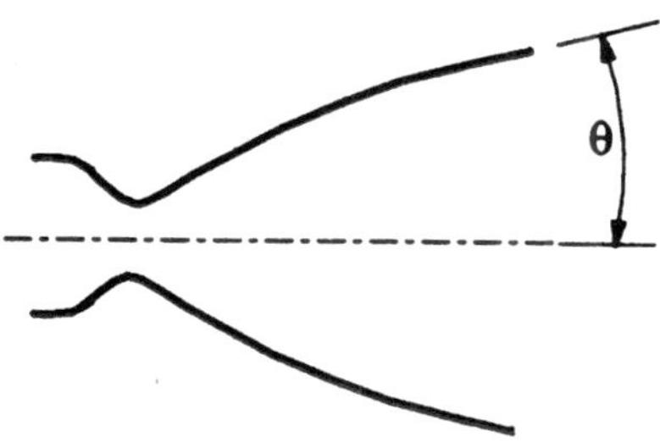

Fig. 3   Simplified nozzle divergence efficiency.

Euler solver is required and a finite rate chemistry solver is preferred. Finite rate chemistry Euler solvers have been available for almost three decades.[12]

The magnitude of the divergence loss is around 2%. For systems without large O/F gradients or very thick boundary layers, good Euler solvers are probably accurate to within 5% of the divergence loss or 0.1% of the Isp loss. Because they tend to use coarser grids than Euler solvers, PNS and FNS solvers are not quite as accurate, and the error is estimated to be 10–15%. PNS codes are less likely to suffer from this problem than FNS codes, because they can be run at higher spatial resolutions. Another problem with higher fidelity models is that the input data are not known to the same order of accuracy of the models. Kushida et al.[13] have reported very good agreement between predictions and measured data when using an MOC/boundary layer method for a very small high area ratio thruster, the Hughes 5 lbf engine in Table 1. The computed boundary-layer thickness for that nozzle was 28% of the radius at the nozzle exit plane.

3. Finite Rate Kinetics

For most systems, the finite rate kinetics loss is less than 2%. Hence, simply assuming a constant value for η'_{kin} of 0.66 will suffice for most systems. There are many systems, however, in which this approximation is very poor, most notably, low pressure amine and fluorinated systems. In the absence of data about a particular system, the minimum acceptable model is a one-dimensional finite rate kinetics calculation. Arrhenius rates are known within an order of magnitude for most of the important reactions in chemical systems of current interest. The difference in calculated kinetics loss between one- and two-dimensional solutions is usually very small. Because we have recommended the use of chemistry models for the divergence loss to handle striations in the flow, however, the use of an Euler solver with finite rate kinetics capability is also recommended.

Although PNS solutions with finite rate kinetics can be achieved in somewhat reasonable computational times, adding kinetics to FNS solutions results in very long execution times for rather poorly resolved spatial meshes. Under normal conditions, we do not recommend the use of FNS solvers to calculate the kinetics loss.

The rates chosen for these calculations have an obvious impact on the magnitude of the loss. In the United States, the rates shown in Table 4 are recommended for CHON systems in the two-dimensional kinetics (TDK) code documentation,[14] with the rate data taken from Baulch et al.[15–17] and Jensen and Jones.[18] The chemical reactions considered are between the species, CO, CO_2, H, H_2, H_2O, N, NO, N_2, O, OH, and O in initial chemical equilibrium. Recommended third body efficiencies for various species are shown in Table 5. Perturbing the $H+H$ and $CO+O$ recombination rates by a factor of 30 downward has the effect on kinetic efficiency shown in Table 6 for an engine with a nozzle expansion ratio of 100, NTO/A50 propellants at a stoichiometric mixture ratio, and a chamber pressure of 7 atm. As can be seen from the table, the changes in performance are minimal.

Table 4 Reaction rate data for the CHON system

Reactions	A^a	N	B	Meas., M	Reference
$H + H + M = H_2 + M$	6.4E17	1.0	0.0	Ar	Baulch[15] 30U
$H + OH + M = H_2O + M$	8.4E21	2.0	0.0	Ar	Baulch[15] 10U
$O + O + M = O_2 + M$	1.9E13	0.0	-1.79	Ar	Baulch[15] 10U
$N + O + M = NO + M$	6.4E16	0.5	0.0	N2	Baulch[17] 10U
$N + N + M = N_2 + M$	3.0E14	0.0	-0.99	N2	Baulch[17] 10U
$CO + O + M = CO_2 + M$	1.0E14	0.0	0.0	Ar	Baulch[16] 30U
$O + H + M = OH + M$	3.62E18	1.0	0.0	Ar	Jensen[18] 30U
$O_2 + H = O + OH$	2.2E14	0.0	16.8		Baulch[15] 1.5U
$H_2 + O = H + OH$	1.8E10	-1.0	8.9		Baulch[15] 1.5U
$H_2 + OH = H_2O + H$	2.2E13	0.0	5.15		Baulch[15] 2U
$OH + OH = H_2O + O$	6.3E12	0.0	1.0		Baulch[15] 3U
$CO + OH = CO_2 + H$	1.5E7	-1.3	-0.765		Baulch[15] 3U
$N_2 + O = NO + N$	7.6E13	0.0	75.5		Baulch[17] 3U
$O_2 + N = NO + O$	6.4E9	-1.0	6.25		Baulch[17] 2U
$CO + O = CO$	2.5E6	0.0	3.18		Baulch[16] 2U
$CO_2 + O = CO + O$	1.7E13	0.0	52.7		Baulch[16] 3U

$^a k = AT^{-N} \exp(-1000B/RT)$, in units of cc, K, mole, s.

Table 5 Third body recombination efficiency ratio (CHON system), as recommended by Kushida[19]

Species	H + H	H + OH	O + O	N + O	N + N	CO + O	O + H
Ar	1.0	1.0	1.0	0.8	1.0	1.0	1.0
CO	1.5	3.0	4.0	1.0	1.0	1.0	4.0
CO_2	6.4	4.0	8.0	3.0	2.0	5.0	5.0
H	25.0	12.5	12.5	10.0	10.0	1.0	12.5
H_2	4.0	5.0	5.0	2.0	2.0	1.0	5.0
H_2O	10.0	17.0	5.0	7.0	3.0	1.0	5.0
N	1.0	1.0	10.0	10.0	10.0	1.0	1.0
NO	1.5	3.0	4.0	1.0	1.0	1.0	4.0
N_2	1.5	3.0	4.0	1.0	1.0	2.0	4.0
O	25.0	12.5	12.5	10.0	10.0	1.0	12.5
OH	25.0	12.5	12.5	10.0	10.0	1.0	12.5
O_2	1.5	6.0	11.0	1.0	1.0	25.0	5.0

C. Poorly Characterized Losses, the Energy Release Efficiency

For the purposes of modeling, we have broken down the injector losses into three categories: vaporization, interzonal mixing, and intrazonal mixing.

1. Vaporization Efficiency

The vaporization loss is that part of the energy release performance degradation due to either the fuel or oxidizer not completely vaporizing in time to be completely burned. The major obstacle in modeling the vaporization efficiency is obtaining reasonable estimates for the sizes and distributions of the liquid drops. If such estimates are known, then the methods outlined by Priem and Heidmann[20] and Nickerson and Johnson[21] can be applied to subcritical droplet vaporization in one- or two-dimensional flows. Issues related to supercritical droplet vaporization and combustion are addressed in Chapter 7.

2. Interzonal Losses

Interzonal variations in mixture ratio are caused by decisions made in the design of the thrust chamber. The most common cause of interzonal striations is the use of a fuel (or oxidizer) film to keep the chamber walls from exceeding

Table 6 Variation in kinetic efficiency with rate data

Reaction	η'_{kin}	η_{kin}
Reference/No changes	0.552	0.9513
$H + H + M = H_2 + M$	0.546	0.9507
$CO + O + M = CO_2 + M$	0.551	0.9512
Change both rates	0.545	0.9505

their maximum design temperature. Other striations can be caused by the presence of baffles used to suppress acoustic waves. The simplest model for interzonal mixing is a simple mass average of the theoretical *Isp* for each mixture ratio. That is,

$$\eta_{xz} = \sum_{i=1}^{\#zones} \left[\frac{\dot{m}_{(O/F)_i}}{\dot{m}_t} \right] \cdot Isp[(O/F)_i]/Isp_{th}(\overline{O/F}) \tag{6}$$

The design values of mass flow for the fuel and oxidizer can be used for initial studies. Cold flow data can supply updated values once testing has begun. Except in rare instances, striations can only be inferred from hot flow heat transfer data.

The moderately complex modeling approach uses the same information as the simple model, but the data are used to generate striation profiles, which are then run though an Euler solver. In such cases the boundary between the striations is a slip line or contact discontinuity. For most cases, the differences between the simple and moderately complex methods are small in terms of performance of the core flow. The effect on the boundary-layer edge condition, however, can be very significant. Also, one must be very careful not to include this loss more than once.

The complex approach can either try to model the injection process or it can use the same data as the moderately complex approach. For the purposes of performance assessment, the use of measured data is probably much more accurate than trying to model the O/F variations directly. However, such modeling can supply a qualitative insight into what is happening in the engine and can be invaluable in understanding problems.

3. Intrazonal Mixing Loss

The major problem with modeling intrazonal or micromixing loss is that it cannot be measured directly in either a rocket engine or a reasonable simulation device. The micromixing losses are always inferred by first subtracting out other losses, such as finite rate kinetics, vaporization, and macromixing losses. Both theoretical and empirical micromixing models exist. Spalding,[22] Elghobashi and Pun,[23] and Tamanini[24] assume a two-parameter probability density function to describe the local variation in fuel mixture fraction. The average fuel mixture fraction and other parameters of interest can then be calculated. On the empirical side, C^* correlations based on similar engines are used to estimate the total energy release efficiency loss.

The only recommendation that we can make with respect to modeling the micromixing loss is to use engine-specific empiricisms to estimate the total energy release efficiency loss, subtract out the vaporization and macromixing loss, and then adjust the input enthalpy to match the measured or estimated performance.

V. Approaches

In the United States, the practices used at each engine manufacturer and cognizant analysis organization can vary significantly. However, for engines employing standard bell nozzles, the JANNAF procedures as outlined in Ref. 2 are generally followed. These procedures are essentially those outlined in this chapter as the moderately complex modeling approach. That is, the TDK computer program is used to model all of the losses that we have termed "well-characterized losses." The TDK code consists of several modules that compute the *Isp* for a variety of input mixture ratios and enthalpies. The code uses a finite rate kinetics MOC solver to compute the core flow and a finite rate kinetics, finite difference boundary-layer module employing a Cebeci-Smith eddy viscosity turbulence model to compute the boundary-layer loss. The macromixing loss is treated by inputting to the code the O/F and energy content of each striation considered. Gas turbine exhaust dumps can be treated as being injected into either the boundary layer or the core flow. Both cold wall, radiation cooled, and adiabatic wall heat transfer treatments are allowed. Furthermore, the solutions of the core flow and boundary layer can be iterated by displacing the potential wall either inward or outward by the boundary-layer displacement thickness.

One curiosity of the JANNAF procedure is the method of computing the boundary-layer loss. The standard JANNAF equation for the boundary-layer loss is

$$\Delta Isp_{\mathrm{BL}} = 2\pi r \rho_e u_e^2 \theta \cos(\alpha)\left[1 - \frac{\delta^*}{\theta} P_e/\rho_e u_e^2\right]\bigg/\dot{m} \tag{7}$$

which is a combination of both the inner and outer solutions applied all at the same time. Kehtarnavaz et al.[25] have extended the derivation of Eq. (7) to thick boundary layers.

The MOC has been shown to be an accurate and efficient flow solver, although there have been many advances in the numerical solutions to rocket nozzle flow. The use of PNS and FNS flow solvers has increased substantially in recent years, especially for very small and very high area ratio engines. Table 7 shows a comparison of results from the TDK and VIPER PNS codes.[26]

Table 7　Comparisons of TDK and PNS nozzle performance predictions

	Isp prediction		
Engine name	TDK/BLM[22]	VIPER[27]	Measured
Adv. Space Engine	473.58	470.20	477.9
Hughes 5 lbf	216.65	218.05	214.52[21]
NASA/LeRC Hi-E	480.31	488.47	468.9
	(458.7)[a]	(466.49)[a]	
SSME	457.7	455.15	452.6[22]
RL 10	463.03	462.29	458.7
XLR-134	468.68	462.29	——

[a]Corrected for 95.5% measured C^* efficiency.

The energy release efficiency is usually estimated by using C^* efficiencies from sub- or full-scale testing, from data on similar engines, or backed out from measured *Isp* data. Each engine manufacturer has its own procedures and analytical methods for estimating this loss. Once the loss has been established, it is applied by decreasing the propellant enthalpy until the desired decrease in *Isp* is achieved.

VI. Conclusions

The assessment of thrust chamber performance has been discussed in terms of both the physical phenomena and the modeling of the phenomena. Ideal performance has been defined and deviations from that ideal have also been discussed. Performance losses have been broken down into two categories, well-characterized and poorly characterized losses. The former involve mainly the flow in the nozzle, and the latter are associated with injector performance.

Three levels of modeling, simple, moderately complex, and complex, were discussed. The moderately complex approach has been used successfully in the United States and is more than adequate for performance prediction. The complex approach is recommended only when the fidelity of the input is commensurate with the fidelity of the modeling.

The state of the art in nozzle loss prediction is much better than that of injector performance. There are, however, still issues that need to be resolved. The most important of the issues for nozzle losses is the establishment of an adequate turbulence model for the boundary-layer calculations. A unified model that is applicable for all speed regimes and includes finite rate chemistry is required. In general, the use of empirical data is required to predict the energy release efficiency.

References

[1]Pieper, J. L., "ICRPG Liquid Propellant Thrust Chamber Performance Evaluation Manual," Chemical Propulsion Information Agency, Pub. 178, Sept. 1968.

[2]Evens, S., "JANNAF Rocket Engine Performance Prediction and Evaluation," Chemical Propulsion Information Agency, Pub. 246, April 1975.

[3]Sutton, G. P., *Rocket Propulsion Elements*, 6th ed., Wiley, New York, 1992.

[4]Zucrow, M. J., and Hoffman, J. D., *Gas Dynamics*, Wiley, New York, 1976.

[5]Gordon, S., and McBride, B. J., "Computer Program for Calculation of Complex Chemical Equilibrium Compositions, Rocket Performance, Incident and Reflected Shocks, and Chapman Jouguet Detonations," NASA SP-273, 1971.

[6]Gordon, S., and McBride, B. J., "Computer Program for Calculation of Complex Chemical Equilibrium Compositions and Applications," NASA RP-1311, Oct. 1994.

[7]Coats, D. E., Berker, D. R., and Kawasaki, A. H., "Boundary Layer Loss Models in Nozzle Performance Predictions," Chemical Propulsion Information Agency, Pub. 529, Oct. 1989.

[8]Cebeci, T., and Smith, A. M. O., *Analysis of Turbulent Boundary Layers*, Academic Press, New York, 1974.

[9]Hall, I. M., "Transonic Flow in Two-Dimensional and Axially-Symmetric Nozzles," *Quarterly Journal of Mechanics and Applied Mechanics*, Vol. XV, Pt. 4, 1962, pp. 487–508.

[10]Kliegel, J. R., and Levine, J. N., "Transonic Flow in Small Throat Radius of Curvature Nozzles," *AIAA Journal*, Vol. 7, No. 7, July 1969, pp. 1375–1378.

[11]Weingold, H. D., "The ICRPG Turbulent Boundary Layer (TBL) Reference Program," ICRPG Perf. Std. Working Group, Pratt and Whitney Aircraft, East Hartford, CT, July 1968.

[12]Kliegel, J. R., Nickerson, G. R., Frey, H. M., Quan, V., and Melde, J. E., "Two-Dimensional Kinetics Nozzle Analysis Computer Program-TDK," ICRPG Performance Standardization Working Group, Dynamic Science Corp., Monrovia, CA, July 1968.

[13]Kushida, R., Hermal, J., Apfel, S., and Zydowicy, M., "Performance of High-Area Ratio Nozzle for a Small Rocker Thruster," *Journal of Propulsion and Power*, Vol. 3, No. 4, p. 329.

[14]Nickerson, G. R., Berker, D. R., Coats, D. E., and Dunn, S. S., "Two-Dimensional Kinetics (TDK) Nozzle Performance Computer Program Volume II, Users Manual," Software and Engineering Associates, Inc., Contract NAS8-39048, Carson City, NV, March 1993.

[15]Baulch, D. L., Drysdale, D. D., Horne, D. G., and Lloyd, A. C., *Evaluated Kinetic Data for High Temperature Reactions, Vol. II Homogeneous Gas Phase Reactions for the H_2-O_2-N_2 System*, CRC Press, Cleveland, OH, 1972.

[16]Baulch, D. L., Drysdale, D. D., Duxbury, J., and Grant, S., *Evaluated Kinetic Data for High Temperature Reactions, Vol. III Homogeneous Gas Phase Reactions of the O_2-O_3 System, The CO-O_2-H_2 System, and of Sulfur-Containing Species*, Butterworths, London, 1976.

[17]Baulch, D. L., Drysdale, D. D., Horne, D. G., and Lloyd, A. C., *Evaluated Kinetic Data for High Temperature Reactions, Vol. II Homogeneous Gas Phase Reactions for the H_2-O_2-N_2 System*, CRC Press, Cleveland, OH, 1973.

[18]Jensen, D. E. and Jones, G. A. "Reaction Rate Coefficients for Flame Calculations," *Combustion and Flame*, Vol. 32, 1978, pp. 1–34.

[19]Kushida, R., "Revision of CPIA 246, Section 6.2, Reaction Rate Data," Jet Propulsion Lab., JPL 383CR-76-211, Pasadena, CA, March 1976.

[20]Priem, R. J., and Heidmann, M. F., "Propellant Vaporization as a Design Criterion for Rocket-Engine Combustion Chambers," NASA TR R-67, 1964.

[21]Nickerson, G. R., and Johnson, C. W., "SCAP-Spray Combustion Analysis Program," Phillips Lab., PL-TR-94-3032, Edwards AFB, CA, Aug. 1993.

[22]Spalding, D. B., "Concentration Fluctuations in a Round Turbulent Jet," *Chemical Engineering Science*, Vol. 26, 95, 1971.

[23]Elghobashi, S. E., and Pun, W. M., "A Theoretical and Experimental Study of Turbulent Diffusion Flames in Cylindrical Furnaces," *Fifteenth Symposium (International) on Combustion*, Combustion Inst., Pittsburgh, PA, 1975, p. 1353.

[24]Tamanini, F., "On the Numerical Prediction of Turbulent Diffusion Flames," Eastern States Section/Combustion Inst., Pittsburgh, PA, April 1976.

[25]Kehtarnavaz, H., Coats, D. E., and Dang, A. L., "Viscous Loss Assessment in Rocket Engines," *Journal of Propulsion and Power*, Vol. 6, No. 6, Nov.–Dec. 1990, pp. 713–717.

[26]Kawasaki, A. H., Dunn, S. S., Coats, D. E., Nickerson, G. R., and Berker, D. R., "Viscous Interaction Performance Evaluation Routine for Two-Phase Nozzle Flows with Finite Rate Chemistry, VIPER 3.0, Volume I: Computer Reference Manual," Software and Engineering Associates, Inc., Contract F29601-91-C-0099, Carson City, NV, May 1994.

[27]Coats, D. E., Berker, D. R., and Dunn, S. S., "Boundary Layer Study," U.S. Air Force Astronomical Lab. AL-TR-90-040, Contract F04611-86-C-0055, Nov. 1990.

Color reproductions for Chapter 12

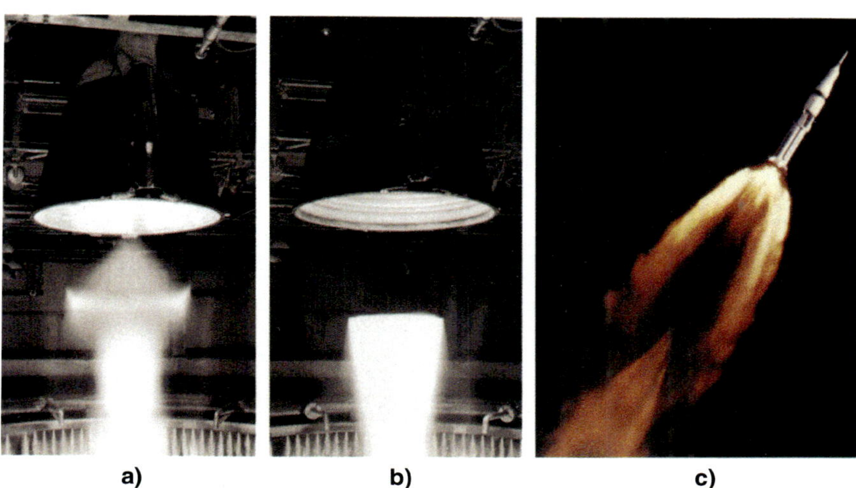

a)   b)   c)

Fig. 2   Rocket nozzle flowfield during off-design operation: a) overexpanded flow with cap-shock pattern Vulcain engine, b) overexpanded flow with Mach disk Vulcain engine, and c) underexpanded flow Saturn-1B, Apollo-7 (photographs Dasa, SNECMA, NASA).

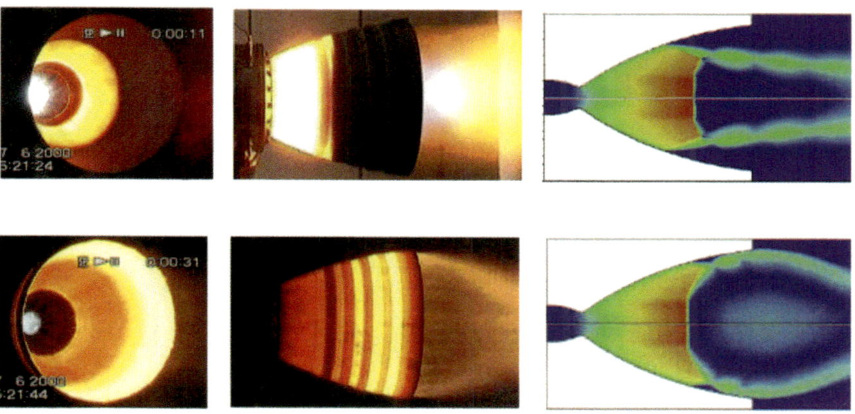

GOX /GH$_2$ thrust chamber operating at 40% power level featuring restricted shock separation

Fig. 6   Hot-firing test with GOX/GH$_2$ thrust chamber designed for 40-kN vacuum thrust level at 40% power level featuring free shock separation (top figures) and restricted shock separation (bottom figures). CFD results for flow visualization (Mach number distribution, from blue to red for increasing Mach number).

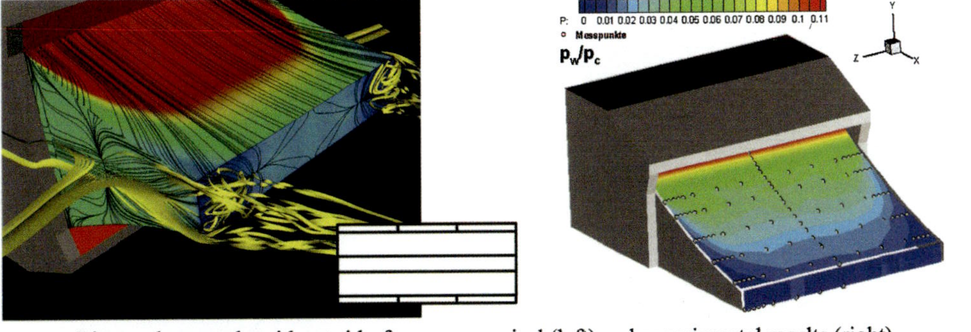

Linear plug nozzle without side-fences, numerical (left) and experimental results (right).

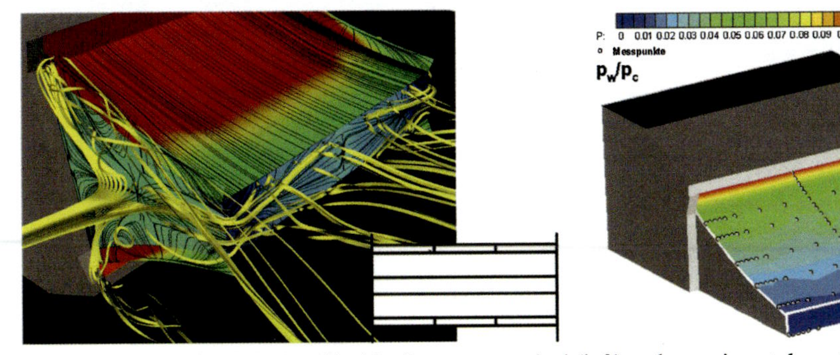

Linear plug nozzle with side-fences, numerical (left) and experimental results (right).

Fig. 19   Flow phenomena for a linear plug nozzle without side fences (top figure) and with side fences (bottom figure): numerical results, Mach number distribution and streamlines (left, from red to blue for increasing Mach number), and experimental results, pressure ratio p_w/p_c distribution (right, from red to blue for increasing pressure ratio). (CFD images courtesy of NASA Marshall Space Flight Center; see also Ref. 56.)

Color reproductions for Chapter 14

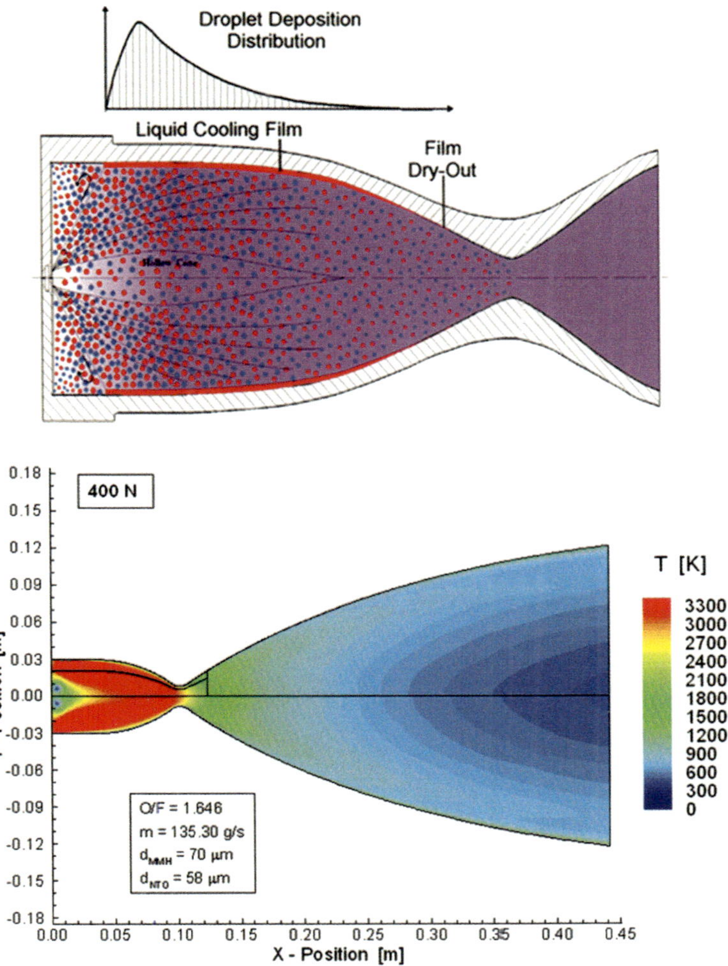

Fig. 3 Sketch of a liquid film cooled 400-N apogee thruster illustrating the droplet flowfield and annular cooling film evolution (top) and temperature contours of a corresponding CFD simulation (bottom).

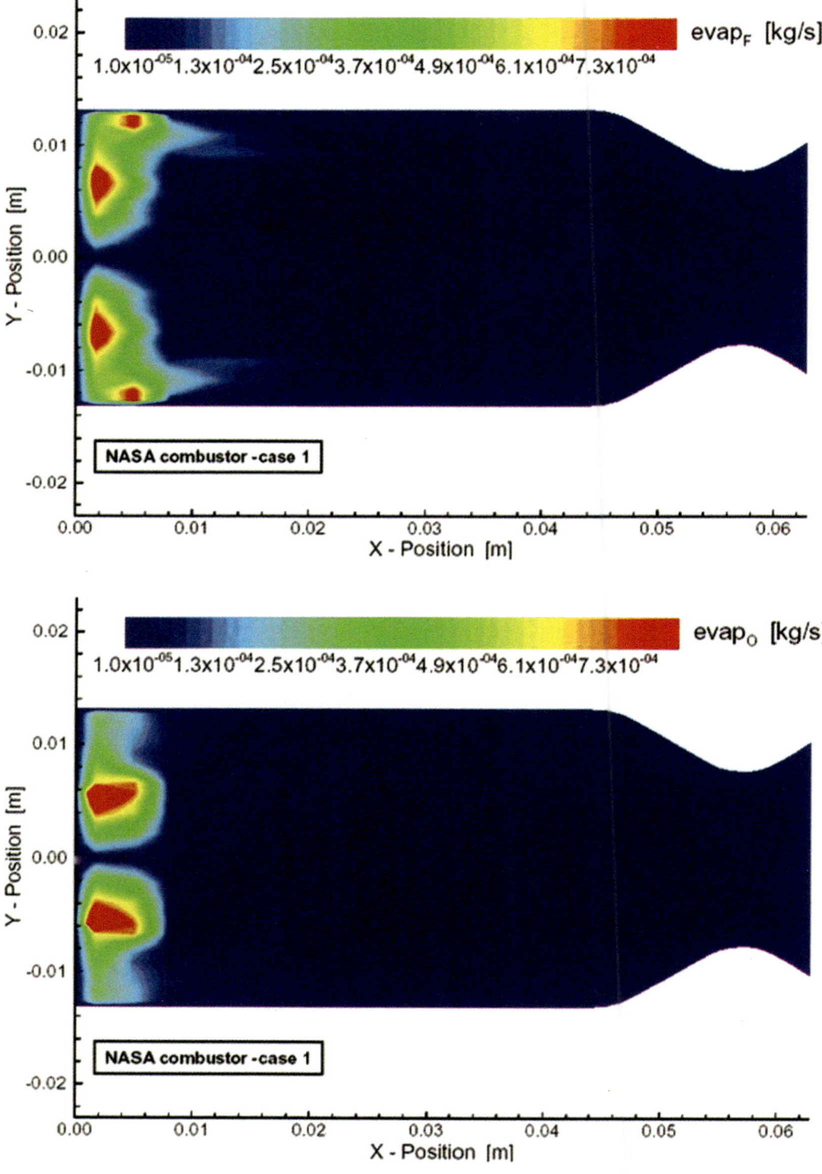

Fig. 4   Calculated hydrazine (top) and NTO (bottom) evaporation rates for combustor test case 1 (Ref. 19).

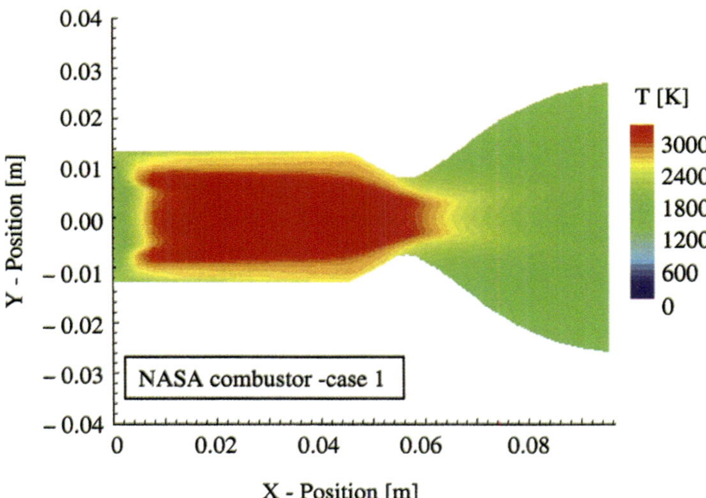

Fig. 5   Calculated gas temperature contours for combustor and nozzle test case 1 (Ref. 19).

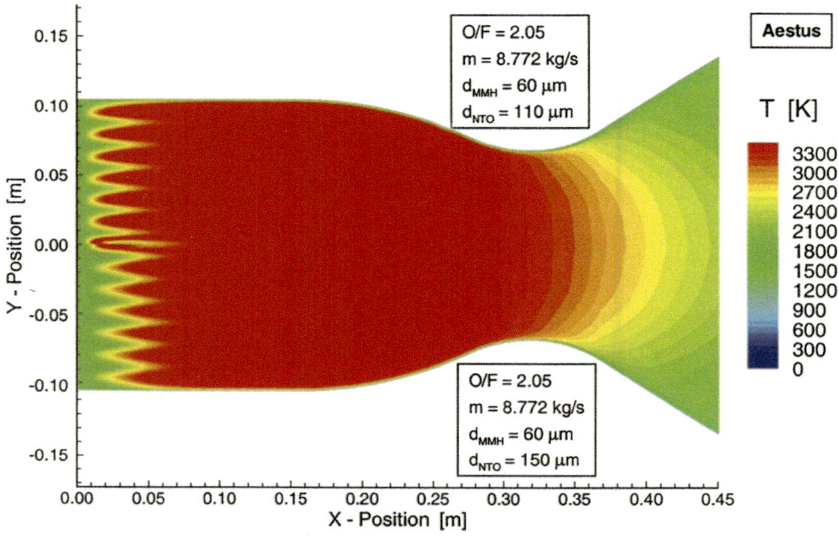

Aestus/OMS Post Test Condition

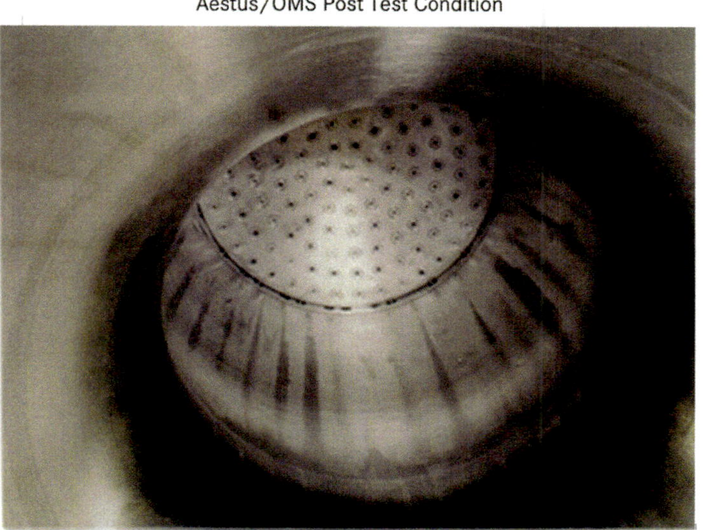

Fig. 8 Comparison of computational temperature stratifications close to the injector plate resulting from two different oxidizer spray patterns: upper half $D_{NTO} = 100\ \mu m$, lower half $D_{NTO} = 150\ \mu m$ (top), and photography taken from tested Aestus hardware illustrating the effect of temperature stratification (bottom).

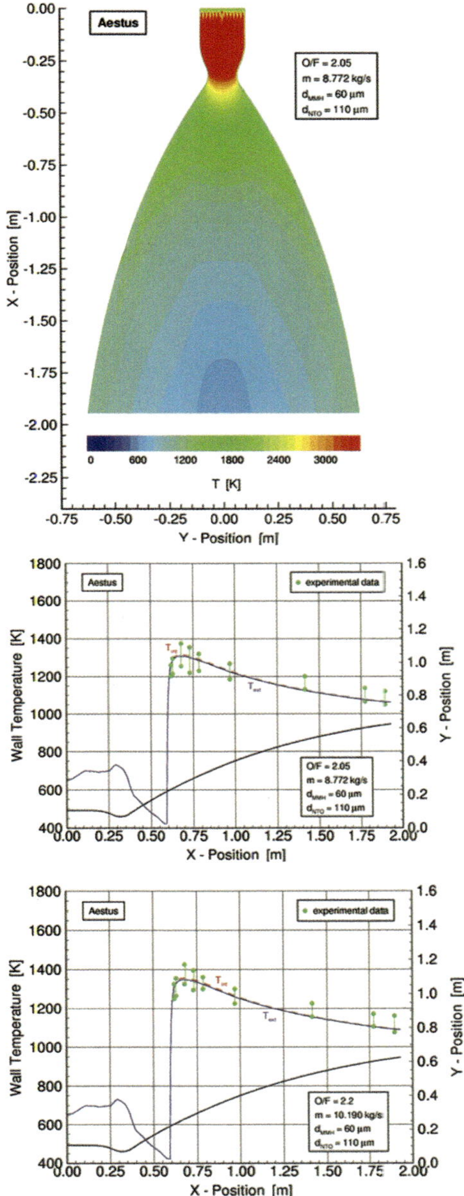

Fig. 10   Temperature contours predicted for the reference load point *R* (top) and external and internal wall temperature distribution for load point *R* (middle) and *Q5′* (bottom).

Color reproductions for Chapter 15

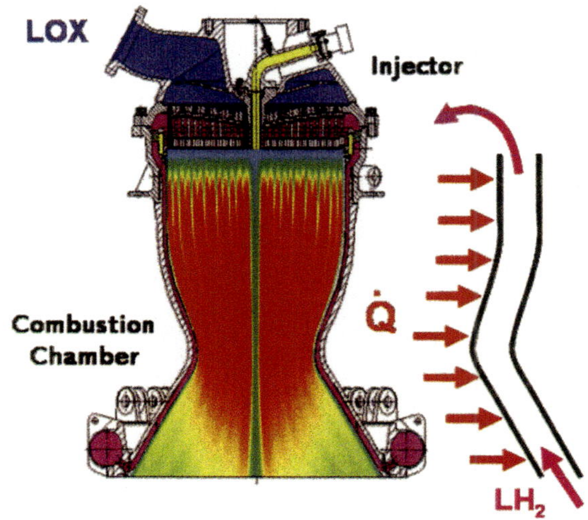

Fig. 2   Flow schematic for regeneratively cooled combustion chamber.

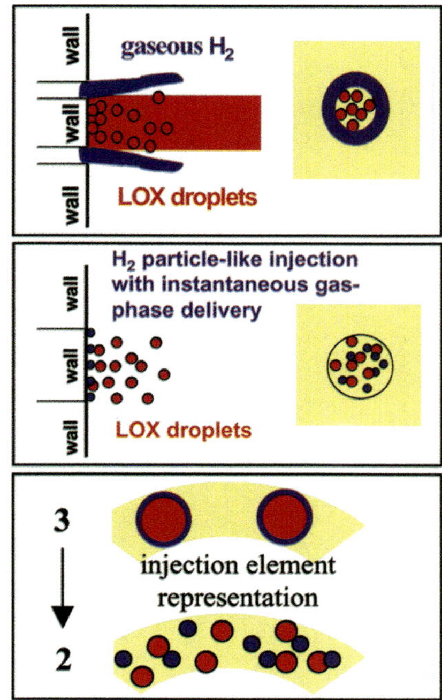

Fig. 4   Schematic of non-premixed, coaxial LOX/H$_2$ injection employing a liquid jet cone (top), well-premixed injection (center), and the principal representation of an injection element row through an axisymmetrical treatment (bottom).

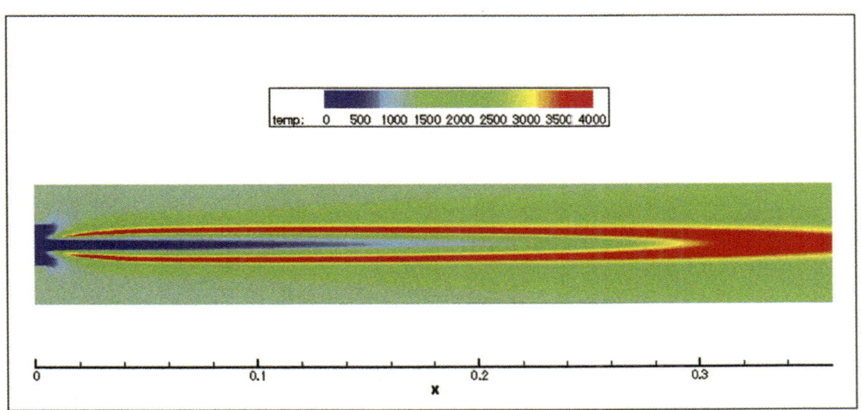
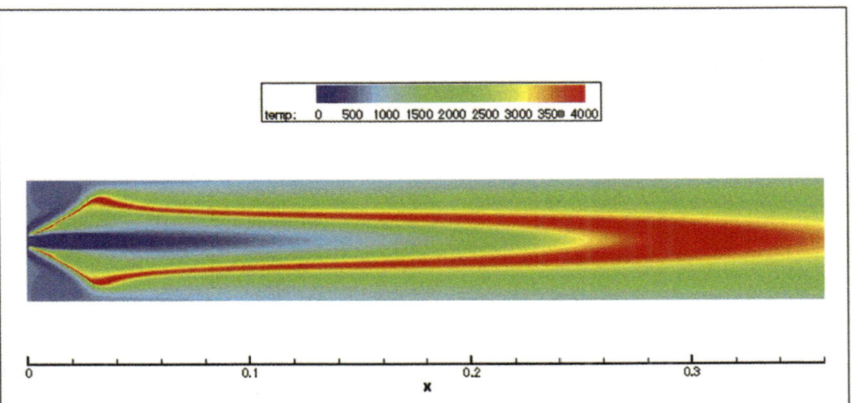

Fig. 5   Temperature distribution for single-element combustor test case ($P_c = 10$ bar) employing a standard grid resolution applicable to a multi-element injection system (top) and a strongly refined grid resolution (bottom).

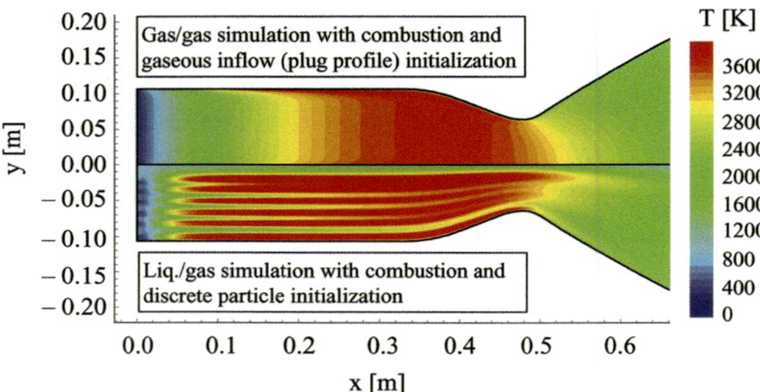

Fig. 9 **Hot gas temperature distribution inside a combustion chamber obtained for a gas/gas simulation with plug-inflow profile (upper half) and liquid/gas discrete particle (lower half) initialization, both with turbulent combustion and prescribed wall temperature boundary conditions (boundary-layer model: logarithmic wall functions).**

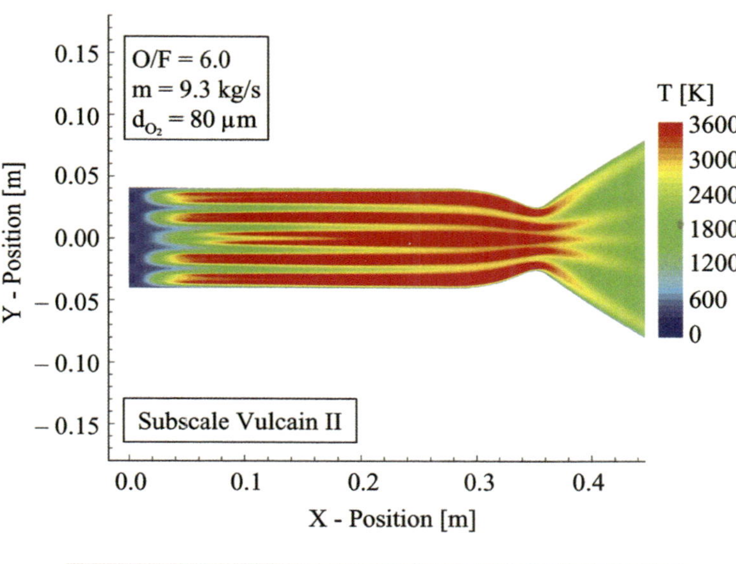

Fig. 11 **LOX/H$_2$ temperature distribution and flowfield stratification inside subscale chamber employing the discrete particle initialization method (top); "footprints" of near-wall injection elements observed on divergent part of nozzle throat wall (bottom) during hot firing; operational load point: P_c = 100 bar; O/F = 6.**

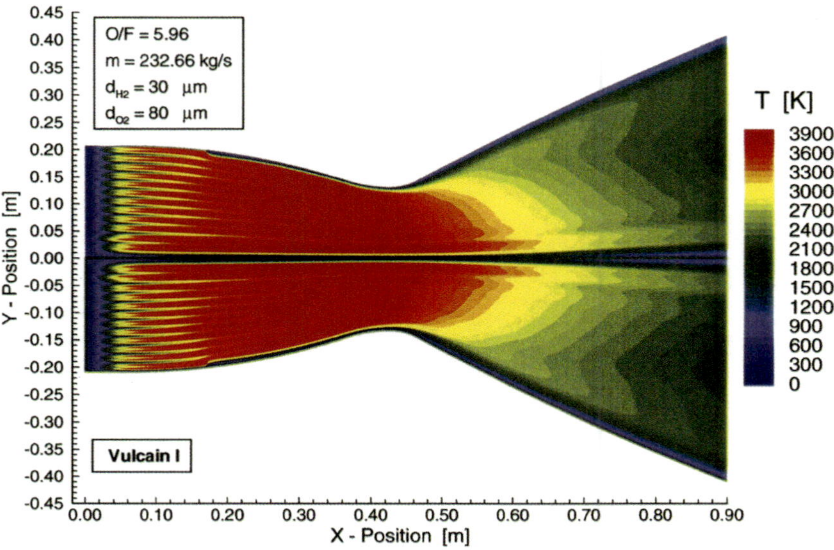

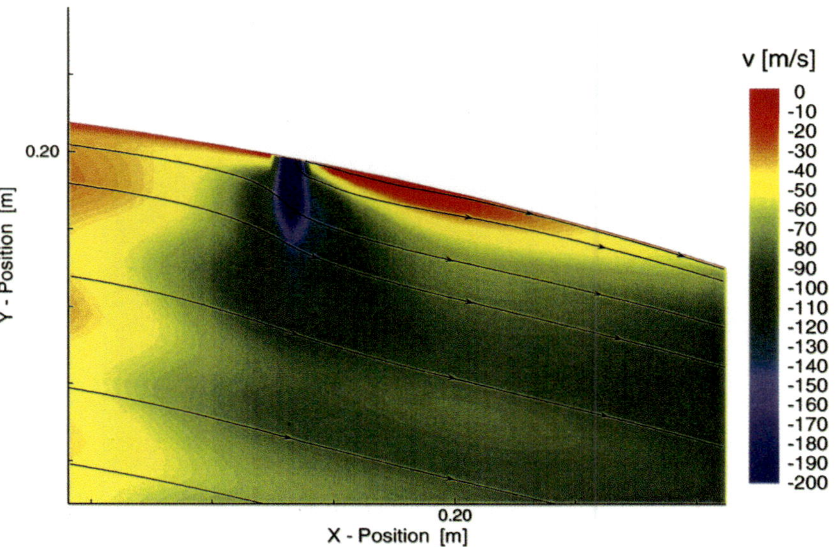

Fig. 14   Temperature distribution inside a rocket combustion chamber with an artificial crack 170 mm downstream of the injector (top), and local velocity profile with streamlines around the crack (bottom).

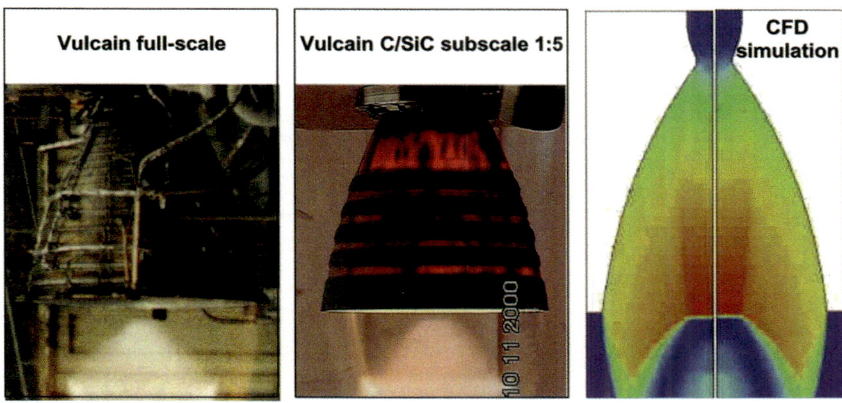

Fig. 18 Comparison of Vulcain-type full-flowing nozzles at nearly identical operational conditions: full-scale (left), long C/SiC subscale (middle), and numerical simulation for full-scale (right)).

Color reproductions for Chapter 20

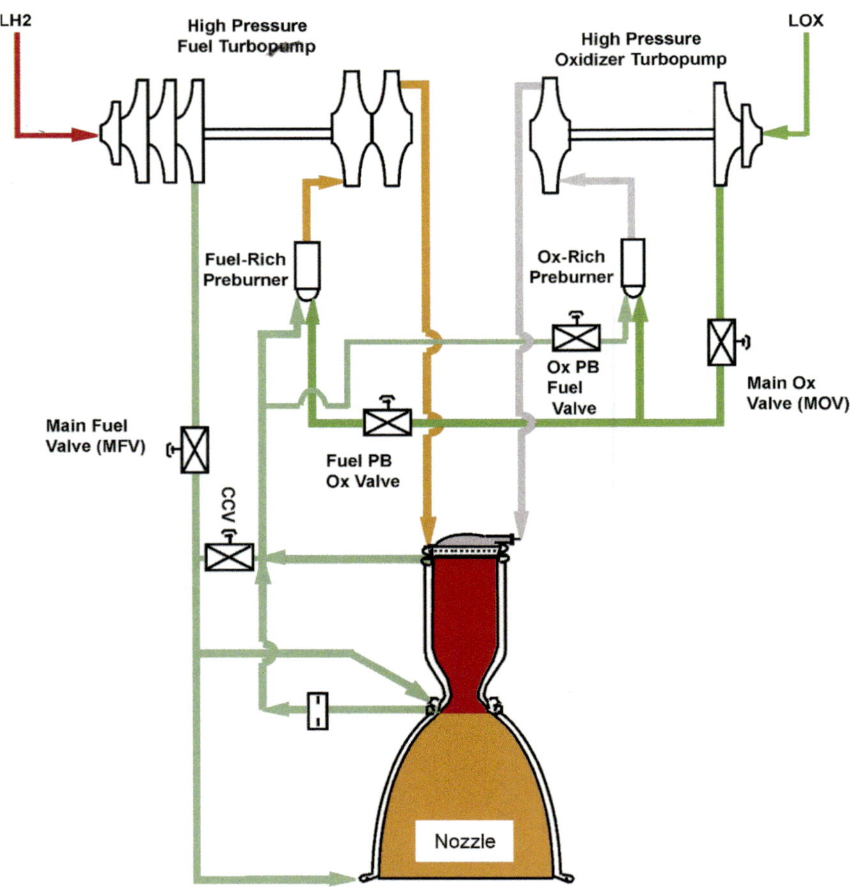

Fig. 1   Full-flow staged combustion booster engine schematic.

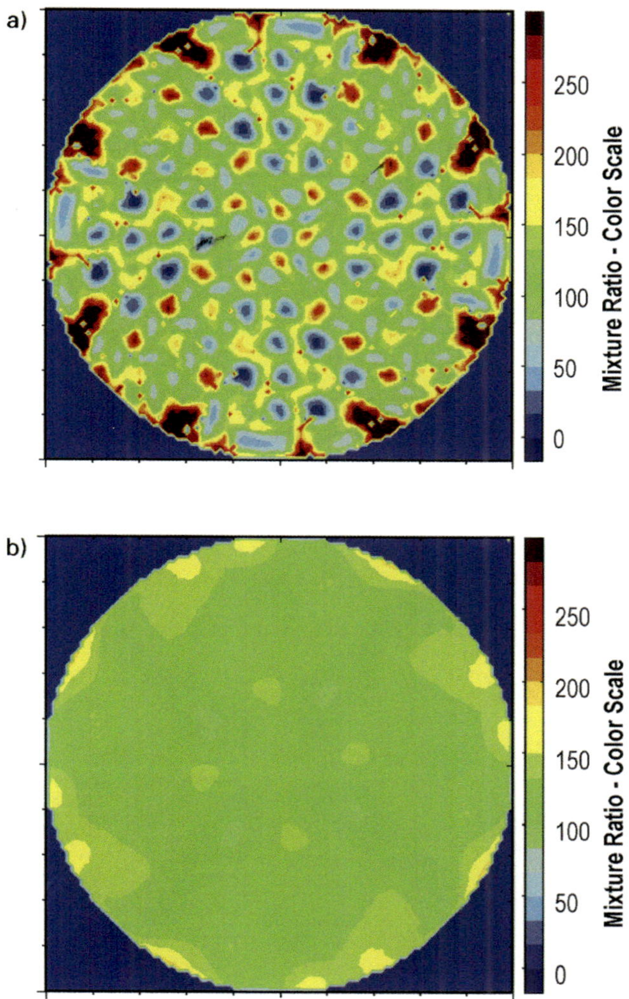

Fig. 7   Predicted mixture ratio distribution at a) 1.27 cm and b) 3.81 cm from the injector for a P_c of 24.1 MPa and MR of 135.

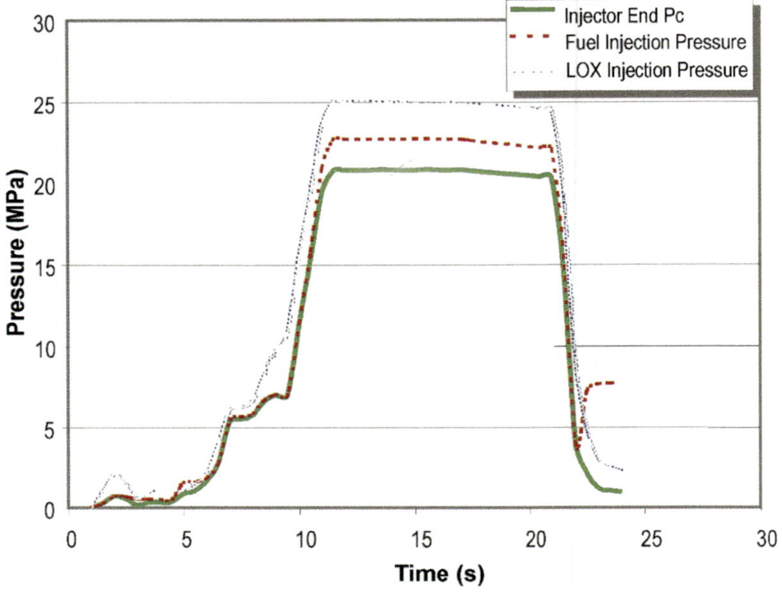

Fig. 11   Test 034 injection and chamber pressures.

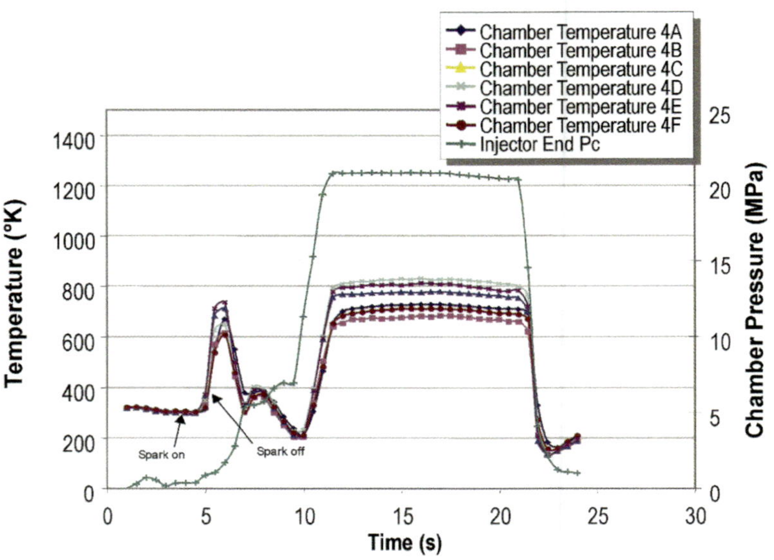

Fig. 12   Test 034 temperature rake data.

Chapter 18

Thermodynamic Power Cycles for Pump-Fed Liquid Rocket Engines

Randy C. Parsley*
Pratt and Whitney, United Technologies Corporation, West Palm Beach, Florida

and

Baojiong Zhang[†]
Shanghai Bureau of Astronautics, Shanghai, People's Republic of China

I. Introduction

THE objective of this chapter is to provide an overview of the thermo-dynamics that influence the configuration selection of pump-fed liquid propellant rocket cycles. The intent is to provide insight into the fundamental differences and inherent advantages of different cycle approaches. To simplify the explanations, first-order calculations are presented with many secondary influences neglected or simplified by assumption. Expander, gas generator, and staged combustion cycles are explained and compared with special emphasis on the thermodynamic implications of including oxidizer-rich combustion for the turbopump drive cycle. Also discussed are the interactions of the cycle ther-modynamics with the engine component stress limitations, thermal limitations, and efficiency trends.

A. Cycle Types and Configurations

Pump-fed liquid rocket cycles are defined by two configuration variables. The first cycle configuration variable is the energy source for the turbine drive. Energy for the turbine can come from an auxiliary combustion device such as

---

*Manager, Propulsion System Analysis and Integration. Associate Fellow AIAA.
[†]Research Professor.

a preburner or a gas generator, or from the main combustion chamber, either directly by the extraction of combusted propellants or indirectly by heat transfer through the chamber walls. The second cycle configuration variable is the turbine discharge location. Historically, there are two options for the turbine discharge flow. If the turbine discharge is to a high-pressure source, specifically the main combustion chamber, the cycle is referred to as a closed cycle. If the turbine discharge flow is to a low-pressure source, generally overboard or into the nozzle skirt, the cycle is referred to as an open cycle. A third turbine discharge source exists but has no development history—turbine discharge flow into an intermediate-pressure combustion device referred to as an afterburner.

Figure 1 summarizes eight possible cycle configuration options based on these two configuration variables and identifies the common names associated with each cycle. Also included are the turbine gas compositional options, propellant limitations, and examples of operational engines associated with each cycle type. All of the eight cycle options, except the afterburning cycles, have been developed into operational engines. This chapter examines advantages, disadvantages, and fundamental performance limitations of each cycle. For simplicity, the supporting engine schematics do not include boost pumps and are examined with separate turbopumps for the fuel and oxidizer. The schematics include the minimum valve complement required for engine startup and control.

In an open cycle, a minimum fraction of the engine propellant is expanded through the turbines to provide power to the pumps. The turbine flow is discharged either overboard or into the divergent section of the primary nozzle, which allows for a turbine pressure ratio of 5 or greater. Because the propellants for an open cycle are pressurized to only slightly above chamber pressure, the pump work is minimized. For a closed cycle, the turbine drive flow is discharged into the main chamber, which is at a relatively high pressure. This generally

		Turbine Energy Source		
		Main Combustor		Auxiliary Combustor
		Chamber Coolant Heat Limited to Hydrogen, Methane, Propane Fuels	Chamber Combustion Gas All Propellants are Compatible	Gas Generator or Preburner All Propellants are Compatible
Turbine Discharge Pressure Sink	High Pressure: Main Combustor	Closed Expander Fuel Cooling Fuel Cooling with Regeneration Fuel & Oxidizer Cooling (Full Flow) Operational Engine Examples: (RL10)	(No Flow Potential) (None Possible)	Staged Combustion Fuel-Rich Oxidizer-Rich Fuel & Oxidizer-Rich (Full Flow) (SSME, LE-7, RD-170, RD-0120, NK-33)
	Intermediate Pressure: A/B Combustor	Afterburning Open Expander Fuel Cooling (None)	Afterburning Tapoff Fuel-Rich (None)	Afterburning Gas Generator Oxidizer-Rich Fuel-Rich (None)
	Low Pressure: Overboard or in Nozzle Skirt	Open Expander Fuel Cooling (LE-5A)	Tapoff Fuel-Rich (J-2S)	Gas Generator Oxidizer-Rich Fuel-Rich Fuel & Oxidizer-Rich (Full Flow) (F-1, J-2, Vulcain, RS-27, LR87-AJ, YF-20)

Fig. 1   Clarification of liquid rocket power cycles by turbopump power options.

limits the turbine pressure ratio to 2 or less to avoid excessive pump discharge pressures.

For either the open or closed cycle approach, energy must be provided to the turbine working fluid prior to its expansion through the turbine. Depending on the option selected to provide this turbine energy, the cycle definition is different. The three common thermodynamic cycles for liquid rocket engines are *expander*, *gas generator*, and *staged combustion*.

The expander cycle is an open or closed cycle in which the energy to drive the turbine comes from the thermal energy absorbed by propellant used to regeneratively cool the thrust chamber and nozzle. In the closed cycle, this turbine flow is discharged into the main chamber. Consequently, all the pump propellants are combusted in the main chamber and expanded through the primary nozzle. For an open cycle the turbine flow is discharged overboard or into the divergent sector of the nozzle. The energy available for the expander cycle is limited by the thrust chamber and nozzle heat transfer, which limits potential chamber pressure to ~ 10 MPa.

The gas generator cycle is an open cycle in which the energy to drive the turbine is supplied by combustion of a minimal fraction of propellants in a gas generator combustion device. Because of the high expansion ratio, the pressure of the turbine discharge flow is below the main combustion chamber pressure, and therefore the discharge must bypass main chamber combustion. The chemical energy released during combustion in the gas generator is influenced by the temperature limit of the turbine. Chamber pressure for a gas generator cycle is selected to optimize total engine performance, which includes both the higher performance main engine flow and the lower performance turbine discharge flow. This performance optimum generally occurs at $10-15$ MPa of chamber pressure, depending on propellant selection, with an overboard flow generally less than 4% of the total engine flow. For simplification, the turbine exhaust gases are usually expanded through a separate nozzle to provide some thrust. A more complex but higher performing option is to dump this gas into the divergent section of the primary nozzle. A third option would be an intermediate pressure afterburner downstream of the gas generator and turbine discharge. The afterburning option has been investigated because of the potential to offset the performance loss due to the main chamber mixture ratio shift that occurs for open cycle configurations. Although this option increases performance significantly relative to the conventional gas generator, the weight and complexity of the afterburner must be considered and optimized for the intended application.

The staged combustion cycle is a closed cycle in which major portions of the propellants are burned, in preburner combustion devices, to provide energy to drive the turbines. The energy released in the preburners minimizes turbine pressure ratio, allowing chamber pressure to be maximized. The energy released in the preburners is influenced by the turbine temperature limit and the percentage of the propellants included in the combustion process. The performance of a staged combustion cycle generally begins to become hardware limited between 20 and 25 MPa chamber pressure.

Once the propellants and engine mixture ratio have been selected, the performance of the various cycles is influenced by only a few parameters. For the closed cycles (expander and staged combustion), these parameters are engine impulse efficiency, engine chamber pressure, and engine nozzle exit pressure. Engine impulse efficiency is the product of the main chamber combustion energy release efficiency and the primary nozzle expansion efficiency. For oxygen and hydrogen propellants, the main chamber combustion energy release efficiency for a well-designed system is generally 98–99%, whereas for oxygen and kerosene propellants the practical limit has been 95–96%. Typical nozzle expansion efficiencies are generally from 98 to 99%, depending on several factors, including operating nozzle pressure ratio and the design area ratio.

First-order performance trends, as functions of chamber pressure and nozzle exit pressure, are shown in Fig. 2a, for hydrogen and Fig. 2b for kerosene fuels. The theoretical specific impulse was generated using NASA's chemical equilibrium program for rocket performance.[1] This theoretical performance was adjusted according to the indicated constant values of combustion energy release efficiency and nozzle expansion efficiency. The performance is shown for three nozzle discharge pressures p_{exit}. The $p_{\text{exit}} = 1.0$ bar line represents the nozzle expansion ratios generally required to maximize sea-level thrust, which is desirable and typical for booster applications. The $p_{\text{exit}} = 0.3$ bar line represents the maximum sea-level expansion ratio that can typically be sustained without nozzle separation, used for engines that must start at sea level but also operate at high altitudes. The $p_{\text{exit}} = 0.1$ bar line is representative of upper-stage expansion ratios that balance performance, weight, and engine geometric size. These performance estimates are presented for initial screening activities only. The secondary effects of changes in combustion energy release efficiency and nozzle efficiency depending on each individual design are important and should be investigated and optimized for each individual application.

B. Pump-Fed Powerhead Power Balance

In a powerhead device, propellants are increased in pressure using single- or multiple-stage pumps. Power to drive the pumps is supplied by single- or multiple-stage turbines. The work relationship for pumping propellants is given by Eq. (1). The required work is related to flowrate $\dot{m}_p$, liquid propellant density ρ, pressure rise Δp, and pump efficiency η_p:

$$W_p = \frac{\dot{m}_p \Delta p}{\rho \eta_p} \tag{1}$$

The delivered turbine work is given by Eq. (2). The delivered power is related to flowrate $\dot{m}_t$, turbine pressure ratio TPR, turbine inlet temperature T, gas constant R, gas specific heat ratio γ, and turbine efficiency η_t:

$$W_t = \dot{m}_t \left(\frac{\gamma}{\gamma - 1} \right) RT \left[1 - \left(\frac{1}{TPR} \right)^{\frac{\gamma-1}{\gamma}} \right] \eta_t \tag{2}$$

Rocket cycle performance is generally improved by increasing the cycle operating pressure (i.e., pump discharge pressure). This trend continues until component thermal or structural limits are reached or until component efficiencies

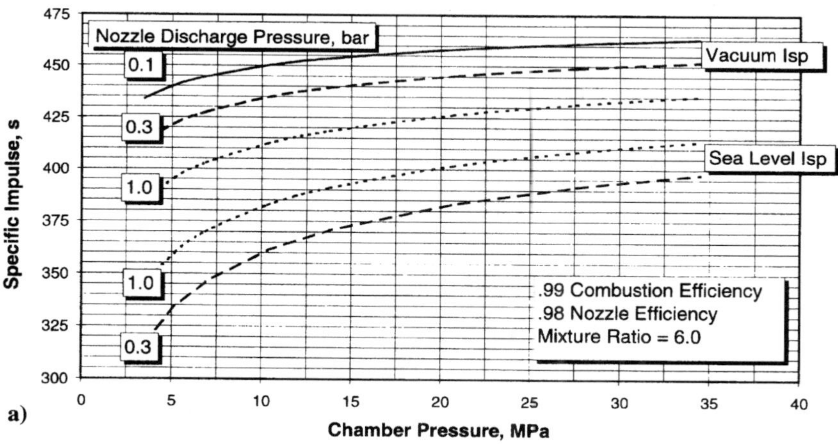

a)

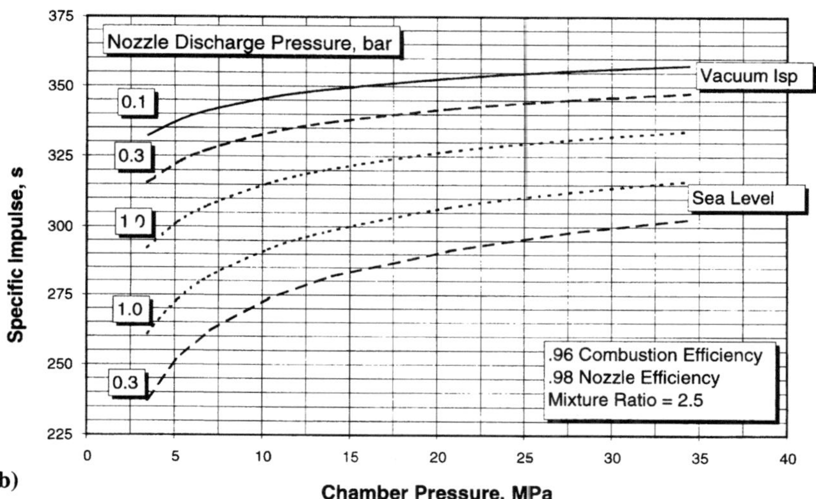

b)

Fig. 2 **Performance trends for a) hydrogen/oxygen and b) kerosene/oxygen.**

begin to drop. Pump and turbine efficiencies generally decrease as cycle pressure is increased, because the components are forced to operate at increased work levels. This component efficiency loss can be mitigated by increasing the number of pump and turbine stages, but only at the expense of engine hardware complexity, engine cost, and engine weight. Although neglected in the simplified discussion in this chapter, turbine power must be greater than the pumping power requirements by the amount of the connecting shaft friction losses (windage) and the turbine power control margin needed to achieve practical steady-state operation. The sum of these real-world effects can be as much as 10% of the total turbine power.

C.   Thermodynamic and Hardware Interactions

Optimization of liquid rocket engines requires closure between the cycle thermodynamics and the component hardware configuration. The engine hardware must be startable and controllable, and weight must be minimized. The thermodynamic cycle and configuration must be compatible with a development program that considers cost, technical risk, and schedule. For expendable applications the acquisition cost of the hardware is important, whereas for reusable applications the hardware durability is the primary consideration. Four of the most common hardware limitations that influence the overall cycle optimization are turbine temperature limits, thrust chamber cooling requirements, turbopump rotational speed limits, and pump discharge pressure limits.

1.   Turbine Temperature Limits

For most rocket engine cycles, it is desirable to maximize turbine temperature within practical limits. For the gas generator cycle, this minimizes the overboard flow and maximizes performance. For staged combustion cycles, increasing temperature allows a reduced turbine pressure ratio at a given pump discharge pressure, enabling an increase in chamber pressure, and therefore, as shown in Fig. 2, an increase in performance.

However, increased turbine temperatures can significantly increase the cost of the hardware and can have an adverse effect on hardware durability. For most applications it is desirable to maintain turbine temperatures at or below 900 K. However, with an increased degree of difficulty and cost, turbine temperatures up to 1200 K can be achieved, using state-of-the-art turbine materials, while maintaining durability and reliability margin.[2]

2.   Thrust Chamber Cooling

For most propellant combinations, the fuel has superior heat-transfer characteristics relative to the oxidizer; therefore, a portion of the fuel is used to cool the main combustion chamber. For fuel-rich closed staged combustion cycles, the chamber coolant flow rate should be minimized, which allows more fuel to be available for preburner combustion to drive the turbopumps. The percentage of the fuel required for most cycles is generally less than 20%. For oxidizer-rich staged combustion gas generator, and closed expander cycles, most or all of the fuel is used in the chamber cooling process. Because it is desirable to maximize the chamber heat transfer, yet minimize the chamber coolant pressure drop, the optimum trade will depend on the chosen cycle details and on the size of the chamber and the engine. For the gas generator cycle and open expander cycle, minimum pressure drop will minimize pump discharge pressure, improving the cost and weight characteristics of the engine.

For high-pressure kerosene cycles, significantly more than 20% cooling flow may be required. This is a contributing factor to the use of oxidizer-rich staged combustion cycles for this propellant. Because only a small amount of kerosene is needed for the preburner supply, over 90% of the kerosene is available for chamber cooling.

3. Turbopump Rotational Speed

Turbopump rotational speed must be optimized to minimize the pump diameter and improve the pump and turbine efficiencies. In many cases, especially for oxygen and hydrogen propellants, the rotational speed should be maximized. Turbopump rotational speed is generally limited by one of four parameters: pump inlet cavitation, bearing rotational speeds, rotordynamics, or turbine blade root-stress.

Turbopump cavitation occurs when the local static pressure of the propellant, as it accelerates around the pump impeller inlet blading, drops below the vaporization pressure. The result is the formation of small pockets of vapor in the low-pressure regions. For a constant inlet flow rate, increased pump speed will increase these local accelerations due to increasingly severe inlet incident angles between the propellant and the blading. Cavitation causes loss of pumping capability and can result in physical damage to the pump hardware. Low-speed boost pumps are often used to increase the main pump inlet pressures to avoid cavitation and allow main pump rotational speed to be increased.

The bearing rotational speed for conventional rotor support systems becomes limited as internal stress in the bearing increases. The bearing stress level is proportional to the product of the inner diameter of the bearing D in mm, and the bearing rotational speed N in revolutions per min. The practical DN limit for state-of-the-art rolling element bearings is $DN \sim 2.0 \times 10^6$. Attempts to reduce the DN by decreasing the diameter will result in increased flexibility of the shaft, which can lead to rotodynamic vibrations. Fluid film bearings are being developed to eliminate this constraint.[3,4] Fluid film bearings support the rotor on high-pressure propellant, eliminating internal surface-to-surface contact during high-speed operation.

Rotordynamic vibrations can occur when the natural bending or natural vibration frequency of the pump rotor falls within the operating speed range. If these natural vibration frequencies "couple" with the rotational energy of the rotor, the pump can fail mechanically. To avoid these vibrations, the pump must operate at a rotational speed below the natural vibration frequency or must transition quickly through the natural frequency modes at a very low speed, where the rotational energy is low. High-speed rotor damping, developed in Russia over the last two decades, has been shown to preclude many of the historical problems with rotordynamic vibration.

The turbine blade root-stress is proportional to the product of the turbine rotor exit annulus area A in cm^2, and the turbine rotational speed squared N^2, in revolutions per min. The practical limit for this parameter is influenced by the operating temperature because the turbine blade material strength drops as temperature increases. For temperatures of 1000 K the practical limit is $AN^2 \sim 3.0 \times 10^{11}$ mm^{2}rpm^2.

4. *Pump Discharge Pressure Limitations*

Staged combustion cycle performance is generally limited by pump discharge pressure. The fuel is generally the limiting propellant because of the lower density of fuel, relative to oxidizer. Lower density results in increased power requirements. Current state-of-the-art limits for pumping hydrogen is $\sim$50 MPa, and for kerosene $\sim$100 MPa. These limits result from the combination of allowable impeller tip speed and allowable number of pump stages. The allowable tip speed of $\sim$700 m/s allows acceptable stresses to be maintained for the impeller. The maximum number of stages is generally limited to three to avoid pump integration concerns such as rotor vibrations, rotor thrust balance, and total power transmitted between the turbine and impellers.

D. Fuel-Rich vs Oxidizer-Rich Combustion for Turbine Drive

Material temperatures are the limiting features in turbines when two propellants are burned to release chemical energy for powerhead work requirements. These temperature limits can be achieved by selecting propellant mixture ratios that are either fuel rich or oxidizer rich. Figure 3 illustrates the relationship of combustion temperature to mixture ratio for oxygen with either hydrogen or kerosene fuel.

Figure 3 shows that 900 K can be achieved for kerosene and hydrogen at fuel-rich mixture ratios of 0.055 and 0.775, respectively, but also at oxidizer-rich mixture ratios of 41.5 and 115.0, respectively. Selection of fuel-rich or oxidizer-rich preburner combustion for a particular application depends on the cycle configuration and propellant selections. There are fundamental thermodynamic differences in energy release potential and net powerhead work potential

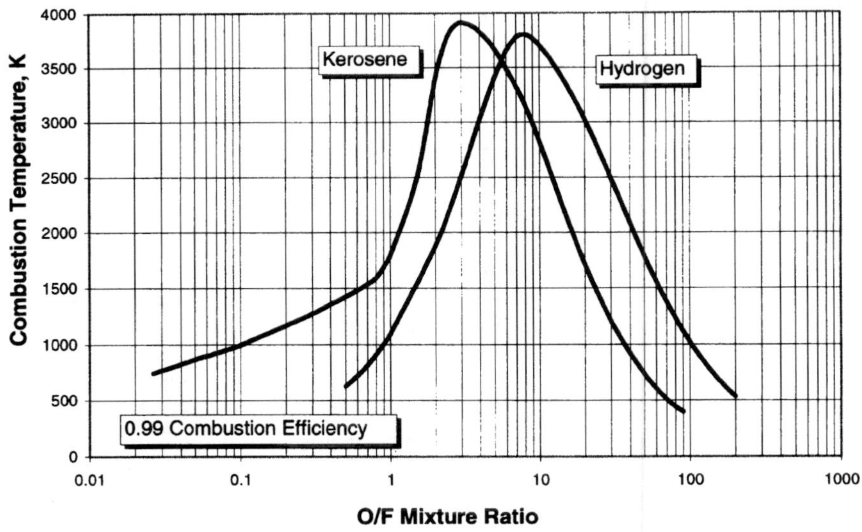

Fig. 3 Combustion temperature vs mixture ratio.

between the various fuel-rich and oxidizer-rich approaches, as discussed in the following sections.

1. Energy Release Potential

The propellant combustion process is a complicated series of mostly exothermic chemical reactions. When fuel and oxidizer are burned, there is a net positive enthalpy release, the magnitude of which is represented by the lower heating value (LHV) of the fuel. More specifically, LHV is the net positive energy, per unit mass of fuel, that is released at constant temperature when the fuel is completely combined with oxygen. This net positive enthalpy release is then absorbed by the combustion products, which increases the resulting energy state of the combustion mixture.

Depending on the selected propellants and mixture ratios, the energy will be absorbed in different ways. For a very oxidizer-rich combustion process, the combustion products are nearly 90% oxygen. In this case the majority of the net positive enthalpy will be used to vaporize and heat the excess oxygen that surrounds the local combustion processes in particular abundance. For a fuel-rich hydrocarbon mixture, the net positive enthalpy will be mostly absorbed by the "cracking" of the excess unburned fuel molecules and then heating of the resulting constituents. For a fuel-rich hydrogen mixture, the positive enthalpy will vaporize and heat the excess flow. In either case the absorption of the liberated LHV increases the net energy state of the combustion products, making them more useful to provide turbine work.

For closed cycle engines with specified inlet flow rates of fuel and oxidizer, the total absolute level of energy that is released and absorbed by the combustion products should be maximized while an acceptable turbine inlet temperature is maintained. This creates the maximum energy state for the turbine drive fluid. In some cases, as for the oxidizer-rich cycles, extra flow must be pressurized and delivered to the preburner to achieve the maximum level of energy release. The penalty of the extra horsepower required to accomplish this increased preburner flow rate is addressed in Section I.D.2.

Table 1 Hypothetical oxygen/kerosene staged combustion engine parameters

Engine parameters	Valve
Engine inlet kerosene flow, kg/s	100
Engine inlet oxygen flow, kg/s (O/F mixture ratio = 2.5)	250
Powerhead turbine temperature limit, K (Either fuel-rich or oxidizer-rich)	900
Kerosene LHV, MJ/kg	43.15
Engine main chamber pressure, MPa	25
Preburner pressure, MPa (Simplified turbine pressure ratio = 2.0)	50

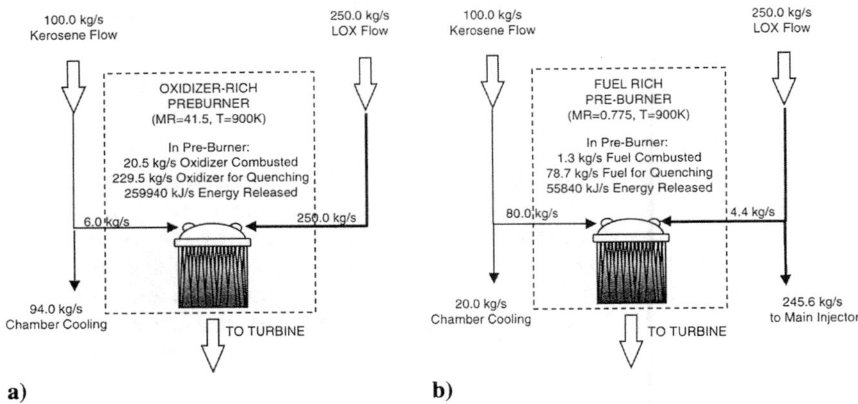

Fig. 4   Comparison of a) oxidizer-rich and b) fuel-rich powerhead propellant distribution.

The following simplified example compares the energy release potential for hypothetical fuel-rich and oxidizer-rich oxygen/kerosene staged combustion engines with the characteristics shown in Table 1. The simplified propellant distribution schematics for both powerhead options are shown in Fig. 4.

Assume that the engine is configured with an oxygen-rich powerhead (Fig. 4a) and that all of the oxygen (250 kg/s) is available for combustion in the preburners. To maintain 900 K temperature, the preburner kerosene flow is limited to 6 kg/s because of the mixture ratio limit of 41.5 (from Fig. 3). Because the stoichiometric mixture ratio for kerosene and oxygen is approximately 3.4, the 6 kg/s of preburner kerosene flow can combine chemically with only 20.4 kg/s of the preburner oxygen flow to release its stored chemical energy. The remaining 229.6 kg/s of preburner oxygen flow effectively serves only to quench the entire combustion mixture down to the 900 K temperature limit. The total energy release is, therefore, the 6 kg/s of kerosene multiplied by 43.15 MJ/kg LHV, equaling 259 MJ/s. The remaining 94 kg/s of kerosene flow, not used in the preburner, is available for chamber cooling.

For the fuel-rich powerhead option (Fig. 4b), a very aggressive level of 20 kg/s of fuel is assumed to be required for chamber cooling. This leaves 80 kg/s of kerosene available for combustion in the preburners. To maintain 900 K temperature, the preburner oxygen flow is limited to 4.4 kg/s because of the mixture ratio limit of 0.055 (from Fig. 3). Because the stoichiometric mixture ratio for kerosene and oxygen is 3.4, only 1.3 kg/s of the preburner kerosene flow can chemically combine with the oxygen to release its stored chemical energy. The remaining kerosene does not release chemical energy but effectively serves only to quench the entire preburner combustion mixture down to the 900 K temperature limit. Therefore, the total energy release is 1.3 kg/s of kerosene multiplied by 43.15 MJ/kg LHV, which equals 56 MJ/s. The remaining 245.6 kg/s of the oxygen, not used in the preburners, is delivered to the main injector.

For oxygen and kerosene propellants, the energy release of the oxidizer-rich approach is over 400% greater than for the fuel-rich approach, within the same turbine temperature limits. The important factor, however, is the potential for this energy release to be converted into useful turbine work, which is discussed next.

2. Net Powerhead Work Potential

To estimate the work potential previously discussed, the thermodynamic characteristics of the propellants and the resulting products of combustion must be understood. These characteristics, which are consistent with the previous example, are given in Table 2. The component efficiencies are assumed to be conservatively low, because this simplified example ignores all of the system pressure losses, including the chamber coolant pressure loss.

The work required by the pumps is different for the oxidizer-rich and the fuel-rich approaches. The oxidizer-rich approach requires that 250 kg/s of oxygen and 6.0 kg/s of kerosene be pumped to preburner pressures of 50 MPa. The remaining 94 kg/s of kerosene needs to be pumped to only chamber pressure, which is 25 MPa. By using Eq. (1) for the individual propellant discharge pressures and flow rates, the estimated total work required for the oxygen-rich pumping combination is $\sim$28.2 MW. The fuel-rich powerhead approach requires that 80 kg/s of kerosene and 4.4 kg/s of oxygen be pumped to preburner pressure of 50 MPa. The remaining 245.6 kg/s of oxygen flow and 20 kg/s of kerosene cooling flow needs to be pumped only to chamber pressure, which is 25 MPa. Again, by using Eq. (1), the fuel-rich pumping combination will require $\sim$23.3 MW. This is less than the oxidizer-rich approach, but the available turbine work must also be compared.

The work delivered by the turbines for each case can be calculated using the information in Tables 1 and 2, and substituting into Eq. (2). Most of the parameters used in Eq. (2) are similar except for the turbine flow rate, which is three times larger for the oxygen-rich approach. Equation (2) indicates a delivered turbine work for the oxygen-rich approach of $\sim$25 MW and for the fuel-rich approach of $\sim$9 MW. For the oxidizer-rich cycle, the turbine work available

Table 2 **Powerhead parameters for hypothetical oxygen/ kerosene staged combustion engine**

Powerhead parameter	Oxygen-rich	Fuel-rich
Powerhead mixture ratio	41.5	0.055
Powerhead temperature, K[a]	900	900
Powerhead total flow rate, kg/s	256	84.4
Gas constant, R, kJ/kg K[a]	0.261	0.275
Ratio of specific heats[a]	1.304	1.107
Assumed pump efficiency, %	50	50
Assumed turbine efficiency, %	65	65
Oxygen density, kg/m^3	1140	1140
Kerosene density, kg/m^3	810	810

[a]As given per Ref. 1 for respective mixture ratios.

is only slightly less than the required pump work, indicating that chamber pressure will drop slightly below the 25 MPa level. For the fuel-rich cycle, however, the turbine work available falls far short of that required by the pumps; therefore, chamber pressure must be reduced significantly.

Iterating these calculations for each approach shows that the chamber pressure for the oxidizer-rich cycle balances at ~23 MPa, whereas the fuel-rich cycle balances at ~12.3 MPa. Therefore, considering equal system temperatures and peak system pressures, the oxygen-rich cycle provides a chamber pressure that is 87% higher. In practice, this ratio could be even higher because the selection of 20% of the kerosene to cool the fuel-rich cycle main chamber was very aggressive.

3.   Propellant Selection Influences

The previous example illustrates the fundamental difference of oxidizer-rich and fuel-rich thermodynamics for oxygen and kerosene propellants. The 87% increase in chamber pressure, assuming equal turbine temperatures and pump discharge pressures, provides the rationale for the selection of the oxidizer-rich cycle as the preferred approach for oxygen and kerosene staged combustion engines. However, for oxygen and hydrogen propellants, a similar iterative investigation would show that a hydrogen-rich cycle is superior to the oxygen-rich cycle approach, largely because of the differences in thermodynamic parameters that strongly influence the pump and turbine work calculations.

For a hydrogen cycle, the flow rate of hydrogen is a smaller percentage of the total flow compared with that in kerosene engines. Although the fuel percentage is lower (~15% for hydrogen vs ~30% for kerosene), the reduced density of hydrogen (71 kg/m^3 for hydrogen vs 810 kg/m^3 for kerosene) results in a substantial increase in the required pump work. However, this increased pump work is more than offset by the profound increase in available turbine work, because of the thermodynamic properties of hydrogen fuel.

The energy release potential of hydrogen, represented by the LHV of hydrogen at 117.7 MJ/kg, is more than 2.5 times larger than that of kerosene, at 43.15 MJ/kg. The turbine work potential is directly influenced by the gas constant of the combustion products, which for oxidizer-rich combustion using hydrogen fuel is 0.275 kJ/kg-K. For hydrogen-rich combustion the gas constant of the combustion products is nearly an order of magnitude higher at 2.32 kJ/kg-K.

The combined effect of improved energy release and improved turbine work potential more than offsets a moderate increase in required pump work, and results in a fuel-rich preference for staged combustion cycles if hydrogen is the fuel. An alternative hydrogen-based staged combustion cycle that operates with two preburners (one fuel-rich and one oxygen-rich) is addressed in Section VIII.B.

II. Expander Cycles

As described in Fig. 1, expander engines can use open or closed cycles. All of the energy to support the cycle is supplied by heat transfer to the propellant, generally the fuel, accomplished by using the propellant to regeneratively cool the thrust chamber and nozzle. The heat added to the propellant is used as work potential to power the turbines. An operative example of a closed expander cycle is the RL10.[5]

A. General Cycle Discussion

A schematic of the basic closed expander cycle engine is shown in Fig. 5. The fuel enters a single- or multiple-stage pump and is increased in pressure before cooling the thrust chamber and nozzle. The separate chamber and nozzle flows are vaporized by the energy that is added during this cooling function, and then mixed before being directed to the turbines. For a hydrogen expander engine, the mixed coolant flow temperature is typically between 200 and 300 K. The gaseous fuel flow is expanded through the turbines before being directed to the main injector and chamber for combustion and expansion in the primary nozzle.

The selection of fuels that are compatible with the expander cycle is limited. The fuel must have a high heat capacity and adequate heat-transfer properties, and it must vaporize easily. Generally, fuels are limited to hydrogen, methane, or propane. The selection of oxidizers for an expander engine is not limited, but it must be compatible with the selected fuel.

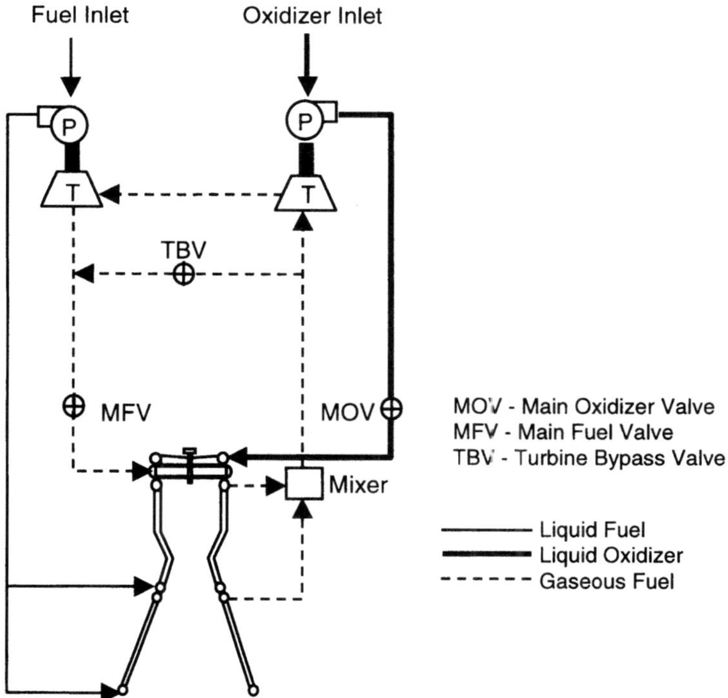

Fig. 5   Basic closed expander cycle.

The fuel side flow rate is controlled by the main fuel valve (MFV) located just upstream of the injector. Configuring the cycle with the MFV at this location has several advantages. The first advantage is that prior to start, with the MFV closed, the coolant passages and turbopump turbine housings will fill with hydrogen. This trapped hydrogen is no longer cryogenic as it will have been vaporized by the latent heat of the metal and the ambient surroundings of the metal. The gaseous fuel will provide an initial acceleration to the turbopumps, prior to main chamber ignition, that allows for a fast, smooth start or restart to begin immediately upon opening of the MFV. The dynamics and stability concerns of powerhead combustion are minimized. A second advantage is that upon engine shutdown, closing the MFV results in immediate main chamber fuel interruption while maintaining trapped propellant in the coolant passages for additional heat capacity to accommodate shutdown thermal transients. Finally, a third advantage is that prior to launch and during launch, the purges are minimized because the MFV isolates the fuel side of the powerhead from ambient conditions.

The power level of the expander cycle is controlled by the turbine bypass valve (TBV). To decrease engine power, the TBV is opened to allow flow to bypass both turbines. This reduces pump speed and flow, and therefore reduces chamber pressure. The oxidizer flow rate is controlled by the main oxidizer valve (MOV), located just upstream of the main injector. This valve is used to control engine mixture ratio and provide shutdown of the oxidizer at engine cutoff. Loading the MOV just upstream of the injector isolates the oxidizer powerhead from ambient contamination prior to engine start or restart, in a manner similar to the MFV on the fuel side.

The influence of cycle operating pressure (i.e., fuel pump discharge pressure) and component efficiencies is important for the expander cycle. Figure 6 shows an example of predicted chamber pressure for an expander cycle configured with different numbers of fuel pump stages, from a detailed oxygen/hydrogen expander engine design study at 1100 kN thrust level.[6] This study also included the influence of a copper chamber liner, which is significantly better than a steel chamber because of the higher thermal conductivity of copper. For the single-stage configuration, the chamber pressure increases as pump discharge pressure increases before reaching a peak. This peak results from the interaction between decreasing main fuel pump efficiency and cycle efficiency improvements with increased cycle pressures. By increasing the number of fuel stages to two, the pump efficiency is improved because of the reduced work per stage, which allows chamber pressure to increase. With a three-stage configuration, there is a small but diminishing gain.

Although not used in this manner historically, expander cycle engines can be considered for booster applications. The reduced cost and weight of the powerhead can offset the reduced performance from the moderately low chamber pressure. Because there are no preburners and the turbine temperatures are low, the engine cost can be affordable for expendable applications. Expander cycle engines are currently used for upper stages and space transfer applications, in which ease of starting and multiple restarts are required and the engine can restart on space-ambient heat capacity. Also, upon restart the concern of

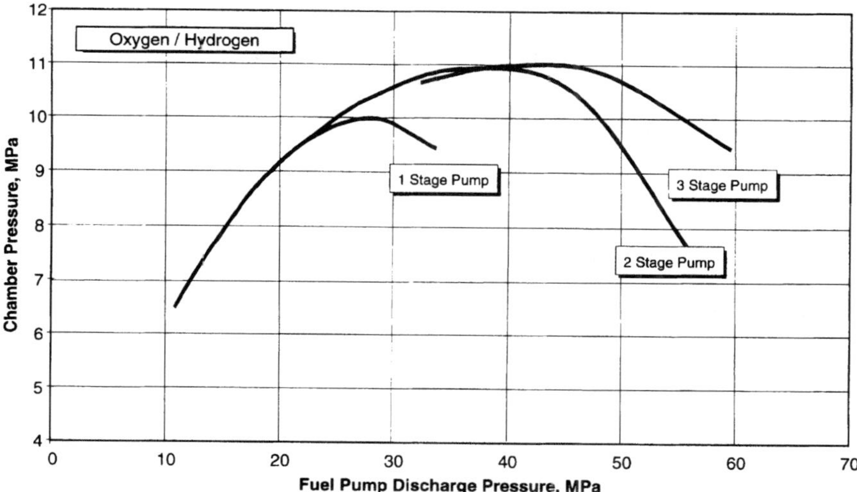

Fig. 6 Typical expander cycle chamber pressure vs fuel pump discharge pressure.

freezing the residual moisture that can result from propellant precombustion is avoided.

B. Configuration Options

There are many possible configurations for expander cycle engines. The configuration selection, however, does very little to change the total energy available to the cycle. Generally the energy available is determined by the chamber and nozzle surface area and combustion-side heat transfer conditions, and it is relatively unaffected by different coolant-side configurations or by the coolant-side pressure drop. The primary objective of the coolant-side design is to maintain adequate chamber and nozzle material temperatures, required for durability, and to minimize coolant pressure drop, which will maximize chamber pressure and cycle performance.

1. Expander Cycle with Regenerator

The basic expander cycle can be enhanced by the addition of a regenerator (heat exchanger) between the turbine discharge and main injector inlet. The regenerator recaptures energy from the turbine discharge flow to provide additional turbine work capability. However, because the turbine discharge flow is approximately 250 K while the liquid hydrogen temperature is approximately 20 K, only moderate levels of energy can be recaptured before the weight of the regenerator becomes excessive.

2. Full-Flow Expander Cycle

In a full-flow expander cycle, the fuel is used to cool the thrust chamber and only a portion of the nozzle, and then used exclusively to drive the fuel pump. The oxidizer is used to cool the remainder of the nozzle, and the energy absorbed is used to provide power exclusively to the oxidizer pump. This cycle requires an additional coolant manifold section to the nozzle and an oxidizer turbine bypass valve (OTBV) on the oxidizer turbine line to control and throttle the oxidizer side of the cycle.

The performance for the full-flow expander cycle will be less than for the conventional expander. The total energy available to the cycle remains the same because the total chamber and nozzle heat transfer is unaffected by this alternate pump configuration, but the total pump horsepower requirement is increased. This occurs because the oxidizer pump discharge pressure must be increased to provide the oxidizer turbine pressure ratio potential that is now required. However, one advantage is that this cycle would eliminate the need for an interpropellant seal (IPS) purge. The IPS purge is required for booster applications to protect against mixing and possible ignition of propellants in the oxygen turbopump. The IPS purge is not required for upper-stage applications because the interpropellant location can be vented to a low ambient pressure that is below the pressure required to allow propellant ignition and combustion. In general there are not many advantages for this full-flow expander cycle as compared with the traditional approach.

C. Expander Thrust Scaling Trends and Issues

All expander cycles are dependent on chamber and nozzle heat transfer to provide energy to the cycle. For expander cycles, as with all cycles, the required propellant pumping power scales proportionally to engine thrust level (neglecting efficiency differences, etc.). The energy available to drive the expander cycle does not scale proportionally to thrust level, however, which results in a reduced chamber pressure as thrust size increases. The following discussion will provide insight into these scaling trends.

Chamber heat transfer does not scale proportionally with thrust. If engine thrust size is increased, then required pump work and throat area will increase proportionally. If combustion chamber length is approximately constant, however, the chamber surface area, and therefore the chamber heat transfer, scales proportionally to the square root of the throat area. This relative loss of available energy can be partially offset by increasing chamber length, but this will increase engine length and weight and increase the chamber coolant pressure losses, which will limit chamber pressure increase.

Unlike chamber heat transfer, nozzle heat transfer does scale proportionally with thrust. For a constant nozzle expansion ratio, the nozzle surface area is proportional to the throat area. Therefore, as throat area increases proportional to thrust, the nozzle heat transfer area also increases proportional to thrust.

Therefore, because total heat transfer (chamber plus nozzle) does not scale proportionally with increased thrust, the energy available is increasing more slowly than the work requirement, and the chamber pressure decreases as thrust size is increased. This trend is illustrated in Fig. 7, indicating the

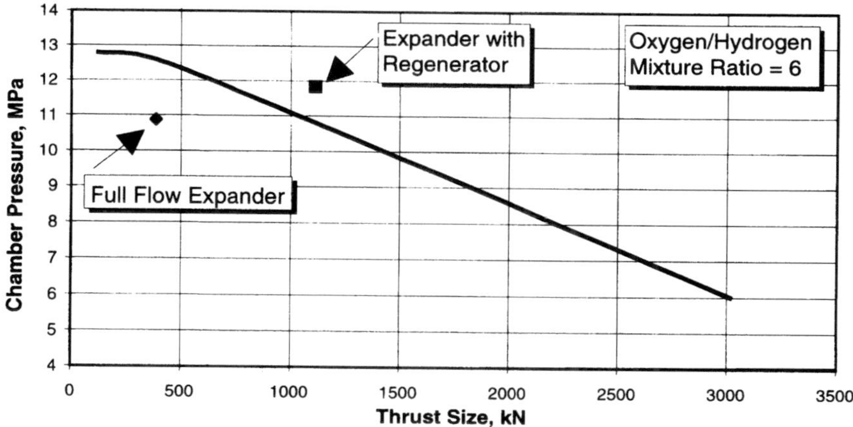

Fig. 7   Thrust scaling trends for oxygen/hydrogen expander engines.

chamber pressure drops to $\sim$8 MPa at a thrust level of 2225 kN. Figure 7 also shows the approximate performance benefit of adding a regenerator and the approximate performance penalty for the full flow expander configuration.

Because expander cycles operate at lower chamber pressures than gas generator or staged combustion cycles, the geometric size of the engine can be an issue, as thrust size is increased. Size issues can be mitigated, however, depending on the application. For upper stages the important dimension is generally the interstage length required for engine packaging. Packaged engine length can be shortened by use of a retracted nozzle skirt that is stowed at liftoff and translates prior to engine start. For booster applications, the overall engine geometry is nearly unaffected by chamber pressure because reduced chamber pressure forces the selection of a reduced nozzle expansion ratio to maximize sea-level thrust or to avoid nozzle flow separation. However, a smaller nozzle expansion ratio will reduce engine performance.

III.   Gas Generator Cycles

Gas generator cycle engines are open cycle engines. An operative example of a gas generator cycle is the Vulcain.[7,8] The energy to support the cycle is supplied by the combustion of a portion of the propellants in a "gas-generator" combustion device. This turbine drive flow is configured in parallel with the main chamber flow and is discharged overboard or into the divergent section of the primary nozzle, providing a minimal contribution to thrust. For a fuel-rich configuration, the mixture ratio of the gas generator flow, selected to maintain acceptable turbine temperature limits, is much lower than the mixture ratio of the main chamber. To maintain a nominal engine inlet mixture ratio, the main chamber must operate above the nominal mixture ratio to compensate for the lower mixture ratio in the gas generator. Similarly, for an oxidizer-rich configuration, the mixture ratio of the main chamber must operate below the nominal mixture ratio, because the mixture ratio of the gas generator is higher. Thus, for both configurations main chamber performance is reduced because the mixture ratio is further from the peak performance value. The gas generator flow should be minimized to avoid the thrust loss of the outer performance gas generator stream and to minimize the main chamber mixture ratio shift. To minimize the impact of this mixture ratio shift, the vehicle tanks can be designed to operate at a "system" or engine inlet mixture ratio that is somewhat reduced. This works better for nonhydrogen fueled engines, where the fuel density is nearer to the density of the oxidizer and mixture ratio selection has minimal impact on the stage tank weight and volume.

Another remedy for the mixture ratio shift would be to include an afterburner located downstream of the turbines but prior to expansion through a secondary expansion nozzle. For the afterburning approach the performance loss of an overboard gas generator flow is mitigated, and the mixture ratio shift of the primary combustion chamber is avoided, although now a portion of the main chamber flow is combusted and expanded from a lower chamber pressure. The gas generation afterburning cycle is discussed further in Section VIIB.

A. General Cycle Discussion

A conventional fuel-rich gas generator cycle is shown in Fig. 8. The fuel enters a single- or multiple-stage pump and is increased to a pressure adequate to cool

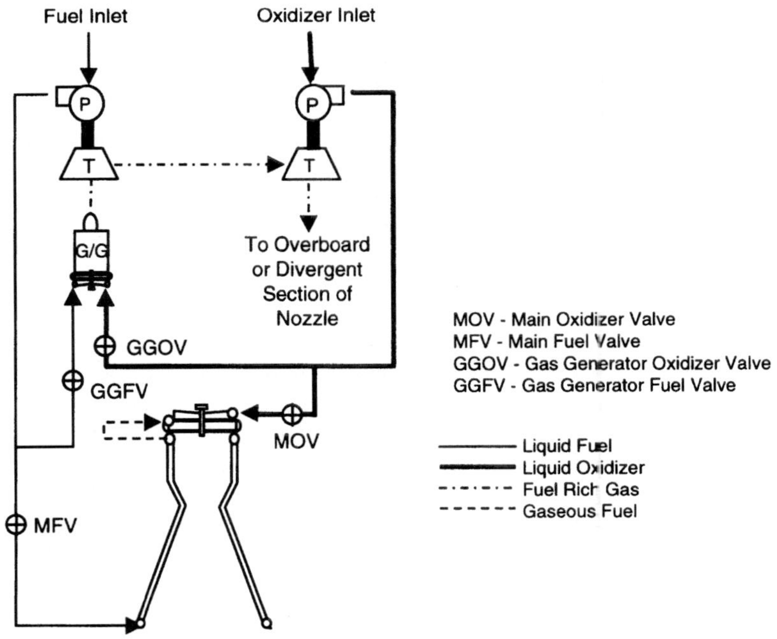

Fig. 8   Fuel-rich gas generator cycle.

the thrust chamber and nozzle. A small percentage of the fuel is directed to the gas generator combustion device. The pump discharge pressure must be adequate to overcome coolant pressure losses prior to fuel entry into the main injector and chamber. The fuel flow is controlled by the MFV located just upstream of the chamber and nozzle coolant circuit. The gas generator fuel flow is controlled by the gas generator fuel valve (GGFV).

The oxidizer flow is pressurized to moderate pressures by a single- or multiple-stage oxidizer pump. The oxidizer flow is controlled by the MOV located just upstream of the main injector, and the gas generator oxidizer valve (GGOV) upstream of the gas generator. The desired engine power level is achieved and maintained by using the GGFV and GGOV to control the gas generator mixture ratio. Engine mixture ratio is controlled by the MOV.

To minimize the overboard flow, the gas generator temperature should be maximized. A moderate temperature limit of 900 K is typical for a low-cost turbine because gas generator cycles are generally configured for expendable applications.

B. Configuration Options

To recover performance from the use of a gas generator, the turbine discharge flow can be directed to an afterburner combustion device for additional combustion and expansion through a secondary nozzle. Although either a fuel-rich or oxidizer-rich gas generator could be considered upstream of the afterburner, the oxidizer-rich option was selected for two reasons. First, the oxidizer-rich combustion products contain much less carbon and are less likely to result in soot buildup in the turbines and the injector passages of the afterburner injector. Second, the fuel dedicated to the afterburner is a more appropriate propellant to use in cooling of the afterburner combustion liner.

An oxidizer-rich gas generator cycle with an afterburner is shown in Fig. 9. The oxidizer-rich turbine discharge is burned with fuel diverted from the fuel pump discharge or the chamber coolant discharge. The afterburner mixture ratio is maintained at the nominal engine level by using the afterburner fuel valve (ABFV); therefore the performance loss due to main chamber mixture ratio shift is avoided. The performance penalty normally associated with overboard gas generator flow is mitigated by the additional combustion in the afterburner, although at a lower pressure than in the main chamber.

The performance of the oxidizer-rich gas generator cycle with afterburning is shown in Fig. 10 for oxygen and kerosene propellants and is compared with a conventional fuel-rich gas generator cycle and an oxidizer-rich staged combustion cycle. For this comparison the specific impulse is plotted vs gas generator (or preburner) pressure, not chamber pressure as is the conventional comparison. This is a more valid comparison, as the staged combustion cycle would require significantly higher pump discharge pressure than the gas generator cycles for the same chamber pressure. Figure 10 shows that the performance of the afterburning cycle is higher than that of the conventional gas generator cycle and of the staged combustion cycle for low to moderate pump discharge pressures. Only at higher pump discharge pressures does the staged combustion cycle begin to provide performance benefit.

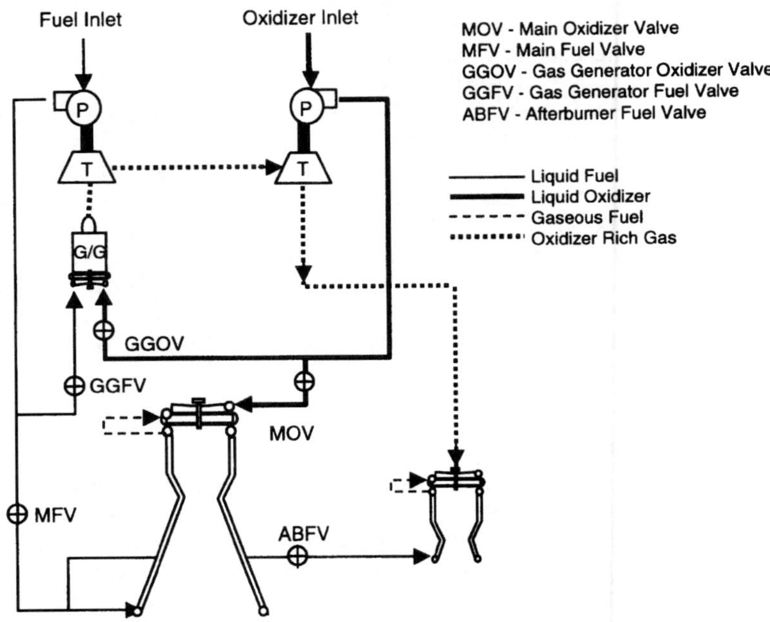

Fig. 9 Oxidizer-rich gas generator cycle with afterburner.

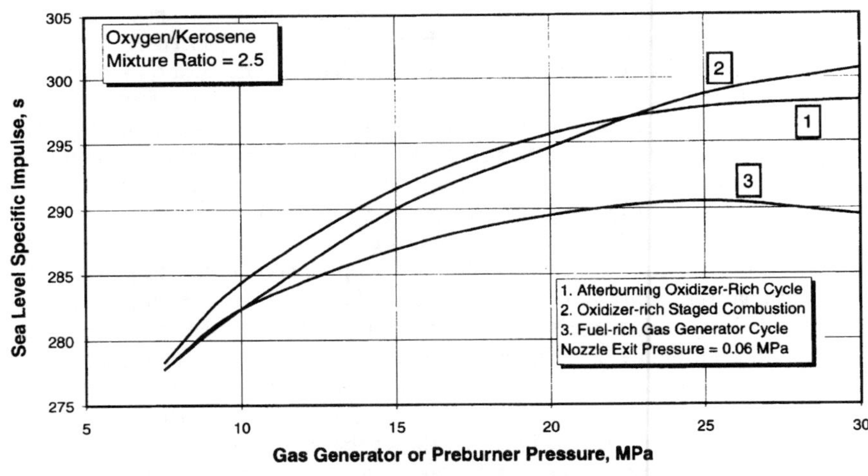

Fig. 10 Performance comparison of afterburning oxidizer-rich gas generator cycle.

The addition of the afterburning hardware will increase complexity as compared with a single chamber design. However, large booster applications may require multiple individually gimbaled thrust chambers, the complexity could be reduced as compared with a staged combustion cycle by dedicating one of the multiple chambers as the afterburner.

C. Gas Generator Thrust Scaling Trends and Issues

The thermodynamics of gas generator cycles generally scale very easily to high and low thrust levels. There are some secondary effects on the thermodynamics because of main chamber cooling and component efficiencies, both of which improve the thermodynamics as thrust size is increased. Because normalized chamber heat transfer is reduced as thrust size increases (see Section VI.B), the chamber coolant flow pressure drop can be decreased. Also, as the pump size is increased the relative internal clearances can be improved, which provides a secondary improvement in component efficiency.

IV. Staged Combustion Cycles

Staged combustion engines are closed-cycle engines. Nearly all of the energy to support the cycle is provided by chemical energy released in a preburning combustion device. Hence, these cycles are sometimes referred to as preburner cycles. The turbine drive flow is configured in series with the main chamber, so that all propellant is combusted in the main chamber and expanded in the primary nozzle. Staged combustion cycles can be used for either booster or upper-stage applications depending on the cost vs performance trades for the particular application.

A. General Cycle Discussion

The flow schematic of a fuel-rich staged combustion cycle engine with dual preburners representative of the space shuttle main engine[8] is shown in Fig. 11. The chemical energy to support the cycle is transported to the preburners by the fuel; the amount of available energy is represented by the LHV of the fuel. To maximize the energy release, the fuel flow provided to the preburners should be maximized. Some of the energy to the staged combustion cycle is supplied by nozzle heat transfer similar to the expander cycles described in Section VI. Unlike in the expander cycle, however, chamber coolant flow cannot be used in the preburner because the increased chamber pressures for the staged combustion cycles result in higher chamber heat flux and higher coolant pressure drop, and therefore chamber coolant discharge pressure is too low to return to the preburners.

For the staged combustion cycle, the fuel enters a single- or multiple-stage pump and is pressurized to high pressure. The current state-of-the-art limit for pumping hydrogen is approximately 70 MPa, because of the high power requirements resulting from the low density of hydrogen. A minimum amount of the high pressure fuel is diverted for main chamber cooling, generally less than 20% of the total fuel flow. The remainder of the fuel is vaporized while

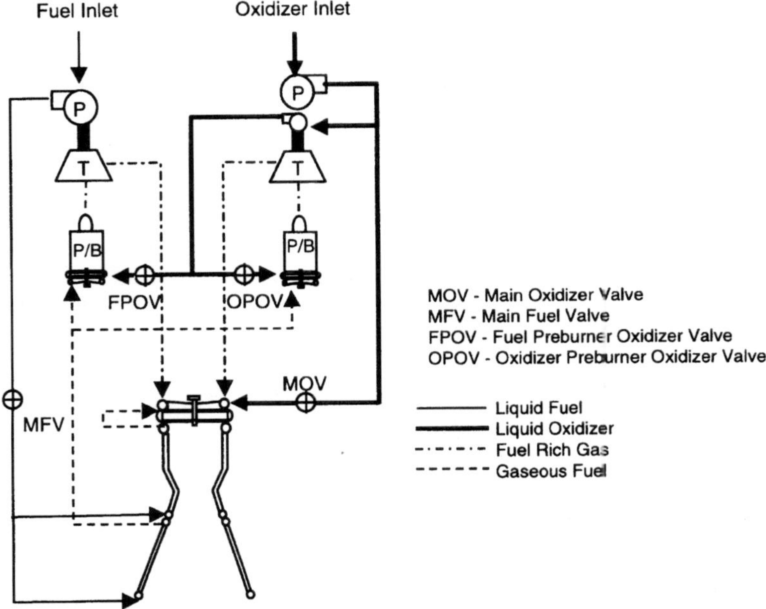

Fig. 11   Fuel-rich staged combustion.

cooling the nozzle, then delivered to the preburners for precombustion to provide power to the turbines. The total engine fuel flow is controlled by the MFV. The fuel distribution between the preburners is set by the respective preburner flow areas to ensure a split relative to the work requirement of each turbopump.

The oxidizer flow is pressurized to moderate pressure by a single- or multiple-stage main oxidizer pump. The pressure must be sufficient to deliver oxidizer to the main chamber but not to the preburners; a kick-stage is used to pressurize a fraction of the oxidizer to preburner pressure levels. The total engine oxidizer flow is controlled by the MOV. A fuel preburner oxidizer valve (FPOV) and an oxidizer preburner oxidizer valve (OPOV) are required to control the mixture ratio for each preburner. The desired engine power level is achieved and maintained by controlling the preburner mixture ratios.

The use of separate preburners has advantages for increased flexibility for component matching and throttling, but it can be a severe disadvantage during starting. The dynamic interaction of the preburners must be tailored to the inertia of each pump during the transition to full power, which must be accomplished without overspeeding the pumps and while avoiding transient preburner mixture ratios that may damage the turbines. Staged combustion cycles can also be configured with a single preburner. In this configuration, the preburner discharge flow must either power the turbines in series or be split to the respective turbines, or a single turbine must be configured to power both the fuel and oxidizer pumps on a single shaft. An operative example for a single preburner staged combustion oxygen/hydrogen engine is the Russian RD-0120.[9] The differences in the potential performance of this thermodynamic cycle are not strongly influenced by the alternate turbine or preburner configurations.

B. Configuration Options

Configuration options for staged combustion cycles are primarily influenced by the selection of a fuel-rich or an oxidizer-rich preburner configuration. As previously discussed in Section V, the fuel-rich powerhead approach is preferred for hydrogen-fueled cycles, and the oxygen-rich powerhead approach is preferred for hydrocarbon-fueled cycles. An additional option that can be considered for a dual preburner configuration is to operate one preburner fuel rich to drive the fuel pump and the other preburner oxidizer rich to drive the oxidizer pump. This is commonly referred to as a full-flow staged combustion cycle because nearly all of the flow is used in the powerhead combustion process. This configuration is also referred to as a gas-gas cycle because both propellants are gaseous entering the main injector. A technology demonstrator for this full-flow staged combustion cycle approach, is Integrated Powerhead Demonstrator.[10]

1. Full-Flow Powerhead

The use of both an oxygen-rich and a hydrogen-rich preburner for a staged combustion cycle has a profound effect on the cycle thermodynamics. A schematic of this approach is shown in Fig. 12. The fuel supply configuration is

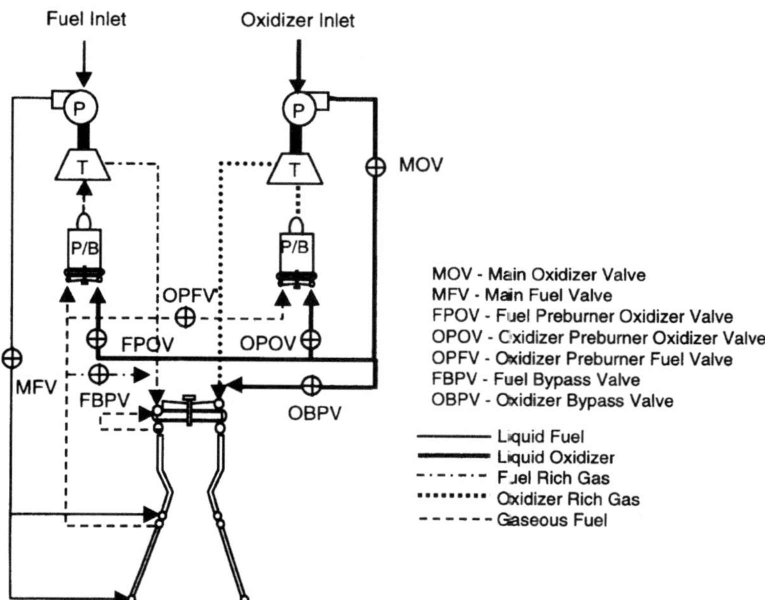

Fig. 12 Full-flow staged combustion.

similar to the conventional fuel-rich approach except that the majority of the fuel is directed to the fuel-rich preburner while a small portion of the fuel flow is directed to the oxidizer-rich preburner through the oxidizer preburner fuel valve (OPFV). The oxidizer side of this configuration requires that all of the oxygen be pumped to the preburner pressure level. Although this increases the total pump work, the pump configuration is simplified by the elimination of the oxidizer kick-stage. The majority of the main oxidizer flow is directed to the oxidizer-rich preburner, while a small portion is directed to the fuel-rich preburner through the FPOV.

A significant control constraint differentiates the conventional fuel rich and full-flow cycle approaches. The conventional fuel-rich approach has the flexibility to freely split the preburner flow rate to match the work requirements for the fuel and oxidizer pumps during both steady-state and transient operation. Also, the mixture ratio of the separate preburners can be controlled in a relatively independent fashion, especially during original design point selection. For the full-flow cycle, there are constraints to allocate only fuel-rich preburner flow to the fuel pump and only oxidizer-rich preburner flow to the oxidizer pump, as well as to hold total powerhead flow equal to the total flow being delivered by the pump. Also, as with the conventional fuel-rich cycle, the turbine pressure ratio of the fuel and oxidizer sides must be equal. The net result of these constraints for the full-flow cycle is that, without bypassing propellant around the preburners, only one of the preburner temperatures can be specified; the other preburner temperature will be determined by the remaining available propellants. This constraint applies for both steady-state and transient conditions. These constraints result in extra valves to control the full-flow cycle during startup, shutdown, and throttling, as indicated by comparison of Figs. 11 and 12.

The thermodynamic advantage of the full-flow staged combustion cycle is that it provides a significant increase in powerhead energy release within the same turbine temperature limits as the conventional fuel-rich approach. The proof of this increase can be shown by examining the energy release potential and the net powerhead work potential in a simplified example similar to that described in Section V. Rather than repeat the calculation, the results from a hypothetical oxygen/hydrogen staged combustion engine with characteristics as shown in Table 3 will be summarized. The simplified propellant distribution schematics

**Table 3   Hypothetical oxygen/hydrogen staged combustion
engine parameters**

Engine parameters	Valve
Engine inlet hydrogen flow, kg/s	100
Engine inlet oxygen flow, kg/s	600
Powerhead turbine temperature limit, K	900 (both fuel-rich and oxidizer-rich)
Hydrogen LHV, MJ/kg	117.7
Engine main chamber pressure, MPa	20
Preburner pressure, MPa	35

for both the conventional fuel-rich and full-flow powerhead approaches are shown in Fig. 13.

From the propellant distributions shown in Fig. 13a, the preburner in the conventional fuel-rich cycle will produce an energy release of 912 MJ/s, given a hydrogen LHV of 117.7 MJ/kg. From Fig. 13b, the fuel-rich preburner in the full-flow cycle will provide an energy release of 859 MJ/s, while the oxidizer-rich preburner will produce 553 MJ/s, for a total of 1412 MJ/s. Thus, the full-flow cycle provides a 55% increase in total energy release that can be used to increase turbine work potential.

Using the characteristics shown in Table 4, the net powerhead work potential is examined next. The component efficiencies in Table 4 are assumed conservatively low because this simplified example ignores all system pressure losses including the chamber coolant pressure loss. Pump work is calculated using Eq. (1), and work delivered by the turbines is calculated using Eq. (2). Results of the noniterated power balance are shown in Table 5.

Table 5 shows that, although the total turbopump work requirement for the full-flow cycle is ~13% higher than the conventional fuel-rich cycle (37 MW + 98 MW vs 120 MW), the available turbine work is more than 40% greater (46 MW + 95 MW vs 100 MW). This comparison alone indicates a higher performance potential for the full-flow cycle. For the conventional fuel-rich cycle the delivered turbine work (100 MW) is less than the required pump work (120 MW), therefore, the chamber pressure will be less than the assumed 20 MPa. An iterative calculation indicates that chamber pressure for the fuel-rich conventional cycle would drop to 14.5 MPa.

For the full-flow cycle, the turbine work for the hydrogen side turbine (95 MW) is nearly equal to the requirement for the hydrogen pumps, indicating that the hydrogen side of the full-flow cycle is approximately balanced at the 20 MPa chamber pressure level. The turbine work for the oxidizer side turbine (46 MW) is greater than the oxidizer pump requirement (37 MW). Because the

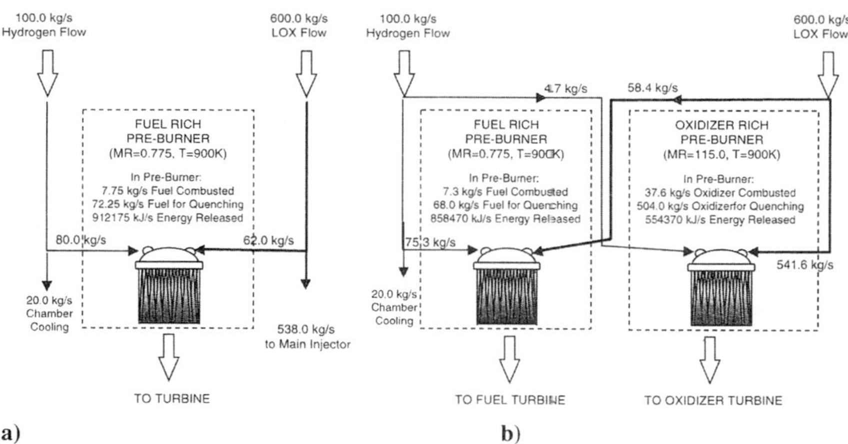

Fig. 13 Powerhead propellant distribution: a) conventional vs b) full-flow option.

Table 4   Powerhead parameters for hypothetical oxygen/hydrogen staged combustion engine

	Conventional	Full-flow	
Powerhead parameters	Fuel-rich	Oxidizer-rich	Fuel-rich
Preburner mixture ratios	0.775	115	0.775
Preburner temperatures, K[a]	900	900	900
Preburner flow rates, kg/s	142	546.3	133.7
Preburner gas constant, R, kJ/kg K[a]	2.323	0.275	2.323
Preburner ration of specific heats[a]	1.368	1.309	1.368
Assumed pump efficiency, %	50	50	50
Assumed turbine efficiency, %	65	65	65
Oxygen density, kg/m^3	1140	1140	1140
Hydrogen density, kg/m^3	71	71	71

[a]As given per Ref. 1 for respective mixture ratios.

chamber pressure is limited by the hydrogen side work balance, the oxidizer side must be balanced by reducing the temperature to ~700 K. Thus, the full-flow cycle can deliver the full assumed 20 MPa. The 38% improvement in chamber pressure over the conventional dual fuel-rich preburner cycle provides the rationale for consideration of the full-flow cycle approach for oxygen and hydrogen staged combustion rocket engines.

Both the conventional and full-flow cycles are strongly influenced by component efficiencies. As pump discharge pressure is increased, chamber pressure for both cycles will increase until component structural or thermal limitations are reached or until component efficiencies begin to decrease more rapidly. An example is shown in Fig. 14. In much the same fashion as shown previously in Fig. 6, local peaks in chamber pressure result from increasing cycle efficiency being offset by accelerating reductions in component efficiencies as their work

Table 5   Power balance parameters for hypothetical oxygen/hydrogen staged combustion engine

	Conventional	Full-flow	
Power balance parameters	Total	Oxidizer-side	Fuel-side
Preburner energy release, MJ/s	912	553	859
Delivered turbine work, MW	100	46	95
Fuel flow pumped to 35 MPa, kg/s	100	0	100
Oxidizer flow pumped to 35 MPa, kg/s	62	600	0
Fuel flow pumped to 20 MPa, kg/s	0	0	0
Oxidizer flow pumped to 20 MPa, kg/s	538	0	0
Turbopump work requirement, MW	120	37	98

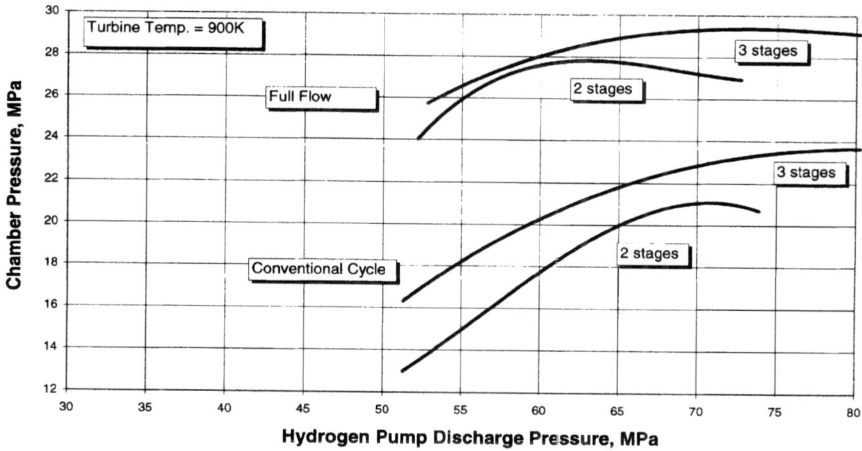

Fig. 14 Staged combustion cycle hydrogen pump discharge pressure.

levels are increased. The effect of reduced fuel pump work load, by increasing from two to three stages, is evident in the increased chamber pressure for a given fuel pump discharge pressure.

2. *Single-Shaft Oxidizer-Rich Powerhead*

As previously discussed in Section V, an oxidizer-rich powerhead is thermodynamically desirable for staged combustion cycles using kerosene propellants. The simplification of a common shaft for the fuel and oxidizer turbomachinery is feasible because the combination of propellant densities and discharge pressures results in similar desired rotational speeds of the fuel and oxidizer impellers. This configuration with a single preburner and single shaft turbopump allows the engine to be controlled with a minimum complement of valves. The starting and throttling of the engine is more controllable because the dynamic complications of separate preburners and separate turbopumps is avoided. Operative examples of oxidizer-rich, oxygen/kerosene, single-shaft staged combustion engines are the NK-33[11] and RD-120.[12]

C. Staged Combustion Thrust Scaling Trends and Issues

The thermodynamics of staged combustion cycles generally scale very easily, in the same fashion as gas generator cycles. As thrust size increases, the required percentage of main chamber coolant flow can be decreased, instead of or in addition to a reduction in coolant flow pressure drop.

V. Summary

This chapter provides an overview of the thermodynamics that influence the cycle configuration selection of liquid rocket engines. An overview of the interaction of the cycle thermodynamics with the engine hardware stress limitations, thermal limitations, and component efficiency trends is addressed. Thermodynamic cycle comparisons are presented for expander cycle, gas generator cycle, and staged combustion cycles. For the expander cycles the thermodynamic issues are presented relative to operation with a regenerator as well as scaling trends. For the staged combustion cycles, the thermodynamic implications of oxidizer-rich combustion and the influence on liquid rocket engine cycles are specifically considered. For the gas generator cycle, the thermodynamic implications of using an afterburner downstream of an oxygen-rich gas generator and turbine drive are addressed.

References

[1]McBride, B. J., and Gordon, S., "Computer Program for Calculation of Complex Chemical Equilibrium Compositions and Applications," NASA RP-1311, 1996.

[2]Beckman, B., Chilton, J., and Jue, F., "Block II SSME Operability Improvements," AIAA Paper 97-2819, June 1997.

[3]Minick, A., and Peery, S., "Design and Development of an Advanced Liquid Hydrogen Turbopump," AIAA Paper 98-3681, July 1998.

[4]Ohta, T., Kitamura, A., and Ogata, H., "LH_2 Turbopump Test with Hydrostatic Bearing," AIAA Paper 99-2195, June 1999.

[5]Santiago, J., "Evolution of the RL10 Liquid Rocket Engine for a New Upper Stage Application," AIAA Paper 96-3013, July 1996.

[6]Parsley, R., and Crocker, A., "The RL200: Incorporating A Balanced Approach for Reusable Propulsion Safety," AIAA Paper 2000-3287, July 2000.

[7]Brossel, P., et al., "Development Status of the Vulcain Engine," AIAA Paper 94-2751, June 1994.

[8]Bradley, M., "Space Shuttle Main Engine Off-Nominal Low Power Level Operation," AIAA Paper 97-2685, July 1997.

[9]Rachuk, V., Martinenko, Y. A., Gontcharov, N. S., and Sciorelli, F., "Design, Development, and History of the CADB RD-0120 Engine," AIAA Paper 95-2540, 1995.

[10]Vilja, J., "Rocketdyne Advanced Propulsion Systems Overview," AIAA Paper 97-3309, July 1997.

[11]Kuznetsov, N. D., "Closed Cycle Liquid Propellant Rocket Engines," AIAA Paper 93-1956, June 1993.

[12]Fatuev, I., "RD-120M High Performance Engine for Booster Applications," AIAA Paper 95-3144, July 1995.

Chapter 19

Tripropellant Engine Technology for Reusable Launch Vehicles

N. S. Gontcharov,[*] V. A. Orlov,[†] V. S. Rachuk,[‡] M. A. Rudis,[§] and A. V. Shostak[¶]
Chemical Automatics Design Bureau, Voronezh, Russia

and

R. G. Starke[**] and J. R. Hulka[††]
Aerojet General Corporation, Sacramento, California

Nomenclature

c^* = characteristic exhaust velocity, m/s
E = modulus of elasticity, MPa
H = pump head, m
H_2 = hydrogen
k = kerosene
l_{ak} = amplitude of elasto-plastic deformation
l_f = ultimate deformation in stress concentration area
$\dot{m}$ = mass flow rate, kg/s
$\bar{\dot{m}}$ = relative mass flow rate
m_o = low cycle fatigue curve exponent
m_e = low cycle fatigue curve exponent

---

*Vice President.
†Doctor of Technical Science, Head of Department.
‡Doctor of Technical Science, President.
§Doctor of Technical Science, Head of Structural Analysis Department.
¶Head of Design Bureau.
**Chief Project Engineer.
††Engineering Specialist; currently with Jacobs/Sverdrup Technology. Senior Member AIAA.

N_e = number of operational cycles or engine cycles
N_o = number of cycles before crack initiation
O_2 = oxygen
p = pressure, MPa
r = mixture ratio
R = gas constant, (kgf · m)/(kg · def K)
T = temperature, K
ΔV = vehicle velocity increment, m/s

Greek

β = stress loading cycle asymmetry factor
β_* = strain loading cycle asymmetry factor
Ψ = relative cross-sectional area reduction
σ_B = ultimate strength of material, MPa
σ_{-1} = fatigue limit of material, MPa
η_N = durability margin factor
ρ = density, kg/m^3

Subscripts

e = experimental
g = gas generator (preburner)
H = hydrogen
k = kerosene
m = main (main combustion chamber)
nom = nominal
o = oxidizer
p = predicted

Superscripts

H = (hydrogen) fuel-rich
O = oxidizer-rich

I. Introduction

AFTER 70 years of development, rocket vehicles have reached a level of sophistication where cost is the deciding factor. Designers of the next generation of launch vehicles must dramatically reduce the costs for transportation of payloads to orbit, while improving vehicle reliability, operability, and efficiency. Cost drivers include the size of the vehicle, the size and weight of the payload, the inclusion of human pilots and passengers, the number of stages, the number of propellants on board, the landing mode, the materials of construction, the reusability of the vehicle, and the propulsion system.

A NASA space transportation architecture study[1,2] examined medium payload class, one-stage and two-stage launch vehicles where different takeoff and landing criteria resulted in vehicles of varying designs. A reusable launch vehicle (RLV) was selected as the baseline vehicle that, used with an interim expendable launch fleet, would offer significant reductions in annual operating costs over existing vehicles.[1,2] In theory, reusable vehicles have the lowest life cycle costs, although they may also have the highest initial costs and therefore are the most sensitive to variable fiscal policies.

Most space transportation operation cost studies, including the NASA study, also favor the one-stage, or single-stage-to-orbit (SSTO) rocket vehicle, for the simple reason that multiple stages require multiple pieces of hardware that need procurement, servicing, assembly, integration, and maintenance. These cost factors generally outweigh the advantages of two-stage systems that may be smaller and lighter, and also less sensitive to weight growth.[3] In addition, SSTO vehicles have no expendable hardware components that would add to the debris in space, no additional stages required to be returned to the launch site, and in theory the most minimal servicing, resulting in the fastest response and shortest turnaround time to flight.

Design features of this ultimate SSTO vehicle have been the subject of debate over the past 40 years, however. Although many concepts have been generated, until recently none has been shown to be technically feasible with the technologies available at the time, due to a combination of issues such as insufficient engine performance, excessive engine and vehicle size and weight, and high development costs. Many early SSTO studies concluded that vehicles designed then could not theoretically deliver a payload into orbit. A primary difficulty is the propulsion—that is, of configuring a lightweight system that emulates booster engine characteristics at low altitude and upper-stage engine characteristics for ascent and orbit insertion.[4,5]

One of the most important theoretical advancements to improve feasibility of SSTO vehicles was the introduction of mixed-mode propulsion by Salkeld,[6] in which different modes of propulsion with different propellant densities were combined in the same stage. Salkeld initially concluded that the optimum burn profile to maximize ideal ΔV for a two-mode, single-stage rocket vehicle was purely sequential, with the higher specific-density impulse mode operating first in the boost portion of the flight, and the higher specific-impulse mode operating during the ascent portion of the flight.[6] Many rocket vehicles already operated this way, in fact, by using the different modes in different stages, although the individual propulsion modes were optimized for each stage.

However, carrying hydrogen engines or engine components without using them at liftoff appeared to create a penalty on vehicle or engine mass.[7] Studies showed that various engine types on a vehicle with parallel burn (i.e., simultaneous operation of hydrocarbon and hydrogen at liftoff) could provide vehicle dry mass and gross liftoff weight at least as good as that of a vehicle with series burn.[7-9] One of the important advantages of a parallel burn vehicle is that the hydrogen engine or engine components can be ignited on the ground and thus can be monitored and evaluated before liftoff, an obvious increase in vehicle reliability.

Since the introduction of the mixed-mode principle to SSTO vehicle propulsion, there have been numerous studies that showed that a mixed-mode or dual-fuel SSTO vehicle has many distinct advantages over reference single-fuel SSTO vehicles.[3,7,9-14] The main benefit of the mixed-mode operation results from the fact that most of the propellant is burned in achieving a relatively small percentage of the velocity to orbit and is maximized by burning the high-density fuel early in the flight. When a high-density propellant combination is burned in this initial phase of the flight, the resultant vehicle size and dry weight are less for a fixed payload mass. The vehicle size and dry weight are historically considered to be roughly representative of both the development, test, and engineering (DT&E) costs and the life cycle costs (LCC) of a vehicle. Dry weight is the weight of the vehicle without propellant, fluids, payload, or crew.

The tripropellant engine concept is in fact a self-contained mixed-mode engine, where the first operating mode, for liftoff on the ground, is tripropellant (hydrocarbon, liquid hydrogen, and liquid oxygen), and the second operating mode, for sustain performance at altitude, is bipropellant (liquid hydrogen and liquid oxygen). A tripropellant engine such as this retains all of the benefits of the mixed-mode principle for the launch vehicle and has additional engine and vehicle benefits. The availability of hydrogen for cooling significantly increases the allowable chamber pressure for hydrocarbon fuel-rich turbine drive systems, and hence increases performance and reduces engine size, since there is no reliance on the hydrocarbon fuel or the oxygen for chamber wall or nozzle cooling. By shutting off the kerosene flow, the tripropellant engine is transformed into a bipropellant liquid oxygen/hydrogen engine, achieving the mixed-mode benefit with a single barrel engine, which will reduce the rear panel or "boat tail" area of SSTO launch vehicles, reducing vehicle weight and improving aerodynamic efficiency by moving the vehicle center of gravity forward. By initiating the hydrogen flow and ignition on the launch pad, there is no reliance on upper atmosphere chilling and ignition of a staged or serial burn engine, which would add system weight for the storage of purge gases and may reduce vehicle reliability.

Mixed-mode vehicle studies performed with separate hydrocarbon-fueled and hydrogen-fueled engines showed that vehicle dry weight could be reduced 30% over an all-hydrogen reference vehicle.[7,11] With dual fuel engines, vehicle dry weight could be reduced 35–50%.[3,7,11] Specific tripropellant engine studies[3,15] considered many advanced concepts available at the time, all integrating the turbopumps and using hydrogen for cooling, but using different drive gas and piping schemes and nozzle concepts. All these studies found they could reduce the vehicle dry weight of a reference all-hydrogen SSTO vehicle, ranging from a 21% reduction with a hydrogen gas generator tripropellant engine design, to over 30% reduction for dual bell and dual expansion engine designs. Tripropellant in this instance refers to a dual fuel scheme because the two fuels, hydrogen and a hydrocarbon, share a common oxidizer, oxygen.

However, in one study, a clean-sheet LO_2/LH_2 engine provided more dry weight reduction (primarily by increasing mixture ratio from 6.0 to 6.9) than a clean-sheet tripropellant engine.[16] With advances in other areas of SSTO vehicle technology, or if clean-sheet engines are considered, the use of tripropellant engines may not be critical to the development of an SSTO vehicle.[1,16,17] The large weight penalties on the vehicle due to the low density of hydrogen, because of larger tanks and feed structures, can be reduced by the proposed use of lightweight advanced composite materials, so that performance from single fuel engines becomes feasible. Preliminary designs have been developed of an SSTO vertical-takeoff, horizontal-landing (VTHL) vehicle that delivers a 9100-kg payload to a Space Station Freedom orbit, using only single mixture ratio, all-hydrogen propulsion, with dry weight growth margin of 15%.[17]

Whether this vehicle dry weight growth margin is adequate is debatable, given that the space shuttle weight growth was 25%.[1] Many vehicle and engine performance issues still remain that create doubts about vehicle dry weight gain, and cost issues remain because of engine reusability. Consequently, continued studies of additional technologies, including a tripropellant engine, were recommended that would provide additional vehicle dry weight reduction.[1] The use of tripropellant engines, by reducing the dry weight of the vehicle, provides for an increase in the dry weight growth margin.[1]

In the NASA study, an all-hydrogen vehicle was chosen based on an evolved engine because of the potential savings in engine development costs.[1] However, continued evaluation of tripropellant engines was recommended due to vehicle dry weight savings and weight growth margin increase. One tripropellant engine considered was also an evolved engine, based on the existing Russian liquid oxygen/kerosene propellant RD-170 engine. The RD-170 engine is a flight-qualified engine that supplied the main propulsion for the booster core stage of the Energia heavy-lift launch vehicle.[18,19] A tripropellant engine concept based on the RD-170 engine, the RD-701, was developed in Russia for the Multipurpose Aerospace System (MAKS).[20] This two-stage-to-orbit space plane was to be air launched from a Russian AN-225. The RD-701 went through complete mechanical design and analysis, including systems integration with the MAKS space plane, before the program was postponed in 1991.[20]

The RD-701 engine uses a significant portion of the RD-170, including the same oxidizer-rich preburner cycle. The preburner is unchanged, while all the hydrogen is used in the main combustion chamber for coolant. The main injector is modified to include injection of the hydrogen from the main combustion chamber. Initial subscale development testing of the tripropellant main injector shows that high performance can be achieved with three propellants,[21] which is corroborated by numerous other studies.[22–32]

A tripropellant engine can also be developed with a fuel-rich preburner cycle. In this chapter, an evolved tripropellant engine is discussed based on another existing Russian engine, the liquid oxygen/hydrogen propellant RD-0120 engine, which operates with a fuel-rich turbine drive gas. The RD-0120 engine is a flight-qualified engine that supplied the main propulsion for the sustainer core stage of the Energia heavy-lift launch vehicle.[33] Under a typical flight profile, it was ignited at sea level and operated 460 s, producing a nominal vacuum thrust of 200 metric tons and a vacuum delivered specific impulse of 455.5 s.[33] The RD-0120 engine completed extensive qualification testing, with more than 163,000 s of testing accumulated on more than 90 engines prior to the first flight, and has a demonstrated reliability of 0.992 at 90% confidence.[33] A tripropellant engine based on this highly evolved engine also would include substantial development savings.

In this chapter, the design and development issues of a tripropellant liquid rocket engine using a closed (or staged combustion) power cycle and a fuel-rich propellant turbine drive gas scheme are discussed. The first section reviews issues for selection of the tripropellant engine turbine drive gas. Next, the optimum configuration of the turbine drive gas engine is reviewed, followed by discussion of the main technical issues this engine will face to be used on an SSTO RLV. Finally, a demonstration tripropellant engine program using the liquid oxygen/liquid hydrogen RD-0120 as a test bed is briefly described, illustrating that many of these issues can be investigated in a cost-effective fashion.

II. Selection of Tripropellant Engine Cycle for Reusable
SSTO Application

The SSTO RLV application places many significant requirements on the engines, of which the most important is reusability. To be cost effective in a reusable vehicle, the engine must withstand typically 25 starts over an operating duration of 12,500 s between refurbishment. Total operating life would be typically 125 starts over a total operating duration of 62,500 s. In addition, the engine must also be high performing and low weight, otherwise its inclusion on the SSTO vehicle would result in pound-for-pound payload deductions and reduced vehicle stability.

There are several variants of engine schemes to be considered that may meet these requirements. The most widely recognized is the staged combustion cycle, with full or partial secondary combustion of the gas from the preburner. In this section, the schemes for two different partial secondary combustion tripropellant schemes are compared. One has an oxidizer-rich turbine drive generated by a bipropellant preburner, along with hydrogen cooling of the chamber and nozzle, and a main chamber injector with liquid/gas/gas propellant injection. The other has a fuel-rich turbine drive generated by a tripropellant preburner, along with hydrogen cooling of the chamber and nozzle, and a main chamber injector with liquid/gas propellant injection. Full secondary combustion, using both fuel-rich and oxidizer-rich gas to drive separate turbines, offers potential advantages over either of these schemes, such as the elimination of unlike propellant pump and drive gas schemes, but also has disadvantages, such as increased control complexity. For brevity, this scheme is not considered in this chapter. However, the advantages and disadvantages of each partial secondary combustion scheme may be weighed for the full combustion scheme.

All the tripropellant engines are dual mode: one mode with hydrocarbon/hydrogen flow for high specific density impulse during liftoff, and the other mode with hydrogen-only flow for high specific impulse for ascent into orbit. Based on one vehicle optimization in Russia for maximizing payload, the ratio of the thrust level of the first mode to the second mode is approximately 2.5. For the example applied in this chapter, vacuum thrust levels per engine at these modes are estimated as 200 and 80 metric tons, respectively, noting that the sea-level thrust per engine must exceed 150 metric tons for typical vehicles.

A. Turbine Drive Power of Preburner Gas

The first important comparison is the ratio of available power or work capability of the turbine drive gas. Table 1 shows the parameters used to calculate the relative power (or work capability) $\dot{m}_g R_g T_g$ of the following three preburner gas compositions: 1) bipropellant oxidizer-rich preburner gas (kerosene and liquid oxygen), 2) tripropellant fuel-rich preburner gas (kerosene, liquid hydrogen, and liquid oxygen), and 3) bipropellant fuel-rich preburner gas (liquid hydrogen and liquid oxygen). The parameters shown are for various gas temperatures and engine propellant mixture ratios r_m, where $\dot{m}_g = \dot{m}_g/\dot{m}$. To calculate

Table 1 Calculation of turbine power for three preburner gas compositions

Propellant composition	Parameter	Units	H_2, 6%; kerosene, 12.6%; O_2, 81.4%			H_2, 5%; kerosene, 14%; O_2, 81%			H_2, 4%; kerosene, 15.6%; O_2, 80.4%		
Tripropellant fuel-rich	r_m	——	4.376			4.263			4.102		
	$\bar{\dot{m}}_g$	——	0.294	0.310	0.321	0.286	0.310	0.320	0.279	0.296	0.307
	R_g	$\dfrac{\mathrm{kg_f - m}}{\mathrm{kg - K}}$	122.5	121.5	119.5	107.3	107.1	107.0	91.4	93.5	94.5
	r_g	——	0.581	0.667	0.726	0.505	0.632	0.684	0.423	0.510	0.564
	T_g	K	850	950	1000	850	950	1000	850	950	1000
	$\bar{\dot{m}}_g R_g T_g$	$\dfrac{\mathrm{kg_f - m}}{\mathrm{kg}}$	30610	35780	38360	26090	31520	34240	21680	26310	28960
Bipropellant fuel-rich	$\bar{\dot{m}}_g$	——	0.093	0.099	0.102	0.078	0.083	0.085	0.062	0.066	0.068
	R_g	$\dfrac{\mathrm{kg_f - m}}{\mathrm{kg - K}}$	273	256	249	273	256	249	273	256	249
	r_g	——	0.55	0.65	0.70	0.55	0.65	0.70	0.55	0.65	0.70
	T_g	K	850	950	1000	850	950	1000	850	950	1000
	$\bar{\dot{m}}_g R_g T_g$	$\dfrac{\mathrm{kg_f - m}}{\mathrm{kg}}$	21580	24080	25400	17980	20060	21170	14387	16050	16930
Bipropellant oxidizer-rich	$\bar{\dot{m}}_g$	——	0.833	0.835	0.836	0.829	0.831	0.832	0.822	0.825	0.826
	R_g	$\dfrac{\mathrm{kg_f - m}}{\mathrm{kg - K}}$	26.6	26.6	26.6	26.6	26.6	26.6	26.6	26.6	26.6
	r_g	——	43.63	38.67	36.70	43.63	38.67	36.7	43.63	38.67	36.70
	T_g	K	850	950	1000	850	950	1000	850	950	1000
	$\bar{\dot{m}}_g R_g T_g$	$\dfrac{\mathrm{kg_f - m}}{\mathrm{kg}}$	18830	21100	22240	18740	20100	22130	18590	20840	21970

the preburner gas flow rates, the following equation was applied:

$$r_m = \bar{\dot{m}}_o / (\bar{\dot{m}}_k + \bar{\dot{m}}_h) \tag{1}$$

where $\bar{\dot{m}}_h$, $\bar{\dot{m}}_k$, and $\bar{\dot{m}}_o$ are the relative flow rates of liquid hydrogen, kerosene, and liquid oxygen, respectively, defined as $\bar{\dot{m}}_i = \dot{m}_i / \dot{m}$, and $\dot{m}$ is the sum of $\dot{m}_h$, $\dot{m}_k$, and $\dot{m}_o$. The propellant flow rates through the preburner are defined by the following equations:

$$\bar{\dot{m}}_g = \frac{r_m}{1 + r_m} \cdot \frac{1 + r_g}{r_g} \tag{2}$$

where $r_g = \dot{m}_o / \dot{m}_{kg}$, for the bipropellant preburner with the oxygen-rich gas;

$$\bar{\dot{m}}_g = \frac{1 + r_g}{1 + r_m} \tag{3}$$

where $r_g = \dot{m}_{og} / (\dot{m}_k + \dot{m}_h)$, for the tripropellant preburner with the fuel-rich gas; and

$$\bar{\dot{m}}_g = \bar{\dot{m}}_h \cdot (1 + r_g) \tag{4}$$

where $r_g = \dot{m}_{og} / \dot{m}_h$, for the bipropellant preburner with the fuel-rich gas, and where $\bar{\dot{m}}_h = \dot{m}_h / \dot{m}$ is the relative flow rate of hydrogen.

Comparison of the $\bar{\dot{m}}_g R_g T_g$ parameters given in Table 1 shows that the relative power of the preburner gas for the fuel-rich scheme applied to the tripropellant preburner, over a gas temperature range of 850–1000 K and at different mixture ratios (i.e., different hydrogen percentages for tripropellant), is 1.72 to 1.17 times higher than for the oxidizer-rich preburner scheme with a tripropellant engine application, and 1.42 to 1.71 times higher than for the bipropellant fuel-rich preburner scheme. These ratios are illustrated in Fig. 1. However, if the main thrust chamber pressures for all the schemes are equal, the tripropellant or the bipropellant fuel-rich scheme requires 1.1 times more turbomachinery power than that for the oxidizer-rich scheme, because the hydrogen must be pumped to a higher pressure to supply the preburner. Therefore, from consideration of the overall utilization of energy in the engine, the fuel-rich scheme for the tripropellant preburner has more relative power than the oxidizer-rich scheme by 1.48 to 1.56 times for the fuel that has 6% hydrogen, and 1.06 to 1.20 times for the fuel with 4% hydrogen, over the preburner gas temperature range of 850–1000 K.

Consequently, considering the energy utilization for the tripropellant engine in the first mode, the tripropellant fuel-rich preburner scheme has advantages in lower preburner gas temperature at the same chamber pressure, or higher chamber pressure at the same preburner gas temperature, than the oxidizer-rich scheme, as shown in Fig. 2. To reach the same chamber pressure of 24.5 MPa for the propellants with 6% hydrogen, the preburner gas temperature with the oxidizer-rich scheme must increase by 200 K. Conversely, at the same preburner gas

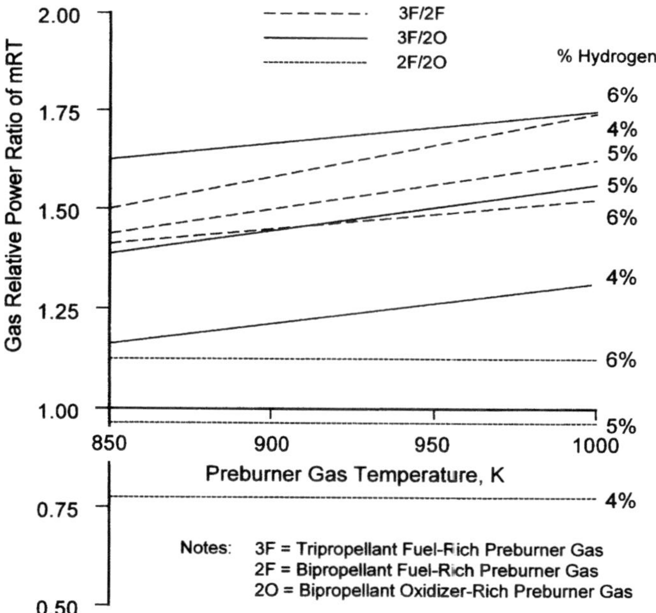

Fig. 1   Comparison of relative turbine drive gas power capability.

temperature of 850 K, the fuel-rich scheme attains a chamber pressure approximately 5 MPa higher. These calculations illustrate that a tripropellant engine with a tripropellant preburner will either 1) provide an increase of specific impulse by 3–4 s in the tripropellant mode, assuming a fixed nozzle exit diameter (or engine envelope), and hence reduce the engine weight as the chamber

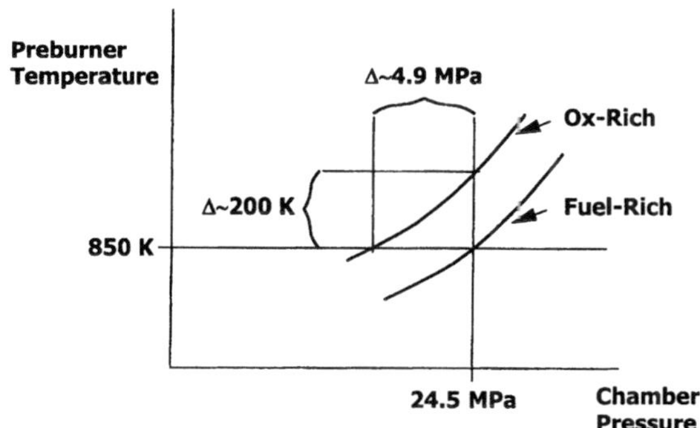

Fig. 2   Comparison of fuel-rich and oxidizer-rich turbine drive gas schemes for generation of turbine power, with 6% hydrogen propellant in engine.

pressure will be higher with the same nozzle exit diameter, at the same preburner temperature; or, 2) at the same chamber pressure, will lower the internal stresses on ducts and turbopump assemblies due to the lower gas temperatures and thus provide higher reliability and reusability.

The engine scheme with the tripropellant fuel-rich preburner provides 1.42 to 1.71 times more relative power compared to the bipropellant fuel-rich preburner (hydrogen and oxygen). The required turbomachinery powers for these two cases are practically the same. Compared to the oxidizer-rich preburner scheme, the bipropellant fuel-rich scheme provides 1.14 more to 0.77 times less relative power, as shown in Fig. 1.

B. Ignition Hazard of Metal Parts in Hot Gas Flow

Experience from liquid rocket engine development demonstrates that the assemblies and piping that contain flowing gaseous and liquid oxidizer provide the most significant fire hazard in the engine. Nearly every metal has some finite kindling temperature with oxygen or oxidizer-rich gases. The fire hazard is significantly influenced by the heat and pressure levels in the engine components, which result in increasing stressed-strained (deformed) characteristics of the assemblies, especially the oxidizer turbopump assembly. There have been cases of burning resulting from not only high temperatures but also rubbing of the metallic parts of the turbopump in the oxidizer environment.

Unfortunately, burning in oxidizer-rich gas ducts or turbines cannot be satisfactorily predicted because many unpredictable factors may cause ignition, including: 1) the appearance of microcracks in the material; 2) the appearance of metallic particles from the engine cavities and turbine cavities and gaps between the movable and unmovable parts of the turbopump assembly, which under conditions of high temperature and pressure in the oxidizing environment will ignite and spur ignition of the environment; and 3) rubbing of movable assemblies and unmovable housings because of e.g., deformation of rotor housings. The use of oxidizer-rich turbine gas for reusable engine applications presents a loss of reliability due to the increase of fire hazard in the engine. Also, the development program for the engine designed with the oxygen-rich staged combustion cycle may be more costly because test failures do not provide enough data on defects and deviations, as the corresponding hardware elements are usually completely burnt and fused. All these factors increase the cost and lengthen the time required for engine design and development.

To prevent metal ignition, nonignitable coatings have been used to protect those parts that may be most sensitive to ignition. However, these coatings have been developed for expendable liquid rocket engine components. For a reusable application, use of these coatings is not desired because they may be pitted or removed during repeated or extensive operation, requiring inspection and repairs that will increase the operation costs of the vehicle.

C. Preburner Temperature for Reusability Requirements

The reusability requirement is the basis for establishment of the engine operating parameters, most especially for the selection of the proper temperature of the preburner gas. Numerous hot-fire tests conducted on engines with long life, as well as studies of the material properties at cyclic loading, provided for the development of the philosophy and principles of durability estimations for the most stressed assemblies of the engine.[34] The temperature of the preburner gas is a determining factor for turbine blade durability and for engine life because the turbine blades are the most sensitive engine components. This section evaluates the maximum temperatures required to satisfy the reusability and life requirements (but do not take into account the additional fire hazard of the oxidizer-rich gas piping).

Studies have been conducted for turbine blades made of heat-resistant chrome-nickel alloy, with the ultimate strength of $\sigma_B \geq 1225$ MPa, and relative area reduction of $\Psi \geq 15\%$ at ambient temperatures.[34] Reusable engine assemblies operate under cyclic loading resulting from thermal variations and mechanical vibrations during the startup and shutdown transients of the engine. Such cyclic loading leads to material low-cycle fatigue, crack initiation in stress concentration areas, and, subsequently, assembly failure. At a low number of loading cycles, the loads may exceed the yield strength, and so the estimations of the elasto-plastic deformation range Δl_k and the amplitude of the elasto-plastic deformations $l_{ak} = \Delta l_k/2$ are important. The amplitude of elasto-plastic deformation l_{ak} relates to the number of cycles until crack initiation, N_o, as

$$l_{ak} = \frac{l_f^T}{2(4N_o)^{m_o} + \beta_*} + \frac{\sigma_B^T / E^T}{(4N_o)^{m_e} + \beta} \tag{5}$$

In this dependence, l_f is an ultimate (destructive) deformation of the material in the stress concentration area and is determined considering the three-dimensional stressed state and the environmental effect (particularly, hydrogen) on the material mechanical properties. The low cycle fatigue curve exponents m_o and m_e reflect the strength and durability characteristics, σ_B^T and σ_{-1}^T, respectively, as

$$m_o = 0.36 + 0.002\sigma_B^T \tag{6}$$

$$m_e = 0.15 \log(\sigma_B^T / \sigma_{-1}^T) \tag{7}$$

We can assume that $\beta_* \cong 0$ and $\beta \cong 1.0$. Note that the dependence $l_{ak}(N_o)$ corresponds to the "overstressed" loading conditions ($\Delta l_k \cong$ constant), which as a rule occurs in stress concentration areas.

The calculations just described were conducted for a preburner gas temperature range of 973–1123 K. The formations of the stress concentration areas on the blades as a result of the blade design and manufacturing peculiarities were taken into account. The blades are exposed to the transient heat loads, which are of primary importance in these studies.

As the preburner gas temperature and consequently the blade temperature rises, the elasto-plastic deformations magnify significantly, especially in stress concentration areas. This reduces the material strength properties and shortens blade durability. Based on the structural analysis calculations at a preburner gas temperature of 973 K, the number of cycles prior to the crack initiation in the stress concentration areas is $N_o \sim 180$. At a preburner gas temperature of

1123 K, the number of cycles prior to the crack initiation is $N_o \sim 19$. The most significant durability decrease is observed at the temperature range of the preburner gas from 1023 to 1173 K. Figure 3 presents the results of these calculations. The number of cycles to crack initiation, N_o, is considered to be the limit. Thus, to determine the number of the operational cycles N_e, the durability margin factor η_N is introduced as follows:

$$\eta_N = N_o/N_e \tag{8}$$

Though currently the durability margin factor is taken as $\eta_N = 4$, the durability margin is considered to decrease as the number of load cycles rises.

Experience has shown that taking $\eta_N = 4$ substantially overestimates durability margin.[34] Furthermore, the constant value of η_N, which does not depend on the number of operating cycles N_e, may result in high durability requirements that can hardly be realized in practice.

The durability margin factor η_N should be a function of N_e, such that increasing N_e should decrease η_N. The approach to the determination of η_N that connects the number of loading cycles until the crack initiation and the number of loading cycles until the complete failure is introduced next.

The corresponding analysis leads to the following dependence:

$$N_e = \frac{15.625 \cdot \eta_N}{(\eta_N - 1)^3} \tag{9}$$

Table 2 shows the values of η_N as a function of N_e. For N_e between 10 and 30, the durability factor margins η_N can be reduced from 4 down to 2.5 to 2.0.

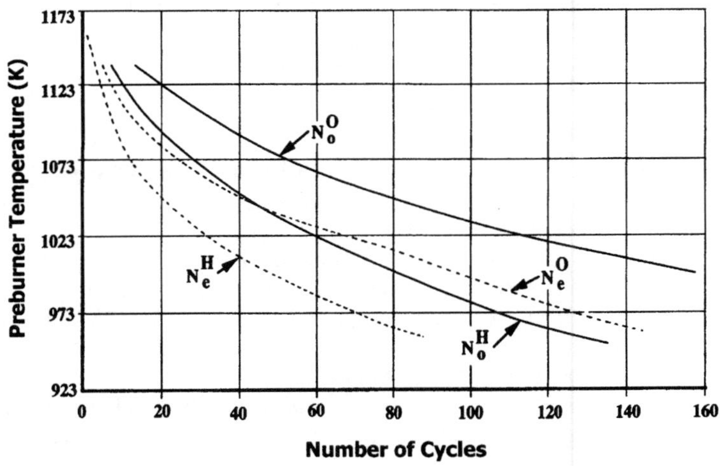

Fig. 3　Number of cycles to crack initiation (N_o) and operational cycles (N_e) as functions of preburner gas temperature and fuel-rich (H) and oxidizer-rich (O) turbine drive gas.

Table 2 Comparison of durability margin factor and number of loading cycles (or engine starts)

η_N	6.0	4.0	3.0	2.5	2.0	1.8	1.5
N_e	~1.0	~2.0	~6	~12	~31	~55	~187

With these concepts defined, the allowable values of operational use of the engine depending on the preburner gas temperature can now be analyzed. The dependence of preburner gas temperature T_g on N_e was shown on Fig. 3. The maximum allowable preburner gas temperature in the oxidizer-rich scheme for an engine with 25 starts is 1075 K, where no account was made toward the possibility of burning of the metal.

When the fuel-rich scheme in the tripropellant engine with tripropellant or bipropellant preburner is used, the preburner gas contains hydrogen, which can cause a reduction of the strength properties of the turbine blade material because of the problem of hydrogen embrittlement. The level of embrittlement is known to depend on the temperature, pressure, and the combination of gases present including hydrogen. High hydrogen pressure in lines and turbopump assemblies does not necessarily lead to problems with hydrogen embrittlement. In preburner gas lines, although the hazard of hydrogen embrittlement is present, the effect of hydrogen embrittlement is reduced because of both the high temperatures and the presence and influence of other components in the gas (such as water vapor). Engine design and manufacture procedures can account for the reduction of the plastic properties of the materials because of hydrogen embrittlement, since the extent of the plastic properties reduction is known.

During engine operation in flight, the temperature of the turbine blades changes from ambient temperature to the maximum during startup, and from maximum to ambient temperature during shutdown. To estimate the durability, material strength properties of the blades over the whole range of temperatures are required, taking into account the influence of embrittlement. For the tripropellant engine that transfers operational modes during flight from tripropellant to bipropellant, the hydrogen environment has influence on both modes of the engine operation.

Using the same concepts of the estimation of the durability of turbine blades operating in environments containing hydrogen, for 25 engine starts the allowable maximum preburner gas temperature is calculated to be 1033 K.

After calculation of the maximum allowable preburner gas temperature for the reusable engine, the nominal gas temperature can be calculated. From the maximum preburner gas temperature are subtracted the possible temperature changes ΔT that are connected with changes of external factors (variations of inlet pressures and temperatures of hydrogen, kerosene, and oxygen), changes of internal factors (variations of pump efficiencies, pump heads $H = \Delta p/\rho$), hydraulic characteristics of the units, and changes of the engine operational modes concerning the thrust and mixture ratio. In this case the nominal temperature of the preburner gas can be estimated by

$$T_{g,\text{nom}} = T_{g,\text{max}} - \Delta T_{\text{ext}} - \Delta T_{\text{int}} - \Delta T_F - \Delta T_{r_m} \tag{10}$$

where

$T_{g,\text{max}}$ = maximum allowable temperature of preburner gas,
ΔT_{ext} = temperature variations caused by inlet conditions variations,
ΔT_{int} = temperature variations caused by engine-to-engine variability,
ΔT_F = temperature variations caused by changes of thrust F,
ΔT_{r_m} = temperature variations caused by changes of mixture ratio r_m.

The calculations of the preburner gas temperature variations in each group can be made to take into account the occasional and systematic laws of changes of these influencing factors. An engine designer must decide according to the technical requirements and the technology of the unit fabrication, especially important for such units as turbopump assemblies, what kind of occasional or systematic laws to use to take this into account. At the present time, the technical requirements for the tripropellant and multi-usable full-scale engine are not yet defined. An estimate of the nominal preburner gas temperature can be made based on temperature uncertainties ΔT_i from RD-0120 engine experience, for which the total sum of the possible preburner gas temperature variations is 149 K.

Using that estimation for the variations, the maximum allowable nominal preburner gas temperature for the fuel-rich scheme is $T_{g,\text{nom}} = 884$ K, and for the oxidizer-rich scheme is $T_{g,\text{nom}} = 926$ K. This difference is less than 5%, which is not significant when choosing the preburner gas combination for development of the tripropellant engine.

D.   Soot Formation in Fuel-Rich Preburner

One problem associated with a fuel-rich hydrocarbon preburner is the possibility of creating solid carbon condensate, or soot, in the combustion products. Soot is a common byproduct in the turbine drive gas produced by fuel-rich oxygen/kerosene gas generators used in expendable open-cycle liquid rocket engines.[35] In simplest form, the predominant chemical reaction that creates soot in hydrocarbon reactions with oxygen is the recombination of carbon monoxide, CO, as follows:

$$2CO = CO_2 + C_s \tag{11}$$

where C_s is solid carbon, or soot. This is a non-equilibrium process not related to the total amount of CO, but it is a strong function of combustion gas pressure and temperature.

To avoid excessive component replacement or maintenance to remove soot from the internal passages of a reusable engine, the production of soot in the preburner must be minimized or eliminated. Preliminary analytical and experimental research in Russia suggested that organizing the burning processes in the tripropellant preburner with the correct application of propellant properties, injection distribution, and gas constituents precluded the possibility of soot formation. The addition of hydrogen to the fuel-rich, sooting hydrocarbon reaction with oxygen changes the amount of soot in two ways: 1) by eliminating the soot by reacting with it to create methane, CH_4,

$$C_s + 2H_2 = CH_4 \qquad (12)$$

and 2) by precluding its creation by reacting with the CO,

$$2CO + 2H_2 = CO_2 + CH_4 \qquad (13)$$

However, because the injection and mixing schemes in full-scale designs are dependent on size and can considerably influence the processes of combustion, more development experimentation was required to demonstrate soot-free operation. The design task was additionally complicated by the requirement for the preburner to satisfactorily operate in both tripropellant and bipropellant modes with fixed hardware geometry, and to transition without problems from one mode to the other.

For these reasons, single-element and multi-element preburners (called models in Russia) were developed for testing various preburner mixing schemes and injection element designs prior to testing in the full-scale preburner.[36,37] These model preburners used the same injection element geometries and propellants as in the full-scale preburner, but in reduced-size combustion chamber hardware. Two variants of mixing schemes—a one-zone process and a two-zone process—were tested in the model hardware using different variants of injection elements. In the one-zone mixing process, all the propellants were injected at the head end of the chamber, while in the two-zone mixing process, the hydrogen was injected at a downstream location.

The primary objectives of the model hot-fire tests were to 1) choose the optimum mixing scheme and injection element for soot-free tripropellant operation, 2) verify the capability to sample and measure the composition of the combustion products (including soot, if present), 3) examine the nonuniformity of the combustion gas temperature field, and 4) investigate the conditions of ignition, providing for startup and shutdown transients without temperature excursions or soot generation. Additional objectives from the unique requirements of dual-mode tripropellant engine operation included: 1) transition from tripropellant to bipropellant operation without soot generation or temperature and pressure excursions, 2) operate in bipropellant mode with minimum temperature nonuniformity, and 3) shutdown in bipropellant mode without temperature excursions.

1. Model Preburner Testing

Four variants of injection elements in each of the one-zone and two-zone mixing schemes were tested in single-element and eight-element model tripropellant preburners.[36,37] Chamber pressures ranged up to 13 MPa, and test durations were typically 20 s.

Test conditions were selected to provide operational modes of the preburner for the standard tripropellant engine flow splits (6% hydrogen, 12.6% kerosene, and 81.4% oxygen). Test conditions also included varying the overall mixture ratio over a wide range, $r_g = 0.20 - 0.90$, and varying hydrogen-to-kerosene ratios at the same overall mixture ratio r_g.

To evaluate the combustion process in the preburners and verify soot-free operation, the gas composition and soot concentration were sampled during hot-fire testing. The gas sampling system is described in detail in Ref. 36. It was designed to discourage further reactions among chemically active components after sampling, and to avoid distorting the phase composition (i.e., soot) of the sampled gas. The sampled gas was initially cooled in a water-cooled sample probe, and then further cooled to $10-20$ C in a water-cooled heat exchanger until it reached chemical equilibrium. The sampled products were then directed into a centrifugal separator where soot fell to the bottom of a sample cup. After a test, this cup was weighed and compared to the weight before the test to determine soot quantity collected in the cup during the sample time.

In a different version of the sample gas system, the centrifugal separator was removed and the gas directed through a filter to measure soot quantity. The filter was made from titanium powder by powder metallurgy and had a filter rating equal to 10 μm and a thickness of $3-4$ mm, with the labyrinth cross section of the filter capable of capturing soot. Pressure sensors were installed in the line to monitor pressure drop across the filter. With no soot in the sample gas, the pressure drop across the filter remained constant. During an operational mode where soot was present, the pressure drop across the filter began to increase, indicating that the filter was being contaminated and filling with solid matter. The intensity of this filling provided a means to determine exactly when sooting was occurring in the preburner chamber, and how strongly. When the operational mode initially began to change to one where sooting occurred, the pressure drop across the filter slowly increased, but when the mode was completely achieved, the pressure drop increased intensely, indicating the presence of soot. With this method, the time and conditions of soot appearance were clearly established. The filters were also weighed before and after the test, with the net weight increase attributed to soot in the gas stream.

The presence of soot in the model testing was thus determined by three quantitative techniques: 1) mass change in the sample cup and filter in the gas sample system, 2) increase in pressure drop across the filter in the gas sample system, and 3) differences between measured and predicted equilibrium combustion gas temperature and pressure. Soot was also shown qualitatively by review of video of the exhaust plume, where a transparent exhaust was a verification of soot-free operation.

According to mass changes of the sample cup and the filter, the combustion gas at nominal operating modes produced only trace amounts of soot. At conditions outside the range of normal operation, near the boundaries where soot began to occur, measured soot content in the combustion gas, given in percent of the total mass of the combustion products gas, ranged from 1 to 4%. No measurements were attempted where severe sooting occurred. These measurements clearly defined the operating conditions required for soot formation and determined the boundaries for soot-free operation. Analyses of these test data showed that operation of the preburner in nominal tripropellant mode without soot formation was possible up to a chamber pressure of 50 MPa.

The absence of soot was also verified, in addition to sampling the combustion products, by comparing measured preburner chamber pressure p_e with calculated pressure p_p, and measured preburner gas temperature T_e with calculated temperature T_p. The comparison between experimental and analytical chamber pressures in the preburner is one way to determine the completeness of combustion and whether soot has been generated in the preburner combustion gas. Calculated pressure was determined using the measured gas temperature, propellant flow rates, and the results of the combustion product composition analyses. If the experimental pressure was less than the calculated value ($p_e < p_p$), then either some of the preburner products did not participate in creating pressure in the preburner chamber (i.e., soot exists) or the combustion was incomplete. Under nominal tripropellant operating conditions as previously defined, high efficiencies of combustion were obtained, with measured pressure 95–100% of calculated pressure, suggesting combustion was mostly complete and soot was not present. However, a drastic reduction in efficiency (less than 85%) was observed during portions of tests where sooting was suspected, suggesting that soot was present at those moments, which agreed with the measurements of soot made in the gas sample systems. Comparison of experimental and analytical preburner chamber pressures showed that, to operate without generating soot in the preburner products, increasing the overall mixture ratio r_m was required.

The uniformity of the combustion gas temperature field is an indication of the mixing level in the injection design, which can also influence the creation of soot at local levels. For all tests conducted, the temperature field was measured by thermocouple rakes at two circumferential locations. Each rake contained either four or eight thermocouples located radially at even intervals along the rake lengths. The maximum temperature non-uniformity did not exceed $+8/$ $-30°C$ at average T_e of 700–750 K. This is a high level of uniformity ($+1/$ -4%), indicating a satisfactory injection process design for preburner development.

The results of the model preburner tests showed that the basis for development of a soot-free kerosene/hydrogen/oxygen preburner was correct, and that operating without soot was possible over a satisfactory range of operating conditions necessary for power generation in the tripropellant engine.

2. Full-Scale Preburner Testing

Following model testing, a one-zone full-scale preburner was fabricated and tested. A cross-sectional detail of this preburner is shown in Fig. 4. The one-zone mixing scheme, with all propellants injected at the head end of the chamber, simplified the design and fabrication. The injection element chosen for the full-scale preburner, based on results of the model testing, is shown in Fig. 5. The element includes oxidizer swirl and both swirled and angled kerosene injection. Hydrogen is injected from the faceplate. Additional kerosene orifices, as shown in Fig. 4, were included on the injector face to provide for the balance of kerosene injection area.

For full-scale preburner testing, a test bench was constructed that included an assemblage of valves and pipes exactly as would appear on the tripropellant engine. With such a configuration, the start transient, mode transfer, and shut-

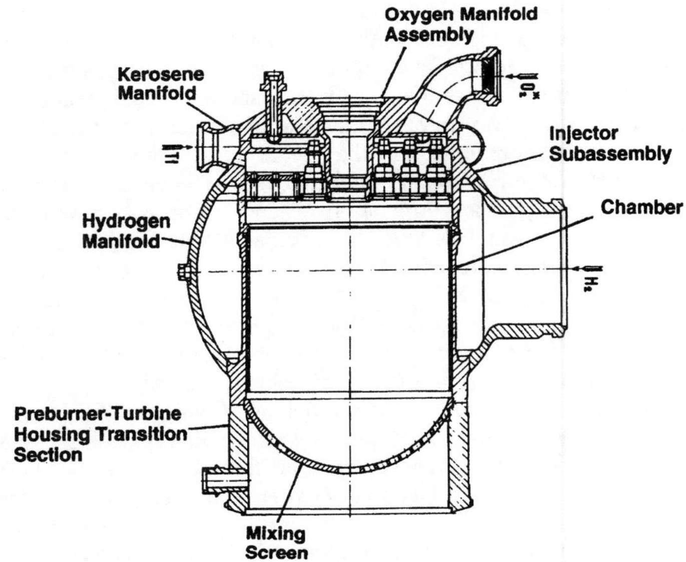

Fig. 4   Cross-section of one-zone full-scale tripropellant preburner.

down transient of the preburner can be examined and developed in a much more realistic fashion. An orifice simulating the turbine and main chamber resistance was installed at the preburner chamber outlet.

Five tests of the full-scale preburner were conducted and were reported in Ref. 37. During tripropellant operation, no soot was detected in the sampled gas by filter pressure drop or filter weight change, or in the exhaust plume by examination with video. Although the preburner was tested at low chamber pressures—to the limits of the test stand—predicted pressures matched model

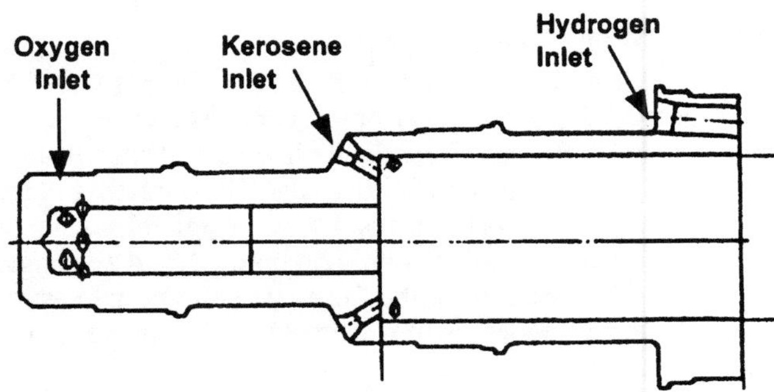

Fig. 5   Cross-section of injection element used in one-zone full-scale tripropellant preburner.

preburner test data, indicating soot-free operation is expected at higher pressures by similarity to the model data.

The temperature fields of the preburner chamber for bipropellant and tripropellant modes were evaluated using two seven-thermocouple rakes. The maximum temperature nonuniformity for the tripropellant mode, including the low temperature at the boundary layer near the wall, ranged from $+48$ to -40 K for a total temperature differential of 88 K based on a mean temperature of 833 K, or $\pm \sim 5\%$. The maximum temperature nonuniformity for the bipropellant mode, including the low temperature at the boundary layer near the wall, ranged from $+43$ to -46 K for a total temperature differential of 89 K based on a mean temperature of 850 K, or $\pm \sim 5\%$. At these low-power operating levels, both of these temperature nonuniformities were acceptable.

The operational stability was evaluated based on the pressure oscillations measured in the fuel and oxidizer manifolds. Low-frequency pressure oscillations were completely absent during the bipropellant mode, while operating at about 25% of nominal power. During the tripropellant mode, there were low-frequency pressure oscillations between 40 Hz and 60 Hz with amplitudes reaching about 9–12% of the nominal preburner chamber pressure in the oxidizer manifold and about 4–7% of the nominal chamber pressure in the fuel manifold. These oxidizer manifold oscillations were mild and did not influence the preburner gas temperature or temperature uniformity, so that the preburner operation was still considered stable. The oscillations were caused by operating at a tripropellant preburner chamber pressure of only about 10% of nominal. At this chamber pressure, the injection element pressure drops were very low (less than 2% of preburner chamber pressure).

Within the limits of the test stand capability, the full-scale preburner was found to operate without soot and with acceptable thermal and stability variations. Comparison with the model testing showed that the full-scale preburner should operate satisfactorily at higher chamber pressures.

III.   Tripropellant Engine Using Fuel-Rich Closed-Power Cycle

A.   Optimum Engine Schematic

The schematic of the optimum closed-cycle tripropellant engine with a tripropellant fuel-rich preburner is shown in Fig. 6. The operating and design parameters of this engine at a main chamber pressure of 24.5 MPa are listed in Table 3, which were calculated for a propellant combination consisting of 6% liquid hydrogen, 12.6% kerosene, and 81.4% liquid oxygen, and an overall $r_m = 4.376$. The vacuum thrust and specific impulse were calculated with an exit nozzle diameter of 2300 mm and geometrical expansion ratio of 104.5 : 1.

This engine contains three groups of turbopumps: 1) booster and main liquid oxygen pumps with gas turbines, 2) booster and main liquid hydrogen pumps with gas turbines, and 3) liquid hydrogen kick pump and main kerosene pump with gas turbine, which are on the same axle, and the booster kerosene pump with gas turbine. The use of two hydrogen pumps is a peculiarity of the engine due to the dual-mode operation. The hydrogen flow rates for tripropellant mode and bipropellant mode are practically equal, as shown in Table 3, but the discharge pressure after the main hydrogen pump drops 2 or 2.5 times in the

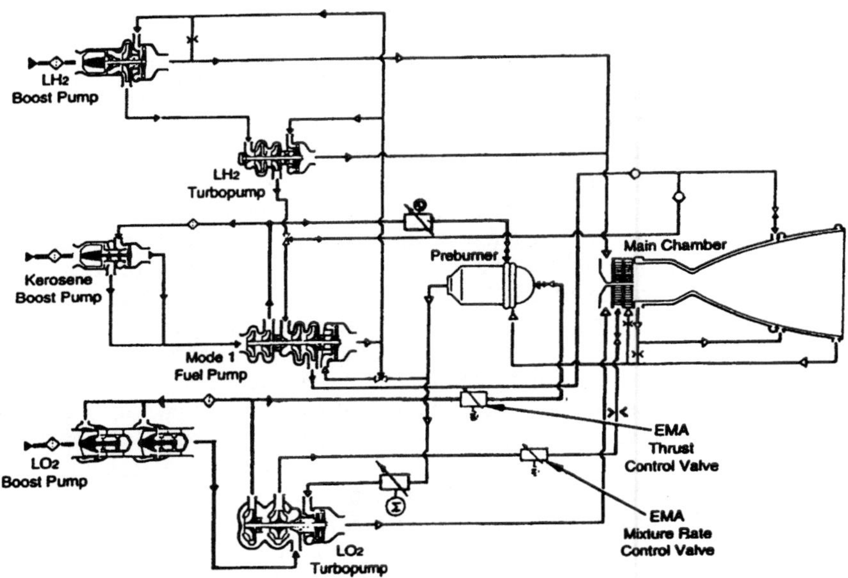

Fig. 6   Schematic of optimum closed-cycle tripropellant engine with fuel-rich turbine drive.

bipropellant mode. Such a range requires two hydrogen pumps for tripropellant mode and one pump for bipropellant mode. One hydrogen pump (along with the kerosene pump) is switched off during transition from tripropellant to bipropellant mode.

To adjust and control the engine operational modes, there are four controlling valves: 1) the liquid oxygen regulator in the preburner feed supply duct, for control of thrust, 2) the kerosene regulator in the preburner feed supply duct, 3) the gas throttle in the hot gas duct between the main liquid oxygen turbopump turbine and the main liquid hydrogen turbopump turbine, and 4) the liquid oxygen throttle in the main injector feed duct, for control of mixture ratio.

One unique feature of this tripropellant engine scheme is that all of the hydrogen flow is supplied to the combustion chamber for chamber wall cooling. The use of heat resistant coatings on the hot copper wall of the chamber is not considered because of the requirements that the tripropellant engine be reusable. The use of different coatings can reduce the reliability because these coatings may not be adequately bonded to the metal chamber walls. Therefore, to maintain a nominal hot-gas copper wall temperature at 800 K, provision of a cold film near the wall is required to reduce the heat transfer to the chamber wall. Because of its superior cooling capabilities, hydrogen can cool the copper chamber wall with a specific heat flux of $90-100 \times 10^6$ kcal/(m$^2 \cdot$ h), and with acceptable hydraulic losses in the cooling ducts assuming coolant flow speed of 250–300 m/s. When the chamber pressure is increased, the specific heat flux also increases, and to reduce the heat flux to allowable values, the mixture ratio in

Table 3 Main parameters of optimum fuel-rich tripropellant engine for reusable vehicle with 25 cycles between refurbishments

Parameter	Units	Tripropellant	Bipropellant
Propellant combination	——	6.0% H_2, 12.6% K, 81.4% O_2	14.3% H_2, 85.7% O
Vacuum thrust	mT	200	80
Sea-level thrust	mT	157 1	——
Sea-level thrust with insert	mT	171 2	——
Vacuum specific impulse	s	409.1	454.0
Sea-level specific impulse	s	321.3	——
Sea-level specific impulse with insert	s	350.2	——
Main chamber pressure	MPa	24.5	10.3
Overall mixture ratio	——	4.376	6.0
Total propellant flow rate	kg/s	488.9	176.2
Liquid oxygen flow rate	kg/s	398	151
Liquid hydrogen flow rate	kg/s	29.3	25.2
Kerosene flow rate	kg/s	61.6	——
Preburner gas temperature	K	850	815
Preburner gas constant	$\dfrac{kg_f - m}{kg - K}$	122.5	280
Preburner mixture ratio	——	0.581	0.515
Preburner total propellant flow rate	kg/s	143.7	38.2
Preburner liquid oxygen flow rate	kg/s	52.8	13.0
Preburner liquid hydrogen flow rate	kg/s	29.3	25.2
Preburner kerosene flow rate	kg/s	6.6	——
Nozzle expansion ratio	——	32 : 1	104.5 : 1
Nozzle exit diameter	mm	1400	2300

the film by the hot gas wall must also be reduced, which causes a reduction of the specific impulse of thrust. Figure 7 shows the change of the mixture ratio of the combustion products in the wall boundary-layer film with changes in chamber pressure according to 1) an expendable chamber with nickel-chrome heat resistant coating on the combustion chamber wall, where the number of thermal cycles (start plus stop) is 5 or less, and 2) a reusable chamber where the number of thermal cycles is 25. Then, based on these dependences, the vacuum specific thrust impulse with a fixed nozzle exit diameter equal to 2300 mm (or a fixed engine envelope) with different chamber pressure can be calculated, as shown on Fig. 8. Thus, for expendable combustion chambers, the optimum vacuum specific impulse occurs with a chamber pressure of about 34.5 MPa, while for reusable combustion chambers the optimum occurs at a chamber pressure not higher than 24.5 MPa.

As a rule, the hydrogen is first supplied to the throat section to provide reliable cooling of this critical section, and then through the chamber to the

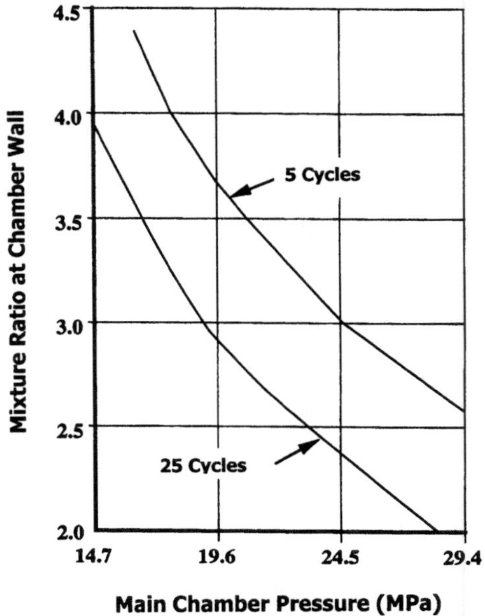

Fig. 7 Effect of number of operating cycles and main chamber pressure on the mixture ratio required at main combustion chamber wall.

head end, and finally in a series path through the nozzle. To decrease the overall hydraulic resistance of the cooling duct flow path, the nozzle can also be cooled by oxygen flow, due to the low level of the specific heat flux in that region, although that is not used on the current schematic. The coolant supply of oxygen is then mixed with the rest of the oxygen, which then proceeds to the main injector.

B. Engine Characteristics with Dual-Mode Operation

1. Main Injector Performance

For the tripropellant engine shown in Fig. 6, the mixing elements in the main injector operate with gaseous and liquid fluid states in both modes. Consequently, the development of the mixing elements and the combustion chamber head of existing O_2/H_2 staged combustion engines (such as the RD-0120 engine) can be used for design of the main injector for the tripropellant engine.

Results of uni-element mixing and combustion process experiments conducted during the development of the RD-0120 engine main injector are shown in Fig. 9. The experimental dependence is shown between completeness of combustion in the combustion chamber for coaxial atomization injector elements operating with fuel-rich preburner gas and liquid oxygen, and the

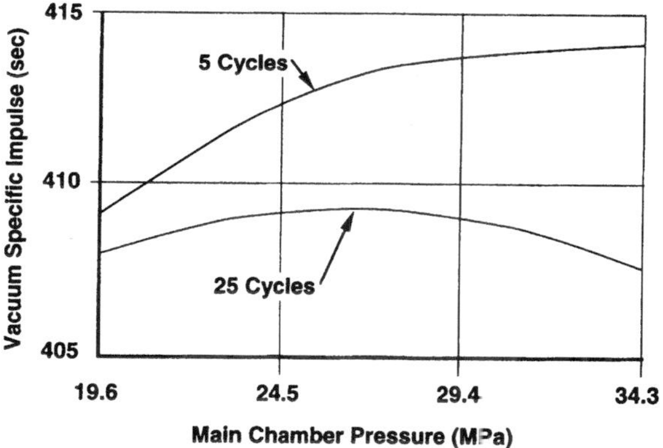

Fig. 8 Effect of main chamber pressure and number of operational cycles on vacuum specific impulse.

ratio of gas injection momentum to liquid oxygen injection momentum. Increasing this momentum ratio clearly increases the efficiency of combustion of the propellants in the chamber. The data shown in Fig. 9 provide a means to analyze the potential performance of these mixing injector elements when operating in tripropellant and bipropellant modes. The momentum ratio for the injector element in the tripropellant mode (i.e., with tripropellant fuel-rich gas for fuel) is about 0.3, and the combustion efficiency is higher than 0.995. The momentum ratio for the injector element in the bipropellant mode (i.e., with bipropellant fuel-rich gas for fuel) is about 0.2, so that the combustion efficiency of the elements would

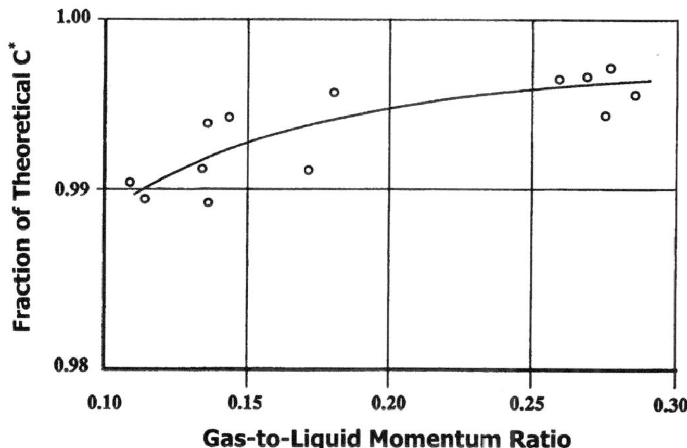

Fig. 9 Characteristic velocity efficiency of RD-0120 injection elements as function of gas-to-liquid element momentum ratio—single-element hot-fire data.

be about 0.995. For both modes, then, the element mixing performance remains high.

2. Preburner Injector Performance

The requirements for the mixing system and elements of the preburner injector are different than for the main combustion chamber. Combustion efficiency is less important, but uniform gas temperature fields are required for both operating cases, and there can be no temperature spikes during the startup and shutdown transients and the mode transfer transient. Because the combustion of the propellants for tripropellant and bipropellant modes can be organized with independent systems of propellant supply into the reaction zone, these requirements can be satisfied. Results of model and autonomous full-scale preburner testing, presented in Section II.D, showed acceptable preburner performance.

3. Mode Transfer Transients

During the transition of operation of the engine from the tripropellant to bipropellant mode, the kerosene pump and the high-pressure hydrogen pump are turned off. Because the three propellants are supplied independently to the preburner (with the oxygen as a cryogenic liquid, the kerosene as a room temperature liquid, and the hydrogen as a room temperature gas), during the transition between modes the kerosene manifold must be filled with hydrogen. This operation does not freeze the kerosene, because the hydrogen supplied to the preburner is at room temperature. However, for proper timing, the hydraulic resistance of the hydrogen and kerosene feed ducts must be considered for both tripropellant and bipropellant conditions. During the process of switching modes, liquid kerosene and warm hydrogen are mixed together in the preburner kerosene line for some time, until the kerosene pump is turned off. During this time, the engine control system switches all engine systems over to bipropellant mode operation. Satisfactory mode transfer operation was demonstrated in testing of the model and autonomous full-scale preburners, described in detail elsewhere.[36,37] Final development of the mode transfer transient must wait for testing in a complete engine.

4. Altitude Performance Compensation

RLVs require a combination of high sea-level takeoff thrust and high vacuum specific impulse performance. To increase both with an SSTO vehicle, some type of altitude compensation in the engine is necessary to provide the required mission-average performance. Modifications of the supersonic portion of the standard bell nozzle are required. One simple variant under development provides an insert in the supersonic portion of the nozzle, as shown in Fig. 10, that is removed at some appropriate moment in flight.[38]

The ejectable insert involves fixing a carbon-carbon insert in the nozzle to reduce the expansion ratio at low altitudes and prevent separation of the exhaust plume. At the appropriate time in the trajectory, the insert is released and guided cleanly out of the nozzle, increasing the expansion ratio for efficient high-altitude operation. The insert is made of carbon-carbon composite material

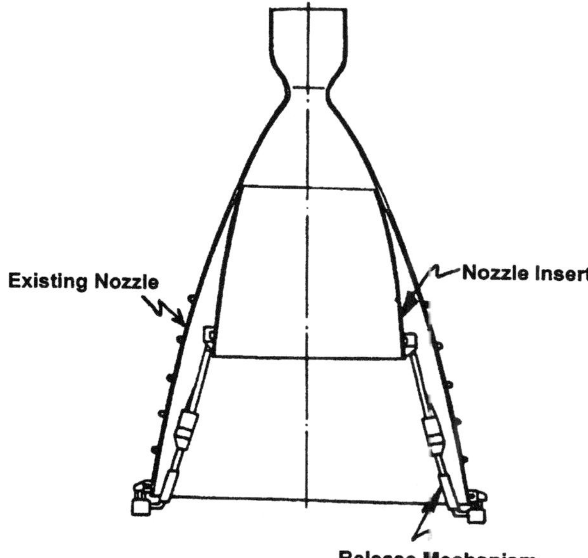

Fig. 10   Altitude compensating nozzle insert for tripropellant engine with single bell nozzle.

to minimize weight and eliminate the necessity for active cooling, and has a profiled contour to provide the highest performance characteristics. The insert is constructed to reduce the geometrical area of the nozzle extension, as shown in Fig. 10, and hence eliminate the appearance of shock waves during the startup and shutdown transients. Consequently, the vibration loads on the nozzle structure are decreased. The geometrical dimensions of the insert and the moment when it is removed are defined by the gas dynamic characteristics of the combustion products flows and by the performance requirements of the rocket flight trajectory.

Full-scale hot-fire experiments with the RD-0120 engine have been conducted with the ejectable insert and have demonstrated the durability of materials, seals, and release mechanisms with gas temperature equal or less than 2200 K.[38] With one of the variants of the inserts installed in the nozzle, the sea-level specific thrust impulse in a 50% operational thrust mode increased by 21.8%, providing for a prediction of an increase of sea-level specific thrust impulse of 9.2% at the 100% operational thrust mode.[38] Vibration levels at sea level were decreased by 16–36 G^2/HZ.[38]

The use of an insert for altitude compensation has several advantages. No changes to an existing nozzle are required to use the insert for altitude compensation. Because the insert is a separate, bolt-on component, adaptations are easily made to re-optimize specific parameters for other missions. The low replacement costs, short fabrication lead times, and three-hour replacement time are very compatible with scenarios for reusable vehicles.

C.   Requirements for Reusability and Operability

1.   RLV Engine Life Improvements

A key factor in making an RLV cost effective is the ability to amortize the purchase price of high-cost items, such as liquid rocket engines, over a large number of uses. Consequently, increasing the life of the liquid rocket engines is critical. A life prediction methodology that relates engine durability (operating cycles and duration) to engine operating conditions has been developed.[34] This methodology is based primarily on structural analysis techniques, accounting for both low and high cycle fatigue, but is also anchored to test data of the RD-0120 engine. The methodology can be used to analyze specific design modifications, predicting life as a function of engine operating parameters.

The methodology has already been used to assess increasing the life of the existing RD-0120 engine. The Energia mission profile required that the RD-0120 operate primarily at 106% of nominal thrust. The engine qualification program demonstrated that the standard production configuration under simultaneous conditions of 106% thrust and maximum mixture ratio can deliver six hot-fire cycles, without infringing on required margins or affecting the 0.992 reliability. The life-limiting feature demonstrated by the engine in serial production configuration was the appearance of cracks in the trailing edges of the turbine blades on the main turbopump. Implementing an identified modification to the turbine blades and the shroud is predicted to increase life by an order of magnitude.[34] Further identified design modifications on the next life-limiting features, such as main combustion chamber and main turbine nozzle cracks, will increase the life limit of the engine further.[38] The effects of these changes on the predicted existing RD-0120 engine life are shown in Fig. 11.[34,38] Such improvements as already identified on an existing engine strongly support that RLV life requirements can be achieved on the tripropellant engine as well.

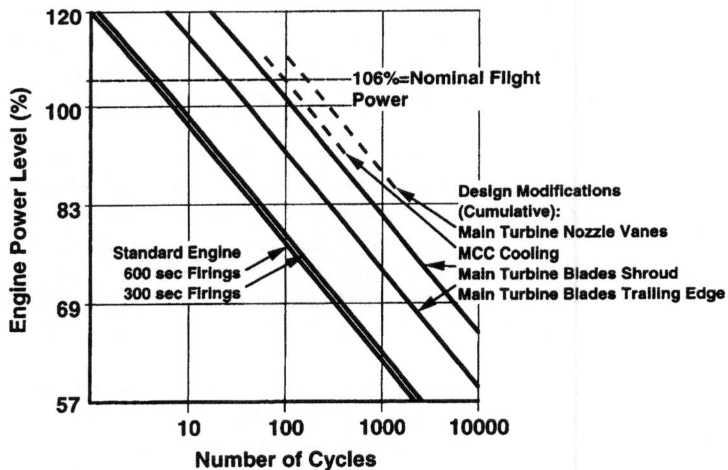

Fig. 11   Improvements in flight RD-0120 engine life due to identified engine modifications.

2. RLV Operability Issues

Labor, equipment, and facilities for initial launch preparation and recycling between flights are a major factor in the life-cycle cost of any reusable vehicle. Quick, low-cost ground operations are absolutely essential to a cost-effective SSTO RLV and will determine the economic feasibility of the concept.

The Russian approach to operations has always emphasized rapid launch preparations with minimal equipment and simple procedures.[38] The RD-0120 engine had the advantage of a legacy of dozens of previous engine developments and thousands of flights that have refined this basic approach. The design of the engine system and the designs of individual components incorporated lessons learned in automated checkout, elimination of maintenance, tolerance of extreme conditions, and management of conditions, such as leakage, which are difficult and costly to completely eliminate. RD-0120 engine operations have been well characterized in the course of preparation for approximately a thousand engine firings (for both ground and flight tests).

A test program was conducted with emphasis on measuring the operability parameters of the RD-0120 engine.[39] Service time on the engine of 48– 72 hours and 250 manhours was all that was required to return the engines to test.[39] These are representative numbers desired for RLV operation.[38]

3. Weight Reduction

High thrust-to-weight at sea level is an absolutely essential feature for the RLV main engine. The nozzle program discussed in Section III.B is a large contributor toward reaching an RLV weight requirement, a synergistic effect of the higher sea-level thrust available from the altitude compensating design and of the lower weight.

Additional weight reductions are necessary, however, and a review of the existing RD-0120 engine, for example, shows that some easily implemented solutions are available. These include changes in the size, configuration, and materials of the propellant inlet ducts. Inlet ducts are relatively massive, and modifications offer the possibility of significant improvements without disturbing the arrangement or functioning of the engine. Removal of external insulation on hot and cold ducts is another relatively high payoff modification. More intrusive weight reductions for the RD-0120 engine require longer-term solutions, such as shifting to higher-strength materials for certain components.[38]

4. Controls and Health Monitoring

For a reliable and reusable vehicle, health monitoring and properly responsive control systems of the engines are important aspects of operability. The assessment of engine health during ignition and flight and the correct response to problems will be critical to avoid sacrificing payloads. The level of development required may not be as large as may appear, however. Sophisticated and effective control laws and safety system algorithms have already been developed and proven for the RD-0120 engine for operation on the Energia vehicle,[33] and their effectiveness is shown by the high demonstrated reliability. One important new requirement is that the control system be autonomous. Engine control system autonomy permits independent development of the vehicle systems and avoids conflicting schedules, requirements, and priorities. With software programmability, operating modes can be rapidly reconfigured and optimized to reduce development testing and operational timelines. Functional redundancy, automated checkout, and integrated condition monitoring and problem diagnosis will improve reliability and reduce life cycle costs.

IV. Use of RD-0120 Engine for Development of Tripropellant Engine

The development cost of a rocket engine, especially for booster applications, is a substantial investment and can be prohibitive. This cost can redefine features of the design or be the ultimate deciding factor between paper and reality. Consequently, the use of previously developed and qualified engines and components is a great advantage that can save money and make new or "next-generation" designs feasible. Because of significant similarities to the RD-0120 engine, as suggested by previous sections of this chapter, a fuel-rich tripropellant engine can use this evolved approach for development.

Tripropellant engine development issues that could be considered already developed because of the design, fabrication, and test history of the RD-0120 engine include: 1) achieving highly efficient combustion processes in the main combustion chamber with injection of liquid/gas cryogenic propellants, 2) development of transpiration cooling of the main chamber injector face, 3) development of the main chamber cooling techniques, 4) creation of a milled, jointless supersonic nozzle with large dimensions, 5) use of powder metallurgical technology for fabrication of turbine blades, 6) creation of efficient axial unloading in the turbopump unit, 7) development of high-frequency rotor balancing methodology for the turbopumps, 8) selection and development of the experimental materials in regard to the hydrogen influence on their mechanical strength properties, 9) design of the units and engine overhaul concerning reusability and long life of operation, and 10) creation of the safety system and monitoring the technical state of the engine after hot-fire tests.[33]

Engine development problems solved during the design, fabrication, and test of the RD-0120 engine whose solutions could be used for tripropellant engine development include: 1) achieving uniform temperature distributions in the preburner with injection of cryogenic propellants, 2) development of engine startup and shutdown transients, and 3) creation of the system to control the engine modes.[33]

Obviously, many of the most difficult technical issues of the fuel-rich tripropellant engine development have already been solved. In addition, the methods of development and experimental verification that were previously developed can be used during the development of the tripropellant engine.

For these reasons, the RD-0120 engine can be used to provide a rapidly integrated and low-cost technology demonstration of a tripropellant kerosene-hydrogen-oxygen engine with a fuel-rich turbine drive.[38,40] Most of the existing bipropellant engine hardware can be incorporated directly into the tripropellant engine demonstration, with no changes whatsoever to approximately 95% of the flight-qualified component designs. A comparison of the bipropellant RD-0120 engine and a tripropellant engine demonstration, shown in Fig. 12,

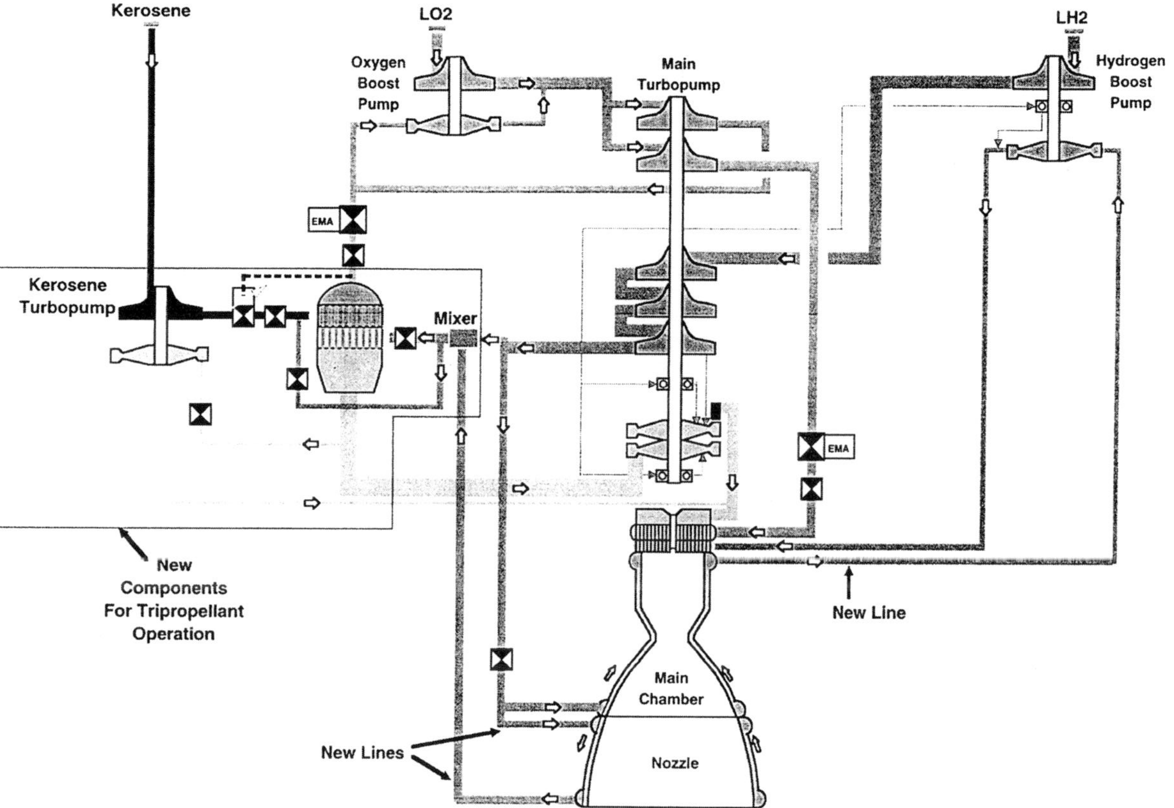

Fig. 12 Comparison of flight RD-0120 engine schematic and tripropellant demonstration engine based on RD-0120 engine.

Table 4   Comparison of flight RD-0120 engine and tripropellant demonstration engine based on RD-0120 engine

Operating parameters	Units	RD-0120 tripropellant demonstrator		Standard RD-0120, 106%
		Mode 1	Mode 2	
Engine performance				
Vacuum thrust	mT	134.3	79	200
Sea-level thrust	mT	94.5	——	155.6
Vacuum specific impulse	s	419	452	455.5
Sea-level specific impulse	s	295	——	354.0
Propellants				
Hydrogen	%	9.1	14.6	14.3
Kerosene	%	10.4	——	——
Oxygen	%	80.5	85.4	85.7
Propellant flow rates				
Total	kg/s	320.4	174.8	439.7
Hydrogen	kg/s	29.2	25.5	62.8
Kerosene	kg/s	33.2	——	——
Oxygen	kg/s	258.0	149.3	376.8
Engine operating parameters				
Main chamber pressure	MPa	14.7	8.1	21.9
Overall main mixture ratio	——	4.13	5.85	6.0
Preburner gas temperature	K	800	800	800
Nozzle expansion ratio	——	85.7 : 1	85.7 : 1	85.7 : 1

illustrates the high degree of commonality. Operating parameters of the RD-0120 and tripropellant demonstration engines are summarized in Table 4. The existing main turbopump, boost pumps, thrust chamber assembly, and most valves and lines can be used without changes. New components include only a tripropellant preburner, kerosene turbopump, and additional valves and lines for kerosene supply and kerosene turbopump turbine drive gas supply and exhaust. The tripropellant preburner, discussed in Section II.D, is itself an evolution of the RD-0120 design, incorporating modifications in the propellant manifolding and injector elements and adding a kerosene inlet, but retaining all the previous propellant inlet and structural interfaces.

Risk for the tripropellant engine demonstration can be minimized by an approach that emphasizes a step-by-step addition of capabilities. For the first step, the tripropellant preburner and associated control systems are installed, while kerosene is supplied by high-pressure facility tanks. Following demonstration of operating characteristics and performance with this engine, a kerosene turbopump is added and a fully pump-fed tripropellant engine is demonstrated. Development of new turbomachinery is avoided by adapting an existing kerosene unit taken from other rocket systems. Any substantial technical risks of tripropellant engines, along with life and durability margins, can be verified by this demonstration engine before commitment to a completely new engine is required.

V. Conclusions

Advanced liquid propellant rocket propulsion will be a requirement for the eventual realization of an SSTO RLV. Dual-mode tripropellant engines have been considered both in the United States and in Russia as important competitive technologies for creating this vehicle. These engines provide a high bulk density fuel for boost phase and a high specific performance fuel for ascent phase, but eliminate the need for separate hydrocarbon and hydrogen engines on the vehicle. Vehicle dry weight is reduced due to smaller tank size for the boost fuel and less vehicle base area to package engines and nozzles.

One of the most critical factors for performance, reliability, and operability of a tripropellant engine in an SSTO RLV is the selection of the power cycle. In this chapter, the rationale for selection of a partial staged combustion cycle with fuel-rich turbine drive gas was presented. First, the capability of the drive gas to produce turbine power was examined. Fuel-rich tripropellant turbine drive gas possesses 6–58% more capability to generate turbine power than oxidizer-rich bipropellant turbine drive gas, over a range of turbine drive gas temperature of 850 to 1000 K, and a percent of hydrogen to total propellant in the engine from 4 to 6%. Because of thrust chamber cooling requirements and typical leakage rates in advanced hydrogen turbomachinery, it is unlikely that the percent of hydrogen in the engine will be less than 4%. For the propellants with 6% hydrogen, the preburner gas temperature with the oxidizer-rich scheme must increase by 200 K (or 14%) to reach the same chamber pressure of 24.5 MPa as the fuel-rich scheme. Conversely, at the same preburner gas temperature of 850 K, the fuel-rich scheme attains a chamber pressure approximately 5 MPa (or 25%) higher. These are substantial numbers when considering the life requirements for a reusable engine.

Second, the effects of the gas on the engine material reliability were discussed. Oxidizer-rich gas has the capability to ignite the metal ducting given unpredictable hazards as can happen in oxidizer turbines. Coatings on the exposed metals are not desirable for use on a highly reusable engine that would retain low operational costs. This type of ignition is not possible with fuel-rich systems. On the other hand, the problem of hydrogen embrittlement on metals is understood enough to be analyzed and avoided in the design of the structure, while providing the necessary reusability.

Third, analyses of the effects of the turbine drive gas on the turbine blades, which are currently the components with the least durability in these engines, showed that the maximum allowable fuel-rich turbine gas temperature to meet an engine life of 25 cycles between refurbishment was only 5% less than the turbine gas temperature of the oxidizer-rich gas, without discounting the metal ignition hazard of the oxidizer-rich gas. This difference is much less significant at the required operating temperatures than the difference in turbine power.

Finally, few technical issues remain to be solved with this engine, mainly because of its similarity with other high-pressure, fuel-rich, staged combustion engines already developed and qualified. The most critical issue for fuel-rich tripropellant operation, sooting in the preburner, has been addressed. Soot-free fuel-rich tripropellant gas generation has been demonstrated in model and autonomous full-scale preburner testing. The tests showed that soot formation was precluded at preburner mixture ratios where soot would otherwise occur, and a sufficient soot-free operating range existed, so that sooting in the preburner would not limit engine operability. The model and full-scale preburner testing also successfully demonstrated the transient between tripropellant and bipropellant modes.

An optimal fuel-rich tripropellant engine scheme was presented. A review of this scheme suggests that issues about engine life, durability, and operability need to be addressed, along with perceived issues of excessive complexity and reduced reliability due to additional engine components with an additional propellant. However, because of the similarity to an existing engine, some of these issues can be assessed now. A life methodology has been developed and shows that the existing RD-0120 engine can reach the life required for RLV. Additionally, because of the Russian approach to operations, emphasizing rapid preparations, minimal equipment, and simple procedures, engine operability requirements are also at acceptable levels for RLV. For the other questions, an engine test demonstration is required. A demonstration tripropellant engine can be quickly brought to test using the RD-0120 engine as the basis. This demonstration would emphasize that the tripropellant engine would not be a new development but an evolution from an existing, flight-qualified engine—probably the only way liquid rocket engines will be developed in the future.

Acknowledgments

The work performed by Aerojet and portions of the work performed by the Chemical Automatics Design Bureau were funded under the NASA Cooperative Agreement NCC8-44.

References

[1]"Access to Space—Advanced Technology Team: Final Report," Office of Space Systems Development, NASA Headquarters, Washington, DC, July 1993.

[2]Aldrich, A. D., "Access to Space Study, Summary Report," Office of Space Systems Development, NASA Headquarters, Washington, DC, Jan. 1994.

[3]Martin, J. A., "Effects of Tripropellant Engines on Earth-to-Orbit Vehicles," *Journal of Spacecraft and Rockets*, Vol. 22, No. 6, 1985, pp. 620–625.

[4]Bekey, I., "SSTO Rockets: A Practical Possibility," *Aerospace America*, July 1994, pp. 32–37.

[5]Martin, J. A., "History of Propulsion for SSTO and Multiple Stage Vehicles," AIAA Paper 93-1942, June 1993.

[6]Salkeld, R., "Mixed-Mode Propulsion for the Space Shuttle," *Astronautics and Aeronautics*, Vol. 9, No. 8, 1971, pp. 52–58.

[7]Wilhite, A. W., "Optimization of Rocket Propulsion Systems for Advanced Earth-to-orbit Shuttles," *Journal of Spacecraft and Rockets*, Vol. 17, No. 2, 1980, pp. 99–104.

[8]Martin, J. A., "A Method for Determining Optimum Phasing of a Multiphase Propulsion System for a Single-Stage Vehicle with Linearized Inert Weight," NASA TN-7792, 1974.

[9]Colasurdo, G., Pastrone, D., and Casalino, L., "Optimal Performance of a Dual-Fuel Single-Stage Rocket," *Journal of Spacecraft and Rockets*, Vol. 35, No. 5, 1998, pp. 667–671.

[10]Henry, B. V., and Decker, J. P., "Future Earth Orbit Transportation Systems/Technology Implications," *Astronautics and Aeronautics*, Vol. 14, No. 9, 1976, pp. 18–28.

[11]Wilhite, A. W., "Advanced Rocket Propulsion Technology Assessment for Future Space Transportation," *Journal of Spacecraft and Rockets*, Vol. 19, No. 4, 1982, pp. 314–319.

[12]Martin, J. A., "Hydrocarbon Rocket Engines for Earth-to-Orbit Vehicles," *Journal of Spacecraft and Rockets*, Vol. 20, No. 3, 1983, pp. 249–256.

[13]Martin, J. A., "Selecting Hydrocarbon Rocket Propulsion Technology," IAF-86-167, 37th Congress of the International Astronautical Federation, Innsbruck, Austria, 1986.

[14]Lepsch, R. A., Jr., Stanley, D. O., and Unal, R., "Dual-fuel Propulsion in Single-Stage Advanced Manned Launch System Vehicle," *Journal of Spacecraft and Rockets*, Vol. 32, No. 3, 1995, pp. 417–425.

[15]Kozlov, A. A., Hinckel, J. N., Koreeda, J., and Comiran, A., "Payload Evaluation of a Tripropellant Carrier Rocket," *Journal of Propulsion and Power*, Vol. 15, No. 2, 1999, pp. 304–309.

[16]Goracke, B. D., Levack, D. J. H., and Johnson, G. W., "Tripropellant Engine Option Comparison for Single Stage to Orbit," *Journal of Spacecraft and Rockets*, Vol. 34, No. 5, 1997, pp. 636–641.

[17]Stanley, D. O., Engelund, W. C., and Lepsch, R. A., "Propulsion System Requirements for Reusable Single-Stage-To-Orbit Rocket Vehicles," AIAA Paper 92-3504, 1992.

[18]Tkachenko, J. N., and Limerick, C. D., "Powerful Liquid Rocket Engine (LRE) Created by NPO Energomash for Up to Date Space Rockets," AIAA Paper 93-1957, 1993.

[19]Katorgin, B. I., Chelkis, F. J., and Limerick, C. D., "The RD-170, A Different Approach to Launch Vehicle Propulsion," AIAA Paper 93-2415, 1993.

[20]Lozino-Lozinsky, G. E., Shkadov, L. M., and Plokhikh, V. P., "Reusable Aerospace System with Horizontal Launch," *Space Technology*, Vol. 13, No. 1, 1993, pp. 11–23.

[21]Lozino-Lozinskaya, I. G., Chelkis, F. J., and Tanner, L. G., "The Current Status of Tripropellant Combustion Technology," *Proceedings of the Second International Symposium on Liquid Rocket Propulsion*, Chatillon, France, June 19–21, 1995.

[22]Visek, W. A., "LOX/Hydrocarbon Booster Engine Concepts," AIAA Paper 86-1687, 1986.

[23]Visek, W. A., "Design Concept for LOX/Hydrocarbon Tripropellant Booster Engine," AIAA Paper 87-1854, 1987.

[24]Yatsuyanagi, N., Sato, K., Sakamoto, H., Ono, F., and Tamura, H., "Stabilizing Effect of Hydrogen Injection on LOX/Kerosene Unstable Combustion," *Proceedings of the Seventeenth International Symposium on Space Technology and Science*, Tokyo, Japan, 1990.

[25]Yatsuyanagi, N., Sato, K., Sakamoto, H., and Ono, F., "Improvement of LOX-Kerosene Combustion Efficiency with Hydrogen Injection," *Proceedings of the Eighteenth International Symposium on Space Technology and Science*, Kagoshima, Japan, 1992.

[26]Ono, F., Tamura, H., Kumakawa, S., Sakamoto, H., Sato, K., Sasaki, M., and Yatsuyanagi, N., "Effects of Hydrogen Addition on Combustion Performance of a LOX/Kerosene Rocket," National Aerospace Laboratory, TR-1177, 1992.

[27]Kumakawa, S., Ono, F., and Yatsuyanagi, N., "Combustion and Heat Transfer of LO_2/HC/Hydrogen Tripropellant," AIAA Paper 95-2501, 1995.

[28]Clemons, K., Schmidt, M., Mouis, A. G. F., Ni, T., Pal, S., and Santoro, R. J., "Tripropellant Combustion using $RP1/GO_2/GH_2$: Drop and Injector Studies," *Proceedings of the 32nd JANNAF Combustion Subcommittee Meeting*, 1995, pp. 199–207.

[29]Rhys, N., and Hawk, C. W., "Tripropellant Combustion: Chemical Kinetics and Combustion Instability," AIAA Paper 95-3151, 1995.

[30]Ramamurthi, K., and Nair, G. M., "Coaxial Swirl Injection Element for a Tripropellant Engine," AIAA Paper 98-3514, 1998.

[31]Schmidt, M. G., and Micci, M. M., "Combustion Performance of $RP-1/O_2/H_2$ Tripropellants," AIAA Paper 98-3686, 1998.

[32]Qinglian, L., Yuhui, H., and Zhenguo, W., "The Numerical Simulation of the Hot Flow Field in Tripropellant LRE," AIAA Paper 01-3715, 2001.

[33]Rachuk, V., Matinenko, Y. A., Gontcharov, N. S., and Sciorelli, F., "Design, Development, and History of the CADB RD-0120 Engine," AIAA Paper 95-2540, 1995.

[34]Rudis, M. A., Orlov, V. A., Rachuk, V. S., Nikitin, L. N., and Campbell, W. E., "A Universal Methodology for Predicting Liquid Rocket Engine Durability Based on Russian RD-0120 Engine Operating Experience," AIAA Paper 95-2963, 1995.

[35]Lauston, M. F., Rousar, D. C., and Bucella, S., "Carbon Deposition with LOX/RP-1 Propellants," AIAA Paper 85-1164, 1985.

[36]Turtushov, V. A., Orlov, V. A., Yefremov, Y. A., Gontcharov, N. S., Hulka, J., and Dexter, C., "Development Status of a Soot-Free Fuel-Rich Kerosene Tripropellant Preburner for Reusable Liquid Rocket Engine Applications," AIAA Paper 95-3002, 1995.

[37]Turtushov, V. A., Orlov, V. A., Gontcharov, N. S., Yefremov, Y. A., Veremyenko, N. A., and Hulka, J., "Final Model Development and Full-Scale Testing of a Soot-Free Fuel-Rich Kerosene Tripropellant Preburner," AIAA Paper 96-2630, 1996.

[38]Orlov, V., Rachuk, V., Gontcharov, N. S., and Starke, R. G., "Reusable Launch Vehicle Propulsion Based on the RD-0120 Engine," AIAA Paper 95-3003, 1995.

[39]Rachuk, V. S., Shostak, A. V., Dmitrenko, A. I., Goncharov, G. I., Hernandez, R., Starke, R. G., and Hulka, J., "Benchmark Testing of an Enhanced Operability LO_2/LH2 RD-0120 Engine," AIAA Paper 96-2609, 1996.

[40]Rachuk, V., Orlov, V., Plis, A., Gontcharov, N., and Fanciullo, T. J., "The Low Risk Development of a Fuel-Rich Preburner Tripropellant Engine Using the RD-0120 Engine," AIAA Paper 94-9465, 1994.

Chapter 20

Oxidizer-Rich Preburner Technology for Oxygen/ Hydrogen Full Flow Cycle Applications

Shahram Farhangi,* Robert J. Jensen,[†] Ken Hunt,[‡]
Linda Tuegel,[†] and Tai Yu[†]
The Boeing Company, Canoga Park, California

Nomenclature

A_t = combustor throat area
c^* = characteristic velocity, $\equiv P_c A_t / m_T$
E_m = Rupe mixing efficiency
MR = mixture ratio, $\equiv$ oxidizer mass flow/fuel mass flow
m_T = total propellant mass flow
P_c = chamber pressure
η_{c^*} = JANNAF methodology c^* efficiency; ratio of measured to theoretical
$\quad c^*$, %
$\eta_{c^*\mathrm{mix}}$ = JANNAF methodology c^* mixing efficiency, %

I. Introduction

OXYGEN-RICH combustion technology is of considerable interest to liquid rocket engine developers because it can provide, for certain thermodynamic cycles, significant performance improvements along with the advantage of lower cost, higher reliability turbomachinery. The performance benefits have been discussed in a previous chapter in this volume.[1] For some cycles, an oxidizer-rich turbine drive offers the benefit of higher chamber pressure at the same turbine gas temperature, or lower temperature at the same chamber pressure.

---

*Member of the Technical Staff, Rocketdyne Propulsion and Power. Senior Member AIAA.
†Member of the Technical Staff, Rocketdyne Propulsion and Power.
‡Process Leader, Rocketdyne Propulsion and Power.

The turbomachinery advantages arise from the lower temperature turbine gas, which improves turbine life, and simplified turbine dynamic seals for high-pressure oxidizer pumps.[2]

For oxygen/hydrogen propellants, an engine cycle with oxidizer-rich gases that drive the oxidizer turbopump and fuel-rich gases that drive the fuel turbopump would provide these benefits relative to a fuel-rich staged combustion cycle.[1] This dual preburner cycle, which is termed the full flow staged combustion cycle (FFSC), differs in several key areas from the fuel-rich staged combustion cycle. First, all of the propellants flow through the preburners. The required enthalpy flux to the turbines can thus be obtained at lower preburner operating temperatures. As a result, turbine inlet temperatures are dramatically lowered for the same turbine power output, and the critical dynamic sealing requirements for separating turbine gases from the oxidizer pump are eliminated. Lower turbine inlet temperatures at a given pump horsepower lead to improved turbopump life. In addition, both the fuel and oxidizer are injected as gases into the main chamber.[3,4] This permits throttling of the main chamber over a wide range while maintaining injection-coupled combustion stability margins, because the injection pressure drop ratio is relatively insensitive to operating conditions for gaseous propellants. Another potential benefit is that, if the vehicle requires gaseous oxygen for a tank pressurization system (as on the space shuttle), the heat exchanger can be located in an oxygen-rich environment. This eliminates a potential failure mechanism should the heat exchanger leak.

In Europe and the United States, gas generator and preburner cycle engines have featured fuel-rich combustion because of a concern for materials compatibility. Thus, oxidizer-rich combustion systems have not received extensive study or development, and the published data are not extensive. In 1965 an oxidizer-rich staged liquid oxygen (LOX)/kerosene engine of 50 kN (11,200 lbf) thrust was developed and tested in Germany. This engine featured oxidizer-rich gas film cooling in the main chamber and the use of LOX as the channel wall coolant.[5] The engine operating pressure was at 8.5 MPa (1200 psia) and featured a unique engine layout with the pumps, preburner, turbine, and main chamber all integrated on a common axis. NASA-MSFC conducted an oxidizer-rich test series with a subscale combustor in 1966.[6] The combustor was operated with O_2/H_2 propellants at mixture ratios between 20 and 150, but chamber pressures were limited to 6.9 MPa (1000 psia). Ignition was achieved using an electric spark harness that was ejected during each test. In 1967 a parametric study of the features of a highly integrated oxidizer-rich drive storable propellant engine was conducted. A range of thrust levels was considered including booster size concepts.[7] As part of a NASA contract, a subscale LOX/methane combustor was tested at mixture ratios up to 50 and at chamber pressures up to 23.8 MPa (3450 psia).[8] Similar component tests on oxidizer-rich LOX/RP-1 combustors were conducted under NASA funding at pressures up to 16.6 MPa (2400 psia).[9,10] More recently Rahman et al.[11] demonstrated ignition and flame stabilization in a uni-element swirl coaxial injector using oxygen and hydrogen propellants over a wide range of mixture ratios at chamber pressures between 1 and 3.4 MPa (140 to 500 psia).

In the past decade, information about oxidizer-rich combustion systems used on engines from the former Soviet Union has begun to become available. The Russian NK-33 is an operational oxygen/kerosene staged combustion engine developed in the late 1960s with an oxidizer-rich turbine drive.[12] The preburner uses multiple axial zones for oxidizer injection and mixing and operates at a chamber pressure of 32 MPa (4670 psia) and a mixture ratio (ratio of oxidizer to fuel mass flows) of 58.[13] The Russian RD-170 is an operational 8000 kN (1.8 Mlbf) vacuum thrust LOX and kerosene engine, which also features an oxidizer-rich drive. A similar engine of one-half that thrust, the RD-180, is also operational.[14] Based on limited published data, engine balance models estimate that the RD-170 oxidizer preburner operates at a mixture ratio of 58 and a chamber pressure of 56.5 MPa (8200 psia).[15]

None of these studies or published data matches the operating conditions typical of current concepts for FFSC booster engine systems with LOX/hydrogen propellants. System studies[16,17] conducted in support of single-stage-to-orbit (SSTO) concept engines have defined an operational envelope for a family of oxidizer-rich preburner components. Figure 1 shows a typical schematic of such a booster class engine. These studies indicate that a preburner design concept capable of operating at mixture ratios of 120–175 and chamber pressures in the range of 34.5 to 69 MPa (5000 to 10,000 psia) is required to achieve thrust-to-weight goals for these engines. These operating conditions lie outside the previously available database for oxygen-rich combustion with oxygen/hydrogen propellants. Hence development and demonstration of an oxidizer-rich preburner operating as close as possible to booster relevant conditions has been considered a key technology milestone.

The primary objective of the present work was to design, fabricate, and hot-fire test a subscale oxidizer-rich preburner article exhibiting reliable ignition and sustained combustion over the desired range of conditions. Associated objectives were to achieve stable, high-performance combustion with a high degree of temperature and compositional uniformity at the preburner exit. Additionally, demonstration of a flight-type ignition system was desired.

This chapter discusses the effort to develop and demonstrate the subscale hydrogen/oxygen oxidizer-rich preburner. A significant pretest analysis effort was conducted to preclude turbine temperature spikes, material compatibility issues, and flow uniformity problems as well as to optimize preburner performance. The traditional measure of performance, c^* efficiency, while important, is shown to be insufficient to guarantee turbine inlet plane temperature and mass flow uniformity. Ignition, flame propagation, and flame holding characteristics of preburner injectors for which all the propellants enter at the face (full-face injection) are also addressed.

II. Oxidizer-Rich Combustion Issues

A key criterion for selecting oxidizer-rich preburner operating conditions is the capability for low turbine inlet temperatures. Reduced thermal loads improve turbine life, which is a key goal for reusable booster engines. Higher oxidizer-rich gas mixture ratios also tend to increase the density of the product gas. This increase in density may reduce the required turbine size and blade tip speed of a given FFSC engine turbopump, in turn reducing weight and radial stresses, although the blade spanwise loading may become substantial relative to radial loads. Hence a high oxidizer preburner mixture ratio is desirable, subject to combustion related constraints (such as flammability, flame stabilization, and required product gas enthalpy and uniformity) and blade mechanical

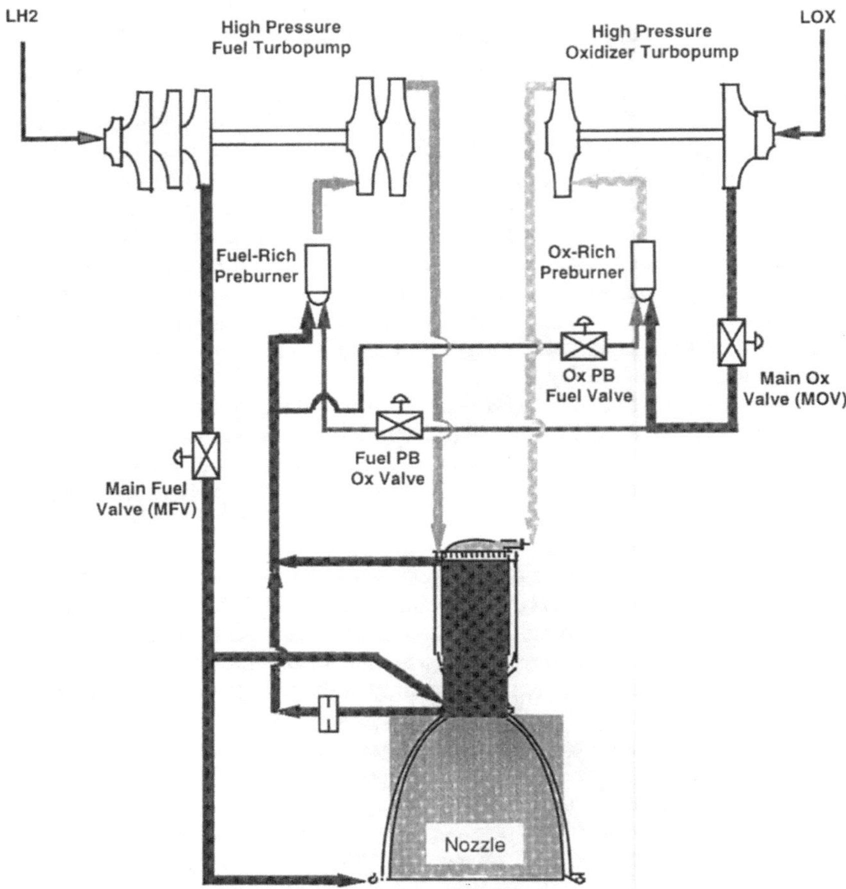

Fig. 1   Full-flow staged combustion booster engine schematic. (See also the color section of figures following page 620.)

loading limits. Product gas temperature uniformity is needed to avoid the potential for thermally induced high-cycle fatigue on the turbine blades.

A homogeneous mixture of hydrogen and oxygen has a wide flammability range, encompassing mixture ratios between approximately 0.84 and 384 (fuel rich and lean limits of 95 and 4% by volume) at standard conditions (i.e., 1 atm and 298 K). Flammability at standard conditions is particularly pertinent to combustor ignition design issues because ignition occurs typically at low pressure. The fuel lean and rich flammability limits both increase linearly with increased temperature. They also vary with pressure. However, the variation of these limits at high pressures (greater than 10.3 MPa) is insignificant. Although mixture ratios of interest to the oxidizer-rich injector design task fall within the flammable region, greater ignition energy and longer residence time (in the area where the flame holding occurs) relative to lower mixture ratio operation are required to sustain combustion and maintain flame propagation. For nonpre-mixed propellants with widely varying mixture ratio regions across the injector face, the task of designing an injector that will promote ignition, flame propagation, and flame stabilization can present a challenge. These issues are resolved by managing both the near face and downstream mixing through proper injector design.

An evaluation of typical oxidizer-rich preburner performance operating at 24.1 MPa (3500 psia) is useful. Figure 2 shows that the oxygen/hydrogen theoretical c^* is sharply reduced below typical main chamber values when operating at high mixture ratios. Note that this curve slope varies significantly over the mixture ratio range of importance to oxidizer-rich combustors. This range is determined by both the system optimized nominal mixture ratio and the mixing efficiency of a given injector pattern. As the mixing efficiency improves, the preburner operates over a narrower mixture ratio range. As the mixture ratio increases, the equilibrium adiabatic flame temperature (chamber total temperature) drops, as illustrated in Fig. 3.

Parametric calculations based on Nurick's mixing distribution model[18] were performed to illustrate the effects of propellant mixing on chamber performance at high mixture ratios. The results of several multiple (50) stream tube calculations are presented in Fig. 4, which shows the variation of characteristic velocity efficiency η_{c^*} with Rupe mixing efficiency[19] E_m for different mixture ratios. Rupe mixing efficiency is an index of the departure of the predicted or measured fuel and oxidizer distribution from a fully mixed flow at the nominal

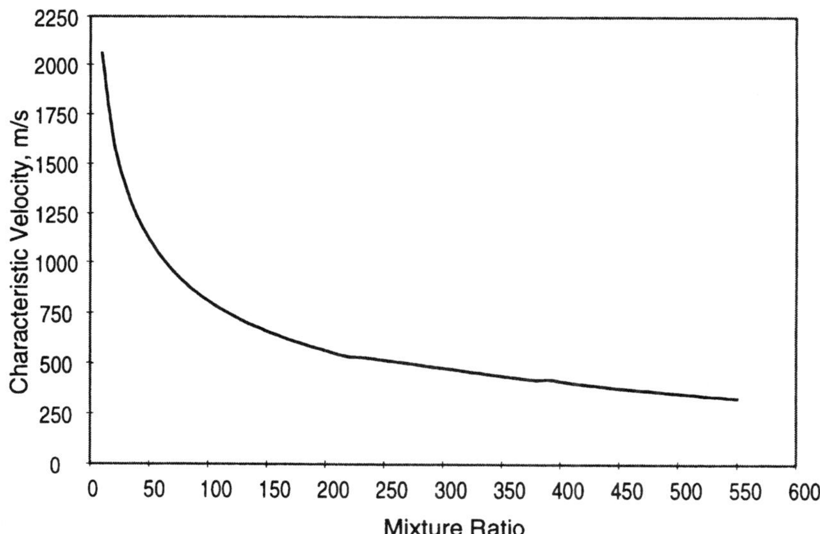

Fig. 2 Characteristic velocity vs mixture ratio for LOX/hydrogen combustion at a P_c of 24.1 MPa.

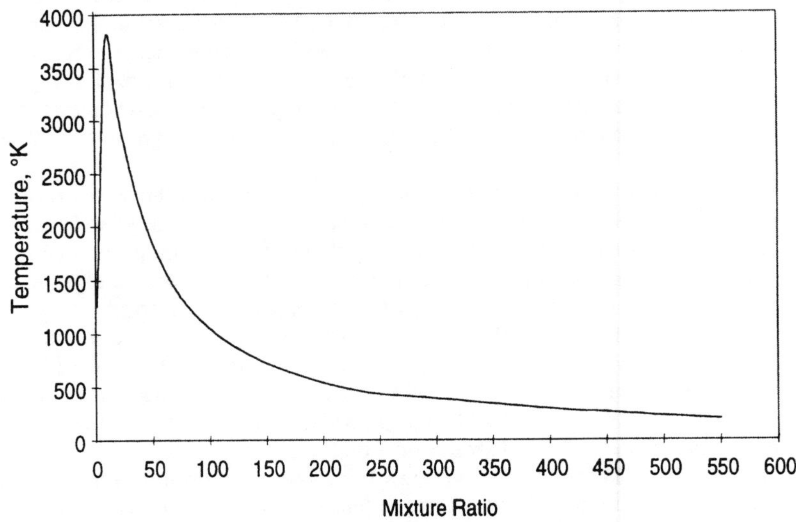

Fig. 3 Combustion gas temperature vs mixture ratio for LOX/hydrogen combustion at a P_c of 24.1 MPa.

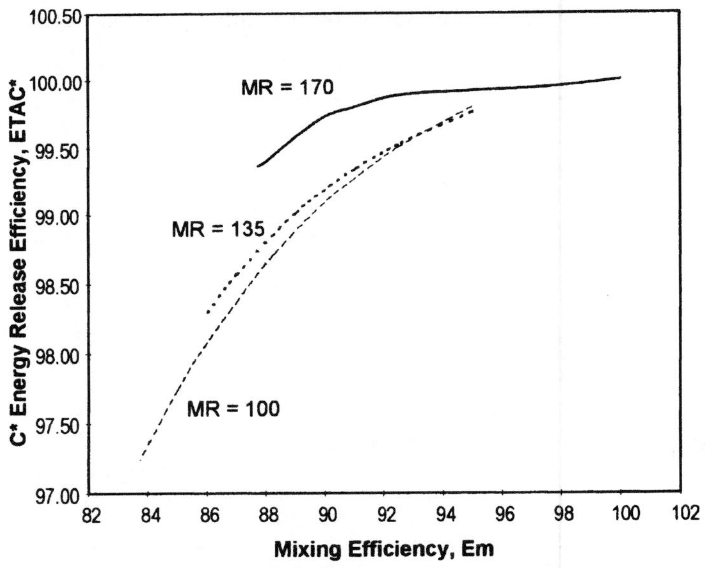

Fig. 4 The c^* efficiency vs Rupe mixing efficiency for LOX/hydrogen combustion at a P_c of 24.1 MPa.

operating conditions. It should not be confused with the JANNAF c^* mixing efficiency $\eta_{c^*\text{mix}}$, which also tends to rise toward a value of 100% as propellant mixing is improved.[20]

These curves illustrate that at the oxidizer-rich mixture ratios of interest, the c^* efficiency remains almost unchanged at mixing efficiencies of 92% and greater. Therefore, it could be very difficult to estimate the mixing efficiency of an oxidizer-rich injector simply based on c^* efficiencies computed from hot-fire test data. In addition, simple two-stream performance analysis shows that the nonlinear nature of the theoretical c^* curve itself shown in Fig. 2 may lead to 100% c^* efficiency values for incompletely mixed flows, assuming no other losses. Naturally, an experimental result of this sort would mask the lack of mixing uniformity required, if no other data were available.

To avoid inducing excessive thermal strain on the turbine blades, uniformity of the turbine gas temperature is required. The preburner chamber total temperature distribution vs normalized cumulative mass flow is presented in Fig. 5 as a function of injector E_m. The figure provides an indication of the degree of temperature uniformity in the product gases. For the indicated operating conditions (chamber pressure of 24.1 MPa and mixture ratio of 135), each curve indicates the cumulative mass fraction of product gas flow exhibiting temperature at or below the given temperature. The temperature uniformity is displayed in this figure in terms of mass-weighted deviations in temperature from the nominal equilibrium adiabatic flame temperature. The flatness of the S-shaped curve represents the level of temperature uniformity. For a uniformly mixed flow

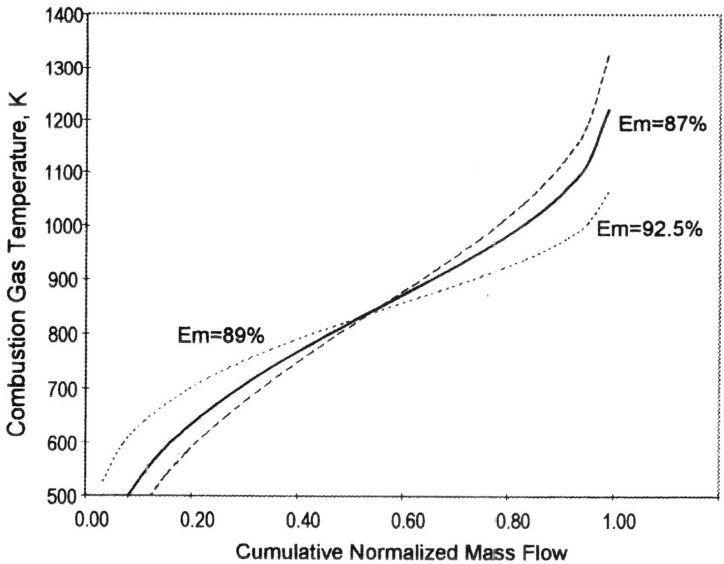

Fig. 5 Effect of Rupe mixing efficiency on temperature uniformity for LOX/ hydrogen combustion at a P_c of 24.1 MPa and a mixture ratio of 135.

($E_m = 100\%$), the curve would be a horizontal line passing through the nominal total temperature of $\sim$833 K (1500°R) corresponding to a mixture ratio of 135. The various curves are for Rupe mixing efficiencies E_m ranging from 87 to 92.5%, which correspond to an η_{c*} range, based on mixing losses only, of 98.6 to 99.5%. This figure indicates that to obtain flow temperature uniformity, a relatively high mixing efficiency is required even for a nominally high performance injector.

For the sake of the longevity of turbomachinery, the liquid fraction in the product gas, whether liquid water or liquid oxygen, should ideally be zero. Because less water is produced as part of the combustion products at high mixture ratios, the partial pressure of water vapor is low. The decrease in water vapor pressure reduces the saturation temperature. However, the reduction (with increasing MR) in chamber total temperature promotes condensation.

The maximum relative amount of condensed water that could form in the chamber as a function of mixing efficiency for various mixture ratios is presented in Fig. 6. These calculations are based on phase equilibrium conditions (saturation temperatures) and are therefore conservative with respect to actual oxidizer-rich combustor operating conditions. As seen from this figure, the percentage of liquid water in the total product mass flow over the range of operating conditions of interest in the current work is predicted to be very low. For a mixture ratio of 135, if the mixing efficiency exceeds 93%, no condensation is predicted to occur. For a mixing efficiency of 90%, less than 0.1% of the total

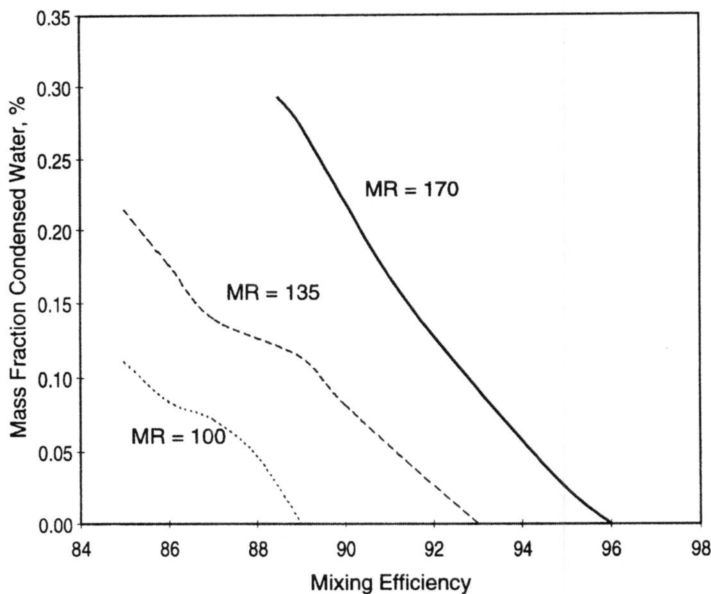

Fig. 6 Liquid condensation as a fraction of the total mass for LOX/hydrogen combustion at a P_c of 24.1 MPa.

flow (by mass) is predicted to be condensed water. If engine restart capability is stipulated as a mission requirement, water condensation or freezing in flow passages could be an issue for injectors with low mixing efficiencies.

III. Preburner Design

To achieve the objective of demonstrating oxidizer-rich preburner operation at targeted conditions, two basic design concepts were considered for the test article. The first concept injected all propellant mass through the preburner injector face. The second concept injected all the fuel mass plus a fraction of the oxidizer mass through the injector face at near stoichiometric conditions, with downstream addition of the remaining oxidizer mass. This second approach would likely require a chamber with an actively cooled zone near the face. A brief trade study comparing the two concepts was performed, resulting in the selection of the former (full face injection) concept for hot-fire demonstration. Key discriminators favoring this concept included reduced combustion chamber design complexity, cost, and weight. These advantages are a result of the simplicity in both the injector and the uncooled chamber configurations, the potential for lower overall component pressure drop (if a hot gas mixer is required with the multizone mixing concept), and a potentially shorter overall combustor length. Minimizing weight of all components is critical to obtain the high thrust-to-weight ratios required for SSTO vehicles.

A. Injector Element Selection

Complete vaporization of the liquid propellants and high mixing efficiency are required to ensure flow uniformity at the turbine inlet. Compliance with maximum turbine inlet temperature constraints must also be ensured. In addition, ignition and flame propagation, flame holding, and sustained combustion can become significant issues at extreme—in this case high—mixture ratios. Good performance and chamber mass flow uniformity are achieved with an injector that provides high vaporization rates, while producing a uniform mixture ratio distribution across the chamber cross section. On the other hand, ignition and flame holding characteristics can be enhanced by limiting propellant mixing near the injector face and promoting initial local nonuniformities. This results in local regions of lower mixture ratio that will sustain combustion. However, incomplete mixing can also lead to nonuniformities in downstream flow properties. Thus, judicious selections of injector element type and pattern layout are required to attain a workable balance between overall performance issues and ignition and flame propagation constraints for high mixture ratio preburner injector operation.

These general considerations and the parametric analyses described earlier were used to develop a set of design criteria for the oxidizer-rich preburner under consideration. Because this development effort was started well in advance of detailed vehicle-specific design trade studies, general rather than vehicle-specific criteria were utilized. Thus, an 8.9-cm-diam (3.5-in.-diam) chamber for which ancillary hardware was available was deemed acceptable for this level of the evaluation. This decision was based in part on favorable comparison of the calculated mass flux, 3660 kg/s-m^2 (748 lbm/s-ft^2), with that of previous oxidizer-rich combustor articles.[6,8,15] Complete vaporization within a reasonable combustor length was a fundamental requirement. Flame holding and flame ignition propagation at the injector face were also issues of concern for the planned mixture ratio operating range. Turbomachinery requirements limited the product gas liquid water content to less than 2% mass fraction. Acoustic cavities that could be blocked to verify the combustion stability margin were part of the baseline design. The combustor assembly included an instrumentation spool equipped with a thermocouple rake to measure temperatures at various radial locations at a given axial position representative of a turbine inlet. Provisions for both hypergolic and spark ignition systems were also required.

Consistent with the limitations of the chosen facility, the nominal operating conditions for the subscale demonstrator design effort were $P_c = 20.7-24.1$ MPa (3000–3500 psia), $MR = 135$, total propellant mass flow rate = 22.7 kg/s (50 lbm/s), and propellant injection pressure drop = 4.14 MPa (600 psi). This represents a pressure drop to P_c ratio of 20%, for P_c of $\sim$20 MPa. However, it was understood that as new engine balance evolved, test conditions would be set accordingly.

Selection of the injector element type (e.g., coaxial or impinging) was the initial issue to be resolved. The coaxial injector element has proven flame holding capability and an extensive development and test history with oxygen/hydrogen propellants. However, flame propagation characteristics across a coaxial element injector face at high mixture ratio are uncertain. Previous design experience and a review of the relative flow rates suggests coaxial element injectors would likely require a relatively long chamber length for complete vaporization at high mixture ratios. In addition, relatively low coaxial injector mixing efficiencies are expected at high mixture ratios, in comparison with impinging element patterns. Finally, a coaxial injector incorporating small diameter elements packaged in a fine pattern could easily entail higher fabrication costs, as influenced by a typically higher part count, than an impinging element injector design for the same conditions. For high mixture ratio applications, impinging injector element patterns should require much shorter vaporization distances, and hence shorter chamber lengths, than coaxial element patterns, due to better atomization characteristics. Therefore, based on mixing, performance, and potential cost issues, a small orifice impinging element pattern was baselined for the oxidizer-rich preburner test article development effort.

B. Injector Design and Analysis

Detailed injector design analyses were performed to develop a complete injector face pattern and to predict resultant performance and stability characteristics.[20,21] Nominal operating conditions assumed for the detailed analyses were as noted previously, $P_c = 20.7-24$ MPa and $MR = 135$. Detailed analyses of performance and stability at off-design conditions (i.e., different mixture ratios and chamber pressures) were not performed. Rather, a design was optimized for performance at nominal conditions, with a plan to investigate off-design characteristics during the hot-fire test program. This goal would be met through mixture ratio and throttling excursions directed by predictions from simple parametric studies.

After consideration of several element pattern concepts, a "box" pattern, where elements are arranged in an orthogonal grid pattern rather than concentric rings, with impinging elements was selected as the baseline for the oxidizer-rich injector. The selection was based on advantageous near-face and far-field mixing characteristics as well as fabrication simplicity. Parametric analyses were performed to establish the number of elements (element density), orifice size, impingement angle, and impingement distance parameters that could be packaged into an 8.9-cm-diam (3.5-in.-diam) face pattern. As the pattern evolved through successive iterations, the liquid injection spray pattern (LISP) module of the standardized distributed energy release (SDER) code[22] was used to characterize near-face flowfield characteristics by computing simplified actual drop size, mass flux, and mixture ratio distributions. E_m and $\eta_{c^*\text{mix}}$ were also predicted. These codes are JANNAF standard codes. This process produced a final element pattern with impinging elements in the core and additional oxidizer elements in the outer zone for enhanced mass flow uniformity and boundary layer cooling (BLC). The BLC contributes to radial temperature and mixture ratio nonuniformities, especially near the injector face. However, the amount selected was small because the BLC orifice area was less than 10% of the total LOX injection area. Wall cooling was required for thermal compatibility between the strongly varying mixture ratio regions near the injector face and the uncooled walls.

Contour plots generated using the LISP module of the SDER code, which show mixture ratio distributions at axial planes 1.27 and 3.81 cm (0.5 and 1.5 in.) downstream of the injector face, are presented in Fig. 7. At axial distances up to 1.27 cm from the injector face, local regions with mixture ratios in the highly flammable range up to $MR = 30$ are in abundance, suggesting good flame-holding characteristics. At the 3.81-cm distance, the MR distribution has become much more uniform. SDER single-stream tube performance calculations, initialized with the LISP module, indicate that the oxidizer-rich injector pattern achieves complete LOX vaporization in a chamber length of approximately 30.5 cm (12 in.), with 90% of the LOX droplets vaporized in approximately 12.7 cm (5 in.). The SDER-predicted LOX vaporization profile is shown in Fig. 8. At normal operating conditions and 30.5 cm (12 in.) downstream distance, E_m and $\eta_{c^*\text{mix}}$ were predicted to be 96 and 99.8%, respectively. It must be noted that SDER models have not been validated for this range of operation. However, as mentioned, until multiphase CFD codes are matured and validated to provide a better option, codes like SDER are routinely used to provide preliminary design guidance.

Intrinsic stability characteristics of the oxidizer-rich injector pattern were evaluated using the three-dimensional oscillatory rocket combustion code,[23] a sensitive time-lag stability analysis. This code was used with the Reardon correlation[24,25] to define the critical frequency range over which the acoustic modes of the chamber might be susceptible to coupling. No intrinsic instabilities were predicted for the oxidizer-rich injector design at nominal operating conditions.

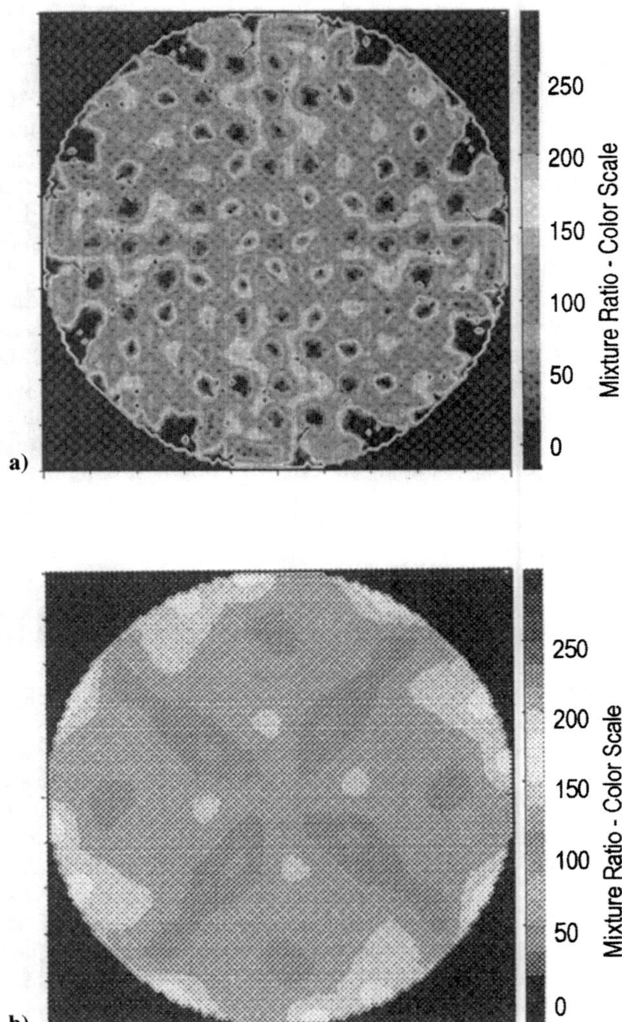

Fig. 7 Predicted mixture ratio distribution at a) 1.27 cm and b) 3.81 cm from the injector for a P_c of 24.1 MPa and MR of 135. (See also the color section of figures following page 620.)

C. Preburner Hardware Design

The layout of the hardware used for the subscale oxidizer-rich preburner demonstration is shown in Fig. 9. The configuration was a bolted assembly, which included an injector insert fabricated as a separate component from the fuel and oxidizer manifolds, an acoustic cavity ring, chamber spool pieces, and a throat section. The chamber and throat diameters were 8.9 and 3.2 cm (3.5 and

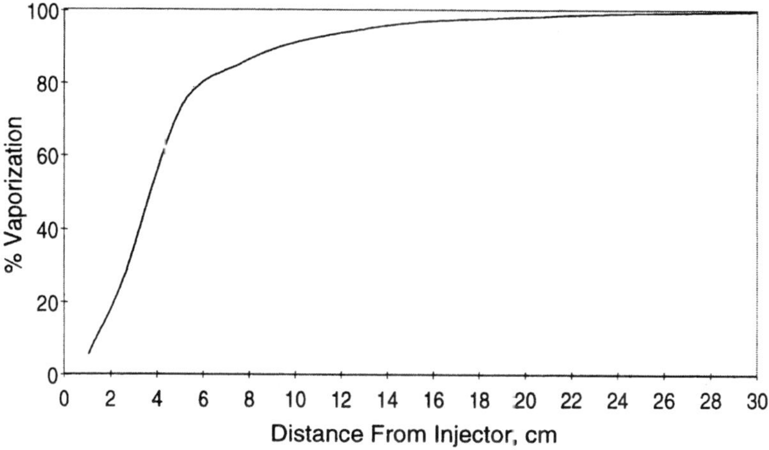

Fig. 8 Predicted LOX vaporization vs axial distance from injector face for LOX/ hydrogen combustion at a P_c of 24.1 MPa and a MR of 135.

1.25 in.), respectively, and the overall injector face-to-throat length with all chamber spools installed as shown was 46 cm (18.1 in.). Flight-type versions of such hardware would feature lighter weight, integrated injector manifold and chamber assemblies.

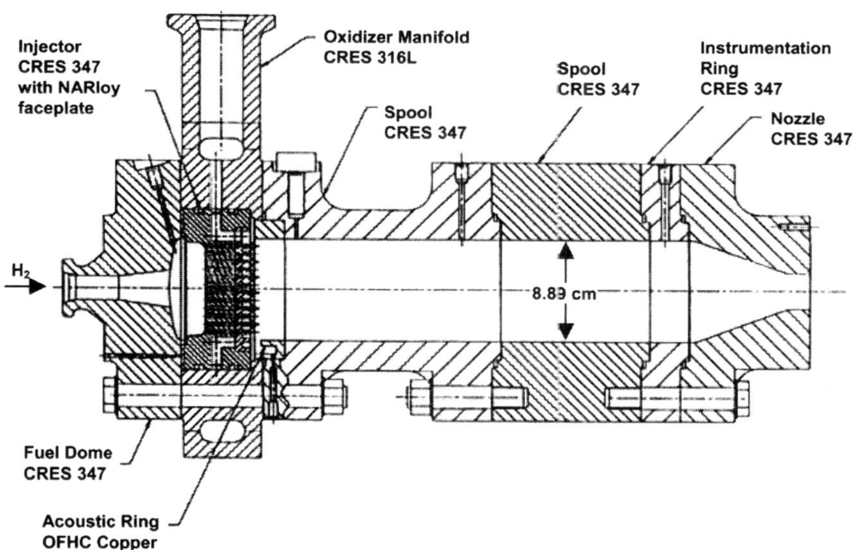

Fig. 9 Oxidizer-rich combustor assembly.

Fabrication materials for all components are shown in Fig. 9. CRES 347 stainless steel was used for all new components except the injector face and acoustic cavity ring. The existing LOX manifold was made from CRES 316L and fuel dome was fabricated from CRES 347. Because of the low (relative to nominal) mixture ratio zones near the face predicted by the LISP analysis, the faceplate was made from NARloy-Z, a high-conductivity copper alloy. The faceplate with laser drilled orifices was brazed to a 347 CRES body. The replaceable acoustic cavity ring was made of OFHC copper, although for all tests a blank ring was used. All of the stainless steel chamber sections were nickel plated on the hot gas wall for improved compatibility with the oxygen-rich environment. On the throat section, a gradated zirconia-NiCrAlY coating was provided for additional hot gas wall thermal protection. None of the hardware was actively cooled.

Product gas temperature was measured by thermocouples installed in an instrumentation ring just upstream of the throat section. Thermocouples were radially inserted to varying depths through six ports. The thermocouples were type K with 0.635 cm (0.25 in.) nominal diameter and reduced tips that were 0.254 cm (0.1 in.) in diameter. High-frequency pressure was measured in each of the propellant manifolds and in the chamber just downstream of the acoustic cavity ring.

Initially hypergolic ignition was achieved by injecting a mixture of triethyl aluminum and triethyl borane (TEA/TEB) into the thrust chamber through a single ignition port just downstream of the acoustic cavity ring, after a LOX lead. Subsequently spark ignition was achieved using a direct spark plug igniter with a staged, multi-setpoint sequence described in the next section. Other details of the hardware and test facility are available.[21,26]

IV. Hot-Fire Test Results

Following fabrication of the hardware, updated booster engine studies[16] defined a different oxygen-rich preburner operating condition from the initial design point. Therefore, most tests were performed over a *MR* range of 150–170 to closely match those operating conditions. A minimum number of tests were conducted at or near the initial design point.

A total of 36 successful main-stage hot-fire tests have been conducted using this oxidizer-rich preburner. Six of these tests were conducted with the preburner installed in a subscale staged combustion assembly featuring a gas-gas main injector.[3] Because the preburner was stable, the blank acoustic cavity ring remained installed for all tests. Chamber pressures ranged from 11.0 to 21.4 MPa (1600–3100 psia) and the mixture ratio was varied from 117 to 272, as shown in Fig. 10. Initial main-stage tests used hypergolic ignition, whereas later tests, including the *MR* = 272 test, were conducted using a spark igniter. Main-stage tests duration typically ranged from 0.9 to 10 s.

The axial plane of the temperature measurements was approximately 33 cm (13 in.) downstream of the injector face and just upstream of the nozzle convergence. The thermocouples used to measure gas temperature had insertion depths of 0.58 and 1.88 cm (0.23 and 0.74 in.). Not all tests included this instrumentation. In tests for which temperature data were available, near chamber wall (0.58 cm) product gas temperatures measured at this plane were found to range

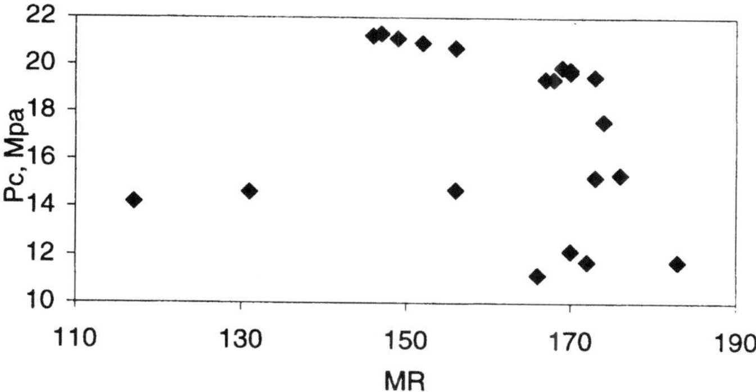

Fig. 10   Chamber pressure and mixture ratio range tested.

from about 406 K (270°F) at $MR = 272$ to 829 K (1032°F) at $MR = 151.7$. The measured product gas temperatures fell within an average range of ± 19 K (± 33°F) for the early tests. In later 10-s duration tests, the average range increased to ± 68 K (± 122°F). The 5- and 10-s test data may be more reliable because hypergolic ignition propellant continued to drain into the main chamber for a substantial part of the main-stage portion of the 1-s duration tests. A comparison between measured temperatures and theoretical combustion (adiabatic flame) temperatures at test conditions during the 10-s tests indicated that the average measured temperature was about 22 K (40°F) higher than the theoretical temperature. This trend was consistently observed in all main-stage tests and may be due to predicted high temperature zones in the combustion gas. Because inspection of the injector face later revealed that some minor erosion of the LOX orifices had occurred, the higher than theoretical temperatures may also be due in part to orifice erosion. Such erosion could cause larger than predicted mixture ratio variations. The minor face erosion was due to a braze process and did not have any noticeable effect on the test program. Moreover, the temperatures and temperature variations measured were well within the range to ensure compatibility with the combustor walls and turbine inlet temperature requirements.

A spark ignition system was tested to demonstrate the feasibility of a flight-weight reusable ignition system. The ignition sequence was staged to provide ignition of gaseous oxygen and hydrogen at low flow rates, followed by increasing LOX and main fuel rates up to full main-stage conditions. A two set-point servo valve schedule with ramp profiles was used for each propellant. Although transient conditions were difficult to establish with the set-point driven sequence, sustained ignition was demonstrated during the spark ignition tests for a range of conditions at or well above (fully mixed) oxygen-hydrogen flammability limits. Combustion was sustained at global mixture ratios as high as 500–1000, based on transient mass flow estimates. The spark plug was located approximately 1.3 cm (0.5 in.) from the face and showed no signs of thermal distress throughout the testing. Based on this success, the spark igniter was used for all subsequent testing.

By way of example, Fig. 11 shows propellant injection and chamber pressure traces for test 034 (chamber pressure of 20.9 MPa and mixture ratio of 152), which was a spark ignition test. Figure 12 shows the test 034 trace of chamber pressure and the product gas temperature measured by the six thermocouples just upstream of the nozzle convergence. The decreases in injection and chamber pressures after about 7 s of main-stage operation were due to the limited gaseous hydrogen run tank capacity.

Some characteristics of the oxidizer-rich preburner operating range can be seen from Fig. 12. During the high mixture ratio excursion, which was part of the start transient described earlier, product gas temperatures were observed to drop as low as 200 K ($-100°F$) at the thermocouple rake before increasing with main propellant flow rates. At the time of that minimum temperature in test 034, the spark igniter had been shut down for approximately 5 s and the chamber pressure was approximately 7 MPa (1000 psia). Although optimization of the start sequence would eliminate this low temperature extreme, the fact that combustion was sustained, as evidenced by the subsequent rise to main-stage conditions, demonstrated the robust flame holding characteristics of this injector. These results suggest that near the injector face, the local mixture ratio remained low enough to sustain combustion, irrespective of the overall MR and the degree of mixing measured downstream. This was consistent with the original goal to limit mixing near the face, creating localized regions of relatively low MR, followed by higher mixing efficiency further downstream.

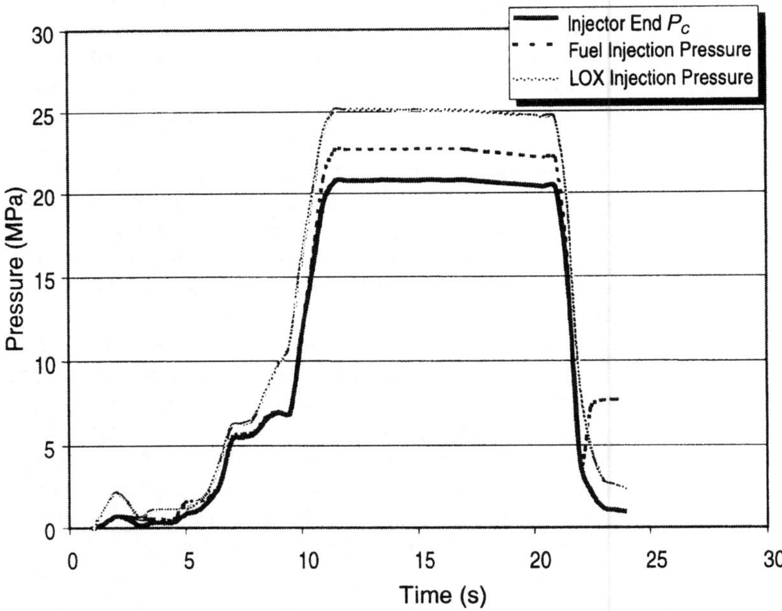

Fig. 11   Test 034 injection and chamber pressures. (See also the color section of figures following page 620.)

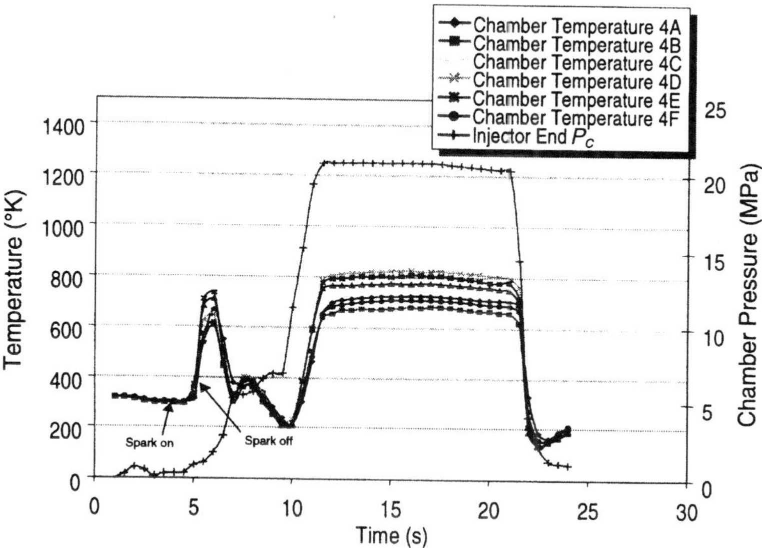

Fig. 12   Test 034 temperature rake data. (See also the color section of figures following page 620.)

As part of a separate contract effort, a derivative oxidizer-rich preburner designed for 9.1 kg/s (20 lbm/s) was succesfully tested at Marshall Space Flight Center using a hypergolic ignition system.[27] This lower flow rate injector used essentially the same pattern as the 22.7 kg/s injector but with smaller orifices. The chamber pressures (14.8–16.9 MPa) and mixture ratio range (147–152) were similar to that demonstrated with the current hardware.

In general, the oxidizer-rich preburner satisfied all test objectives. Combustion was stable for all tests. In addition, the narrow range of measured product gas temperatures suggests that the mixing efficiency is relatively good, even considering the potential BLC influence and limited thermocouple rake coverage. With reference to Fig. 5, a high mixing efficiency would be required to achieve the measured level of thermal uniformity, although an exact estimate of the mixing efficiency is not feasible without more temperature data.

V.   Summary and Conclusions

An oxidizer-rich oxygen/hydrogen preburner has been designed, fabricated, and successfully hot-fire tested over a main-stage operating range of 11.0 to 21.4 MPa (1600–3100 psia) chamber pressure over a series of 36 main-stage tests. The range of oxidizer-to-fuel mass mixture ratio (MR) was 117 to 272 at nominal main-stage conditions. Significantly higher mixture ratio operation was achieved during ignition start transients. Both hypergolic and spark ignition techniques were successfully demonstrated.

This oxidizer-rich preburner design successfully achieved all test objectives of the hot-fire test program. The full-face flow injector design demonstrated successful ignition, flame propagation, and flame holding over a wide operating range, including typical flight hardware start sequence conditions. Stable combustion was achieved on all tests. Relatively uniform product gas temperatures were obtained. The results demonstrate that O_2/H_2 oxidizer-rich preburners can be made compatible with practical engine systems.

The test program represents an enabling technology milestone and a significant design database, permitting resources to be directed toward other critical technology demonstrations. Spark ignition as implemented in the testing is compatible with launch vehicle requirements, allowing engine restart if necessary. In addition, the demonstrated wide range of start transient operating conditions, particularly with respect to mixture ratio, will simplify the development of the overall engine start sequence and greatly reduce development test costs.

Acknowledgments

Acknowledgments are gratefully extended to M. Schuman and F. Dodd for their technical support, and R. Mills, R. Metzner, and the Rocketdyne test crew in preparing for and conducting the hot-fire tests. Thanks also go to D. Matthews for his design efforts.

References

[1]Parsley, R. C., and Zhang, B., "Thermodynamic Power Cycles for Pump-Fed Liquid Rocket Engines," *Liquid Rocket Thrust Chambers: Aspects of Modeling, Analysis, and Design*, Progress in Astronautics and Aeronautics, Vol. 200, AIAA, Reston, VA, 2004, Chap. 18.

[2]Knuth, W. K., and Crawford, R. C., "Oxygen-Rich Combustion Process Applications and Benefits," AIAA Paper 91-2042, June 1991.

[3]Farhangi, S., Yu, T., Rojas, L., Sprouse, K., and McKinon, J., "Gas-Gas Injector Technology for Full Flow Stage Combustion Cycle Application," AIAA Paper 99-2757, June 1999.

[4]Tucker, P. K., Klem, M. D., Smith, T. D., Farhangi, S., Fisher, S. C., and Santoro, R. J., "Design of Efficient GO_2/GH_2 Injectors: A NASA, Industry, and University Cooperative Effort," AIAA Paper 97-3350, July 1997.

[5]Stökel, K., "History of the Development of the Staged Combustion Rocket-Engine in Germany," *Zeitschrift Fr Flugwissenschaften und Weltraumforschung* (in German), Vol. 9, No. 1, 1985, pp. 1–14.

[6]Bailey, C. R., "A Preliminary Investigation of Oxidizer-Rich Oxygen-Hydrogen Combustion Characteristics," NASA TN D-3729, Dec. 1966.

[7]Andrus, S. R., Bishop, H. L., Duckering, R. E., Gibb, J. A., Nelson, A. W., and Ransom, V. H., "Throttling and Scaling Study for Advanced Storable Engine," Aerojet, Rept. 68-C-0008-F, Parts 1 and 2, Sacramento, CA, 1968.

[8]Huebner, A. W., "High Pressure LOX/Hydrocarbon Preburners and Gas Generators— Final Report," Rocketdyne, Rockwell Rept. RI/RD81-129, Contract NAS8-33243, Canoga Park, CA, 1981.

[9]Schoenman, L., "Fuel/Oxidizer-Rich High Pressure Preburners," NASA CR-165404, 1981.

[10]Lawver, B. R., "Testing of Fuel/Oxidizer-Rich, High-Pressure Preburners," NASA CR-165609, 1982.

[11]Rahman, S. A., Pal, S., and Santoro, R. J., "Swirl Coaxial Atomization: Cold-Flow and Hot-Fire Experiments," AIAA Paper 95-0381, Jan. 1995.

[12]Kuznetsov, N. D., "Closed Cycle Liquid Propellant Rocket Engines," AIAA Paper 93-1956, June 1993.

[13]Hulka, J., Forde, J. S., Werling, R. E., Anisimov, V. S., Kozlov, V. A., and Nositsin, I. P., "Modification and Verification Testing of a Russian NK-33 Rocket Engine for Reusable and Restartable Applications," AIAA Paper 98-3361, July 1998.

[14]Ford, R. N., Pipes, W. E., and Josef, J. F., "The Next Generation in Rocket Engines— The RD-180," Space Technology and Applications Internationl Forum, 1998, pp. 1051–1054.

[15]Tkachenko, J. N., and Limerick, C. D., "Powerful Liquid Rocket Engine (LRE) Created by NPO Energomash for Up to Date Space Rockets," AIAA Paper 93-1957, June 1993.

[16]Goracke, B. D., Levack, J. H. D., and Nixon, F. R., "Margin Considerations in SSTO O_2/H_2 Engines," AIAA Paper 94-4676, Sept. 1994.

[17]Davis, J., and Campbell, R., "Advantages of a Full-Flow Staged Combustion Cycle Engine System," AIAA Paper 97-3318, July 1997.

[18]Nurick, W. H., "Dropmix-A PC Based Program for Rocket Engine Injector Design," Chemical Propulsion Information Agency, Pub. 557, Vol. 2, 1990, pp. 435–468.

[19]Rupe, J. H., "The Liquid Phase Mixing of a Pair of Impinging Streams," Jet Propulsion Laboratory, ORDCIT Project, Contract No. DA-04-495-Ord 18, Rept. 20-195, Pasadena, CA, 1953, pp. 18–19.

[20]"JANNAF Rocket Engine and Performance Prediction Manual," Johns Hopkins Univ., Chemical Propulsion Information Agency, CPIA Publ. 246, April 1975.

[21]Farhangi, S., Hunt, K., Tuegel, L., Matthews, D., and Fisher, S., "Oxidizer-Rich Preburner for Advanced Rocket Engine Application," AIAA Paper 94-3260, June 1994.

[22]Schuman, M. D., and Beshore, D. G., "Standardized Distributed Energy Release (SDER) Computer Program Final Report," Rocketdyne, Rept. APRPL-TR-78-7, Canoga Park, CA, 1978.

[23]Mitchell, C. E., "Final Report: Stability Design Methodology," Air Force Contract F04611-86-K0020, AF-TR-89-041, Vols. 1 and 2, 1989.

[24]Smith, A. J., Reardon, F. H., Crocco, L. M., Sirignano, W. A., and Harrje, D. T., "The Sensitive Time Lag Theory and Its Application to Rocket Combustion Instability Problems," Aerojet, AFRPL-TR-67-314, Sacramento, CA. 1968.

[25]Harrje, D. T. (ed.), *Liquid Propellant Rocket Combustion Instability*, NASA SP-194, 1972, pp. 67–68.

[26]Farhangi, S., Hunt, K., Tuegel, L., and Matthews, D., "Design and Testing of an Oxidizer-Rich Preburner," Chemical Propulsion Information Agency, Pub. 620, Vol. 1, Oct. 1994, pp. 401–407.

[27]Elam, S. K., and Strickland, D. E., "LOX Rich Oxygen/Hydrogen Preburner Test Program," George C. Marshall Spaceflight Center, TR-EP87-95-02, Huntsville, AL, April 1995.

Subject Index

Author Index

Supporting Materials

A complete listing of titles in Progress in Astronautics and Aeronautics and other AIAA publications is available at http://www.aiaa.org.